W9-AAZ-614

SURVEY OF IMAGINATIVE LEAPS

> **Reality leaves a lot to the imagination.**
> ***—John Lennon***

> **Always have a vivid imagination, for you never know when you might need it.**
> ***—J. K. Rowling***

When you were younger or possibly even very recently, someone may have suggested that you do something you thought was incredible, or impossible, so you responded, "How am I supposed to do that?!" Chances are, the answer you got was, "Use your imagination." In this familiar exchange, the message is, "Be resourceful, be creative." Most of us imagine circumstances or events fairly regularly: what it would be like to have a child or get a new car, what we would say if we got to meet a celebrity we admire, what it will be like when we achieve a long-term goal. Using your sociological imagination is not so different.

What is the sociological imagination?

The mid-twentieth-century sociologist C. Wright Mills used the term sociological imagination to describe sociological reasoning—what he considered the ability to see the relationship between individual experiences and the larger society. When we use our sociological imagination, we are thinking sociologically, viewing our and others' personal experiences in the social contexts in which they occur. To think sociologically, then, we first have to recognize that we are members of a society, that we are not completely autonomous individuals making choices independent of outside influences.

That means your relationship to society does not look like this:

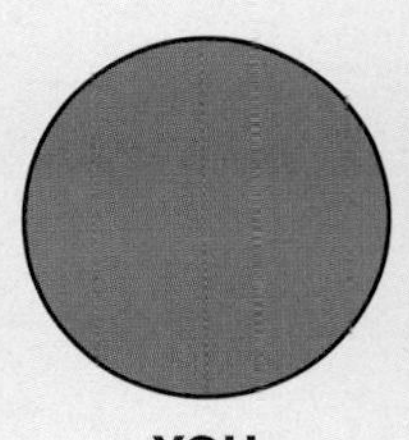

YOU

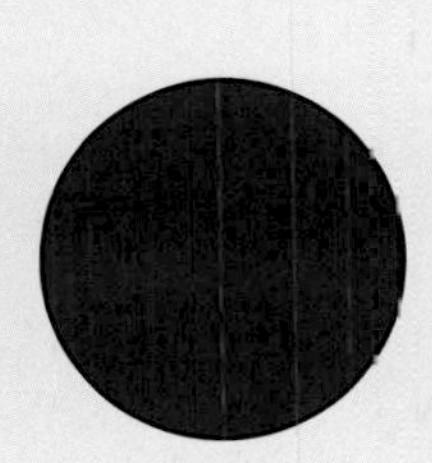

SOCIETY

It looks like this:

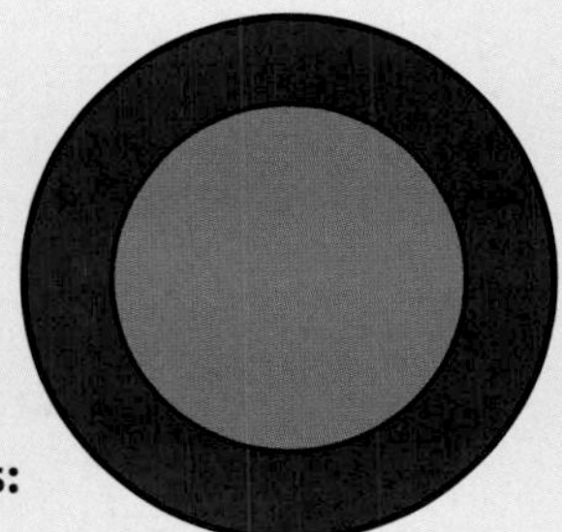

Why is the sociological imagination important?

The sociological imagination is at the center of the course you're taking and this textbook. As a result, the earlier you develop an awareness of it and how it can be applied, the greater edge you'll have as you work through the sociological concepts and theories presented in each chapter you're assigned. That edge will speed up your understanding and help you retain what you learn. Better yet, the sooner you start using your sociological imagination, the sooner you'll be able to move beyond established ways of thinking to gain new insights into yourself and to develop a greater awareness of the connection between your own world and that of other people. You will see that, as another mid-twentieth-century sociologist, Peter Berger, put it, "things are not what they seem." As a result, you will develop new ways of approaching problems and making decisions in everyday life.

UNFOLD PAGE TO CONTINUE

CHAPTER 7 Deviance and Crime

- How do peer groups sometimes encourage inappropriate behavior and even criminal conduct among individuals?
- Do you engage in any type of behavior or possess something that might harm or undermine someone else? If yes, what is it, and how is it a threat to others?
- Under what circumstances do you think you would discard this item or stop this behavior?

CHAPTER 8 Class and Stratification in the United States

- Have you experienced an increase in enrollment of immigrant students at your school? If so, has this influenced your close-knit peer group?
- When you envision your ideal life, what does it look like, and do you think it will be possible for you to achieve it?
- What obstacles do you anticipate might limit your ability to realize your ideal life?

CHAPTER 9 Global Stratification

- Have you lived in or visited a country other than the one in which you were born?
- Have you spoken to a customer service representative working in a call center in India or another middle-income country?
- How do you think the representative's ambitions might be different from those of his or her elder relatives when they were younger?

CHAPTER 10 Race and Ethnicity

- Do you think race/ethnicity plays a part in career goals? Why or why not?
- How many friends or family members do you have whose racial or ethnic background is different from your own?
- If you have friends or family from different backgrounds, what benefits and challenges have you experienced as a result of these relationships?

CHAPTER 11 Sex and Gender

- How does our body image coincide with our level of self-confidence?
- Depending on your sex, where would you place yourself on a scale of traditional femininity or masculinity, with 1 indicating "barely" and 5 "very"? Why?
- Have you ever tried to move yourself up or down this scale? If yes, why, and how did you do it?

CHAPTER 12 Aging and Inequality Based on Age

- At what age do you think someone becomes "old"?
- Do you think there should be a mandatory retirement age for workers in the United States? Why or why not?
- If you were to live to be 90, what would you like to be doing at that age?

CHAPTER 13 The Economy and Work in Global Perspective

- Why are so many people unemployed who would like to have a job?
- What are an employer's responsibilities to its employees, and why do you think what you do?
- Overall, do you think unions affect workers' lives positively or negatively?

Turn to back of page two to continue.

SPRING INTO SOCIAL, IMAGINATIVE THINKING

INSTRUCTIONS: Try answering the questions below to get yourself thinking sociologically. Each set of questions aligns with the content of each chapter. While responding, the more often you find yourself considering other people's experiences, whether those experiences are the same as your own or different, and the more you find yourself considering possible relationships between your and others' experience and the social circumstances in which our lives occur, the more you're using your sociological imagination.

CHAPTER 1 The Sociological Perspective

- Has credit card usage gotten out of hand in this country? Do you think credit hinders or empowers people?
- If you don't use credit cards, what's your reason for not using them?
- How, if at all, has the recession affected your spending (or saving) habits?

CHAPTER 2 Sociological Research Methods

- Do you think bullying is a more serious problem than in years past, or has the media blown it out of proportion?
- How have job insecurity and unemployment negatively affected workers?
- What societal factors influence high rates of suicide?

CHAPTER 3 Culture

- How does the food we eat become part of our personal and cultural memory?
- How does culture influence our philosophical and moral beliefs?
- Do you believe that cultural differences can lead to misunderstandings between people? Why or why not?

CHAPTER 4 Socialization

- Who was your primary caregiver when you were a small child—a parent or guardian, a grandparent, staff at a daycare center, a babysitter or nanny, or someone else?
- What childhood experiences did you have that you think had positive or negative effects on who you are as an adult?
- If you wanted to do so, do you believe that you could change the person you've become? Why or why not?

CHAPTER 5 Society, Social Structure, and Interaction in Everyday Life

- How is where you are currently living similar to and different from the place where you primarily lived when you were growing up?
- Explain how clothing choices represent or misrepresent who you are.
- How has clothing told stories of race, ethnicity, and class throughout history?

CHAPTER 6 Groups and Organizations

- Think of a club, organization, or other group you belong to: On a scale of 1 (very little) to 5 (a lot), how much do the other members know about you?
- What is your favorite method of communicating with other people?
- What would cause you to start using a different method of communication?

CHAPTER 14 Politics and Government in Global Perspective

- What have you learned from the media about politics and the U.S. economy? How might this information be different from what you've learned elsewhere—through word of mouth or in classes or an organization to which you belong?
- In what ways, if any, has the U.S. economy influenced your decision to attend college or to choose a specific major?
- What relationships do you think exists between the political and economic systems in a nation and that nation's domestic turmoil—or the wars it wages against other nations?

CHAPTER 15 Families and Intimate Relationships

- Describe how the individuals who raised you have influenced your relationships with other people.
- Were the persons you lived with while growing up raised in a household that was similar to or different from the arrangement in which you grew up?
- When you were growing up, how "normal" did your family's living arrangement feel to you, and to what did you attribute those feelings?

CHAPTER 16 Education

- In what kind of schools and classroom settings did you receive your education for kindergarten through high school?
- Do you think that your early childhood education shaped the person you are today? If so, how?
- If you could change one thing about your education thus far, or if you have one hope for your education going forward, what is it, and what accounts for it?

CHAPTER 17 Religion

- When is religion a divisive force in society? When is it most likely to produce harmony and stability?
- In what ways are or aren't all religions equal?
- Why do you think people change churches or drop their religious affiliation altogether?

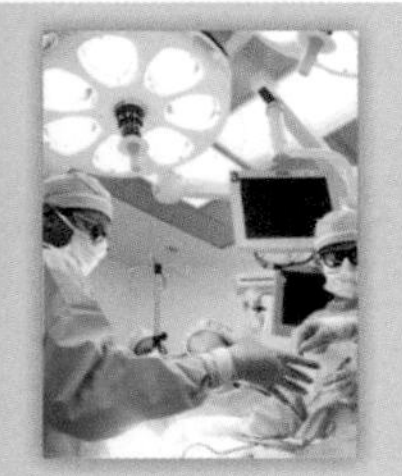

CHAPTER 18 Health, Health Care, and Disability

- Is affordable health care a human right?
- Where do you currently receive the health care you need, and how is it paid for?
- Do you know anyone with, or do you yourself have, a disability? If so, what are the challenges associated with it?

CHAPTER 19 Population and Urbanization

- Did either you or your parents emigrate to the United States from another country? How about your grandparents or an earlier generation?
- What are some of the primary benefits and challenges of residing where you do?
- Do you think that you might want to live in a different kind of community some day? Why or why not?

CHAPTER 20 Collective Behavior, Social Movements, and Social Change

- Do you recycle? If yes, why did you start doing it? How is recycling related to larger social concerns such as the environment?
- Have you ever marched in the street or gone to a rally with other people for a cause of any kind?
- If yes, for what cause and why? If not, why not?

is committed to
vailable, with the support
ormat you choose.

Learning Solutions – Training & Support

We're your partner in the classroom – online and off.

CengageCourse

CengageCourse delivers dynamic, interactive ways to teach and learn using relevant, engaging content with an accent on flexibility and reliability. When you select the **CengageCourse** solution that matches your needs, you'll discover ease and efficiency in teaching and course management.

CourseCare

CourseCare is a revolutionary program providing you with exceptional services and support to integrate your Cengage Learning Digital Solution into your course. Our dedicated team of digital solutions experts will partner with you to design and implement a program built around your course needs. We offer in-depth, one-on-one professional training of our programs and access to a 24/7 Technical Support website. **CourseCare** provides one-on-one service every step of the way—from finding the right solution for your course to training to ongoing support—helping you to drive student engagement.

TEAMUP

For more than a decade, **TeamUP Faculty Programs** have helped faculty reach and engage students through peer-to-peer consultations, workshops, and professional development conferences. Our team of **Faculty Programs Consultants and Faculty Advisors** provide implementation training and professional development opportunities customized to your needs.

Access, Rent, Save, and Engage.

Save up to 60%

CENGAGEbrain.com

At CengageBrain.com students will be able **to save up to 60%** on their course materials through our full spectrum of options. Students will have the option to **Rent** their textbooks, purchase print **textbooks, eTextbooks,** or individual **eChapters** and **Audio Books** all for substantial savings over average retail prices.

CengageBrain.com also includes single sign-on access to Cengage Learning's broad range of homework and study tools, and features a selection of free content.

Diana Kendall

Sociology in Our Times, 9th Edition

ISBN 13: 978-1-111-83157-8
Copyright 2013

reflects **our commitment** to you and your students:

CENTERING ON
Value

BBC Motion Gallery, ABC® videos, PowerLecture™ for easy lecture preparation, and other teaching and learning supplements are available to enhance your course.

CENTERING ON
Choice

Choose the format that best suits the needs of you and your students:

- Loose-leaf format
- Cengage Learning eBook
- Chapter-by-chapter purchase
- Textbook rental

CENTERING ON
Engagement

Innovative, interactive media resources such as **CengageNOW™** and **CourseMate** engage students in the excitement of sociology.

9 edition

SOCIOLOGY in Our Times

Diana Kendall

Baylor University

Australia • Brazil • Japan • Korea • Mexico • Singapore • Spain • United Kingdom • United States

***Sociology in Our Times:* Ninth Edition**
Diana Kendall

Sponsoring Editor: Erin Mitchell
Developmental Editor: Renee Deljon/Kristin Makarewycz
Freelance Development Editor: Tricia Louvar
Assistant Editor: Linda Stewart
Editorial Assistant: Mallory Ortberg
Media Editor: Mary Noel
Marketing Manager: Andrew Keay
Marketing Assistant: Jack Ward
Marketing Communications Manager: Laura Localio
Content Project Manager: Cheri Palmer
Design Director: Rob Hugel
Art Director: Caryl Gorska
Print Buyer: Judy Inouye
Rights Acquisitions Specialist: Dean Dauphinais
Production Service: Greg Hubit Bookworks
Text Designer: Diane Beasley
Photo Researcher: Julie Caruso, PreMedia Global
Text Researcher: Pablo D'Stair
Copy Editor: Donald Pharr
Proofreader: Debra Nichols
Illustrator: MPS Limited, a Macmillan Company
Cover Designer: Design is Play
Cover Image: Design is Play
Compositor: MPS Limited, a Macmillan Company

Box-Feature Icon Credits

Sociology and Everyday Life (Questions): Doug Menuez/Photodisc/Getty Images; (Answers): Jack Hollingsworth/ Photodisc/Getty Images

Sociology Works!: blue jean images/Getty Images

Sociology and Social Policy: Sami Sarkis Photodisc/Getty Images

Media Framing: Nick Koudis/Photodisc/Getty Images

Sociology in Global Perspective: Artifacts Images/Photodisc/Getty Images

You Can Make a Difference: Odilon Dimier/PhotoAlto Agency RF Collections/Getty Images

Census Profiles: © Andrew Johnson/istockphoto.com

Library of Congress Control Number: 2011925980

Student Edition:

ISBN-13: 978-1-111-83157-8

ISBN-10: 1-111-83157-2

Loose-leaf Edition:

ISBN-13: 978-1-111-83247-6

ISBN-10: 1-111-83247-1

Wadsworth
20 Davis Drive
Belmont, CA 94002-3098
USA

Cengage Learning is a leading provider of customized learning solutions with office locations around the globe, including Singapore, the United Kingdom, Australia, Mexico, Brazil, and Japan. Locate your local office at **www.cengage.com/global.**

Cengage Learning products are represented in Canada by Nelson Education, Ltd.

To learn more about Wadsworth, visit **www.cengage.com/wadsworth**
Purchase any of our products at your local college store or at our preferred online store **www.CengageBrain.com**

Printed in the United States of America
1 2 3 4 5 6 7 15 14 13 12 11

brief contents

contents

FEATURES

4 Socialization 90

FEATURES

part 2 Social Groups and Social Control

5 Society, Social Structure, and Interaction in Everyday Life 118

FEATURES

PRICE REDUCED
FORECLOSURE
BANK OWNED

FEATURES

11 Sex and Gender 310

FEATURES

12 Aging and Inequality Based on Age 346

15 Families and Intimate Relationships 432

FEATURES

16 Education 460

FEATURES

17 Religion 488

Love
Thy
Neighbor!

features

sociology and everyday life

sociology in global perspective

sociology and social policy

media framing

ing a meaningful, concrete context within which to learn. Specifically, it presents the stories—the *lived experiences*—of real individuals and the social issues they face while discussing a diverse array of classical and contemporary theory and examining interesting and relevant research. The first-person commentaries that open and are revisited throughout each chapter show students how sociology can help them understand the important questions and social issues that these other individuals face and that they themselves may face as well.

only a collection of concepts and theories but also a field that can make a difference in their lives, their communities, and the world at large.

What's New to the Ninth Edition?

The ninth edition builds on the best of previous editions while offering new insights, learning tools, and opportunities to apply the content of each chapter to relevant sociological issues and major concerns

sociology works!

census profiles

this text helps students consider the significance of

Focus on Making a Difference

the classroom, methods for incorporating film into the classroom, and learning outcomes. The authors give a synopsis of various films and a description of what sociological concept each one demonstrates. Accompanying each feature film is an activity for students to complete.

CourseReader for Sociology. CourseReader for Sociology allows you to create a fully customized online reader in minutes. You can access a rich collection of thousands of primary and secondary sources, readings, and audio and video selections from multiple disciplines. Each selection includes a descriptive introduction that puts it into context, and every selection is further supported by both critical-thinking and multiple-choice questions designed to reinforce key points. This easy-to-use solution allows you to select exactly the content you need for your courses, and it is loaded with convenient pedagogical features, such as highlighting, printing, note taking, and downloadable MP3 audio files for each reading. You have the freedom to assign and customize individualized content at an affordable price. CourseReader is the perfect complement to any class.

Careers in Sociology Module. Written by leading author Joan Ferrante, Northern Kentucky University, the Careers in Sociology module offers the most extensive and useful information on careers that is available. This module provides six career tracks, each of which has a "featured employer," a job description, and a letter of recommendation (written by a professor for a sociology student) or application (written by a sociology student). The module also includes résumé-building tips on how to make the most out of being a sociology major and offers specific course suggestions along with the transferable skills gained by taking them. As part of Wadsworth's Add-a-Module Program, Careers in Sociology can be purchased separately, bundled, or customized with any of our introductory texts. The modules present topics not typically covered in most introductory texts but often requested by instructors.

Sociology of Sports Module. The Sociology of Sports module, authored by Jerry M. Lewis, Kent State University, examines why sociologists are interested in sports, mass media and sports, popular culture and sports (including feature-length films on sports), sports and religion, drugs and sports, and violence and sports. As part of Wadsworth's Add-a-Module Program, Sociology of Sports can be purchased separately, bundled, or customized with any of our introductory texts. The modules present topics not typically covered in most introductory texts but often requested by instructors.

Rural Sociology Module. The Rural Sociology module, authored by Carol A. Jenkins, Glendale Community College–Arizona, presents the realities of life in rural America. Many people imagine a rural America characterized by farming, similar cultures, and close-knit communities. However, rural Americans and rural communities are extremely diverse—demographically, culturally, socially, economically, and environmentally. The module presents these characteristics of rural life in a comprehensive and accessible format for introductory sociology students. As part of Wadsworth's Add-a-Module program, Rural Sociology can be purchased separately, bundled, or customized with any of our introductory sociology texts. The modules present topics not typically covered in most introductory texts but often requested by instructors.

Supplements for Students

Sociology CourseMate. The more you study, the better the results. Make the most of your study time by accessing everything you need to succeed in one place. Read your textbook, take notes, review flash cards, watch videos, and take practice quizzes—online with CourseMate. Go to **www.cengagebrain.com** and search for your book.

CengageNOW™. CengageNOW offers all of your learning resources in one intuitive program. CengageNOW's intuitive "tabbed" design allows you to navigate to all key functions with a single click, and a unique home page tells you just what needs to be done and when. CengageNOW provides you access to an integrated eBook, interactive tutorials, videos, and animations that help you get the most out of your course. Go to **www.cengagebrain.com** for more information.

Study Card for Intro Sociology. This handy card, created by Matisa Wilbon, Bellarmine University, provides all the important sociological concepts covered in introductory sociology courses, broken down into sections. Providing a large amount of information at a glance, this study card is an invaluable tool for a quick review.

Study Guide. Prepared by Mark Melder, Louisiana Tech University, this student study tool contains both brief and detailed chapter outlines, chapter summaries, learning objectives, a list of key terms and key people with page references to the text, questions to guide student reading, student projects/activities, Internet and InfoTrac exercises, and practice tests consisting of more than 65 questions per chapter. All multiple-choice, true/false, and fill-in-the-blank and short-answer questions include answer explanations and page references to the text. This edition also features integration of media through podcasts, video links, and social media.

Acknowledgments

Sociology in Our Times, ninth edition, would not have been possible without the insightful critiques of these colleagues, who have reviewed some or all of this book. My profound thanks to each one for engaging in this time-consuming process:

David Eller, Community College of Denver
Mary Grisby, University of Missouri–Columbia
Christine Janis, Northern Illinois University
Minu Mather, College of San Mateo
Erin Niclaus, Bucks County Community College
Scott Powell, Ivy Tech Community College–Wabash Valley
Marie Sheneman, Itawamba Community College
Jennifer Sumerel, Itawamba Community College

I deeply appreciate the energy, creativity, and dedication of the many people responsible for the development and production of this edition of *Sociology in Our Times.* I wish to thank Wadsworth Publishing Company's Linda Schreiber-Ganster, Erin Mitchell, Renee Deljon, Kristin Makarewycz, Andrew Keay, Jack Ward, Laura Localio, Linda Stewart, Melanie Cregger, and Mallory Ortberg for their enthusiasm and insights throughout the development of this text. I am also deeply appreciative of the assistance and encouragement I received from freelance development editor Tricia Louvar. Many other people worked hard on the production of this ninth edition, especially Cheri Palmer, Greg Hubit, and Donald Pharr, without whom this book would not be possible. I am extremely grateful to them.

I invite you to send your comments and suggestions about this book to me in care of:

Wadsworth/Cengage Learning
20 Davis Drive
Belmont, CA 94002

about the author

DIANA KENDALL is currently Professor of Sociology at Baylor University, where she was awarded the title of Outstanding Professor. Dr. Kendall has taught a variety of courses, including Introduction to Sociology; Sociological Theory (undergraduate and graduate); Sociology of Medicine; Sociology of Law; and Race, Class, and Gender. Previously, she enjoyed many years of teaching sociology and serving as chair of the Social and Behavioral Science Division at Austin Community College.

Diana Kendall received her Ph.D. from the University of Texas at Austin, where she was invited to membership in Phi Kappa Phi Honor Society. Her areas of specialization and primary research interests are sociological theory and the sociology of medicine. In addition to *Sociology in Our Times,* ninth edition, she is the author of *Sociology in Our Times: The Essentials.* She is also the author of *The Power of Good Deeds: Privileged Women and the Social Reproduction of the Upper Class* (Rowman & Littlefield, 2002); *Framing Class: Media Representations of Wealth and Poverty in America,* second edition (Rowman & Littlefield, 2011); and *Members Only: Elite Clubs and the Process of Exclusion* (Rowman & Littlefield, 2008). Professor Kendall is actively involved in national and regional sociological associations, including the American Sociological Association, Sociologists for Women in Society, the Society for the Study of Social Problems, the Southwestern Sociological Association, and the Law and Society Association.

SOCIOLOGY in Our Times

1 The Sociological Perspective

What I enjoyed about college was that I was able to walk away with a degree and go find a job, but what I regret most is getting a credit card, racking it up and getting multiple credit cards and doing the same thing, 'cause now I have to deal with it and I'm paying it off now and it's kinda hard to deal with. Things that I charged on my credit card in college were those spring break vacations, going out to eat with friends numerous times. Other things were like materialistic things like clothes, accessories, makeup—all that good stuff—trying to keep up with everyone else. [Slight laugh.] I wish I could do those things now. Now, I can't have those things; I have to do with what I've got. . . . I can't enjoy the things I enjoyed in college because I enjoyed them in college. I guess when I was making the purchases in college with my credit card saying, "Oh, I can just pay that off later," I figured I would be making more money than what I was given through financial aid and through my parents, [but] in reality, you're not. You have to compensate for other things like tax being taken out of your salary, groceries, gas is something I didn't even think about because my parents always paid it. I mean, all those little things: They will add up!

—Robyn Beck (featured in YouTube, 2011), a college graduate who is struggling to pay off $7,000 in credit card debt with high interest rates, explains why she must use 20 percent of her postgraduation take-home salary to try to reduce this debt (to see her interview, go to http://www.youtube.com/watch?v=7U6pmkTC8i0).

Digital Vision/Getty Images

▲ Young people who run up credit card debt may eventually find that paying off this debt can take decades.

Chapter Focus Question:

How does sociology add to our knowledge of human societies and of social issues such as consumerism?

Like millions of college students in the United States and other high-income nations, Robyn Beck quickly learned both the liberating and constraining aspects of living in a "consumer society" where many of us rely on our credit cards to pay for items we want to purchase or services we need. Many college students continue to increase their credit card purchases to pay for expenses and online shopping, which has become more popular than going to the mall. For many years companies targeted college students, trying to get them to apply for a credit card regardless of whether they had the ability to pay off their balances or not. However, in 2009 Congress passed the Credit Card Accountability, Responsibility, and Disclosure (CARD) Act, making it illegal for companies to offer credit cards to individuals under twenty-one years of age unless they can prove that they can pay their bills or unless an adult (age twenty-one or over) cosigns the application.

Why are sociologists interested in studying consumerism? Sociologists study the *consumer society*—a society in which discretionary consumption is a mass phenomenon among people across diverse income categories—because it provides interesting and important insights into many aspects of social life and our world. In the consumer society, for example, purchasing goods and services is not limited to the wealthy or even the middle class; people in all but the lowest income brackets spend time, energy, and money shopping, and some amass large debts in the process (see Baudrillard, 1998/1970; Ritzer, 1995; Schor, 1999). According to sociologists, shopping and consumption—in this instance, the money that people spend on goods and services—are processes that extend beyond our individual choices and are rooted in larger structural conditions in the social, political, and economic order in which we live. In the 2000s many people have had financial problems not only because of their own consumerism but also because of national and global economic instability.

Why have shopping, spending, credit card debt, and bankruptcy become major problems for some people? How are social relations and social meanings shaped by what people in a given society produce and how they consume? What national and worldwide social processes shape the production and consumption of goods, services, and information? In this chapter we see how the sociological perspective helps us examine complex questions such as these, and we wrestle with some of the difficulties of attempting to study human behavior. Before reading on, take the Sociology and Everyday Life quiz, which lists a number of commonsense notions about consumption and credit card debt.

Highlighted Learning Objectives

- Explain why the sociological imagination is important for studying society.
- Discuss why early social thinkers were concerned with social order and stability.
- Identify reasons why later social thinkers were concerned with social change.
- Describe the key assumptions behind each of the contemporary theoretical perspectives.

Putting Social Life into Perspective

***Sociology* is the systematic study of human society and social interaction.** It is a *systematic* study because sociologists apply both theoretical perspectives and research methods (or orderly approaches) to examinations of social behavior. Sociologists study human societies and their social interactions to develop theories of how human behavior is shaped by group life and how, in turn, group life is affected by individuals. Sociological studies range in size from a focus on entire nations and large-scale organizations and institutions to an analysis of small groups and individual interactions. To better understand the scope of sociology, you might compare it to other social sciences, such as anthropology, psychology, economics, and political science. Like anthropology, sociology studies many aspects of human behavior; however, sociology is particularly interested in contemporary social organization, relations, and social change. Anthropology primarily concentrates on human existence over geographic space and evolutionary time, meaning that it focuses more on traditional societies and the development of diverse cultures. Cultural anthropology most closely overlaps sociology. Unlike psychology, sociology examines the individual in relation to *external* factors, such as the effects of groups, organizations, and social institutions on individuals and social life; psychology primarily focuses on *internal* factors relating to the individual in explanations of human behavior and mental processes—what occurs in the mind. Social psychology is similar to sociology in that it emphasizes how social conditions affect individual behavior. Although sociology examines all major social institutions, including the economy and politics, the fields of economics and political science concentrate primarily on a single institution—the economy or the political system. Topics of mutual interest to economics and sociology include issues such as consumerism and debt, which can be analyzed at global, national, and individual levels. Topics of mutual interest to political science and sociology are how political systems are organized and how power is distributed in society. As you can see, sociology shares similarities with other social sciences but offers a comprehensive approach to understanding many aspects of social life.

Why Should You Study Sociology?

Sociology helps us gain a better understanding of ourselves and our social world. Sociology offers us new insights into our lives as well as an opportunity to learn about other people. It enables us to see how behavior is largely shaped by the groups to which we belong and the society in which we live. By studying sociology, you can gain valuable new tools that will help you in daily life and provide important insights into your interactions with others.

Most of us take our social world for granted and view our lives in very personal terms. Because of our culture's emphasis on individualism, we often do not consider the complex connections between our lives and the larger, recurring patterns of the society and world in which we live. Sociology helps us look beyond our personal experiences and gain insights into society and the larger world order. A ***society* is a large social grouping that shares the same geographical territory and is subject to the same political authority and dominant cultural expectations,** such as the United States, Mexico, or Nigeria. Examining the world order helps us understand that each of us is affected by *global interdependence*—a relationship in which the lives of all people are intertwined closely and any one nation's problems are part of a larger global problem.

Individuals can make use of sociology on a more personal level. Sociology enables us to move beyond established ways of thinking, thus allowing us to gain new insights into ourselves and to develop a greater awareness of the connection between our own "world" and that of other people. According to the sociologist Peter Berger (1963: 23), sociological inquiry helps us see that "things are not what they seem." Sociology provides new ways of approaching problems and making decisions in everyday life. Sociology also promotes understanding and tolerance by enabling each of us to look beyond our personal experiences (see ▶ Figure 1.1).

Many of us rely on intuition or common sense gained from personal experience to help us understand our daily lives and other people's behavior. *Commonsense knowledge* guides ordinary conduct in everyday life. We often rely on common sense—or "what everybody knows"—to answer key questions about behavior: Why do people behave the way they do? Who makes the rules? Why do some people break rules and other people follow rules?

Many commonsense notions are actually myths. A *myth* is a popular but false notion that may be used, either intentionally or unintentionally, to perpetuate certain beliefs or "theories" even in the light of conclusive evidence to the contrary. For example, one widely held myth is that "money can buy happiness." By contrast, sociologists strive to use scientific standards, not popular myths or hearsay, in studying society and social interaction. They use systematic research techniques and are accountable to the scientific community for their methods and the presentation of their findings. Although some

sociology and everyday life

How Much Do You Know About Consumption and Credit Cards?

True	False	
T	F	1. The average U.S. household owes more than $10,000 in credit card debt.
T	F	2. The average debt owed on undergraduate college students' credit cards is less than $1,000.
T	F	3. Fewer than half of all undergraduate students at four-year colleges have at least one credit card.
T	F	4. College students spend more money online than people in any other age category.
T	F	5. Consumer activist groups have been successful in getting Congress to pass a law requiring people under age 21 to get parental approval or show that they have sufficient income prior to obtaining a credit card.
T	F	6. More than one million people in this country file for bankruptcy each year.
T	F	7. If we added up all consumer debt in the United States, we would find that the total amount owed is more than $1.5 trillion.
T	F	8. Overspending is primarily a problem for people in the higher-income brackets in the United States and other affluent nations.

Answers on page 6.

Health and Human Services	Business	Communication	Academia	Law
Counseling Education Medicine Nursing Social Work	Advertising Labor Relations Management Marketing	Broadcasting Public Relations Journalism	Anthropology Economics Geography History Information Studies Media Studies/ Communication Political Science Psychology Sociology	Law Criminal Justice

▲ **FIGURE 1.1 FIELDS THAT USE SOCIAL SCIENCE RESEARCH**
In many careers, including jobs in health and human services, business, communication, academia, and law, the ability to analyze social science research is an important asset.
Source: Based on Katzer, Cook, and Crouch, 1991.

sociologists argue that sociology must be completely value free—without distorting subjective (personal or emotional) bias—others do not think that total objectivity is an attainable or desirable goal when studying human behavior. However, all sociologists attempt to discover patterns or commonalities in human behavior. For example, when they study shopping behavior or credit card abuse, sociologists look for recurring patterns of behavior and for larger, structural factors that contribute to people's behavior. Women's studies scholar Juliet B. Schor, who wrote *The Overspent American* (1999: 68), refers to consumption as the "see–want–borrow–buy" process, which she believes is a comparative process in which desire is structured by what we see around us. As sociologists examine patterns such as these, they begin to use the sociological imagination.

The Sociological Imagination

How can we make a connection between our personal experiences and what goes on in the larger society? Sociologist C. Wright Mills (1959b) described the process of making this linkage the ***sociological imagination*—the ability to see the relationship**

sociology the systematic study of human society and social interaction.

society a large social grouping that shares the same geographical territory and is subject to the same political authority and dominant cultural expectations.

sociological imagination C. Wright Mills's term for the ability to see the relationship between individual experiences and the larger society.

sociology and everyday life

ANSWERS to the Sociology Quiz on Consumption and Credit Cards

1. True. The credit card debt owed by the average U.S. household (that had credit card debt) in 2009 was $15,788.

2. False. The average debt on undergraduate college students' credit cards in 2009 was about $3,173.

3. False. About 84 percent of undergraduate college students have at least one credit card.

4. True. College students are the biggest spenders online, and most of the purchases are for clothing.

5. True. Aggressive marketing of credit cards to college students is illegal. The 2009 Credit Card Accountability, Responsibility, and Disclosure (CARD) Act made it illegal for banks and credit card companies to continue the business practice of routinely sending out mailings and engaging in campus solicitations for new cardholders.

6. True. In the United States, 1.4 million people filed for bankruptcy in 2009.

7. True. The U.S. total was about $2.5 trillion when all consumer debt was taken into account in 2009, amounting to nearly $8,100 in debt for every person living in this country.

8. False. People in all income brackets have problems with overspending, including excessive use of credit and not paying off debts in a timely manner, at least partly due to difficult economic times.

Sources: Based on SallieMae.com, 2009; and Woolsey and Schulz, 2010.

between individual experiences and the larger society. This sociological awareness enables us to understand the link between our personal experiences and the social contexts in which they occur. The sociological imagination helps us distinguish between personal troubles and social (or public) issues. *Personal troubles* are private problems that affect individuals and the networks of people with which they regularly associate. As a result, those problems must be solved by individuals within their immediate social settings. For example, one person being unemployed or running up a high credit card debt could be identified as a personal trouble. *Public issues* are problems that affect large numbers of people and often require solutions at the societal level. Widespread unemployment and massive, nationwide consumer debt are examples of public issues. The sociological imagination helps us place seemingly personal troubles, such as losing one's job or overspending on credit cards, into a larger social context, where we can distinguish whether and how personal troubles may be related to public issues.

Overspending as a Personal Trouble Although individual behavior can contribute to social problems, our individual experiences are largely beyond the individual's control. They are influenced and in some situations determined by the society as a whole—by its historical development and its organization. In everyday life we often blame individuals for "creating" their own problems. If a person sinks into debt due to overspending or credit card abuse, many people consider it to be the result of his or her own personal failings. However, this approach overlooks debt among people who are in low-income brackets, having no way other than debt to gain the basic necessities of life. By contrast, at middle- and upper-income levels, overspending takes on a variety of other meanings.

At the individual level, people may accumulate credit cards and spend more than they can afford, thereby affecting all aspects of their lives, including

© altrendo images/Getty Images

▲ Because of an over-reliance on credit, many Americans now owe more than they can pay back. This couple is signing up for debt consolidation, a somewhat controversial process that may help them avoid bankruptcy.

health, family relationships, and employment stability. Sociologist George Ritzer (1999: 29) suggests that people may overspend through a gradual process in which credit cards "lure people into consumption by easy credit and then entice them into still further consumption by offers of 'payment holidays,' new cards, and increased credit limits."

Overspending as a Public Issue We can use the sociological imagination to look at the problem of overspending and credit card debt as a public issue—a societal problem. For example, Ritzer (1998) suggests that the relationship between credit card debt and the relatively low *savings rate* in the United States constitutes a public issue. Between 2000 and 2010, credit card debt continued to grow rapidly in the United States while savings diminished. Because savings is money that governments, businesses, and individuals can borrow for expansion, a lack of savings may create problems for future economic growth. The rate of bankruptcies in this country is a problem both for financial institutions and the government. As corporations "write off" bad debt from those who declare bankruptcy or simply do not pay their bills, all consumers pay either directly or indirectly for that debt. Finally, poverty is forgotten as a social issue when more-affluent people are having a spending holiday and consuming all, or more than, they can afford to purchase. Some practices of the credit card industry are also a public issue because they harm consumers. Companies may encourage overspending and then substantially increase interest rates and other fees, making it more difficult for consumers to pay off debts. Mills's *The Sociological Imagination* (1959b) is useful for examining issues because it helps integrate microlevel (individual and small-group) troubles with compelling public issues of our day. Recently, his ideas have been applied at the global level as well.

The Importance of a Global Sociological Imagination

Although existing sociological theory and research provide the foundation for sociological thinking, we must reach beyond past studies that have focused primarily on the United States to develop a more comprehensive *global* approach for the future. In the twenty-first century, we face important challenges in a rapidly changing nation and world. The world's ***high-income countries*** **are nations with highly industrialized economies; technologically advanced industrial, administrative, and service occupations; and relatively high levels of national and personal income.** Examples include the United States, Canada, Australia, New Zealand, Japan, and the countries of Western Europe (see ▶ Map 1.1).

As compared with other nations of the world, many high-income nations have a high standard of living and a lower death rate due to advances in nutrition and medical technology. However, everyone living in a so-called high-income country does not necessarily have a high income or an outstanding quality of life. Even among middle-income and upper-income people, problems such as personal debt may threaten economic and social stability.

In contrast, ***middle-income countries*** **are nations with industrializing economies, particularly in urban areas, and moderate levels of national and personal income.** Examples of middle-income countries include the nations of Eastern Europe and many Latin American countries, where nations such as Brazil and Mexico are industrializing rapidly. ***Low-income countries*** **are primarily agrarian nations with little industrialization and low levels of national and personal income.** Examples of low-income countries include many of the nations of Africa and Asia, particularly the People's Republic of China and India, where people typically work the land and are among the poorest in the world. However, generalizations are difficult to make because there are wide differences in income and standards of living within many nations (see Chapter 9, "Global Stratification").

The global expansion of credit cards and other forms of consumerism, including the proliferation of "big-box" retail establishments such as Wal-Mart, shows the influence of U.S.-based megacorporations on other nations of the world. Consider Wal-Mart, for example. Sam Walton opened his first Wal-Mart store in Rogers, Arkansas, in 1962, and the company's home office was established in Bentonville, Arkansas, in the early 1970s. From a small-scale, regional operation in Arkansas, the Wal-Mart chain has now built a worldwide empire. Although the global expansion of credit cards and Wal-Mart Supercenters has produced benefits for some people, it has also affected the everyday lives of many individuals around the world (see "Sociology in Global Perspective").

high-income countries (sometimes referred to as **industrial countries**) nations with highly industrialized economies; technologically advanced industrial, administrative, and service occupations; and relatively high levels of national and personal income.

middle-income countries (sometimes referred to as **developing countries**) nations with industrializing economies and moderate levels of national and personal income.

low-income countries (sometimes referred to as **underdeveloped countries**) primarily agrarian nations with little industrialization and low levels of national and personal income.

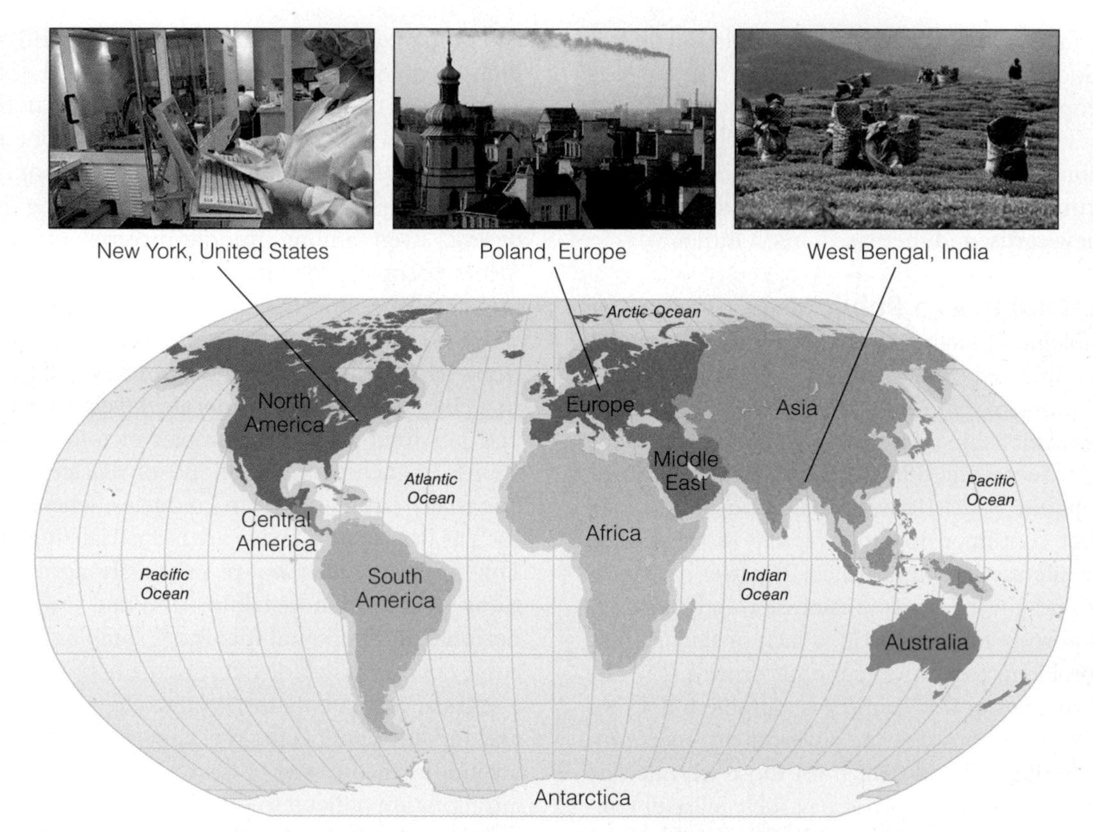

▲ **MAP 1.1 THE WORLD'S ECONOMIES IN THE EARLY TWENTY-FIRST CENTURY**
High-income, middle-income, and low-income countries.
Photos: Left, © Syracuse Newspapers/John Berry/The Image Works; center, © Andrew Ward/Life File/Getty Images; right, © DEA/M. BORCHI/Getty Images

Throughout this text we will continue to develop our sociological imaginations by examining social life in the United States and other nations. The future of our nation is deeply intertwined with the future of all other nations of the world on economic, political, environmental, and humanitarian levels. Whatever your race/ethnicity, class, sex, or age, are you able to include in your thinking the perspectives of people who are quite different from you in experiences and points of view? Before you answer this question, a few definitions are in order. *Race* is a term used by many people to specify groups of people distinguished by physical characteristics such as skin color; in fact, there are no "pure" racial types, and the concept of race is considered by most sociologists to be a social construction that people use to justify existing social inequalities. *Ethnicity* refers to the cultural heritage or identity of a group and is based on factors such as language or country of origin. *Class* is the relative location of a person or group within the larger society, based on wealth, power, prestige, or other valued resources. *Sex* refers to the biological and anatomical differences between females and males. By contrast, *gender* refers to the meanings, beliefs, and practices associated with sex differences, referred to as *femininity* and *masculinity*.

In forming your own global sociological imagination and in seeing the possibilities for sociology in the twenty-first century, it will be helpful for you to understand the development of the discipline.

The Origins of Sociological Thinking

Throughout history, social philosophers and religious authorities have made countless observations about human behavior, but the first systematic analysis of society is found in the philosophies of early Greek philosophers such as Plato (c. 427–347 B.C.E.) and Aristotle (384–322 B.C.E.). For example, Aristotle was concerned with developing a system of knowledge, and he engaged in theorizing and the empirical analysis of data collected from people in Greek cities regarding their views about social life when ruled by kings or aristocracies or when living in democracies (Collins, 1994). However, early thinkers such as Plato and Aristotle provided thoughts on what they believed society *ought* to be like, rather than describing how society actually *was*.

Social thought began to change rapidly in the seventeenth century with the scientific revolution. Like their predecessors in the natural sciences, social thinkers sought to develop a scientific understanding of social life, believing that their work might enable people to reach their full potential. The contributions of Isaac Newton (1642–1727) to modern science,

sociology in global perspective

The Global Wal-Mart Effect? Big-Box Stores and Credit Cards

Did you know that:

- More than half of all Wal-Mart stores worldwide are located outside the United States?
- Wal-Mart operates nearly 300 stores, including supercenters, neighborhood markets, and Sam's Clubs, in China?
- Wal-Mart is a major player in the credit card business in China, where people in the past were opposed to buying anything on credit?

Although most of us are aware that Wal-Mart stores are visible in virtually every city in the United States, we are less aware of the extent to which Wal-Mart and other "big-box" stores are changing the face of the world economy as the megacorporations that own them expand their operations into other nations and into the credit card business.

The strategic placement of Wal-Mart stores both here and abroad accounts for part of the financial success of this retailing giant, but another U.S. export—credit cards—is also part of the company's business plan. Credit cards are changing the way that people shop and how they think about spending money in emerging nations such as China. For example, Wal-Mart China is aggressively seeking both shoppers and credit card holders. By encouraging people to spend money now rather than save it for later, corporations such as Wal-Mart that issue "co-branded" credit cards gain in two ways: (1) people buy more goods than they would otherwise acquire, thus increasing sales; and (2) the corporation whose "brand" is on the credit card increases its earnings as a result of the interest the cardholder pays on credit card debt. Co-branded credit cards are issued jointly by a bank (which provides the credit) and a business (such as an airline or a retailer) that offers some sort of reward for using its credit card.

The motto for the Wal-Mart credit card in China is "Maximizing value, enjoying life," and this idea encourages a change in attitude from the past, when—regardless of income level—most residents of that country did not possess credit cards. This has brought a corresponding surge in credit card debt, which can be partly attributed to aggressive marketing by transnational retailers but also to credit card companies encouraging consumers to buy now, pay later.

reflect & analyze

If theorists such as Karl Marx were alive today, how might they analyze consumerism and debt in the twenty-first century? What new insights on the global economy does our sociological imagination give us?

Sources: Based on Walmart Corporate, 2010.

© AP Images/Eugene Hoshiko

▲ An exciting aspect of studying sociology is comparing our own lives with those of people around the world. Global consumerism, as evidenced by the opening of the new Wal-Mart Supercenter in Shanghai, China, provides a window through which we can observe how issues such as shopping and credit affect all of us. Which aspects of this photo reflect local culture? Which aspects reflect a global cultural phenomenon?

including the discovery of the laws of gravity and motion and the development of calculus, inspired social thinkers to believe that similar advances could be made in the systematic study of human behavior. As Newton advanced the cause of physics and the natural sciences, he was viewed by many as the model of a true scientist. Moreover, his belief that the universe is an orderly, self-regulating system strongly influenced the thinking of early social theorists.

Sociology and the Age of Enlightenment

The origins of sociological thinking as we know it today can be traced to the scientific revolution in the late seventeenth and mid-eighteenth centuries and to the Age of Enlightenment. In this period of European thought, emphasis was placed on the individual's possession of critical reasoning and experience. There was also widespread skepticism regarding the primacy of religion as a source of knowledge and heartfelt opposition to traditional authority. A basic assumption of the Enlightenment was that scientific laws had been designed with a view to human happiness and that the "invisible hand" of either Providence or the emerging economic system of capitalism would ensure that the individual's pursuit of enlightened self-interest would always be conducive to the welfare of society as a whole.

In France, the Enlightenment (also referred to as the *Age of Reason*) was dominated by a group of thinkers referred to collectively as the *philosophes*. The philosophes included such well-known intellectuals as Charles Montesquieu (1689–1755), Jean-Jacques Rousseau (1712–1778), and Jacques Turgot (1727–1781). For the most part, these men were optimistic about the future, believing that human society could be improved through scientific discoveries. The philosophes believed that if people were free from the ignorance and superstition of the past, they could create new forms of political and economic organization such as democracy and capitalism, which would eventually produce wealth and destroy aristocracy and other oppressive forms of political leadership. Although the women of that day were categorically excluded from much of public life in France, some women strongly influenced the philosophes and their thinking through their participation in the *salon*—an open house held to stimulate discussion and intellectual debate. Salons provided a place for intellectuals and authors to discuss ideas and opinions and for people to engage in witty repartee regarding current events. However, the idea of observing how people lived in order to find out what they thought, and doing so in a systematic manner that could be verified, did not take hold until sweeping political and economic changes in the late eighteenth and early nineteenth centuries caused many people to realize that the ideas of some philosophers and theologians no longer seemed relevant. Many of these questions concerned the social upheaval brought about by the age of revolution, particularly the American Revolution of 1776 and the French Revolution of 1789, and the rapid industrialization and urbanization that occurred first in Britain, then in Western Europe, and later in the United States.

Sociology and the Age of Revolution, Industrialization, and Urbanization

Several types of revolution that took place in the eighteenth century had a profound influence on the origins of sociology. The Enlightenment produced an *intellectual revolution* in how people thought about social change, progress, and critical thinking. The optimistic views of the philosophes and other social thinkers regarding progress and equal opportunity (at least for some people) became part of the impetus for *political revolutions* and *economic revolutions*, first in America and then in France. The Enlightenment thinkers had emphasized a sense of common purpose and hope for human progress; the French Revolution and its aftermath replaced these ideals with discord and overt conflict (see Schama, 1989; Arendt, 1973).

During the nineteenth and early twentieth centuries, another form of revolution occurred: the *Industrial Revolution*. ***Industrialization* is the process by which societies are transformed from dependence on agriculture and handmade products to an emphasis on manufacturing and related industries.** This process first occurred during the Industrial Revolution in Britain between 1760 and 1850, and was soon repeated throughout Western Europe. By the mid-nineteenth century, industrialization was well under way in the United States. Massive economic, technological, and social changes occurred as machine technology and the factory system shifted the economic base of these nations from agriculture to manufacturing. A new social class of industrialists emerged in textiles, iron smelting, and related industries. Many people who had labored on the land were forced to leave their tightly knit rural communities and sacrifice well-defined social relationships to seek employment as factory workers in the emerging cities, which became the centers of industrial work.

Urbanization accompanied modernization and the rapid process of industrialization. ***Urbanization* is the process by which an increasing proportion of a population lives in cities rather than in rural areas.** Although cities existed long before the Industrial Revolution, the development of the factory system led to a rapid increase in both the number of cities and the size of their populations. People from very diverse backgrounds worked together in the same factory. At the same time, many people shifted from being *producers* to being *consumers*. For example, families living in the cities had to buy food with their wages because they could no longer grow their own crops to consume or to barter for other resources. Similarly, people had to pay rent for their lodging because they could no longer exchange their services for shelter.

These living and working conditions led to the development of new social problems: inadequate housing, crowding, unsanitary conditions, poverty, pollution, and crime. Wages were so low that entire families—including very young children—were forced to work, often under hazardous conditions and with no job security. As these conditions became more visible, a new breed of social thinkers turned their attention to trying to understand why and how society was changing.

© Hulton Archive/Getty Images

◀ As the Industrial Revolution swept through the United States beginning in the nineteenth century, sights like this became increasingly common. This early automobile assembly line is emblematic of the factory system that shifted the base of the U.S. economy from agriculture to manufacturing. What new technologies are transforming the U.S. economy in the twenty-first century?

The Development of Modern Sociology

At the same time that urban problems were growing worse, natural scientists had been using reason, or rational thinking, to discover the laws of physics and the movement of the planets. Social thinkers started to believe that by applying the methods developed by the natural sciences, they might discover the laws of human behavior and apply these laws to solve social problems. Historically, the time was ripe for such thoughts because the Age of Enlightenment had produced a belief in reason and humanity's ability to perfect itself.

Early Thinkers: A Concern with Social Order and Stability

Early social thinkers—such as Auguste Comte, Harriet Martineau, Herbert Spencer, and Emile Durkheim—were interested in analyzing social order and stability, and many of their ideas had a dramatic influence on modern sociology.

Auguste Comte The French philosopher Auguste Comte (1798–1857) coined the term *sociology* from the Latin *socius* ("social, being with others") and the Greek *logos* ("study of") to describe a new science that would engage in the study of society. Even though he never actually conducted sociological research, Comte is considered by some to be the "founder of sociology." Comte's theory that societies contain *social statics* (forces for social order and stability) and *social dynamics* (forces for conflict and change) continues to be used in contemporary sociology. He stressed that the methods of the natural sciences should be applied to the objective study of society.

Auguste Comte (1798–1857) (oil on canvas), Etex, Louis Jules (1810–1889)/Temple de la Religion de l'Humanité, Paris, France/The Bridgeman Art Library International

▲ Auguste Comte

industrialization the process by which societies are transformed from dependence on agriculture and handmade products to an emphasis on manufacturing and related industries.

urbanization the process by which an increasing proportion of a population lives in cities rather than in rural areas.

Comte's philosophy is known as ***positivism*—a belief that the world can best be understood through scientific inquiry.** Comte believed that objective, bias-free knowledge was attainable only through the use of science rather than religion. However, scientific knowledge was "relative knowledge," not absolute and final. Comte's positivism had two dimensions: (1) methodological—the application of scientific knowledge to both physical and social phenomena—and (2) social and political—the use of such knowledge to predict the likely results of different policies so that the best one could be chosen.

The ideas of Comte are deeply embedded in the discipline of sociology. Of particular importance is his idea that the nature of human thinking and knowledge passed through several stages as societies evolved from simple to more complex. Comte described how the idea systems and their corresponding social structural arrangements changed in what he termed the *law of the three stages:* the theological, metaphysical, and scientific (or positivistic) stages. Comte believed that knowledge began in the *theological stage*—explanations were based on religion and the supernatural. Next, knowledge moved to the *metaphysical stage*—explanations were based on abstract philosophical speculation. Finally, knowledge would reach the *scientific* or *positive stage*—explanations are based on systematic observation, experimentation, comparison, and historical analysis. Shifts in the forms of knowledge in societies were linked to changes in the structural systems of society. In the theological stage, kinship was the most prominent unit of society; however, in the metaphysical stage, the state became the prominent unit, and control shifted from small groups to the state, military, and law. In the scientific or positive stage, industry became the prominent structural unit in society, and scientists became the spiritual leaders, replacing in importance the priests and philosophers of the previous stages of knowledge. For Comte, this progression through the three stages constituted the basic law of social dynamics, and, when coupled with the laws of statics (which emphasized social order and stability), the new science of sociology could bring about positive social change.

Harriet Martineau Comte's works were made more accessible for a wide variety of scholars through the efforts of the British sociologist Harriet Martineau (1802–1876). Until recently, Martineau received no recognition in the field of sociology, partly because she was a woman in a male-dominated discipline and society. Not only did she translate and condense Comte's work, but she was also an active sociologist in her own right. Martineau studied the social customs

▲ Harriet Martineau

of Britain and the United States and analyzed the consequences of industrialization and capitalism. In *Society in America* (1962/1837), she examined religion, politics, child rearing, slavery, and immigration in the United States, paying special attention to social distinctions based on class, race, and gender. Her works explore the status of women, children, and "sufferers" (persons who are considered to be criminal, mentally ill, handicapped, poor, or alcoholic).

Based on her reading of Mary Wollstonecraft's *A Vindication of the Rights of Women* (1974/1797), Martineau advocated racial and gender equality. She was also committed to creating a science of society that would be grounded in empirical observations and widely accessible to people. She argued that sociologists should be impartial in their assessment of society but that it is entirely appropriate to compare the existing state of society with the principles on which it was founded (Lengermann and Niebrugge-Brantley, 1998).

Some scholars believe that Martineau's place in the history of sociology should be as a founding member of this field of study, not just as the translator of Auguste Comte's work (Hoecker-Drysdale, 1992; Lengermann and Niebrugge-Brantley, 1998). Others have highlighted her influence in spreading the idea that societal progress could be brought about by the spread of democracy and the growth of industrial capitalism (Polanyi, 1944). Martineau believed that a better society would emerge if women and men were treated equally, enlightened reform occurred, and cooperation existed among people in all social classes (but led by the middle class).

In keeping with the sociological imagination, Martineau not only analyzed large-scale social structures in society, but she also explored how these factors influenced the lives of people, particularly women, children, and those who were marginalized by virtue of being criminal, mentally ill, disabled, poor, or alcoholic (Lengermann and Niebrugge-Brantley, 1998). She remained convinced that sociology, the "true science of human nature," could bring about new knowledge and understanding, enlarging people's capacity to create a just society and live heroic lives (Hoecker-Drysdale, 1992).

Herbert Spencer Unlike Comte, who was strongly influenced by the upheavals of the French Revolution, the British social theorist Herbert Spencer (1820–1903) was born in a more peaceful and optimistic period in his country's history. Spencer's major contribution to sociology was an evolutionary perspective on social order and social change. Although the term *evolution* has various meanings, evolutionary theory should be taken to mean "a theory to explain the mechanisms of organic/social change" (Haines, 1997: 81). According to Spencer's Theory of General Evolution, society, like a biological organism, has various interdependent parts (such as the family, the economy, and the government) that work to ensure the stability and survival of the entire society.

Spencer believed that societies developed through a process of "struggle" (for existence) and "fitness" (for survival), which he referred to as the "survival of the fittest." Because this phrase is often attributed to Charles Darwin, Spencer's view of society is known as ***social Darwinism*—the belief that those species of animals, including human beings, best adapted to their environment survive and prosper, whereas those poorly adapted die out.** Spencer equated this process of *natural selection* with progress because only the "fittest" members of society would survive the competition, and the "unfit" would be filtered out of society. Based on this belief, he strongly opposed any social reform that might interfere with the natural selection process and, thus, damage society by favoring its least-worthy members.

Critics have suggested that many of Spencer's ideas contain serious flaws. For one thing, societies are not the same as biological systems; people are able to create and transform the environment in which they live. Moreover, the notion of the survival of the fittest can easily be used to justify class, racial–ethnic, and gender inequalities and to rationalize the lack of action to eliminate harmful practices that contribute to such inequalities. Not surprisingly, Spencer's "hands-off" view was applauded by many wealthy industrialists of his day. John D. Rockefeller, who gained monopolistic control of much of the U.S. oil industry early in the twentieth century, maintained that the growth of giant businesses was merely the "survival of the fittest" (Feagin, Baker, and Feagin, 2006).

Social Darwinism served as a rationalization for some people's assertion of the superiority of the white race. After the Civil War, it was used to justify the repression and neglect of African Americans as well as the policies that resulted in the annihilation of Native American populations. Although some social reformers spoke out against these justifications, "scientific" racism continued to exist (Turner, Singleton, and Musick, 1984). In both positive and negative ways, many of Spencer's ideas and concepts have been deeply embedded in social thinking and public policy for more than a century.

© Hulton Archive/Getty Images

▲ Herbert Spencer

Emile Durkheim French sociologist Emile Durkheim (1858–1917) was an avowed critic of some of Spencer's views while incorporating others into his own writing. Durkheim stressed that people are the product of their social environment and that behavior cannot be fully understood in terms of *individual* biological and psychological traits. He believed that the limits of human potential are *socially* based, not *biologically* based. As Durkheim saw religious traditions evaporating in his society, he searched for a scientific, rational way to provide for societal integration and stability (Hadden, 1997).

In *The Rules of Sociological Method* (1964a/1895), Durkheim set forth one of his most important contributions to sociology: the idea that societies are built

positivism a term describing Auguste Comte's belief that the world can best be understood through scientific inquiry.

social Darwinism Herbert Spencer's belief that those species of animals, including human beings, best adapted to their environment survive and prosper, whereas those poorly adapted die out.

sociology works!

Ahead of His Time: Marx and Alienation

The texture of discontent (or lack thereof) can say a lot about a nation, and that Americans today are less likely to rebel may not be an entirely positive sign.

> It certainly doesn't mean we have more love, patience or tolerance for one another. Indeed, it may mean just the opposite, that we tend not to trust one another and that we are more alienated from our neighbors than ever before. The lack of direct action could signal the weakening of a social contract that keeps people meaningfully invested in the fate of our country—which may, in turn, be hindering our ability to resolve this [financial] crisis.
>
> —Sudhir Venkatesh (2009: WK10), a contemporary sociologist who sees alienation as one possible explanation of why many people are not more actively protesting against the actions of high-powered individuals and organizations that have contributed to our current financial plight

Social scientists have long been fascinated by alienation. This concept is often attributed to the economist and philosopher Karl Marx. As further discussed in Chapter 8, *alienation* is a term used to refer to an individual's feeling of *powerlessness* and *estrangement* from other people and from oneself. Marx specifically linked alienation to social relations that are inherent in capitalism; however, more-recent social thinkers have expanded his ideas to include social psychological feelings of powerlessness, meaninglessness, and isolation. These may be present because people experience social injustice and vast economic inequalities in contempo-

In Venkatesh's explanation of why people are not more rebellious when they are grappling with issues of corporate malfeasance and their own economic hardship, Venkatesh states that we hear a lot about "populist rage" but actually see very few protests or acts of outright rebellion: People may feel alienated, but they do not sense that there is much that they can do about the problem. They also do not feel strong social ties with other individuals that might lead them to bond together for joint action. According to Venkatesh (2009), rather than coming together for social action, we often express our individual frustrations on social networks, radio or TV call-in shows, and our cell phones. According to Venkatesh, "With headsets on and our hands busily texting, we are less aware of one another's behavior in public space." And this lack of awareness contributes to, rather than reduces, our alienation from one another and from the larger society of which we are a part.

If we apply the earlier theorizing of Marx regarding alienation to the contemporary views of Venkatesh on the weakening social contract, we gain new insights on how sociology works. Today, we can view pressing social issues as important problems that we must all work together to solve. We should come together to talk about how we might solve problems rather than continuing to live in our own isolated social worlds, where many individuals feel alienated from others.

reflect & analyze

Why are we often more concerned about trivial matters, such as who will be the big winner in a sporting event or on a reality TV show, than we are about how to address our most pressing social and economic concerns? What can we learn from Marx's concept of alienation to show us that

▲ Emile Durkheim

on social facts. ***Social facts* are patterned ways of acting, thinking, and feeling that exist *outside* any one individual but that exert social control over each person.** Durkheim believed that social facts must be explained by other social facts—by reference to the social structure rather than to individual attributes.

Durkheim was concerned with social order and social stability because he lived during the period of rapid social changes in Europe resulting from industrialization and urbanization. His recurring question was this: How do societies manage to hold together? In *The Division of Labor in Society* (1933/1893), Durkheim concluded that preindustrial societies were held together by strong traditions and by members' shared moral beliefs and values. As societies industrialized, more specialized economic activity became the basis of the social bond because people became dependent on one another.

Durkheim observed that rapid social change and a more specialized division of labor produce *strains* in society. These strains lead to a breakdown in traditional organization, values, and authority and to a dramatic increase in ***anomie*—a condition in which social control becomes ineffective as a result of the loss of shared values and of a sense of purpose in society.** According to Durkheim, anomie is most likely to occur during a period of rapid social change. In *Suicide* (1964b/1897) he explored the relationship between anomic social conditions and suicide, as discussed in Chapter 2.

Durkheim's contributions to sociology are so significant that he has been referred to as "*the* crucial figure in the development of sociology as an academic discipline [and as] one of the deepest roots of the sociological imagination" (Tiryakian, 1978: 187). He has long been viewed as a proponent of the scientific approach to examining social facts that lie outside individuals. He is also described as the founding figure of the functionalist theoretical tradition. Recently, scholars have acknowledged Durkheim's influence on contemporary social theory, including the structuralist and postmodernist schools of thought. Like Comte, Martineau, and Spencer, Durkheim emphasized that sociology should be a science based on observation and the systematic study of social facts rather than on individual characteristics or traits.

Can Durkheim's ideas be applied to our ongoing analysis of consumerism and credit cards? Durkheim was interested in examining the "social glue" that could hold contemporary societies together and provide people with a "sense of belonging." Ironically, shopping and the credit card industry have created what we might call a "pseudo-sense of belonging" through "membership" in Sam's Club–like businesses and the creation of affinity credit cards designed to encourage members of an organization (such as a fraternity or university alumni association) or people who share interests and activities (such as dog owners and skydiving enthusiasts) to possess a particular card. In later chapters we examine Durkheim's theoretical contributions to diverse subjects ranging from suicide and deviance to education and religion.

Differing Views on the Status Quo: Stability Versus Change

Together with Karl Marx, Max Weber, and Georg Simmel, Durkheim established the course for modern sociology. We will look first at Marx's and Weber's divergent thoughts about conflict and social change in societies, and then at Georg Simmel's analysis of society.

Karl Marx In sharp contrast to Durkheim's focus on the stability of society, German economist and philosopher Karl Marx (1818–1883) stressed that history is a continuous clash between conflicting

▲ Karl Marx

skills from one generation to the next; a latent function is the establishment of social relations and networks. Merton noted that all features of a social system may not be functional at all times; *dysfunctions* are the undesirable consequences of any element of a society. A dysfunction of education in the United States is the perpetuation of gender, racial–ethnic, and class inequalities. Such dysfunctions may threaten the capacity of a society to adapt and survive (Merton, 1968).

Applying a Functional Perspective to Shopping and Consumption How might functionalists analyze shopping and consumption? When we examine the part-to-whole relationships of contemporary society in high-income nations, it immediately becomes apparent that each social institution depends on the others for its well-being. For example, a booming economy benefits other social institutions, including the family (members are gainfully employed), religion (churches, mosques, synagogues, and temples receive larger contributions), and education (school taxes are higher when property values are higher). A strong economy also makes it possible for more people to purchase more goods and services. By contrast, a weak economy has a negative influence on people's opportunities and spending patterns. Because of the significance of the economy in other aspects of social life, the U.S. Census Bureau conducts surveys (for the Bureau of Labor Statistics) to determine how people are spending their money (see "Census Profiles"). If people have "extra" money to spend and can afford leisure time away from work, they are more likely to dine out, take trips, and purchase things they might otherwise forgo. However, in difficult economic times, people are more likely to curtail family outings and some purchases.

Clearly, a manifest function of shopping and consumption is purchasing necessary items such as food, clothing, household items, and sometimes transportation. But what are the latent functions of shopping? Consider, shopping malls, for example: Many young people go to the mall to "hang out," visit with friends, and eat lunch at the food court. People of all ages go shopping for pleasure, relaxation, and perhaps to enhance their feelings of self-worth. ("If I buy this product, I'll look younger/beautiful/handsome/sexy, etc.!") However, shopping and consuming may also produce problems or dysfunctions. Some people are "shopaholics" or "credit card junkies" who cannot

Consumer Spending

The U.S. Census Bureau provides a wealth of data that helps sociologists and other researchers answer questions about the characteristics of the U.S. population. Although the decennial census occurs only once every ten years, the Census Bureau conducts surveys and produces reports on many topics throughout every year. For example, the Census Bureau conducts surveys that the Bureau of Labor Statistics uses to compute the Consumer Price Index, which is a measure of the average change over time in the prices paid by urban consumers for consumer goods and services. This index is used to measure how the nation's economy is performing.

The Consumer Price Index is based on a survey of 7,500 randomly selected households in which people keep a diary of all expenditures they make over a period of time and record whether those expenditures occur on a regular basis (such as for food or rent) or involve relatively large purchases (such as a house or a car). The Census Bureau also conducts interviews to obtain additional data about people's expenditures.

According to the most recent survey, here is the distribution of how we spend our money on an annual basis:

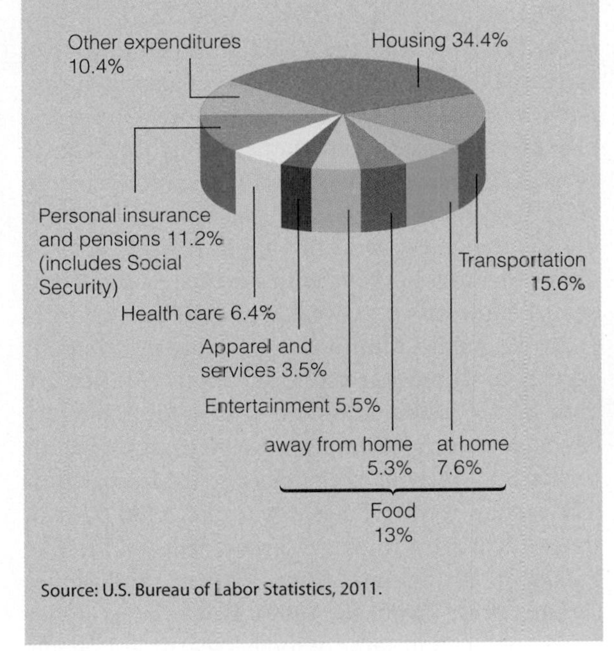

Source: U.S. Bureau of Labor Statistics, 2011.

stop spending money; others are kleptomaniacs, who steal products rather than pay for them.

The functionalist perspective is useful in analyzing consumerism because of the way in which it examines the relationship between part-to-whole relationships. How the economy is doing affects individuals' consumption patterns, and when the economy is not doing well, political leaders often encourage us to spend more to help the national economy and keep other people employed.

Conflict Perspectives

According to ***conflict perspectives*, groups in society are engaged in a continuous power struggle for control of scarce resources**. Conflict may take the form of politics, litigation, negotiations, or family discussions about financial matters. Simmel, Marx, and Weber contributed significantly to this perspective by focusing on the inevitability of clashes between social groups. Today, advocates of the conflict perspective view social life as a continuous power struggle among competing social groups.

Max Weber and C. Wright Mills As previously discussed, Karl Marx focused on the exploitation and oppression of the proletariat (the workers) by the bourgeoisie (the owners or capitalist class).

© Walter McBride/Corbis

▲ As one of the wealthiest and most-beloved entertainers in the world, Oprah Winfrey is an example of Max Weber's concept of prestige—a positive estimate of honor. In 2011 Ms. Winfrey moved beyond her very popular television show and has started a cable channel, OWN.

conflict perspectives the sociological approach that views groups in society as engaged in a continuous power struggle for control of scarce resources.

© Hulton Archive/Getty Images

▲ C. Wright Mills

Max Weber recognized the importance of economic conditions in producing inequality and conflict in society but added *power* and *prestige* as other sources of inequality. Weber (1968/1922) defined *power* as the ability of a person within a social relationship to carry out his or her own will despite resistance from others, and *prestige* as a positive or negative social estimation of honor (Weber, 1968/1922).

C. Wright Mills (1916–1962), a key figure in the development of contemporary conflict theory, encouraged sociologists to get involved in social reform. Mills encouraged everyone to look beneath everyday events in order to observe the major resource and power inequalities that exist in society. He believed that the most important decisions in the United States are made largely behind the scenes by the *power elite*—a small clique composed of the top corporate, political, and military officials. Mills's power elite theory is discussed in Chapter 14 ("Politics and Government in Global Perspective").

The conflict perspective is not one unified theory but rather encompasses several branches. One branch is the neo-Marxist approach, which views struggle between the classes as inevitable and as a prime source of social change. A second branch focuses on racial–ethnic inequalities and the continued exploitation of members of some racial–ethnic groups. A third branch is the feminist perspective, which focuses on gender issues.

The Feminist Approach A feminist approach (or "feminism") directs attention to women's experiences and the importance of gender as an element of social structure. This approach is based on the belief that "women and men are equal and should be equally valued as well as have equal rights" (Basow, 1992). According to feminist theorists, we live in a *patriarchy*, a system in which men dominate women and in which things that are considered to be "male" or "masculine" are more highly valued than those considered to be "female" or "feminine." The feminist perspective assumes that gender is socially created, rather than determined by one's biological inheritance, and that change is essential in order for people to achieve their human potential without limits based on gender. Some feminists argue that women's subordination can end only after the patriarchal system becomes obsolete. However, note that feminism is not one single, unified approach. Rather, there are several feminist perspectives, which are discussed in Chapter 11 ("Sex and Gender").

Applying Conflict Perspectives to Shopping and Consumption How might advocates of a conflict approach analyze the process of shopping and consumption? A contemporary conflict analysis of consumption might look at how inequalities based on racism, sexism, and income differentials affect people's ability to acquire the things they need and want. It might also look at inequalities regarding the issuance of credit cards and access to "cathedrals of consumption" such as megashopping malls and tourist resorts (see Ritzer, 1999: 197–214). However, one of the earliest social theorists to discuss the relationship between social class and consumption patterns was the U.S. social scientist Thorstein Veblen (1857–1929). In *The Theory of the Leisure Class* (1967/1899), Veblen described early wealthy U.S. industrialists as engaging in *conspicuous consumption*—the continuous public display of one's wealth and status through purchases such as expensive houses, clothing, motor vehicles, and other consumer goods. According to Veblen, the leisurely lifestyle of the upper classes typically does not provide them with adequate opportunities to show off their wealth and status. In order to attract public admiration, the wealthy often engage in consumption and leisure activities that are both highly visible and highly wasteful. Examples of conspicuous consumption range from Cornelius Vanderbilt's 8 lavish mansions and 10 major summer estates in the Gilded Age to the 2,400 pairs of shoes owned by Imelda Marcos, wife of the late President Ferdinand Marcos of the Philippines (Frank, 1999; Twitchell, 1999). However, as Ritzer (1999) points out, some of today's wealthiest people engage in *inconspicuous consumption*, perhaps to maintain a low public profile or out of fear for their own safety.

Conspicuous consumption has become more widely acceptable at all income levels, and many families live on credit in order to purchase the goods and services that they would like to have. According to conflict theorists, the economic gains of the wealthiest people are often at the expense of those in the lower classes, who may have to struggle (sometimes unsuccessfully) to have adequate food, clothing, and shelter for themselves and their children. Chapter 8 ("Class and Stratification in the United States") and Chapter 9 ("Global Stratification") discuss contemporary conflict perspectives on class-based inequalities.

However, some people reject the idea of overconsumption and decide not to overspend, instead seeking to make changes in their lives and encouraging others to do likewise (see "You Can Make a Difference").

Symbolic Interactionist Perspectives

The conflict and functionalist perspectives have been criticized for focusing primarily on macrolevel analysis. A ***macrolevel analysis*** **examines whole societies, large-scale social structures, and social systems** instead of looking at important social dynamics in individuals' lives. Our third perspective, symbolic interactionism, fills this void by examining people's day-to-day interactions and their behavior in groups. Thus, symbolic interactionist approaches are based on a ***microlevel analysis*****, which focuses on small groups rather than on large-scale social structures.**

We can trace the origins of this perspective to the Chicago School, especially George Herbert Mead and Herbert Blumer (1900–1986), who is credited with coining the term *symbolic interactionism*. According to ***symbolic interactionist perspectives*****, society is the sum of the interactions of individuals and groups.** Theorists using this perspective focus on the process of *interaction*—defined as immediate reciprocally oriented communication between two or more people—and the part that *symbols* play in giving meaning to human communication. A *symbol* is anything that meaningfully represents something else. Examples of symbols include signs, gestures, written language, and shared values. Symbolic interaction occurs when people communicate through the use of symbols; for example, a gift of food—a cake or a casserole—to a newcomer in a neighborhood is a symbol of welcome and friendship. But symbolic communication occurs in a variety of forms, including facial gestures, posture, tone of voice, and other symbolic gestures (such as a handshake or a clenched fist).

Symbols are instrumental in helping people derive meanings from social situations. In social encounters each person's interpretation or definition of a given situation becomes a *subjective reality* from that person's viewpoint. We often assume that what we consider to be "reality" is shared by others; however, this assumption is often incorrect. Subjective reality is acquired and shared through agreed-upon symbols, especially language. If a person shouts "Fire!" in a crowded movie theater, for example, that language produces the same response (attempting to escape) in all of those who hear and understand it. When people in a group do not share the same meaning for a given symbol, however, confusion results; for example, people who did not know the meaning of the word *fire* would not know what the commotion was about. How people *interpret* the messages they receive and the situations they encounter becomes their subjective reality and may strongly influence their behavior.

Symbolic interactionists attempt to study how people make sense of their life situations and the way they go about their activities, in conjunction with others, on a day-to-day basis (Prus, 1996). How do people develop the capacity to think and act in socially prescribed ways? According to symbolic interactionists, our thoughts and behavior are shaped by our social interactions with others. Early theorists such as Charles H. Cooley and George Herbert Mead explored how individual personalities are developed from social experience and concluded that we would not have an identity, a "self," without communication with other people. This idea is developed in Cooley's notion of the "looking-glass self" and Mead's "generalized other," as discussed in Chapter 4 ("Socialization"). From this perspective the attainment of language is essential not only for the development of a "self" but also for establishing common understandings about social life.

How do symbolic interactionists view social organization and the larger society? According to symbolic interactionists, social organization and society are possible only through people's everyday interactions. In other words, group life takes its shape as people interact with one another (Blumer, 1986/1969). Although macrolevel factors such as

macrolevel analysis an approach that examines whole societies, large-scale social structures, and social systems.

microlevel analysis sociological theory and research that focus on small groups rather than on large-scale social structures.

symbolic interactionist perspectives the sociological approach that views society as the sum of the interactions of individuals and groups.

▲ Sporting events are a prime location for seeing how college students use symbols to convey shared meanings. The colors of clothing and the display of fraternity and sorority signs emphasize these students' pride in their school.

© Jeff Greenberg/PhotoEdit

economic and political institutions constrain and define the forms of interaction that we have with others, the social world is dynamic and always changing. Chapter 5 ("Society, Social Structure, and Interaction") explores two similar approaches—rational choice and exchange theories—that focus specifically on how people rationally try to get what they need by exchanging valued resources with others.

As we attempt to present ourselves to others in a particular way, we engage in behavior that the our own subjective reality. This theoretical viewpoint applies to shopping and consumption just as it does to other types of conduct. For example, when a customer goes into a store to make a purchase and offers a credit card to the cashier, what meanings are embedded in the interaction process that takes place between the two of them? The roles that the two people play are based on their histories of interaction in previous situations. They bring to the present encounter symbolically charged ideas

you can make a difference

Thinking Less About Things and More About People

We must rapidly begin the shift from a thing-oriented society to a person-oriented society. When machines and computers, profit motives and property rights are considered more important than people, the giant triplets of racism, militarism and economic exploitation are incapable of being conquered.

—Dr. Martin Luther King, Jr., April 1967 (qtd. in Postman, 2011)

More than forty years ago, Dr. King encouraged people to find fulfillment in social relationships with other people rather than in new technologies or in making more money. Since King's era, the United States has had periods of economic boom and bust, accompanied by unparalleled consumerism. Many analysts believe that consumerism and constant pressures to "buy, buy, buy more!" have created financial havoc for many individuals and families. We are continually surrounded by advertisements, shopping malls, and online buying opportunities that set in front of us a veritable banquet of merchandise to buy. However, shopping that gets out of hand is a serious habit that may have lasting psychological and economic consequences. If we are aware of these problems, we may be able to help ourselves or others avoid hyperconsumerism.

Do you know the symptoms of compulsive overspending and debt dependency? Consider these questions:

- Do you or someone you know spend large amounts of time shopping or thinking about going shopping?
- Do you or someone you know rush to the store or to the computer for online shopping when feeling frustrated, depressed, or otherwise "out of sorts"?
- Do you or someone you know routinely argue with parents, friends, or partners about spending too much money or overcharging on credit cards?
- Do you or someone you know hide purchases or make dishonest statements—such as "It was a gift from a friend"—to explain where new merchandise came from?

According to economist Juliet Schor (1999), who has extensively studied the problems associated with excessive spending and credit card debt, each of us can empower ourselves and help others as well if we follow simple steps in our consumer behavior. Among these steps are *controlling desire* by gaining knowledge of the process of consumption and its effect on people, *helping to make exclusivity uncool* by demystifying the belief that people are "better" simply because they own excessively expensive items, and *discouraging competitive consumption* by encouraging our friends and acquaintances to spend less on presents and other purchases. Finally, Schor suggests that we should *become educated consumers* and *avoid use of shopping as a form of therapy*. By following Schor's simple steps and encouraging our friends and relatives to do likewise, we may be able to free ourselves from the demands of a hyperconsumer society that continually bombards us with messages indicating that we should spend more and go deeper in debt on our credit cards.

How might we think more about people? Some analysts suggest that we should make a list of things that are more important to us than money and material possessions. These might include our *relationships* and *experiences* with family, friends, and others whom we encounter in daily life. Are we so engrossed in our own life that we fail to take others into account? Around school, are we so busy texting or talking on our cell phone that we fail to speak to others? During holidays and special occasions, do we make time for friends and loved ones even if we think we have "better" things to do? Other suggestions for thinking about others might include *looking for ways, small and large, to help others*. Small ways to help others might be opening a door for someone whose hands are full, letting someone go before us in a line or while driving in traffic, or any one of a million small kindnesses that might brighten someone else's day as well as our own. Large ways of helping others would include joining voluntary organizations that assist people in the community, including older individuals, persons with health problems, children who need a tutor or mentor, or many others you might learn of from school organizations, social service agencies, or churches in your area.

Are you up to the challenge? Many who have tried thinking less about things and more about people highly recommend this as a life-affirming endeavor for all involved.

Source: Schor, 1999.

© Bob Daemmrich / Alamy

As part of Project 2009, these University of Texas at Austin students spent a day of service, painting and sprucing up this house. What similar projects could you and your peers undertake?

may affect the cashier's interactions with subsequent customers. Likewise, the next time the customer uses a credit card, he or she may say something like "I hope this card isn't over its limit; sometimes I lose track," even if the person knows that the card's credit limit has not been exceeded. This is only one of many ways in which the rich tradition of symbolic interactionism might be used to examine shopping and consumption. Other areas of interest might include the social nature of the shopping experience, social interaction patterns in families regarding credit card debts, and why we might spend money to impress others.

Postmodern Perspectives

According to ***postmodern perspectives*, existing theories have been unsuccessful in explaining social life in contemporary societies that are characterized by postindustrialization, consumerism, and global communications**. Postmodern social theorists reject the theoretical perspectives we have previously discussed, as well as how those thinkers created the theories (Ritzer, 2011).

Postmodern theories are based on the assumption that the rapid social change that occurs as societies move from modern to postmodern (or postindustrial) conditions has a harmful effect on people. One evident change is the significant decline in the influence of social institutions such as the family, religion, and education on people's lives. Those who live in postmodern societies typically pursue individual freedom and do not want the structural constraints that are imposed by social institutions.

factory workers in industrial economies, to enhance their profits and to keep everyday people from rebelling against social inequality (1998/1970). How does this work? When consumers are encouraged to purchase more than they need or can afford, they often sink deeper in debt and must keep working to meet their monthly payments. Instead of consumption being related to our needs, it is based on factors such as our "wants" and the need we feel to distinguish ourselves from others. We will look at this idea in more detail in the next section, where we apply a postmodern perspective to shopping and consumption. We will also return to Baudrillard's general ideas on postmodern societies in Chapter 3 ("Culture").

Postmodern theory opens up broad new avenues of inquiry by challenging existing perspectives and questioning current belief systems. However, postmodern theory has also been criticized for raising more questions than it answers.

Applying Postmodern Perspectives to Shopping and Consumption According to some social theorists, the postmodern society is a consumer society. The focus of the capitalist economy has shifted from production to consumption: The emphasis is on getting people to consume more and to own a greater variety of things. As previously discussed, credit cards may encourage people to spend more money than they should, and often more than they can afford (Ritzer, 1998). Television networks and online shopping make it possible for people to shop around the clock without having to leave home or encounter "real" people. As Ritzer

[concept quick review]

The Major Theoretical Perspectives

Perspective	Analysis Level	View of Society
Functionalist	Macrolevel	Society is composed of interrelated parts that work together to maintain stability within society. This stability is threatened by dysfunctional acts and institutions.
Conflict	Macrolevel	Society is characterized by social inequality; social life is a struggle for scarce resources. Social arrangements benefit some groups at the expense of others.
Symbolic Interactionist	Microlevel	Society is the sum of the interactions of people and groups. Behavior is learned in interaction with other people; how people define a situation becomes the foundation for how they behave.
Postmodernist	Macrolevel/ Microlevel	Societies characterized by postindustrialization, consumerism, and global communications bring into question existing assumptions about social life and the nature of reality.

The Concept Quick Review reviews all four of these perspectives. Throughout this book we will be using these perspectives as lenses through which to view our social world.

postmodern perspectives the sociological approach that attempts to explain social life in modern societies that are characterized by postindustrialization, consumerism, and global communications.

chapter review Q & A

Use these questions and answers to check how well you've achieved the learning objectives set out at the beginning of this chapter.

- **What is sociology, and how can it help us to understand ourselves and others?**

Sociology is the systematic study of human society and social interaction. We study sociology to understand how human behavior is shaped by group life and, in turn, how group life is affected by individuals. Our culture tends to emphasize individualism, and sociology pushes us to consider more-complex connections between our personal lives and the larger world.

- **What is the sociological imagination, and why is it important to have a global sociological imagination?**

According to C. Wright Mills, the sociological imagination helps us understand how seemingly personal troubles, such as suicide, are actually related to larger social forces. It is the ability to see the relationship between individual experiences and the larger society. It is important to have a global sociological imagination because the future of this nation is deeply intertwined with the future of all nations of the world on economic, political, and humanitarian levels.

- **What factors contributed to the emergence of sociology as a discipline?**

Industrialization and urbanization increased rapidly in the late eighteenth century, and social thinkers began to examine the consequences of these powerful forces. Auguste Comte coined the term *sociology* to describe a new science that would engage in the study of society.

- **What are the major contributions of early sociologists such as Durkheim, Marx, and Weber?**

The ideas of Emile Durkheim, Karl Marx, and Max Weber helped lead the way to contemporary sociology. Durkheim argued that societies are built on social facts, that rapid social change produces strains in society, and that the loss of shared values and purpose can lead to a condition of anomie. Marx stressed that within society there is a continuous clash between the owners of the means of production and the workers, who have no choice but to sell their labor to others. According to Weber, sociology should be value free, and people should become more aware of the role that bureaucracies play in daily life.

- **How did Simmel's perspective differ from that of other early sociologists?**

Whereas other sociologists primarily focused on society as a whole, Simmel explored small social groups and argued that society was best seen as a web of patterned interactions among people.

- **What are the major contemporary sociological perspectives?**

Functionalist perspectives assume that society is a stable, orderly system characterized by societal consensus. Conflict perspectives argue that society is a continuous power struggle among competing groups, often based on class, race, ethnicity, or gender. Interactionist perspectives focus on how people make sense of their everyday social interactions, which are made possible by the use of mutually understood symbols. From an alternative perspective, postmodern theorists believe that entirely new ways of examining social life are needed and that it is time to move beyond functionalist, conflict, and interactionist approaches.

key terms

anomie 14
conflict perspectives 23
functionalist perspectives 21
high-income countries 7
industrialization 10
latent functions 21
low-income countries 7
macrolevel analysis 25
manifest functions 21
microlevel analysis 25
middle-income countries 7
positivism 12
postmodern perspectives 28
social Darwinism 13
social facts 14
society 4
sociological imagination 5
sociology 4
symbolic interactionist perspectives 25
theory 21
urbanization 10

questions for critical thinking

1. What does C. Wright Mills mean when he says the sociological imagination helps us "to grasp history and biography and the relations between the two within society?" (Mills, 1959b: 6). How might this idea be applied to today's consumer society?
2. As a sociologist, how would you remain objective yet see the world as others see it? Would you make subjective decisions when trying to understand the perspectives of others?
3. Early social thinkers were concerned about stability in times of rapid change. In our more global world, is stability still a primary goal? Or is constant conflict important for the well-being of all humans? Use the conflict and functionalist perspectives to bolster your analysis.
4. Some social analysts believe that college students relate better to commercials and advertising culture than they do to history, literature, or probably anything else (Twitchell, 1996). How would you use the various sociological perspectives to explore the validity of this assertion in regard to students on your college campus?

turning to video

Watch the CBS video *Economic Meltdown* (running time 2:22), available through **CengageBrain.com**. This video asks the question, "Will the new economic realities change the way Americans look at money and possessions?" As you watch the video, think about your own attitudes toward money and possessions before the recession—was there anything "excessive" about the way you lived then? After you've watched the video, consider these questions: Do you believe Americans have undergone a cultural shift in regard to personal finances? Why or why not?

online study resources

Go to CENGAGE brain.com to access online study resources, including the Sociology CourseMate for this text as well as special features such as video, an interactive sociology time line and interactive maps, General Social Survey (GSS) data, and U.S. Census 2010 data.

CourseMate brings course concepts to life with interactive learning, study, and exam-preparation tools that support the printed textbook. A textbook-specific website, **Sociology CourseMate** includes an integrated interactive eBook and other interactive learning tools, including quizzes, flash cards, and videos.

Visit **www.cengagebrain.com** to access your account and purchase materials.

2 Sociological Research Methods

Jumping off the gw [George Washington] bridge. Sorry.

—Tyler Clementi, a freshman at Rutgers University, posted this terse good-bye message on his Facebook page before jumping to his death after his roommate secretly filmed him during a "sexual encounter" in his dorm room and posted it live on the Internet (*People*, 2010: 58)

Tyler was a fine young man, and a distinguished musician. The family is heartbroken beyond words.

—a statement issued by Tyler's family (qtd. in Foderaro, 2010: A4)

I will always remember everything from our preschool's Halloween party to your amazing musical talents. When you picked up the violin and began to play, it was as if everything just paused until you put it down. We will never forget you Tyler. May you rest in peace.

—a posting on a social network by Courtney Ayukawa, a friend of Tyler (qtd. in Friedman, 2010)

© AFP/Getty Images/Newscom

▲ Tyler Clementi, a gifted violinist and first-year student at Rutgers University, committed suicide after becoming the victim of online bullying. Studying sociology provides us with new insights on problems such as suicide by making us aware that much more goes on in social life than we initially observe.

He was always by himself with his iPod in his ears. . . . There was deep emotion when he put that bow to his violin. That's how he expressed himself. . . . When I felt isolated, he showed me an immense amount of compassion.

—various acquaintances describing Tyler (qtd. in Schwartz, 2010)

As a parent of two daughters, it breaks my heart. It's something that just shouldn't happen in this country. It's time for Americans to dispel the myth that bullying is "just a normal rite of passage." I don't know what it's like to be picked on for being gay. But I do know what it's like to grow up feeling that sometimes you don't belong. It's tough. But what I want to say is this: You are not alone. You didn't do anything wrong. You didn't do anything to deserve being bullied.

—President Barack Obama commenting on the death of Tyler Clementi, as well as on the recent suicides of other young people who experienced bullying or taunting for being gay (qtd. in Powers, 2010)

Chapter Focus Question:

How do sociological theory and research add to our knowledge of human societies and social issues such as suicide?

Clearly, the internationally publicized suicide of Tyler Clementi deeply touched his family, friends, and millions of people who did not know him. It also raised interesting sociological questions about the individual and social causes of behavior such as suicide. Although we will never know the full story of Tyler's life, this tragic occurrence brings us to larger sociological questions: Why does anyone commit suicide? Is suicide purely an individual phenomenon, or is it related to our social interactions and the social environment and society in which we live? How have newer technologies such as cell phones, webcams, and the Internet affected our communications—both positively and negatively—with others?

In this chapter we examine how sociological theories and research help us understand social life, including seemingly individualistic acts such as taking one's own life. We will see how sociological theory and research methods might be used to answer complex questions, and we will wrestle with some of the difficulties that sociologists experience as they study human behavior.

Highlighted Learning Objectives

- Explain the relationship between theory and research and the benefits of studying theory.
- Identify and describe the steps in the conventional research process.
- Discuss several benefits of qualitative methods that can add to our understanding of human behavior.
- Explain why it is important to have a variety of research methods available.
- Explain what research has contributed to our understanding of suicide.
- Identify reasons why a code of ethics is necessary for sociological research.

sociology and everyday life

How Much Do You Know About Suicide?

True	False	
T	F	1. For people thinking of suicide, it is difficult, if not impossible, to see the bright side of life.
T	F	2. People who talk about suicide don't do it.
T	F	3. Suicide rates in the United States are highest for Asian/Pacific Islanders because of pressure to achieve.
T	F	4. Rates of suicide are highest in the intermountain states located in the western and northwestern regions of the United States.
T	F	5. Females complete suicide (take their own life) at a much higher rate than that of males.
T	F	6. Over half of all suicides occur in adult women between the ages of 25 and 65.
T	F	7. Suicide rates are higher for African Americans than for white Americans.
T	F	8. More teenagers and young adults die from suicide than from cancer, heart disease, AIDS, birth defects, stroke, pneumonia, influenza, and chronic lung disease combined.

Answers on page 36.

Why Is Sociological Research Necessary?

Most of us rely on our own experiences and personal knowledge to help us form ideas about what happens in everyday life and how the social world works. However, there are many occasions when this personal knowledge is not enough to provide a thorough understanding of what is going on around us. This is why sociologists and other social scientists learn to question ordinary assumptions and to use specific research methods to find out more about the social world.

Sociologists obtain their knowledge of human behavior through research, which results in a body of information that helps us move beyond guesswork and common sense in understanding society. The sociological perspective incorporates theory and research to arrive at a more accurate understanding of the "hows" and "whys" of human social interaction. Once we have an informed perspective about social issues, we are in a better position to find solutions and make changes. Social research, then, is a key part of sociology.

Common Sense and Sociological Research

How does sociological research challenge our commonsense beliefs about an issue? Consider our discussion of suicide, for example. Most of us have commonsense ideas about suicide. Common sense may tell us that people who threaten suicide will not commit suicide. However, sociological research indicates that this assumption is frequently incorrect: People who threaten to kill themselves are often sending messages to others and may indeed attempt suicide. Common sense may also tell us that suicide is caused by despair or depression. However, research suggests that suicide is sometimes used as a means of lashing out at friends and relatives because of real or imagined wrongs. Research also shows that some younger people commit suicide because they believe there is no way out of their problems, particularly when they are continually harassed or bullied by individuals whom they encounter daily. Before reading on, take the Sociology and Everyday Life quiz to find out more about suicide.

Historically, the commonsense view of suicide was that it was a sin, a crime, and a mental illness (Evans and Farberow, 1988). Emile Durkheim refused to accept these explanations. In what is probably the first sociological study to use scientific research methods, he related suicide to the issue of cohesiveness (or lack of cohesiveness) in society instead of viewing suicide as an isolated act that could be understood only by studying individual personalities or inherited tendencies. In *Suicide* (1964b/1897), Durkheim documented his contention that a high suicide rate was symptomatic of large-scale societal problems. In the process, he

sociology works!

Durkheim's Sociology of Suicide and Twenty-First-Century India

> The bond attaching [people] to life slackens because the bond which attaches [them] to society is itself slack.
>
> —Emile Durkheim, *Suicide* (1964b/1897)

Although this statement described social conditions accompanying the high rates of suicide found in late-nineteenth-century France, Durkheim's words ring true today as we look at contemporary suicide rates for cities such as Bangalore, which some refer to as "India's Suicide City" (Guha, 2004).

At first glance, we might think that the outsourcing of jobs in the technology sector—from high-income nations such as the United States to India—would provide happiness and job satisfaction for individuals in cities such as Bangalore and New Delhi who have gained new opportunities and higher salaries in recent years as a result of outsourcing. News stories have focused on the wealth of opportunities that these outsourced jobs have brought to millions of men and women in India, most of whom are in their twenties and thirties and who now earn larger incomes than do their parents and many of their contemporaries. However, the underlying story of what is really going on in these cities stands in stark contrast: Rapid urbanization and fast-paced changes in the economy and society are weakening social ties that have been very important to individuals. Social bonds have been weakened or dissolved as people move away from their families and their community. Life in the cities moves at a much faster pace than in the rural areas, and many individuals experience loneliness, sleep disorders, family discord, and major health risks such as heart disease and depression (Mahapatra, 2007). In the words of Ramachandra Guha (2004), a historian residing in India, Durkheim's sociology of suicide remains highly relevant to finding new answers to this challenging problem: "The rash of suicides in city and village is a qualitatively new development in our history. We sense that tragedies are as much social as they are individual. But we know very little of what lies behind them. What we now await, in sum, is an Indian Durkheim."

reflect & analyze

How does sociology help us to examine seemingly private acts such as suicide within a larger social context? Why are some people more inclined to commit suicide if they are not part of a strong social fabric?

Ross Anania/Photographer's Choice/Getty Images

▲ Although scientific studies about human behavior are readily available to most of us, and trained professionals can help us with our personal and social problems, many individuals rely—at least in part—on the advice of psychics such as the one shown here, who provides her services on line. What problems might occur as a result of relying on psychics for counseling and advice?

developed an approach to research that influences researchers to this day (see "Sociology Works!"). As we discuss sociological research, we will use the problem of suicide to demonstrate the research process.

Because much of sociology deals with everyday life, we might think that common sense, our own personal experiences, and the media are the best sources of information. However, our personal experiences are subjective, and much of the information provided by the media comes from sources seeking support for a particular point of view. The content of the media is also influenced by the necessity of selling advertisements based on readership, audience ratings, or the number of online hits a website receives.

We need to be able to evaluate the information we receive. This is especially true because the quantity—but, in some instances, not the quality—of information available has grown dramatically as a result of the information explosion brought about by computers and by the telecommunications industry.

sociology and everyday life

ANSWERS to the Sociology Quiz on Suicide

1. True. To people thinking of suicide, an acknowledgment that there is a bright side only confirms and conveys the message that they have failed; otherwise, they, too, could see the bright side of life.

2. False. Some people who talk about suicide do kill themselves. Warning signals of possible suicide attempts include talk of suicide, the desire not to exist anymore, despair, and hopelessness.

3. False. Asian/Pacific Islanders have the lowest rates of suicide, and Native Americans (American Indians/Alaskan Natives) have the highest suicide rates per 100,000 in the United States.

4. True. Suicide rates are highest in the western and northwestern regions of the United States. What sociological factors might help explain this trend?

5. False. Males *complete* suicide at a rate four times that of females. However, females *attempt* suicide three times more often than males.

6. False. Just the opposite is true. Over half of all suicides occur in adult men aged 25 to 65.

7. False. Suicide rates are much higher among white Americans than African Americans.

8. False. Suicide is a leading cause of death among teenagers and young adults. It is the third leading cause of death among young people between 15 and 24 years of age, following accidents (unintentional injuries) and homicide.

Sources: Based on American Association of Suicidology, 2009; and National Centers for Disease Control and Prevention, 2009b.

Sociology and Scientific Evidence

In taking this course, you will be studying social science research and may be asked to write research reports or read and evaluate journal articles. If you attend graduate or professional school in fields that use sociological research, you will be expected to evaluate existing research and perhaps do your own. Hopefully, you will find that social research is relevant to the practical, everyday concerns of the real world.

Sociology involves *debunking*—the unmasking of fallacies (false or mistaken ideas or opinions) in the everyday and official interpretations of society (Mills, 1959b). Because problems such as suicide involve threats to existing societal values, we cannot analyze these problems without acknowledging the values involved. For example, should assisted suicide for terminally ill patients who wish to die be legal? We often answer questions like this by using either the normative or the empirical approach. The *normative approach* uses religion, customs, habits, traditions, and law to answer important questions. It is based on strong beliefs about what is right and wrong and what "ought to be" in society. Issues such as assisted suicide are often answered by the normative approach.

Although these issues are immediate and profound, some sociologists discourage the use of the normative approach in their field and advocate the use of the empirical approach instead. The *empirical approach* attempts to answer questions through systematic collection and analysis of data. This approach is referred to as the conventional model, or the "scientific method," and is based on the assumption that knowledge is best gained by direct, systematic observation. Many sociologists believe that two basic scientific standards must be met: (1) scientific beliefs should be supported by good evidence or information, and (2) these beliefs should be open to public debate and critiques from other scholars, with alternative interpretations being considered (Cancian, 1992).

▲ Why do older African American men have a lower rate of suicide than white males of similar ages? Questions such as this often serve as the foundation for an explanatory study as sociologists attempt to understand and describe certain cause-and-effect relationships.

The Theory and Research Cycle

The relationship between theory and research has been referred to as a continuous cycle, as shown in ▶ Figure 2.1 (Wallace, 1971). You will recall that a *theory* is a set of logically interrelated statements that attempts to describe, explain, and (occasionally) predict social events. A theory attempts to explain why something is the way it is. *Research* is the process of systematically collecting information for the purpose of testing an existing theory or generating a new one. The theory and research cycle consists of deductive and inductive approaches. In the *deductive approach* the researcher begins with a theory and uses research to test the theory. This approach proceeds as follows: (1) theories generate hypotheses, (2) hypotheses lead to observations (data gathering), (3) observations lead to the formation of generalizations, and (4) generalizations are used to support the theory, to suggest modifications to it, or to refute it. To illustrate, if we use the deductive method to determine why people commit suicide, we start by formulating a theory about the "causes" of suicide and then test our theory by collecting and analyzing data (for example, vital statistics on suicides or surveys to determine whether adult church members view suicide differently from nonmembers).

In the *inductive approach* the researcher collects information or data (facts or evidence) and then generates theories from the analysis of that data.

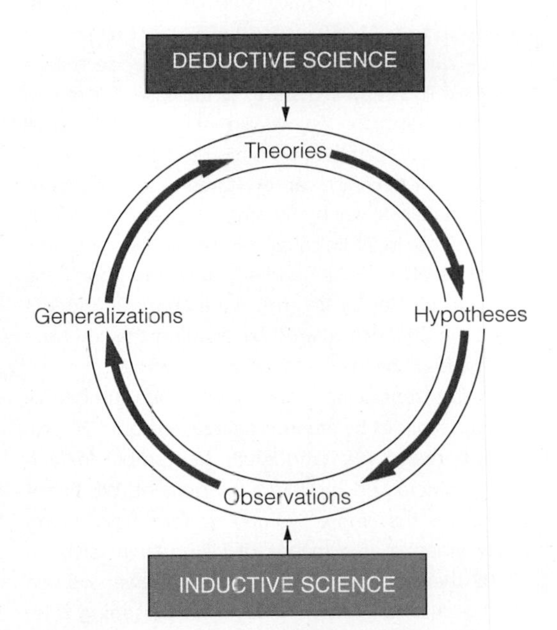

▲ **FIGURE 2.1 THE THEORY AND RESEARCH CYCLE**

The theory and research cycle can be compared to a relay race; although all participants do not necessarily start or stop at the same point, they share a common goal—to examine all levels of social life.

Sources: Adapted from Walter Wallace, *The Logic of Science in Sociology*. New York: Aldine de Gruyter, 1971.

Under the inductive approach, we would proceed as follows: (1) specific observations suggest generalizations, (2) generalizations produce a tentative theory, (3) the theory is tested through the formation of hypotheses, and (4) hypotheses may provide suggestions for additional observations. Using the inductive approach to study suicide, we might start by simultaneously collecting and analyzing data related to suicidal behavior and then generate a theory (see Glaser and Strauss, 1967; Reinharz, 1992). Researchers may break into the cycle at different points depending on what they want to know and what information is available.

Theory gives meaning to research; research helps support theory. Sociologists suggest that a healthy skepticism (a feature of science) is important in research because it keeps us open to the possibility of alternative explanations. Some degree of skepticism is built into each step of the research process.

The Sociological Research Process

Not all sociologists conduct research in the same manner. Some researchers primarily engage in quantitative research, whereas others engage in qualitative research. With *quantitative research,* the goal is scientific objectivity, and the focus is on data that can be measured numerically. Quantitative research typically emphasizes complex statistical techniques. Most sociological studies on suicide have used quantitative research. They have compared rates of suicide with almost every conceivable variable, including age, sex, race/ethnicity, education, and even sports participation (see Lester, 1992). For example, researchers in one study examined the effects of church membership, divorce, and migration on suicide rates in the United States and concluded that suicide rates are typically higher where divorce and migration rates are higher and church membership is lower (Breault, 1986). (The "Understanding Statistical Data Presentations" box explains how to read numerical tables, how to interpret the data and draw conclusions, and how to calculate ratios and rates.)

With *qualitative research,* interpretive description (words) rather than statistics (numbers) is used to analyze underlying meanings and patterns of social relationships. An example of qualitative research is a study in which the researcher systematically analyzed the contents of the notes of suicide victims to determine recurring themes, such as a feeling of despair or failure. Through this study, the researcher hoped to determine if any patterns could be found that would help in understanding why people might kill themselves.

understanding statistical data presentations

Are men or women more likely to commit suicide? Are suicide rates increasing or decreasing? These questions can be answered in numerical terms. Sociologists often use statistical tables as a concise way to present data because such tables convey a large amount of information in a relatively small space; ■ Table 1 gives an example. To understand a table, follow these steps:

table 1

U.S. Suicides, by Sex and Method Used, 1984 and 2005[a]

Method	Males		Females	
	1984	2005	1984	2005
Total	22,689	25,907	6,597	6,730
Firearms	14,504	14,916	2,609	2,086
(% of total)[b]	(64.0)	(57.6)	(39.5)	(30.9)
Poisoning[c]	3,203	3,112	2,406	2,632
(% of total)[b]	(14.1)	(12.0)	(36.5)	(39.1)
Suffocation[d]	3,478	5,887	863	1,361
(% of total)[b]	(15.3)	(22.7)	(13.0)	(20.2)
Other	1,504	1,992	719	651
(% of total)[b]	(6.6)	(7.7)	(10.9)	(9.8)

[a]Excludes deaths of nonresidents of the United States.

[b]Because of rounding, the percentages in a column may not add up to 100.0%.

[c]Includes solids, liquids, and gases.

[d]Includes hanging and strangulation.

Source: National Center for Injury Prevention and Control, 2006.

1. *Read the title*. The title indicates the topic. From the title, "U.S. Suicides, by Sex and Method Used, 1984 and 2005," we learn that the table shows relationships between two variables: sex and method of suicide used. It also indicates that the table contains data for two different time periods: 1984 and 2005.
2. *Check the source and other explanatory notes*. In this case, the source is *National Center for Injury Prevention and Control, 2006*. (Because newer reports combine data for the period from 2002 to 2006, I have used 2005, the latest year available for a one-year comparison.) Checking the source helps determine its reliability and timeliness. The first footnote indicates that the table includes only people who reside in the United States. The next footnote reflects that, because of rounding, the percentages in a column may not total 100.0%. The final two footnotes provide more information about exactly what is included in each category.
3. *Read the headings for each column and each row*. The main column headings in Table 1 are "Method," "Males," and "Females." These latter two column headings are divided into two groups: 1984 and 2005. The columns present information (usually numbers) arranged vertically. The rows present information horizontally. Here, the row headings indicate suicide methods.
4. *Examine and compare the data*. To examine the data, determine what units of measurement have been used. In Table 1 the figures are numerical counts (for example, the total number of reported female suicides by poisoning in 2005 was 2,632) and percentages (for example, in 2005, poisoning accounted for 39.1 percent of all female suicides reported). A *percentage*, or proportion, shows how many of a given item there are in every one hundred. Percentages allow us to compare groups of different sizes. For example, percentages show the proportion of people who used each method, thus giving a more meaningful comparison.
5. *Draw conclusions*. By looking for patterns, some conclusions can be drawn from Table 1.

   a. *Determining the increase or decrease*. Between 1984 and 2005, reported male suicides by firearms increased from 14,504 to 14,916—an increase of 412—while female suicides by firearms decreased by 523. This represents a *total* increase (for males) and decrease (for females) in suicides by firearms for the two years being compared. The *amount* of increase or decrease can be stated as a percentage: Total male suicides by firearms were about 2.8 percent higher in 2005, calculated by dividing the total increase (412) by the earlier (lower) number. Total female suicides by firearms were about 20 percent lower in 2005, calculated by dividing the total decrease (523) by the earlier (higher) number.

   b. *Drawing appropriate conclusions*. The number of female suicides by firearms decreased about 20 percent between 1984 and 2005; the number for poisoning increased by about 9.4 percent. We might conclude that more women preferred poisoning over firearms as a means of killing themselves in 2005 than in 1984. Does that mean fewer women had access to guns in 2005? That poisoning oneself became more acceptable? Such generalizations do not take into account that we are only looking at data from two years and that the difference in statistics for those two years may not really represent a trend.

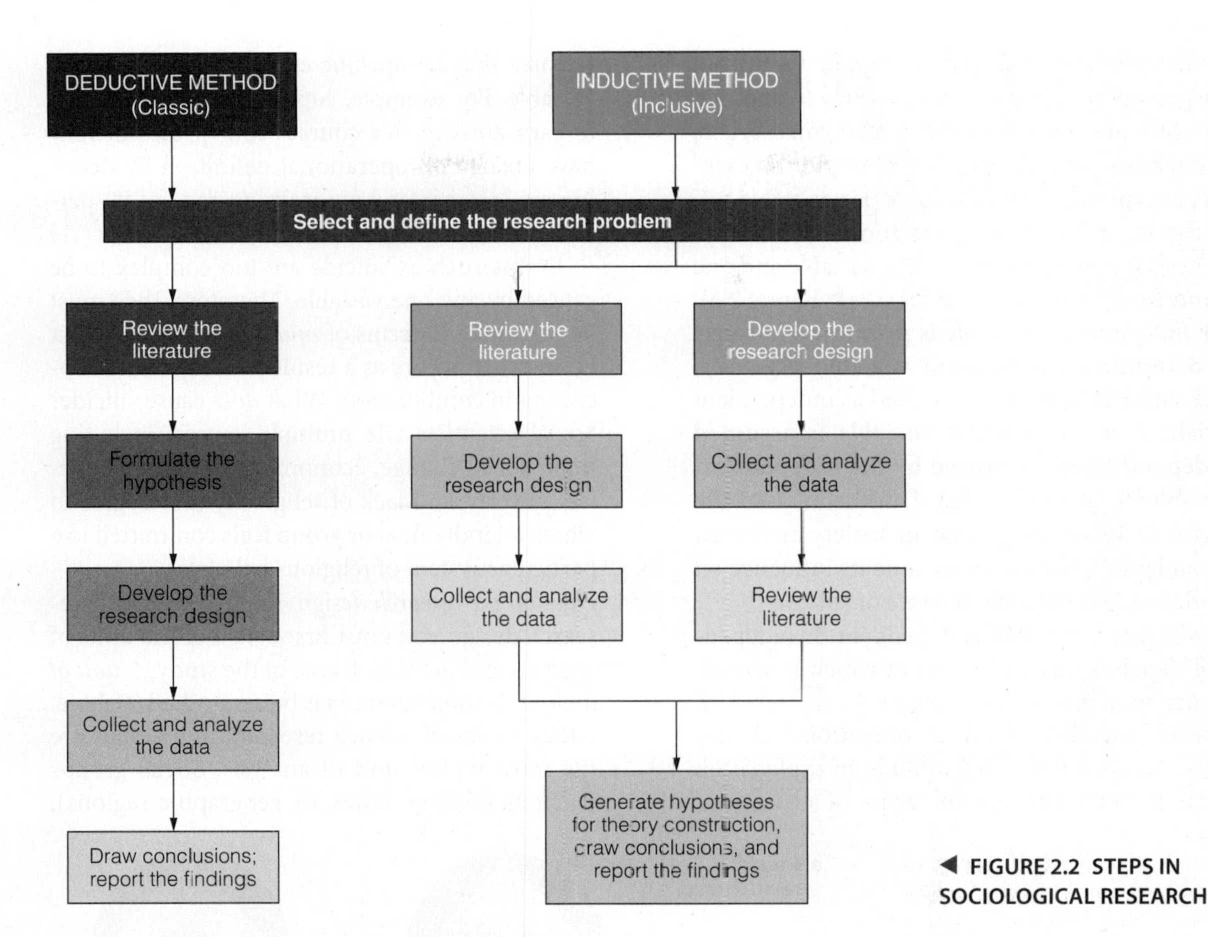

◀ **FIGURE 2.2 STEPS IN SOCIOLOGICAL RESEARCH**

The "Conventional"—Quantitative—Research Model

Research models are tailored to the specific problem being investigated and the focus of the researcher. Both quantitative research and qualitative research contribute to our knowledge of society and human social interaction, and both involve a series of steps, as shown in ▶ Figure 2.2. We will now trace the steps in the "conventional" research model, which focuses on quantitative research. Then we will describe an alternative model that emphasizes qualitative research.

1. *Select and define the research problem.* When you engage in research, the first step is to select and clearly define the research topic. Sometimes, a specific experience such as having known someone who committed suicide can trigger your interest in a topic. Other times, you might select topics to fill gaps or challenge misconceptions in existing research or to test a specific theory (Babbie, 2010). Emile Durkheim selected suicide because he wanted to demonstrate the importance of *society* in situations that might appear to be arbitrary acts by individuals: In his time, suicide was widely believed to be a uniquely individualistic act. Durkheim emphasized that *suicide rates* provide better explanations for suicide than do *individual acts* of suicide. He reasoned that if suicide were purely an individual act, then the rate of suicide (the relative number of people who kill themselves each year) should be the same for every group regardless of culture and social structure. Durkheim wanted to know why there were different rates of suicide—whether factors such as religion, marital status, sex, and age had an effect on social cohesion.
2. *Review previous research.* Before you begin your research, it is important to review the literature to see what others have written about the topic. Analyzing what previous researchers have found helps to clarify issues and focus the direction of your own research. But when Durkheim began his study, very little sociological literature existed for him to review other than the works of Henry Morselli (1975/1881), who concluded that suicide was a part of an evolutionary process whereby "weak-brained" individuals were sorted out by insanity and voluntary death.
3. *Formulate the hypothesis (if applicable).* You may formulate a ***hypothesis*—a tentative statement of the relationship between two or more concepts.** Concepts are the abstract elements representing some aspect of the world in simplified form (such as "social integration" or "loneliness"). As you formulate your hypothesis about suicide, you may need to convert concepts to variables. A *variable* is any concept with measurable traits

hypothesis in research studies, a tentative statement of the relationship between two or more concepts.

or characteristics that can change or vary from one person, time, situation, or society to another. Variables are the observable and/or measurable counterparts of concepts. For example, "suicide" is a concept; the "rate of suicide" is a variable.

The most fundamental relationship in a hypothesis is between a dependent variable and one or more independent variables (see ▶ Figure 2.3). The ***independent variable* is presumed to cause or determine a dependent variable**. Age, sex, race, and ethnicity are often used as independent variables. The ***dependent variable* is assumed to depend on or be caused by the independent variable(s)** (Babbie, 2010). Durkheim used the degree of social integration in society as the independent variable to determine its influence on the dependent variable, the rate of suicide.

Whether a variable is dependent or independent depends on the context in which it is used. To use variables in the contemporary research process, sociologists create operational definitions. An *operational definition* is an explanation of an abstract concept in terms of observable features that are specific enough to measure the variable. For example, suppose that your goal is to earn an *A* in this course. Your professor may have created an operational definition by defining an *A* as earning an exam average of 90 percent or above (Babbie, 2010).

Events such as suicide are too complex to be caused by any one variable. Therefore, they must be explained in terms of *multiple causation*—that is, an event occurs as a result of many factors operating in combination. What *does* cause suicide? Social scientists cite multiple causes, including rapid social change, economic conditions, hopeless poverty, and lack of religiosity (the degree to which an individual or group feels committed to a particular system of religious beliefs).

4. *Develop the research design.* In developing the research design you must first consider the units of analysis and the time frame of the study. A *unit of analysis* is *what* or *whom* is being studied (Babbie, 2010). In social science research, individuals are the most typical unit of analysis. Social groups (such as families, cities, or geographic regions),

a. Causal relationship

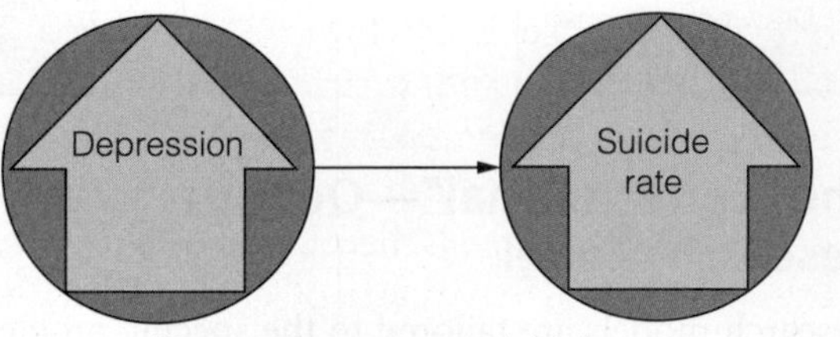

"Depression causes suicide."

b. Inverse causal relationship (Durkheim)

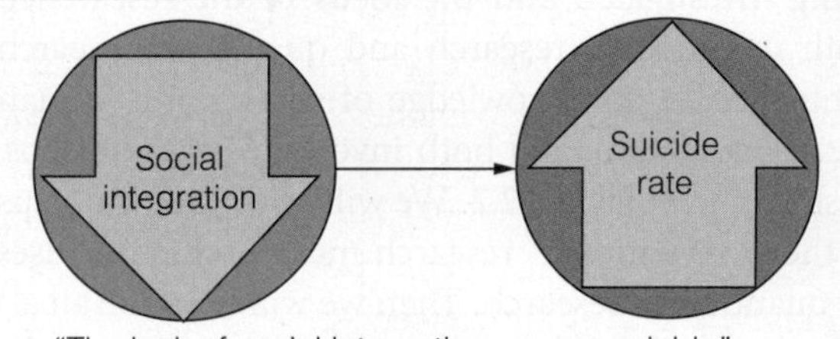

"The lack of social integration causes suicide."

c. Multiple-cause explanation

Rate of social change
Poverty
Religiosity
Suicide rate

"Many factors interact to cause suicide."

▶ FIGURE 2.3 HYPOTHESIZED RELATIONSHIPS BETWEEN VARIABLES

A causal hypothesis connects one or more independent (causal) variables with a dependent (affected) variable. The diagram illustrates three hypotheses about the causes of suicide. To test these hypotheses, social scientists would need to operationalize the variables (define them in measurable terms) and then investigate whether the data support the proposed explanation.

© David R. Frazier Photolibrary, Inc./Alamy

▲ An operational definition is an explanation of an abstract concept in terms of observable features that are specific enough to measure the variable. For example, the operational definition of an *A* may be an exam average of 90 percent or above. After college professors have established the grading requirements for a course, students seek to meet those expectations by performing well on examinations.

organizations (such as clubs, labor unions, or political parties), and social artifacts (such as books, paintings, or weddings) may also be units of analysis. Durkheim's unit of analysis was social groups, not individuals, because he believed that the study of individual cases of suicide would not explain the rates of suicide.

After determining the unit of analysis for your study, you must select a time frame for study: cross-sectional or longitudinal. *Cross-sectional studies* are based on observations that take place at a single point in time; these studies focus on behavior or responses at a specific moment. *Longitudinal studies* are concerned with what is happening over a period of time or at several different points in time; they focus on processes and social change. Some longitudinal studies are designed to examine the same set of people each time, whereas others look at trends within a general population. Using longitudinal data, Durkheim was able to compare suicide rates over a period of time in France and other European nations.

5. *Collect and analyze the data.* Your next step is to collect and analyze data. You must decide which population—persons about whom we want to be able to draw conclusions—will be observed or questioned. Then it is necessary to select a sample of people from the larger population to be studied. It is important that the sample accurately represent the larger population. For example, if you arbitrarily selected five students from your class to interview about the problem of bullying, they probably would not be representative of your school's total student body. However, if you selected five students from the total student body by a random sample, they might be closer to being representative (although a random sample of five students would be too small to yield much useful data). In ***random sampling*, every member of an entire population being studied has the same chance of being selected.** You would have a more representative sample of the total student body, for example, if you placed all the students' names in a rotating drum and conducted a drawing. By contrast, in ***probability sampling*, participants are deliberately chosen because they have specific characteristics, possibly including such factors as age, sex, race/ethnicity, and educational attainment.**

For his study of suicide, Durkheim collected data from vital statistics for approximately 26,000 suicides. He classified them separately according to age, sex,

independent variable a variable that is presumed to cause or determine a dependent variable.

dependent variable a variable that is assumed to depend on or be caused by one or more other (independent) variables.

random sampling a study approach in which every member of an entire population being studied has the same chance of being selected.

probability sampling choosing participants for a study on the basis of specific characteristics, possibly including such factors as age, sex, race/ethnicity, and educational attainment.

sociology in global perspective

Contemporary Problems in France: Worker Stress and Suicide Rates

When I started as a psychiatrist, 35 years ago, my patients were talking about their personal lives. Now it's all about their jobs. People are suffering in the workplace. They shouldn't be, from the logic of management. After all, they have a good job, a nice vacation. But they are suffering.

—Marie-France Hirigoyen, a psychiatrist in France, describing how workers who many believe should feel secure are complaining about rapid economic changes and job-related stress (qtd. in Jolly and Saltmarsh, 2009)

Although social scientists and health care professionals believe that many factors contribute to an increase in the number of suicides in a nation, stress in the workplace is one factor that often brings about an increasing rate of suicide in countries such as France. In 2010, for example, France was listed eighth in the world in the suicide rate per 100,000 people living in that country: Specifically, the rate of suicide was 26.4 per 100,000 for men and 9.2 per 100,000 for women. Some suicides have drawn widespread media attention and brought up pressing questions such as this one: What is going on in a nation that is assumed by much of the rest of the world to have it made when it comes to work because workers in France have had excellent job security, a 35-hour workweek, and lengthy, paid vacations?

What contributed to the increase in suicides in France? Is it possible to envision that people might have a steady job and good benefits but still be fearful enough about the future to end their life? When there is a high level of uncertainty, some people may experience an increase in feelings of depression, fear, and ambivalence, particularly if they are exposed to growing levels of personal risk. In the words of a labor union official, "Stress has become a national sport. We need employers to modify the way that they organize work, but we don't have the impression that anything will happen soon" (qtd. in Jolly and Saltmarsh, 2009: B3). Instead, numerous companies, facing global competition, are cutting jobs through voluntary departures, early retirements, and other mechanisms that fall short of actually giving workers "pink slips" to terminate their employment. But as current workers see all of this going on around them, many become concerned about their own future. Despite keeping their jobs for now, some employees are being asked to move from one job to another within the company or relocate to other jobsites. Workers who see constant change and upheaval going on around them often become insecure despite reassurances that they have job security.

When we look at economic issues from a global perspective, we can see that rapid social change (as suggested by Durkheim) may be a factor in why nations such as France have higher suicide rates than some others. However, we can also see that factors such as the increasingly fragmented nature of contemporary life may produce stress and ambivalence, even if the workers have nothing to fear. However, according to Xavier Darcos, France's employment minister, it is better to feel insecure than to be unemployed: "We are in a transforming economy. . . . For us, unemployment is the absolute failure. We prefer to have people who don't feel totally happy at work, or to work part time, rather than people being unemployed" (qtd. in Jolly and Saltmarsh, 2009: B3).

reflect & analyze

Why do some workers in France feel as much stress when they are asked to move from one position to another as workers feel in the United States when they are laid off or fired? How might sociological research be put to work to study the problem of an increasing rate of suicide in France?

MIGUEL MEDINA/AFP/Getty Images

Rapid social change can cause tension and insecurity. These French public employees were part of a strike to protest against the government's plan to change working hours and to increase the minimum retirement age from 60 to 62.

marital status, presence or absence of children in the family, religion, geographic location, calendar date, method of suicide, and a number of other variables. As Durkheim analyzed his data, four distinct categories of suicide emerged: egoistic, altruistic, anomic, and fatalistic. *Egoistic suicide* occurs among people who are isolated from any social group. For example, Durkheim concluded that suicide rates were relatively high in Protestant countries in Europe because Protestants believed in individualism and were more loosely tied to the church than were Catholics. Single people had proportionately higher suicide rates than married persons because they had a low degree of social integration, which contributed to their loneliness. In contrast, *altruistic suicide* occurs among individuals who are excessively integrated into society. An example is military leaders who kill themselves after defeat in battle because they have so strongly identified themselves with their cause that they believe they cannot live with defeat. According to Durkheim, people are more likely to kill themselves when social cohesion is either very weak or very strong, and/or when nations experience rapid social change (see "Sociology in Global Perspective").

We have traced the steps in the "conventional" research process (based on quantitative research). But what steps might be taken in an alternative approach based on qualitative research?

A Qualitative Research Model

Although the same underlying logic is involved in both quantitative and qualitative sociological research, the *styles* of these two models are very different (King, Keohane, and Verba, 1994). As previously stated, qualitative research is more likely to be used when the research question does not easily lend itself to numbers and statistical methods. As compared to a quantitative model, a qualitative approach often involves a different type of research question and a smaller number of cases.

Although the qualitative approach follows the conventional research approach in presenting a problem, asking a question, collecting and analyzing data, and seeking to answer the question, it also has several unique features (based on Creswell, 1998, and Kvale, 1996):

1. *The researcher begins with a general approach rather than a highly detailed plan.* Flexibility is necessary because of the nature of the research question. The topic needs to be explored so that we can know "how" or "what" is going on, but we may not be able to explain "why" a particular social phenomenon is occurring.
2. *The researcher has to decide when the literature review and theory application should take place.* Initial work may involve redefining existing concepts or reconceptualizing how existing studies have been conducted. The literature review may take place at an early stage, before the research design is fully developed, or it may occur after the development of the research design, and after the data collection has already occurred.
3. *The study presents a detailed view of the topic.* Qualitative research usually involves a smaller number of cases and many variables, whereas quantitative researchers typically work with a few variables and many cases.
4. *Access to people or other resources that can provide the necessary data is crucial.* Unlike the quantitative researcher, who often uses existing databases, many qualitative researchers generate their own data. As a result, it is necessary to have access to people and build rapport with them.
5. *Appropriate research method(s) are important for acquiring useful qualitative data.* Qualitative studies are often based on field research such as observation, participant observation, case studies, ethnography, and unstructured interviews, as discussed in the Research Methods section of this chapter.

How might qualitative research be used to study suicidal behavior? In studying different rates of suicide among women and men, for example, the social psychologist Silvia Canetto (1992) questioned whether existing theories and quantitative research provided an adequate explanation for gender differences in suicidal behavior and decided that she

▲ Sociological research on suicide has begun to look at issues such as what social factors might motivate suicide bombers. Some researchers might ask why suicide bomber Raed Abdel-Hameed Mesk (shown here with his children) would take his own life in the process of committing a terrorist attack.

would explore alternate explanations. Analyzing previous research, Canetto learned that most studies linked suicidal behavior in women to problems in their personal relationships, particularly with members of the opposite sex. By contrast, most studies of men's suicides focused on their performance and found that men are more likely to be suicidal when their self-esteem and independence are threatened. According to Canetto's analysis, gender differences in suicidal behavior are more closely associated with beliefs about and expectations for men and women rather than purely interpersonal crises.

As in Canetto's study, researchers using a qualitative approach may engage in *problem formulation* to clarify the research question and to develop questions of concern and interest to the research participants (Reinharz, 1992). To create a research design for Canetto's study, we might start with the proposition that most studies may have attributed women's and men's suicidal behavior to the wrong causes. Next, we might decide to interview people who have attempted suicide by using a collaborative approach in which the participants suggest avenues of inquiry that the researcher should explore (Reinharz, 1992).

Although Canetto did not gather data in her study, she reevaluated existing research, concluding that alternative explanations of women's and men's suicidal behavior are justified from existing data.

In a qualitative approach the next step is collecting and analyzing data to assess the validity of the starting proposition. Data gathering is the foundation of the research. Researchers pursuing a qualitative approach tend to gather data in natural settings, such as where the person lives or works, rather than in a laboratory or other research setting. Data collection and analysis frequently occur concurrently, and the analysis draws heavily on the language of the persons being studied, not the researcher.

Research Methods

How do sociologists know which research method to use? Are some approaches better than others? Which method is best for a particular problem? ***Research methods* are specific strategies or techniques for systematically conducting research**. We will look at four of these methods.

Survey Research

A *survey* is a poll in which the researcher gathers facts or attempts to determine the relationships among facts. Surveys are the most widely used research method in the social sciences because they make it possible to study things that are not directly observable—such as people's attitudes and beliefs—and to describe a population too large to observe directly (Babbie, 2010). Researchers frequently select a representative sample (a small group of respondents) from a larger population (the total group of people) to answer questions about their attitudes, opinions, or behavior. ***Respondents* are persons who provide data for analysis through interviews or questionnaires**. The Gallup, Harris, Roper, and Pew polls are among the most widely known large-scale surveys; however, government agencies such as the U.S. Census Bureau conduct a variety of surveys as well.

Unlike many polls that use various methods of gaining a representative sample of the larger population, the Census Bureau attempts to gain information from all persons in the United States. The decennial census occurs every 10 years, in the years ending in "0." The purpose of this census is to count the population and housing units of the entire United States. The population count determines how seats in the U.S. House of Representatives are apportioned; however, census figures are also used in formulating public policy and in planning and decision making in the private sector. The Census Bureau attempts to survey the *entire* U.S. population by using two forms—a "short form" of questions asked of *all* respondents and a "long form" that contains additional questions asked of a *representative sample* of about one in six respondents. Statistics from the Census Bureau provide information that sociologists use in their research. An example is shown in the Census Profiles feature: "How People in the United States Self-Identify Regarding Race." Note that because of recent changes in the methods used to collect data by the Census Bureau, information on race from the 2000 census is not directly comparable with data from earlier censuses.

Surveys are the most widely used research method in the social sciences because they make it possible to study things that are not directly observable—such as people's attitudes and beliefs—and to describe a population too large to observe directly (Babbie, 2010). Let's take a brief look at the most frequently used types of surveys.

Types of Surveys Survey data are collected by using self-administered questionnaires, face-to-face interviews, and/or telephone interviews. **A *questionnaire* is a printed research instrument containing a series of items to which subjects respond.** Items are often in the form of statements with which the respondent is asked to "agree" or "disagree." Questionnaires may be administered by interviewers in face-to-face encounters or by telephone, but the most commonly used technique is the *self-administered questionnaire*. The questionnaires are typically mailed or delivered to the respondents'

How People in the United States Self-Identify Regarding Race

Beginning with Census 2000 and continuing with Census 2010, the U.S. Census Bureau has made it possible for people responding to census questions regarding their race and Hispanic origin to indicate more than one category. If you look at the figures set forth below, they total more than 100 percent of the total population. How can this be? Simply stated, some individuals are counted at least twice, based on the number of racial categories they listed.

Race	Percentage of Total Population
White alone or in combination with one or more other races	74.8
African American alone or in combination with one or more other races	13.6
Asian American alone or in combination with one or more other races	5.6
Native American and Alaska Native alone or in combination with one or more other races	1.7
Native Hawaiian and other Pacific Islander alone or in combination with one or more other races	0.4
Some other race alone or in combination with one or more other races	7.0

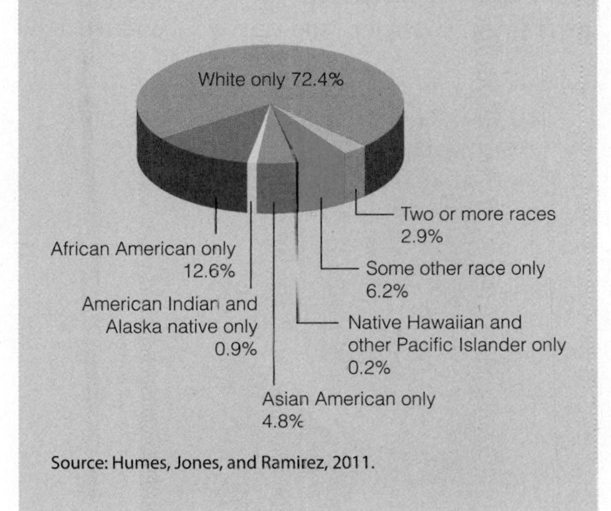

Source: Humes, Jones, and Ramirez, 2011.

homes; however, they may also be administered to groups of respondents gathered at the same place at the same time. For example, the sociologist Kevin E. Early (1992) used survey data collected through questionnaires to test his hypothesis that suicide rates are lower among African Americans than among white Americans because of the influence of the black church. Data from questionnaires filled out by members of six African American churches in Florida supported Early's hypothesis that the church buffers some African Americans against harsh social forces, such as racism, that might otherwise lead to suicide.

Survey data may also be collected by interviews. An ***interview* is a data-collection encounter in which an interviewer asks the respondent questions and records the answers.** Survey research often uses *structured interviews*, in which the interviewer asks questions from a standardized questionnaire. Structured interviews tend to produce uniform or replicable data that can be elicited time after time by different interviews. For example, in addition to surveying congregation members, Early (1992) conducted interviews with pastors of African American churches to determine the pastors' opinions about the extent to which the African American church reinforces values and beliefs that discourage suicide.

Interviews have specific advantages. They are usually more effective in dealing with complicated issues and provide an opportunity for face-to-face communication between the interviewer and the respondent. Although interviews provide a wide variety of useful information, a major disadvantage is the cost and time involved in conducting the interviews and analyzing the results. A quicker method of administering questionnaires is the *telephone* or *computer survey*. Telephone and computer surveys give greater control over data collection and provide greater personal safety for respondents and researchers than do personal encounters. In *computer-assisted telephone interviewing* (sometimes called CATI), the interviewer uses a computer to dial random telephone numbers, reads the questions shown on the video monitor to the respondent, and then

research methods specific strategies or techniques for systematically conducting research.

survey a poll in which the researcher gathers facts or attempts to determine the relationships among facts.

respondents persons who provide data for analysis through interviews or questionnaires.

questionnaire a printed research instrument containing a series of items to which subjects respond.

interview a research method using a data-collection encounter in which an interviewer asks the respondent questions and records the answers.

© Masterfile-RF

◀ Computer-assisted telephone interviewing is an easy and cost-efficient method of conducting research. However, the widespread use of answering machines, voice mail, and caller ID may make this form of research more difficult in the twenty-first century.

types the responses into the computer terminal. The answers are immediately stored in the central computer, which automatically prepares them for data analysis. However, the respondent must answer the phone before the interview can take place, and many people screen their phone calls. In the 2000s online survey research has increased dramatically as software packages and online survey services have made this type of research easier to conduct. Online research makes it possible to study virtual communities, online relationships, and other types of computer-mediated communications networks around the world.

Strengths and Weaknesses of Surveys Survey research is useful in describing the characteristics of a large population without having to interview each person in that population. In recent years, computer technology has enhanced researchers' ability to do *multivariate analysis*—research involving more than two independent variables. For example, to assess the influence of religion on suicidal behavior among African Americans, a researcher might look at the effects of age, sex, income level, and other variables all at once to determine which of these independent variables influences suicide the most or least and how influential each variable is relative to the others. However, a weakness of survey research is the use of standardized questions; this approach tends to force respondents into categories in which they may or may not belong. Moreover, survey research relies on self-reported information, and some people may be less than truthful, particularly on emotionally charged issues such as suicide. Some scholars have also criticized the way survey data are used. They believe that survey data do not always constitute the "hard facts" that other analysts may use to justify changes in public policy or law. For example, survey statistics may over- or underestimate the extent of a problem and work against some categories of people more than others, as shown in ■ Table 2.1.

Secondary Analysis of Existing Data

In ***secondary analysis*****, researchers use existing material and analyze data that were originally collected by others.** Existing data sources include public records, official reports of organizations and government agencies, and surveys conducted by

© Angela Hampton Picture Library/Alamy

▲ Conducting surveys and polls is an important means of gathering data from respondents. Some surveys take place on street corners; increasingly, however, such surveys are done by telephone, Internet, or other means.

table 2.1

Statistics: What We Know (and Don't Know)

Topic	Homelessness in the United States	Suicide in the United States
Research Finding	At least 250,000 people in this country are homeless.	More than 34,000 suicides occurred in this country in 2007.
Possible Problem	Does that badly underestimate the total number of homeless people?	Are suicide rates different for some categories of U.S. residents?
Explanation	The homeless are difficult to count, frequently attempting to avoid interviews with census takers. Critics of the census figures assert that the actual number may be 3 million and that the government intentionally undercounts the homeless.	It is difficult to know about racial and ethnic disparities in suicide rates: Census data tend to shift categories over time, making comparisons difficult, if not impossible.

As the examples in this table show, statistics provide certain insights into the prevalence of social issues such as homelessness and suicide but do not always provide the *answer* regarding the nature and extent of the problem. What difficulties do researchers encounter when gathering data on people?

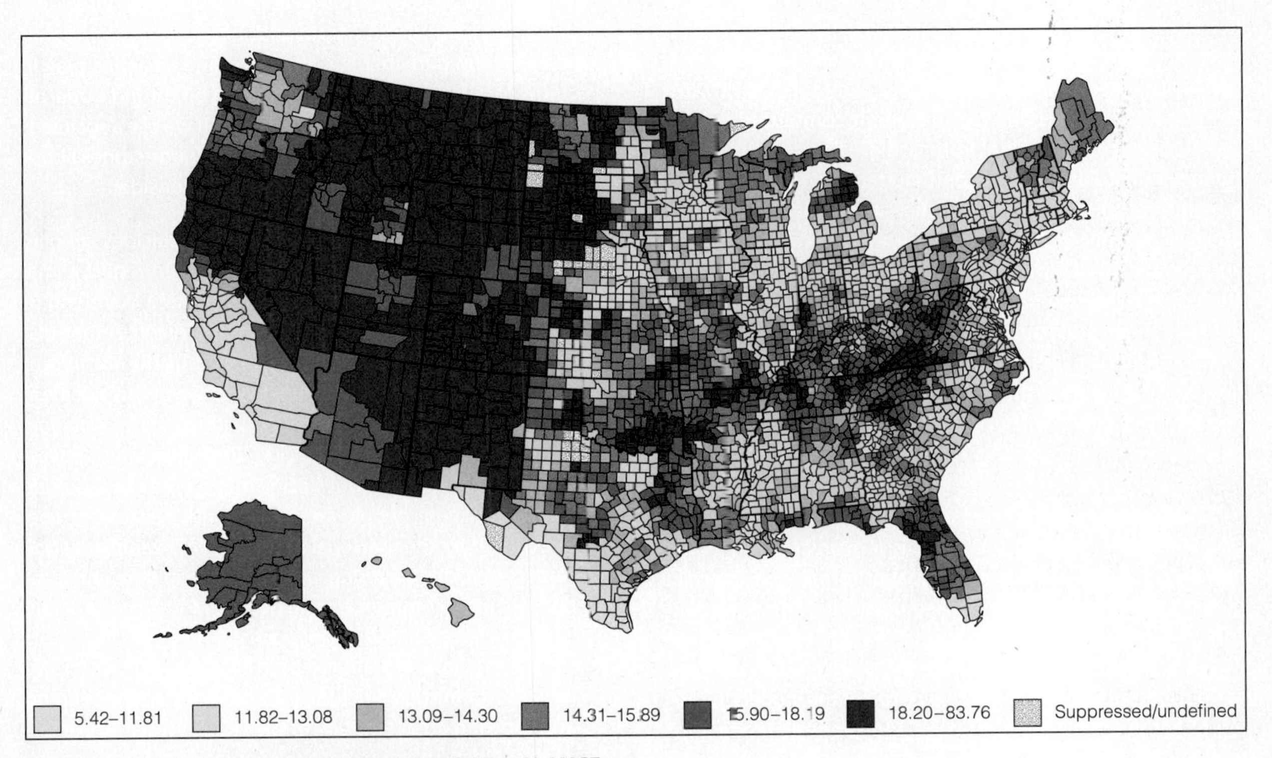

▲ **MAP 2.1 NATIONAL SUICIDE STATISTICS AT A GLANCE**

Source: www.cdc.gov/violenceprevention/suicide/statistics/suicide_map.html

researchers in universities and private corporations. Research data gathered from studies are available in data banks, such as the Inter-University Consortium for Political and Social Research, the National Opinion Research Center (NORC), and the Roper Public Opinion Research Center. Today, many researchers studying suicide use data compiled by the National Centers for Disease Control and Prevention (CDC) and the National Center for Injury Prevention and Control. For example, look at ▶ Map 2.1, "National Suicide Statistics at a Glance," based on data compiled by the CDC. Other sources of data for secondary analysis are books, magazines, newspapers, radio and television programs, websites, and personal documents. Secondary analysis is referred to as *unobtrusive research* because it has no impact on the people being studied. In Durkheim's study of suicide, for example, his analysis of existing statistics on suicide did nothing to increase or decrease the number of people who *actually* committed suicide.

Analyzing Existing Statistics Secondary analysis may involve obtaining *raw data* collected by other researchers and undertaking a statistical analysis of the data, or it may involve the use of other

secondary analysis a research method in which researchers use existing material and analyze data that were originally collected by others.

framing suicide in the media

Sociology Versus Sensationalism

Front Page: *New York Post*, March 10, 2004:

. . . New York University student to die in a plunge this academic year. [Name of student,] 19, jumped . . . from the roof of her boyfriend's 24-story apartment building Saturday after having a fight with him. (*New York Post*, 2004: 1)

New York Times, March 10, 2004:

The suicide of a New York University student who fell to her death from a rooftop off campus on Saturday chilled students, who learned about it in an e-mail message from university officials on Monday. It was the fourth N.Y.U. student death this year. One N.Y.U. student committed suicide last semester by jumping from a high floor of the Elmer Holmes Bobst Library. A second student also jumped from a high floor in the library, but the city medical examiner's office found that he had been on drugs and ruled his death an accident. They have not yet ruled on the death of a third student, a young woman who fell to her death from the window of a friend's apartment near Washington Square Park. . . .
—Karen Arenson, "Suicide of N.Y.U. Student, 19, Brings Sadness and Questions," the New York Times, March 10, 2004.

Compare these two news accounts of the same suicide: One source is a tabloid; the other is a mainstream national newspaper. Tabloids have a newspaper format but typically provide readers with a condensed version of the news and contain illustrated, often sensational material that they hope will encourage people to buy that day's paper. In this case the tabloid features a large, four-color photo of the student jumping to her death on the front page and suggests that a fight with her boyfriend was the cause of her suicide. By contrast, the *New York Times* article begins with a description of how this student's death might affect other students and explains how school officials notified students of the tragedy. No picture is included with the coverage.

As this comparison shows, the media offer us different vantage points from which to view a given social event based on how they *frame* the information they provide to their audience. The term *media framing* refers to the process by which information and entertainment are packaged by the mass media (newspapers, magazines, radio and television networks and stations, and the Internet) before being presented to an audience. This process includes factors such as the amount of exposure given to a story, where it is placed, the positive or negative tone it conveys, and its accompanying headlines, photographs, or other visual or auditory effects (if any). Through framing, the media emphasize some beliefs and values over others and manipulate salience by directing people's attention to some ideas while ignoring others. As such, a frame constitutes a story line or an unfolding narrative about an issue. These narratives are organizations of experience that bring order to events. Consequently, such narratives wield power because they influence how we make sense of the world (Kendall, 2011).

Thinking sociologically, what problems exist in how the media frame stories about a social issue such as suicide? As discussed in this chapter, the early sociologist Emile Durkheim believed that we should view even seemingly individual actions—such as suicide—from a *sociological perspective* that focuses on the part that social groups and societies play in patterns of behavior rather than focusing on the *individual* attributes of people who commit such acts. If we use a sociological approach to analyzing how the media frame stories about suicide, here are three questions for you to consider:

- Do television and newspaper reports of suicides simplify the *reasons* for the suicide? The media often report a final precipitating situation (such as a fight with a boyfriend or girlfriend, losing one's job, or getting a divorce) that distressed the individual prior to suicide but do not inform audiences that this was not the *only* cause of this suicide.
- Are readers and viewers provided with *repetitive, ongoing, and excessive reporting* on a suicide? Repeated sensationalistic framing of stories about suicide may contribute to *suicide contagion*. Some people are more at risk because of age, stress, and/or other personal problems. For example, *suicide clusters* (suicides occurring close together in time and location) are most common among people 15 to 24 years of age who may have learned details from the media.
- Do the media use dramatic photographs (such as of the person committing suicide or friends and family weeping and showing great emotion at the victim's funeral) primarily to sell papers or increase viewer ratings? Photographs and other sensational material are potentially most damaging for people who have long-standing mental health problems or who have limited social networks to provide them with hope, encouragement, and guidance.

In our mass-mediated culture, many sociologists agree that the media do much more than simply *mirror* society: The media help shape our society and our cultural perceptions on many issues, including seemingly individual behavior such as suicide.

reflect & analyze

Have you ever read two media accounts of the same event that were examples of very different approaches to framing? Are we influenced by how the media frame news and entertainment stories even when we think we are unaffected by such coverage?

researchers' existing statistical analyses. In analysis of existing statistics, the unit of analysis is often *not* the individual. Most existing statistics are *aggregated:* They describe a group. For example, Durkheim used vital statistics (death records) that were originally collected for other purposes to examine the relationship among variables such as age, marital status, and the circumstances surrounding the suicides of a large number of individuals.

In a contemporary study of suicide, K. D. Breault (1986) analyzed secondary data collected by government agencies to test Durkheim's hypothesis that religion and social integration provide protection from suicide. Using suicide as the dependent variable and church membership, divorce, unemployment, and female labor-force participation as several of his independent variables, Breault performed a series of sophisticated statistical analyses and concluded that the data supported Durkheim's views on social integration and his theory of egoistic suicide.

Analyzing Content ***Content analysis*** **is the systematic examination of cultural artifacts or various forms of communication to extract thematic data and draw conclusions about social life.** *Cultural artifacts* are products of individual activity, social organizations, technology, and cultural patterns (Reinharz, 1992). Among the materials studied are written records (such as diaries, love letters, poems, books, and graffiti), narratives and visual texts (such as movies, television programs, websites, advertisements, and greeting cards), and material culture (such as music, art, and even garbage). Researchers may look for regular patterns, such as frequency of suicide as a topic on television talk shows. They may also examine subject matter to determine how it has been represented, such as how the mass media frame presentations on suicide (see "Framing Suicide in the Media" for an example).

Content analysis provides objective coding procedures for analyzing written material (see Berg, 1998; Manning and Cullum-Swan, 1994). It also allows for the counting and arranging of data into clearly identifiable categories (manifest coding) and provides for the creation of analytically developed categories (latent or open coding). Using latent or open coding, it is possible to identify general themes, create generalizations, and develop "grounded theoretical" explanations (Glaser and Strauss, 1967). As this explanation suggests, researchers use both qualitative and quantitative procedures in content analysis.

How might a social scientist use content analysis in research on why people commit suicide? Suicide notes and diaries are useful forms of cultural artifacts. Suicide notes have been subjected to extensive analysis because they are "ultrapersonal documents" that are not solicited by others and are frequently written just before the person's death (Leenaars, 1988: 34). Many notes provide new levels of meaning regarding the *individuality* of the person who committed or attempted suicide. For example, some notes indicate that people may want to get revenge and make other people feel guilty or responsible for their suicide (Leenaars, 1988). Thus, suicide notes may be a valuable starting point for finding patterns of suicidal behavior and determining the characteristics of people who are most likely to commit suicide (Leenaars, 1988).

Strengths and Weaknesses of Secondary Analysis One strength of secondary analysis is that data are readily available and inexpensive. Another is that, because the researcher often does not collect the data personally, the chances of bias may be reduced. In addition, the use of existing sources makes it possible to analyze longitudinal data (things that take place over a period of time or at several different points in time) to provide a historical context within which to locate original research. However, secondary analysis has inherent problems. For one thing, the researcher does not always know if the data are incomplete, unauthentic, or inaccurate.

Field Research

Field research **is the study of social life in its natural setting: observing and interviewing people where they live, work, and play.** Some kinds of behavior can be studied best by "being there"; a fuller understanding can be developed through observations, face-to-face discussions, and participation in events. Researchers use these methods to generate *qualitative* data: observations that are best described verbally rather than numerically.

Participant Observation Sociologists who are interested in observing social interaction as it occurs may use ***participant observation*****—the process of collecting systematic observations while being part**

content analysis the systematic examination of cultural artifacts or various forms of communication to extract thematic data and draw conclusions about social life.

field research the study of social life in its natural setting: observing and interviewing people where they live, work, and play.

participant observation a research method in which researchers collect data while being part of the activities of the group being studied.

of the activities of the group that the researcher is studying. Participant observation generates more "inside" information than simply asking questions or observing from the outside. For example, to learn more about how coroners make a ruling of "suicide" in connection with a death and to analyze what (if any) effect such a ruling has on the accuracy of official suicide statistics, the sociologist Steve Taylor (1982) engaged in participant observation at a coroner's office over a six-month period. As he followed a number of cases from the initial report of death through the various stages of investigation, Taylor learned that it was important to "be around" so that he could listen to discussions and ask the coroners questions because intuition and guesswork play a large part in some decisions to rule a death as a suicide.

Case Studies Most participant observation research takes the form of a *case study,* which is often an in-depth, multifaceted investigation of a single event, person, or social grouping (Feagin, Orum, and Sjoberg, 1991). However, a case study may also involve multiple cases and is then referred to as a *collective case study* (Stake, 1995). Whether the case is single or collective, most case studies require detailed, in-depth data collection involving multiple sources of rich information such as documents and records and the use of methods such as participant observation, unstructured or in-depth interviews, and life histories (Creswell, 1998). As they collect extensive amounts of data, the researchers seek to develop a detailed description of the case, to analyze the themes or issues that emerge, and to interpret or create their own assertions about the case (Stake, 1995).

When do social scientists decide to do case studies? Initially, some researchers have only a general idea of what they wish to investigate. In other cases, they literally "back into" the research. They may find themselves close to interesting people or situations. For example, the anthropologist Elliot Liebow "backed into" his study of single, homeless women living in emergency shelters by becoming a volunteer at a shelter. As he got to know the women, Liebow became fascinated with their lives and survival strategies. Prior to Liebow's research, most studies of the homeless focused primarily on men. By contrast, Liebow wanted to know more about homeless women, wondering such things as "What are they carrying in those [shopping] bags?" (Coughlin, 1993: A8). Liebow spent the next four years engaged in participant observation research that culminated in his book *Tell Them Who I Am* (1993). Liebow's findings are discussed in Chapter 5 ("Society, Social Structure, and Interaction in Everyday Life").

▲ Field research takes place in a wide variety of settings. For example, how might sociologists study the ways in which parents and their college-age children cope with change when the students first leave home and move into college housing?

Ethnography An *ethnography* **is a detailed study of the life and activities of a group of people by researchers who may live with that group over a period of years** (Feagin, Orum, and Sjoberg, 1991). Unlike participant observation, ethnographic studies usually take place over a longer period of time. For example, in a classic community study, *Middletown* and *Middletown in Transition,* the sociologists Robert Lynd and Helen Lynd (1929, 1937) lived in Muncie, Indiana, for a number of years and conducted ethnographic research on the daily lives of residents and the composition of the local power structure. The Lynds' study concluded that people in Muncie had strong beliefs about the importance of religion, hard work, self-reliance, and civic pride; that one dominant family "ruled" the city; and that a working class emerged in the community when companies opened factories in Muncie.

In research reported in the book *Code of the Street,* the sociologist Elijah Anderson (1990) conducted a study in two areas of a major city—one African American and low-income, the other racially mixed but becoming increasingly middle- to upper-income and white. As Anderson spent numerous hours on the streets, talking and listening to the people, he was

© Addison Geary

▲ Elijah Anderson (at left in photo) conducted an ethnographic study of two very different Philadelphia neighborhoods that became the basis for his landmark study *Code of the Street*. What can researchers learn from ethnographic research that might be less apparent if they used other methods to study human behavior?

Cover used by permission of W. W. Norton & Company, Inc. Cover photo © Camilo Jose Vergara.

able to document the changes in residents' everyday lives brought about by increasing drug abuse, loss of jobs, decreases in city services despite increases in taxes, and the eventual exodus of middle-income people from the central city.

Unstructured Interviews An ***unstructured interview* is an extended, open-ended interaction between an interviewer and an interviewee.** This type of interview is referred to as *unstructured* because few predetermined or standardized procedures are established for conducting it. Because many decisions have to be made during the interview, this approach requires that the researcher have a high level of skill in interviewing and extensive knowledge regarding the interview topic (Kvale, 1996). Unstructured interviews are essentially conversations in which interviewers establish the general direction by asking open-ended questions, to which interviewees may respond flexibly, and then interviewers may "shift gears" to pursue specific topics raised by interviewees.

Sociologist Joe Feagin's (1991) study of middle-class African Americans is an example of research that used in-depth interviews to examine public discrimination and victims' coping strategies. No specific questions were asked regarding discrimination in public accommodations or other public places. Rather, discussion of discrimination was generated by answers to general questions about barriers to personal goals or in digressions in answers to specific questions about employment, education, and housing.

Before conducting in-depth interviews, researchers must make a number of decisions, including how the people to be interviewed will be selected. Respondents are often chosen by "snowball sampling"—a method in which the researcher interviews a few individuals who possess a certain characteristic and then these interviewees are asked to supply the names of others with the same characteristic (such as persons who are members of the same social organization). This process continues until the sample has "snowballed" into an acceptable size and no new information of any significance is being gained by the researchers.

Using unstructured, open-ended interviews does not mean that the researcher simply walks into a room, has a conversation with someone, and the research is complete. Planning and preparation are essential. Similarly, the follow-up, analysis of data, and write-up of the study must be carefully designed and carried out.

Interviews and Theory Construction In-depth interviews, along with participant observation and case studies, are frequently used to develop theories through observation. The term *grounded theory* was developed by sociologists Barney Glaser and Anselm Strauss (1967) to describe this inductive

ethnography a detailed study of the life and activities of a group of people by researchers who may live with that group over a period of years.

unstructured interview an extended, open-ended interaction between an interviewer and an interviewee.

method of theory construction. Researchers who use grounded theory collect and analyze data simultaneously. For example, after in-depth interviews with 106 suicide attempters, researchers in one study concluded that half of the individuals who attempted suicide wanted *both* to live *and* to die at the time of their attempt. From these unstructured interviews it became obvious that ambivalence led about half of "serious" suicidal attempters to "literally gamble with death" (Kovacs and Beck, 1977, qtd. in Taylor, 1982: 144). After asking their initial unstructured questions of the interviewees, Kovacs and Taylor decided to widen the research question from "Why do people kill themselves?" to a broader question: "Why do people engage in acts of self damage which may result in death?" In other words, uncertainty of outcome is a common feature of most suicidal acts.

Strengths and Weaknesses of Field Research Participant observation research, case studies, ethnography, and unstructured interviews provide opportunities for researchers to view from the inside what may not be obvious to an outside observer. They are useful when attitudes and behaviors can be understood best within their natural setting or when the researcher wants to study social processes and change over a period of time. They provide a wealth of information about the reactions of people and give us an opportunity to generate theories from the data collected (Whyte, 1989).

A weakness of field research is the inability to generalize what is learned from a specific group or community to a larger population. Data collected in natural settings are descriptive and do not lend themselves to precise measurement. To counteract these criticisms, some qualitative researchers use computer-assisted qualitative data analysis programs that make it easier for researchers to enter, organize, annotate, code, retrieve, count, and analyze data (Dohan and Sanchez-Jankowski, 1998).

Experiments

An ***experiment*** **is a carefully designed situation in which the researcher studies the impact of certain variables on subjects' attitudes or behavior.** Experiments are designed to create "real-life" situations, ideally under controlled circumstances, in which the influence of different variables can be modified and measured.

Types of Experiments Conventional experiments require that subjects be divided into two groups: an experimental group and a control group. The ***experimental group*** **contains the subjects who are exposed to an independent variable (the experimental condition) to study its effect on them.** The ***control group*** **contains the subjects who are not exposed to the independent variable.** The

▲ Do extremely violent video games cause an increase in violent tendencies in their users? Experiments are one way to test this hypothesis.

members of the two groups are matched for similar characteristics so that comparisons may be made between the groups. The experimental and control groups are then compared to see if they differ in relation to the dependent variable, and the hypothesis stating the relationship of the two variables is confirmed or rejected.

In a *laboratory experiment,* subjects are studied in a closed setting so that researchers can maintain as much control as possible over the research. For example, the sociologist Arturo Biblarz and colleagues (1991) examined the effects of media violence and depictions of suicide on attitudes toward suicide by showing one group of subjects (an experimental group) a film about suicide, while a second (another experimental group) saw a film about violence, and a third (the control group) saw a film containing neither suicide nor violence. The research found some evidence that people exposed to suicidal acts or violence in the media may be more likely to demonstrate an emotional state favorable to suicidal behavior, particularly if they are already "at risk" for suicide.

Researchers may use experiments when they want to demonstrate that a cause-and-effect relationship exists between variables. In order to show that a change in one variable causes a change in another, these three conditions must be satisfied:

1. *You must show that a correlation exists between the two variables.* ***Correlation*** **exists when two variables are associated more frequently than could be expected by chance** (Hoover, 1992). For example, suppose that you wanted to test the hypothesis that the availability of a crisis intervention center with a twenty-four-hour counseling "hotline" on your campus causes a change in students' attitudes toward suicide (see ▶ Figure 2.4). To demonstrate

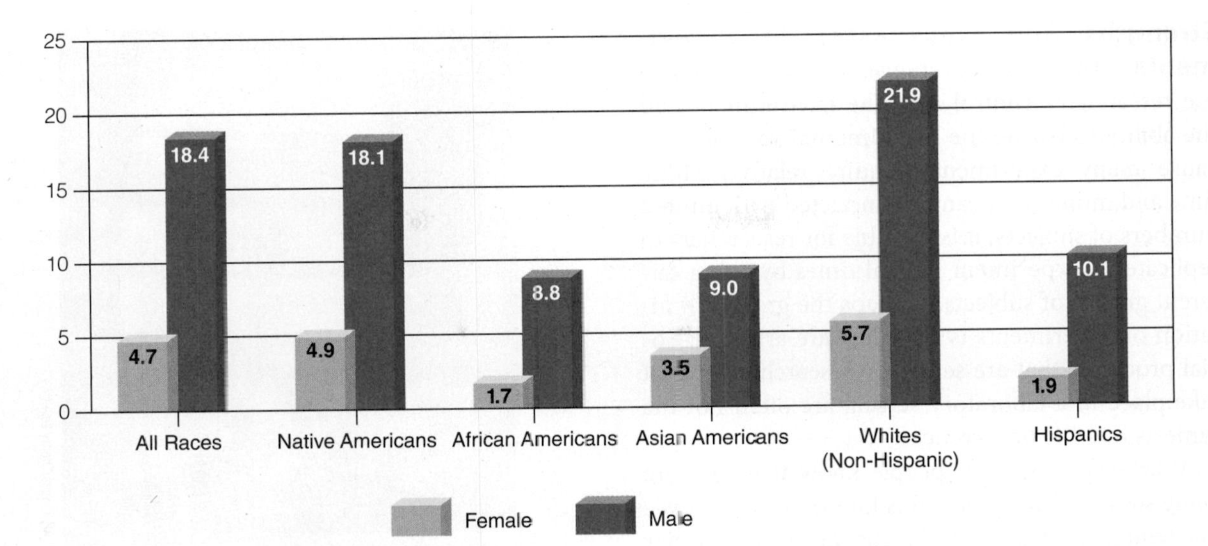

▲ **FIGURE 2.4 SUICIDE RATES BY RACE AND SEX**
Rates are for U.S. suicides and indicate the number of deaths by suicide for every 100,000 people in each category for 2006.
Source: National Center for Health Statistics, 2008.

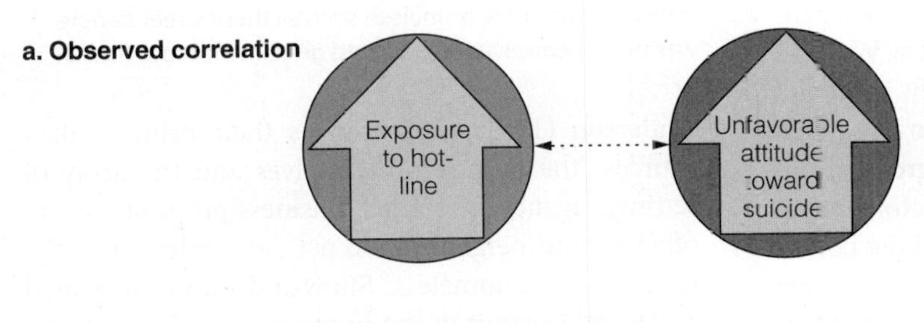

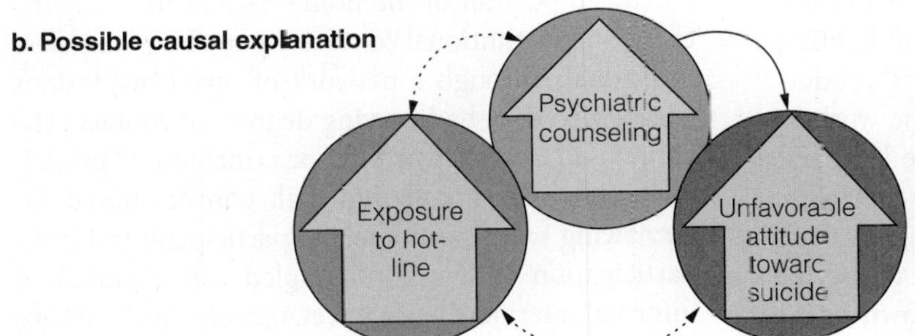

◀ **FIGURE 2.5 CORRELATION VERSUS CAUSATION**
A study might find that exposure to a suicide hotline is associated (correlated) with a change in attitude toward suicide. But if some of the people who were exposed to the hotline also received psychiatric counseling, the counseling may be the "hidden" cause of the observed change in attitude. In general, correlations alone do not prove causation.

correlation you would need to show that the students had different attitudes toward committing suicide depending on whether they had any experience with the crisis intervention center.

2. *You must ensure that the independent variable preceded the dependent variable.* If differences in students' attitudes toward suicide were evident before the students were exposed to the intervention center, exposure to the center could not be the cause of these differences.
3. *You must make sure that any change in the dependent variable was not because of an extraneous variable*—one outside the stated hypothesis. If some of the students receive counseling from off-campus psychiatrists, any change in attitude that they experience could be because of this third variable, not the hotline. This is referred to as a *spurious correlation*—the association of two variables that is actually caused by a third variable and does not demonstrate a cause-and-effect relationship (see ▶ Figure 2.5).

experiment a research method involving a carefully designed situation in which the researcher studies the impact of certain variables on subjects' attitudes or behavior.

experimental group in an experiment, the group that contains the subjects who are exposed to an independent variable (the experimental condition) to study its effect on them.

control group in an experiment, the group containing the subjects who are not exposed to the independent variable.

correlation a relationship that exists when two variables are associated more frequently than could be expected by chance.

Strengths and Weaknesses of Experiments The major advantage of an experiment is the researcher's control over the environment and the ability to isolate the experimental variable. Because many experiments require relatively little time and money and can be conducted with limited numbers of subjects, it is possible for researchers to replicate an experiment several times by using different groups of subjects. Perhaps the greatest limitation of experiments is that they are artificial. Social processes that are set up by researchers or that take place in a laboratory setting are often not the same as real-life occurrences.

What happens when people know that they are being studied? This problem is known as *reactivity*—the tendency of subjects to change their behavior in response to the researcher or to the fact that they know they are being studied. Social psychologist Elton Mayo first noticed this problem in a study conducted between 1927 and 1932 to determine how worker productivity and morale could be improved at Western Electric's Hawthorne plant. To identify variables that increase worker productivity, Mayo separated one group of women (the experimental group) from other workers and systematically varied factors in that group's work environment, while the working conditions for the other workers (the control group) were not changed. The researchers tested a number of factors, including an increase in the amount of lighting, to see if more light would raise the workers' productivity. Much to the researchers' surprise, the workers' productivity increased not only when the light was brighter but also when it was dimmed. In fact, all of the changes increased productivity, leading Mayo to conclude that the subjects were trying to please the researchers because interest was being shown in the workers (Roethlisberger and Dickson, 1939). Thus, the ***Hawthorne effect*** (based on the name of the plant where the research took place) **refers to changes in a subject's behavior caused by the researcher's presence or by the subject's awareness of being studied.**

Multiple Methods: Triangulation

What is the best method for studying a topic? The Concept Quick Review compares the various social research methods. There is no one best research method because social reality is complex and all research methods have limitations. Many sociologists believe that *triangulation*—the use of multiple methods in one study—is the solution to this problem. Triangulation refers not only to research methods but also to multiple data sources, investigators, and theoretical perspectives in a study. For example, in a study of more than 700 homeless people in Austin, Texas, in the mid-1980s, the sociologists David Snow and Leon

▲ Multiple research methods are often used to gain information about important social concerns. Which methods might be most effective in learning more about the problems of the homeless, such as these street people warming themselves on a heated grate in Moscow, Russia?

Anderson (1991: 158) used as their primary data sources "the homeless themselves and the array of settings, agency personnel, business proprietors, city officials, and neighborhood activities relevant to the routines of the homeless." Snow and Anderson gained a detailed portrait of the homeless and their experiences and institutional contacts by tracking homeless individuals through a network of seven institutions with which they had varying degrees of contact. The study used a variety of methods, including "participant observation and informal, conversational interviewing with the homeless; participant and nonparticipation observation, coupled with formal and informal interviewing in street agencies and settings; and a systematic survey of agency records" (Snow and Anderson, 1991: 158–169). This study is discussed in depth in Chapter 5 ("Society, Social Structure, and Interaction in Everyday Life").

Multiple methods and approaches provide a wider scope of information and enhance our understanding of critical issues. Many researchers also use multiple methods to validate or refine one type of data by use of another type.

Ethical Issues in Sociological Research

The study of people ("human subjects") raises vital questions about ethical concerns in sociological research. Researchers are required to obtain written "informed consent" statements from the persons they study—but what constitutes "informed consent"? And how do researchers protect the identity and confidentiality of their sources?

[concept quick review]

Strengths and Weaknesses of Social Research Methods

Research Method	Strengths	Weaknesses
Experiments (Laboratory, Field, Natural)	Control over research	Artificial by nature
	Ability to isolate experimental factors	Frequent reliance on volunteers or captive audiences
	Relatively little time and money required	Ethical questions of deception
	Replication possible, except for natural experiments	
Survey Research (Questionnaire, Interview, Telephone Survey)	Useful in describing features of a large population without interviewing everyone	Potentially forced answers
	Relatively large samples possible	Respondent untruthfulness on emotional issues
	Multivariate analysis possible	Data that are not always "hard facts" presented as such in statistical analyses
Secondary Analysis of Existing Data (Existing Statistics, Content Analysis)	Data often readily available, inexpensive to collect	Difficulty in determining accuracy of some of the data
	Longitudinal and comparative studies	Failure of data gathered by others to meet goals of current research
	Replication possible	Questions of privacy when using diaries, other personal documents
Field Research (Participant Observation, Case Study, Ethnography, Unstructured)	Opportunity to gain insider's view	Problems in generalizing results to a larger population
	Useful for studying attitudes and behavior in natural settings	Imprecise data measurements
	Longitudinal/comparative studies possible	Inability to demonstrate cause/effect relationships or test theories
	Documentation of important social problems of excluded groups possible	Difficult to make comparisons because of lack of structure
	Access to people's ideas in their words	Not a representative sample
	Forum for previously excluded groups	
	Documentation of need for social reform	

The ASA *Code of Ethics*

The American Sociological Association (ASA) *Code of Ethics* (1997) sets forth certain basic standards that sociologists must follow in conducting research:

1. Researchers must endeavor to maintain objectivity and integrity in their research by disclosing their research findings in full and including all possible interpretations of the data (even those interpretations that do not support their own viewpoints).
2. Researchers must safeguard the participants' right to privacy and dignity while protecting them from harm.

Hawthorne effect a phenomenon in which changes in a subject's behavior are caused by the researcher's presence or by the subject's awareness of being studied.

you can make a difference

Responding to a Cry for Help

Chad felt that he knew Frank quite well. After all, they had been roommates for two years at State U. As a result, Chad was taken aback when Frank became very withdrawn, sleeping most of the day and mumbling about how unhappy he was. One evening, Chad began to wonder whether he needed to do something because Frank had begun to talk about "ending it all" and saying things like "the world will be better off without me." If you were in Chad's place, would you know the warning signs that you should look for? Do you know what you might do to help someone like Frank?

The American Foundation for Suicide Prevention, a national nonprofit organization dedicated to funding research, education, and treatment programs for depression and suicide prevention, suggests that each of us should be aware of these warning signs of suicide:

- *Talking about death or suicide.* Be alert to such statements as "Everyone would be better off without me." Sometimes, individuals who are thinking about suicide speak as if they are saying good-bye.
- *Making plans.* The person may do such things as giving away valuable items, paying off debts, and otherwise "putting things in order."
- *Showing signs of depression.* Although most depressed people are not suicidal, most suicidal people are depressed. Serious depression tends to be expressed as a loss of pleasure or withdrawal from activities that a person has previously enjoyed. It is especially important to note whether five of the following symptoms are present almost every day for several weeks: change in appetite or weight, change in sleeping patterns, speaking or moving with unusual speed or slowness, loss of interest in usual activities, decrease in sexual drive, fatigue, feelings of guilt or worthlessness, and indecisiveness or inability to concentrate.

The possibility of suicide must be taken seriously: Most people who commit suicide give some warning to family members or friends. Instead of attempting to argue the person out of suicide or saying "You have so much to live for," let the person know that you care and understand, and that his or her problems can be solved. Urge the person to see a school counselor, a physician, or a mental health professional immediately. If you think the person is in imminent danger of committing suicide, you should take the person to an emergency room or a walk-in clinic at a psychiatric

3. Researchers must protect confidential information provided by participants, even when this information is not considered to be "privileged" (legally protected, as is the case between doctor and patient and between attorney and client) and legal pressure is applied to reveal this information.
4. Researchers must acknowledge research collaboration and assistance they receive from others and disclose all sources of financial support.

Sociologists are obligated to adhere to this code and to protect research participants; however, many ethical issues arise that cannot be easily resolved. Ethics in sociological research is a difficult and often ambiguous topic. But ethical issues cannot be ignored by researchers, whether they are sociology professors, graduate students conducting investigations for their dissertations, or undergraduates conducting a class research project. Sociologists have a burden of "self-reflection"—of seeking to understand the role they play in contemporary social processes while at the same time assessing how these social processes affect their findings (Gouldner, 1970).

How honest do researchers have to be with potential participants? Let's look at two specific cases in point. Where does the "right to know" end and the "right to privacy" begin in these situations?

The Zellner Research

Sociologist William Zellner (1978) wanted to look at fatal single-occupant automobile accidents to determine if some drivers were actually committing suicide. To examine this issue further, he sought to interview the family, friends, and acquaintances of persons killed in single-car crashes to determine if the deaths were possibly intentional. To recruit respondents, Zellner told them that he hoped the research would reduce the number of automobile accidents in the future. He did not mention that he suspected "autocide" might have occurred in the case of their friend or loved one. From his data, Zellner concluded that at least 12 percent of the fatal single-occupant crashes were suicides—and that those crashes sometimes also killed or critically injured other people as well. However, Zellner's research raised important research questions: Was his research unethical? Did he misrepresent the reasons for his study?

The Humphreys Research

As a sociology graduate student, Laud Humphreys (1970) decided to study homosexual conduct for his doctoral dissertation and chose to focus on homosexual acts between strangers in the public

© Gari Engberg/The Image Works

Can a suicide crisis center prevent a person from committing suicide? People who understand factors that contribute to suicide may be able to better counsel those who call for help.

hospital. It is best to remain with the person until help is available.

For more information about suicide prevention, contact the following organizations:

- American Foundation for Suicide Prevention (**http://www.afsp.org**) is a leading not-for-profit organization dedicated to understanding and preventing suicide through research and education.
- Suicide Awareness Voices of Education (**http://www.save.org**) is a resource index with links to other valuable resources, such as "Questions Most Frequently Asked on Suicide," "Symptoms of Depression and Danger Signs of Suicide," and "What to Do If Someone You Love Is Suicidal."
- Befrienders Worldwide (**http://www.befrienders.org**) is a website providing information for anyone feeling depressed or suicidal or who is worried about a friend or relative who feels that way. Includes a directory of suicide and crisis help lines.

restrooms (referred to as "tearooms") of city parks. Before beginning his observations, Humphreys did not ask his subjects' permission or inform them that they were being studied. Instead, he showed up at public restrooms offering to be the lookout for police while others engaged in homosexual acts as he systematically recorded the various encounters that took place. To learn more about their everyday lives away from this setting, Humphreys tracked down their names and addresses from their auto license numbers and invited them to participate in a medical survey. Because he did not want the men to recognize him, he wore various disguises and drove a different car to their homes. From these interviews he collected personal information and learned most of the men were married and lived very conventional lives apart from these closeted encounters with gay men in public restrooms.

Would Humphreys have gained access to these subjects if he had identified himself as a researcher? Probably not—nevertheless, the fact that he did not do so produced widespread criticism from sociologists and journalists. Despite the fact that his study, *Tearoom Trade* (1970), won an award for its scholarship, the controversy surrounding his project was never fully resolved.

In this chapter we have looked at the research process and the methods used to pursue sociological knowledge. The important thing to remember is that research is the "lifeblood" of sociology: Theory provides the framework for analysis, and research takes us beyond common sense and provides opportunities for us to use our sociological imagination to generate new knowledge. As we have seen in this chapter, for example, suicide cannot be explained by common sense or a few isolated variables. We have to take into account many aspects of personal choice and social structure that are related to one another in extremely complex ways. Research can help us unravel the complexities of social life if sociologists observe, talk to, and interact with people in real-life situations (Feagin, Orum, and Sjoberg, 1991).

Our challenge today is to find new ways to integrate knowledge and action and to include all people in the research process in order to help fill the gaps in our existing knowledge about social life and how it is shaped by gender, race, class, age, and the broader social and cultural contexts in which everyday life occurs (Cancian, 1992). Each of us can and should find new ways to integrate knowledge and action into our daily lives (see "You Can Make a Difference").

chapter review Q&A

Use these questions and answers to check how well you've achieved the learning objectives set out at the beginning of this chapter.

- **How does sociological research differ from commonsense knowledge?**

Sociological research provides a factual and objective counterpoint to commonsense knowledge and ill-informed sources of information. It is based on an empirical approach that answers questions through a direct, systematic collection and analysis of data.

- **What is the relationship between theory and research?**

Theory and research form a continuous cycle that encompasses both deductive and inductive approaches. With the deductive approach, the researcher begins with a theory and then collects and analyzes research to test it. With the inductive approach, the researcher collects and analyzes data and then generates a theory based on that analysis.

- **How does quantitative research differ from qualitative research?**

Quantitative research focuses on data that can be measured numerically (comparing rates of suicide, for example). Qualitative research focuses on interpretive description (words) rather than statistics to analyze underlying meanings and patterns of social relationships.

- **What are the key steps in the conventional research process?**

A conventional research process based on deduction and the quantitative approach has these key steps: (1) selecting and defining the research problem; (2) reviewing previous research; (3) formulating the hypothesis, which involves constructing variables; (4) developing the research design; (5) collecting and analyzing the data; and (6) drawing conclusions and reporting the findings.

- **What steps are often taken by researchers using the qualitative approach?**

A researcher taking the qualitative approach might (1) formulate the problem to be studied instead of creating a hypothesis, (2) collect and analyze the data, and (3) report the results.

- **What are the major types of research methods?**

The main types of research methods are surveys, secondary analysis of existing data, field research, and experiments. Surveys are polls used to gather facts about people's attitudes, opinions, or behaviors; a representative sample of respondents provides data through questionnaires or interviews. In secondary analysis, researchers analyze existing data, such as a government census, or cultural artifacts, such as a diary. In field research, sociologists study social life in its natural setting through participant observation, case studies, unstructured interviews, and ethnography. Through experiments, researchers study the impact of certain variables on their subjects.

- **What ethical issues are involved in sociological research?**

Because sociology involves the study of people ("human subjects"), researchers are required to obtain the informed consent of the people they study; however, in some instances what constitutes "informed consent" may be difficult to determine.

key terms

questions for critical thinking

1. The agency that funds the local suicide clinic has asked you to study the clinic's effectiveness in preventing suicide. What would you need to measure? What can you measure? What research method(s) would provide the best data for analysis?

2. Recent studies have suggested that groups with high levels of *suicide acceptability* (holding the belief that suicide is an acceptable way to end one's life under certain circumstances) tend to have a higher than average suicide risk (Stack and Wasserman, 1995; Stack, 1998). What implications might such findings have on public policy issues such as the legalization of physician-assisted suicide and euthanasia? What implications might the findings have on an individual who is thinking about committing suicide? Analyze your responses using a sociological perspective.
3. In high-income nations, computers have changed many aspects of people's lives. Thinking about the various research methods discussed in this chapter, which approaches do you believe would be most affected by greater reliance on computers for collecting, organizing, and analyzing data? What are the advantages and limitations of conducting sociological research via the Internet?

turning to video

Watch the CBS video *Census 2010* (running time 2:25), available through **CengageBrain.com**. This video reports the purpose of the census and the efforts to count everyone. (See page xiv at the front of this book for a list of Census Profiles included within select chapters.) As you watch the video, think about the saying, "Stand up and be counted," and consider the ways in which your community may be affected by the results of the census. After you've watched the video, consider these questions: Were you counted in the 2010 Census? Why were or weren't you?

online study resources

Go to CENGAGE brain.com to access online study resources, including the Sociology CourseMate for this text as well as special features such as video, an interactive sociology time line and interactive maps, General Social Survey (GSS) data, and U.S. Census 2010 data.

CourseMate brings course concepts to life with interactive learning, study, and exam-preparation tools that support the printed textbook. A textbook-specific website, **Sociology CourseMate** includes an integrated interactive eBook and other interactive learning tools, including quizzes, flash cards, and videos.

Visit **www.cengagebrain.com** to access your account and purchase materials.

3 Culture

At home, I kept opening the refrigerator and cupboards, wishing for American foods to magically appear. I wanted what the other kids had: Bundt cakes and casseroles, Cheetos and Doritos.... The more American foods I ate, the more my desires multiplied, outpacing my interest in Vietnamese food. I had memorized the menu at Dairy Cone, the sugary options in the cereal aisle at Meijer's [grocery], and every inch of the candy display at Gas City: the rows of gum, the rows of chocolate, the rows without chocolate....I knew Reese's peanut butter cups, Twix, Heath Crunch, Nestlé Crunch, Baby Ruth, Bar None, Oh Henry!, Mounds and Almond Joy, Snickers, Mr. Goodbar[,] ... Milk Duds, [and] Junior Mints. I dreamed of taking it all, plus the freezer full of popsicles and nutty, chocolate-coated ice cream drumsticks. I dreamed of Little Debbie, Dolly Madison, Swiss Miss, all the bakeries presided over by prim and proper girls.

—Bich Minh Nguyen (2007: 50–51), an English professor at Purdue University, describing how food served as a powerful cultural symbol in her childhood as a Vietnamese American

© Bob Daemmrich/The Image Works

▲ How is the food that we consume linked to our identity and the larger culture of which we are a part? Do people who identify with more than one culture face more-complex issues when it comes to food preferences?

Growing up in Oakland ... I came to dislike Chinese food. That may have been, in part, because I was Chinese and desperately wanted to be American. I *was* American, of course, but being born and raised in Chinatown—in a restaurant my parents operated, in fact—I didn't feel much like the people I saw outside Chinatown, or in books and movies.

It didn't help that for lunch at school, my mother would pack—*Ai ya!*—Chinese food. Barbecued pork sandwiches, not ham and cheese; Chinese pears, not apples. At home—that is, at the New Eastern Café—it was Chinese food night after night. No wonder I would sneak off, on the way to Chinese school, to Hamburger Gus for a helping of thick-cut French fries.

—author Ben Fong-Torres (2007: 11) describing his experiences as a Chinese American who desired to "Americanize" his eating habits

Chapter Focus Question:

What part does culture play in shaping people and the social relations in which they participate?

Why are these authors concerned about the food they ate as children? For all of us, the food we consume is linked to our identity and to the larger culture of which we are a part. For people who identify with more than one culture, food and eating patterns may become a very complex issue. To some people, food consumption is nothing more than how we meet a basic biological need; however, many sociologists are interested in food and eating because of their cultural significance in our lives (see Mennell, 1996; Mennell, Murcott, and van Otterloo, 1993).

What is culture? ***Culture* is the knowledge, language, values, customs, and material objects that are passed from person to person and from one generation to the next in a human group or society.** As previously defined, a *society* is a large social grouping that occupies the same geographic territory and is subject to the same political authority and dominant cultural expectations. Whereas a society is composed of people, a culture is composed of ideas, behavior, and material possessions. Society and culture are interdependent; neither could exist without the other.

In this chapter we examine society and culture, with special attention to how our material culture, including the food we eat, is related to our beliefs, values and actions. We also examine cultural diversity and why diversity matters at college and throughout our lives. Before reading on, test your knowledge of food and culture by answering the questions in the Sociology and Everyday Life quiz.

Highlighted Learning Objectives

- Identify and define the essential components of culture.
- Explain how subcultures and countercultures reflect diversity within a society.
- Discuss the degree to which you are influenced by popular culture.
- Describe how the various sociological perspectives view culture.

sociology and everyday life

How Much Do You Know About Global Food and Culture?

True	False	
T	F	1. Cheese is a universal food enjoyed by people of all nations and cultures.
T	F	2. Giving round-shaped foods to the parents of new babies is considered to be lucky in some cultures.
T	F	3. Wedding cakes are made of similar ingredients in all countries, regardless of culture or religion.
T	F	4. Food is an important part of religious observance for many different faiths.
T	F	5. In authentic Chinese cuisine, cooking methods are divided into "yin" and "yang" qualities.
T	F	6. Because of the fast pace of life in the United States, virtually everyone relies on mixes and instant foods at home and fast foods when eating out.
T	F	7. Potatoes are the most popular mainstay in the diet of first- and second-generation immigrants who have arrived in the United States over the past 40 years.
T	F	8. According to sociologists, individuals may be offended when a person from another culture does not understand local food preferences or the cultural traditions associated with eating, even if the person is obviously an "outsider" or a "tourist."

Answers on page 64.

Culture and Society in a Changing World

How important is culture in determining how people think and act on a daily basis? Simply stated, culture is essential for our individual survival and our communication with other people. We rely on culture because we are not born with the information we need to survive. We do not know how to take care of ourselves, how to behave, how to dress, what to eat, which gods to worship, or how to make or spend money. We must learn about culture through interaction, observation, and imitation in order to participate as members of the group. Sharing a common culture with others simplifies day-to-day interactions. However, we must also understand other cultures and the worldviews therein.

Just as culture is essential for individuals, it is also fundamental for the survival of societies. Culture has been described as "the common denominator that makes the actions of individuals intelligible to the group" (Haviland, 1993: 30). Some system of rule making and enforcing necessarily exists in all societies. What would happen, for example, if *all* rules and laws in the United States suddenly disappeared? At a basic level, we need rules in order to navigate our bicycles and cars through traffic. At a more abstract level, we need laws to establish and protect our rights.

In order to survive, societies need rules about civility and tolerance. We are not born knowing how to express certain kinds of feelings toward others. When a person shows kindness or hatred toward another individual, some people may say "Well, that's just human nature" when explaining this behavior. Such a statement is built on the assumption that what we do as human beings is determined by *nature* (our biological and genetic makeup) rather than *nurture* (our social environment)—in other words, that our behavior is instinctive. An *instinct* is an unlearned, biologically determined behavior pattern common to all members of a species that predictably occurs whenever certain environmental conditions exist. For example, spiders do not learn to build webs. They build webs because of instincts that are triggered by basic biological needs such as protection and reproduction.

Culture is similar to instinct in animals because it helps us deal with everyday life. Although people may have some instincts, what we most often think of as instinctive behavior can actually be attributed to reflexes and drives. A *reflex* is an unlearned, biologically determined, involuntary response to some physical stimuli (such as sneezing after breathing some pepper in through the nose or the blinking of an eye when a speck of dust gets in it). *Drives* are unlearned, biologically determined impulses common to all members of a species that satisfy needs such as those for sleep, food, water, or sexual gratification. Reflexes and drives do not determine how people will behave in human societies; even the expression of these biological characteristics is channeled by culture. For example, we may be taught that the "appropriate" way to sneeze (an involuntary response) is to use a tissue or turn our head away from others (a learned response). Similarly, we may learn to sleep on mats or

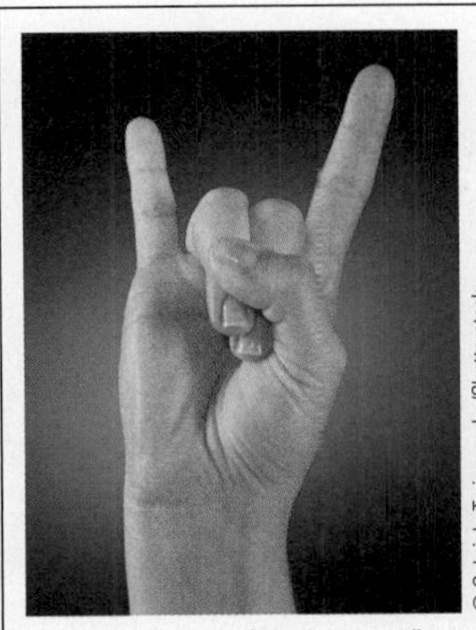

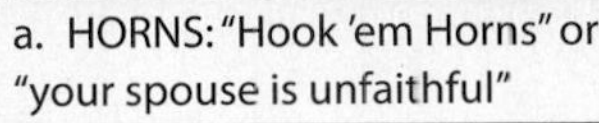
a. HORNS: "Hook 'em Horns" or "your spouse is unfaithful"

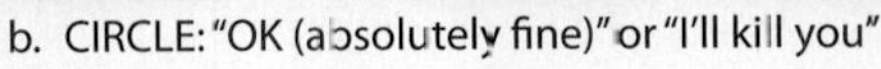
b. CIRCLE: "OK (absolutely fine)" or "I'll kill you"

c. THUMBS UP: "Great" or an obscenity

▲ **FIGURE 3.1 HAND GESTURES WITH DIFFERENT MEANINGS**
As international travelers and businesspeople have learned, hand gestures may have very different meanings in different cultures.

in beds. Most contemporary sociologists agree that culture and social learning, not nature, account for virtually all of our behavior patterns.

Because humans cannot rely on instincts in order to survive, culture is a "tool kit" for survival. According to the sociologist Ann Swidler (1986: 273), culture is a "tool kit of symbols, stories, rituals, and world views, which people may use in varying configurations to solve different kinds of problems." The tools we choose will vary according to our personality and the situations that we face. We are not puppets on a string; we make choices from among the items in our own "tool box." Why is this important to us? Because you and I make decisions about what we are going to do and how we are going to do it based on the culture of which we are a part.

Material Culture and Nonmaterial Culture

Our cultural tool box is divided into two major parts: material culture and nonmaterial culture (Ogburn, 1966/1922). ***Material culture*** **consists of the physical or tangible creations that members of a society make, use, and share.** Initially, items of material culture begin as raw materials or resources such as ore, trees, and oil. Through technology, these raw materials are transformed into usable items (ranging from books and computers to guns and tanks). Sociologists define *technology* as the knowledge, techniques, and tools that make it possible for people to transform resources into usable forms, and the knowledge and skills required to use them after they are developed. From this standpoint, technology is both concrete and abstract. For example, technology includes a pair of scissors and the knowledge and skill necessary to make them from iron, carbon, and chromium (Westrum, 1991). At the most basic level, material culture is important because it is our buffer against the environment. For example, we create shelter to protect ourselves from the weather and to give ourselves privacy. Beyond the survival level, we make, use, and share objects that are interesting and important to us. Why are you wearing the particular clothes you have on today? Perhaps you're communicating something about yourself, such as where you attend school, what kind of music you like, or where you went on vacation.

Nonmaterial culture **consists of the abstract or intangible human creations of society that influence people's behavior.** Language, beliefs, values, rules of behavior, family patterns, and political systems are examples of nonmaterial culture. Even the gestures that we use in daily conversation are part of the nonmaterial culture in a society. As many international travelers and businesspeople have learned, it is important to know what gestures mean in various nations (see ▶ Figure 3.1). Although the "hook 'em Horns" sign—the pinky and index finger raised up and the middle two fingers folded down—is used by fans to express their support for University of Texas at Austin sports teams, for millions of Italians the same gesture means "Your spouse is being

culture the knowledge, language, values, customs, and material objects that are passed from person to person and from one generation to the next in a human group or society.

material culture a component of culture that consists of the physical or tangible creations that members of a society make, use, and share.

nonmaterial culture a component of culture that consists of the abstract or intangible human creations of society that influence people's behavior.

sociology and everyday life

ANSWERS to the Sociology Quiz on Global Food and Culture

1. False. Although cheese is a popular food in many cultures, most of the people living in China find cheese very distasteful and prefer delicacies such as ducks' feet.

2. True. Round foods such as pears, grapes, and moon cakes are given to celebrate the birth of babies because the shape of the food is believed to symbolize family unity.

3. False. Although wedding cakes are a tradition in virtually all nations and cultures, the ingredients of the cake—as well as other foods served at the celebration—vary widely at this important family celebration. The traditional wedding cake in Italy is made from biscuits, for example, whereas in Norway the wedding cake is made from bread topped with cream, cheese, and syrup.

4. True. Many faiths, including Christianity, Judaism, Islam, Hinduism, and Buddhism, have dietary rules and rituals that involve food; however, these practices and beliefs vary widely among individuals and communities. For some people, food forms an integral part of religion in their life; for others, food is less relevant.

5. True. Just as foods are divided into yin foods (e.g., bean sprouts, cabbage, and carrots) and yang foods (beef, chicken, eggs, and mushrooms), cooking methods are also referred to as having yin qualities (e.g., boiling, poaching, and steaming) or yang qualities (deep-frying, roasting, and stir-frying). For many Chinese Americans, yin and yang are complementary pairs that should be incorporated into all aspects of social life, including the ingredients and preparation of foods.

6. False. Although more people now rely on fast foods, there is a "slow-food" movement afoot to encourage people to prepare their food from scratch for a healthier lifestyle. Also, some cultural and religious communities—such as the Amish of Ohio, Pennsylvania, and Indiana—encourage families to prepare their food from scratch and to preserve their own fruits, vegetables, and meats. Rural families are more likely to grow their own food or prepare it from scratch than are families residing in urban areas.

7. False. Rice is a popular mainstay in the diets of people from diverse cultural backgrounds who have arrived in the United States over the past four decades. Groups ranging from the Hmong and Vietnamese to Puerto Ricans and Mexican Americans use rice as a central ingredient in their diets. Among some in the younger generations, however, food choices have become increasingly Americanized, and items such as french fries and pizza have become very popular.

8. True. Cultural diversity is a major issue in eating, and people in some cultures, religions, and nations expect that even an "outsider" will have a basic familiarity with, and respect for, their traditions and practices. However, social analysts also suggest that we should not generalize or imply that certain characteristics apply to *all* people in a cultural group or nation.

Sources: Based on Better Health Channel, 2007; Ohio State University, 2007; and PBS, 2005a.

unfaithful." In Argentina rotating one's index finger around the front of the ear means "You have a telephone call," but in the United States it usually suggests that a person is "crazy" (Axtell, 1991). Similarly, making a circle with your thumb and index finger indicates "OK" in the United States, but in Tunisia it means "I'll kill you!" (Samovar and Porter, 1991).

As the example of hand gestures shows, a central component of nonmaterial culture is ***beliefs*—the mental acceptance or conviction that certain things are true or real.** Beliefs may be based on tradition, faith, experience, scientific research, or some combination of these. Faith in a supreme being and trust in another person are examples of beliefs. We may also have a belief in items of material culture. When we travel by airplane, for instance, we believe that it is possible to fly at 33,000 feet and to arrive at our destination even though we know that we could not do this without the airplane itself.

Cultural Universals

Because all humans face the same basic needs (such as for food, clothing, and shelter), we engage in similar activities that contribute to our survival. Anthropologist George Murdock (1945: 124) compiled a list of more than seventy ***cultural universals*—customs and practices that occur across all societies.** His categories included appearance (such as bodily adornment and hairstyles), activities (such as sports, dancing, games, joking,

© Celia Peterson/Getty Images

© Frans Lemmens/Getty Images

© Mark Henley/Impact/HIP/The Image Works

▲ Food is a universal type of material culture, but what people eat and how they eat it vary widely, as shown in these cross-cultural examples from the United Arab Emirates (top), Holland (middle), and China (bottom). What might be some reasons for the similarities and differences that you see in these photos?

and visiting), social institutions (such as family, law, and religion), and customary practices (such as cooking, folklore, gift giving, and hospitality). Whereas these general customs and practices may be present in all cultures, their specific forms vary from one group to another and from one time to another within the same group. For example, although telling jokes may be a universal practice, what is considered to be a joke in one society may be an insult in another.

How do sociologists view cultural universals? In terms of their functions, cultural universals are useful because they ensure the smooth and continual operation of society (Radcliffe-Brown, 1952). A society must meet basic human needs by providing food, shelter, and some degree of safety for its members so that they will survive. Children and other new members (such as immigrants) must be taught the ways of the group. A society must also settle disputes and deal with people's emotions. All the while, the self-interest of individuals must be balanced with the needs of society as a whole. Cultural universals help fulfill these important functions of society.

From another perspective, however, cultural universals are not the result of functional necessity; these practices may have been *imposed* by members of one society on members of another. Similar customs and practices do not necessarily constitute cultural universals. They may be an indication that a conquering nation used its power to enforce certain types of behavior on those who were defeated (Sargent, 1987). Sociologists might ask questions such as "Who determines the dominant cultural patterns?" For example, although religion is a cultural universal, traditional religious practices of indigenous peoples (those who first live in an area) have often been repressed and even stamped out by subsequent settlers or conquerors who hold political and economic power over them. However, many people believe there is cause for optimism in the United States because the democratic ideas of this nation provide more guarantees of religious freedom than do some other nations.

Components of Culture

Even though the specifics of individual cultures vary widely, all cultures have four common nonmaterial cultural components: symbols, language, values, and norms. These components contribute to both harmony and strife in a society.

Symbols

A ***symbol*** **is anything that meaningfully represents something else.** Culture could not exist without symbols because there would be no shared meanings among people. Symbols can simultaneously produce loyalty and animosity, and love and hate. They help us communicate ideas such as love

beliefs the mental acceptance or conviction that certain things are true or real.

cultural universals customs and practices that occur across all societies.

symbol anything that meaningfully represents something else.

© AMIT DAVE/Reuters/Corbis

© Myimagefiles/Alamy

© Michael Greenlar/The Image Works

▲ The customs and rituals associated with weddings are one example of nonmaterial culture. What can you infer about beliefs and attitudes concerning marriage in the societies represented by these photographs?

or patriotism because they express abstract concepts with visible objects. For example, flags can stand for patriotism, nationalism, school spirit, or religious beliefs held by members of a group or society. Symbols can stand for love (a heart on a valentine), peace (a dove), or hate (a Nazi swastika), just as words can be used to convey these meanings. Symbols can also transmit other types of ideas. A siren is a symbol that denotes an emergency situation and sends the message to clear the way immediately. Gestures are also a symbolic form of communication—a movement of the head, body, or hands can express our ideas or feelings to others. For example, in the United States, pointing toward your chest with your thumb or finger is a symbol for "me."

Symbols affect our thoughts about class. For example, how a person is dressed or the kind of car that he or she drives is often at least subconsciously used as a measure of that individual's economic standing or position. With regard to clothing, although many people wear casual clothes on a daily basis, where the clothing was purchased is sometimes used as a symbol of social status. Were the items purchased at Wal-Mart, Old Navy, Abercrombie & Fitch, or Saks Fifth Avenue? What indicators are there on the items of clothing—such as the Nike *swoosh,* some other logo, or a brand name—that point out something about the status of the product? Automobiles and their logos are also symbols that have cultural meaning beyond the shopping environment in which they originate.

Finally, symbols may be specific to a given culture and have special meaning to individuals who share that culture but not necessarily to other people. Consider, for example, the use of certain foods to celebrate the Chinese New Year: Bamboo shoots and black-moss seaweed both represent wealth, peanuts and noodles symbolize a long life, and tangerines represent good luck. What foods in other cultures represent "good luck" or prosperity?

Language

***Language* is a set of symbols that expresses ideas and enables people to think and communicate with one another.** Verbal (spoken) language and nonverbal (written or gestured) language help us describe reality. One of our most important human attributes is the ability to use language to share our experiences, feelings, and knowledge with others. Language can create visual images in our head, such as "the kittens look like little cotton balls" (Samovar and Porter, 1991). Language also allows people to distinguish themselves from outsiders and maintain group boundaries and solidarity (Farb, 1973).

Language is not solely a human characteristic. Other animals use sounds, gestures, touch, and smell to communicate with one another, but they use signals with fixed meanings that are limited to the immediate situation (the present) and that cannot encompass past or future situations. For example, chimpanzees can use elements of Standard American Sign Language and manipulate physical objects to make "sentences," but they are not physically endowed with the vocal apparatus needed to form the consonants required for oral language. As a result, nonhuman animals cannot transmit the

© George Frey/Bloomberg via Getty Images

▲ Would you expect the user of this device to be impoverished or affluent? What do possessions indicate about their owner's social class?

more complex aspects of culture to their offspring. Humans have a unique ability to manipulate symbols to express abstract concepts and rules and thus to create and transmit culture from one generation to the next.

Language and Social Reality Does language *create* or simply *communicate* reality? Anthropological linguists Edward Sapir and Benjamin Whorf have suggested that language not only expresses our thoughts and perceptions but also influences our perception of reality. According to the ***Sapir–Whorf hypothesis,* language shapes the view of reality of its speakers** (Whorf, 1956; Sapir, 1961). If people are able to think only through language, then language must precede thought. If language actually shapes the reality that we perceive and experience, then some aspects of the world are viewed as important and others are virtually neglected because people know the world only in terms of the vocabulary and grammar of their own language.

If language does create reality, are we trapped by our language? Many social scientists agree that the Sapir–Whorf hypothesis overstates the relationship between language and our thoughts and behavior patterns. Although they acknowledge that language has many subtle meanings and that words used by people reflect their central concerns, most sociologists contend that language may *influence* our behavior and interpretation of social reality but does not *determine* it.

Language and Gender What is the relationship between language and gender? What cultural assumptions about women and men does language reflect? Scholars have suggested several ways in which language and gender are intertwined:

- The English language ignores women by using the masculine form to refer to human beings in general (Basow, 1992). For example, the word *man* is used generically in words such as *chairman* and *mankind,* which allegedly include both men and women.
- Use of the pronouns *he* and *she* affects our thinking about gender. Pronouns show the gender of the person we *expect* to be in a particular occupation. For instance, nurses, secretaries, and schoolteachers are usually referred to as *she,* but doctors, engineers, electricians, and presidents are usually referred to as *he* (Baron, 1986).
- Words have positive connotations when relating to male power, prestige, and leadership; when related to women, they carry negative overtones of weakness, inferiority, and immaturity (Epstein, 1988: 224). ■ Table 3.1 shows how gender-based language reflects the traditional acceptance of men and women in certain positions, implying that the jobs are different when filled by women rather than men.
- A language-based predisposition to think about women in sexual terms reinforces the notion that women are sexual objects. Women are often described by terms such as *fox, broad, bitch, babe,* or *doll,* which ascribe childlike or even pet-like characteristics to them. By contrast, men have performance pressures placed on them by being defined in terms of their sexual prowess, such as *dude, stud,* and *hunk* (Baker, 1993).

Gender in language has been debated and studied extensively in recent years, and some changes have occurred. The preference of many women to be called *Ms.* (rather than *Miss* or *Mrs.* in reference to their marital status) has received a degree of acceptance in public life and the media. Many

language a set of symbols that expresses ideas and enables people to think and communicate with one another.

Sapir–Whorf hypothesis the proposition that language shapes the view of reality of its speakers.

table 3.1
Language and Gender

Male Term	Female Term	Neutral Term
Teacher	Teacher	Teacher
Chairman	Chairwoman	Chair, chairperson
Congressman	Congresswoman	Representative
Policeman	Policewoman	Police officer
Fireman	Lady fireman	Firefighter
Airline steward	Airline stewardess	Flight attendant
Race car driver	Woman race car driver	Race car driver
Professor	Teacher/female professor	Professor
Doctor	Lady/woman doctor	Doctor
Bachelor	Old maid	Single person
Male prostitute	Prostitute	Prostitute
Welfare recipient	Welfare mother	Welfare recipient
Worker/employee	Working mother	Worker/employee
Janitor/maintenance man	Maid/cleaning lady	Custodial attendant

Sources: Adapted from Korsmeyer, 1981: 122; and Miller and Swift, 1991.

© Jim West/The Image Works

▲ Certain jobs are stereotypically considered to be "men's jobs"; others are "women's jobs." Is your perception of a male flight attendant the same as your perception of a female flight attendant?

organizations and publications have established guidelines for the use of nonsexist language and have changed titles such as *chairman* to *chair* or *chairperson*. "Men Working" signs in many areas have been replaced with "People Working." Some occupations have been given "genderless" titles, such as *firefighter* or *flight attendant*. Yet many people resist change, arguing that the English language is being ruined (Epstein, 1988). To develop a more inclusive and equitable society, many scholars suggest that a more inclusive language is needed (see Basow, 1992).

Language, Race, and Ethnicity Language may create and reinforce our perceptions about race and ethnicity by transmitting preconceived ideas about the superiority of one category of people over another. Let's look at a few images conveyed by words in the English language in regard to race/ethnicity:

- Words may have more than one meaning and create and/or reinforce negative images. Terms such as *blackhearted* (malevolent) and expressions such as *a black mark* (a detrimental fact) and *Chinaman's chance of success* (unlikely to succeed) associate the words *black* and *Chinaman* with negative connotations and derogatory imagery. By contrast, expressions such as *that's white of you* and *the good guys wear white hats* reinforce positive associations with the color white.
- Overtly derogatory terms such as *nigger, kike, gook, honkey, chink, spic,* and other racial–ethnic slurs have been "popularized" in movies, music, comic routines, and so on. Such derogatory terms are often used in conjunction with physical threats against persons.
- Words are frequently used to create or reinforce perceptions about a group. For example, Native Americans have been referred to as "savage" and "primitive," and African Americans have been described as "uncivilized," "cannibalistic," and "pagan."
- The "voice" of verbs may minimize or incorrectly identify the activities or achievements of people of color. For example, the use of the passive voice in the statement "African Americans *were given* the right to vote" ignores how African Americans *fought* for that right. Active-voice verbs may also inaccurately attribute achievements to people or groups. Some historians argue that cultural bias is shown by the very notion that "Columbus discovered America"—given that America was already inhabited by people who later became known as Native Americans (see Stannard, 1992; Takaki, 1993).

In addition to these concerns about the English language, problems also arise when more than one language is involved. Across the nation, the question of whether or not the United States should have an "official" language continues to arise. Some people believe that there is no need to designate an official language; other people believe that English should be designated as the official language and that the

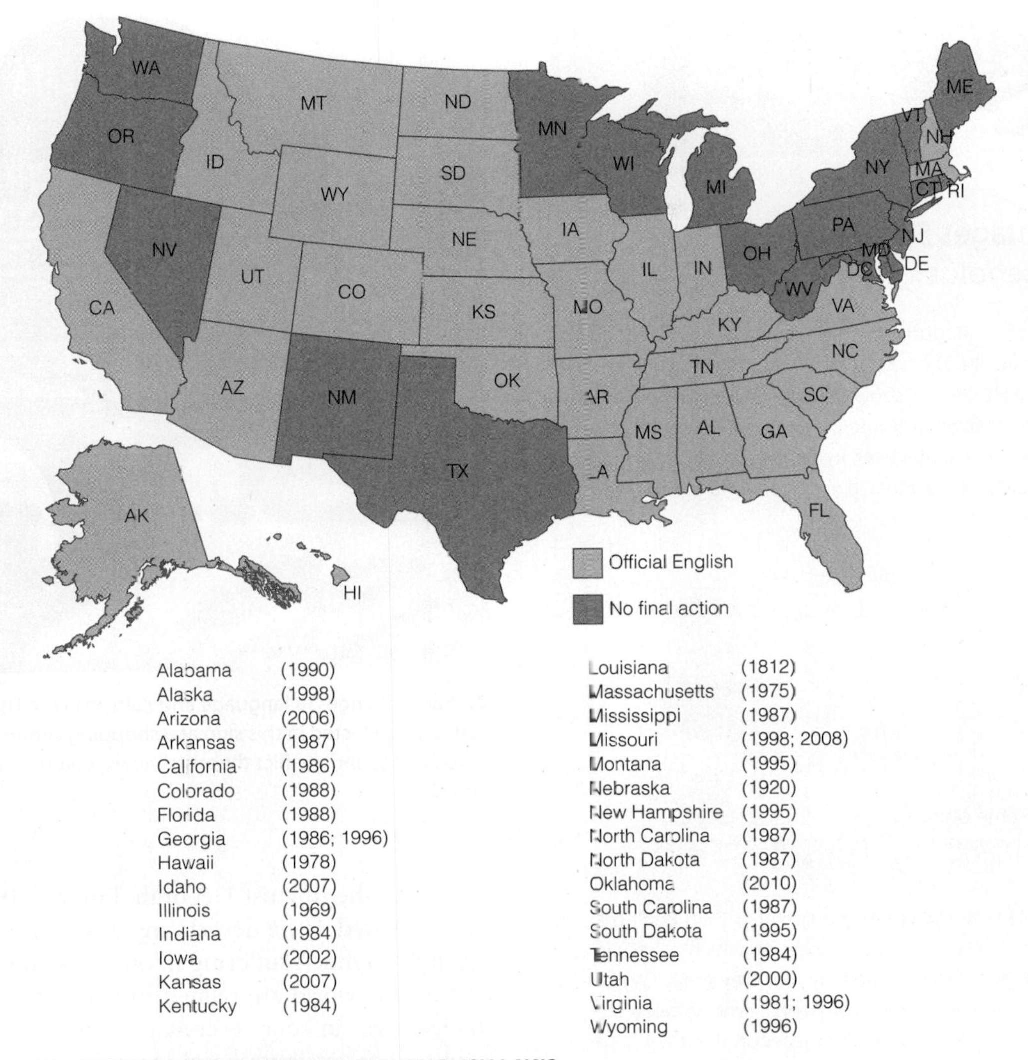

▲ **MAP 3.1 STATES WITH OFFICIAL ENGLISH LAWS**

Do you see any similarities in the states that have official English laws versus those that don't? What conclusions can you draw about the populations in those states?

Source: U.S. English, 2011.

use of any other language in official government business should be discouraged or negatively sanctioned. By 2011 thirty-one states (see ▶ Map 3.1) had passed laws that require all public documents, records, legislation, and regulations, as well as hearings, official ceremonies, and public meetings, to be conducted solely in English.

Are deep-seated social and cultural issues embedded in social policy decisions such as these? Although the United States has always been a nation of immigrants, in recent decades this country has experienced rapid changes in population that have brought about greater diversity in languages and cultures. Recent data gathered by the U.S. Census Bureau (see "Census Profiles: Languages Spoken in U.S. Households") indicate that although more than 80 percent of the people in this country speak only English at home, almost 20 percent speak a language other than English. The largest portion (over 10 percent of the U.S. population) of non-English speakers speak Spanish at home.

Language is an important means of cultural transmission. Through language, children learn about their cultural heritage and develop a sense of personal identity in relationship to their group. For example, Latinos/as in New Mexico and south Texas use *dichos*—proverbs or sayings that are unique to the Spanish language—as a means of expressing themselves and as a reflection of their cultural heritage. Examples of *dichos* include *Anda tu camino sin ayuda de vecino* ("Walk your own road without the help of a neighbor") and *Amor de lejos es para pendejos* ("A long-distance romance is for fools"). *Dichos* are passed from generation to generation as a priceless verbal tradition whereby people can give advice or teach a lesson (Gandara, 1995).

Language is also a source of power and social control; language perpetuates inequalities between people and between groups because words are used (whether or not intentionally) to "keep people in their

Languages Spoken in U.S. Households

Among the categories of information gathered by the U.S. Census Bureau is data on the languages spoken in U.S. households. As shown below, English is the only language spoken at home in more than 80 percent of U.S. households; however, in almost 20 percent of U.S. households, some other language is the primary language spoken at home.

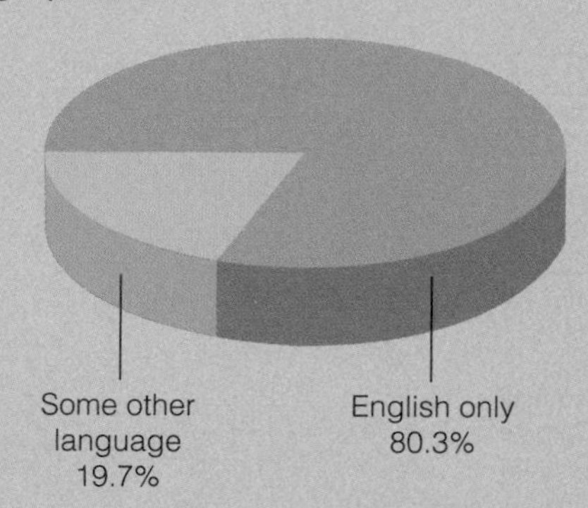

People who speak a language other than English at home are asked not only to indicate which other languages they speak but also how well they speak English. Approximately 44 percent of people who speak a language other than English at home report that they speak English "less than well." The principal languages other than English that are most frequently spoken at home are shown in the following chart. Do you think that changes in the languages spoken in this country will bring about other significant changes in U.S. culture? Why or why not?

Languages Spoken at Home Other Than English, by Percentage

Spanish 62.0%
Other 15.0%
Chinese 4.4%
French 2.7%
Tagalog 2.7%
German 2.2%
Vietnamese 2.2%
Korean 1.9%
Russian 1.6%
Italian 1.5%
Portuguese 1.3%
Arabic 1.3%
Polish 1.2%

Source: U.S. Census Bureau, 2010b.

© AP Images/Ann Johansson

▲ Rapid changes in language and culture in the United States are reflected in this sign at a shopping center. How do functionalist and conflict theorists' views regarding language differ?

place." As the linguist Deborah Tannen (1993: B5) has suggested, "The devastating group hatreds that result in so much suffering in our own country and around the world are related in origin to the small intolerances in our everyday conversations—our readiness to attribute good intentions to ourselves and bad intentions to others." Language, then, is a reflection of our feelings and values.

Values

***Values* are collective ideas about what is right or wrong, good or bad, and desirable or undesirable in a particular culture** (Williams, 1970). Values do not dictate which behaviors are appropriate and which ones are not, but they provide us with the criteria by which we evaluate people, objects, and events. Values typically come in pairs of positive and negative values, such as being brave or cowardly, hardworking or lazy. Because we use values to justify our behavior, we tend to defend them staunchly.

Core American Values Do we have shared values in the United States? Sociologists disagree about the extent to which all people in this country share a core set of values. Functionalists tend to believe that shared values are essential for the maintenance of a society, and scholars using a functionalist approach have conducted most of the research on core values. Analysts who focus on the importance of core values maintain that the following ten values, identified more than forty years ago by sociologist Robin M. Williams, Jr. (1970), are still very important to people in the United States:

1. *Individualism*. People are responsible for their own success or failure. Individual ability and hard work are the keys to success. Those who do not succeed have only themselves to blame because of their lack of ability, laziness, immorality, or other character defects.
2. *Achievement and success*. Personal achievement results from successful competition with others. Individuals are encouraged to do better than others in school and to work in order to gain wealth, power, and prestige. Material possessions are seen as a sign of personal achievement.
3. *Activity and work*. People who are industrious are praised for their achievement; those perceived as lazy are ridiculed. From the time of the early Puritans, work has been viewed as important. Even during their leisure time, many people "work" in their play. For example, think of all the individuals who take exercise classes, run in marathons, garden, repair or restore cars, and so on in their spare time.
4. *Science and technology*. People in the United States have a great deal of faith in science and technology. They expect scientific and technological advances ultimately to control nature, the aging process, and even death.
5. *Progress and material comfort*. The material comforts of life include not only basic necessities (such as adequate shelter, nutrition, and medical care) but also the goods and services that make life easier and more pleasant.
6. *Efficiency and practicality*. People want things to be bigger, better, and faster. As a result, great value is placed on efficiency ("How well does it work?") and practicality ("Is this a realistic thing to do?").
7. *Equality*. Since colonial times, overt class distinctions have been rejected in the United States. However, *equality* has been defined as "equality of *opportunity*"—an assumed equal chance to achieve success—not as "equality of *outcome*."
8. *Morality and humanitarianism*. Aiding others, especially following natural disasters (such as floods or hurricanes), is seen as a value. The notion of helping others was originally a part of religious teachings and tied to the idea of morality. Today, people engage in humanitarian acts without necessarily perceiving that it is the "moral" thing to do.
9. *Freedom and liberty*. Individual freedom is highly valued in the United States. The idea of freedom includes the right to private ownership of property, the ability to engage in private enterprise, freedom of the press, and other freedoms that are considered to be "basic" rights.
10. *Racism and group superiority*. People value their own racial or ethnic group above all others. Such feelings of superiority may lead to discrimination; slavery and segregation laws are classic examples. Many people also believe in the superiority of their country and that "the American way of life" is best.

Do you think that these values are still important today? Are there core values that you believe should be added to this list? Although sociologists have not agreed upon a specific list of emerging core values, various social analysts have suggested that some additional shared values in the United States today include the following:

- Ecological sensitivity, with an increased awareness of global problems such as overpopulation and global warming.
- Emphasis on developing and maintaining relationships through honesty and with openness, fairness, and tolerance of others.
- Spirituality and a need for meaning in life that reaches beyond oneself.

Value Contradictions All societies, including the United States, have value contradictions. ***Value contradictions* are values that conflict with one another or are mutually exclusive** (achieving one makes it difficult, if not impossible, to achieve another). Core values of morality and humanitarianism may conflict with values of individual achievement and success. For example, humanitarian values reflected in welfare and other government-aid programs for people in need come into conflict with values emphasizing hard work and personal achievement.

Ideal Culture Versus Real Culture What is the relationship between values and human behavior? Sociologists stress that a gap always exists between ideal culture and real culture in a society. *Ideal culture* refers to the values and standards of behavior that people in a society profess to hold. *Real culture* refers to the values and standards of behavior

values collective ideas about what is right or wrong, good or bad, and desirable or undesirable in a particular culture.

value contradictions values that conflict with one another or are mutually exclusive.

© Allen Eyestone/ZUMA Press/Corbis

▲ Basketball star LeBron James is widely considered to be one of the greatest players in the history of the NBA. Which core American values are reflected in sports such as basketball?

that people actually follow. For example, we may claim to be law-abiding (ideal cultural value) but smoke marijuana (real cultural behavior), or we may regularly drive over the speed limit but think of ourselves as "good citizens."

Most of us are not completely honest about how well we adhere to societal values. In a University of Arizona study known as the "Garbage Project," household waste was analyzed to determine the rate of alcohol consumption in Tucson. People were asked about their level of alcohol consumption, and in some areas of the city, they reported very low levels of alcohol use. However, when these people's garbage was analyzed, researchers found that over 80 percent of those households consumed some beer, and more than half discarded eight or more empty beer cans a week (Haviland, 1993). Obviously, this study shows a discrepancy between ideal cultural values and people's actual behavior.

Norms

Values provide ideals or beliefs about behavior but do not state explicitly how we should behave. Norms, on the other hand, do have specific behavioral expectations. ***Norms* are established rules of behavior or standards of conduct.** *Prescriptive norms* state what behavior is appropriate or acceptable. For example, persons making a certain amount of money are expected to file a tax return and pay any taxes they owe. Norms based on custom direct us to open a door for a person carrying a heavy load. By contrast, *proscriptive norms* state what behavior is inappropriate or unacceptable. Laws that prohibit us from driving over the speed limit and "good manners" that preclude you from texting during class are examples. Prescriptive and proscriptive norms operate at all levels of society, from our everyday actions to the formulation of laws.

Formal and Informal Norms Not all norms are of equal importance; those that are most crucial are formalized. *Formal norms* are written down and involve specific punishments for violators. Laws are the most common type of formal norms; they have been codified and may be enforced by sanctions. ***Sanctions* are rewards for appropriate behavior or penalties for inappropriate behavior.** Examples of *positive sanctions* include praise, honors, or medals for conformity to specific norms. *Negative sanctions* range from mild disapproval to the death penalty.

Norms considered to be less important are referred to as *informal norms*—unwritten standards of behavior understood by people who share a common identity. When individuals violate informal norms, other people may apply informal sanctions. *Informal sanctions* are not clearly defined and can be applied by any member of a group (such as frowning at someone or making a negative comment or gesture).

Folkways Norms are also classified according to their relative social importance. ***Folkways* are informal norms or everyday customs that may be violated without serious consequences within a particular culture** (Sumner, 1959/1906). They provide rules for conduct but are not considered to be essential to society's survival. In the United States,

© David Falconer/SuperStock

▲ Crowded conditions exist around the world, yet certain norms prevail in everyday life. Is the behavior of the people in this Osaka, Japan, train station a reflection of formal or informal norms?

folkways include using underarm deodorant, brushing our teeth, and wearing appropriate clothing for a specific occasion. Often, folkways are not enforced; when they are enforced, the resulting sanctions tend to be informal and relatively mild.

Mores Other norms are considered to be highly essential to the stability of society. ***Mores*** **are strongly held norms with moral and ethical connotations that may not be violated without serious consequences in a particular culture.** Because mores (pronounced MOR-ays) are based on cultural values and are considered to be crucial to the well-being of the group, violators are subject to more-severe negative sanctions (such as ridicule, loss of employment, or imprisonment) than are those who fail to adhere to folkways. The strongest mores are referred to as taboos. ***Taboos*** **are mores so strong that their violation is considered to be extremely offensive and even unmentionable.** Violation of taboos is punishable by the group or even, according to certain belief systems, by a supernatural force. The incest taboo, which prohibits sexual or marital relations between certain categories of kin, is an example of a nearly universal taboo.

Laws ***Laws*** **are formal, standardized norms that have been enacted by legislatures and are enforced by formal sanctions.** Laws may be either civil or criminal. *Civil law* deals with disputes among persons or groups. Persons who lose civil suits may encounter negative sanctions such as having to pay compensation to the other party or being ordered to stop certain conduct. *Criminal law,* on the other hand, deals with public safety and well-being. When criminal laws are violated, fines and prison sentences are the most likely negative sanctions, although in some states the death penalty is handed down for certain major offenses.

Technology, Cultural Change, and Diversity

Cultures do not generally remain static. There are many forces working toward change and diversity. Some societies and individuals adapt to this change, whereas others suffer culture shock and succumb to ethnocentrism.

Cultural Change

Societies continually experience cultural change at both the material and nonmaterial levels. Changes in technology continue to shape the material culture of society. ***Technology*** **refers to the knowledge, techniques, and tools that allow people to transform resources into usable forms and the knowledge and skills required to use what is developed.** Although most technological changes are primarily modifications of existing technology, *new technologies* refers to changes that make a significant difference in many people's lives. Examples of new technologies include the introduction of the printing press more than 500 years ago and the advent of computers and electronic communications in the twentieth century. The pace of technological change has increased rapidly in the past 150 years, as contrasted with the 4,000 years prior to that, during which humans advanced from digging sticks and hoes to the plow.

All parts of culture do not change at the same pace. When a change occurs in the material culture of a society, nonmaterial culture must adapt to that

▲ In 1999 Napster began as an online service that allowed its users to share music for free. However, the music industry claimed that Napster was violating its copyrights and sued the company. Napster was in business until 2001, an example of cultural lag, as the courts took time coming to grips with the implications of the new technology.

norms established rules of behavior or standards of conduct.

sanctions rewards for appropriate behavior or penalties for inappropriate behavior.

folkways informal norms or everyday customs that may be violated without serious consequences within a particular culture.

mores strongly held norms with moral and ethical connotations that may not be violated without serious consequences in a particular culture.

taboos mores so strong that their violation is considered to be extremely offensive and even unmentionable.

laws formal, standardized norms that have been enacted by legislatures and are enforced by formal sanctions.

technology the knowledge, techniques, and tools that allow people to transform resources into a usable form and the knowledge and skills required to use what is developed.

change. Frequently, this rate of change is uneven, resulting in a gap between the two. Sociologist William F. Ogburn (1966/1922) referred to this disparity as ***cultural lag*—a gap between the technical development of a society and its moral and legal institutions**. In other words, cultural lag occurs when material culture changes faster than nonmaterial culture, thus creating a lag between the two cultural components. For example, at the material cultural level, the personal computer and electronic coding have made it possible to create a unique health identifier for each person in the United States. Based on available technology (material culture), it would be possible to create a national data bank that included everyone's individual medical records from birth to death. Using this identifier, health providers and insurance companies could rapidly transfer medical records around the globe, and researchers could access unlimited data on people's diseases, test results, and treatments. However, the availability of this technology does not mean that it will be accepted by people who believe (nonmaterial culture) that such a national data bank would constitute an invasion of privacy and could easily be abused by others. The failure of nonmaterial culture to keep pace with material culture is linked to social conflict and societal problems. As in the above example, such changes are often set in motion by discovery, invention, and diffusion.

Discovery is the process of learning about something previously unknown or unrecognized. Historically, discovery involved unearthing natural elements or existing realities, such as "discovering" fire or the true shape of the Earth. Today, discovery most often results from scientific research. For example, the discovery of a polio vaccine virtually eliminated one of the major childhood diseases. A future discovery of a cure for cancer or the common cold could result in longer and more productive lives for many people.

As more discoveries have occurred, people have been able to reconfigure existing material and nonmaterial cultural items through invention. *Invention* is the process of reshaping existing cultural items into a new form. Guns, video games, airplanes, and First Amendment rights are examples of inventions that positively or negatively affect our lives today.

When diverse groups of people come into contact, they begin to adapt one another's discoveries, inventions, and ideas for their own use. *Diffusion* is the transmission of cultural items or social practices from one group or society to another through such means as exploration, war, the media, tourism, and immigration. Today, cultural diffusion moves at a very rapid pace in the global economy.

▲ In heterogeneous societies such as the United States, people from diverse cultures encourage their children to learn about their heritage. This East Indian mother and daughter in California dance with flower petals.

Cultural Diversity

Cultural diversity refers to the wide range of cultural differences found between and within nations. Cultural diversity between countries may be the result of natural circumstances (such as climate and geography) or social circumstances (such as level of technology and composition of the population). Some nations—such as Sweden—are referred to as *homogeneous societies,* meaning that they include people who share a common culture and who are typically from similar social, religious, political, and economic backgrounds. By contrast, other nations—including the United States—are referred to as *heterogeneous societies,* meaning that they include people who are dissimilar in regard to social characteristics such as religion, income, or race/ethnicity (see ► Figure 3.2).

Immigration contributes to cultural diversity in a society. Throughout its history, the United States has been a nation of immigrants (see ► Map 3.2). Over the past 175 years, more than 55 million "documented" (legal) immigrants have arrived here; innumerable people have also entered the country as

Religious Affiliation

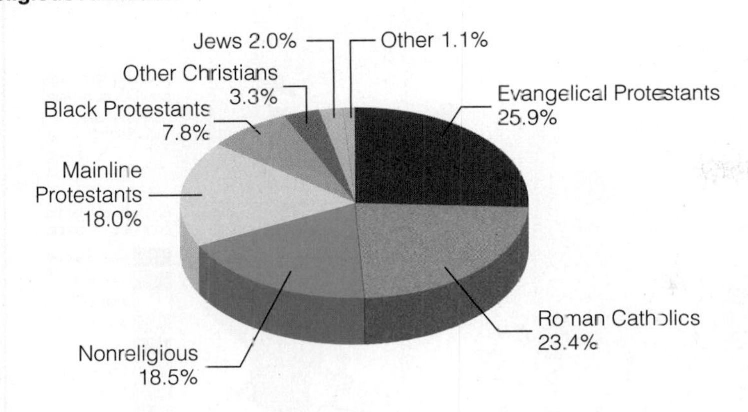

Household Income[a]

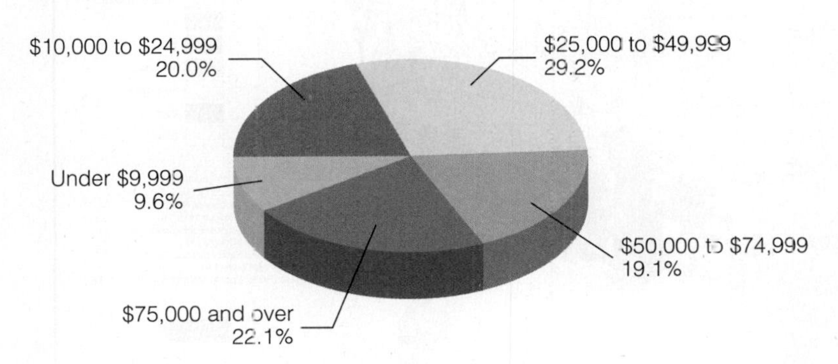

Race and Ethnic Distribution

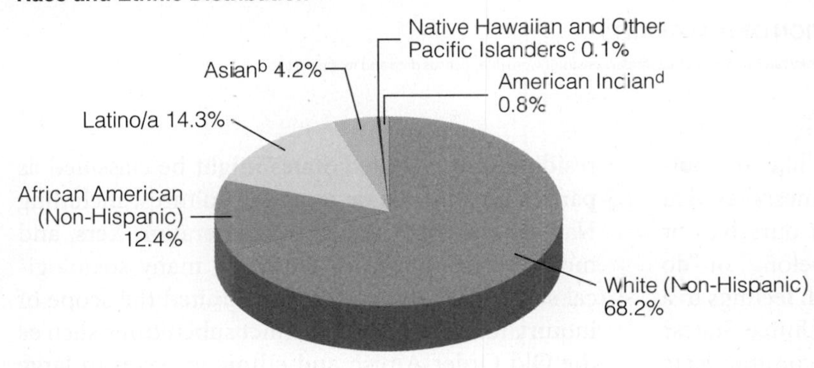

◀ **FIGURE 3.2 HETEROGENEITY OF U.S. SOCIETY**

Throughout history, the United States has been heterogeneous. Today, we represent a wide diversity of social categories, including our religious affiliations, income levels, and racial/ethnic categories.

[a]In Census Bureau terminology, a household consists of people who occupy a housing unit.
[b]Includes Chinese, Filipino, Japanese, Asian, Indian, Korean, Vietnamese, and other Asians.
[c]Includes Native Hawaiian, Guamanian or Chamorro, Samoan, and other Pacific Islanders.
[d]Includes American Indians, Eskimos, and Aleuts.

Source: U.S. Census Bureau, 2008.

undocumented immigrants. Immigration can cause feelings of frustration and hostility, especially in people who feel threatened by the changes that large numbers of immigrants may produce (Mydans, 1993). Often, people are intolerant of those who are different from themselves. When societal tensions rise, people may look for others on whom they can place blame—or single out persons because they are the "other," the "outsider," the one who does not "belong." Ronald Takaki, an ethnic studies scholar, described his experience of being singled out as an "other":

> I had flown from San Francisco to Norfolk and was riding in a taxi to my hotel to attend a conference on multiculturalism.... My driver and I chatted about the weather and the tourists.... The rearview mirror reflected a white man in his forties. "How long have you been in this country?" he asked. "All my life," I replied, wincing. "I was born in the United States." With a strong southern drawl, he remarked: "I was wondering because your English is excellent!" Then, as I had many times before, I explained: "My grandfather came here from Japan in the 1880s. My family has been here, in America, for over a hundred years." He glanced at me in the mirror. Somehow I did not look "American" to him; my eyes and complexion looked foreign. (Takaki, 1993: 1)

cultural lag William Ogburn's term for a gap between the technical development of a society and its moral and legal institutions.

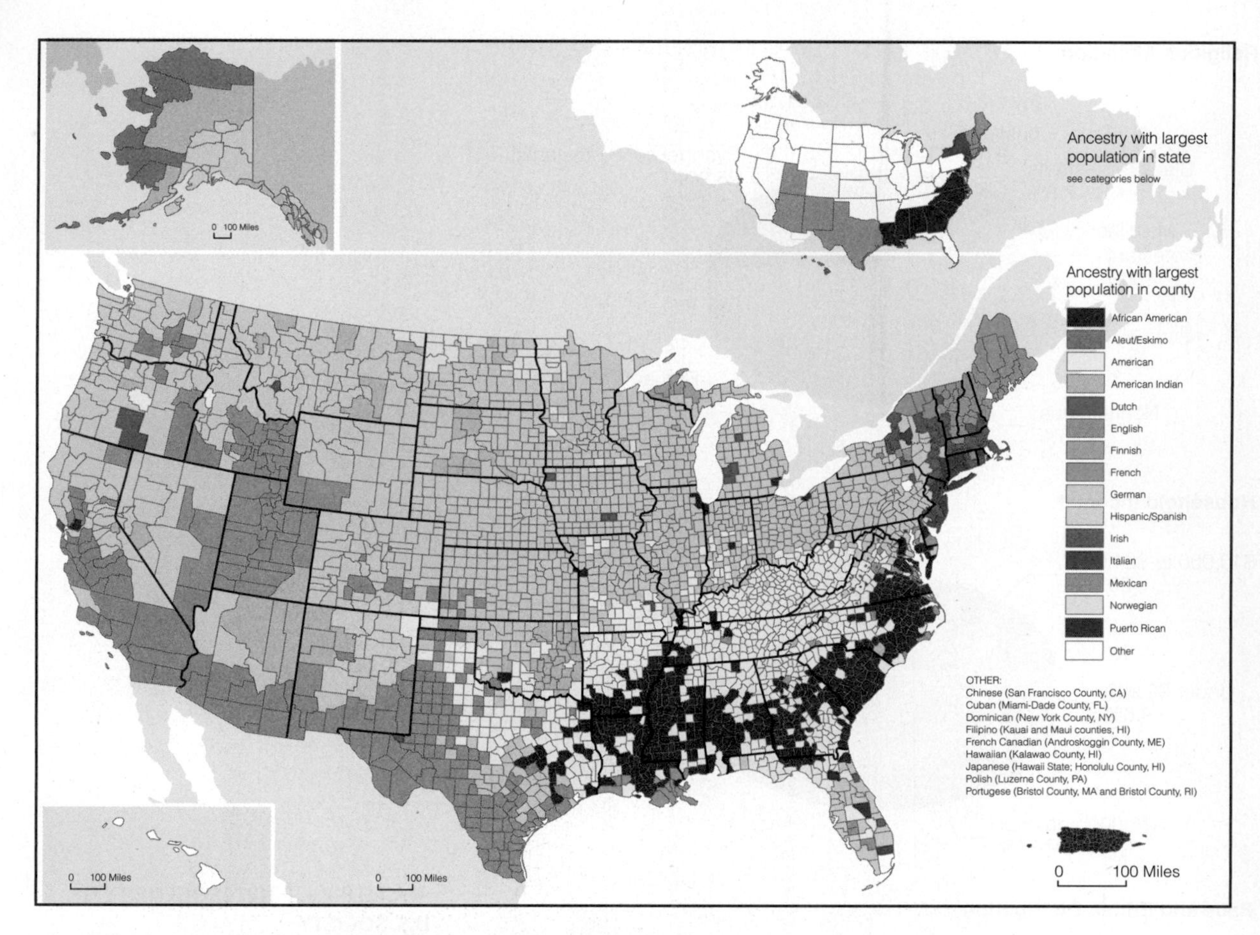

▲ **MAP 3.2 CULTURAL DIVERSITY: A NATION OF IMMIGRANTS**

Source: U.S. Census Bureau, Census 2000 special tabulation. American Factfinder at factfinder.census.gov provides census data and mapping tools.

Have you ever been made to feel like an "outsider"? Each of us receives cultural messages that may make us feel good or bad about ourselves or may give us the perception that we "belong" or "do not belong." Can people overcome such feelings in a culturally diverse society such as the United States? Some analysts believe it is possible to communicate with others despite differences in race, ethnicity, national origin, age, sexual orientation, religion, social class, occupation, leisure pursuits, regionalism, and so on (see "You Can Make a Difference"). People who differ from the dominant group may also find reassurance and social support in a subculture or a counterculture.

Subcultures A ***subculture*** **is a category of people who share distinguishing attributes, beliefs, values, and/or norms that set them apart in some significant manner from the dominant culture.** Emerging from the functionalist tradition, this concept has been applied to distinctions ranging from ethnic, religious, regional, and age-based categories to those categories presumed to be "deviant" or marginalized from the larger society. In the broadest use of the concept, thousands of categories of people residing in the United States might be classified as participants in one or more subcultures, including Native Americans, Muslims, Generation Xers, and motorcycle enthusiasts. However, many sociological studies of subcultures have limited the scope of inquiry to more-visible, distinct subcultures such as the Old Order Amish and ethnic enclaves in large urban areas to see how subcultural participants interact with the dominant U.S. culture.

The Old Order Amish Having arrived in the United States in the early 1700s, members of the Old Order Amish have fought to maintain their distinct identity. Today, over 75 percent of the more than 100,000 Amish live in Pennsylvania, Ohio, and Indiana, where they practice their religious beliefs and remain a relatively closed social network. According to sociologists, this religious community is a subculture because its members share values and norms that differ significantly from those of people who primarily identify with the dominant culture. The Amish have a strong faith in God and reject worldly concerns. Their core values include the joy of work, the primacy of the home, faithfulness, thriftiness, tradition, and humility. The Amish

you can make a difference

Bonding with Others Through Food and Conversation

[A reader recently said] how meeting for lunch was central to her relationship with her friends. I think that's true for a lot of us.... [Food is a] medium through which connections [are] built. My new friends and I started getting together once a month for dinner to celebrate birthdays, with lots of food involved. Most of them were non-native English speakers, so deep discussion of intellectual topics was difficult. Functional English only goes so far, and my smattering of Arabic wasn't up to the challenge, either. So we'd spend hours over dinner in various restaurants, exchanging food off our plates....There's something intimate about sharing food....When we share food, we share ourselves. It bonds us in a way other things don't.

—From Terry Karago's blog **www.dailytroll.com.** (2006) Reprinted by permission

Terry's blog describes how sharing food (an important component of culture) can help us in bonding with and learning more about people from diverse cultures. However, simply meeting with people from other cultural backgrounds or sharing a meal with them is not the same thing as really getting to understand them and helping them to understand us. Daisy Kabagarama, a U.S. college professor who was born in Uganda, suggests in *Breaking the Ice* (1993) that the following techniques can help each of us in communicating across cultures:

- *Get acquainted.* Show genuine interest, have a sense of curiosity and appreciation, feel empathy for others, be nonjudgmental, and demonstrate flexibility.
- *Ask the right questions.* Ask general questions first and specific ones later, making sure that questions are clear and simple and are asked in a relaxed, nonthreatening manner.
- *Consider visual images.* Use compliments carefully; it is easy to misjudge other people based on their physical appearance alone, and appearance norms differ widely across cultures.
- *Deal with stereotypes.* Overcome stereotyping and myths about people from other cultures through sincere self-examination, searching for knowledge, and practicing objectivity.
- *Establish trust and cooperation.* Be available when needed. Give and accept criticism in a positive manner and be spontaneous in interactions with others, but remember that rules regarding spontaneity are different for each culture.

Electronic systems now link people around the world, making it possible for us to communicate with people from diverse racial–ethnic backgrounds and cultures without even leaving home or school. Try this website for interesting information on multicultural issues and cultural diversity:

- Multicultural Education Pavilion provides resources on racism, sexism, and classism in the United States, as well as access to multicultural newsgroups, essays, and a large list of multicultural links on the Web: **http://www.edchange.org/multicultural.**

AP Images/Tony Dejak

▲ Although modernization and consumerism have changed the way of life of some subcultures, groups such as the Old Order Amish have preserved some of their historical practices, including traveling by horse-drawn buggy.

hold a conservative view of the family, believing that women are subordinate to men, birth control is unacceptable, and wives should remain at home. Children (about seven per family) are cherished and seen as an economic asset: They help with the farming and other work. Many of the Old Order Amish speak Pennsylvania Dutch (a dialect of German) as well as English. They dress in traditional clothing, live on farms, and rely on the horse and buggy for transportation.

The Amish are aware that they share distinctive values and look different from other people; these differences provide them with a collective identity

subculture a category of people who share distinguishing attributes, beliefs, values, and/or norms that set them apart in some significant manner from the dominant culture.

and make them feel close to one another (Schaefer and Zellner, 2007). The belief system and group cohesiveness of the Amish remain strong despite the intrusion of corporations and tourists, the vanishing farmlands, and increasing levels of government regulation in their daily lives (Schaefer and Zellner, 2007).

Ethnic Subcultures Some people who have unique shared behaviors linked to a common racial, language, or nationality background identify themselves as members of a specific subculture, whereas others do not. Examples of ethnic subcultures include African Americans, Latinos/Latinas (Hispanic Americans), Asian Americans, and Native Americans. Some analysts include "white ethnics" such as Irish Americans, Italian Americans, and Polish Americans. Others also include Anglo Americans (Caucasians).

Although people in ethnic subcultures are dispersed throughout the United States, a concentration of members of some ethnic subcultures is visible in many larger communities and cities. For example, Chinatowns, located in cities such as San Francisco, Los Angeles, and New York, are one of the more visible ethnic subcultures in the United States. By living close to one another and clinging to their original customs and language, first-generation immigrants can survive the abrupt changes they experience in material and nonmaterial cultural patterns. In New York City, for example, Korean Americans and Puerto Rican Americans constitute distinctive subcultures, each with its own food, music, and personal style. In San Antonio, Mexican Americans enjoy different food and music than do Puerto Rican Americans or other groups. Subcultures provide opportunities for expression of distinctive lifestyles, as well as sometimes helping people adapt to abrupt cultural change. Subcultures can also serve as a buffer against the discrimination experienced by many ethnic or religious groups in the United States. However, some people may be forced by economic or social disadvantage to remain in such ethnic enclaves.

Countercultures Some subcultures actively oppose the larger society. A ***counterculture* is a group that strongly rejects dominant societal values and norms and seeks alternative lifestyles** (Yinger, 1960, 1982). Young people are most likely to join countercultural groups, perhaps because younger persons generally have less invested in the existing culture. Examples of countercultures include the beatniks of the 1950s, the flower children of the 1960s, the drug enthusiasts of the 1970s, and contemporary members of nonmainstream religious sects, or cults.

Culture Shock

***Culture shock* is the disorientation that people feel when they encounter cultures radically different from their own and believe that they cannot depend on their own taken-for-granted assumptions about life.** When people travel to another society, they may not know how to respond to that setting. For example, Napoleon Chagnon (1992) described his initial shock at seeing the Yanomamö (pronounced yah-noh-MAH-mah) tribe of South America on his first trip in 1964.

The Yanomamö (also referred to as the "Yanomami") are a tribe of about 20,000 South American Indians who live in the rain forest. Although Chagnon traveled in a small aluminum motorboat for three days to reach these people, he was not prepared for the sight that met his eyes when he arrived:

> I looked up and gasped to see a dozen burly, naked, sweaty, hideous men staring at us down the shafts of their drawn arrows. Immense wads of green tobacco were stuck between their lower teeth and lips, making them look even more hideous, and strands of dark-green slime dripped from their nostrils—strands so long that they reached down to their pectoral muscles or drizzled down their chins and stuck to their chests and bellies. We arrived as the men were blowing ebene, a hallucinogenic drug, up their noses.... I was horrified. What kind of welcome was this for someone who had come to live with these people and learn their way of life—to become friends with them? But when they recognized Barker [a guide], they put their weapons down and returned to their chanting, while keeping a nervous eye on the village entrances. (Chagnon, 1992: 12–14)

The Yanomamö have no written language, system of numbers, or calendar. They lead a nomadic lifestyle, carrying everything they own on their backs. They wear no clothes and paint their bodies; the women insert slender sticks through holes in the lower lip and through the pierced nasal septum. In other words, the Yanomamö—like the members of thousands of other cultures around the world—live in a culture very different from that of the United States.

Ethnocentrism and Cultural Relativism

When observing people from other cultures, many of us use our own culture as the yardstick by which we judge their behavior. Sociologists refer to this approach as ***ethnocentrism*—the practice of judging**

© Victor Englebert/Time Life Pictures/Getty Images

▲ Even as global travel and the media make us more aware of people around the world, the distinctiveness of the Yanomamö in South America remains apparent. Are people today more or less likely than those in the past to experience culture shock upon encountering diverse groups of people such as these Yanomamö?

all other cultures by one's own culture (Sumner, 1959/1906). Ethnocentrism is based on the assumption that one's own way of life is superior to all others. For example, most schoolchildren are taught that their own school and country are the best. The school song, the pledge to the flag, and the national anthem are forms of *positive ethnocentrism*. However, *negative ethnocentrism* can also result from constant emphasis on the superiority of one's own group or nation. Negative ethnocentrism is manifested in derogatory stereotypes that ridicule recent immigrants whose customs, dress, eating habits, or religious beliefs are markedly different from those of dominant-group members. Long-term U.S. residents who are members of racial and ethnic minority groups, such as Native Americans, African Americans, and Latinas/os, have also been the target of ethnocentric practices by other groups.

An alternative to ethnocentrism is ***cultural relativism*—the belief that the behaviors and customs of any culture must be viewed and analyzed by the culture's own standards.** For example, the anthropologist Marvin Harris (1974, 1985) uses cultural relativism to explain why cattle, which are viewed as sacred, are not killed and eaten in India, where widespread hunger and malnutrition exist. From an ethnocentric viewpoint, we might conclude that cow worship is the cause of the hunger and poverty in India. However, according to Harris, the Hindu taboo against killing cattle is very important to their economic system. Live cows are more valuable than dead ones because they have more important uses than as a direct source of food. As part of the ecological system, cows consume grasses of little value to humans. Then they produce valuable resources—oxen (the neutered offspring of cows) to power the plows and manure (for fuel and fertilizer)—as well as milk, floor covering, and leather. As Harris's study reveals, culture must be viewed from the standpoint of those who live in a particular society.

Cultural relativism also has a downside. It may be used to excuse customs and behavior (such as cannibalism) that may violate basic human rights. Cultural relativism is a part of the sociological imagination; researchers must be aware of the customs and norms of the society they are studying and then spell out their background assumptions so that others can spot possible biases in their studies. However, according to some social scientists, issues surrounding ethnocentrism and cultural relativism may become less distinct in the future as people around the globe increasingly share a common popular culture. Others, of course, disagree with this perspective. Let's see what you think.

A Global Popular Culture?

Before taking this course, what was the first thing you thought about when you heard the term *culture*? In everyday life, culture is often used to describe the fine arts, literature, and classical music. When people say that a person is "cultured," they may mean that the individual has a highly developed sense of style or aesthetic appreciation of the "finer" things.

High Culture and Popular Culture

Some sociologists use the concepts of high culture and popular culture to distinguish between different cultural forms. ***High culture* consists of**

counterculture a group that strongly rejects dominant societal values and norms and seeks alternative lifestyles.

culture shock the disorientation that people feel when they encounter cultures radically different from their own and believe that they cannot depend on their own taken-for-granted assumptions about life.

ethnocentrism the practice of judging all other cultures by one's own culture.

cultural relativism the belief that the behaviors and customs of any culture must be viewed and analyzed by the culture's own standards.

high culture classical music, opera, ballet, live theater, and other activities usually patronized by elite audiences.

classical music, opera, ballet, live theater, and other activities usually patronized by elite audiences, composed primarily of members of the upper-middle and upper classes, who have the time, money, and knowledge assumed to be necessary for its appreciation. In the United States, high culture is often viewed as being international in scope, arriving in this country through the process of diffusion, because many art forms originated in European nations or other countries of the world. By contrast, much of U.S. popular culture is often thought of as "homegrown" in this country. ***Popular culture*** **consists of activities, products, and services that are assumed to appeal primarily to members of the middle and working classes.** These include rock concerts, spectator sports, movies, and television soap operas and situation comedies. Although we will distinguish between "high" and "popular" culture in our discussion, it is important to note that some social analysts believe that the rise of a consumer society in which luxury items have become more widely accessible to the masses has substantially reduced the great divide between activities and possessions associated with wealthy people or a social elite (see Huyssen, 1984; Lash and Urry, 1994). Other analysts use terms such as *highbrow, middlebrow,* and *lowbrow* to identify the relationship between culture and class (Gans, 1999).

In fact, most sociological examinations of high culture and popular culture primarily focus on the link between culture and social class. French sociologist Pierre Bourdieu's (1984) *cultural capital theory* views high culture as a device used by the dominant class to exclude the subordinate classes. According to Bourdieu, people must be trained to appreciate and understand high culture. Individuals learn about high culture in upper-middle-class and upper-class families and in elite education systems, especially higher education. Once they acquire this trained capacity, they possess a form of cultural capital. Persons from poor and working-class backgrounds typically do not acquire this cultural capital. Because knowledge and appreciation of high culture are considered a prerequisite for access to the dominant class, its members can use their cultural capital to deny access to subordinate-group members and thus preserve and reproduce the existing class structure.

According to the sociologist Herbert J. Gans (1999), class explains only a part of why people choose the forms of culture they do: Age, gender, race, and religion are other important factors in how and why people make cultural choices, and many individuals do not limit their choices to one culture. Gans identifies *taste publics* as people who share similar artistic, recreational, and intellectual interests but are not necessarily members of an organized group. By contrast, *taste cultures* are made up of people who not only share similar tastes but also participate in the same cultural groups or organizations. For example, some of us enjoy the music of a particular recording artist but do not join that person's fan club; others join a real or virtual fan club to discuss the artist's latest release or participate in celebrity gossip. Do you identify with a taste public? Are you part of a taste culture?

▲ The meteoric rise of Lady Gaga raises some interesting questions. Are her fans more interested in her music and musical performances, or are they more interested in Lady Gaga as a phenomenon? How does this question relate to Herbert Gans's distinction between taste publics and taste cultures?

Forms of Popular Culture

What are the most common forms of popular culture? Three prevalent forms of popular culture are fads, fashions, and leisure activities. A *fad* is a temporary but widely copied activity followed enthusiastically by large numbers of people. Most fads are short-lived novelties, but some can endure for surprisingly long periods of time. According to the sociologist John Lofland (1993), fads are divided into four major categories. First, *object fads* are items that people purchase even though they have little use or intrinsic value. Past and present examples include Webkinz, Harry Potter wands, SpongeBob

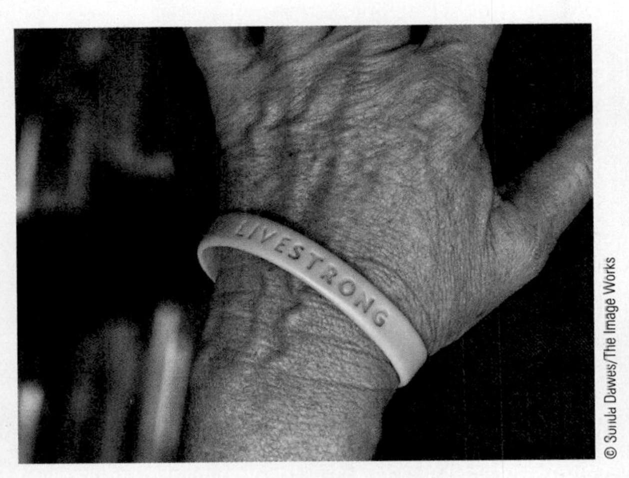

© Sunda Dawes/The Image Works

▲ Is this wristband an example of a fad or a fashion?

SquarePants items, and Zhu Zhu pets (erratically moving robotic hamsters). Second, *activity fads* include everyone you know playing games of Angry Birds on their cell phones, posting Facebook pictures of celebrity look-alikes (a "doppelganger") in place of their own photo, and 24/7 texting or Tweeting friends. Third are *idea fads,* such as New Age ideologies, the "Go Green" movement, and various locally grown food movements and the resurgence of farmers' markets. Fourth are *personality fads*—for example, Lady Gaga, Beyoncé, Oprah, Sarah Palin, and Matthew Morrison or other characters on *Glee,* a hit Fox television series about an unusual high school glee club.

A *fashion* is a currently valued style of behavior, thinking, or appearance that is longer lasting and more widespread than a fad. Examples of fashion are found in many areas, including child rearing, education, arts, clothing, music, and sports. In regard to clothing, for example, consider women's shoes, where recent fashion trends have ranged from very casual flip-flops and boots to extremely dressy, very high-heeled shoes that are first worn by media celebrities. In the 2000s fashion is an international phenomenon as both street and designer fashions are quickly embraced by people around the world.

Will the spread of popular culture produce a homogeneous global culture? Critics argue that the world is not developing a global culture; rather, other cultures are becoming Westernized. Political and religious leaders in some nations oppose this process, which they view as ***cultural imperialism*—the extensive infusion of one nation's culture into other nations.** For example, some view the widespread infusion of the English language into countries that speak other languages as a form of cultural imperialism. On the other hand, the concept of cultural imperialism may fail to take into account various cross-cultural influences. For example, the cultural diffusion of literature, music, clothing, and food has occurred on a global scale. A global culture, if it comes into existence, will most likely include components from many societies and cultures.

Sociological Analysis of Culture

Sociologists regard culture as a central ingredient in human behavior. Although all sociologists share a similar purpose, they typically see culture through somewhat different lenses as they are guided by different theoretical perspectives in their research. What do these perspectives tell us about culture?

Functionalist Perspectives

As previously discussed, functionalist perspectives are based on the assumption that society is a stable, orderly system with interrelated parts that serve specific functions. Anthropologist Bronislaw Malinowski (1922) suggested that culture helps people meet their *biological needs* (including food and procreation), *instrumental needs* (including law and education), and *integrative needs* (including religion and art). Societies in which people share a common language and core values are more likely to have consensus and harmony.

How might functionalist analysts view popular culture? According to many functionalist theorists, popular culture serves a significant function in society in that it may be the "glue" which holds society together. Regardless of race, class, sex, age, or other characteristics, many people are brought together (at least in spirit) to cheer teams competing in major sporting events such as the Super Bowl and the Olympic Games. Television and the Internet help integrate recent immigrants into the mainstream culture, whereas longer-term residents may become more homogenized as a result of seeing the same images and being exposed to the same beliefs and values (Gerbner et al., 1987).

popular culture activities, products, and services that are assumed to appeal primarily to members of the middle and working classes.

cultural imperialism the extensive infusion of one nation's culture into other nations.

framing culture in the media

You Are What You Eat?

The agonizing decision to pick Yale over Harvard didn't come down only to academics for Philip Gant....It also came down to his tummy. And his eco-savvy....When he chose Yale last year, Gant wasn't swayed by its running tab of presidential alumni: President Bush, George H. W. Bush, Bill Clinton, Gerald Ford and William Howard Taft. He was more impressed by Yale's leading-edge dedication to serving "sustainable" food.... In addition to wanting sustainable food, students such as Gant want it to be organic: grown without pesticides, herbicides, antibiotics or hormones.

—From Bruce Horovitz, "More university students call for organic, 'sustainable' food; Campuses nationwide buy more food from local farms." *USA Today*, Sept. 27th, 2006. Copyright © 2006 USA Today. Reprinted by permission.

You may ask "What does a newspaper article about university cafeterias and organic food have to do with culture?" The answer is simple: Food is very much a part of all cultures. What we eat and how it is grown and prepared are a product of the culture of the society in which we live. Fads and fashions in food may come and go, but we often become aware of them as a result of mass media such as television, magazines, newspapers, and the Internet. Our ideas may be influenced by media framing of stories about food and eating habits. The term *media framing* refers to the process by which information and entertainment are packaged by the mass media (newspapers, magazines, radio and television networks and stations, and the Internet) before being presented to an audience. This process includes factors such as the amount of exposure given to a story, where it is placed, the positive or negative tone it conveys, and its accompanying headlines, photographs, or other visual or auditory effects (if any). Through framing, the media emphasize some beliefs and values over others and manipulate salience by directing people's attention to some ideas while ignoring others. As such, a frame constitutes a story line or an unfolding narrative about an issue. These narratives are organizations of experience that bring order to events. Consequently, such narratives wield power because they influence how we make sense of the world (Kendall, 2011).

Media framing of food takes place in television networks and magazines, some of which are devoted solely to the topic of food. There are stories and articles about food on an almost daily basis in the other forms of mass media.

AP Photo/David Stluka

▲ According to many functionalist theorists, popular culture serves a significant function in society in that it may be the "glue" which holds society together. This NFL game is a good example of the wide diversity of people who come together to support a common interest.

However, functionalists acknowledge that all societies have dysfunctions which produce a variety of societal problems. When a society contains numerous subcultures, discord results from a lack of consensus about core values. In fact, popular culture may undermine core cultural values rather than reinforce them. For example, movies may glorify crime, rather than hard work, as the quickest way to get ahead. According to some analysts, excessive violence in music videos, movies, and television programs may be harmful to children and young people. From this perspective, popular culture can be a factor in antisocial behavior as seemingly diverse as hate crimes and fatal shootings in public schools.

A strength of the functionalist perspective on culture is its focus on the needs of society and the fact that stability is essential for society's continued survival. A shortcoming is its overemphasis on harmony and cooperation. This approach also fails to fully account for factors embedded in the structure of society—such as class-based inequalities, racism, and sexism—that may contribute to conflict among people in the United States or to global strife.

Conflict Perspectives

Conflict perspectives are based on the assumption that social life is a continuous struggle in which members of powerful groups seek to control scarce resources. According to this approach, values and norms help create and sustain the privileged position of the powerful in society while excluding others. As early conflict theorist Karl Marx stressed,

So why did *USA Today* report on Philip Gant's concerns about campus food? At least in part, the article resulted from a recent emphasis on organic food reporting in the media. Organic food refers to crops that are grown without the use of artificial fertilizers or most pesticides and that are processed without ionizing radiation or food additives, and to meat that is raised without antibiotics or growth hormones.

What most of us know about the "good" and "bad" sides of organic foods comes from the media. According to some media analysts, organic foods contain some nutrients that are not present in commercial foods, and organic foods do not have certain toxins that may be present in commercial foods (Crinnion, 1995). As one journalist stated, organic food methods "honor the fragile complexity of our ecosystem, the health of those who work the land, and the long-term well-being of customers who enjoy [the] harvest ..." (Shapin, 2006). However, not all media reports agree on this issue: Some sources note that pesticides *are* used on organic farms (Idaho Association of Soil Conservation Districts, 2004) and that organic foods typically cost the consumer more money.

How a story about food is framed has a major effect on how each of us feels about the subject of that story. When the media report that some type of food—spinach, packaged salads, or some brand of peanut butter—is being recalled by the manufacturer because of health concerns, for example, we may quit buying that particular product for a while.

Thinking specifically about the production of food, the media often use the term "Big Agra" to describe the major corporations around the globe that grow and market much of the food that we eat. These megacorporations own giant cultivated tracts that use procedures intended to maximize the crop yield, harvest that crop (whether plants or animals) at the lowest price possible, and distribute the crop to markets in the United States and other countries. Maximizing crop yield involves the use of chemicals and pesticides—and cheap labor. "Big Agra" obviously has a stake in the battle over how the media frame stories that compare its foods with organic foods; the organic food industry also has a stake in the battle. Accordingly, both sides attempt to influence how stories about their products are framed in the media because that can make a big difference in their respective profits. Whether it is fast food or fresh spinach, they want to have an impact on what you buy and where you buy it.

reflect & analyze

What factors affect your perceptions of what is "good" food and "bad" food? Are these factors influenced by advertising and media framing?

ideas are *cultural creations* of a society's most powerful members. Thus, it is possible for political, economic, and social leaders to use *ideology*—an integrated system of ideas that is external to, and coercive of, people—to maintain their positions of dominance in a society. As Marx stated,

> The ideas of the ruling class are in every epoch the ruling ideas, i.e., the class which is the ruling material force in society, is at the same time, its ruling intellectual force. The class, which has the means of material production at its disposal, has control at the same time over the means of mental production.... The ruling ideas are nothing more than the ideal expression of the dominant material relationships, the dominant material relationships grasped as ideas. (Marx and Engels, 1970/1845–1846: 64)

Many contemporary conflict theorists agree with Marx's assertion that ideas, a nonmaterial component of culture, are used by agents of the ruling class to affect the thoughts and actions of members of other classes. The role of the mass media in influencing people's thinking about the foods that they should—or should not—eat is an example of ideological control (see "Framing Culture in the Media").

How might conflict theorists view popular culture? Some conflict theorists believe that popular culture, which originated with everyday people, has been largely removed from their domain and has become nothing more than a part of the capitalist economy in the United States (Gans, 1999; Cantor, 1980, 1987). From this approach, media conglomerates such as Time Warner, Disney, and Viacom create popular culture, such as films, television shows, and amusement parks, in the same way that they would produce any other product or service. Creating new popular culture also promotes consumption of *commodities*—objects outside ourselves that we purchase to satisfy our human needs or wants (Fjellman, 1992). According to contemporary social analysts, consumption—even of things that we do not necessarily need—has become prevalent at all social levels, and some middle- and lower-income individuals and families now use as their frame of reference the lifestyles of the more affluent in their communities. As a result, many families live on credit in order to purchase the goods and services that they would like to have or that keep them on the competitive edge with their friends, neighbors, and coworkers (Schor, 1999).

Other conflict theorists examine the intertwining relationship among race, gender, and popular culture. According to the sociologist K. Sue Jewell (1993), popular cultural images are often linked to negative stereotypes of people of color, particularly African American women. Jewell believes that cultural images depicting African American women as mammies or domestics—such as those previously used in Aunt Jemima Pancake ads and recent resurrections of films such as *Gone with the Wind*—affect contemporary black women's economic prospects in profound ways (Jewell, 1993).

A strength of the conflict perspective is that it stresses how cultural values and norms may perpetuate social inequalities. It also highlights the inevitability of change and the constant tension between those who want to maintain the status quo and those who desire change. A limitation is its focus on societal discord and the divisiveness of culture.

Symbolic Interactionist Perspectives

Unlike functionalists and conflict theorists, who focus primarily on macrolevel concerns, symbolic interactionists engage in a microlevel analysis that views society as the sum of all people's interactions. From this perspective, people create, maintain, and modify culture as they go about their everyday activities. Symbols make communication with others possible because they provide us with shared meanings.

According to some symbolic interactionists, people continually negotiate their social realities. Values and norms are not independent realities that automatically determine our behavior. Instead, we reinterpret them in each social situation we encounter. However, the classical sociologist Georg Simmel warned that the larger cultural world—including both material culture and nonmaterial culture—eventually takes on a life of its own apart from the actors who daily re-create social life. As a result, individuals may be more controlled by culture than they realize. Simmel (1990/1907) suggested that money is an example of how people may be controlled by their culture. According to Simmel, people initially create money as a means of exchange, but then money acquires a social meaning that extends beyond its purely economic function. Money becomes an end in itself, rather than a means to an end. Today, we are aware of the relative "worth" not only of objects but also of individuals. Many people revere wealthy entrepreneurs and highly paid celebrities, entertainers, and sports figures for the amount of money they make, not for their intrinsic qualities. According to Simmel (1990/1907), money makes it possible for us to *relativize* everything, including our relationships with other people. When social life can be reduced to money, people become cynical, believing that anything—including people, objects, beauty, and truth—can be bought if we can pay the price. Although Simmel acknowledged the positive functions of money, he believed that the social interpretations that people give to money often produce individual feelings of cynicism and isolation.

A symbolic interactionist approach highlights how people maintain and change culture through their interactions with others. However, interactionism does not provide a systematic framework for analyzing how we shape culture and how it, in turn, shapes us. It also does not provide insight into how shared meanings are developed among people, and it does not take into account the many situations in which there is disagreement on meanings. Whereas the functional and conflict approaches tend to overemphasize the macrolevel workings of society, the interactionist viewpoint often fails to take these larger social structures into account.

Postmodernist Perspectives

Postmodernist theorists believe that much of what has been written about culture in the Western world is Eurocentric—that it is based on the uncritical assumption that European culture (including its dispersed versions in countries such as the United States, Australia, and South Africa) is the true, universal culture in which all the world's people ought to believe (Lemert, 1997). By contrast, postmodernists believe that we should speak of *cultures,* rather than *culture.*

However, Jean Baudrillard, one of the best-known French social theorists and a key figure in postmodern theory, believes that the world of culture today is based on *simulation,* not reality. According to Baudrillard, social life is much more a spectacle that simulates reality than reality itself. Many U.S. children, upon entering school for the first time, have already watched more hours of television than the total number of hours of classroom instruction they will encounter in their entire school careers (Lemert, 1997). Add to this the number of hours that some children will have spent playing computer games or using the Internet, where they often find that it is more interesting to deal with imaginary heroes and villains than to interact with "real people" in real life. Baudrillard refers to this social creation as *hyperreality*—a situation in which the *simulation* of reality is more real than experiencing the event itself and having any actual connection with what is taking place. For Baudrillard, everyday life has been captured by the signs and symbols generated to represent it, and we ultimately relate to simulations and models as if they were reality.

▲ People of all ages are spending many hours each week using computers, playing video games, and watching television. How is this behavior different from the ways in which people enjoyed popular culture in previous generations?

Baudrillard (1983) uses Disneyland as an example of a simulation—one that conceals the reality that exists outside rather than inside the boundaries of the artificial perimeter. According to Baudrillard, Disney-like theme parks constitute a form of seduction that substitutes symbolic (seductive) power for real power, particularly the ability to bring about social change. From this perspective, amusement park "guests" may feel like "survivors" after enduring the rapid speed and gravity-defying movements of the roller-coaster rides or see themselves as "winners" after surviving fights with hideous cartoon villains on the "dark rides." In reality, they have been made to *appear* to have power, but they do not actually possess any real power.

In their examination of culture, postmodernist social theorists thus make us aware of the fact that no single perspective can grasp the complexity and diversity of the social world. There is no one, single, universal culture. They also make us aware that reality may not be what it seems. According to the postmodernist view, no one authority can claim to know social reality, and we should deconstruct—take apart and subject to intense critical scrutiny—existing beliefs and theories about culture in hopes of gaining new insights (Ritzer, 1997).

Although postmodern theories of culture have been criticized on a number of grounds, we will mention only three. One criticism is postmodernism's lack of a clear conceptualization of ideas. Another is the tendency to critique other perspectives as being "grand narratives," whereas postmodernists offer their own varieties of such narratives. Finally, some analysts believe that postmodern analyses of culture lead to profound pessimism about the future.

This chapter's Concept Quick Review summarizes the components of culture as well as how the four major perspectives view it.

Culture in the Future

As we have discussed in this chapter, many changes are occurring in the United States. Increasing cultural diversity can either cause long-simmering racial and ethnic antagonisms to come closer to a boiling point or result in the creation of a truly "rainbow culture" in which diversity is respected and encouraged.

In the future the issue of cultural diversity will increase in importance, especially in schools. Multicultural education that focuses on the contributions of a wide variety of people from different backgrounds will continue to be an issue of controversy from kindergarten through college. In the Los Angeles school district, for example, students speak more than 114 different languages and dialects. Schools will face the challenge of embracing widespread cultural diversity while conveying a sense of community and national identity to students (see "Sociology Works!").

Technology will continue to have a profound effect on culture. Television and radio, films and videos, and electronic communications will continue to accelerate the flow of information and expand cultural diffusion throughout the world. Global communication devices will move images of people's lives, behavior, and fashions instantaneously among almost all nations. Increasingly, computers and cyberspace will become people's window on the world and, in the process, promote greater integration or fragmentation among nations. Integration occurs when there is a widespread acceptance of ideas and items—such as

© David McNew/Getty Images

© AP Images/John Raoux

▲ In recent years there has been a significant increase in the number of immigrants who have become U.S. citizens. However, an upsurge in anti-immigrant sentiment has put pressure on the Border Patrol and the U.S. Citizenship and Immigration Service, which are charged with enforcing immigration laws.

democracy, rock music, blue jeans, and McDonald's hamburgers—among cultures. By contrast, fragmentation occurs when people in one culture disdain the beliefs and actions of other cultures. As a force for both cultural integration and fragmentation, technology will continue to revolutionize communications, but most of the world's population will not participate in this revolution.

From a sociological perspective, the study of culture helps us not only understand our own "tool kit" of symbols, stories, rituals, and worldviews but also expand our insights to include those of other people of the world, who also seek strategies for enhancing their own lives. If we understand how culture is used by people, how cultural elements constrain or further certain patterns of action, what aspects of our cultural heritage have enduring effects on our actions, and what specific historical changes undermine the validity of some cultural patterns and give rise to others, we can apply our sociological imagination not only to our own society but to the entire world as well (see Swidler, 1986).

[concept quick review]

Analysis of Culture

Components of Culture	Symbol	Anything that meaningfully represents something else.
	Language	A set of symbols that expresses ideas and enables people to think and communicate with one another.
	Values	Collective ideas about what is right or wrong, good or bad, and desirable or undesirable in a particular culture.
	Norms	Established rules of behavior or standards of conduct.
Sociological Analysis of Culture	Functionalist Perspectives	Culture helps people meet their biological, instrumental, and expressive needs.
	Conflict Perspectives	Ideas are a cultural creation of society's most powerful members and can be used by the ruling class to affect the thoughts and actions of members of other classes.
	Symbolic Interactionist Perspectives	People create, maintain, and modify culture during their everyday activities; however, cultural creations can take on a life of their own and end up controlling people.
	Postmodern Perspectives	Much of culture today is based on simulation of reality (e.g., what we see on television) rather than reality itself.

sociology works!

Schools as Laboratories for Getting Along

Sociology makes us aware of the importance of culture in daily life. Research in sociology has also shown the significance of schools and friendship groups in exposing children and young people to cultures that are different from their own. Recent studies have shown that it may be easier for children to set aside their differences and get to know one another than it is for adults to do so. Consider what is happening among some children at International Community School, an innovative Decatur, Georgia, school where some students were born in the United States, but most are refugees from as many as forty war-torn countries: This school has become a "laboratory for getting along," particularly because some of the children have taken the initiative to befriend and help others (St. John, 2007). An excellent example is the friendship that developed between nine-year-old Dante Ramirez and Soung Oo Hlaing, an eleven-year-old Burmese refugee who spoke no English:

> The two boys met on the first day of school this year. Despite the language barrier, Dante managed to invite the newcomer to sit with him at lunch.
>
> "I didn't think he'd make friends at the beginning because he didn't speak that much English," Dante said. "So I thought I should be his friend."
>
> In the next weeks, the boys had a sleepover. They trick-or-treated on Soung's first Halloween. Soung, a gifted artist, gave Dante pointers on how to draw. And Dante helped Soung with his English. "I use simple words that are easy to know and sometimes hand movements," Dante explained. "For 'huge,' I would make my hands bigger. And for 'big,' I would make my hands smaller than for huge." (St. John, 2007: A14)

Over time, as the boys got to know each other better, their mothers also developed a friendship and began to celebrate ethnic holidays together even though they largely relied on gestures (a form of nonverbal communication) to communicate with each other. Only time will tell how successful this "laboratory" will be in helping people from diverse cultures get along, but from a sociological perspective, community efforts such as this are clearly a good start.

Sociologists believe that it is important for cross-cultural communications and cooperation to develop among individuals from diverse cultural backgrounds who now share common spaces. If a chance exists for greater understanding and cooperation in the twenty-first century, it may well originate in the small-group interactions of children in settings such as this school.

reflect & analyze

What examples can you provide that show how sociology works in regard to culture on your own college campus or in the community where you reside?

chapter review Q & A

Use these questions and answers to check how well you've achieved the learning objectives set out at the beginning of this chapter.

- **What is culture?**

Culture is the knowledge, language, values, and customs passed from one generation to the next in a human group or society. Culture can be either material or nonmaterial. Material culture consists of the physical creations of society. Nonmaterial culture is more abstract and reflects the ideas, values, and beliefs of a society.

- **What are cultural universals?**

Cultural universals are customs and practices that exist in all societies and include activities and institutions such as storytelling, families, and laws. However, specific forms of these universals vary from one cultural group to another.

- **What are the four nonmaterial components of culture that are common to all societies?**

These components are symbols, language, values, and norms. Symbols express shared meanings; through them, groups communicate cultural ideas and abstract concepts. Language is a set of symbols through which groups communicate. Values are a culture's collective ideas about what is acceptable or not acceptable. Norms are the specific behavioral expectations within a culture.

- **What are the main types of norms?**

Folkways are norms that express the everyday customs of a group, whereas mores are norms with strong moral and ethical connotations, and are essential to the stability of a culture. Laws are formal, standardized norms that are enforced by formal sanctions.

- **What are high culture and popular culture?**

High culture consists of classical music, opera, ballet, and other activities usually patronized by elite audiences. Popular culture consists of the activities, products, and services of a culture that appeal primarily to members of the middle and working classes.

- **How is cultural diversity reflected in society?**

Cultural diversity is reflected through race, ethnicity, age, sexual orientation, religion, occupation, and so forth. A diverse culture also includes subcultures and countercultures. A subculture has distinctive ideas and behaviors that differ from the larger society to which it belongs. A counterculture rejects the dominant societal values and norms.

- **What are culture shock, ethnocentrism, and cultural relativism?**

Culture shock refers to the anxiety that people experience when they encounter cultures radically different from their own. Ethnocentrism is the assumption that one's own culture is superior to others. Cultural relativism views and analyzes another culture in terms of that culture's own values and standards.

- **How do the major sociological perspectives view culture?**

A functionalist analysis of culture assumes that a common language and shared values help produce consensus and harmony. According to some conflict theorists, culture may be used by certain groups to maintain their privilege and exclude others from society's benefits. Symbolic interactionists suggest that people create, maintain, and modify culture as they go about their everyday activities. Postmodern thinkers believe that there are many cultures within the United States alone. In order to grasp a better understanding of how popular culture may simulate reality rather than be reality, postmodernists believe that we need a new way of conceptualizing culture and society.

key terms

beliefs 64
counterculture 78
cultural imperialism 81
cultural lag 74
cultural relativism 79
cultural universals 64
culture 61
culture shock 78
ethnocentrism 78
folkways 72
high culture 79
language 66
laws 73
material culture 63
mores 73
nonmaterial culture 63
norms 72
popular culture 80
sanctions 72
Sapir–Whorf hypothesis 67
subculture 76
symbol 65
taboos 73
technology 73
value contradictions 71
values 70

questions for critical thinking

1. Would it be possible today to live in a totally separate culture in the United States? Could you avoid all influences from the mainstream popular culture or from the values and norms of other cultures? How would you be able to avoid any change in your culture?
2. Do fads and fashions reflect and reinforce or challenge and change the values and norms of a society? Consider a wide variety of fads and fashions: musical styles; computer and video games and other technologies; literature; and political, social, and religious ideas.
3. You are doing a survey analysis of recent immigrants to the United States to determine the effects of popular culture on their views and behavior. What are some of the questions you would use in your survey?

turning to video

Watch the ABC video *Longevity: Does It Matter Where You Live?* (running time 3:00), available through **CengageBrain.com**. This video presents research findings showing that although more people in the United States live to be 100 than in any other country, depending on where you live, your life expectancy could be getting shorter. As you watch the video, think about your own lifestyle choices, including diet and exercise. After you've watched the video, consider these questions: In what ways is your diet and activity level influenced by cultural factors? How would you rate your overall healthfulness, especially those aspects of it that may be culturally influenced?

online study resources

Go to CENGAGE brain.com to access online study resources, including the Sociology CourseMate for this text as well as special features such as video, an interactive sociology time line and interactive maps, General Social Survey (GSS) data, and U.S. Census 2010 data.

CourseMate brings course concepts to life with interactive learning, study, and exam-preparation tools that support the printed textbook. A textbook-specific website, **Sociology CourseMate** includes an integrated interactive eBook and other interactive learning tools, including quizzes, flash cards, and videos.

Visit **www.cengagebrain.com** to access your account and purchase materials.

4 Socialization

Nothing could have gone more wrong for me on my first day [as a student] at Penn State. I didn't know who Joe Paterno [the school's long-time, championship-winning football coach] is or why the library is named after him. I hadn't bought any of my books yet because I thought they would be passed out to me as was customary in high school, and, worst of all, I had never really read a map on my own before so finding the classrooms in the huge main campus was really hard. Avoid looking like an idiot on your first day of higher education by following a few pieces of advice, all of which revolve around the two most important things I learned about being a college student: stay organized and make friends whenever possible.... Good luck on your first day!

—As a student, Mdmse. Amelie (2010) found that she was completely unprepared for the first day of college, so she now helps other students with the socialization process by providing Web tips on how to survive in college.

© John Powell Photographer/Alamy

▲ For students attending college for the first time, the socialization process is complex and immediate. What socialization issues did you face during your first term in higher education?

The white coat ceremony... was intended to herald our introduction into the [Harvard Medical School] community on our first day of medical school. While not the long coat of a physician or resident, the white coat signaled our medical affiliation and differentiated us from the civilian visitors and volunteers.

This was not an affiliation I was ready to claim as a first-year medical student. Over the course of the year, after taking courses in anatomy, pharmacology, physiology, genetics, and embryology, I was more deeply impressed by how little I knew than by how much I had learned. Yet every Monday in our Patient–Doctor course I found myself in my white coat interviewing still another patient.

Despite the uncertainty of my place in the medical world, my white coat ushered me into the foreign world of the patient–doctor dynamic.... These weekly interviews as part of our Patient–Doctor course were about learning the important questions, the right mannerisms, and the appropriate responses to our patients. Our instructors taught us to take a careful, methodical history, which I more or less skillfully replicated every week with a different patient. Although the goal of these weekly patient interactions was to discover a person's experience with illness, these interviews were more about my learning process than about the patient's story....

When I interviewed patients, they saw my white coat.... The white coat masked my youth. It masked my inexperience. It masked my nervousness. Yet in the medical world my white coat did not offer the solace of anonymity but forced me to take on power that I was not ready to accept.

— Ellen Lerner Rothman, M.D. (1999: 2–3), describing the professional socialization process that she and most other medical students encounter in the early years of their training as they learn what is expected of them as doctors in training and how to communicate most effectively with patients

Highlighted Learning Objectives

- Discuss the purpose of socialization.
- Explain what problems develop when children receive inadequate socialization.
- Identify the key agents of socialization.
- Describe socialization throughout various stages in the life course.
- Define resocialization and explain what categories of people are most likely to go through this process.

Chapter Focus Question:

How does socialization occur throughout our lives, including our college years?

What do the comments by Mdmse. Amelie and Dr. Ellen Lerner Rothman have in common? Sociologically speaking, their statements express concern about the socialization process and how we learn to adapt when we join a new social organization. Many of us experience stress when we take on new and seemingly unfamiliar roles and find that we must learn the appropriate norms regarding how persons in a specific role should think, act, and communicate with others.

Look around in your classes at the beginning of each semester, and you will probably see other students who are trying to find out what is going to be expected of them as a student in a particular course. What is the course going to cover? What are the instructor's requirements? How should students communicate with the instructor and other students in the class? Some information of this type is learned through formal instruction, such as in a classroom, but much of what we know about school is learned informally through our observations of other people, by listening to what they say when we are in their physical presence, or through interacting with them by cell phone, e-mail, or text messaging when we are apart. Sociologists use the term *socialization* to refer to both the formal and informal processes by which people learn a new role and find out how to be a part of a group or organization. As we shall see in this chapter, this process takes place throughout our life.

In this chapter we examine the process of socialization and identify reasons why socialization is crucial to the well-being of individuals, groups, and societies. We discuss both sociological and social psychological theories of human development. We look at the dynamics of socialization—how it occurs and what shapes it. Throughout the chapter we focus on positive and negative aspects of the socialization process, including the daily stresses that may be involved in this process. Before reading on, test your knowledge about socialization and the college experience by taking the Sociology and Everyday Life quiz.

Why Is Socialization Important Around the Globe?

***Socialization* is the lifelong process of social interaction through which individuals acquire a self-identity and the physical, mental, and social skills needed for survival in society.** It is the essential link between the individual and society because it helps us to become aware of ourselves as a member of the larger groups and organizations of which we are a part. Socialization also helps us to learn how to communicate with other people and to have knowledge of how other people expect us to behave in a variety of social settings. Briefly stated, socialization enables us to develop our human potential and to learn the ways of thinking, talking, and acting that are necessary for social living.

Socialization is most crucial during childhood because it is essential for the individual's survival and for human development. The many people who met the early material and social needs of each of us were central to our establishing our own identity. During the first three years of our life, we begin to develop both a unique identity and the ability to manipulate things and to walk. We acquire sophisticated cognitive tools for thinking and for analyzing a wide variety of situations, and we learn effective communication skills. By so doing, we begin a socialization process that takes place throughout our lives and through which we also have an effect on other people who watch us.

Socialization is also essential for the survival and stability of society. Members of a society must be socialized to support and maintain the existing social structure. From a functionalist perspective, individual conformity to existing norms is not taken for granted; rather, basic individual needs and desires must be balanced against the needs of the social structure. The socialization process is most effective when people conform to the norms of society because they believe that doing so is the best course of action. Socialization enables a society to "reproduce" itself by passing on its culture from one generation to the next.

Although the techniques used to teach newcomers the beliefs, values, and rules of behavior are somewhat similar in many nations, the *content* of socialization differs greatly from society to society. How people walk, talk, eat, make love, and wage war are all functions of the culture in which they are raised. At the same time, we are also influenced by our exposure to subcultures of class, race, ethnicity, religion, and gender. In addition, each of us has unique experiences in our family and friendship groupings. The kind of human being that we become depends greatly on the particular society and social groups that surround us at birth and during early childhood. What we believe about ourselves, our society, and the world does not spring full-blown from inside ourselves; rather, we learn these things from our interactions with others.

Human Development: Biology and Society

What does it mean to be "human"? To be human includes being conscious of ourselves as individuals with unique identities, personalities, and

sociology and everyday life

How Much Do You Know About Socialization and the College Experience?

True	False	
T	F	1. Professors are the primary agents of socialization for college students.
T	F	2. In recent studies few students report that they spend time studying with other students.
T	F	3. Many students find that taking college courses is stressful because it is an abrupt change from high school.
T	F	4. Law and medical students often report that they experience a high level of academic pressure because they know that their classmates were top students during their undergraduate years.
T	F	5. Academic stress may be positive for students; it does not necessarily trigger psychological stress.
T	F	6. College students typically find the socialization process in higher education to be less stressful than the professional socialization process they experience when they enter an occupation or profession.
T	F	7. Students who have paid employment (outside of school) experience higher levels of stress than students who are not employed during their college years.
T	F	8. Getting good grades and completing schoolwork are the top sources of stress reported by college students.

Answers on page 94.

relationships with others. As humans, we have ideas, emotions, and values. We have the capacity to think and to make rational decisions. But what is the source of "humanness"? Are we born with these human characteristics, or do we develop them through our interactions with others?

When we are born, we are totally dependent on others for our survival. We cannot turn ourselves over, speak, reason, plan, or do many of the things that are associated with being human. Although we can nurse, wet, and cry, most small mammals can also do those things. As discussed in Chapter 3, we humans differ from nonhuman animals because, for the most part, we lack instincts and must rely on learning for our survival. Human infants have the potential to develop human characteristics if they are exposed to an adequate socialization process.

Every human being is a product of biology, society, and personal experiences—that is, of heredity and environment or, in even more basic terms, "nature" and "nurture." How much of our development can be explained by socialization? How much by our genetic heritage? Sociologists focus on how humans design their own culture and transmit it from generation to generation through socialization. By contrast, sociobiologists assert that nature, in the form of our genetic makeup, is a major factor in shaping human behavior. ***Sociobiology*** **is the systematic study of how biology affects social behavior** (Wilson, 1975). According to the zoologist Edward O. Wilson, who pioneered sociobiology, genetic inheritance underlies many forms of social behavior such as war and peace, envy of and concern for others, and competition and cooperation. Most sociologists disagree with the notion that biological principles can be used to explain all human behavior. Obviously, however, some aspects of our physical makeup—such as eye color, hair color, height, and weight—are largely determined by our heredity.

How important is social influence ("nurture") in human development? There is hardly a single behavior that is not influenced socially. Except for simple reflexes, most human actions are social, either in their causes or in their consequences. Even solitary actions such as crying or brushing our teeth are ultimately social. We cry because someone has hurt us. We brush our teeth because our parents (or dentist) told us it was important. Social environment probably has a greater effect than heredity on the way we develop and the way we act. However, heredity does provide the basic material from which other people help to mold an individual's human characteristics.

Our biological and emotional needs are related in a complex equation. Children whose needs are met in settings characterized by affection, warmth, and closeness see the world as a safe and

socialization the lifelong process of social interaction through which individuals acquire a self-identity and the physical, mental, and social skills needed for survival in society.

sociobiology the systematic study of how biology affects social behavior.

sociology and everyday life

ANSWERS to the Sociology Quiz on Socialization and the College Experience

1. False. Numerous studies have concluded that although professors are important in helping students learn about the academic side of the college experience, our friends and acquaintances help us adapt to higher education.

2. False. A recent study reported in the *Chronicle of Higher Education* found that 87.7 percent of first-year students at four-year colleges stated that they studied with other students. Similar data are not available for students at two-year schools. Would this percentage be higher, lower, or about the same at two-year and community colleges?

3. True. The college environment is stressful for many students, who find that it is an abrupt change from high school because workloads increase, students are expected to manage their time independently and effectively, and grades are increasingly important for a person's career goals and other future endeavors.

4. True. The competitive nature of the admission process in law schools and medical schools virtually guarantees that new students will be surrounded by classmates who were exceptional students during their undergraduate years. However, this level of achievement may be a source of stimulation for some students rather than a source of discomfort and stress.

5. True. Some amount of academic stress may be positive in helping students reach their academic and career goals; however, excessive stress may be detrimental if it results in high levels of psychological stress or problematic behaviors such as alcohol abuse.

6. False. Recent studies have found that stress levels among college students are higher than those of people entering a new occupation or profession. For this reason, students are encouraged to develop good coping skills and build support networks of friends, family, and other individuals in the college community so that they have someone they can turn to if they believe that the pressure has become excessive.

7. False. Although numerous studies have been conducted to determine whether or not paid employment (outside of school) contributes to higher stress levels among college students, most research has not shown a significant relationship between the number of hours worked and levels of stress among students. Earning more money for school and personal expenses appears to offset additional time and responsibility in the workplace.

8. True. The two top stressors most frequently reported on college campuses are getting good grades and completing schoolwork. However, first-year college students also report that changes in eating and sleeping habits, increased workloads and new responsibilities, and going home for holidays and other breaks are major sources of stress for them.

Sources: *Campus Times*, 2008; *Chronicle of Higher Education*, 2009; Messenger, 2009; Reuters, 2008; Ross, Niebling, and Heckert, 1999; and Whitman, 1985.

comfortable place and see other people as trustworthy and helpful. By contrast, infants and children who receive less-than-adequate care or who are emotionally rejected or abused often view the world as hostile and have feelings of suspicion and fear.

Problems Associated with Social Isolation and Maltreatment

Social environment, then, is a crucial part of an individual's socialization. Even nonhuman primates such as monkeys and chimpanzees need social contact with others of their species in order to develop properly. As we will see, appropriate social contact is even more important for humans.

Isolation and Nonhuman Primates Researchers have attempted to demonstrate the effects of social isolation on nonhuman primates raised without contact with others of their own species. In a series of laboratory experiments, the psychologists Harry and Margaret Harlow (1962, 1977) took infant rhesus monkeys from their mothers and isolated them in separate cages. Each cage contained two nonliving "mother substitutes" made of wire, one with a feeding bottle attached and the other covered with soft terry cloth but without a bottle. The infant monkeys instinctively clung to the cloth "mother" and would not abandon it until hunger drove them to the bottle attached to the wire "mother." As soon as they were full, they went back to the cloth "mother" seeking warmth, affection, and physical comfort.

The Harlows' experiments show the detrimental effects of isolation on nonhuman primates. When the

© Martin Rogers/Getty Images

▲ As Harry and Margaret Harlow discovered, humans are not the only primates that need contact with others. Deprived of its mother, this infant monkey found a substitute.

young monkeys were later introduced to other members of their species, they cringed in the corner. Having been deprived of social contact with other monkeys during their first six months of life, they never learned how to relate to other monkeys or to become well-adjusted adults—they were fearful of or hostile toward other monkeys (Harlow and Harlow, 1962, 1977).

Because humans rely more heavily on social learning than do monkeys, the process of socialization is even more important for us.

Isolated Children Of course, sociologists would never place children in isolated circumstances so that they could observe what happened to them. However, some cases have arisen in which parents or other caregivers failed to fulfill their responsibilities, leaving children alone or placing them in isolated circumstances. From analysis of these situations, social scientists have documented cases in which children were deliberately raised in isolation. A look at the lives of two children who suffered such emotional abuse provides important insights into the importance of a positive socialization process and the negative effects of social isolation.

Anna Born in 1932 to an unmarried, mentally impaired woman, Anna was an unwanted child. She was kept in an attic-like room in her grandfather's house. Her mother, who worked on the farm all day and often went out at night, gave Anna just enough care to keep her alive; she received no other care. Sociologist Kingsley Davis (1940) described Anna's condition when she was found in 1938:

> [Anna] had no glimmering of speech, absolutely no ability to walk, no sense of gesture, not the least capacity to feed herself even when the food was put in front of her, and no comprehension of cleanliness. She was so apathetic that it was hard to tell whether or not she could hear. And all of this at the age of nearly six years.

When she was placed in a special school and given the necessary care, Anna slowly learned to walk, talk, and care for herself. Just before her death at the age of ten, Anna reportedly could follow directions, talk in phrases, wash her hands, brush her teeth, and try to help other children (Davis, 1940).

Genie About three decades later, Genie was found in 1970 at the age of thirteen. She had been locked in a bedroom alone, alternately strapped down to a child's potty chair or straitjacketed into a sleeping bag, since she was twenty months old. She had been fed baby food and beaten with a wooden paddle when she whimpered. She had not heard the sounds of human speech because no one talked to her and there was no television or radio in her room (Curtiss, 1977; Pines, 1981). Genie was placed in a pediatric hospital, where one of the psychologists described her condition:

> At the time of her admission she was virtually unsocialized. She could not stand erect, salivated continuously, had never been toilet-trained and had no control over her urinary or bowel functions. She was unable to chew solid food and had the weight, height and appearance of a child half her age (Rigler, 1993: 35).

In addition to her physical condition, Genie showed psychological traits associated with neglect, as described by one of her psychiatrists:

> If you gave [Genie] a toy, she would reach out and touch it, hold it, caress it with her fingertips, as though she didn't trust her eyes. She would rub it against her cheek to feel it. So when I met her and she began to notice me standing beside her bed, I held my hand out and she reached out and took my hand and carefully felt my thumb and fingers individually, and then put my hand against her cheek. She was exactly like a blind child (Rymer, 1993: 45).

Extensive therapy was used in an attempt to socialize Genie and develop her language abilities (Curtiss,

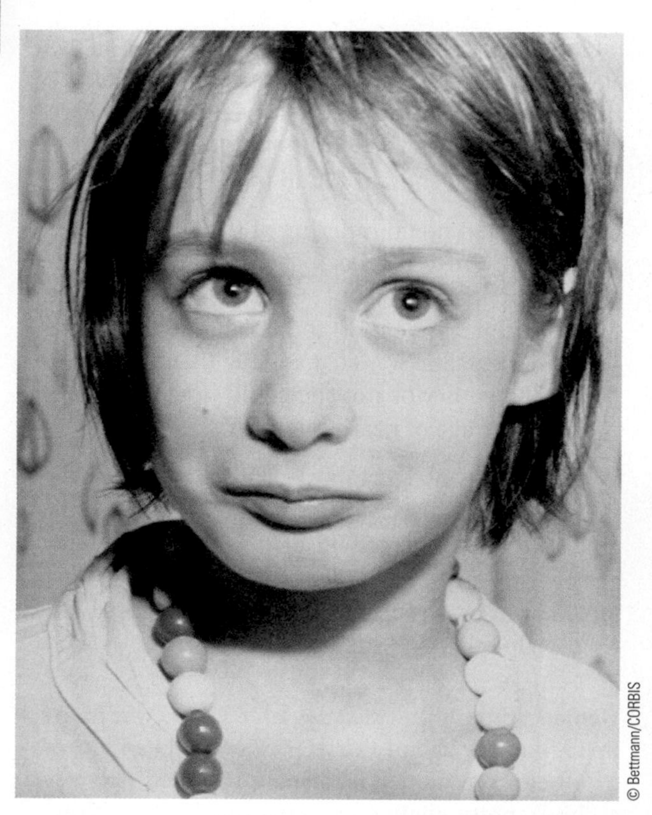

© Bettmann/CORBIS

▲ A victim of extreme child abuse, Genie was isolated from human contact and tortured until she was rescued at the age of thirteen. Subsequent attempts to socialize her were largely unsuccessful.

1977; Pines, 1981). These efforts met with limited success: In the 1990s Genie was living in a board-and-care home for retarded adults (see Angier, 1993; Rigler, 1993; Rymer, 1993).

Why do we discuss children who have been the victims of maltreatment in a chapter that looks at the socialization process? The answer lies in the fact that such cases are important to our understanding of the socialization process because they show the importance of this process and reflect how detrimental social isolation and neglect can be to the well-being of people.

Child Maltreatment What do the terms *child maltreatment* and *child abuse* mean to you? When asked what constitutes child maltreatment, many people first think of cases that involve severe physical injuries or sexual abuse. However, neglect is the most frequent form of child maltreatment (Dubowitz et al., 1993). Child neglect occurs when children's basic needs—including emotional warmth and security, adequate shelter, food, health care, education, clothing, and protection—are not met, regardless of cause (Dubowitz et al., 1993: 12). Neglect often involves acts of omission (where parents or caregivers fail to provide adequate physical or emotional care for children) rather than acts of commission (such as physical or sexual abuse). Of course, what constitutes child maltreatment differs from society to society.

Social Psychological Theories of Human Development

Over the past hundred years, a variety of psychological and sociological theories have been developed not only to explain child abuse but also to describe how a positive process of socialization occurs. Let's look first at several social psychological theories that focus primarily on how the individual personality develops.

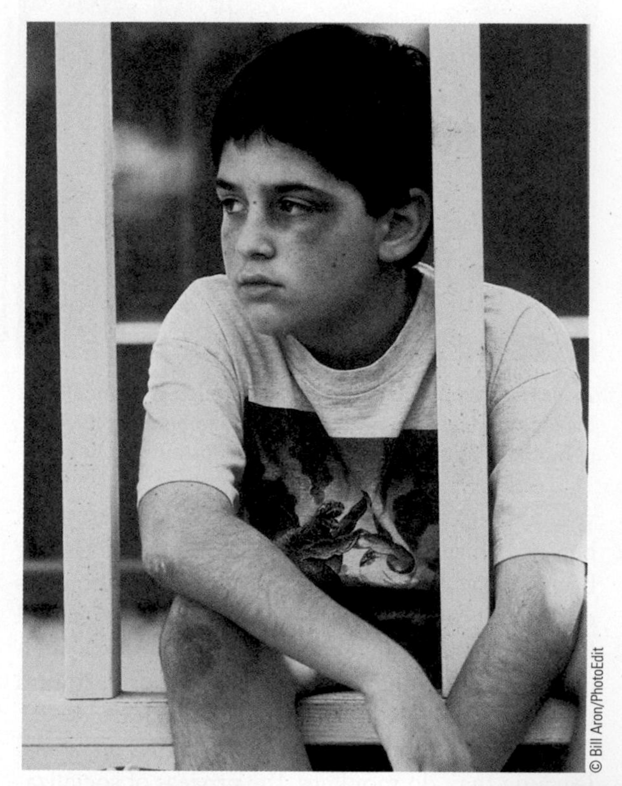

© Bill Aron/PhotoEdit

© Jose Luis Pelaez, Inc./Getty Images

▲ What are the consequences to children of isolation and physical abuse, as contrasted with social interaction and parental affection? Sociologists emphasize that social environment is a crucial part of an individual's socialization.

© Mary Evans/The Image Works

▲ Sigmund Freud, founder of the psychoanalytic perspective.

Freud and the Psychoanalytic Perspective

The basic assumption in Sigmund Freud's (1924) psychoanalytic approach is that human behavior and personality originate from unconscious forces within individuals. Freud (1856–1939), who is known as the founder of psychoanalytic theory, developed his major theories in the Victorian era, when biological explanations of human behavior were prevalent. It was also an era of extreme sexual repression and male dominance when compared to contemporary U.S. standards. Freud's theory was greatly influenced by these cultural factors, as reflected in the importance he assigned to sexual motives in explaining behavior. For example, Freud based his ideas on the belief that people have two basic tendencies: the urge to survive and the urge to procreate.

According to Freud (1924), human development occurs in three states that reflect different levels of the personality, which he referred to as the *id*, *ego*, and *superego*. The ***id* is the component of personality that includes all of the individual's basic biological drives and needs that demand immediate gratification.** For Freud, the newborn child's personality is all id, and from birth the child finds that urges for self-gratification—such as wanting to be held, fed, or changed—are not going to be satisfied immediately. However, id remains with people throughout their life in the form of *psychic energy*, the urges and desires that account for behavior. By contrast, the second level of personality—the ego—develops as infants discover that their most basic desires are not always going to be met by others. The ***ego* is the rational, reality-oriented component of personality that imposes restrictions on the innate pleasure-seeking drives of the id.** The ego channels the desire of the id for immediate gratification into the most advantageous direction for the individual. The third level of personality—the superego—is in opposition to both the id and the ego. The ***superego*, or conscience, consists of the moral and ethical aspects of personality.** It is first expressed as the recognition of parental control and eventually matures as the child learns that parental control is a reflection of the values and moral demands of the larger society. When a person is well adjusted, the ego successfully manages the opposing forces of the id and the superego. ▶ Figure 4.1 illustrates Freud's theory of personality.

Although subject to harsh criticism, Freud's theory made people aware of the importance of early childhood experiences, including abuse and neglect. His theories have also had a profound influence on contemporary mental health practitioners and on other human development theories.

Piaget and Cognitive Development

Jean Piaget (1896–1980), a Swiss psychologist, was a pioneer in the field of cognitive (intellectual) development. Cognitive theorists are interested in how people obtain, process, and use information—that is, in how we think. Cognitive development relates to changes over time in how we think.

Piaget (1954) believed that in each stage of development (from birth through adolescence), children's activities are governed by their perception of the world around them. His four stages of cognitive development are organized around specific tasks that, when mastered, lead to the acquisition of new mental capacities, which then serve as the basis for the next level of development. Piaget emphasized that all children must go through each stage in sequence before moving on to the next one, although some children move through them faster than others.

1. *Sensorimotor stage* (birth to age two). During this period, children understand the world only through sensory contact and immediate action because they cannot engage in symbolic thought or use language. Toward the end of the second year, children comprehend *object permanence;* in other words, they start to realize that objects continue to exist even when the items are out of sight.

id Sigmund Freud's term for the component of personality that includes all of the individual's basic biological drives and needs that demand immediate gratification.

ego Sigmund Freud's term for the rational, reality-oriented component of personality that imposes restrictions on the innate pleasure-seeking drives of the id.

superego Sigmund Freud's term for the conscience, consisting of the moral and ethical aspects of personality.

2. *Preoperational stage* (age two to seven). In this stage, children begin to use words as mental symbols and to form mental images. However, they still are limited in their ability to use logic to solve problems or to realize that physical objects may change in shape or appearance while still retaining their physical properties. For example, Piaget showed children two identical beakers filled with the same amount of water. After the children agreed that both beakers held the same amount of water, Piaget poured the water from one beaker into a taller, narrower beaker and then asked them about the amounts of water in each beaker. Those still in the preoperational stage believed that the taller beaker held more water because the water line was higher than in the shorter, wider beaker.
3. *Concrete operational stage* (age seven to eleven). During this stage, children think in terms of tangible objects and actual events. They can draw conclusions about the likely physical consequences of an action without always having to try the action out. Children begin to take the role of others and start to empathize with the viewpoints of others.
4. *Formal operational stage* (age twelve through adolescence). By this stage, adolescents are able to engage in highly abstract thought and understand places, things, and events they have never seen. They can think about the future and evaluate different options or courses of action.

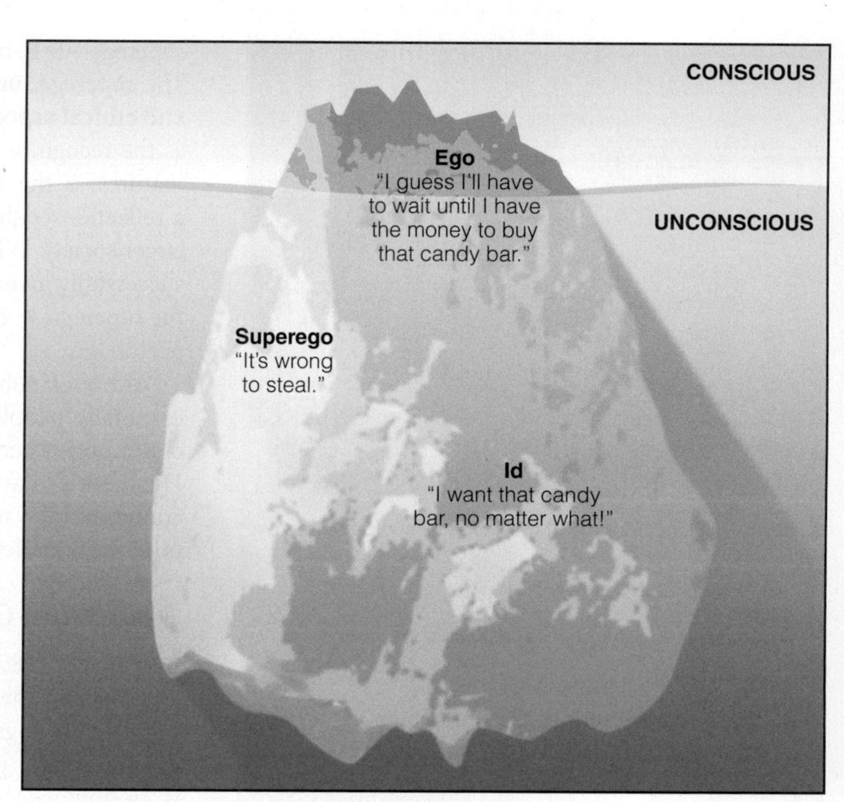

▲ **FIGURE 4.1 FREUD'S THEORY OF PERSONALITY**
This illustration shows how Freud might picture a person's internal conflict over whether to commit an antisocial act such as stealing a candy bar. In addition to dividing personality into three components, Freud theorized that our personalities are largely unconscious—hidden from our normal awareness. To dramatize his point, Freud compared conscious awareness (portions of the ego and superego) to the visible tip of an iceberg. Most of personality—including the id, with its raw desires and impulses—lies submerged in our subconscious.

▲ Jean Piaget, a pioneer in the field of cognitive development.

Piaget provided useful insights into the emergence of logical thinking as the result of biological maturation and socialization. However, critics have noted several weaknesses in Piaget's approach to cognitive development. For one thing, the theory says little about individual differences among children, nor does it provide for cultural differences. For another, as the psychologist Carol Gilligan (1982) has noted, Piaget did not take into account how gender affects the process of social development.

Kohlberg and the Stages of Moral Development

Lawrence Kohlberg (1927–1987) elaborated on Piaget's theories of cognitive reasoning by conducting

© Tony Freeman/PhotoEdit

© Tony Freeman/PhotoEdit

▲ Psychologist Jean Piaget identified four stages of cognitive development, including the preoperative stage, in which children have limited ability to realize that physical objects may change in shape or appearance. Piaget poured liquid from one beaker into a taller, narrower beaker and then asked children about the amounts of liquid in each beaker.

a series of studies in which children, adolescents, and adults were presented with moral dilemmas that took the form of stories. Based on his findings, Kohlberg (1969, 1981) classified moral reasoning into three sequential levels:

1. *Preconventional level* (age seven to ten). Children's perceptions are based on punishment and obedience. Evil behavior is that which is likely to be punished; good conduct is based on obedience and avoidance of unwanted consequences.
2. *Conventional level* (age ten through adulthood). People are most concerned with how they are perceived by their peers and with how one conforms to rules.
3. *Postconventional level* (few adults reach this stage). People view morality in terms of individual rights; "moral conduct" is judged by principles based on human rights that transcend government and laws.

Although Kohlberg presents interesting ideas about the moral judgments of children, some critics have challenged the universality of his stages of moral development. They have also suggested that the elaborate "moral dilemmas" he used are too abstract for children. In one story, for example, a husband contemplates stealing for his critically ill wife medicine that he cannot afford. When questions are made simpler, or when children and adolescents are observed in natural (as opposed to laboratory) settings, they often demonstrate sophisticated levels of moral reasoning (Darley and Shultz, 1990; Lapsley, 1990).

Gilligan's View on Gender and Moral Development

Psychologist Carol Gilligan (b. 1936) is one of the major critics of Kohlberg's theory of moral development. According to Gilligan (1982), Kohlberg's model was developed solely on the basis of research with male respondents, and women and men often have divergent views on morality based on differences in socialization and life experiences. Gilligan believes that men become more concerned with law and order but that women tend to analyze social relationships and the social consequences of behavior. For example, in Kohlberg's story about the man who is thinking about stealing medicine for his wife, Gilligan argues that male respondents are more likely to use *abstract standards* of right and wrong, whereas female respondents are more likely to be concerned about what *consequences* his stealing the drug might have for the man and his family. Does this constitute a "moral deficiency" on the part of either women or men? Not according to Gilligan.

Subsequent research that directly compared women's and men's reasoning about moral dilemmas has supported some of Gilligan's assertions but not others. For example, some other researchers have not found that women are more compassionate than men (Tavris, 1993). Overall, however, Gilligan's argument that people make moral decisions according to both abstract principles of justice and principles of compassion and care is an important contribution to our knowledge about moral reasoning.

Sociological Theories of Human Development

Although social scientists acknowledge the contributions of psychoanalytic and psychologically based explanations of human development, sociologists believe that it is important to bring a sociological perspective to bear on how people develop an awareness of self and learn about the culture in which they live. According to a sociological perspective, we cannot form a sense of self or personal identity without intense social contact with others. The self represents the sum total of perceptions and feelings that an individual has of being a distinct, unique person—a sense of who and what one is. When we speak of the "self," we typically use words such as *I, me, my, mine,* and *myself* (Cooley, 1998/1902). This sense of self (also referred to as *self-concept*) is not present at birth; it arises in the process of social experience. ***Self-concept*** **is the totality of our beliefs and feelings about ourselves.** Four components make up our self-concept: (1) the physical self ("I am tall"), (2) the active self ("I am good at soccer"), (3) the social self ("I am nice to others"), and (4) the psychological self ("I believe in world peace"). Between early and late childhood a child's focus tends to shift from the physical and active dimensions of self toward the social and psychological aspects. Self-concept is the foundation for communication with others; it continues to develop and change throughout our lives.

Our *self-identity* is our perception about what kind of person we are. As we have seen, socially isolated children do not have typical self-identities because they have had no experience of "humanness." According to symbolic interactionists, we do not know who we are until we see ourselves as we believe that others see us. The perspectives of symbolic interactionists Charles Horton Cooley and George Herbert Mead help us understand how our self-identity is developed through our interactions with others.

Cooley and the Looking-Glass Self

According to the sociologist Charles Horton Cooley (1864–1929), the ***looking-glass self*** **refers to the way in which a person's sense of self is derived from the perceptions of others.** Our looking-glass self is based on our perception of *how* other people think of us (Cooley, 1998/1902). As ▶ Figure 4.2 shows, the looking-glass self is a self-concept derived from a three-step process:

▲ **FIGURE 4.2 HOW THE LOOKING-GLASS SELF WORKS**

Source: Based on Katzer, Cook, and Crouch, 1991.

1. We imagine how our personality and appearance will look to other people.
2. We imagine how other people judge the appearance and personality that we think we present.
3. We develop a self-concept. If we think the evaluation of others is favorable, our self-concept is enhanced. If we think the evaluation is unfavorable, our self-concept is diminished. (Cooley, 1998/1902)

The self develops only through contact with others, just as social institutions and societies do not exist independently of the interaction of individuals (Schubert, 1998).

© PhotoAlto/Alamy

▲ How do these teenagers' perceptions of the world differ from their perceptions ten years earlier, according to Piaget?

Mead and Role-Taking

George Herbert Mead (1863–1931) extended Cooley's insights by linking the idea of self-concept to ***role-taking*—the process by which a person mentally assumes the role of another person or group in order to understand the world from that person's or group's point of view.** Role-taking often occurs through play and games, as children try out different roles (such as being mommy, daddy, doctor, or teacher) and gain an appreciation of them. First, people come to take the role of the other (role-taking). By taking the roles of others,

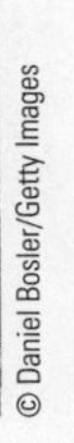

© Digital Vision/Alamy

▲ According to sociologist George Herbert Mead, the self develops through three stages. In the preparatory stage, children imitate others; in the play stage, children pretend to take the roles of specific people; and in the game stage, children become aware of the "rules of the game" and the expectations of others.

© Daniel Bosler/Getty Images

© Stella/Getty Images

the individual hopes to ascertain the intention or direction of the acts of others. Then the person begins to construct his or her own roles (role-making) and to anticipate other individuals' responses. Finally, the person plays at her or his particular role (role-playing).

According to Mead (1934), in the early months of life children do not realize that they are separate from others. However, they do begin early on to see a mirrored image of themselves in others. Shortly after birth, infants start to notice the faces of those around them, especially the significant others, whose faces start to have meaning because they are associated with experiences such as feeding and cuddling. ***Significant others* are those persons whose care, affection, and approval are especially desired and who are most important in the development of the self.** Gradually, we distinguish ourselves from our caregivers and begin to perceive ourselves in contrast to them. As we develop language skills and learn to understand symbols, we begin to develop a self-concept. When we can represent ourselves in our minds as objects distinct from everything else, our self has been formed.

Mead (1934) divided the self into the "I" and the "me." The "I" is the subjective element of the self and represents the spontaneous and unique traits of each person. The "me" is the objective element of the self, which is composed of the internalized attitudes and demands of other members of society and the individual's awareness of those demands. Both the "I" and the "me" are needed to form the social self. The unity of the two constitutes the full development of the individual. According to Mead, the "I" develops first, and the "me" takes form during the three stages of self-development:

1. During the *preparatory stage,* up to about age three, interactions lack meaning, and children largely imitate the people around them. At this stage, children are preparing for role-taking.
2. In the *play stage,* from about age three to five, children learn to use language and other symbols, thus enabling them to pretend to take the roles of specific people. At this stage, they begin to see themselves in relation to others, but they do not see role-taking as something they have to do.
3. During the *game stage,* which begins in the early school years, children understand not only their own social position but also the positions of others around them. In contrast to play, games are structured by rules, are often competitive, and

self-concept the totality of our beliefs and feelings about ourselves.

looking-glass self Charles Horton Cooley's term for the way in which a person's sense of self is derived from the perceptions of others.

role-taking the process by which a person mentally assumes the role of another person or group in order to understand the world from that person's or group's point of view.

significant others those persons whose care, affection, and approval are especially desired and who are most important in the development of the self.

sociology works!

"Good Job!": Mead's Generalized Other and the Issue of Excessive Praise

> Hang out at a playground, visit a school, or show up at a child's birthday party, and there's one phrase you can count on hearing repeatedly: "Good job!" Even tiny infants are praised for smacking their hands together ("Good clapping!"). Many of us blurt out these judgments of our children to the point that it has become almost a verbal tic. (Kohn, 2001)

Educational analyst Alfie Kohn describes the common practice of praising children for practically everything they say or do. According to Kohn, excessive praise or unearned compliments may be problematic for children because, rather than bolstering their self-esteem, such praise may increase a child's dependence on adults. As children increasingly rely on constant praise and on significant others to identify what is good or bad about their performance, they may not develop the ability to make meaningful judgments about what they have done. As Kohn suggests (2001), "Sadly, some of these kids will grow into adults who continue to need someone to pat them on the head and tell them whether what they did was OK."

Kohn's ideas remind us of the earlier sociological insights of George Herbert Mead, who described how children learn to take into account the expectations of the larger society and to balance the "I" (the subjective element of the self: the spontaneous and unique traits of each person) with the "me" (the objective element of the self: the internalized attitudes and demands of other members of society and the individual's awareness of those demands). As Mead (1934: 160) stated, "What goes on in the game goes on in the life of the child at all times. He is continually taking the attitudes of those about him, especially the roles of those who in some sense control him and on whom he depends." According to Mead, role-taking is vital to the formation of a mature sense of self as each individual learns to visualize the intentions and expectations of other people and groups. Excessively praising children may make it more difficult for them to develop a positive self-concept and visualize an accurate picture of what is expected of them as they grow into young adulthood.

Does this mean that children should not be praised? Definitely not! It means that we should think about when and how to praise children. What children may need sometimes is not praise, but encouragement. As child development specialist Docia Zavitkovsky has stated,

> I sometimes say that praise is fine "when praise is due." We get into the habit of praising when it isn't praise that is appropriate but encouragement. For example, we're always saying to young children: "Oh, what a beautiful picture," even when their pictures aren't necessarily beautiful. So why not really look at each picture? Maybe a child has painted a picture with many wonderful colors. Why don't we comment on that—on the reality of the picture? (qtd. in *Scholastic Parent & Child*, 2007)

From this perspective, positive feedback can have a very important influence on a child's self-esteem because he or she can learn how to do a "good job" when engaging in a specific activity or accomplishing a task rather than simply being praised for any effort expended. Mead's concept of the generalized other makes us aware of the importance of other people's actions in how self-concept develops.

reflect & analyze

What effect does receiving praise when we are young have on us when we are college students? Also, when we are dealing with our peers, how might we thoughtfully use the phrase "Good job!" without making it into an overworked expression?

involve a number of other "players." At this time, children become concerned about the demands and expectations of others and of the larger society. Mead used the example of a baseball game to describe this stage because children, like baseball players, must take into account the roles of all the other players at the same time. Mead's concept of the ***generalized other*** **refers to the child's awareness of the demands and expectations of the society as a whole or of the child's subculture.**

How useful are symbolic interactionist perspectives in enhancing our understanding of the socialization process? Certainly, this approach contributes to our understanding of how the self develops (see "Sociology Works!"). However, this approach has certain limitations. Sociologist Anne Kaspar (1986) suggests that Mead's ideas about the social self may be more applicable to men than to women because women are more likely to experience inherent conflicts between the meanings they derive from their personal experiences and those they take from the culture, particularly in regard to balancing the responsibilities of family life and paid employment. (This chapter's Concept Quick Review summarizes the major theories of human development.)

Recent Symbolic Interactionist Perspectives

The symbolic interactionist approach emphasizes that socialization is a collective process in which children are active and creative agents, not just passive recipients of the socialization process. From this view,

[concept quick review]

Psychological and Sociological Theories of Human Development

Social Psychological Theories of Human Development	Freud's psychoanalytic perspective	Children first develop the id (drives and needs), then the ego (restrictions on the id), and then the superego (moral and ethical aspects of personality).
	Piaget's cognitive development	Children go through four stages of cognitive (intellectual) development, going from understanding only through sensory contact to engaging in highly abstract thought.
	Kohlberg's stages of moral development	People go through three stages of moral development, from avoidance of unwanted consequences to viewing morality based on human rights.
	Gilligan: gender and moral development	Women go through stages of moral development from personal wants to the greatest good for themselves and others.
Sociological Theories of Human Development	Cooley's looking-glass self	A person's sense of self is derived from his or her perception of how others view him or her.
	Mead's three stages of self-development	In the preparatory stage, children imitate the people around them; in the play stage, children pretend to take the roles of specific people; and in the game stage, children learn the demands and expectations of roles.

childhood is a *socially constructed* category (Adler and Adler, 1998). Children are capable of actively constructing their own shared meanings as they acquire language skills and accumulate interactive experiences (Qvortrup, 1990). According to the "orb web model" of the sociologist William A. Corsaro (1985, 1997), children's cultural knowledge reflects not only the beliefs of the adult world but also the unique interpretations and aspects of the children's own peer culture. Corsaro (1992: 162) states that *peer culture* is "a stable set of activities or routines, artifacts, values, and concerns that children produce and share." This peer culture emerges through interactions as children "borrow" from the adult culture but transform it so that it fits their own situation. In fact, Corsaro (1992) believes that the peer group is the most significant arena in which children and young people acquire cultural knowledge.

Agents of Socialization

Agents of socialization **are the persons, groups, or institutions that teach us what we need to know in order to participate in society.** We are exposed to many agents of socialization throughout our lifetime; in turn, we have an influence on those socializing agents and organizations. Here, we look at the most pervasive ones in childhood—the family, the school, peer groups, and mass media.

The Family

The family is the most important agent of socialization in all societies. From our infancy onward, our families transmit cultural and social values to us. As discussed later in this book, families vary in size and structure. Some families consist of two parents and their biological children, whereas others consist of a single parent and one or more children. Still other families reflect changing patterns of divorce and remarriage, and an increasing number are made up of same-sex partners and their children. Over time, patterns have changed in some two-parent families so that fathers, rather than mothers, are the primary daytime agents of socialization for their young children.

Theorists using a functionalist perspective emphasize that families serve important functions in society because they are the primary locus for the procreation and socialization of children. Most of us form an emerging sense of self and acquire most of our beliefs and values within the family context. We also learn about the larger dominant culture (including language, attitudes, beliefs, values, and norms) and the primary subcultures to which our parents and other relatives belong.

Families are also the primary source of emotional support. Ideally, people receive love, understanding, security, acceptance, intimacy, and companionship

generalized other George Herbert Mead's term for the child's awareness of the demands and expectations of the society as a whole or of the child's subculture.

agents of socialization the persons, groups, or institutions that teach us what we need to know in order to participate in society.

© Michael Newman/PhotoEdit

▲ As this celebration attended by three generations of family members illustrates, socialization enables society to "reproduce" itself.

within families. The role of the family is especially significant because young children have little social experience beyond the family's boundaries; they have no basis for comparing or evaluating how they are treated by their own family.

To a large extent, the family is where we acquire our specific social position in society. From birth, we are a part of the specific racial, ethnic, class, religious, and regional subcultural grouping of our family. Studies show that families socialize their children somewhat differently based on race, ethnicity, and class (Kohn, 1977; Kohn et al., 1990; Harrison et al., 1990). For example, sociologist Melvin Kohn (1977; Kohn et al., 1990) has suggested that social class (as measured by parental occupation) is one of the strongest influences on what and how parents teach their children. On the one hand, working-class parents, who are closely supervised and expected to follow orders at work, typically emphasize to their children the importance of obedience and conformity. On the other hand, parents from the middle and professional classes, who have more freedom and flexibility at work, tend to give their children more freedom to make their own decisions and to be creative. Kohn concluded that differences in the parents' occupations are a better predictor of child-rearing practices than is social class itself.

Whether or not Kohn's findings are valid today, the issues he examined make us aware that not everyone has the same family experiences. Many factors—including our cultural background, nation of origin, religion, and gender—are important in determining how we are socialized by family members and others who are a part of our daily life.

Conflict theorists stress that socialization contributes to false consciousness—a lack of awareness and a distorted perception of the reality of class as it affects all aspects of social life. As a result, socialization reaffirms and reproduces the class structure in the next generation rather than challenging the conditions that presently exist. For example, children in low-income families may be unintentionally socialized to believe that acquiring an education and aspiring to lofty ambitions are pointless because of existing economic conditions in the family. By contrast, middle-income and upper-income families typically instill ideas of monetary and social success in children while encouraging them to think and behave in "socially acceptable" ways.

The School

As the amount of specialized technical and scientific knowledge has expanded rapidly and as the amount of time that children are in educational settings has increased, schools continue to play an enormous role in the socialization of young people. For many people the formal education process is an undertaking that lasts twenty years or more. Colleges and universities play an extensive role in socializing young adults for good citizenship, marriage, and careers.

The number of one-parent families and families in which both parents work outside the home has increased dramatically, and the number of children in day-care and preschool programs has also grown rapidly. Currently, about 60 percent of all U.S. preschool children are in day care, either in private homes or institutional settings, and this percentage continues to climb (Children's Defense Fund, 2009). Generally, studies have found that quality day-care and preschool programs have a positive effect on the overall socialization of children. These programs provide children with the opportunity to have frequent interactions with teachers and to learn how to build their language and literacy skills. High-quality programs also have a positive effect on the academic performance of children, particularly those from low-income families. Today, however, the cost of child-care programs has become a major concern for many families.

Although schools teach specific knowledge and skills, they also have a profound effect on children's self-image, beliefs, and values. As children enter school for the first time, they are evaluated and systematically compared with one another by the teacher. A permanent, official record is kept of each

© Tony Freeman/PhotoEdit

▲ Students are sent to school to be educated. However, what else will they learn in school beyond the academic curriculum? Sociologists differ in their responses to this question.

child's personal behavior and academic activities. From a functionalist perspective, schools are responsible for (1) socialization, or teaching students to be productive members of society; (2) transmission of culture; (3) social control and personal development; and (4) the selection, training, and placement of individuals on different rungs in the society (Ballantine and Hammack, 2009).

In contrast, conflict theorists assert that students have different experiences in the school system depending on their social class, their racial–ethnic background, the neighborhood in which they live, their gender, and other factors. According to the sociologists Samuel Bowles and Herbert Gintis (1976), much of what happens in school amounts to teaching a hidden curriculum in which children learn to be neat, to be on time, to be quiet, to wait their turn, and to remain attentive to their work. Thus, schools do not socialize children for their own well-being but rather for their later roles in the workforce, where it is important to be punctual and to show deference to supervisors. Students who are destined for leadership or elite positions acquire different skills and knowledge than those who will enter working-class and middle-class occupations (see Cookson and Persell, 1985).

Peer Groups

As soon as we are old enough to have acquaintances outside the home, most of us begin to rely heavily on peer groups as a source of information and approval about social behavior. A ***peer group* is a group of people who are linked by common interests, equal social position, and (usually) similar age.** In early childhood, peer groups are often composed of classmates in day care, preschool, and elementary school. Recent studies have found that preadolescence—the latter part of the elementary school years—is an age period in which children's peer culture has an important effect on how children perceive themselves and how they internalize society's expectations (Adler and Adler, 1998). In adolescence, peer groups are typically made up of people with similar interests and social activities. As adults, we continue to participate in peer groups of people with whom we share common interests and comparable occupations, income, and/or social position.

Peer groups function as agents of socialization by contributing to our sense of "belonging" and our feelings of self-worth. As early as the preschool years, peer groups provide children with an opportunity for successful adaptation to situations such as gaining access to ongoing play, protecting shared activities from intruders, and building solidarity and mutual trust during ongoing activities (Corsaro, 1985; Rizzo and Corsaro, 1995). Unlike families and schools, peer groups provide children and adolescents with some degree of freedom from parents and other authority figures (Corsaro, 1992). Although peer groups afford children some degree of freedom, they also teach cultural norms such as what constitutes "acceptable" behavior in a specific situation. Peer groups simultaneously reflect the larger culture and serve as a conduit for passing on culture to young people. As a result, the peer group is both a product of culture and one of its major transmitters (Elkin and Handel, 1989).

Is there such a thing as "peer pressure"? Individuals must earn their acceptance with their peers by conforming to a given group's norms, attitudes, speech patterns, and dress codes. When we conform to our peer group's expectations, we are rewarded; if we do not conform, we may be ridiculed or even expelled

peer group a group of people who are linked by common interests, equal social position, and (usually) similar age.

www.Beepstock.com/Robinbeckham/Alamy Limited

▲ For decades, analysts have been concerned about the effects of television viewing on the young. However, the relatively recent advent of video games, the Internet, cell phones, and texting devices has exacerbated the problem. Today, it is possible—*and common*—for every spare minute of a young person's day to be spent on audiovisual and digital entertainment.

people watching so much television. Television has been praised for offering numerous positive experiences to children. Some scholars suggest that television (when used wisely) can enhance children's development by improving their language abilities, concept-formation skills, and reading skills and by encouraging prosocial development (Winn, 1985). However, other studies have shown that children and adolescents who spend a lot of time watching television often have lower grades in school, read fewer books, exercise less, and are overweight (American Academy of Child and Adolescent Psychiatry, 1997). Of special concern to many people is the issue of television violence. It is estimated that the typical young person who watches 28 hours of television per week will have seen 16,000 simulated murders and 200,000 acts of violence by the time he or she reaches age 18. A report by the American Psychological Association states that about 80 percent of all television programs contain acts of violence and that commercial television for children is 50 to 60 times more violent than prime-time television for adults. For example, some cartoons average more than 80 violent acts per hour (APA Online, 2000). The violent content of media programming and the marketing and advertising practices of mass media industries that routinely target children under age 17 have come under the scrutiny of government agencies such as the Federal Trade Commission because of concerns raised by parents and social analysts.

In addition to concerns about social media and television programming, video games have been criticized for absorbing excessive amounts of time among younger people, particularly boys. Survey data from a nationally represented sample of 10- to 19-year-olds found that boys are much more likely to play video games than girls are. Male gamers average 58 minutes of play per weekday and 90 minutes per weekend day; female gamers average 44 minutes of play per weekday and 64 minutes per weekend day (ScienceDaily, 2007). We do not have an accurate assessment of how much time young people spend in social networking. Some analysts believe that social networking enhances a person's ability to interact with others; however, critics believe that this type of communication limits the ability of young people to get to know other individuals and work well with them.

Gender and Racial–Ethnic Socialization

***Gender socialization* is the aspect of socialization that contains specific messages and practices concerning the nature of being female or male in a specific group or society.** Through the process of gender socialization we learn about what attitudes and behaviors are considered to be appropriate for girls and boys, men and women, in a particular society. Different sets of gender norms are appropriate for females and males in the United States and most other nations.

One of the primary agents of gender socialization is the family. In some families this process begins even before the child's birth. Parents who learn the sex of the fetus through ultrasound or amniocentesis often purchase color-coded and gender-typed clothes, toys, and nursery decorations in anticipation of their daughter's or son's arrival. After birth, parents may respond differently toward male and female infants; they often play more roughly with boys and talk more lovingly to girls. Throughout childhood and adolescence, boys and girls are typically assigned different household chores and given different privileges (such as how late they may stay out at night).

When we look at the relationship between gender socialization and social class, the picture becomes more complex. Although some studies have found less-rigid gender stereotyping in higher-income families (Seegmiller, Suter, and Duviant, 1980; Brooks-Gunn, 1986), others have found more (Bardwell, Cochran, and Walker, 1986). One study found that higher-income families are more likely than lower-income families to give "male-oriented" toys (which develop visual spatial and problem-solving skills) to children of both sexes (Serbin et al., 1990). Working-class families tend to adhere to more-rigid gender expectations than do middle-class families (Canter and Ageton, 1984; Brooks-Gunn, 1986).

We are limited in our knowledge about gender socialization practices among racial–ethnic groups because most studies have focused on white, middle-class families. In a study of African American families, the sociologist Janice Hale-Benson (1986) found that children typically are not taught to think of gender strictly in "male–female" terms. Both daughters and sons are socialized toward autonomy, independence, self-confidence, and nurturance of children (Bardwell, Cochran, and Walker, 1986). Sociologist Patricia Hill Collins (1990) has suggested that "othermothers" (women other than a child's biological mother) play an important part in the gender socialization and motivation of African American children, especially girls. Othermothers often serve as gender role models and encourage women to become activists on behalf of their children and community (Collins, 1990). By contrast, studies of Korean American and Latino/a families have found more-traditional gender socialization (Min, 1988), although some evidence indicates that this pattern may be changing (Jaramillo and Zapata, 1987).

Like the family, schools, peer groups, and the media also contribute to our gender socialization. From kindergarten through college, teachers and peers reward gender-appropriate attitudes and behavior. Sports reinforce traditional gender roles through a rigid division of events into male and female categories. The media are also a powerful source of gender socialization; starting very early in childhood, children's books, television programs, movies, and music provide subtle and not-so-subtle messages about how boys and girls should act (see Chapter 11, "Sex and Gender").

In addition to gender-role socialization, we receive racial socialization throughout our lives. ***Racial socialization* is the aspect of socialization that contains specific messages and practices concerning the nature of one's racial or ethnic status** as it relates to one's identity, interpersonal relationships, and location in the social hierarchy. Racial socialization includes direct statements regarding race, modeling behavior (wherein a child imitates the behavior of a parent or other caregiver), and indirect activities such as exposure to an environment that conveys a specific message about a racial or ethnic group ("We are better than they are," for example).

The most important aspects of racial identity and attitudes toward other racial–ethnic groups are passed down in families from generation to generation. As the sociologist Martin Marger (1994: 97) notes, "Fear of, dislike for, and antipathy toward one group or another is learned in much the same way that people learn to eat with a knife or fork rather than with their bare hands or to respect others' privacy in personal matters." These beliefs can be transmitted in subtle and largely unconscious ways; they do not have to be taught directly or intentionally. Scholars have found that ethnic values and attitudes begin to crystallize among children as young as age four (Van Ausdale and Feagin, 2001). By this age, the society's ethnic hierarchy has become apparent to the child. Some minority parents feel that racial socialization is essential because it provides children with the skills and abilities they will need to survive in the larger society.

Socialization Through the Life Course

Why is socialization a lifelong process? Throughout our lives, we continue to learn. Each time we experience a change in status (such as becoming a college student or getting married), we learn a new set of rules, roles, and relationships. Even before we achieve a new status, we often participate in ***anticipatory socialization*—the process by which knowledge and skills are learned for future roles.** Many societies organize social activities according to age and gather data regarding the age composition of the people who live in that society. Some societies have distinct *rites of passage*, based on age or other factors, that publicly dramatize and validate changes in a person's status. In the United States and other industrialized societies, the most common categories of age are childhood, adolescence, and adulthood (often subdivided into young adulthood, middle adulthood, and older adulthood).

Childhood

Some social scientists believe that a child's sense of self is formed at a very early age and that it is difficult to change this self-perception later in life. Symbolic interactionists emphasize that during infancy and early childhood, family support and guidance are crucial to a child's developing self-concept. In some families children are provided with emotional warmth, feelings of mutual trust, and a sense of security. These families come closer to our ideal cultural belief that childhood should be a time of carefree play, safety, and freedom from economic, political, and sexual responsibilities. However, other families reflect the discrepancy between cultural ideals and reality—children grow up in a setting characterized by fear, danger, and risks that are created by parental neglect, emotional maltreatment, or premature economic and sexual demands (Knudsen, 1992). Abused children often experience low self-esteem, an inability to trust others, feelings of isolationism and powerlessness, and denial of their feelings.

Adolescence

In industrialized societies the adolescent (or teenage) years represent a buffer between childhood and

gender socialization the aspect of socialization that contains specific messages and practices concerning the nature of being female or male in a specific group or society.

racial socialization the aspect of socialization that contains specific messages and practices concerning the nature of one's racial or ethnic status.

anticipatory socialization the process by which knowledge and skills are learned for future roles.

© AP Images

▲ Do you believe that what this child is learning here will have an influence on her actions in the future? What other childhood experiences might offset early negative racial socialization?

adulthood. In the United States no specific rites of passage exist to mark children's move into adulthood; therefore, young people have to pursue their own routes to self-identity and adulthood (Gilmore, 1990). Anticipatory socialization is often associated with adolescence, during which many young people spend much of their time planning or being educated for future roles they hope to occupy. Rites of passage may be used to mark the transition between childhood and adolescence or adolescence and adulthood. A celebration known as a Bar Mitzvah is held for some Jewish boys on their thirteenth birthday, and a Bat Mitzvah is held for some Jewish girls on their twelfth birthday; these events mark the occasion upon which young people accept moral responsibility for their own actions and the fact that they are now old enough to own personal property. Similarly, some Latinas are honored with the *quinceañera*—a celebration of their fifteenth birthday that marks their passage into young womanhood. Although it is not officially designated as a rite of passage, many of us think of the time when we get our first driver's license or graduate from high school as another way in which we mark the transition from one period of our life to the next.

Adolescence is often characterized by emotional and social unrest. In the process of developing their own identities, some young people come into conflict with parents, teachers, and other authority figures who attempt to restrict their freedom. Adolescents may also find themselves caught between the demands of adulthood and their own lack of financial independence and experience in the job market. The experiences of individuals during adolescence vary according to race, class, and gender.

Based on their family's economic situation and personal choices, some young people leave high school and move directly into the world of work while others pursue a college education and may continue to receive advice and financial support from their parents. Others are involved in both the world of work and the world of higher education as they seek to support themselves and to acquire more years of formal education or vocational/career training. Whether or not a student works while in college may affect the process of adjusting to college life (see ▶ Figure 4.3, "Time Line for First-Semester College Socialization"). In the second decade of the twenty-first century, more college students are exploring international study programs as part of their adult socialization to help them gain new insights into divergent cultures and the larger world around them (see "Sociology in Global Perspective").

Adulthood

One of the major differences between child socialization and adult socialization is the degree of freedom of choice. If young adults are able to support themselves financially, they gain the ability to make more choices about their own lives. In early adulthood (usually until about age forty), people work toward their own goals of creating meaningful relationships with others, finding employment, and seeking personal fulfillment. Of course, young adults continue to be socialized by their parents, teachers, peers, and the media, but they also learn new attitudes and behaviors. For example, when we marry or have children,

© David Young-Wolff/PhotoEdit

▲ An important rite of passage for many Latinas is the *quinceañera*—a celebration of their fifteenth birthday and their passage into womanhood. Can you see how this occasion might also be a form of anticipatory socialization?

EARLY FALL

© David Young-Wolff/PhotoEdit

- Adapting to new people and new situations
- Anticipation and excitement about studying in a new setting
- Insecurity about academic demands
- Homesickness
- If employed, trying to balance school and work life

MID FALL

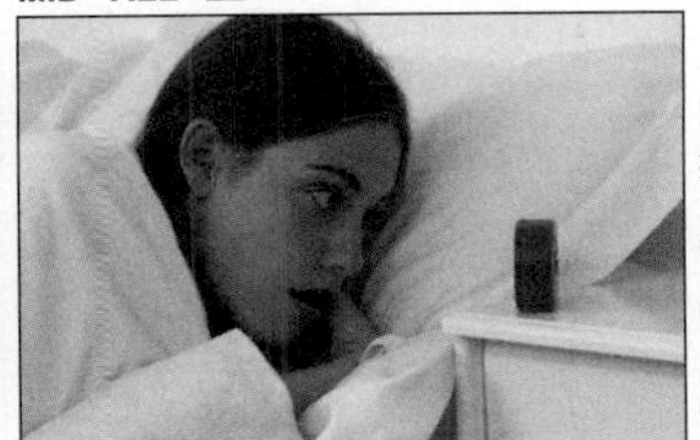

© Bubbles Photolibrary/Alamy

- Social pressures from others: What would my parents think?
- Anticipation (and dread) of midterm exams and major papers
- Time-management problems between school and social life
- Intense need for a break
- Concerns about role conflict between school and work

LATE FALL

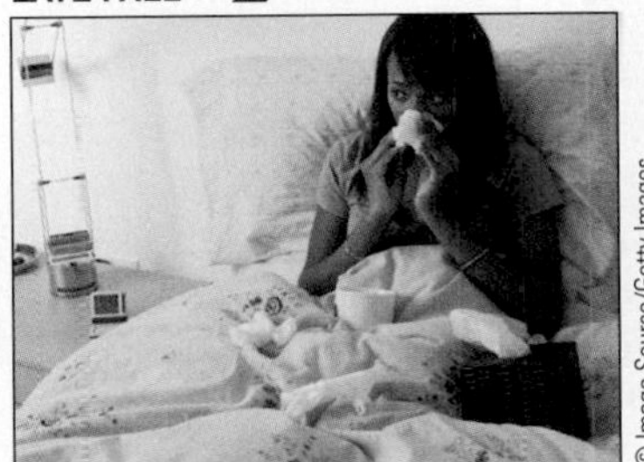

© Image Source/Getty Images

- Positive or negative assessment of grades so far
- Pre-final studying and jitters
- Making up for lost time and procrastination
- First college illnesses likely to occur because of late hours, poor eating habits, and proximity to others who become ill
- Potential problems with roommates or others who make excessive demands on one's time and/or personal space

END OF TERM

© The Copyright Group Super Stock

- Final exams: late nights, extra effort, and stress
- Concerns about leaving new friends and college setting for winter break
- Anticipation (and tension) associated with going home for break for those who have been away
- Reassessment of college choice, major, and career options: Am I on the right track?
- Acknowledgment that growth has occurred and much has been learned, both academically and otherwise, during the first college term

▲ **FIGURE 4.3 TIME LINE FOR FIRST-SEMESTER COLLEGE SOCIALIZATION**

Sources: Based on the author's observations of student life and on Kansas State University, "Timeline for Transition," 2010.

we learn new roles as partners or parents. Adults often learn about fads and fashions in clothing, music, and language from their children.

Workplace (occupational) socialization is one of the most important types of early adult socialization. This type of socialization tends to be most intense immediately after a person makes the transition from school to the workplace; however, this process may continue throughout our years of employment. Many people experience continuous workplace socialization as a result of having more than one career in their lifetime.

In middle adulthood—between the ages of forty and sixty-five—people begin to compare their accomplishments with their earlier expectations. This is the point at which people either decide that they have reached their goals or recognize that they have attained as much as they are likely to achieve.

Late adulthood may be divided into three categories: (1) the "young-old" (ages sixty-five to seventy-four), (2) the "old-old" (ages seventy-five to eighty-five), and (3) the "oldest-old" (over age eighty-five). Although these are somewhat arbitrary divisions, the "young-old" are less likely to suffer from disabling illnesses, whereas some of the "old-old" are more likely to suffer such illnesses. Increasingly, studies in gerontology and the sociology of medicine have come to question these arbitrary categories and show that some persons defy the expectations of their age grouping based on individual genetic makeup, lifestyle choices, and a zest for living. Perhaps "old age" is what we make it!

Late Adulthood and Ageism

In older adulthood, some people are quite happy and content; others are not. Erik Erikson noted that difficult changes in adult attitudes and behavior occur in the last years of life, when people experience decreased physical ability, lower prestige, and the

sociology in global perspective

Open Doors: Study Abroad and Global Socialization

As I had been told, the first month or so of the study abroad experience feels like a vacation in that everything is exciting and new. After this "honeymoon" period, the experience becomes something other than merely a vacation or fleeting visit. You start to relate to the people, the culture, and life in that country not from the eyes of a tourist passing through, but progressively from the eyes of those around you—the citizens who were born and raised there. That is the perspective which is unattainable without actually *living* in another country, and a perspective which I have come to appreciate and understand more fully as I settle back into life here back at home.

—John R. R. Howie (2010), a Boston College economics and Mandarin Chinese major, explaining what studying abroad at Peking University, in Beijing, meant to him

Although we may read and hear about what goes on in other countries, it is quite different to be able to see and experience those cultures firsthand. Perhaps this is why a record number of U.S. students are choosing to study abroad: Studying in another country is an important part of the college socialization process for preparing to live and work in an interconnected world. According to recent reports, more than 260,000 students annually participate in study-abroad programs, and this number continues to increase each year. China, India, and the Middle East have become increasingly popular destinations for study abroad; however, the leading destinations continue to be in Europe, with the United Kingdom, Italy, Spain, and France leading the list (Institute of International Education, 2010). More than half (56.3 percent) of study-abroad students remain in their host country for a short-term stay (summer, January term, or eight weeks or less during the academic year). About 40 percent of all study-abroad students remain for one or two quarters or one semester. Some analysts believe that longer periods of study abroad provide students with greater opportunities to learn the language and way of life of people in other nations (Institute of International Education, 2010).

Sociologists are interested in studying the profile of U.S. study-abroad students because the data provide interesting insights into differences in students' participation by

prospect of death. Older adults in industrialized societies may experience ***social devaluation*—wherein a person or group is considered to have less social value than other persons or groups.** Social devaluation is especially acute when people are leaving roles that have defined their sense of social identity and provided them with meaningful activity.

Negative images regarding older persons reinforce ***ageism*—prejudice and discrimination against people on the basis of age, particularly against older persons.** Ageism is reinforced by stereotypes, whereby people have narrow, fixed images of certain groups. Older persons are often stereotyped as thinking and moving slowly; as being bound to themselves and their past, unable to change and grow; as being unable to move forward and often moving backward.

Negative images also contribute to the view that women are "old" ten or fifteen years sooner than men (Bell, 1989). The multibillion-dollar cosmetics industry helps perpetuate the myth that age reduces the "sexual value" of women but increases it for men. Men's sexual value is defined more in terms of personality, intelligence, and earning power than by physical appearance. For women, however, sexual attractiveness is based on youthful appearance. By idealizing this "youthful" image of women and playing up the fear of growing older, sponsors sell thousands of products and services that claim to prevent or fix the "ravages" of aging.

© Sonda Dawes/The Image Works

▲ Throughout life, our self-concept is influenced by our interactions with others.

Although not all people act on appearances alone, Patricia Moore, an industrial designer, found that many do. At age twenty-seven, Moore disguised herself as an eighty-five-year-old woman by donning age-appropriate clothing and placing baby oil in her eyes to create the appearance of cataracts. With the help of a makeup artist, Moore supplemented the "aging process" with latex wrinkles, stained teeth, and a gray wig. For three years, "Old Pat Moore" went to various locations, including a grocery store, to see how people responded to her:

classification, gender, race, and class. Based on the latest figures available (2007/2008), most students participating in study-abroad programs are classified as juniors (35.9 percent) or seniors (21.3 percent). Women make up 65.1 percent of all study-abroad students while men make up 34.9 percent. White students make up the vast majority of study-abroad students (81.8 percent). Other groups include Asian or Pacific Islander (6.6 percent), Hispanic or Latino(a) (5.9 percent), and black or African American (4.0 percent).

Community college students have the fewest opportunities to study abroad because of a shortage of programs and lack of support for programs that do exist. Studies have found that many community college students would like to participate in study-abroad programs but that institutional barriers and prevailing beliefs by school officials about students' personal barriers (such as inability to afford study abroad, conflicting work and family obligations, and lack of understanding about the importance of possible cultural capital that might be gained from such an experience) would keep students from participating if such programs were offered (Raby, 2010).

Socialization for life in the global community is necessary to all students because of the increasing significance of international understanding and the need to learn how to live and work in a diversified nation and world. Even more important may be the opportunity for each student to gain direction and meaning in his or her own life, as John R. R. Howie (2010), the study-abroad student, explains:

> In a sense, the opportunity to live away from my life as I knew it made my future aspirations more apparent to me. As I came back to my life at Boston College, that clarity gave direction and more meaning to what I was doing now. The months abroad definitely weren't always easy—I remember how hard it was adjusting to the food, being away from my girlfriend, friends, and family, and seemingly being out of place in every way—but it was undoubtedly one of the most rewarding and meaningful experiences I have ever had.

reflect & analyze

What are the positive aspects of study-abroad programs in the college socialization process? What are the limitations of such programs? If you are unable to participate in a study-abroad program, what other methods and resources might you use to gain "global socialization," which could be beneficial in helping you meet your goals for the future?

> When I did my grocery shopping while in character, I learned quickly that the Old Pat Moore behaved—and was treated—differently from the Young Pat Moore. When I was 85, people were more likely to jockey ahead of me in the checkout line. And even more interesting, I found that when it happened, I didn't say anything to the offender, as I certainly would at age 27. It seemed somehow, even to me, that it was okay for them to do this to the Old Pat Moore, since they were undoubtedly busier than I was anyway. And further, they apparently thought it was okay, too! After all, little old ladies have plenty of time, don't they? And then when I did get to the checkout counter, the clerk might start yelling, assuming I was deaf, or becoming immediately testy, assuming I would take a long time to get my money out, or would ask to have the price repeated, or somehow become confused about the transaction. What it all added up to was that people feared I would be trouble, so they tried to have as little to do with me as possible. And the amazing thing is that I began almost to believe it myself.... I think perhaps the worst thing about aging may be the overwhelming sense that everything around you is letting you know that you are not terribly important any more (Moore with Conn, 1985: 75–76).

If we apply our sociological imagination to Moore's study, we find that "Old Pat Moore's" experiences reflect what many older persons already know—it is other people's reactions to their age, not their age itself, that place them at a disadvantage.

Many older people buffer themselves against ageism by continuing to view themselves as being in middle adulthood long after their actual chronological age would suggest otherwise. Other people begin a process of resocialization to redefine their own identity as mature adults.

Resocialization

***Resocialization* is the process of learning a new and different set of attitudes, values, and behaviors from those in one's background and previous experience.** Resocialization may be voluntary or involuntary. In either case, people undergo changes that are much more rapid and pervasive than the gradual adaptations that socialization usually involves.

social devaluation a situation in which a person or group is considered to have less social value than other individuals or groups.

ageism prejudice and discrimination against people on the basis of age, particularly against older persons.

resocialization the process of learning a new and different set of attitudes, values, and behaviors from those in one's background and previous experience.

you can make a difference

Don't Be Stressed Out in College: Helping Yourself and Others

- MTV/Associated Press Survey on What Stresses Out College Students Finds That While Two-Thirds of College Students Say They're Generally Happy, 80% Feel Day-to-Day Stress! (Ypulse.com, 2009)
- A study in the *Journal of Adolescent Health* reports that stresses from the daily routine of college life keep 68 percent of students awake at night, with 20 percent of them reporting being sleepless for some period of time at least once a week. (Messenger, 2009)

The college experience is an important socialization process for young people and adults. The economic and social benefits from achieving additional years of education beyond high school are great; however, some students are concerned that the price (not only in terms of dollars and cents) of such an education is also great because of academic and psychological stress. Is there anything that you can do to reduce stress? Do you have tips for coping that you might pass on to other students?

Here are a few thoughts on how to reduce stress in college:

- *Don't stress about being stressed.* Sometimes we worry even more when we realize that we are feeling pressure to succeed, to get along well with others, and to fit into new surroundings, particularly during our early college years. For students taking classes online, a whole new experience occurs as you learn how to interact with professors and others in the virtual community. Sometimes it is best to quickly admit that we are stressed out and then to set up a plan for handling the problem. Don't wait to seek help from others if the worries seem overwhelming.
- *Get more sleep.* This may sound odd when you are already concerned about there not being enough hours in your day; however, this is sound advice. One of the major stressors of college students (and others!) is a chronic lack of sleep. Although professionals suggest that college students should get a minimum of eight hours of sleep each night, 70 percent of students surveyed in one study reported that they slept far less than eight hours per night. If we are organized, we can often do more work in a shorter period of time, and this gives us more time for rest and relaxation.
- *Stay well.* Getting sick is one of the major ways in which college students get behind in their studies, work, and personal life. Cold and flu are among the key types of illnesses that affect students' studies and class attendance.

Voluntary Resocialization

Resocialization is voluntary when we assume a new status (such as becoming a student, an employee, or a retiree) of our own free will. Sometimes, voluntary resocialization involves medical or psychological treatment or religious conversion, in which case the person's existing attitudes, beliefs, and behaviors must undergo strenuous modification to a new regime

© Journal Courier/The Image Works

▲ New inmates are taught how to order their meals. Two fingers raised means two portions. There is no talking in line. Inmates must eat all of their food. This "ceremony" suggests how much freedom and dignity that an inmate loses when beginning the resocialization process.

As simple as it may seem, Mother's adage about washing your hands regularly is excellent advice for reducing the likelihood of becoming sick. Hand sanitizers have become increasingly popular on college campuses since the outbreak of the H1N1 flu virus. Make use of receptacles dispensing sanitizer on campus and elsewhere. Dress warmly in cold weather, cultivate good eating habits (despite the 24-hour-a-day availability of pizza and junk food), and squeeze in time for some exercise even when you think you definitely don't have time to exercise.

- *Plan some quiet time and some fun time.* Many of us get immersed in our work projects and forget that we need some time to ourselves to think, meditate, and engage in activities we find relaxing. We need some personal space, and if you are attending a brick-and-mortar college, you may have to look around to find a place where you can have a few moments for quality quiet time without other people around. If you are taking classes away from a traditional college campus, you may have to carve out a space in which to do your studies and to spend time without interruptions from other family members or coworkers.
- Gain a new perspective on stress by helping other people cope with their own stress. Sometimes the surest way to learn new information or to develop a new pattern is to share ideas with another individual. If you know someone who appears to be stressing out, pass on positive suggestions about how you have coped with a similar situation. This may be especially helpful if you have been in college for several years and can give insights from your own experience to a first-year student or someone else who is new to your college. Although colleges offer orientation and advising programs, many students like to turn to peers to find out how to cope with problematic situations. Often, the individuals we meet in college—and with whom we share our stresses and coping mechanisms—are the same people we later identify as our best friends (based on Lynn, 2010). Sharing helps us to talk aloud about our problems and coping strategies; it also provides us with an opportunity to learn from other people about their life experiences and strategies for remaining calm even in seemingly stressful circumstances.

What other suggestions do you have for dealing with stress in college? At home? At work?

Interested in learning more online? Use keywords such as "college student stress" and "tips for preventing stress" to search for sources of information and assistance.

and a new way of life. For example, resocialization for adult survivors of emotional/physical child abuse includes extensive therapy in order to form new patterns of thinking and action, somewhat like Alcoholics Anonymous and its twelve-step program, which has become the basis for many other programs dealing with addictive behavior (Parrish, 1990).

Involuntary Resocialization

Involuntary resocialization occurs against a person's wishes and generally takes place within a ***total institution*—a place where people are isolated from the rest of society for a set period of time and come under the control of the officials who run the institution** (Goffman, 1961a). Military boot camps, jails and prisons, concentration camps, and some mental hospitals are total institutions. Resocialization is a two-step process. First, people are totally stripped of their former selves—or depersonalized—through a degradation ceremony (Goffman, 1961a). For example, inmates entering prison are required to strip, shower, and wear assigned institutional clothing. In the process, they are searched, weighed, fingerprinted, photographed, and given no privacy even in showers and restrooms. Their official identification becomes not a name but a number. In this abrupt break from their former existence, they must leave behind their personal possessions and their family and friends. The depersonalization process continues as they are required to obey rigid rules and to conform to their new environment.

The second step in the resocialization process occurs when the staff at an institution attempt to build a more compliant person. A system of rewards and punishments (such as providing or withholding television or exercise privileges) encourages conformity to institutional norms.

Individuals respond to resocialization in different ways. Some people are rehabilitated; others become angry and hostile toward the system that has taken away their freedom. Although the assumed purpose of involuntary resocialization is to reform persons so that they will conform to societal standards of conduct after their release, the ability of total institutions to modify offenders' behavior in a meaningful manner has been widely questioned. In many prisons, for example, inmates may conform to the norms of the prison or of other inmates but have little respect for the norms and the laws of the larger society.

total institution Erving Goffman's term for a place where people are isolated from the rest of society for a set period of time and come under the control of the officials who run the institution.

Socialization in the Future

What will socialization be like in the future? The family is likely to remain the institution that most fundamentally shapes and nurtures people's personal values and self-identity. However, other institutions, including education, religion, and the media, will continue to exert a profound influence on individuals of all ages. A central value-oriented issue facing parents and teachers as they attempt to socialize children is the growing dominance of television, the Internet, and social media such as Facebook, Twitter, and e-mail, which make it possible for children and young people to experience many things outside their own homes and schools and to communicate routinely with people around the world.

The socialization process in colleges and universities will become more diverse as students have an even wider array of options in higher education, including attending traditional classes in brick-and-mortar buildings, taking independent-study courses, enrolling in online courses and degree programs, participating in study-abroad programs, and facing additional options that are unknown at this time. However, it remains to be seen whether newer approaches to socialization in higher education will be more stressful or less stressful than current methods. (See "You Can Make a Difference" to learn how some students are working to reduce stress in their current college environment.)

It is very likely that socialization in the future will be vastly different in the world of global instant communication than it has been in the past. We are already bombarded with massive quantities of information that vary widely in usefulness and quality. If analysts are correct in their assumption that we are moving toward a paperless society in the future, the flow of information will increasingly shift to the Web and intensify the level of data with which we are bombarded. At the same time, we will find it difficult to discern what information is useful and what is entertainment or trivia. One thing remains clear: The socialization process will continue to be a dynamic and important part of our life as we assume various roles throughout our life span.

chapter review Q & A

Use these questions and answers to check how well you've achieved the learning objectives set out at the beginning of this chapter.

- **What is socialization, and why is it important for human beings?**

Socialization is the lifelong process through which individuals acquire their self-identity and learn the physical, mental, and social skills needed for survival in society. The kind of person we become depends greatly on what we learn during our formative years from our surrounding social groups and social environment.

- **How much of our unique human characteristics comes from heredity and how much from our social environment?**

As individual human beings, we have unique identities, personalities, and relationships with others. Each of us is a product of two forces: (1) heredity, referred to as "nature," and (2) the social environment, referred to as "nurture." Whereas biology dictates our physical makeup, the social environment largely determines how we develop and behave.

- **Why is social contact essential for human beings?**

Social contact is essential in developing a self, or self-concept, which represents an individual's perceptions and feelings of being a distinct or separate person. Much of what we think about ourselves is gained from our interactions with others and from what we perceive that others think of us.

- **What are the main social psychological theories on human development?**

According to Sigmund Freud, the self emerges from three interrelated forces: the id, the ego, and the superego. When a person is well adjusted, the three forces act in balance. Jean Piaget identified four cognitive stages of development; each child must go through each stage in sequence before moving on to the next one, although some children move through them faster than others.

- **How do sociologists believe that we develop a self-concept?**

According to Charles Horton Cooley's concept of the looking-glass self, we develop a self-concept as we see ourselves through the perceptions of others. Our initial sense of self is typically based on how families perceive and treat us. George Herbert Mead suggested that we develop a self-concept through role-taking and learning the rules of social interaction. According to Mead, the self is divided into the "I" and the "me." The "I" represents the spontaneous and unique traits of each person. The "me" represents the internalized attitudes and demands of other members of society.

- **What are the primary agents of socialization?**

The agents of socialization include the family, schools, peer groups, and mass media. Our families, which transmit cultural and social values to us, are the most important agents of socialization in all societies, serving these functions: (1) procreating and socializing children, (2) providing emotional support, and (3) assigning social position. Schools primarily teach knowledge and skills but also have a profound influence on the self-image, beliefs, and values of children. Peer groups contribute to our sense of belonging and self-worth, and are a key source of information about acceptable behavior. The media function as socializing agents by (1) informing us about world events, (2) introducing us to a wide variety of people, and (3) providing an opportunity to live vicariously through other people's experiences.

- **When does socialization end?**

Socialization is ongoing throughout the life course. We learn knowledge and skills for future roles through anticipatory socialization. Parents are socialized by their own children, and adults learn through workplace socialization. Resocialization is the process of learning new attitudes, values, and behaviors, either voluntarily or involuntarily.

key terms

questions for critical thinking

1. Consider the concept of the looking-glass self. How do you think others perceive you? Do you think most people perceive you correctly?
2. What are your "I" traits? What are your "me" traits? Which ones are stronger?
3. What are some different ways that you might study the effect of toys on the socialization of children? How could you isolate the toy variable from other variables that influence children's socialization?
4. Is the attempted rehabilitation of criminal offenders—through boot camp programs, for example—a form of socialization or resocialization?

turning to video

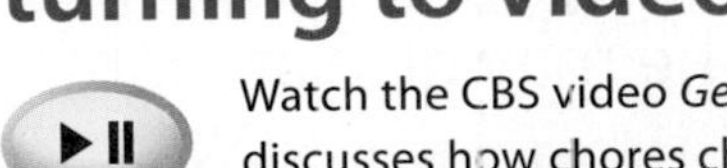

Watch the CBS video *Gender Roles* (running time 3:12), available through **CengageBrain.com**. This video discusses how chores children are assigned contribute to the development of gender roles and could affect beliefs about housework for life. As you watch the video, think about your own experiences growing up and the kinds of chores were routinely assigned. After you've watched the video, consider these questions: What kind of chores were you assigned as a child, and how do these relate to the typical chores that might be attributed to your gender? How has this affected the kinds of chores you do now?

online study resources

Go to CENGAGE brain.com to access online study resources, including the Sociology CourseMate for this text as well as special features such as video, an interactive sociology time line and interactive maps, General Social Survey (GSS) data, and U.S. Census 2010 data.

CourseMate brings course concepts to life with interactive learning, study, and exam-preparation tools that support the printed textbook. A textbook-specific website, **Sociology CourseMate** includes an integrated interactive eBook and other interactive learning tools, including quizzes, flash cards, and videos.

Visit **www.cengagebrain.com** to access your account and purchase materials.

5 Society, Social Structure, and Interaction in Everyday Life

I began Dumpster diving [scavenging in a large garbage bin] about a year before I became homeless. . . . The area I frequent is inhabited by many affluent college students. I am not here by chance; the Dumpsters in this area are very rich. Students throw out many good things, including food. In particular they tend to throw everything out when they move at the end of a semester, before and after breaks, and around midterm, when many of them despair of college. So I find it advantageous to keep an eye on the academic calendar. I learned to scavenge gradually, on my own. Since then I have initiated several companions into the trade. I have learned that there is a predictable series of stages a person goes through in learning to scavenge.

© Bob Collins/The Image Works

▲ **All activities in life—including scavenging in garbage bins and living "on the streets"—are social in nature.**

At first the new scavenger is filled with disgust and self-loathing. He is ashamed of being seen and may lurk around, trying to duck behind things, or he may dive at night. (In fact, most people instinctively look away from a scavenger. By skulking around, the novice calls attention to himself and arouses suspicion. Diving at night is ineffective and needlessly messy.) . . . That stage passes with experience. The scavenger finds a pair of running shoes that fit and look and smell brand-new. . . . He begins to understand: People throw away perfectly good stuff, a lot of perfectly good stuff.

At this stage, Dumpster shyness begins to dissipate. The diver, after all, has the last laugh. He is finding al manner of good things that are his for the taking. Those who disparage his profession are the fools, not he.

—Author Lars Eighner recalls his experiences as a Dumpster diver while living under a shower curtain in a stand of bamboo in a public park. Eighner became homeless when he was evicted from his "shack" after being unemployed for about a year. (Eighner, 1993: 111–119)

Chapter Focus Question:

How is homelessness related to the social structure of a society?

Eighner's "diving" activities reflect a specific pattern of social behavior. All activities in life—including scavenging in garbage bins and living "on the streets"—are social in nature. Homeless persons and domiciled persons (those with homes) live in social worlds that have predictable patterns of social interaction. ***Social interaction* is the process by which people act toward or respond to other people** and is the foundation for all relationships and groups in society. In this chapter we look at the relationship between social structure and social interaction. In the process, homelessness is used as an example of how social problems occur and how they may be perpetuated within social structures and patterns of interaction.

***Social structure* is the complex framework of societal institutions (such as the economy, politics, and religion) and the social practices (such as rules and social roles) that make up a society and that organize and establish limits on people's behavior.** This structure is essential for the survival of society and for the well-being of individuals because it provides a social web of familial support and social relationships that connects each of us to the larger society. Many homeless people have lost this vital linkage. As a result, they often experience a loss of personal dignity and a sense of moral worth because of their "homeless" condition (Snow and Anderson, 1993). This feeling is of concern to everyone in a society, not just those who experience homelessness.

Homeless persons such as Eighner come from all walks of life. They include full-time and part-time workers, parolees, runaway youths and children, veterans, and the elderly. They live in cities, suburbs, and rural areas. Contrary to popular myths, most of the homeless are not on the streets by choice or because they are mentally ill. Before reading on, learn more about homeless persons and the pressing national problem of homelessness by taking the Sociology and Everyday Life quiz.

Why do we need to know about social structure? Social structure provides the framework within which we interact with others. This framework is an orderly, fixed arrangement of parts that together make up the whole group or society (see ▶ Figure 5.1). As defined in Chapter 1, a *society* is a large social grouping that shares the same geographical territory and is subject to the same political authority and dominant cultural expectations. At the macrolevel, the social structure of a society has several essential elements: social institutions, groups, statuses, roles, and norms.

Highlighted Learning Objectives

- Identify the components of social structure.
- Describe how societies change over time.
- Explain why societies have shared patterns of social interaction.
- Describe how daily interactions are similar to being onstage.
- Discuss whether positive changes in society occur through individual or institutional efforts.

social interaction the process by which people act toward or respond to other people.

social structure the complex framework of societal institutions (such as the economy, politics, and religion) and the social practices (such as rules and social roles) that make up a society and that organize and establish limits on people's behavior.

sociology and everyday life

How Much Do You Know About Homelessness and Homeless Persons?

True	False	
T	F	1. A significant increase in family homelessness has occurred in the United States in recent years.
T	F	2. Alcoholism and domestic violence are the primary factors that bring about family homelessness.
T	F	3. Many homeless people have full-time employment.
T	F	4. Most homeless people are mentally ill.
T	F	5. Homeless people typically panhandle (beg for money) so that they can buy alcohol or drugs.
T	F	6. Shelters for the homeless consistently have clients who sleep on overflow cots, in chairs, in hallways, and in other nonstandard sleeping arrangements.
T	F	7. There have always been homeless persons throughout the history of the United States.
T	F	8. In large urban areas such as Los Angeles, many homeless people live in tent cities or other large encampments.

Answers on page 122.

Do social scientists agree about how social structure operates? No, diverse theoretical approaches have different interpretations of how structure operates. For example, functional theorists emphasize that social structure creates order and predictability in a society (Parsons, 1951). Social structure is also important for human development: You and I develop a self-concept as each of us learns the attitudes, values, and behaviors of the people around us. When these attitudes and values are part of a predictable structure, it is easier for us to develop a positive self-concept.

By contrast, conflict theorists maintain that social structure helps determine social relations in a society and may be the source of inequality and injustice. For example, Karl Marx suggested that how economic production is organized is the most important structural aspect of society. In capitalistic societies, few people control the labor of many, and the social structure helps create a system of domination and subordination that affects certain categories of people, including owners and workers, landlords and tenants, and rich celebrities and poor "nobodies."

Whether we look at social structure through the lens of functionalist or conflict theories, this structure creates boundaries that define persons and groups as "insiders," "outsiders," or "marginals." *Social marginality* is the state of being part insider and part outsider in the social structure. Sociologist Robert Park (1928) coined this term to refer to persons (such as immigrants) who simultaneously share the life and traditions of two distinct groups. Social marginality is an important concern for people because it often results

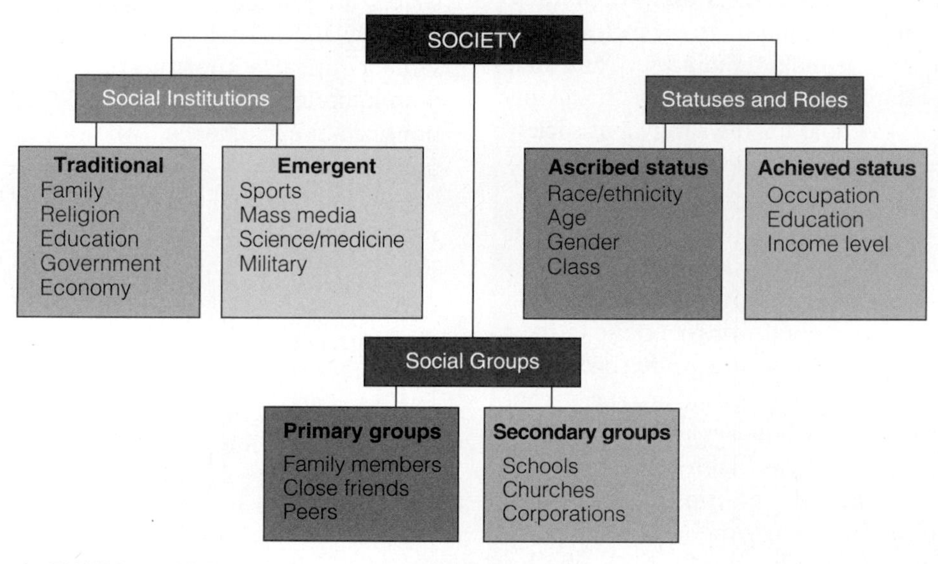

▲ FIGURE 5.1 SOCIAL STRUCTURE FRAMEWORK

in stigmatization. A *stigma* is any physical or social attribute or sign that so devalues a person's social identity that it disqualifies that person from full social acceptance (Goffman, 1963b). A convicted criminal wearing a prison uniform is an example of a person who has been stigmatized; the uniform says that the person has done something wrong and should not be allowed unsupervised outside the prison walls. The stigmatization of homelessness is discussed later in this chapter.

Why is social structure important to you? Social structure gives us the ability to interpret the social situations we encounter. For example, we expect our families to care for us, our schools to educate us, and our police to protect us. When our circumstances change dramatically, most of us feel an acute sense of anxiety because we do not know what to expect or what is expected of us. Consider, for instance, why newly homeless individuals may feel disoriented when they do not know how to function in their new situation. These persons are likely to ask questions such as "How will I survive on the streets?" "Where do I go to get help?" "Should I stay at a shelter?" "Where can I get a job?" Social structure helps people make sense out of their social setting even when they find themselves on the streets.

Components of Social Structure

What is included in the social structure of a society? Most importantly, social structure encompasses social positions, a scheme of how those positions relate to each other, and an idea of the kinds of resources that are attached to each social position. As well as social positions, social structure also includes all the groups that make up society and the relationships among those groups (Smelser, 1988). Let's start our study of the components of social structure by examining the social positions that are closest to us—the individual.

Status

No doubt you have heard the word *status* for many years. Sometimes we use this word to describe a person as "high status" or "low status," but sociologically speaking, what does the term really mean? A ***status* is a socially defined position in a group or society characterized by certain expectations, rights, and duties.** It is important to note that statuses exist *independently* of the specific people who occupy them (Linton, 1936). What are a few of the many statuses that exist? The statuses of professional athlete, rock musician, professor, college student, and homeless person all exist exclusive of the specific individuals who occupy these social positions. For example, although thousands of new students arrive on college campuses each year to occupy the status of first-year student, the status of college student and the expectations attached to that position have remained relatively unchanged for centuries. New technologies, such as computers, smartphones, and iPads, may change college instruction and how students study, but the basic status of student remains similar over time and place.

As we previously mentioned, the term *status* does *not* refer to high-level positions only. Sociologists use it to refer to all socially defined positions—high rank and low rank. For example, both the position of director of the Department of Health and Human Services in Washington, D.C., and that of a homeless person who is paid about five dollars a week (plus bed and board) to clean up the dining room at a homeless shelter are social statuses (see Snow and Anderson, 1993).

A *status set* comprises all the statuses that a person occupies at a given time. For example, Maria may be a psychologist, a professor, a wife, a mother, a Catholic, a school volunteer, a Texas resident, and a Mexican American. All of these socially defined positions constitute her status set.

Ascribed Status and Achieved Status Statuses are distinguished by the manner in which we acquire them. An ***ascribed status* is a social position conferred at birth or received involuntarily later in life, based on attributes over which the individual has little or no control, such as race/ethnicity, age, and gender.** For example, Maria is a female born to Mexican American parents; she was assigned these statuses at birth. She is an adult and—if she lives long enough—will someday become an "older adult," which is an ascribed status received involuntarily later in life. An ***achieved status* is a social position a person assumes voluntarily as a result of personal choice, merit, or direct effort.** Achieved statuses (such as occupation, education, and income) are thought to be gained as a result of personal ability or successful competition. Most occupational positions in modern societies are achieved statuses. For instance, Maria voluntarily assumed the statuses of psychologist, professor, wife, mother, and school

status a socially defined position in a group or society characterized by certain expectations, rights, and duties.

ascribed status a social position conferred at birth or received involuntarily later in life, based on attributes over which the individual has little or no control, such as race/ethnicity, age, and gender.

achieved status a social position that a person assumes voluntarily as a result of personal choice, merit, or direct effort.

sociology and everyday life

ANSWERS to the Sociology Quiz on Homelessness and Homeless Persons

1. **True.** Recently, a significant increase has occurred in the number of homeless families while there has been a decrease or leveling in the number of homeless single adults, partly because of policies aimed at ending chronic homelessness among single adults with disabilities.
2. **False.** Primary causes of family homelessness are as follows: job loss, foreclosures on homes brought about by the recession, and a lack of affordable housing in many cities.
3. **True.** Many homeless people do have full-time employment, but they are among the working poor. The minimum-wage jobs that they hold do not pay enough for them to support their families and pay the high rents that are typical in many cities.
4. **False.** Most homeless people are not mentally ill; estimates suggest that about one-fourth of the homeless are emotionally disturbed.
5. **False.** Many homeless persons panhandle to pay for food, a bed at a shelter, or other survival needs.
6. **True.** Overcrowded shelters throughout the nation often attempt to accommodate as many homeless people as possible on a given night, particularly when the weather is bad. As a result, any available spaces—including offices, closets, and hallways—are used as sleeping areas until the individuals can find another location or weather conditions improve.
7. **True.** Scholars have found that homelessness has always existed in the United States. However, the number of homeless persons has increased or decreased with fluctuations in the national economy.
8. **False.** Although media reports frequently show homeless individuals and families living in tent cities or other large homeless encampments in major U.S. cities, official studies by the U.S. Conference of Mayors (2010) have found that only ten large U.S. cities have tent cities and that these cities hold only a very small percentage of people who are homeless.

Sources: U.S. Conference of Mayors, 2010.

AP Photo/Jason DeCrow

▲ In the past, a person's status was primarily linked to his or her family background, education, occupation, and other sociological attributes. Today, some sociologists believe that celebrity status has overtaken the more traditional social indicators of status. The popular-music star Rihanna, shown here performing at a concert, is an example of celebrity status.

volunteer. However, not all achieved statuses are positions most people would want to attain; for example, being a criminal, a drug addict, or a homeless person is a negative achieved status.

Ascribed statuses have a significant influence on the achieved statuses that we occupy. Race/ethnicity, gender, and age affect each person's opportunity to acquire certain achieved statuses. Those who are privileged by their positive ascribed statuses are more likely to achieve the more prestigious positions in a society. Those who are disadvantaged by their ascribed statuses may more easily acquire negative achieved statuses.

Master Status If we occupy many different statuses, how can we determine which one is the most important? Sociologist Everett Hughes has stated that societies resolve this ambiguity by determining master statuses. A ***master status* is the most important status a person occupies**; it dominates all the individual's other statuses and is the overriding ingredient in determining a person's general social position (Hughes, 1945). Being poor or rich is a master status that influences many other areas of life, including

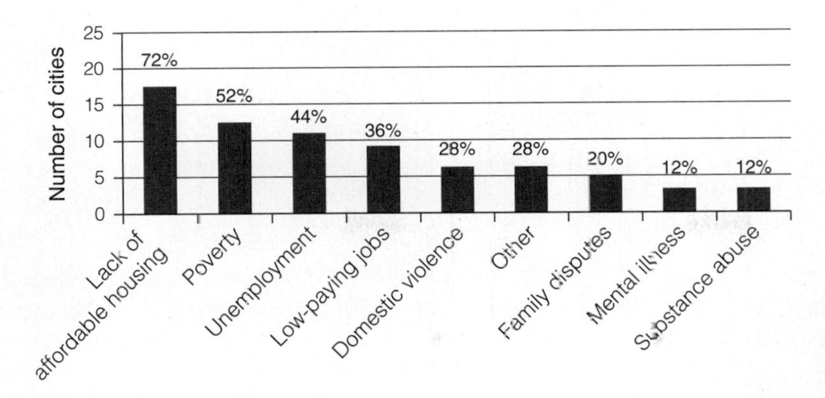

◀ **FIGURE 5.2 CAUSES OF FAMILY HOMELESSNESS IN 25 CITIES**

Source: U.S. Conference of Mayors, 2008.

health, education, and life opportunities. Historically, the most common master statuses for women have related to positions in the family, such as daughter, wife, and mother. For men, occupation has usually been the most important status, although occupation is increasingly a master status for many women as well. "What do you do?" is one of the first questions most people ask when meeting another. Occupation provides important clues to a person's educational level, income, and family background. An individual's race/ethnicity may also constitute a master status in a society in which dominant-group members single out members of other groups as "inferior" on the basis of real or alleged physical, cultural, or nationality characteristics (see Feagin and Feagin, 2003).

Master statuses confer high or low levels of personal worth and dignity on people. These are not characteristics that we inherently possess; they are derived from the statuses we occupy. For those who have no residence, being a homeless person readily becomes a master status regardless of the person's other attributes. Homelessness is a stigmatized master status that confers disrepute on its occupant because domiciled people often believe that a homeless person has a "character flaw." Sometimes this assumption is supported by how the media frame stories about homeless people (see "Framing Homelessness in the Media"). The circumstances under which someone becomes homeless determine the extent to which that person is stigmatized (see ▶ Figure 5.2). Snow and Anderson (1993: 199) observed the effects of homelessness as a master status:

> It was late afternoon, and the homeless were congregated in front of [the Salvation Army shelter] for dinner. A school bus approached that was packed with Anglo junior high school students being bused from an eastside barrio school to their upper-middle and upper-class homes in the city's northwest neighborhoods. As the bus rolled by, a fusillade of coins came flying out the windows, as the students made obscene gestures and shouted, "Get a job." Some of the homeless gestured back, some scrambled for the scattered coins—mostly pennies—others angrily threw the coins at the bus, and a few seemed oblivious to the encounter. For the passing junior high schoolers, the exchange was harmless fun, a way to work off the restless energy built up in school; but for the homeless it was a stark reminder of their stigmatized status and of the extent to which they are the objects of negative attention.

Status Symbols When people are proud of a particular social status that they occupy, they often choose to use visible means to let others know about their position. ***Status symbols* are material signs that inform others of a person's specific status.** For example, just as wearing a wedding ring proclaims that a person is married, owning a Rolls-Royce announces that one has "made it." As we saw in Chapter 3, achievement and success are core U.S. values. For this reason, people who have "made it" tend to want to display symbols to inform others of their accomplishments.

Status symbols for the domiciled and for the homeless may have different meanings. Among affluent persons, a full shopping cart in the grocery store and bags of merchandise from expensive department stores indicate a lofty financial position. By contrast, among the homeless, bulging shopping bags and overloaded grocery carts suggest a completely different status. Carts and bags are essential to street life; there is no other place to keep things, as shown by this description of Darian, a homeless woman in New York City:

> The possessions in her postal cart consist of a whole house full of things, from pots and pans to books, shoes, magazines, toilet articles, personal papers and clothing, most of which she made herself. . . .

master status the most important status that a person occupies.

status symbol a material sign that informs others of a person's specific status.

framing homelessness in the media

Thematic and Episodic Framing

They live—and die—on a traffic island in the middle of a busy downtown street, surviving by panhandling drivers or turning tricks. Everyone in their colony is hooked on drugs or alcohol. They are the harsh face of the homeless in San Francisco.

The traffic island where these homeless people live is a 40-by-75 foot triangle chunk of concrete just west of San Francisco's downtown. . . . The little concrete divider wouldn't get a second glance, or have a name—if not for the colony that lives there in a jumble of shopping carts loaded with everything they own. It's called Homeless Island by the shopkeepers who work near it and the street sweepers who clean it; to the homeless, it is just the Island. The inhabitants live hand-to-mouth, sleep on the cement and abuse booze and drugs, mostly heroin. There are at least 3,000 others like them in San Francisco, social workers say. They are known as the "hard core," the people most visible on the streets, the most difficult to help. . . . (Fagan, 2003)

This news article is an example of typical media framing of stories about homeless people. The full article includes statements about how the homeless of San Francisco use drugs, lack ambition, and present a generally disreputable appearance on the streets. This type of framing of stories about the homeless is not unique. According to the media scholar Eungjun Min (1999: ix), media images typically portray the homeless as "drunk, stoned, crazy, sick, and drug abusers." Such representations of homeless people limit our understanding of the larger issues surrounding the problem of homelessness in the United States.

Most media framing of newspaper articles and television reports about the problem of homelessness can be classified into one of two major categories: *thematic framing* and *episodic framing*. Thematic framing refers to news stories that focus primarily on statistics about the homeless population and recent trends in homelessness. Examples include stories about changes in the U.S. poverty rate and articles about states and cities that have had the largest increases in poverty. Most articles of this type are abstract and impersonal, primarily presenting data and some expert's interpretation of what those data mean. Media representations of this type convey a message to readers that "the poor and homeless are faceless." According to some analysts, thematic framing of poverty is often dehumanizing because it "ignores the human tragedy of poverty—the suffering, indignities, and misery endured by millions of children and adults" (Mantsios, 2003: 101).

By contrast, episodic framing presents public issues such as poverty and homelessness as concrete events, showing them to be specific instances that occur more or less in isolation. For example, a news article may focus on the problems of one homeless family, describing how the parents and kids live in a car and eat meals from a soup kitchen. Often, what is not included is the *big picture of homelessness*: How many people throughout the city or nation are living in their cars or in shelters? What larger structural factors (such as reductions in public and private assistance to the poor, or high rates of unemployment in some regions) contribute to or intensify the problem of homelessness in this country?

The poor have been a topic of interest to journalists and social commentators for many years. How stories about the poor and homeless are framed in the media has been and remains an important concern for each of us because these reports influence how we view the less fortunate in our society. If we come to see the problem of homelessness as nothing more than isolated statistical data or as marginal situations that affect only a few people, then we are unable to make a balanced assessment of the larger social problems involved.

reflect & analyze

How are the poor and homeless represented in the news reports and television entertainment shows that you watch? Are the larger social issues surrounding homelessness discussed within the context of these shows? Should they be?

Because of its weight and size, Darian cannot get the cart up over the curb. She keeps it in the street near the cars. This means that as she pushes it slowly up and down the street all day long, she is living almost her entire life directly in traffic. She stops off along her route to sit or sleep for awhile and to be both stared at as a spectacle and to stare back. Every aspect of her life including sleeping, eating, and going to the bathroom is constantly in public view. . . . [S]he has no space to call her own and she never has a moment's privacy. Her privacy, her home, is her cart with all its possessions. (Rousseau, 1981: 141)

For homeless women and men, possessions are not status symbols as much as they are a link with the past, a hope for the future, and a potential source of immediate cash.

Role A role is the dynamic aspect of a status. Whereas we occupy a status, we play a role. A ***role* is a set of behavioral expectations associated with a given status.** For example, a carpenter (employee) hired to remodel a kitchen is not expected to sit down uninvited and join the family (employer) for dinner.

***Role expectation* is a group's or society's definition of the way a specific role *ought* to be played.** By

© Xinhua/Landov

▲ Sociologists believe that being rich or poor may be a master status in the United States. How do the lifestyles of these two men differ based on their master statuses?

© DUSAN ZIDAR, 2010/Used under license from Shutterstock.com

contrast, ***role performance*** **is how a person *actually* plays the role.** Role performance does not always match role expectation. Some statuses have role expectations that are highly specific, such as that of surgeon or college professor. Other statuses, such as friend or significant other, have less-structured expectations. The role expectations tied to the status of student are more specific than those of being a friend. Role expectations are typically based on a range of acceptable behavior rather than on strictly defined standards.

Our roles are relational (or complementary); that is, they are defined in the context of roles performed by others. We can play the role of student because someone else fulfills the role of professor. Conversely, to perform the role of professor, the teacher must have one or more students.

Role ambiguity occurs when the expectations associated with a role are unclear. For example, it is not always clear when the provider–dependent aspect of the parent–child relationship ends. Should it end at age eighteen or twenty-one? When a person is no longer in school? Different people will answer these questions differently depending on their experiences and socialization, as well as on the parents' financial capability and psychological willingness to continue contributing to the welfare of their adult children.

Role Conflict and Role Strain

Most people occupy a number of statuses, each of which has numerous role expectations attached. For example, Charles is a student who attends morning classes at the university, and he is an employee at a fast-food restaurant, where he works from 3:00 to 10:00 P.M. He is also Stephanie's boyfriend, and she would like to see him more often. On December 7 Charles has a final exam at 7:00 P.M., when he is supposed to be working. Meanwhile, Stephanie is pressuring him to take her to a movie. To top it off, his mother calls, asking him to fly home because his father is going to have emergency surgery. How can Charles be in all these places at once? Such experiences of role conflict can be overwhelming.

Role conflict **occurs when incompatible role demands are placed on a person by two or more statuses held at the same time.** When role conflict occurs, we may feel pulled in different directions. To deal with this problem, we may prioritize our roles and first complete the one we consider to be most important. Or we may compartmentalize our lives and "insulate" our various roles (Merton, 1968). That is, we may perform the activities linked to one role for part of the day and then engage in the activities associated with another role in some other time period or elsewhere. For example, under routine circumstances Charles would fulfill his student role for part of the day and his employee role for another part of the day. In his current situation, however, he is unable to compartmentalize his roles.

Role conflict may also occur as a result of changing statuses and roles in society. Research has found that women who engage in behavior that is

role a set of behavioral expectations associated with a given status.

role expectation a group's or society's definition of the way that a specific role *ought* to be played.

role performance how a person *actually* plays a role.

role conflict a situation in which incompatible role demands are placed on a person by two or more statuses held at the same time.

gender-typed as "masculine" tend to have higher rates of role conflict than those who engage in traditional "feminine" behavior (Basow, 1992). According to the sociologist Tracey Watson (1987), role conflict can sometimes be attributed not to the roles themselves but to the pressures that people feel when they do not fit into culturally prescribed roles. In her study of women athletes in college sports programs, Watson found role conflict in the traditionally incongruent identities of being a woman and being an athlete. Even though the women athletes in her study wore makeup and presented a conventional image when they were not on the basketball court, their peers in school still saw them as "female jocks," thus leading to role conflict.

Whereas role conflict occurs between two or more statuses (such as being homeless and being a temporary employee of a social service agency), role strain takes place within one status. ***Role strain* occurs when incompatible demands are built into a single status that a person occupies** (Goode, 1960). For example, physicians want to provide their patients with the best possible health care but are often required by insurance companies or governmental regulations (such as Medicare or Medicaid programs) to keep costs down. The concepts of role expectation, role performance, role conflict, and role strain are illustrated in ▶ Figure 5.3.

▲ Parents sometimes experience role conflict when they are faced with societal expectations that they will earn a living for their family and that they will also be good parents to their children. Obviously, this father needs to leave for work; however, his son has other needs.

Individuals frequently distance themselves from a role they find extremely stressful or otherwise problematic. *Role distancing* occurs when people consciously foster the impression of a lack of commitment or attachment to a particular role and merely go through the motions of role performance (Goffman, 1961b). People use distancing techniques when they do not want others to take them as the

◀ **FIGURE 5.3 ROLE EXPECTATION, PERFORMANCE, CONFLICT, AND STRAIN**
When playing the role of "student," do you sometimes personally encounter these concepts?

© Gilles Mingasson

© MARIO ANZUONI/Reuters/Landov

▲ *Los Angeles Times* columnist Steve Lopez (left) met a homeless man, Nathaniel Ayers (above), and learned that he had been a promising musician studying at the Juilliard School who had dropped out because of his struggle with mental illness. In his 2008 book titled *The Soloist*, Lopez chronicles the relationship that he developed with Ayers and how he eventually helped get Ayers off the street and treated for his schizophrenia. This story is an example of role exit, and you can see it in the movie version of *The Soloist*, released in 2009.

"self" implied in a particular role, especially if they think the role is "beneath them." While Charles is working in the fast-food restaurant, for example, he does not want people to think of him as a "loser in a dead-end job." He wants them to view him as a college student who is working there just to "pick up a few bucks" until he graduates. When customers from the university come in, Charles talks to them about what courses they are taking, what they are majoring in, and what professors they have. He does not discuss whether the bacon cheeseburger is better than the chili burger. When Charles is really involved in role distancing, he tells his friends that he "works there but wouldn't eat there."

Role Exit ***Role exit*** **occurs when people disengage from social roles that have been central to their self-identity** (Ebaugh, 1988). Sociologist Helen Rose Fuchs Ebaugh studied this process by interviewing ex-convicts, ex-nuns, retirees, divorced men and women, and others who had exited voluntarily from significant social roles. According to Ebaugh, role exit occurs in four stages. The first stage is doubt, in which people experience frustration or burnout when they reflect on their existing roles. The second stage involves a search for alternatives; here, people may take a leave of absence from their work or temporarily separate from their marriage partner. The third stage is the turning point at which people realize that they must take some final action, such as quitting their job or getting a divorce. The fourth and final stage involves the creation of a new identity.

Exiting the "homeless" role is often very difficult. The longer a person remains on the streets, the more difficult it becomes to exit this role. Personal resources diminish over time. Possessions are often stolen, lost, sold, or pawned. Work experience and skills become outdated, and physical disabilities that prevent individuals from working are likely to develop.

Groups

Groups are another important component of social structure. To sociologists, a ***social group*** **consists of two or more people who interact frequently and share a common identity and a feeling of interdependence.** Throughout our lives, most of us participate in groups, from our families and childhood friends, to our college classes, to our work and community organizations, and even to society.

role strain a condition that occurs when incompatible demands are built into a single status that a person occupies.

role exit a situation in which people disengage from social roles that have been central to their self-identity.

social group a group that consists of two or more people who interact frequently and share a common identity and a feeling of interdependence.

Primary and secondary groups are the two basic types of social groups. A ***primary group*** **is a small, less specialized group in which members engage in face-to-face, emotion-based interactions over an extended period of time.** Primary groups include our family, close friends, and school- or work-related peer groups. By contrast, a ***secondary group*** **is a larger, more specialized group in which members engage in more-impersonal, goal-oriented relationships for a limited period of time.** Schools, churches, and corporations are examples of secondary groups. In secondary groups people have few, if any, emotional ties to one another. Instead, they come together for some specific, practical purpose, such as getting a degree or a paycheck. Secondary groups are more specialized than primary ones; individuals relate to one another in terms of specific roles (such as professor and student) and more limited activities (such as course-related endeavors). Primary and secondary groups are further discussed in Chapter 6 ("Groups and Organizations").

Social solidarity, or cohesion, refers to a group's ability to maintain itself in the face of obstacles. Social solidarity exists when social bonds, attractions, or other forces hold members of a group in interaction over a period of time (Jary and Jary, 1991). For example, if a local church is destroyed by fire and congregation members still worship together in a makeshift setting, then they have a high degree of social solidarity.

Many of us build social networks that involve our personal friends in primary groups and our acquaintances in secondary groups. A *social network* is a series of social relationships that links an individual to others. Social networks work differently for men and women, for different races/ethnicities, and for members of different social classes. Social networks particularly do not work effectively for poor and homeless individuals. Snow and Anderson (1993) found that homeless men have fragile social networks that are plagued with instability. Most of the avenues for exiting the homeless role and acquiring housing are intertwined with the large-scale, secondary groups that sociologists refer to as formal organizations.

A ***formal organization*** **is a highly structured group formed for the purpose of completing certain tasks or achieving specific goals.** Many of us spend most of our time in formal organizations such as colleges, corporations, or the government. Chapter 6 ("Groups and Organizations") analyzes the characteristics of bureaucratic organizations; however, at this point we should note that these organizations are a very important component of social structure in all industrialized societies. We expect such organizations to educate us, solve our social problems (such as crime and homelessness), and provide us work opportunities.

Today, formal organizations such as the U.S. Conference of Mayors and the National Law Center on Homelessness and Poverty work with groups around the country to make people aware that homelessness must be viewed within the larger context of poverty and to educate the public on the nature and extent of homelessness among various categories of people in the United States (see ▶ Figure 5.4 for statistics on homelessness).

© Bob Daemmrich/PhotoEdit

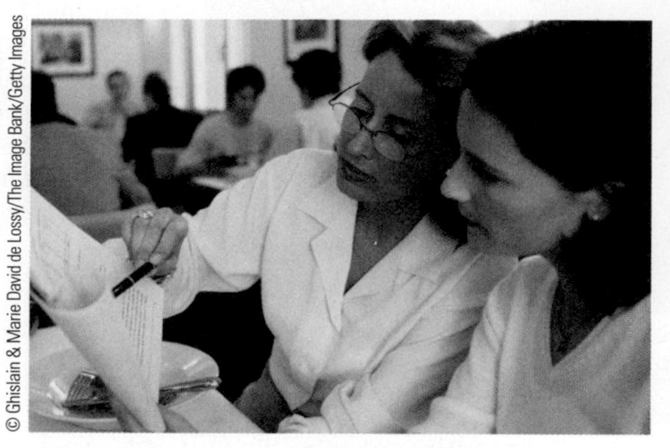

© Ghislain & Marie David de Lossy/The Image Bank/Getty Images

▲ For many years, capitalism has been dominated by powerful "old-boy" social networks. Professional women have increasingly created their own social networks to enhance their business opportunities.

Social Institutions

At the macrolevel of all societies, certain basic activities routinely occur—children are born and socialized, goods and services are produced and distributed, order is preserved, and a sense of purpose is maintained (Aberle et al., 1950; Mack and Bradford, 1979). Social institutions are the means by which these basic needs are met. A ***social institution*** **is a set of organized beliefs and rules that establishes how a society will attempt to meet its basic social needs.** In the past these needs have centered around five basic social institutions: the family, religion, education, the economy, and the government or politics. Today, mass media, sports, science and medicine, and the military are also considered to be social institutions.

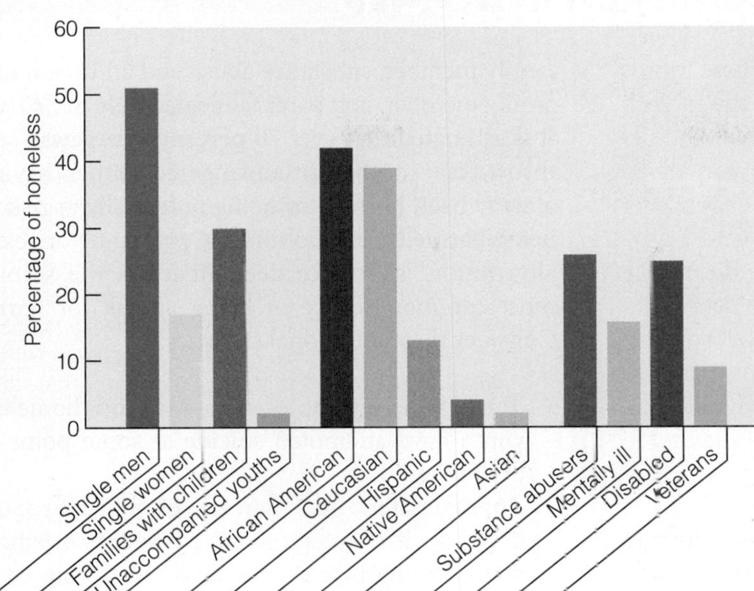

◀ **FIGURE 5.4 WHO ARE THE HOMELESS?**

Source: National Law Center on Homelessness and Poverty, 2004. Reprinted courtesy of HowStuffWorks.com

What is the difference between a group and a social institution? A group is composed of specific, identifiable people; an institution is a standardized way of doing something. The concept of "family" helps to distinguish between the two. When we talk about "your family" or "my family," we are referring to a specific family. When we refer to the family as a social institution, we are talking about ideologies and standardized patterns of behavior that organize family life. For example, the family as a social institution contains certain statuses organized into well-defined relationships, such as husband–wife, parent–child, and brother–sister. Specific families do not always conform to these ideologies and behavior patterns.

Functional theorists emphasize that social institutions exist because they perform five essential tasks:

▲ As a formal organization, the Salvation Army completes certain tasks and achieves certain goals that otherwise might not be fulfilled in contemporary societies, such as providing meals for the homeless.

© David Dow/NBAE via Getty Images

1. *Replacing members.* Societies and groups must have socially approved ways of replacing members who move away or die.
2. *Teaching new members.* People who are born into a society or move into it must learn the group's values and customs.
3. *Producing, distributing, and consuming goods and services.* All societies must provide and distribute goods and services for their members.
4. *Preserving order.* Every group or society must preserve order within its boundaries and protect itself from attack by outsiders.
5. *Providing and maintaining a sense of purpose.* In order to motivate people to cooperate with one another, a sense of purpose is needed.

Although this list of functional prerequisites is shared by all societies, the institutions in each society perform these tasks in somewhat different ways depending on their specific cultural values and norms.

primary group Charles Horton Cooley's term for a small, less specialized group in which members engage in face-to-face, emotion-based interactions over an extended period of time.

secondary group a larger, more specialized group in which members engage in more-impersonal, goal-oriented relationships for a limited period of time.

formal organization a highly structured group formed for the purpose of completing certain tasks or achieving specific goals.

social institution a set of organized beliefs and rules that establishes how a society will attempt to meet its basic social needs.

areas. However, some analysts predict that these groups will soon cease to exist, as food producers with more dominating technologies usurp the geographic areas from which these groups have derived their food supply (Nolan and Lenski, 1999).

Horticultural and Pastoral Societies

The period between 13,000 and 7,000 B.C.E. marks the beginning of horticultural and pastoral societies. During this period there was a gradual shift from *collecting* food to *producing* food, a change that has been attributed to three factors: (1) the depletion of the supply of large game animals as a source of food, (2) an increase in the size of the human population to feed, and (3) dramatic weather and environmental changes that probably occurred by the end of the Ice Age (Ferraro, 1992).

Why did some societies become horticultural while others became pastoral? Water supply, terrain, and soils are three critical factors in whether horticultural activities or pastoral activities became a society's primary mode of food production. ***Pastoral societies*** **are based on technology that supports the domestication of large animals to provide food** and emerged in mountainous regions and areas with low amounts of annual rainfall. Pastoralists—people in pastoral societies—typically remain nomadic as they seek new grazing lands and water sources for their animals. ***Horticultural societies*** **are based on technology that supports the cultivation of plants to provide food.** These societies emerged in more-fertile areas that were better suited for growing plants through the use of hand tools.

The family is the basic unit in horticultural and pastoral societies. Because they typically do not move as often as hunter-gatherers or pastoralists, horticulturalists establish more-permanent family ties and create complex systems for tracing family lineage. Some social analysts believe that the invention of a hoe with a metal blade was a contributing factor to the less nomadic lifestyle of the horticulturalists because this made planting more efficient and productive. As a result, people become more *sedentary,* remaining settled for longer periods in the same location.

Unless there are fires, floods, droughts, or environmental problems, herding animals and farming are more reliable sources of food than hunting and gathering. When food is no longer in short supply, more infants are born, and children have a greater likelihood of surviving. When people are no longer nomadic, children are viewed as an economic asset: They can cultivate crops, tend flocks, or care for younger siblings.

Division of labor increases in horticultural and pastoral societies. As the food supply grows, not everyone needs to be engaged in food production. Some pursue activities such as weaving cloth or carpets, crafting jewelry, serving as priests, or creating the tools needed for building the society's structure. Horticultural and pastoral societies are less egalitarian than hunter-gatherers, and the idea of property rights emerges as people establish more-permanent settlements. At this stage, families with the largest surpluses have an economic advantage and gain prestige and power.

In contemporary horticultural societies women do most of the farming while men hunt game, clear land, work with arts and crafts, make tools, participate in religious and ceremonial activities, and engage in war (Nielsen, 1990). Gender inequality is greater in pastoral societies because men herd the large animals and women contribute relatively little to subsistence production. In some herding societies women's primary value is their ability to produce male offspring so the family lineage is preserved and a sufficient number of males are available to protect against enemy attack (Nielsen, 1990).

Agrarian Societies

About five to six thousand years ago, agrarian (or agricultural) societies emerged, first in Mesopotamia and Egypt and slightly later in China. ***Agrarian societies*** **use the technology of large-scale farming, including animal-drawn or energy-powered plows and equipment, to produce their food supply.** Farming made it possible for people to spend their entire lives in the same location, and food surpluses made it possible for people to live in cities, where they were not directly involved in food production. The use of animals to pull plows made it possible for people to generate a large surplus of food. The land can be used more or less continuously because the plow turns the topsoil, thus returning more nutrients to the soil. In some cases farmers reap several harvests each year from the same plot of land.

In agrarian societies social inequality is the highest of all preindustrial societies in terms of both class and gender. The two major classes are the landlords and the peasants. The landlords own the fields and the harvests produced by the peasants. Inheritance becomes important as families of wealthy landlords own the same land for generations. By contrast, the landless peasants enter into an agreement with the landowners to live on and cultivate a parcel of land in exchange for part of the harvest or other economic incentives. Over

© Steve Satushek/Photographer's Choice/Getty Images

▲ In the twenty-first century, most people around the globe still reside in agrarian societies that are in various stages of industrialization. Open-air markets such as this one in Bali, where people barter or buy their food from one another, are a common sight in agrarian societies.

time, the landlords grow increasingly wealthy and powerful as they extract labor, rent, and taxation from the landless workers. Politics is based on a feudal system controlled by a political–economic elite made up of the ruler, his royal family, and members of the landowning class. Peasants have no political power and may be suppressed through the use of force or military power.

Gender-based inequality grows dramatically in agrarian societies. Men gain control over both the disposition of the food surplus and the kinship system (Lorber, 1994). Because agrarian tasks require more labor and greater physical strength than horticultural ones, men become more involved in food production. Women may be excluded from these tasks because they are seen as too weak for the work or because it is believed that their child-care responsibilities are incompatible with the full-time labor that the tasks require (Nielsen, 1990). Today, gender inequality continues in agrarian societies; the division of labor between women and men is very distinct in areas such as parts of the Middle East. Here, women's work takes place in the private sphere (inside the home), and men's work occurs in the public sphere, providing men with more recognition and greater formal status.

Industrial Societies

***Industrial societies* are based on technology that mechanizes production.** Originating in England during the Industrial Revolution, this mode of production dramatically transformed predominantly rural and agrarian societies into urban and industrial societies. Chapter 1 describes how the revolution first began in Britain and then spread to other countries, including the United States.

Industrialism involves the application of scientific knowledge to the technology of production, thus making it possible for machines to do the work previously done by people or animals. New technologies, such as the invention of the steam engine and fuel-powered machinery, stimulated many changes. Previously, machines were run by natural power sources (such as wind or water mills) or harnessed power (either human or animal power). The steam engine made it possible to produce goods by machines powered by fuels rather than undependable natural sources or physical labor.

As inventions and discoveries build upon one another, the rate of social and technological change increases. For example, the invention of the steam engine brought about new types of transportation, including trains and steamships. Inventions such as electric lights made it possible for people to work around the clock without regard to whether it was daylight or dark outside. Industrialism changes the nature of subsistence production. In countries such as the United States, large-scale agribusinesses have practically replaced small, family-owned farms and ranches. However, large-scale agriculture has produced many environmental problems while providing solutions to the problem of food supply.

In industrial societies a large proportion of the population lives in or near cities. Large corporations and government bureaucracies grow in size and complexity. The nature of social life changes as people come to know one another more as statuses than as individuals. In fact, a person's occupation becomes a key defining characteristic in industrial societies, whereas his or her kinship ties are most important in preindustrial societies.

pastoral societies societies based on technology that supports the domestication of large animals to provide food.

horticultural societies societies based on technology that supports the cultivation of plants to provide food.

agrarian societies societies that use the technology of large-scale farming, including animal-drawn or energy-powered plows and equipment, to produce their food supply.

industrial societies societies based on technology that mechanizes production.

Social institutions are transformed by industrialism. The family diminishes in significance as the economy, education, and political institutions grow in size and complexity. The family is now a consumption unit, not a production unit. Although the influence of traditional religion is diminished in industrial societies, religion remains a powerful institution. Religious organizations are important in determining what moral issues will be brought to the forefront (e.g., unapproved drugs, abortion, and violence and sex in the media) and in trying to influence lawmakers to pass laws regulating people's conduct. Politics in industrial societies is usually based on a democratic form of government. As nations such as South Korea, the People's Republic of China, and Mexico have become more industrialized, many people in these nations have intensified their demands for political participation.

Although the standard of living rises in industrial societies, social inequality remains a pressing problem. As societies industrialize, the status of women tends to decline further. For example, after industrialization occurred in the United States, the division of labor between men and women in the middle and upper classes became much more distinct: Men were responsible for being "breadwinners"; women were seen as "homemakers" (Amott and Matthaei, 1996). This gendered division of labor increased the economic and political subordination of women. In short, industrial societies have brought about some of the greatest innovations in all of human history, but they have also maintained and perpetuated some of the greatest problems.

Postindustrial Societies

A ***postindustrial society*** **is one in which technology supports a service- and information-based economy.** As discussed in Chapter 1, postmodern (or "postindustrial") societies are characterized by an *information explosion* and an economy in which large numbers of people either provide or apply information (IT specialists, for example) or are employed in service jobs (such as fast-food server or health care worker). For example, banking, law, and the travel industry are characteristic forms of employment in postindustrial societies, whereas producing steel or automobiles is representative of employment in industrial societies. There is a corresponding rise of a *consumer society* and the emergence of a *global village* in which people around the world communicate with one another by cell phone, e-mail, social networking, and the Internet.

▲ In postindustrial economies, many service- and information-based jobs are located in countries far removed from where a corporation's consumers actually live. These call-center employees in India are helping customers around the world.

Postindustrial societies produce knowledge that becomes a commodity. This knowledge can be leased or sold to others, or it can be used to generate goods, services, or more knowledge. In postindustrial societies the economy is based on involvement with people and communications technologies such as the mass media, computers, and the World Wide Web. Consider, for example, that more than three-quarters of all U.S. households have at least one computer (see "Census Profiles: Computer and Internet Access in U.S. Households").

Previous forms of production, including agriculture and manufacturing, do not disappear in postindustrial societies. Instead, they become more efficient through computerization and other technological innovations. Work that relies on manual labor is often shifted to less technologically advanced societies, where workers are paid low wages to produce profits for corporations based in industrial and postindustrial societies.

Knowledge is viewed as the basic source of innovation and policy formulation in postindustrial societies. As a result, formal education and other sources of information become crucial to the success of individuals and organizations. Scientific research becomes institutionalized, and newer industries—such as computer manufacturing and software development—come into existence that would not have been possible without the new knowledge and technological strategies. (The features of the different types of societies, distinguished by technoeconomic base, are summarized in ■ Table 5.1.)

Computer and Internet Access in U.S. Households

The U.S. Census Bureau collects extensive data on U.S. households in addition to the questions it used for Census 2000. For example, Current Population Survey data, collected from about 50,000 U.S. households during 2009, show an increase in the percentage of homes with access to the Internet, as the following figure illustrates:

Computers and/or Internet Access in the Home: 1984 to 2009
(civilian noninstitutional population)

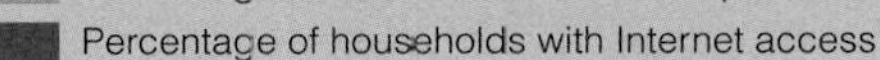

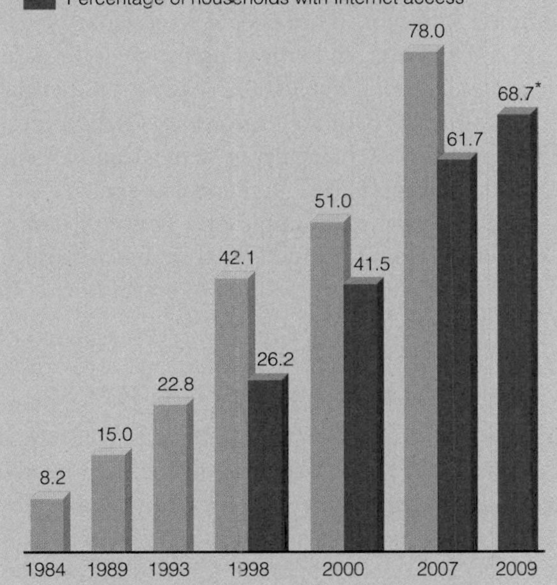

*Note: Data on Internet access were not collected before 1997, and data on computer access or ownership were not collected beginning in 2007. This is an example of why it is difficult to compare certain kinds of data over time.

Between 1998 and 2009, there has been a significant increase in the percentage of households with Internet use at home (from 26 percent to almost 69 percent).

Source: U.S. Census Bureau, 2010a.

Stability and Change in Societies

Changes in social structure have a dramatic impact on individuals, groups, and societies. Social arrangements in contemporary societies have grown more complex with the introduction of new technology, changes in values and norms, and the rapidly shrinking "global village." How do societies maintain some degree of social solidarity in the face of such changes? Sociologists Emile Durkheim and Ferdinand Tönnies developed typologies to explain the processes of stability and change in the social structure of societies. A *typology* is a classification scheme containing two or more mutually exclusive categories that are used to compare different kinds of behavior or types of societies.

Durkheim: Mechanical and Organic Solidarity

Emile Durkheim (1933/1893) was concerned with the question "How do societies manage to hold together?" He asserted that preindustrial societies are held together by strong traditions and by the members' shared moral beliefs and values. As societies industrialized and developed more-specialized economic activities, social solidarity came to be rooted in the members' shared dependence on one another. From Durkheim's perspective, social solidarity derives from a society's social structure, which, in turn, is based on the society's division of labor, which refers to how the various tasks of a society are divided up and performed. People in diverse societies (or in the same society at different points in time) divide their tasks somewhat differently, based on their own history, physical environment, and level of technological development.

How do societies change? To explain social change, Durkheim categorized societies as having either mechanical or organic solidarity. ***Mechanical solidarity*** **refers to the social cohesion of preindustrial societies, in which there is minimal division of labor and people feel united by shared values and common social bonds.** Durkheim used the term *mechanical solidarity* because he believed that people in such preindustrial societies feel a more or less automatic sense of belonging. Social interaction is characterized by face-to-face, intimate, primary-group relationships. Everyone is engaged in similar work, and little specialization is found in the division of labor.

postindustrial societies societies in which technology supports a service- and information-based economy.

mechanical solidarity Emile Durkheim's term for the social cohesion of preindustrial societies, in which there is minimal division of labor and people feel united by shared values and common social bonds.

table 5.1
Technoeconomic Bases of Society

	Hunting and Gathering	Horticultural and Pastoral	Agrarian
Change from Prior Society	—	Use of hand tools, such as digging stick and hoe	Use of animal-drawn plows and equipment
Economic Characteristics	Hunting game, gathering roots and berries	Planting crops, domestication of animals for food	Labor-intensive farming
Control of Surplus	None	Men begin to control societies	Men own land or herds
Inheritance	None	Shared—patrilineal and matrilineal	Patrilineal
Control Over Procreation	None	Increasingly by men	Men—to ensure legitimacy of heirs
Women's Status	Relative equality	Decreasing in move to pastoralism	Low

Organic solidarity **refers to the social cohesion found in industrial (and perhaps postindustrial) societies, in which people perform very specialized tasks and feel united by their mutual dependence.** Durkheim chose the term *organic solidarity* because he believed that individuals in industrial societies come to rely on one another in much the same way that the organs of the human body function interdependently. Social interaction is less personal, more status oriented, and more focused on specific goals and objectives. People no longer rely on morality or shared values for social solidarity; instead, they are bound together by practical considerations.

Tönnies: *Gemeinschaft* and *Gesellschaft*

Another social analyst who examined social change was the sociologist Ferdinand Tönnies (1855–1936), who used the terms *Gemeinschaft* and *Gesellschaft* to characterize the degree of social solidarity and social control found in societies. He was especially concerned about what happens to social solidarity in a society when a "loss of community" occurs.

The ***Gemeinschaft*** **(guh-MINE-shoft) is a traditional society in which social relationships are based on personal bonds of friendship and kinship and on intergenerational stability.** These relationships are based on ascribed rather than achieved status. In such societies people have a commitment to the entire group and feel a sense of togetherness. Tönnies (1963/1887) used the German term *Gemeinschaft* because it means "commune" or "community"; social solidarity and social control are maintained by the community. Members have a strong sense of belonging, but they also have very limited privacy.

By contrast, the ***Gesellschaft*** **(guh-ZELL-shoft) is a large, urban society in which social bonds are based on impersonal and specialized relationships, with little long-term commitment to the group or consensus on values.** In such societies most people are "strangers" who perceive that they have very little in common with most other people. Consequently, self-interest dominates, and little consensus exists regarding values. Tönnies (1963/1887) selected the German term *Gesellschaft* because it means "association"; relationships are based on achieved statuses, and interactions among people are both rational and calculated.

Is our society an example of a *Gemeinschaft* or a *Gesellschaft?* Is it characterized by mechanical or organic solidarity? What kinds of social bonds are still important today?

Social Structure and Homelessness

In *Gesellschaft* societies such as the United States, a prevailing core value is that people should be able to take care of themselves. Thus, many people view the homeless as "throwaways"—as beyond help or as having already had enough done for them by society. Some argue that the homeless made their own bad decisions and should be held responsible for the consequences of their actions. In this sense homeless people serve as a visible example to others to "follow the rules" lest they experience a similar fate.

Alternative explanations for homelessness in *Gesellschaft* societies have been suggested. Elliot Liebow (1993: 224) notes that homelessness is a "social class phenomenon, the direct result of a steady, across-the-board lowering of the standard of living of the working class and lower class." The problem is exacerbated by a lack of jobs and adequate housing. Clearly, there is no simple answer to the question about what should be done to help the homeless. Nor, as discussed in "Sociology and Social Policy," is there any consensus on what rights the homeless have in public spaces, such as parks and sidewalks. The answers we derive as a society and as individuals are often based on our social construction of this reality of life.

table 5.1
(continued)

	Industrial	Postindustrial
Change from Prior Society	Invention of steam engine	Invention of computer and development of "high-tech" society
Economic Characteristics	Mechanized production of goods	Information and service economy
Control of Surplus	Men own means of production	Corporate shareholders and high-tech entrepreneurs
Inheritance	Bilateral	Bilateral
Control Over Procreation	Men—but less so in later stages	Mixed
Women's Status	Low	Varies by class, race, and age

Source: Adapted from Lorber, 1994: 140.

Social Interaction: The Microlevel Perspective

So far in this chapter, we have focused on society and social structure from a macrolevel perspective, seeing how the structure of society affects the statuses we occupy, the roles we play, and the groups and organizations to which we belong. Functionalist and conflict perspectives provide a macrosociological overview because they concentrate on large-scale events and broad social features. By contrast, the symbolic interactionist perspective takes a microsociological approach, asking how social institutions affect our daily lives.

© David Young-Wolff/PhotoEdit

▲ Have you watched other people's reactions to one another in an elevator? How might we explain the lack of eye contact and the general demeanor of the individuals pictured here?

Social Interaction and Meaning

When you are with other people, do you often wonder what they think of you? If so, you are not alone! Because most of us are concerned about the meanings that others ascribe to our behavior, we try to interpret their words and actions so that we can plan how we will react toward them (Blumer, 1969). We know that others have expectations of us. We also have certain expectations about them. For example, if we enter an elevator that has only one other person in it, we do not expect that individual to confront us and stare into our eyes. As a matter of fact, we would be quite upset if the person did so.

organic solidarity Emile Durkheim's term for the social cohesion found in industrial (and perhaps postindustrial) societies, in which people perform very specialized tasks and feel united by their mutual dependence.

Gemeinschaft (guh-MINE-shoft) a traditional society in which social relationships are based on personal bonds of friendship and kinship and on intergenerational stability.

Gesellschaft (guh-ZELL-shoft) a large, urban society in which social bonds are based on impersonal and specialized relationships, with little long-term commitment to the group or consensus on values.

Social interaction within a given society has certain shared meanings across situations. For instance, our reaction would be the same regardless of *which* elevator we rode in *which* building. Sociologist Erving Goffman (1961b) described these shared meanings in his observation about two pedestrians approaching each other on a public sidewalk. He noted that each will tend to look at the other just long enough to acknowledge the other's presence. By the time they are about eight feet away from each other, both individuals will tend to look downward. Goffman referred to this behavior as *civil inattention*—the ways in which an individual shows an awareness that another is present without making this person the object of particular attention. The fact that people engage in civil inattention demonstrates that interaction does have a pattern, or *interaction order*, which regulates the form and processes (but not the content) of social interaction.

Does everyone interpret social interaction rituals in the same way? No. Race/ethnicity, gender, and social class play a part in the meanings we give to our interactions with others, including chance encounters on elevators or the street. Our perceptions about the meaning of a situation vary widely based on the statuses we occupy and our unique personal experiences. For example, sociologist Carol Brooks Gardner (1989) found that women frequently do not perceive street encounters to be "routine" rituals. They fear for their personal safety and try to avoid comments and propositions that are sexual in nature. African Americans may also feel uncomfortable in street encounters. A middle-class African American college student described his experiences walking home at night from a campus job:

As this passage indicates, social encounters have different meanings for men and women, whites and people of color, and individuals from different social classes. Members of the dominant classes regard the poor, unemployed, and working class as less worthy of attention, frequently subjecting them to subtle yet systematic "attention deprivation" (Derber, 1983). The same can certainly be said about how members of the dominant classes "interact" with the homeless.

The Social Construction of Reality

If we interpret other people's actions so subjectively, can we have a shared social reality? Some symbolic interaction theorists believe that there is very little shared reality beyond that which is socially created. Symbolic interactionists refer to this as the ***social construction of reality*—the process by which our perception of reality is largely shaped by the subjective meaning that we give to an experience** (Berger and Luckmann, 1967). This meaning strongly influences what we "see" and how we respond to situations.

Our perceptions and behavior are influenced by how we initially define situations: We act on reality as we see it. Sociologists describe this process as the *definition of the situation*, meaning that we analyze a social context in which we find ourselves, determine what is in our best interest, and adjust our attitudes and actions accordingly. This process can result in a ***self-fulfilling prophecy*—a false belief or prediction that produces behavior that makes the originally false belief come true** (Merton, 1968). An example would be a person who has been told repeatedly that she or he is not a good student; eventually, this person might come to believe it to be true, stop studying, and receive failing grades.

sociology and social policy

Homeless Rights Versus Public Space

I had a bit of a disturbing experience yesterday as I was running errands downtown. First, I was glad to see the south Queen sidewalk east of University [in Toronto, Canada,] open. (Months of construction on the new opera house had blocked it off.) As I continued walking eastward past the acclaimed new structure (where I have enjoyed a performance or two) I wondered why the sidewalk was so narrow. It seems this stretch of Queen should feel a bit grander. When I reached the corner of Queen and Bay, I saw some police officers and city workers "taking action on sidewalk clearance." They were clearing a homeless person's worldly belongings off the sidewalk. Using shovels. And a pickup truck. . . .

I think what I saw yesterday is unacceptable. Sure, the situation is complicated. Yes, there are a lot of stakeholders and stories to appreciate. But it's unfairness I want to see shoveled out of public space. Not people. Not blankets. Not kindness. And I hope I'm not alone. (Sandals, 2007)

"Public space protection" has become an issue in many cities, both in the United States and elsewhere. Record numbers of homeless individuals and families seek refuge on the streets and in public parks because they have nowhere else to go. However, this seemingly individualistic problem is actually linked to larger social concerns, including long-term unemployment, lack of education and affordable housing, and cutbacks in government and social service budgets. The problem of homelessness also raises significant social policy issues, including the extent to which cities can make it illegal for people to remain for extended periods of time in public spaces.

Should homeless persons be allowed to sleep on sidewalks, in parks, and in other public areas? This issue has been the source of controversy. As cities have sought to improve their downtown areas and public spaces, they have taken measures to enforce city ordinances controlling loitering (standing around or sleeping in public spaces), "aggressive panhandling," and disorderly conduct. Advocates for the homeless and civil liberties groups have filed lawsuits claiming that the rights of the homeless are being violated by the enforcement of these laws. The lawsuits assert that the homeless have a right to sleep in parks because no affordable housing is available for them. Advocates also argue that panhandling is a legitimate means of livelihood for some of the homeless and is protected speech under the First Amendment. In addition, they accuse public and law enforcement officials of seeking to punish the homeless on the basis of their "status," a cruel and unusual punishment prohibited by the Eighth Amendment.

The "homeless problem" is not a new one for city governments. Of the limited public funding that is designated for the homeless, most has been spent on shelters that are frequently overcrowded and otherwise inadequate. Officials in some cities have given homeless people a one-way ticket to another city. Still others have routinely run them out of public spaces.

What responsibility does society have to the homeless? Are laws restricting the hours that public areas or parks are open to the public unfair to homeless persons? Some critics have argued that if the homeless and their advocates win these lawsuits, what they have won (at best) is the right for the homeless to live on the street under extremely adverse conditions. Others have disputed this assertion and note that if society does not make available affordable housing and job opportunities, the least it can do is stop harassing homeless people who are getting by as best they can.

reflect & analyze

What do you think? What rights are involved? Whose rights should prevail?

Sources: Based on Kaufman, 1996; Sandals, 2007; and Wood, 2002.

© Mark Ludak/The Image Works

Sidewalk clearance and public space protection are controversial topics in cities where law enforcement officials have been instructed to remove homeless individuals and their possessions from public spaces. What are the central issues in this social policy debate? Why should this problem be of concern to each of us?

© Bruce Ayres/Stone/Getty Images

Contrary to a popular myth that most homeless people are single drifters, an increasing number of families are now homeless.

© SAUL LOEB/AFP/Getty Images

© JIM RUYMEN/UPI/Landov

▲ People can have sharply contrasting perceptions of the same reality.

To uncover people's background expectancies, ethnomethodologists frequently break "rules" or act as though they do not understand some basic rule of social life so that they can observe other people's responses. In a series of *breaching experiments*, Garfinkel assigned different activities to his students to see how breaking the unspoken rules of behavior created confusion.

The ethnomethodological approach contributes to our knowledge of social interaction by making us aware of subconscious social realities in our daily lives. However, a number of sociologists regard ethnomethodology as a frivolous approach to studying human behavior because it does not examine the impact of macrolevel social institutions—such as the economy and education—on people's expectancies, and it fails to look at how social realities are initially created.

© AP Photo/Carolyn Kaster

© AP Photo/Alex Brandon

▲ Erving Goffman believed that people spend a great amount of time and effort managing the impression that they present. How do political candidates use impression management as they seek to accomplish their goal of being elected to public office?

Dramaturgical Analysis

How is everyday life like watching a dramatic presentation? Erving Goffman suggested that day-to-day interactions have much in common with being on stage or in a dramatic production. ***Dramaturgical analysis* is the study of social interaction that compares everyday life to a theatrical presentation.** Members of our "audience" judge our performance and are aware that we may slip and reveal our true character (Goffman, 1959, 1963a). Consequently, most of us attempt to play our role as well as possible and to control the impressions we give to others. ***Impression management (presentation of self)* refers to people's efforts to present themselves to others in ways that are most favorable to their own interests or image.**

For example, suppose that a professor has returned graded exams to your class. Will you discuss the exam and your grade with others in the class? If you are like most people, you probably play your student role differently depending on whom you are talking to and what grade you received on the exam. Your "presentation" may vary depending on the grade earned by the other person (your "audience"). In one study, students who all received high grades ("Ace–Ace encounters") willingly talked with one another about their grades and sometimes engaged in a little bragging about how they had "aced" the test. However, encounters between students who had received high grades and those who had received low or failing grades ("Ace–Bomber encounters") were uncomfortable. The Aces felt as if they had to minimize their own grade. Consequently, they tended to attribute their success to "luck" and were quick to offer the Bombers words of encouragement. On the other hand, the Bombers believed that they had to praise the Aces and hide their own feelings of frustration and disappointment. Students who received low or failing grades ("Bomber–Bomber encounters") were more comfortable when they talked with one another because they could share their negative emotions. They often indulged in self-pity and relied on face-saving excuses (such as an illness or an unfair exam) for their poor performances (Albas and Albas, 1988).

ethnomethodology the study of the commonsense knowledge that people use to understand the situations in which they find themselves.

dramaturgical analysis the study of social interaction that compares everyday life to a theatrical presentation.

impression management (presentation of self) Erving Goffman's term for people's efforts to present themselves to others in ways that are most favorable to their own interests or image.

Can you think of similar situations that you have observed at your school?

In Goffman's terminology *face-saving behavior* refers to the strategies we use to rescue our performance when we experience a potential or actual loss of face. When the Bombers made excuses for their low scores, they were engaged in face-saving; the Aces attempted to help them save face by asserting that the test was unfair or that it was only a small part of the final grade. Why would the Aces and Bombers both participate in face-saving behavior? In most social interactions all role players have an interest in keeping the "play" going so that they can maintain their overall definition of the situation in which they perform their roles.

Goffman noted that people consciously participate in *studied nonobservance,* a face-saving technique in which one role-player ignores the flaws in another's performance to avoid embarrassment for everyone involved. Most of us remember times when we have failed in our role and know that it is likely to happen again; thus, we may be more forgiving of the role failures of others.

Social interaction, like a theater, has a front stage and a back stage. The *front stage* is the area where a player performs a specific role before an audience. The *back stage* is the area where a player is not required to perform a specific role because it is out of view of a given audience. For example, when the Aces and Bombers were talking with one another at school, they were on the "front stage." When they were in the privacy of their own residences, they were in "back stage" settings—they no longer had to perform the Ace and Bomber roles and could be themselves. What are the front stages and back stages in your own life? Where are you the most comfortable?

The need for impression management is most intense when role players have widely divergent or devalued statuses. As we have seen with the Aces and Bombers, the participants often play different roles under different circumstances and keep their various audiences separated from one another. If one audience becomes aware of other roles that a person plays, the impression being given at that time may be ruined. For example, people facing or experiencing homelessness are not only stigmatized but may also find that they lose the opportunity to get a job if their homelessness becomes known (see "Sociology Works!"). However, many homeless individuals do not passively accept the roles into which they are cast. For the most part, they attempt—as we all do—to engage in impression management in their everyday life.

The dramaturgical approach helps us think about the roles we play and the audiences who judge our presentation of self; however, this perspective has also been criticized for focusing on appearances and not the underlying substance. This approach may not place enough emphasis on the ways in which our everyday interactions with other people are influenced by occurrences within the larger society. For example, if some political leaders or social elites in a community deride homeless people by saying they are "lazy" or "unwilling to work," it may become easier for everyday people walking down a street to treat homeless individuals poorly. Overall, however, Goffman's dramaturgical analysis has been highly influential in the development of the sociology of emotions, an important area of contemporary theory and research.

The Sociology of Emotions

Why do we laugh, cry, or become angry? Are these emotional expressions biological or social in nature? To some extent, emotions are a biologically given sense (like hearing, smell, and touch), but they are also social in origin. We are socialized to feel certain emotions, and we learn how and when to express (or not express) those emotions (Hochschild, 1983).

How do we know which emotions are appropriate for a given role? Sociologist Arlie Hochschild (1983) suggests that we acquire a set of *feeling rules* that shapes the appropriate emotions for a given role or specific situation. These rules include how, where, when, and with whom an emotion should be expressed. For example, for the role of a mourner at a funeral, feeling rules tell us which emotions are required (sadness and grief, for example), which are acceptable (a sense of relief that the deceased no longer has to suffer), and which are unacceptable (enjoyment of the occasion expressed by laughing out loud) (see Hochschild, 1983).

Feeling rules also apply to our occupational roles. For example, the truck driver who handles explosive cargos must be able to suppress fear. Although all jobs place some burden on our feelings, *emotional labor* occurs only in jobs that require personal contact with the public or the production of a state of mind (such as hope, desire, or fear) in others (Hochschild, 1983). With emotional labor, employees must display only certain carefully selected emotions. For example, flight attendants are required to act friendly toward passengers, to be helpful and open to requests, and to maintain an "omnipresent smile" in order to enhance the customers' status. By contrast, bill collectors are encouraged to show anger and make threats to customers, thereby supposedly deflating the customers' status and wearing down their presumed resistance to paying past-due bills. In both jobs the employees are expected to show feelings that are often not their true ones (Hochschild, 1983).

Social class and race are determinants in managed expression and emotion management. Emotional labor is emphasized in middle-class and upper-class families. Because middle- and upper-class parents often work with people, they are more likely to teach their children the importance of emotional labor in their own careers than are working-class parents, who tend to work with things, not people (Hochschild, 1983). Race is also an

sociology works!

Goffman's Stigmatization Theory and Contemporary Homelessness

We need as a state and a community here in South Australia to stop blaming and stigmatizing people who are homeless. Any one of us could be homeless if the circumstances turned for us at any time and we need to recognize that.

—Jo Wickes, a representative of Homelessness South Australia, emphasizing the importance of doing something to reduce homelessness in her country rather than merely stigmatizing those who are homeless (ABC.net.au, 2008)

As we have seen in this chapter, homelessness carries a stigma like the ones that sociologist Erving Goffman described in his important book *Stigma: Notes on the Management of Spoiled Identity*, which was originally published in 1963. According to Goffman, a gap may exist between an individual's *virtual social identity*—what a person *ought* to be—and his or her *actual social identity*—what a person actually *is*. Having a permanent residence is considered to be normal, and individuals who fall outside the norm are stigmatized by those who are domiciled and view themselves as the mainstream group.

People who are visibly homeless are more likely to be stigmatized than those who make up the invisible homeless. Goffman identifies two types of stigma: (1) *discredited stigmas* are obvious to other people because the source of the stigma is visible (such as a person with a missing nose or other physical impairment), and (2) *discreditable stigmas* are not obvious to others and are not known or perceived by them. Homelessness is more visible in some cases than in others, and homeless persons must cope with the extent to which their situation is seen and known by others. When homelessness is highly visible, homeless individuals must manage the tension that occurs because others know of their problem. When homelessness is not highly visible (for example, homeless persons who live in their cars or with relatives), the central concern of these individuals is managing information so that others do not find out about their problem (Goffman, 1963b).

Current scholars studying problems associated with homelessness continue to find that Goffman's work on stigma is useful. According to Guy Johnson, an Australian scholar, "For people facing or experiencing homelessness, stigma is real in that they have to deal with the devalued identity attached to homelessness as much as they have to deal with shortages in the supply of affordable housing" (2006: 42). As a result of the stigma that is attached to homelessness, it is difficult for homeless people to reestablish a routine in everyday life. Some individuals who are homeless try to "pass" by acting like they have a place to live and by disassociating themselves with other homeless people. Others respond differently to stigma and instead embrace homelessness by becoming part of a subculture comprising homeless individuals who provide mutual support for one another (Johnson, 2006).

As we look at Goffman's ideas on stigma today, we find rich new opportunities for application of classical sociological insights to our understanding of pressing social problems such as homelessness. We can also hope that political and social leaders in various nations will seek to provide new pathways *out* of homelessness through innovative public policy rather than continuing to reinforce social structures that contribute to long-term homelessness and stigmatization for many people.

reflect & analyze

How might you apply Goffman's ideas about stigma to problems at your college or university? For example, would his analysis be useful in studying issues such as eating disorders or alcohol abuse among college students?

important factor in emotional labor. People of color spend much of their life engaged in emotional labor because racist attitudes and discrimination make it continually necessary to manage one's feelings.

Emotional labor may produce feelings of estrangement from one's "true" self. C. Wright Mills (1956) suggested that when we "sell our personality" in the course of selling goods or services, we engage in a seriously self-alienating process. In other words, the "commercialization" of our feelings may dehumanize our work role performance and create alienation and contempt that spill over into other aspects of our life (Hochschild, 1983; Smith and Kleinman, 1989).

Clearly, the sociology of emotions helps us understand the social context of our feelings and the relationship between the roles we play and the emotions we experience. However, it may overemphasize the cost of emotional labor and the emotional controls that exist outside the individual (Wouters, 1989).

Nonverbal Communication

In a typical stage drama the players not only speak their lines but also use nonverbal communication to convey information. ***Nonverbal communication* is the transfer of information between persons without the use of words.** It includes not only visual cues (gestures, appearances) but also vocal features (inflection, volume, pitch) and environmental factors (use of space, position) that

nonverbal communication the transfer of information between persons without the use of words.

© Tom Prettyman/PhotoEdit

© David Silverman/Reportage/Getty Images

▲ Are there different gender-based expectations in the United States about the kinds of emotions that men, as compared with women, are supposed to show? What feeling rules shape the emotions of the men in these two roles?

affect meanings (Wood, 1999). Facial expressions, head movements, body positions, and other gestures carry as much of the total meaning of our communication with others as our spoken words do (Wood, 1999).

Functions of Nonverbal Communication Why is nonverbal communication important to you? We obtain first impressions of others from various kinds of nonverbal communication, such as the clothing they wear and their body positions. Head and facial movements may provide us with information about other people's emotional states, and others receive similar information from us (Samovar and Porter, 1991). Through our body posture and eye contact we signal that we do or do not wish to speak to someone. For example, we may look down at the sidewalk or off into the distance when we pass homeless persons who look as if they are going to ask for money.

Nonverbal communication establishes the relationship among people in terms of their responsiveness to and power over one another (Wood, 1999). For example, we show that we are responsive toward or like another person by maintaining eye contact and attentive body posture and perhaps by touching and standing close. We can even express power or control over others through nonverbal communication. Goffman (1956) suggested that *demeanor* (how we behave or conduct ourselves) is relative to social power. People in positions of dominance are allowed a wider range of permissible actions than are their subordinates, who are expected to show deference. *Deference* is the symbolic means by which subordinates give a required permissive response to those in power; it confirms the existence of inequality and reaffirms each person's relationship to the other (Rollins, 1985).

Facial Expression, Eye Contact, and Touching Nonverbal communication is symbolic of our relationships with others. Who smiles? Who stares? Who makes and sustains eye contact? Who touches whom? All these questions relate to demeanor and deference; the key issue is the status of the person who is doing the smiling, staring, or touching relative to the status of the recipient (Goffman, 1967).

Facial expressions, especially smiles and eye contact, also reflect patterns of dominance and subordination in society, particularly as they relate to gender. Typically, white women have been socialized to smile and frequently do so even when they are not actually happy (Halberstadt and Saitta, 1987). Jobs held predominantly by women (including flight attendant, secretary, elementary schoolteacher, and nurse) are more closely associated with being pleasant and smiling than are "men's jobs." By contrast, men tend to display less emotion through smiles or other facial expressions and instead seek to show that they are reserved and in control (Wood, 1999).

Women are more likely to sustain eye contact during conversations (but not otherwise) as a means of showing their interest in and involvement with others. By contrast, men are less likely to maintain prolonged eye contact during conversations but are more likely to stare at other people (especially men) in order to challenge them and assert their own status (Pearson, 1985).

Eye contact can be a sign of domination or deference. For example, in a participant observation study of domestic (household) workers and their employers, the sociologist Judith Rollins (1985) found that the domestics were supposed to show deference by averting their eyes when they talked to their employers. Deference also required that they present an "exaggeratedly subservient demeanor" by standing less erect and walking tentatively.

Touching is another form of nonverbal behavior that has many different shades of meaning. Gender and power differences are evident in tactile communication from birth. Studies have shown that touching has variable meanings to parents: Boys are touched more roughly and playfully, whereas girls are handled more gently and protectively (Condry, Condry, and Pogatshnik, 1983). This pattern continues into adulthood, with women touched more frequently than men. Sociologist Nancy Henley (1977) attributed this pattern to power differentials between men and women and to the nature of women's roles as mothers, nurses, teachers, and secretaries. Clearly, touching has a different meaning to women than to men. Women may hug and touch others to indicate affection and emotional support, but men are more likely to touch others to give directions, assert power, and express sexual interest (Wood, 1999). The "meaning" we give to touching is related to its "duration, intensity, frequency, and the body parts touching and being touched" (Wood, 1994: 162).

Personal Space How much space do you like between yourself and other people? Anthropologist Edward Hall (1966) analyzed the physical distance between people speaking to each other and found the amount of personal space that people prefer

© Dana White/PhotoEdit

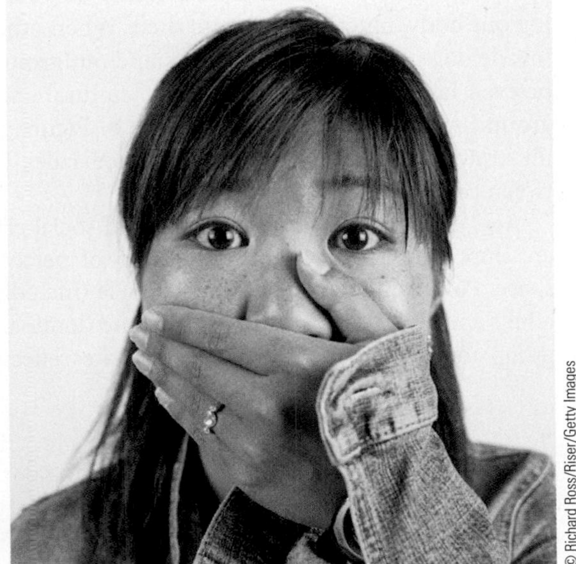

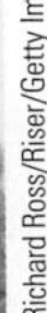

© Richard Ross/Riser/Getty Images

© Krzysztof Mystkowski/AFP/Getty Images

▲ Nonverbal communication may be thought of as an international language. What message do you receive from the facial expression, body position, and gestures of each of these people? Is it possible to misinterpret these messages?

[concept quick review]

Social Interaction: The Microlevel Perspective

Social Interaction and Meaning	In a given society, forms of social interaction have shared meanings, although these may vary to some extent based on race/ethnicity, gender, and social class.
Social Construction of Reality	The process by which our perception of reality is largely shaped by the subjective meaning that we give to an experience.
Ethnomethodology	Studying the commonsense knowledge that people use to understand the situations in which they find themselves makes us aware of subconscious social realities in daily life.
Dramaturgical Analysis	The study of social interaction that compares everyday life to a theatrical presentation. This approach includes impression management (people's efforts to present themselves favorably to others).
Sociology of Emotions	We are socialized to feel certain emotions, and we learn how and when to express (or not express) them.
Nonverbal Communication	The transfer of information between persons without the use of speech, such as by facial expressions, head movements, and gestures.

varies from one culture to another. ***Personal space* is the immediate area surrounding a person that the person claims as private.** Our personal space is contained within an invisible boundary surrounding our body, much like a snail's shell. When others invade our space, we may retreat, stand our ground, or even lash out, depending on our cultural background (Samovar and Porter, 1991). ▶ Figure 5.5 illustrates differences in social distance rules between two contrasting cultures.

Age, gender, kind of relationship, and social class are important factors in the allocation of personal space. Power differentials between people (including adults and children, men and women, and dominant-group members and people of color) are reflected in personal space and privacy issues. With regard to age, adults generally do not hesitate to enter the personal space of a child (Thorne, Kramarae, and Henley, 1983). Similarly, young children who invade the personal space of an adult tend to elicit a more favorable response than do older uninvited visitors (Dean, Willis, and la Rocco, 1976). The need for personal space appears to increase with age (Baxter, 1970; Aiello and Jones, 1971), although it may begin to decrease at about age forty (Heshka and Nelson, 1972).

In sum, all forms of nonverbal communication are influenced by our culture, gender, ethnicity, social class, and the personal contexts in which they occur. Although it is difficult to generalize about people's nonverbal behavior, we still need to think about our own nonverbal communication patterns. Recognizing that differences in social interaction exist is important. Learning to understand and respect alternative styles of social interaction enhances our personal effectiveness by increasing the range of options we have for communicating with different people in diverse contexts and for varied reasons (Wood, 1999). (The Concept Quick Review summarizes the microlevel approach to social interaction.)

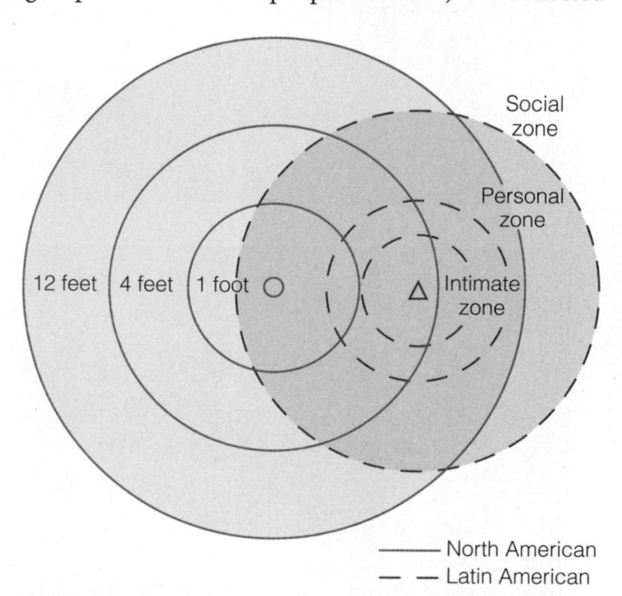

▲ **FIGURE 5.5 NORTH AMERICAN AND LATIN AMERICAN SOCIAL DISTANCE RULES**

Source: From *Cultural Anthropology* by Paul Hiebert. Copyright © 1983 by Baker Academic, a division of Baker Publishing Group. Used by permission.

Changing Social Structure and Interaction in the Future

The social structure in the United States has been changing rapidly in recent decades. Currently, there are more possible statuses for persons to occupy

personal space the immediate area surrounding a person that the person claims as private.

you can make a difference

Offering a Helping Hand to Homeless People

When you pull up at an intersection and see a person holding a torn piece of cardboard with a handwritten sign on it, how do you react? Many of us shy away from chance encounters such as this because we know, without actually looking, that the sign says something like "Homeless, please help." In an attempt to avoid eye contact with the person on the street corner, we suddenly look with newfound interest at something lying on our car seat, or we check our appearance in the rear-view mirror, or we adjust the radio. In fact, we do just about whatever it takes to divert our attention, making eye contact with this person impossible until the traffic light changes and we can be on our way.

Does this scenario sound familiar? Many of us see homeless individuals on street corners and elsewhere as we go about our daily routine. We are uncomfortable in their presence because we don't know what we can do to help them, or even if we should. Frequently, we hear media reports stating that some allegedly homeless people abuse the practice of asking for money on the streets and that many are faking injury or poverty so that they can take advantage of generous individuals. Stereotypes such as this are commonplace when some laypersons, members of the media, and politicians describe the homeless in America. But it is far from the entire picture: Many homeless people are in need of assistance, and many of the homeless are children, persons with disabilities, and people with other problems that make it difficult, if not impossible, for them to earn enough money to pay for housing in many cities.

Do all of these "big-picture" problems in our society mean that we have no individual responsibility to help homeless people? We do not necessarily have to hand money over to the person on the street to help individuals who are homeless. There are other, and perhaps even better, ways in which we can provide help to the homeless through our small acts of generosity and kindness.

In some communities, college students lead the way in helping homeless individuals and families. Some programs help homeless children by providing them with clothing, other basic necessities, and even school supplies so that the children will feel comfortable in a classroom setting. Still other college students work in, or run, homeless shelters in their communities. For example, Harvard University students, along with some city officials and church leaders in Cambridge, Massachusetts, created the Harvard Square Homeless Shelter in 1983 to address the housing needs of the area's poorest residents. Although it was hoped that the shelter would be a temporary project that would be rendered unnecessary when society recognized and dealt with its homeless problem, the shelter was still in existence in 2011.

As organizers of some college groups that seek to help the homeless have suggested, individuals without homes need food, clothing, and shelter, but they also need compassion and caring that extend beyond what most bureaucratic

Courtesy of Gretchen Otto

A unique way that some college students recycle items they no longer want is to conduct a garage sale that benefits a local charity or community organization.

organizations can offer. Here are a few ways in which you and others at your school might help homeless individuals and families in your community:

- *Understand who the homeless are* so that you can help dispel the stereotypes often associated with homeless people. Learn what causes homelessness, and remember that each person's story is unique.
- *Buy a street newspaper sold by homeless persons if you live in an urban area where these newspapers are sold.* Newspapers produced and sold by homeless individuals have grown in number and circulation: Homeless persons receive a small amount for every paper they sell. Examples include *Street Roots* in Portland, Oregon; *Real Change* in Seattle, Washington; and *Street Sense* in Washington, D.C.
- *Give money, clothing, and/or recyclables to organizations that aid the homeless.* In addition to money or clean, usable clothing, recyclable cans and bottles are helpful because they can be turned into small sums of money for living expenses.
- *Volunteer at a shelter, soup kitchen, or battered women's shelter* where you can help staff and other volunteers meet the daily needs of people who are without shelter and food, as well as women and children who need assistance in getting away from abusive relationships with family members.
- *Look for campus organizations that work with the homeless,* or create your own and enlist friends and existing organizations (such as your service organization, sorority, or fraternity) to engage in community service projects that will benefit both the temporarily and permanently homeless.

For additional ways you can help the homeless, check with shelters in your area. You may also want to visit the websites of organizations such as the following:

- Just Give: **www.justgive.org**
- The Doe Fund: **www.doe.org**
- U.S. Department of Housing and Urban Development: **www.hud.gov/homeless**

and roles to play than at any other time in history. Although achieved statuses are considered very important, ascribed statuses still have a significant effect on people's options and opportunities.

Ironically, at a time when we have more technological capability, more leisure activities and types of entertainment, and more quantities of material goods available for consumption than ever before, many people experience high levels of stress, fear for their lives because of crime, and face problems such as homelessness. In a society that can send astronauts into space to perform complex scientific experiments, is it impossible to solve some of the problems that plague us here on Earth?

Individuals and groups often show initiative in trying to solve some of our pressing problems, at least on a local level (see an example in "You Can Make a Difference"). However, the future of this country rests on our collective ability to deal with major social problems at both the macrolevel and the microlevel of society.

chapter review Q&A

Use these questions and answers to check how well you've achieved the learning objectives set out at the beginning of this chapter.

- **How does social structure shape our social interactions?**

The stable patterns of social relationships within a particular society make up its social structure. Social structure is a macrolevel influence because it shapes and determines the overall patterns in which social interaction occurs. Social structure provides an ordered framework for society and for our interactions with others.

- **What are the main components of social structure?**

Social structure comprises statuses, roles, groups, and social institutions. A status is a specific position in a group or society and is characterized by certain expectations, rights, and duties. Ascribed statuses, such as gender, class, and race/ethnicity, are acquired at birth or involuntarily later in life. Achieved statuses, such as education and occupation, are assumed voluntarily as a result of personal choice, merit, or direct effort. We occupy a status, but a role is the set of behavioral expectations associated with a given status. A social group consists of two or more people who interact frequently and share a common identity and sense of interdependence. A formal organization is a highly structured group formed to complete certain tasks or achieve specific goals. A social institution is a set of organized beliefs and rules that establishes how a society attempts to meet its basic needs.

- **What are the functionalist and conflict perspectives on social institutions?**

According to functionalist theorists, social institutions perform several prerequisites of all societies: replace members; teach new members; produce, distribute, and consume goods and services; preserve order; and provide and maintain a sense of purpose. Conflict theorists suggest that social institutions do not work for the common good of all individuals: Institutions may enhance and uphold the power of some groups but exclude others, such as the homeless.

- **How do societies maintain stability in times of social change?**

According to Emile Durkheim, although changes in social structure may dramatically affect individuals and groups, societies manage to maintain some degree of stability. People in preindustrial societies are united by mechanical solidarity because they have shared values and common social bonds. Industrial societies are characterized by organic solidarity, which refers to the cohesion that results when people perform specialized tasks and are united by mutual dependence.

- **How do *Gemeinschaft* and *Gesellschaft* societies differ in social solidarity?**

According to Ferdinand Tönnies, the *Gemeinschaft* is a traditional society in which relationships are based on personal bonds of friendship and kinship and on intergenerational stability. The *Gesellschaft* is an urban society in which social bonds are based on impersonal and specialized relationships, with little group commitment or consensus on values.

- **What is the dramaturgical perspective?**

According to Erving Goffman's dramaturgical analysis, our daily interactions are similar to dramatic productions. Presentation of self refers to efforts to present our own self to others in ways that are most favorable to our interests or self-image.

- **Why are feeling rules important?**

Feeling rules shape the appropriate emotions for a given role or specific situation. Our emotions are not always private, and specific emotions may be demanded of us on certain occasions.

key terms

achieved status 121
agrarian society 134
ascribed status 121
dramaturgical analysis 143
ethnomethodology 142
formal organization 128
Gemeinschaft 138
Gesellschaft 138
horticultural society 134
hunting and gathering society 133
impression management (presentation of self) 143
industrial society 135
master status 122
mechanical solidarity 137
nonverbal communication 145
organic solidarity 138
pastoral society 134
personal space 148
postindustrial society 136
primary group 128
role 124
role conflict 125
role exit 127
role expectation 124
role performance 125
role strain 126
secondary group 128
self-fulfilling prophecy 141
social construction of reality 141
social group 127
social institution 128
social interaction 119
social structure 119
status 121
status symbol 123

questions for critical thinking

1. Think of a person you know well who often irritates you or whose behavior grates on your nerves (it could be a parent, friend, relative, teacher). First, list that person's statuses and roles. Then analyze the person's possible role expectations, role performance, role conflicts, and role strains. Does anything you find in your analysis help to explain the irritating behavior? How helpful are the concepts of social structure in analyzing individual behavior?
2. You are conducting field research on gender differences in nonverbal communication styles. How are you going to account for variations among age, race, and social class?
3. When communicating with other genders, races, and ages, is it better to express and acknowledge different styles or to develop a common, uniform style?

turning to video

Watch the CBS video *Poverty in Greenville, Mississippi* (running time 3:14), available through **CengageBrain.com**. The poverty rate in Greenville, Mississippi, is about two and a half times the national average. This video shows some of those who are struggling to makes ends meet. As you watch the video, think about your own family or people you know who have struggled to find a job or pay the bills. After you've watched the video, consider these questions: How did the stress of being out of work or not being able to pay the bills affect you or someone you know? Were there any larger societal factors contributing to the situation?

online study resources

Go to CENGAGE brain.com to access online study resources, including the Sociology CourseMate for this text as well as special features such as video, an interactive sociology time line and interactive maps, General Social Survey (GSS) data, and U.S. Census 2010 data.

CourseMate brings course concepts to life with interactive learning, study, and exam-preparation tools that support the printed textbook. A textbook-specific website, **Sociology CourseMate** includes an integrated interactive eBook and other interactive learning tools, including quizzes, flash cards, and videos.

Visit **www.cengagebrain.com** to access your account and purchase materials.

6 Groups and Organizations

At my university, professors are divided about whether they should meddle [with students who bring smartphones, iPads, and computers to class]. Our students, some say, are grown-ups. It is not for us to dictate how they take notes or to get involved if they let their attention wander from class-related materials. But when I stand in back of our Wi-Fi enabled lecture halls, students are on Facebook and YouTube, and they are shopping, mostly for music. I want to engage my students in conversation. I don't think they should use class time for any other purpose. One year, I raised the topic for general discussion and suggested using notebooks (the paper kind) for note taking. Some of my students claimed to be relieved. "Now I won't be tempted by Facebook messages," said one sophomore. Others were annoyed, almost surly. . . . I maintained my resolve, but the following year, I bowed to common practice and allowed students to do what they wished.

—Sherry Turkle (2011), a professor at MIT, describing her feelings about students' use of digital technology in classrooms where professors are attempting to form groups and build community among students

AP Images/L. G. Patterson

▲ **Have Facebook and other networking websites influenced our social interactions and group participation? Why are face-to-face encounters in groups and organizations still important in everyday life?**

Chapter Focus Question:

Why is it important for groups and organizations to enhance communication among participants and improve the flow of information while protecting the privacy of individuals?

According to sociologists, we need groups and organizations—just as we need culture and socialization—to live and participate in a society. Historically, the basic premise of groups and organizations was that individuals engage in face-to-face interactions in order to be part of such a group; however, millions of people today communicate with others through the Internet, cell phones, and other forms of digital technology that make it possible for them to "talk" with individuals they have never met and who may live thousands of miles away. A variety of social networking websites, including Facebook, MySpace, LinkedIn, and Twitter, now compete with, or in some cases replace, live, person-to-person communications. Despite the wealth of information and opportunities for new social connections that such websites offer, many of our daily activities require that we participate in social groups and formal organizations where *face time*—time spent interacting with others on a face-to-face basis, rather than via Internet or cell phone—is necessary. However, some people live in a multidimensional world where they seek to interact in both the "real world" and the cyber world at the same time.

What do social groups and formal organizations mean to us in an age of rapid digital communications? What is the relationship between information and social organizations in societies such as ours? How can we balance the information that we provide to other people about us with our own right to privacy and need for security? These questions are of interest to sociologists who seek to apply the sociological imagination to their studies of social groups, bureaucratic organizations, social networking, and virtual communities. Before we take a closer look at groups and organizations, take the quiz in the Sociology and Everyday Life box on issues regarding personal privacy in groups and organizations.

Highlighted Learning Objectives

- Explain what constitutes a social group.
- Discuss how groups and their members are shaped by group size, leadership style, and pressures to conform.
- Describe the relationship between information and social organizations in societies such as ours.
- Discuss the purposes that bureaucracy serves.
- Identify the most widespread and alternative forms of organization that exist today.

sociology and everyday life

How Much Do You Know About Privacy in Groups and Organizations?

True	False	
T	F	1. A college student's privacy is protected when using a school-owned computer as long as he or she deletes from the computer all e-mails or other documents he or she has worked on and thus prevents anyone else from examining those documents.
T	F	2. Parents of students at all U.S. colleges and universities are entitled to obtain a transcript of their children's college grades, regardless of the student's age.
T	F	3. If you work for a business that monitors phone calls with a pen register, your employer has the right to maintain and examine a list of phone numbers dialed by your extension and how long each call lasted.
T	F	4. Members of a high school football team can be required to submit to periodic, unannounced drug testing.
T	F	5. A company has the right to keep its employees under video surveillance anywhere at the company's place of business—even in the restrooms.
T	F	6. A professor can legally post students' grades in public, using a student's Social Security number as an identifier, as long as the student's name does not appear with the number.
T	F	7. Students at a church youth-group meeting who hear one member of the group confess to an illegal act can be required to divulge what that member said.
T	F	8. If you apply for a job at a company that has more than 25 employees, your employer can require that you provide a history of your medical background or take a physical examination prior to offering you a job.

Answers on page 157.

Social Groups

Three strangers are standing at a street corner waiting for a traffic light to change. Do they constitute a group? Five hundred women and men are first-year graduate students at a university. Do they constitute a group? In everyday usage we use the word *group* to mean any collection of people. According to sociologists, however, the answer to these questions is no; individuals who happen to share a common feature or to be in the same place at the same time do not constitute social groups.

Groups, Aggregates, and Categories

As discussed in Chapter 5, a *social group* is a collection of two or more people who interact frequently with one another, share a sense of belonging, and have a feeling of interdependence. Several people waiting for a traffic light to change constitute an ***aggregate*—a collection of people who happen to be in the same place at the same time but share little else in common.** Shoppers in a department store and passengers on an airplane flight are also examples of aggregates. People in aggregates share a common purpose (such as purchasing items or arriving at their destination) but generally do not interact with one another, except perhaps briefly. The first-year graduate students, at least initially, constitute a ***category*—a number of people who may never have met one another but share a similar characteristic (such as education level, age, race, or gender).** Men and women make up categories, as do Native Americans and Latinos/as, and victims of sexual or racial harassment. Categories are not social groups because the people in them do not usually create a social structure or have anything in common other than a particular trait.

Michael Newman/PhotoEdit

▲ These college students are education majors—students who plan to become teachers. Do they constitute a category or a social group?

Occasionally, people in aggregates and categories form social groups. For instance, people within the category known as "graduate students" may become an aggregate when they get together for an orientation to graduate school. Some of them may form social groups as they interact with one another in classes and seminars, find that they have mutual interests and concerns, and develop a sense of belonging to the group. Information technology raises new and interesting questions about what constitutes a group. For example, some people question whether we can form a social group on the Internet (see "Framing 'Community' in the Media").

Types of Groups

As you will recall from Chapter 5, groups have varying degrees of social solidarity and structure. This structure is flexible in some groups and more rigid in others. Some groups are small and personal; others are large and impersonal. We more closely identify with the members of some groups than we do with others.

Cooley's Primary and Secondary Groups Sociologist Charles H. Cooley (1963/1909) used the term *primary group* to describe a small, less specialized group in which members engage in face-to-face, emotion-based interactions over an extended period of time. We have primary relationships with other individuals in our primary groups—that is, with our *significant others*, who frequently serve as role models.

In contrast, you will recall, a *secondary group* is a larger, more specialized group in which the members engage in more impersonal, goal-oriented relationships for a limited period of time. The size of a secondary group may vary. Twelve students in a graduate seminar may start out as a secondary group but eventually become a primary group as they get to know one another and communicate on a more personal basis. Formal organizations are secondary groups, but they also contain many primary groups within them. For example, how many primary groups do you think there are within the secondary-group setting of your college?

Sumner's Ingroups and Outgroups All groups set boundaries by distinguishing between insiders who are members and outsiders who are not. Sociologist William Graham Sumner (1959/1906) coined the terms *ingroup* and *outgroup* to describe people's feelings toward members of their own and other groups. An ***ingroup*** **is a group to which a person belongs and with which the person feels a sense of identity.** An ***outgroup*** **is a group to which a person does not belong and toward which the person may feel a sense of competitiveness or hostility.** Distinguishing between our ingroups and our outgroups helps us establish our individual identity and self-worth. Likewise, groups are solidified by ingroup and outgroup distinctions; the presence of an enemy or a hostile group binds members more closely together (Coser, 1956).

Group boundaries may be formal, with clearly defined criteria for membership. For example, a country club that requires an applicant for membership to be recommended by four current members and to pay a $50,000 initiation fee has clearly set requirements for its members (see "Sociology Works!" on p. 158). However, group boundaries are not always that formal. For example, friendship groups usually do not have clear guidelines for membership; rather, the boundaries tend to be very informal and vaguely defined.

Ingroup and outgroup distinctions may encourage social cohesion among members, but they may also promote classism, racism, sexism, and ageism. Ingroup members typically view themselves positively and members of outgroups negatively. These feelings of group superiority, or *ethnocentrism,* are somewhat inevitable. However, members of some groups feel more free than others to act on their beliefs. If groups are embedded in larger groups or organizations, the larger organization may discourage such beliefs and their consequences (Merton, 1968). Conversely, organizations may covertly foster these ingroup/outgroup distinctions by denying their existence or by failing to take action when misconduct occurs.

Reference Groups Ingroups provide us not only with a source of identity but also with a point of reference. A ***reference group*** **is a group that strongly influences a person's behavior and social attitudes, regardless of whether that individual is an actual member.** When we attempt to evaluate our appearance, ideas, or goals, we automatically refer to the standards of some group. Sometimes, we will refer to our membership groups, such as family

aggregate a collection of people who happen to be in the same place at the same time but share little else in common.

category a number of people who may never have met one another but share a similar characteristic (such as education level, age, race, or gender).

ingroup a group to which a person belongs and with which the person feels a sense of identity.

outgroup a group to which a person does not belong and toward which the person may feel a sense of competitiveness or hostility.

reference group a group that strongly influences a person's behavior and social attitudes, regardless of whether that individual is an actual member.

framing "community" in the media

"Virtual Communities" on the Internet

Meeting new friends,
Imagining smiles . . .
Across the networks
Spanning the miles. . . .

From all walks of life
We come to the net.
A community of friends
Who have never met.
—from "Thoughts of Internet Friendships" by Jamie Wilkerson (1996)

As this excerpt from a poem posted on the Internet suggests, many people believe that they can make new friends and establish a community online. Although chat groups are one of the oldest forms of communication in the online community, they remain popular all around the Internet, from search engines to social networking sites (PerfSpot.com, 2011). One of the advantages of chat groups is that you can have a real-time conversation with another person, and chat groups now provide people with an option to have a seemingly face-to-face conversation by using a webcam. Many popular chat groups are dedicated to a specific topic, such as support groups for people with certain types of chronic illnesses or individuals who like to talk about a special interest or hobby. Social networking sites have contributed to the ongoing popularity of chat groups.

▲ Chatrooms and other forms of communication on the Internet are extremely popular with millions of people; however, some sociologists question whether we can actually form social groups and true communities on the Internet. Is cyber chat that different from our face-to-face interactions with others?

Although chat groups are framed as a way to make friends, get dates, and establish a cyber community, as you study sociology you might ask whether this form of "community" is actually a true community. Because sociologists define a *social group* as a collection of two or more people who interact frequently with one another, share a sense of belonging, and have a feeling of interdependence, this definition suggests that people must have a sense of place (be in the same place at the same time at least part of the time) in order to establish a true social group or community. However, this definition was developed before the Internet provided people with the rapid communications that link them with others around the world today. Are we able to form groups and establish communities with people whom we have never actually met?

Some social scientists believe that virtual communities established on the Internet constitute true communities (see Wellman, 2001). However, the sociologists Robyn Driskell and Larry Lyon examined existing theories and research on this topic and concluded that true communities cannot be established in the digital environment of cyberspace. According to Driskell and Lyon, although the Internet provides us with the opportunity to share interests with others whom we have not met and to communicate with people we already know, the original concept of community, which "emphasized local place, common ties, and social interaction that is intimate, holistic, and all-encompassing," is lacking (Driskell and Lyon, 2002: 6). Virtual communities on the Internet do not have geographic and social boundaries, are limited in their scope to specific areas of interest, are psychologically detached from close interpersonalties, and have only limited concern for their "members" (Driskell and Lyon, 2002). In fact, if we spend many hours in social isolation at our computers or using hand-held devices such as smartphones and iPads, these technologies may *reduce* our sense of community rather than enhance it (see Turkle, 2011). Even so, it may be possible that cyber communication will create a "weak community replacement" for some people who develop specialized ties to others based on their use of these digital technologies (Driskell and Lyon, 2002).

reflect & analyze

Have you had opportunities to gain new friends and build "community" on social networking sites? From your perspective, what are the strengths and limitations of virtual communications?

sociology and everyday life

ANSWERS to the Sociology Quiz on Privacy

1. False. Deleting an e-mail or other document from a computer does not actually remove it from the computer's memory. Until other files are entered that write over the space where the document was located, experts can retrieve the document that was deleted.

2. False. The Family Educational Right to Privacy Act, which allows parents of a student under age 18 to obtain their child's grades, requires the student's consent once he or she has attained age 18; however, that law applies only to institutions that receive federal educational funds.

3. True. Telephone numbers called from a company's phone extensions can be recorded on a pen register, and this information can be used by the employer in evaluating the amount of time you have spent talking with clients—or with other people.

4. True. The U.S. Supreme Court has ruled that schools may require students to submit to random drug testing as a condition of participating in extracurricular activities such as sports teams, the school band, the future homemakers' club, the cheerleading squad, and the choir.

5. False. An employer may not engage in video surveillance of its employees in situations where they have a reasonable right of privacy. At least in the absence of a sign warning of this type of surveillance, employees have such a right in company restrooms.

6. False. The Federal Educational Rights and Privacy Act states that Social Security numbers are considered to be "personally identifiable information" that may not be released without written consent from the student. Posting grades by Social Security number violates this provision unless the student has consented to the number being disclosed to others.

7. True. Although confidential communications made privately to a minister, priest, rabbi, or other religious leader (or to an individual the person reasonably believes to hold such a position) generally cannot be divulged without the consent of the person making the communication, this does not apply when other people are present who are likely to hear the statement.

8. False. The Americans with Disabilities Act prohibits employers in companies with more than 25 employees from asking job applicants about medical information or requiring a physical examination prior to employment.

or friends. Other times, we will rely on groups to which we do not currently belong but that we might wish to join in the future, such as a social club or a profession.

Reference groups help explain why our behavior and attitudes sometimes differ from those of our membership groups. We may accept the values and norms of a group with which we identify rather than one to which we belong. We may also act more like members of a group that we want to join than members of groups to which we already belong. In this case, reference groups are a source of anticipatory socialization. For most of us, our reference-group attachments change many times during our life course, especially when we acquire a new status in a formal organization.

Networks A ***network*** **is a web of social relationships that links one person with other people and, through them, with other people they know.** Frequently, networks connect people who share common interests but who otherwise might not identify and interact with one another. For example, if A is tied to B, and B is tied to C, then a network is more likely to be formed among individuals A, B, and C. If this seems a little confusing at first, let's assume that Alice knows of Dolores and Eduardo only through her good friends Bill and Carolyn. For almost a year, Alice has been trying (without success) to purchase a house she can afford. Because large numbers of people are moving into her community, the real estate market is "tight," and houses frequently sell before a "for sale" sign goes up in the yard. However, through her friends Bill and Carolyn, Alice learns that their friends—Dolores and Eduardo—are about to put their house up for sale. Bill and Carolyn call Dolores and Eduardo to set up an appointment for Alice to see the house before it goes on the real estate market. Thanks to Alice's network, she is able to purchase the house before other people learn

network a web of social relationships that links one person with other people and, through them, with other people they know.

sociology works!

Ingroups, Outgroups, and "Members Only" Clubs

In this country we have a God-given right to associate with whomever we please. And frankly, this includes my right to *not* associate with people I don't want to. If I don't want to be around somebody, why should I have to let them in my club? Let them go start their own club.

—Phil, a white, male attorney who is a member of several prestigious private clubs, explaining why he believes he has the right to establish his own ingroup through private club memberships (qtd. in Kendall, 2008)

A key characteristic of the city clubs and country clubs where Phil is a member is that each organization has formal group boundaries, with people becoming members "by invitation only." In other words, prospective members must be nominated by current members and be voted into the club: They cannot simply decide to join the organization. For this reason, people who are invited to join typically feel special (like "insiders") because they know that club membership is not available to everyone. Club members such as Phil often develop *consciousness of kind*—a term used by sociologists to describe the awareness that individuals may have when they believe that they share important commonalities with certain other people. Consciousness of kind is strengthened by membership in clubs ranging from country clubs to college sororities, fraternities, and other by-invitation-only university social clubs. Members of ingroups typically share strong feelings of consciousness of kind and believe that they have little in common with people in the outgroup.

Recent studies on private clubs and exclusive college social organizations show that the sociological concepts of "ingroup" and "outgroup" remain highly relevant today when we conduct research on the processes of inclusion and exclusion to learn more about how such activities affect individuals and groups (see Kendall, 2008). Most of us are aware that our ingroups are very important to us: They provide us with a unique sense of identity, but they also give us the ability to exclude those individuals whom we do not want in our inner circle of friends and acquaintances. The early sociologist Max Weber captured this idea in his description of the *closed relationship*—a setting in which the "participation of certain persons is excluded, limited, or subjected to conditions" (Gerth and Mills, 1946: 139). Exclusive clubs typically have signs posted on gates, fences, or buildings that state "Members Only." These organizations do not welcome outsiders within their walls, and members are often pledged to loyalty and secrecy about their club's activities. Similarly, many college fraternities and sororities thrive on rituals, secrecy, and the importance of what it means to pledge—to have accepted a bid to join but not having yet been initiated into—the group of one's choice (Robbins, 2004: 342).

reflect & analyze

What areas of sociological research or personal interest can you think of that might benefit from applying the ingroup/outgroup concept to your analysis? How might these concepts be applied to other areas of college life besides invitational social organizations?

© Duncan Hale-Sutton/Alamy

© Jeff Greenberg/Alamy

▲ Sometimes, the distinction between what constitutes an ingroup and an outgroup is subtle. Other times, it is not subtle at all. Would you feel comfortable entering either of these establishments if you were not a member?

that it is for sale. Although Alice had not previously met Dolores and Eduardo, they are part of her network through her friendship with Bill and Carolyn. Scarce resources (in this case, the number of affordable houses available) are unequally distributed, and people often must engage in collaboration and competition in their efforts to deal with this scarcity. Another example of the use of networks to help overcome scarce resources is recent college graduates who seek help from friends and acquaintances in order to find a good job.

What are your networks? For a start, your networks consist of all the people linked to you by primary ties, including your relatives and close friends. Your networks also include your secondary ties, such as acquaintances, classmates, professors, and—if you are employed—your supervisor and coworkers. However, your networks actually extend far beyond these ties to include not only the people that you *know* but also the people that you *know of*—and who know of you—through your primary and secondary ties. In fact, your networks potentially include a pool of between 500 and 2,500 acquaintances if you count the connections of everyone in your networks (Milgram, 1967). Today, the term *networking* is widely used to describe the contacts that people make to find jobs or other opportunities; however, sociologists have studied social networks for many years in an effort to learn more about the linkages between individuals and their group memberships.

Group Characteristics and Dynamics

What purpose do groups serve? Why are individuals willing to relinquish some of their freedom to participate in groups? According to functionalists, people form groups to meet instrumental and expressive needs. *Instrumental,* or task-oriented, needs cannot always be met by one person, so the group works cooperatively to fulfill a specific goal. Groups help members do jobs that are impossible to do alone or that would be very difficult and time-consuming at best. For example, think of how hard it would be to function as a one-person football team or to single-handedly build a skyscraper. In addition to instrumental needs, groups also help people meet their *expressive,* or emotional, needs, especially those involving self-expression and support from family, friends, and peers.

Although not disputing that groups ideally perform such functions, conflict theorists suggest that groups also involve a series of power relationships whereby the needs of individual members may not be equally served. Symbolic interactionists focus on how the size of a group influences the kind of interaction that takes place among members. To many postmodernists, groups and organizations—like other aspects of postmodern societies—are generally characterized by superficiality and depthlessness in social relationships (Jameson, 1984). One postmodern thinker who focuses on this issue is the literary theorist Fredric Jameson, who believes that people experience a waning of emotion in organizations where fragmentation and superficiality are a way of life (cited in Ritzer, 1997). For example, fast-food restaurant employees and customers interact in extremely superficial ways that are largely scripted: The employees follow scripts in taking and filling customers' orders ("Would you like fries and a drink with that?"), and the customers respond with their own "recipied" action. According to the sociologist George Ritzer (1997: 226), "[C]ustomers are mindlessly following what they consider tried-and-true social recipes, either learned or created by them previously, on how to deal with restaurant employees and, more generally, how to work their way through the system associated with the fast-food restaurant."

We will now look at certain characteristics of groups, such as how size affects group dynamics.

Group Size

The size of a group is one of its most important features. Interactions are more personal and intense in a ***small group,* a collectivity small enough for all members to be acquainted with one another and to interact simultaneously.**

Sociologist Georg Simmel (1950/1902–1917) suggested that small groups have distinctive interaction patterns that do not exist in larger groups. According to Simmel, in a ***dyad*—a group composed of two members**—the active participation of both members is crucial to the group's survival. If one member withdraws from interaction or "quits," the group ceases to exist. Examples of dyads include two people who are best friends, married couples, and domestic partnerships. Dyads provide members with an intense bond and a sense of unity not found in most larger groups.

When a third person is added to a dyad, a ***triad,* a group composed of three members,** is formed. The nature of the relationship and interaction patterns changes with the addition of the third person. In a triad, even if one member ignores another or declines to participate, the group can still function.

small group a collectivity small enough for all members to be acquainted with one another and to interact simultaneously.

dyad a group composed of two members.

triad a group composed of three members.

▲ According to the sociologist Georg Simmel, interaction patterns change when a third person joins a dyad—a group composed of two members. How might the conversation between these two women change when another person arrives to talk with them?

In addition, two members may unite to create a coalition that can subject the third member to group pressure to conform. A *coalition* is an alliance created in an attempt to reach a shared objective or goal. If two members form a coalition, the other member may be seen as an outsider or intruder.

As the size of a group increases beyond three people, members tend to specialize in different tasks, and everyday communication patterns change. For instance, in groups of more than six or seven people, it becomes increasingly difficult for everyone to take part in the same conversation; therefore, several conversations will probably take place simultaneously. Members are also likely to take sides on issues and form a number of coalitions. In groups of more than ten or twelve people, it becomes virtually impossible for all members to participate in a single conversation unless one person serves as moderator and guides the discussion. As shown in ▶ Figure 6.1, when the size of the group increases, the number of possible social interactions also increases.

Although large groups typically have less social solidarity than small ones, they may have more power. However, the relationship between size and power is more complicated than it might initially seem. The power relationship depends on both a group's *absolute* size and its *relative* size (Simmel, 1950/1902–1917; Merton, 1968). The absolute size is the number of members the group actually has; the relative size is the number of potential members. For example, suppose that 300 people band together to "march on Washington" and demand enactment of a law on some issue that they believe to be important. Although 300 people is a large number in some contexts, opponents of this group would argue that the low turnout (compared with the number of people in this country) demonstrates that most people don't believe the issue is important. At the same time, the power of a small group to demand change may be based on a "strength in numbers" factor if the group is seen as speaking on behalf of a large number of other people (who are also voters).

Larger groups typically have more formalized leadership structures. Their leaders are expected to

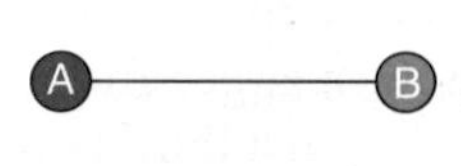

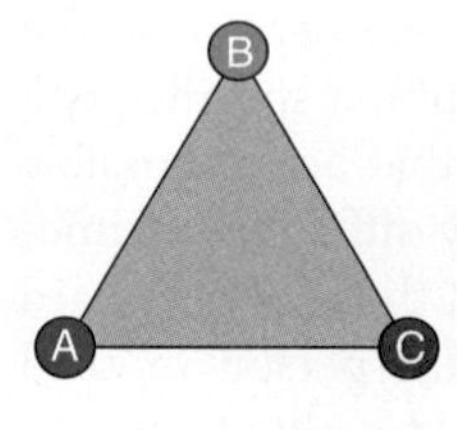

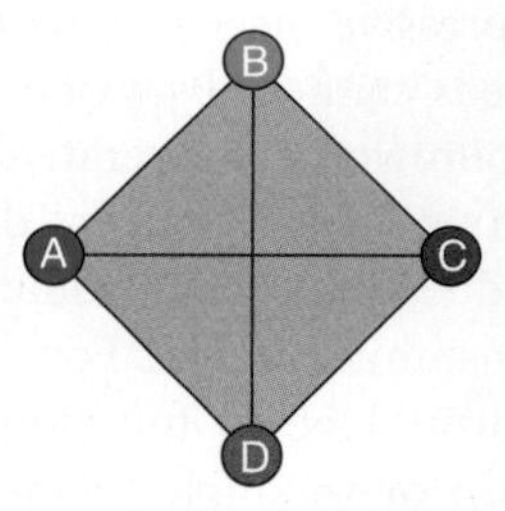

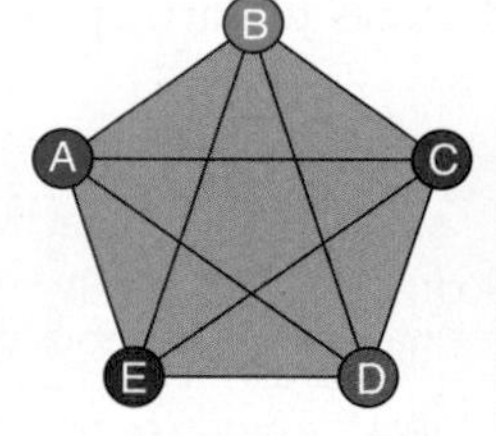

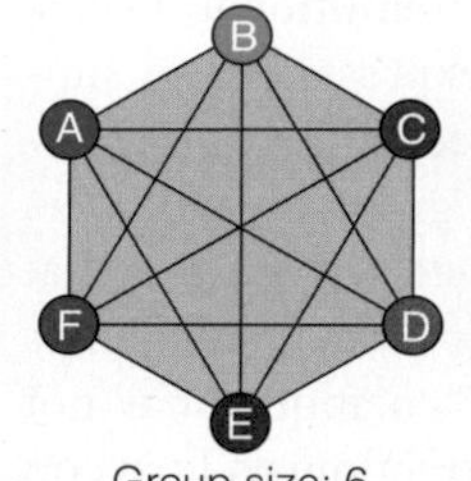

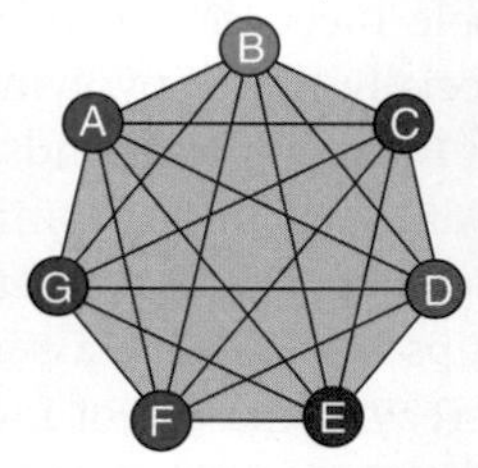

▶ **FIGURE 6.1 GROWTH OF POSSIBLE SOCIAL INTERACTION BASED ON GROUP SIZE**

perform a variety of roles, some related to the internal workings of the group and others related to external relationships with other groups.

Group Leadership

What role do leaders play in groups? Leaders are responsible for directing plans and activities so that the group completes its task or fulfills its goals. Primary groups generally have informal leadership. For example, most of us do not elect or appoint leaders in our own families. Various family members may assume a leadership role at various times or act as leaders for specific tasks. In traditional families, the father or eldest male is usually the leader. However, in today's more diverse families, leadership and power are frequently in question, and power relationships may be quite different, as discussed later in this text. By comparison, larger groups typically have more formalized leadership structures. For example, leadership in secondary groups (such as colleges, governmental agencies, and corporations) involves a clearly defined chain of command, with written responsibilities assigned to each position in the organizational structure.

Leadership Functions Both primary and secondary groups have some type of leadership or positions that enable certain people to be leaders, or at least to wield power over others. From a functionalist perspective, if groups exist to meet the instrumental and expressive needs of their members, then leaders are responsible for helping the group meet those needs. ***Instrumental leadership* is goal or task oriented;** this type of leadership is most appropriate when the group's purpose is to complete a task or reach a particular goal. ***Expressive leadership* provides emotional support for members;** this type of leadership is most appropriate when the group is dealing with emotional issues, and harmony, solidarity, and high morale are needed. Both kinds of leadership are needed for groups to work effectively.

Leadership Styles Three major styles of leadership exist in groups: authoritarian, democratic, and laissez-faire. ***Authoritarian leaders* make all major group decisions and assign tasks to members.** These leaders focus on the instrumental tasks of the group and demand compliance from others. In times of crisis, such as a war or natural disaster, authoritarian leaders may be commended for their decisive actions. In other situations, however, they may be criticized for being dictatorial and for fostering intergroup hostility. By contrast, ***democratic leaders* encourage group discussion and decision making through consensus building.** These leaders may be praised for their expressive, supportive behavior toward group members, but they may also be blamed for being indecisive in times of crisis.

Laissez-faire literally means "to leave alone." ***Laissez-faire leaders* are only minimally involved in decision making and encourage group members to make their own decisions.** On the one hand, laissez-faire leaders may be viewed positively by group members because they do not flaunt their power or position. On the other hand, a group that needs active leadership is not likely to find it with this style of leadership, which does not work vigorously to promote group goals.

▲ Organizations have different leadership styles based on the purpose of the group. How do leadership styles in the military differ from those on college and university campuses?

instrumental leadership goal- or task-oriented leadership.

expressive leadership an approach to leadership that provides emotional support for members.

authoritarian leaders people who make all major group decisions and assign tasks to members.

democratic leaders leaders who encourage group discussion and decision making through consensus building.

laissez-faire leaders leaders who are only minimally involved in decision making and who encourage group members to make their own decisions.

Studies of kinds of leadership and decision-making styles have certain inherent limitations. They tend to focus on leadership that is imposed externally on a group (such as bosses or political leaders) rather than leadership that arises within a group. Different decision-making styles may be more effective in one setting than another. For example, imagine attending a college class in which the professor asked the students to determine what should be covered in the course, what the course requirements should be, and how students should be graded. It would be a difficult and cumbersome way to start the semester; students might spend the entire term negotiating these matters and never actually learn anything.

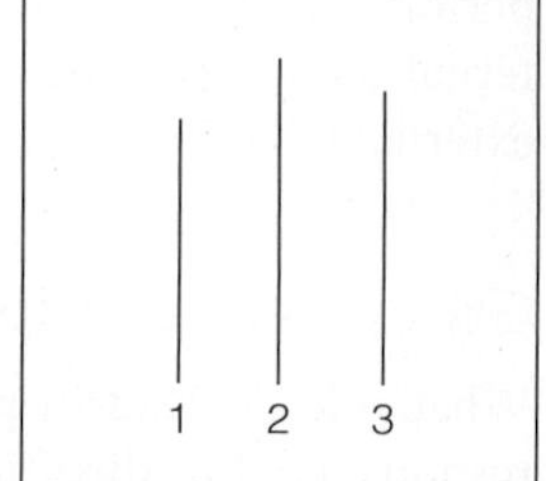

▶ **FIGURE 6.2 ASCH'S CARDS**
Although Line 2 is clearly the same length as the line in the lower card, Solomon Asch's research assistants tried to influence "actual" participants by deliberately picking Line 1 or Line 3 as the correct match. Many of the participants went along rather than risking the opposition of the "group."
Source: Asch, 1955.

Group Conformity

To what extent do groups exert a powerful influence in our lives? Groups have a significant amount of influence on our values, attitudes, and behavior. In order to gain and then retain our membership in groups, most of us are willing to exhibit a high level of conformity to the wishes of other group members. ***Conformity* is the process of maintaining or changing behavior to comply with the norms established by a society, subculture, or other group.** We often experience powerful pressure from other group members to conform. In some situations this pressure may be almost overwhelming.

In several studies (which would be impossible to conduct today for ethical reasons), researchers found that the pressure to conform may cause group members to say they see something that is contradictory to what they are actually seeing or to do something that they would otherwise be unwilling to do. As we look at two of these studies, ask yourself what you might have done if you had been involved in this research.

Asch's Research Pressure to conform is especially strong in small groups in which members want to fit in with the group. In a series of experiments conducted by Solomon Asch (1955, 1956), the pressure toward group conformity was so great that participants were willing to contradict their own best judgment if the rest of the group disagreed with them.

One of Asch's experiments involved groups of undergraduate men (seven in each group) who were allegedly recruited for a study of visual perception. All the men were seated in chairs. However, the person in the sixth chair did not know that he was the only actual subject; all the others were assisting the researcher. The participants were first shown a large card with a vertical line on it and then a second card with three vertical lines (see ▶ Figure 6.2). Each of the seven participants was asked to indicate which of the three lines on the second card was identical in length to the "standard line" on the first card.

In the first test with each group, all seven men selected the correct matching line. In the second trial, all seven still answered correctly. In the third trial, however, the actual subject became very uncomfortable when all the others selected the incorrect line. The subject could not understand what was happening and became even more confused as the others continued to give incorrect responses on eleven out of the next fifteen trials.

Asch (1955) found that about one-third of all subjects chose to conform by giving the same (incorrect) responses as Asch's assistants. In discussing the experiment afterward, most of the subjects who gave incorrect responses indicated that they had known the answers were wrong but decided to go along with the group in order to avoid ridicule or ostracism.

Asch concluded that the size of the group and the degree of social cohesion felt by participants were important influences on the extent to which individuals respond to group pressure (see ▶ Figure 6.3). If you had been in the position of the subject, how would you have responded? Would you have continued to give the correct answer, or would you have been swayed by the others?

Milgram's Research How willing are we to do something because someone in a position of authority has told us to do it? How far are we willing to go to follow the demands of that individual? Stanley Milgram (1963, 1974) conducted a series of controversial experiments to find answers to these questions about people's obedience

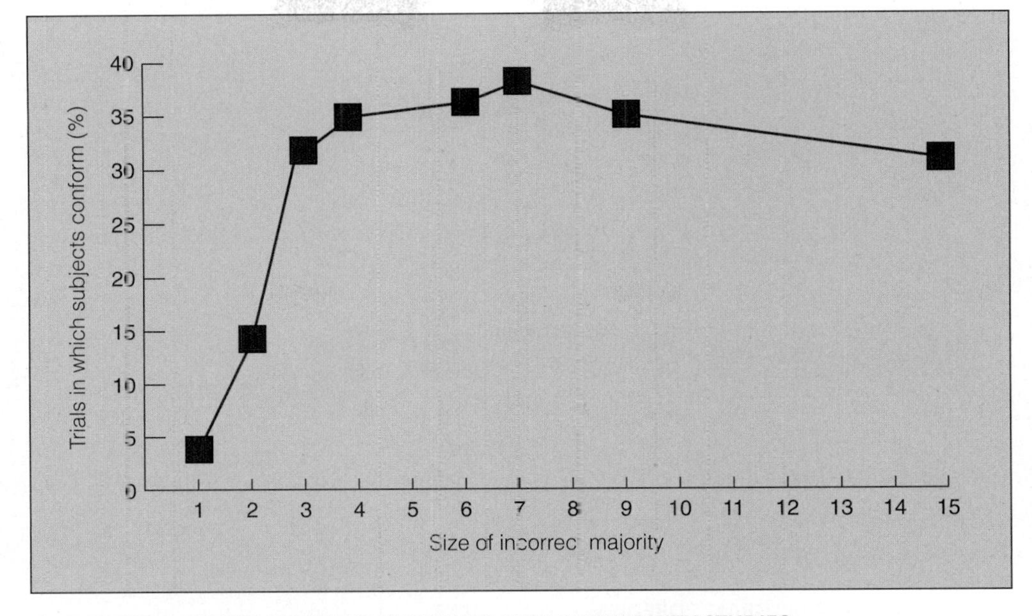

▲ **FIGURE 6.3 EFFECT OF GROUP SIZE IN THE ASCH CONFORMITY STUDIES**
As more people are added to the "incorrect" majority, subjects' tendency to conform by giving wrong answers increases—but only up to a point. Adding more than seven people to the incorrect majority does not further increase subjects' tendency to conform—perhaps because subjects are suspicious about why so many people agree with one another.
Source: Asch, 1955.

to authority. *Obedience* is a form of compliance in which people follow direct orders from someone in a position of authority.

Milgram's subjects were men who had responded to an advertisement for participants in an experiment. When the first (actual) subject arrived, he was told that the study concerned the effects of punishment on learning. After the second subject (an assistant of Milgram's) arrived, the two men were instructed to draw slips of paper from a hat to get their assignments as either the "teacher" or the "learner." Because the drawing was rigged, the actual subject always became the teacher, and the assistant the learner. Next, the learner was strapped into a chair with protruding electrodes that looked something like an electric chair. The teacher was placed in an adjoining room and given a realistic-looking but nonoperative shock generator. The "generator's" control panel showed levels that went from "Slight Shock" (15 volts) on the left, to "Intense Shock" (255 volts) in the middle, to "DANGER: SEVERE SHOCK" (375 volts), and finally "XXX" (450 volts) on the right.

The teacher was instructed to read aloud a pair of words and then repeat the first of the two words. At that time, the learner was supposed to respond with the second of the two words. If the learner could not provide the second word, the teacher was instructed to press the lever on the shock generator so that the learner would be punished for forgetting the word. Each time the learner gave an incorrect response, the teacher was supposed to increase the shock level by 15 volts. The alleged purpose of the shock was to determine if punishment improves a person's memory.

What was the maximum level of shock that a "teacher" was willing to inflict on a "learner"? The learner had been instructed (in advance) to beat on the wall between him and the teacher as the experiment continued, pretending that he was in intense pain. The teacher was told that the shocks might be "extremely painful" but that they would cause no permanent damage. At about 300 volts, when the learner quit responding at all to questions, the teacher often turned to the experimenter to see what he should do next. When the experimenter indicated that the teacher should give increasingly painful shocks, 65 percent of the teachers administered shocks all the way up to the "XXX" (450-volt) level (see ▶ Figure 6.4). By this point in the process, the teachers were frequently sweating, stuttering, or biting on their lip. According to Milgram, the teachers (who were free to leave whenever they wanted to) continued in the experiment because

conformity the process of maintaining or changing behavior to comply with the norms established by a society, subculture, or other group.

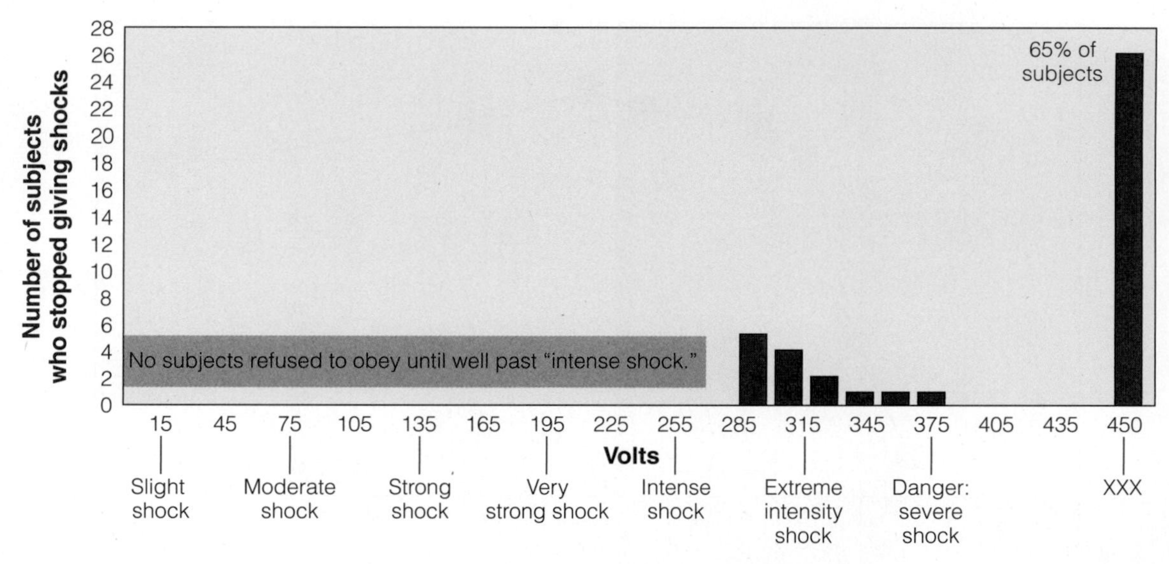

▲ **FIGURE 6.4 RESULTS OF MILGRAM'S OBEDIENCE EXPERIMENT**
Even Milgram was surprised by subjects' willingness to administer what they thought were severely painful and even dangerous shocks to a helpless "learner."
Source: Milgram, 1963.

they were being given directions by a person in a position of authority (a university scientist wearing a white coat).

What can we learn from Milgram's study? The study provides evidence that obedience to authority may be more common than most of us would like to believe. None of the "teachers" challenged the process before they had applied 300 volts. Almost two-thirds went all the way to what could have been a deadly jolt of electricity if the shock generator had been real. For many years, Milgram's findings were found to be consistent in a number of different settings and with variations in the research design (Miller, 1986).

This research once again raises some questions concerning research ethics. As was true of Asch's research, Milgram's subjects were deceived about the nature of the study in which they were asked to participate. Many of them found the experiment extremely stressful. Such conditions cannot be ignored by social scientists because subjects may receive lasting emotional scars from this kind of research. Today, it would be virtually impossible to obtain permission to replicate this experiment in a university setting.

Groupthink

As we have seen, individuals often respond differently in a group context than they might if they were alone. Social psychologist Irving Janis (1972, 1989) examined group decision making among political experts and found that major blunders in U.S. history can be attributed to pressure toward group conformity. To describe this phenomenon, he coined the term ***groupthink*—the process by which members of a cohesive group arrive at a decision that many individual members privately believe is unwise.** Why not speak up at the time? Members usually want to be "team players." They may not want to be the ones who undermine the group's consensus or who challenge the group's leaders. Consequently, members often limit or withhold their opinions and focus on consensus rather than on exploring all of the options and determining the best course of action. ▶ Figure 6.5 summarizes the dynamics and results of groupthink.

The tragic 2003 explosion of the space shuttle *Columbia* while preparing to land has been cited as an example of this process. During takeoff, a chunk of insulated foam fell off the bipod ramp of the external fuel tank, striking and damaging the shuttle's left wing. Although some NASA engineers had previously raised concerns that hardened foam popping off the fuel tank could cause damage to the ceramic tiles protecting the shuttle, and although their concerns were again raised following *Columbia*'s liftoff, these concerns were overruled by NASA officials prior to and during the flight (Glanz and Wong, 2003; Schwartz, 2003). One analyst subsequently described the way that NASA dealt with these concerns as an example of "the ways that smart people working collectively can be dumber than the sum of their brains" (Schwartz and Wald, 2003: WK3).

Process of Groupthink	Example: *Columbia* Explosion
PRIOR CONDITIONS Isolated, cohesive, homogeneous decision-making group Lack of impartial leadership High stress	NASA had previously orchestrated many successful shuttle missions and was under pressure to complete additional space missions that would fulfill agency goals and keep its budget intact.
SYMPTOMS OF GROUPTHINK Closed-mindedness Rationalization Squelching of dissent "Mindguards" Feelings of righteousness and invulnerability	Although *Columbia*'s left wing had been damaged on takeoff when a chunk of insulated foam from the external fuel tank struck it, NASA did not regard this as a serious problem because it had occurred on previous launches. Some NASA engineers stated that they did not feel free to raise questions about problems.
DEFECTIVE DECISION MAKING Incomplete examination of alternatives Failure to examine risks and contingencies Incomplete search for information	The debate among engineers regarding whether the shuttle had been damaged to the extent that the wing might burn off on reentry was not passed on to the shuttle crew or to NASA's top officials in a timely manner because the engineers either harbored doubts about their concerns or were unwilling to believe that the mission was truly imperiled.
CONSEQUENCES Poor decisions	The shuttle *Columbia* was destroyed during reentry into the Earth's atmosphere, killing all seven crew members and strewing debris across large portions of the United States.

© AP Photo/Chris O'Meara

© AP Photo/Dr. Scott Lieberman

NASA Kennedy Space Center (NASA-KSC)

▲ **FIGURE 6.5 JANIS'S DESCRIPTION OF GROUPTHINK**

In Janis's model, prior conditions such as a highly homogeneous group with committed leadership can lead to potentially disastrous "groupthink," which short-circuits careful and impartial deliberation. Events leading up to the tragic 2003 explosion of the space shuttle *Columbia* have been cited as an example of this process.

Sources: Broder, 2003; Glanz and Wong, 2003; Schwartz (with Wald), 2003; Schwartz and Broder, 2003; Schwartz and Wald, 2003.

Formal Organizations in Global Perspective

Over the past century, the number of formal organizations has increased dramatically in the United States and other industrialized nations. Previously, everyday life was centered in small, informal, primary groups, such as the family and the village. With the advent of industrialization and urbanization (as discussed in Chapter 1), people's lives became increasingly dominated by large, formal, secondary organizations. A *formal organization,* you will recall, is a highly structured secondary group formed for the purpose of achieving specific goals in the most efficient manner. Formal organizations (such as corporations, schools, and government agencies)

groupthink the process by which members of a cohesive group arrive at a decision that many individual members privately believe is unwise.

usually keep their basic structure for many years in order to meet their specific goals.

Types of Formal Organizations

We join some organizations voluntarily and others out of necessity. Sociologist Amitai Etzioni (1975) classified formal organizations into three categories—normative, coercive, and utilitarian—based on the nature of membership in each.

Normative Organizations We voluntarily join *normative organizations* when we want to pursue some common interest or gain personal satisfaction or prestige from being a member. Political parties, ecological activist groups, religious organizations, parent–teacher associations, and college sororities and fraternities are examples of normative, or voluntary, associations.

Class, gender, and race are important determinants of a person's participation in a normative association. Class (socioeconomic status based on a person's education, occupation, and income) is the most significant predictor of whether a person will participate in mainstream normative organizations; membership costs may exclude some from joining. Those with higher socioeconomic status are more likely to be not only members but also active participants in these groups. Gender is also an important determinant. Half of the voluntary associations in the United States have all-female memberships; one-fifth are all male. However, all-male organizations usually have higher levels of prestige than do all-female ones (Odendahl, 1990).

Throughout history, people of all racial–ethnic categories have participated in voluntary organizations, but the involvement of women in these groups has largely gone unrecognized. For example, African American women were actively involved in antislavery societies in the nineteenth century and in the civil rights movement in the twentieth century (see Scott, 1990). Other normative organizations focusing on civil rights, self-help, and philanthropic activities in which African American women and men have been involved include the National Association for the Advancement of Colored People (NAACP) and the Urban League. Similarly, Native American women have participated in the American Indian Movement, a group organized

© AP Images/Vincent Thian

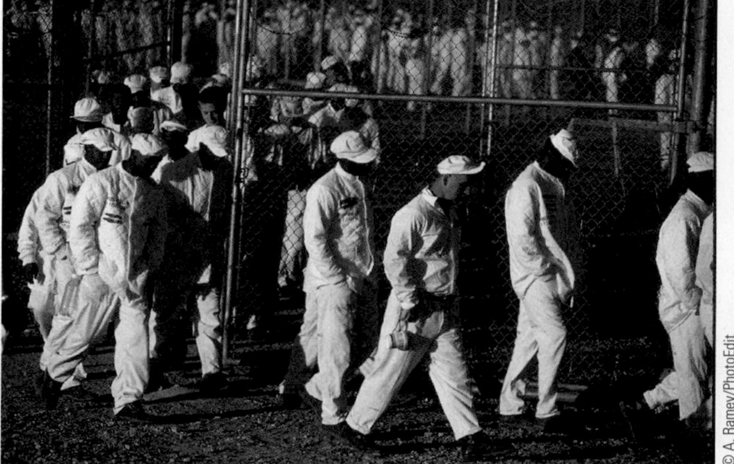

© A. Ramey/PhotoEdit

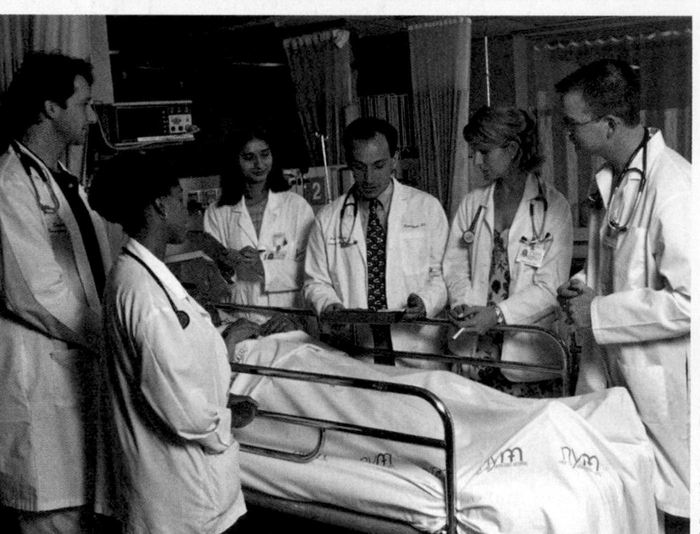

© David Grossman/The Image Works

▲ Normative organizations rely on volunteers to fulfill their goals; for example, Red Cross workers in Sri Lanka aided the relief efforts in that country following a deadly tsunami. Coercive organizations rely on involuntary recruitment; these prison inmates in Alabama are being resocialized in a total institution. Utilitarian organizations provide material rewards to participants; in teaching hospitals such as this one, medical students and patients hope that they may benefit from involvement within the organization.

to fight problems ranging from police brutality to housing and employment discrimination (Feagin and Feagin, 2003). Mexican American women (as well as men) have held a wide range of leadership positions in La Raza Unida Party and the League of United Latin American Citizens, organizations oriented toward civic activities and protests against injustices (Amott and Matthaei, 1996).

Coercive Organizations People do not voluntarily become members of *coercive organizations*—associations that people are forced to join. Total institutions, such as boot camps, prisons, and some mental hospitals, are examples of coercive organizations. As discussed in Chapter 4, the assumed goal of total institutions is to resocialize people through incarceration. These environments are characterized by restrictive barriers (such as locks, bars, and security guards) that make it impossible for people to leave freely. When people leave without being officially dismissed, their exit is referred to as an "escape."

Utilitarian Organizations We voluntarily join *utilitarian organizations* when they can provide us with a material reward that we seek. To make a living or earn a college degree, we must participate in organizations that can provide us these opportunities. Although we have some choice regarding where we work or attend school, utilitarian organizations are not always completely voluntary. For example, most people must continue to work even if the conditions of their employment are less than ideal. (This chapter's Concept Quick Review summarizes the types of groups, sizes of groups, and types of formal organizations.)

Bureaucracies

The bureaucratic model of organization remains the most universal organizational form in government, business, education, and religion. A ***bureaucracy* is an organizational model characterized by a hierarchy of authority, a clear division of labor, explicit rules and procedures, and impersonality in personnel matters.**

Sociologist Max Weber (1968/1922) was interested in the historical trend toward bureaucratization that accelerated during the Industrial Revolution. To Weber, bureaucracy was the most "rational" and efficient means of attaining organizational goals because it contributed to coordination and control. According to Weber, ***rationality* is the process by which traditional methods of social organization, characterized by informality and spontaneity, are gradually replaced by efficiently administered formal rules and procedures.** Bureaucracy can be seen in all aspects of our lives, from small colleges with perhaps a thousand students to multinational corporations employing many thousands of workers worldwide.

In his study of bureaucracies, Weber relied on an ideal-type analysis, which he adapted from the field of economics. An ***ideal type* is an abstract model that describes the recurring characteristics of some phenomenon (such as bureaucracy).** To develop this ideal type, Weber abstracted the most characteristic bureaucratic aspects of religious, educational, political, and business organizations. Weber acknowledged that no existing organization would exactly fit his ideal type of bureaucracy (Blau and Meyer, 1987).

Ideal Characteristics of Bureaucracy Weber set forth several ideal-type characteristics of bureaucratic organizations. His model (see ▶ Figure 6.6) highlights the organizational efficiency and productivity that bureaucracies strive for in these five central elements of the ideal organization:

Division of labor Bureaucratic organizations are characterized by specialization, and each member has highly specialized tasks to fulfill.

Hierarchy of authority In a bureaucracy, each lower office is under the control and supervision of a higher one. Those few individuals at the top of the hierarchy have more power and exercise more control than do the many at the lower levels. Those who are lower in the hierarchy report to (and often take orders from) those above them in the organizational pyramid. Persons at the upper levels are responsible not only for their own actions but also for those of the individuals they supervise.

Rules and regulations Rules and regulations establish authority within an organization. These rules are typically standardized and provided to members in a written format. In theory, written rules

bureaucracy an organizational model characterized by a hierarchy of authority, a clear division of labor, explicit rules and procedures, and impersonality in personnel matters.

rationality the process by which traditional methods of social organization, characterized by informality and spontaneity, are gradually replaced by efficiently administered formal rules and procedures.

ideal type an abstract model that describes the recurring characteristics of some phenomenon (such as bureaucracy).

[concept quick review]

Characteristics of Groups and Organizations

Types of Social Groups	Primary group	Small, less specialized group in which members engage in face-to-face, emotion-based interaction over an extended period of time
	Secondary group	Larger, more specialized group in which members engage in more-impersonal, goal-oriented relationships for a limited period of time
	Ingroup	A group to which a person belongs and with which the person feels a sense of identity
	Outgroup	A group to which a person does not belong and toward which the person may feel a sense of competitiveness or hostility
	Reference group	A group that strongly influences a person's behavior and social attitudes, regardless of whether the person is actually a member
Group Size	Dyad	A group composed of two members
	Triad	A group composed of three members
	Formal organization	A highly structured secondary group formed for the purpose of achieving specific goals
Types of Formal Organizations	Normative	Organizations that we join voluntarily to pursue some common interest or gain personal satisfaction or prestige by joining
	Coercive	Associations that people are forced to join (for example, total institutions such as boot camps and prisons)
	Utilitarian	Organizations that we join voluntarily when they can provide us with a material reward that we seek

and regulations offer clear-cut standards for determining satisfactory performance so that each new member does not have to reinvent the rules.

Qualification-based employment Bureaucracies require competence and hire staff members and professional employees based on specific qualifications. Individual performance is evaluated against specific standards, and promotions are based on merit as spelled out in personnel policies.

Impersonality Bureaucracies require that everyone must play by the same rules and be treated the same. Personal feelings should not interfere with organizational decisions.

Contemporary Applications of Weber's Theory How well do Weber's theory of rationality and his ideal-type characteristics of bureaucracy withstand the test of time? More than 100 years later, many organizational theorists still apply Weber's perspective. For example, the sociologist George Ritzer used Weber's theories to examine fast-food restaurants such as McDonald's. According to Ritzer, the process of "McDonaldization" has become a global phenomenon that can be seen in fast-food restaurants and other "speedy" or "jiffy" businesses (such as Sir Speedy Printing and Jiffy Lube). *McDonaldization* is the term coined by Ritzer to describe the process of rationalization, which means the substitution of logically consistent rules

Left: John Aikins/Corbis. Right: Lovesky Pavel/Shutterstock.com

◄ **FIGURE 6.6 CHARACTERISTICS AND EFFECTS OF BUREAUCRACY**

The very characteristics that define Weber's idealized bureaucracy can create or exacerbate the problems that many people associate with this type of organization. Can you apply this model to an organization with which you are familiar?

for traditional rules. The process of rationalization takes a task and breaks it down into smaller tasks. This process is repeated until all tasks have been broken down into the smallest possible level. The resulting tasks are then rationalized to find the single most efficient method for completing each task. The result is an efficient, logical sequence of methods that can be completed the same way every time to produce the desired outcome. Ritzer (2000a) identifies four main dimensions of McDonaldization:

- *Efficiency*—the optimum method of completing a task that involves a rational determination of the best mode of production. In the fast-food restaurant, the drive-through window is a good example of heightening the efficiency of obtaining a meal.
- *Predictability*—the production process organized to guarantee uniformity of product and standardized outcomes. The Big Mac in Los Angeles is indistinguishable from the one in New York; similarly, the one we consume tomorrow or next year will be just like the one we eat today.
- *Calculability*—the assessment of outcomes based on quantifiable rather than subjective criteria. In other words, rational systems emphasize quantity (usually large quantities) rather than quality. The Big Mac is a good example of this emphasis on quantity rather than quality.
- *Control*—the substitution of more predictable nonhuman labor for human labor, either through automation or the deskilling of the workforce, such as having unskilled cooks follow detailed directions and assembly-line methods for cooking and serving food.

Studying McDonaldization helps us to see the usefulness of Weber's ideas in the twenty-first century; however, as some critics have pointed out, his ideas on bureaucracy and organizational structures largely failed to take into account the informal side of bureaucracy.

The Informal Side of Bureaucracy When we look at an organizational chart, the official, formal structure of a bureaucracy is readily apparent. In practice, however, a bureaucracy has patterns of activities and interactions that cannot be accounted for by its organizational chart. These have been referred to as *bureaucracy's other face* (Page, 1946).

The ***informal side of a bureaucracy* is composed of those aspects of participants' day-to-day activities and interactions that ignore, bypass, or do not correspond with the official rules and procedures of the bureaucracy.** An example is an informal "grapevine" that spreads information (with varying degrees of accuracy) much faster than do official channels of communication, which tend to be slow and unresponsive. The informal structure has also been referred to as *work culture* because it includes the ideology and practices of workers on the job. Workers create this work culture in order to confront, resist, or adapt to the constraints of their jobs, as well as to guide and interpret social relations on the job (Zavella, 1987). Today, computer networks and e-mail offer additional opportunities for workers to enhance or degrade their work culture. Some organizations have sought to control offensive communications so that workers will not be exposed to a hostile work environment brought about by colleagues, but such control has raised significant privacy issues (see "Sociology and Social Policy").

Is the informal side of bureaucracy good or bad? Should it be controlled or encouraged? Two schools of thought have emerged with regard to

▲ Colleges and universities rely a great deal on the use of standardized tests to assess student applications. How do such tests relate to Weber's model of bureaucracy?

informal side of a bureaucracy those aspects of participants' day-to-day activities and interactions that ignore, bypass, or do not correspond with the official rules and procedures of the bureaucracy.

sociology and social policy

Computer Privacy in the Workplace

I think I'm getting paranoid. When I'm at work, I think somebody is watching me. When I send an e-mail or search the Web, I wonder who knows about it besides me. Don't get me wrong. I get all my work done first, but when I have some spare time, I may play a computer game or check out some web site I'm interested in. I know that when I'm at work, my time belongs to the company, but somehow I still feel like it's an invasion of my privacy for some computer to monitor every single thing I do. I mean, I work for a company that makes cardboard boxes, not the CIA!

—a student in one of the author's classes, expressing her irritation over computer surveillance at work

Do employers really have the right to monitor everything that their employees do on company-owned computers? Generally speaking, the answer is yes, and the practice is widespread. Surveys have found that about two-thirds of all U.S. companies monitor their employees' website visits in order to prevent inappropriate surfing. Likewise, 65 percent use software to block connections to websites that are deemed off-limits to employees (American Management Association, 2008).

Employers assert not only that they have the right to engage in such surveillance but also that it may be necessary for them to do so for their own protection. As for their right to do so, they note that they own the computer network and the terminals, pay for the Internet service, and pay the employee to spend his or her time on company business. As for it possibly being necessary for them to monitor their employees' computer use, employers argue that they may be held legally responsible for harassing or discriminatory instant messages or e-mail sent on company computers and that surveillance is the only way to protect against such liability. As a result, many employers take the position that "You leave your First Amendment [privacy] rights at the door when you work for a private employer. That's the way it has always been" (Agonafir, 2002).

In most instances, courts have upheld monitoring by employers (see, for example, *Bourke v. Nissan*, *Smyth v. Pillsbury*, and *Shoars v. Epson*). Yet there are valid arguments against computer surveillance, and invasion of a worker's privacy is certainly one of them. When an employee makes a personal phone call while at work, the employee usually has a reasonable expectation of privacy—a reasonable belief that neither fellow workers nor his or her employer is eavesdropping on that call. How about "snail mail"? An employee has a reasonable expectation of privacy that the employer will not open a personal letter addressed to the employee, read it, and reseal the envelope. Why should an e-mail exchange with friends or relatives not be equally private and protected? If the employer is going to read an employee's e-mail or track the person's Internet activities, shouldn't the employer at least have to make sure that its workers are aware of that policy?

With regard to the employer possibly being held responsible for its employees' actions, Chief Judge Edith H. Jones (qtd. in Gordon, 2001) of the U.S. Fifth Circuit Court has observed that "It seems highly disproportionate to inflict a monitoring program that may invade thousands of people's privacy for the sake of exposing a handful of miscreants." The need to prevent a crime or to protect a company against potential liability must be balanced against each individual's privacy rights.

Ultimately, this balancing must be done by the legislature or the courts. There are no easy answers to this pressing social policy issue, but it should remain a concern for all who live in a democratic society.

reflect & analyze

Are you concerned about computer and cell phone privacy in your own life? Should businesses and colleges have the right to monitor our digital communications? Why or why not?

these questions. One approach emphasizes control of informal groups in order to ensure greater worker productivity. By contrast, the other school of thought asserts that informal groups should be nurtured because such networks may serve as a means of communication and cohesion among individuals. Large organizations would be unable to function without strong informal norms and relations among participants (Blau and Meyer, 1987).

Informal networks thrive in contemporary organizations because e-mail and websites have made it possible for people to communicate throughout the day without ever having to engage in face-to-face interaction. The need to meet at the water fountain or the copy machine in order to exchange information is long gone: Workers now have an opportunity to tell one another—and higher-ups, as well—what they think.

Problems of Bureaucracies

The characteristics that make up Weber's "rational" model of bureaucracy have a dark side that has frequently given this type of organization a bad name. Three of the major problems of bureaucracies are (1) inefficiency and rigidity, (2) resistance to change, and (3) perpetuation of race, class, and gender inequalities.

Inefficiency and Rigidity Bureaucracies experience inefficiency and rigidity at both the upper

© Sean Justice/The Image Bank/Getty Images

▲ How do people use the informal "grapevine" to spread information? Is this faster than the organization's official channels of communication? Is it more or less accurate than official channels?

and lower levels of the organization. The self-protective behavior of officials at the top may render the organization inefficient. One type of self-protective behavior is the monopolization of information in order to maintain control over subordinates and outsiders. Information is a valuable commodity in organizations, and those persons in positions of authority guard information because it is a source of power for them—others cannot "second-guess" their decisions without access to relevant (and often "confidential") information (Blau and Meyer, 1987).

When those at the top tend to use their power and authority to monopolize information, they also fail to communicate with workers at the lower levels. As a result, they are often unaware of potential problems facing the organization and of high levels of worker frustration. Bureaucratic regulations are written in far greater detail than is necessary in order to ensure that almost all conceivable situations are covered. ***Goal displacement* occurs when the rules become an end in themselves rather than a means to an end, and organizational survival becomes more important than achievement of goals** (Merton, 1968).

Inefficiency and rigidity occur at the lower levels of the organization as well. Workers often engage in *ritualism*; that is, they become most concerned with "going through the motions" and "following the rules." According to Robert Merton (1968), the term ***bureaucratic personality* describes those workers who are more concerned with following correct procedures than they are with getting the job done correctly.** Such workers are usually able to handle routine situations effectively but are frequently incapable of handling a unique problem or an emergency. Thorstein Veblen (1967/1899) used the term *trained incapacity* to characterize situations in which workers have become so highly specialized, or have been given such fragmented jobs to do, that they are unable to come up with creative solutions to problems. Workers who have reached this point also tend to experience bureaucratic alienation—they really do not care what is happening around them.

Resistance to Change Once bureaucratic organizations are created, they tend to resist change. This resistance not only makes bureaucracies virtually impossible to eliminate but also contributes to bureaucratic enlargement. Because of the assumed relationship between size and importance, officials tend to press for larger budgets and more staff and office space. To justify growth, administrators and managers must come up with more tasks for workers to perform.

Resistance to change may also lead to incompetence. Based on organizational policy, bureaucracies tend to promote people from within the organization. As a consequence, a person who performs satisfactorily in one position is promoted to a higher level in the organization. Eventually, people reach a level that is beyond their knowledge, experience, and capabilities.

Perpetuation of Race, Class, and Gender Inequalities Some bureaucracies perpetuate inequalities of race, class, and gender because this form of organizational structure creates a specific type of work or learning environment. This structure was typically created for middle-class and upper-middle-class white men, who for many years were the predominant organizational participants.

For people of color, *entry* into dominant white bureaucratic organizations does not equal actual *integration* (Feagin, 1991). Instead, many have experienced an internal conflict between the bureaucratic

goal displacement a process that occurs in organizations when the rules become an end in themselves rather than a means to an end, and organizational survival becomes more important than achievement of goals.

bureaucratic personality a psychological construct that describes those workers who are more concerned with following correct procedures than they are with getting the job done correctly.

ideals of equal opportunity and fairness and the prevailing norms of discrimination and hostility that exist in many organizations. Research has found that people of color are more adversely affected than dominant-group members by hierarchical bureaucratic structures and exclusion from informal networks.

Like racial inequality, social class divisions may be perpetuated in bureaucracies (Blau and Meyer, 1987). The theory of a "dual labor market" has been developed to explain how social class distinctions are perpetuated through different types of employment. Middle-class and upper-middle-class employees are more likely to have careers characterized by higher wages, more job security, and opportunities for advancement. By contrast, poor and working-class employees work in occupations characterized by low wages, lack of job security, and few opportunities for promotion. The "dual economy" not only reflects but may also perpetuate people's current class position.

Gender inequalities are also perpetuated in bureaucracies. Women in traditionally male organizations may feel more visible and experience greater performance pressure. They may also find it harder to gain credibility in management positions.

© Blend Images/Alamy

© Mark Richards/PhotoEdit

▲ Corporate employees in different work settings vary widely in their manner and appearance. How does the environment in which we work affect how we dress and act?

© Jack Kurtz/The Image Works

▲ According to conflict theorists, members of the capitalist class benefit from the work of laborers such as the people shown here, who are harvesting onions on a farm in the Texas Rio Grande Valley. How do low wages and lack of job security contribute to class-based inequalities in the United States?

Inequality in organizations has many consequences. People who lack opportunities for integration and advancement tend to be pessimistic and to have lower self-esteem. Believing that they have few opportunities, they resign themselves to staying put and surviving at that level. By contrast, those who enjoy full access to organizational opportunities tend to have high aspirations and high self-esteem. They feel loyalty to the organization and typically see their job as a means for mobility and growth.

Bureaucracy and Oligarchy

Why do a small number of leaders at the top make all the important organizational decisions? According to the German political sociologist Robert Michels (1949/1911), all organizations encounter the ***iron law of oligarchy*—the tendency to become a bureaucracy ruled by the few.** His central idea was that those who control bureaucracies not only wield power but also have an interest in retaining their power. Michels found that the hierarchical structures of bureaucracies and oligarchies go hand in hand. On the one hand, power may be concentrated in the hands of a few people because rank-and-file members must inevitably delegate a certain amount of decision-making authority to their leaders. Leaders have access to information that other members do not have, and they have "clout," which they may use to protect their own interests. On the other hand, oligarchy may result when individuals have certain outstanding qualities that make it possible for them to manage, if not control, others. The members choose to look to their leaders for direction; the leaders are strongly motivated to maintain the power and privileges that go with their leadership positions.

◀ The Japanese model of organization—including planned group-exercise sessions for employees—has become a part of the workplace in many nations. Would it be a positive change if more workplace settings, such as the one shown here, were viewed as an extension of the family? Why or why not?

Are there limits to the iron law of oligarchy? The leaders in most organizations do not have unlimited power. Divergent groups within a large-scale organization often compete for power, and informal networks can be used to "go behind the backs" of leaders. In addition, members routinely challenge, and sometimes they (or the organization's governing board) remove leaders when they are not pleased with their actions.

Alternative Forms of Organization

Many organizations have sought new and innovative ways to organize work more efficiently than the traditional hierarchical model. In the early 1980s there was a movement in the United States to *humanize bureaucracy*—to establish an organizational environment that develops rather than impedes human resources. More-humane bureaucracies are characterized by (1) less-rigid hierarchical structures and greater sharing of power and responsibility by all participants, (2) encouragement of participants to share their ideas and try new approaches to problem solving, and (3) efforts to reduce the number of people in dead-end jobs, train people in needed skills and competencies, and help people meet outside family responsibilities while still receiving equal treatment inside the organization (Kanter, 1983, 1985, 1993/1977). However, this movement has been overshadowed by globalization and the perceived strengths of systems of organizing work in other nations, such as Japan.

Organizational Structure in Japan

For several decades the Japanese model of organization was widely praised for its innovative structure because it focused on long-term employment and company loyalty. Until recently, many Japanese employees remained with the same company for their entire career, whereas their U.S. counterparts often changed employers every few years. Although the practice of lifetime employment has been replaced by the concept of long-term employment, workers in Japan often have higher levels of job security than do workers in the United States. According to advocates of the Japanese system, this model encourages worker loyalty and a high level of productivity. Managers move through various parts of the organization and acquire technical knowledge about the workings of many aspects of the corporation, unlike their U.S. counterparts, who tend to become highly specialized (Sengoku, 1985). Unlike top managers in the United States who have given themselves pay raises and bonuses even when their companies were financially strapped, many Japanese managers have taken pay cuts under similar circumstances. Japanese management is characterized as being people oriented, taking a long-term view, and having a culture that focuses on *how* work gets done rather than on the result alone.

Quality Circles How work is organized, such as by the use of quality circles, may also affect job satisfaction and worker productivity. Small work groups made up of about five to fifteen workers who meet regularly with one or two managers to discuss the group's performance and working conditions are known as *quality circles*. The purpose of this team approach to management is both to improve product quality and to lower product costs. Workers are motivated to save the corporation money because they,

iron law of oligarchy according to Robert Michels, the tendency of bureaucracies to be ruled by a few people.

in turn, receive bonuses or higher wages for their efforts. Quality circles have been praised for creating worker satisfaction, helping employees develop their potential, and improving productivity (Ishikawa, 1984). Because quality circles focus on both productivity and worker satisfaction, these circles (at least ideally) meet the needs of both the corporation and the workers.

Organizations in the Future

What is the best organizational structure for the future? Of course, this question is difficult to answer because it requires the ability to predict economic, political, and social conditions. Nevertheless, we can make several observations.

First, organizations have been affected by growing social inequality in the United States and other nations because of heightening differences between high- and low-income segments of populations. Having *socially sustainable organizations* is of increasing importance because television, the Internet, and international travel have made people more aware of the wide disparities in the resources and power of "haves" and "have-nots" both within a single country and across nations. As a result, organizations must be developed that are both economically efficient and as equitable as possible. Some organizational and management analysts suggest that more attention must be paid to the "stakeholders" of an organization (MIT 21st Century Manifesto Working Group, 1999). For example, at a college or university, stakeholders would include (but not be limited to) students, faculty and staff, administrators, alumni, major contributors, boards of regents, suppliers, the community where the school is located, and the society as a whole. Goals of the organization should be based on taking into account the interests of these various stakeholders and working toward organizational goals and outcomes that will not only ensure organizational success but also provide the greatest good for the greatest number of stakeholders. Although academic success, winning sports teams, and college financial stability are important in higher education, other criteria should also be used in assessing the effectiveness and overall output of the college community. In other types of organizations, similar stakeholders can be identified and goals established to meet the needs of various constituencies.

Second, *globalization* is the key word for management and change in many organizations, and the use of technology is intricately linked with performing flexible, mobile work anywhere in the world. Based on the assumption that organizations must respond to a rapidly changing environment or they will not thrive, several twenty-first-century organizational models are based on the need to relegate traditional organizational structure to dinosaur status and to move ahead with structures that fully use technology and focus on the need to communicate more effectively. As the pace of communication has increased dramatically and information overload has become prevalent, the leaders of organizations are seeking new ways in which to more efficiently manage their organizations and to be ahead of change, rather than merely adapting to change after it occurs. One recent approach is referred to as "smart working," which is based on the assumption that innovation is crucial and organizational leaders must be able to use the talents and energies of the people who work with them. At one level, "smart working" refers to "anytime, anywhere" ways of work that have become prevalent because of communications technologies such as smartphones and computers. However, another level focuses on the ways in which smart working makes it possible for people to have flexibility and autonomy in where, when, and how they work (chiefexecutive.com, 2010). According to one management specialist,

> It turns out that the sort of collaborative, challenging work with potential for learning and personal development that people find satisfying is exactly the sort of work needed to adapt to current turbulent global operating conditions. Smart working is an outcome of designing organizational systems that are good for business and good for people. (chiefexecutive.com, 2010)

From this perspective, organizations must adapt to change; empower all organizational participants to become involved in collaboration, problem solving, and innovation; and create a work environment that people find engaging and that inspires them to give their best to that organization. Exactly how these organizations might look is not fully clear, although some analysts suggest that corporations such as Google, Microsoft, and other high-tech companies have actively sought to redefine organizational culture and environment by being responsive to employees, customers, and other stakeholders. Although management continues to exist, the distinction between managers and the managed becomes less prevalent, and the idea that management knowledge will be everyone's responsibility becomes more predominant. Emphasis is also placed on the importance of improving communication and on acquiring the latest technologies to make this process even more fast, secure, and efficient. Overall, there is a focus on change and the assumption that people in an organization should be change agents, not individuals who merely respond to change after it occurs.

Ultimately, everyone has a stake in seeing that organizations operate in an effective, humane manner and that opportunities are widely available to all people

you can make a difference

Social Networking, Volunteering, and Becoming a Part of Something Larger Than Ourselves!

"We are the service generation!!!" @BEXwithanX tweeted. And @sjtetreault picked this quote from the first lady to share: "'You didn't think I'd show up here without another challenge, did you? Be yourself, just take it global.' Michelle Obama."

—reactions from two students who (along with about 25,000 other people) listened to First Lady Michelle Obama's commencement address at George Washington University (Johnson, 2010)

Although it is not unusual for political leaders and their spouses to be keynote speakers at university commencements, Michelle Obama's address at George Washington University (GWU) was unique in that it was her payoff in a bet in which she challenged students to do 100,000 hours of community service in exchange for a graduation speech. GWU students easily met the deadline because of social networking and students, such as Christine French, who were highly motivated not only to reach but also to surpass the goal. According to VolunteerMatch.org (2010), a website that links volunteers with community service opportunities, "You can't really major in volunteering, but if you could your schedule might look a lot like Christine French's." In addition to leading the charge to complete the 100,000 service hours by graduation day, Christine was president of the Human Service Student Organization and the Teach for America chapter at GWU.

Like many college students who are actively involved in community service, French traces the beginnings of her volunteer endeavors to her childhood: "I remember my mom driving me to the retirement home when I was in the eighth grade to play Uno and Bingo every week" (qtd. in VolunteerMatch.org, 2010).

What unique factors contribute to the success of college volunteers as they (and you!) make a difference in people's lives? Christine French believes that this is the secret: "I think it's that I listen to people. Often, all people really need is someone to listen to them and validate their feelings. We all just want human connection and to know that we are loved and valuable. This is what I can do for others, and it's more important than the fact that I am a hard worker or a critical thinker."

Given this model for making a difference, how might we connect with individuals and organizations that are in need of our assistance? Online social networks connect people together: people with similar interests, people who may come to know one another. Volunteer organizations use online networks as one way to find a new generation of supporters and activists.

Can social networking and virtual communities successfully inspire us to get active in the real world? It seems that the answer is a resounding "Yes!" Worldwide, a new generation of volunteers is being recruited through the power of the media and social networking, and the success of these endeavors is being reported on Facebook, MySpace, Twitter, YouTube, and similar networks in the United States. Why not explore Facebook, Twitter, your favorite social networking site, and your school's volunteer information system to learn more about available opportunities where you might share your time and resources with other people in your community and around the world?

regardless of race, gender, class, or age. Workers and students alike can benefit from organizational environments that make it possible for people to explore their joint interests without fear of losing their privacy or being pitted against one another in a competitive struggle for advantage. (For an example of students working together on a meaningful activity that benefits others, see "You Can Make a Difference.")

chapter review Q & A

Use these questions and answers to check how well you've achieved the learning objectives set out at the beginning of this chapter.

- **How do sociologists distinguish among social groups, aggregates, and categories?**

Sociologists define a social group as a collection of two or more people who interact frequently, share a sense of belonging, and depend on one another. People who happen to be in the same place at the same time are considered an aggregate. Those who share a similar characteristic are considered a category. Neither aggregates nor categories are considered social groups.

- **How do sociologists classify groups?**

Sociologists distinguish between primary groups and secondary groups. Primary groups are small and personal, and members engage in emotion-based interactions over an extended period. Secondary groups are larger and more specialized, and members have less personal and more formal, goal-oriented relationships. Sociologists also divide groups into ingroups, outgroups, and reference groups. Ingroups are groups to which we belong and with which we identify. Outgroups are groups we do not belong to or perhaps feel hostile toward. Reference groups are groups that strongly influence people's behavior whether or not they are actually members.

- **What is the significance of group size?**

In small groups, all members know one another and interact simultaneously. In groups with more than three members, communication dynamics change, and members tend to assume specialized tasks.

- **What are the major styles of leadership?**

Leadership may be authoritarian, democratic, or laissez-faire. Authoritarian leaders make major decisions and assign tasks to individual members. Democratic leaders encourage discussion and collaborative decision making. Laissez-faire leaders are minimally involved and encourage members to make their own decisions.

- **What do experiments on conformity show us about the importance of groups?**

Groups may have significant influence on members' values, attitudes, and behaviors. In order to maintain ties with a group, many members are willing to conform to norms established and reinforced by group members.

- **What are the strengths and weaknesses of bureaucracies?**

A bureaucracy is a formal organization characterized by hierarchical authority, division of labor, explicit procedures, and impersonality. According to Max Weber, bureaucracy supplies a rational means of attaining organizational goals because it contributes to coordination and control. A bureaucracy also has an informal structure, which includes the daily activities and interactions that bypass the official rules and procedures. The informal structure may enhance productivity or may be counterproductive to the organization. A bureaucracy may be inefficient, resistant to change, and a vehicle for perpetuating class, race, and gender inequalities.

- **What types of organizational structures have been used in recent decades to modify or change typical bureaucracies?**

Some organizations have adopted Japanese management techniques based on long-term employment and company loyalty or the use of quality circles as alternative forms of bureaucratic structures. More recently the focus has been on "smart working," which is based on the needs of the global organizational environment and relies heavily on communication technologies.

key terms

aggregate 154
authoritarian leaders 161
bureaucracy 167
bureaucratic personality 171
category 154
conformity 162
democratic leaders 161
dyad 159
expressive leadership 161
goal displacement 171
groupthink 164
ideal type 167
informal side of a bureaucracy 169
ingroup 155
instrumental leadership 161
iron law of oligarchy 172
laissez-faire leaders 161
network 157
outgroup 155
rationality 167
reference group 155
small group 159
triad 159

questions for critical thinking

1. Who might be more likely to conform in a bureaucracy, those with power or those wanting more power?
2. Do the insights gained from Milgram's research on obedience outweigh the elements of deception and stress that were forced on its subjects?
3. How would you organize a large-scale organization or company for the twenty-first century?

turning to video

Watch the BBC video *Internet Freedom* (running time 1:38), available through **CengageBrain.com.** This video discusses the role of social networking sites such as Facebook and Twitter in the revolutions in the Middle East. As you watch the video, think about how you use the Internet and particularly social networking sites. After you've watched the video, consider these questions: How have you used social networking sites to connect with your peers? What advantages and disadvantages come with it?

online study resources

Go to CENGAGE brain.com to access online study resources, including the Sociology CourseMate for this text as well as special features such as video, an interactive sociology time line and interactive maps, General Social Survey (GSS) data, and U.S. Census 2010 data.

CourseMate brings course concepts to life with interactive learning, study, and exam-preparation tools that support the printed textbook. A textbook-specific website, **Sociology CourseMate** includes an integrated interactive eBook and other interactive learning tools, including quizzes, flash cards, and videos.

Visit **www.cengagebrain.com** to access your account and purchase materials.

7 Deviance and Crime

"What worries me," said Marlene [a Southside Chicago block-club president], "is that there's about seventy children on my block who use that park—and that's not counting the ones who live on the other side. Can't have them around your boys [gang members]."

"You all are something else," Big Cat [a gang leader for the "Black Kings"] said, shaking his head. "I been cooperating with you all for years now, never complaining that I'm losing money.... I don't get no respect for that?"

"If you're in our park, we can't be. It's as simple as that," Marlene replied. "I'll give you the nighttime. Maybe I can convince folks that you all need to work at night, but that's going to be tough. But, bottom line, baby, is we can't have you all there during the day...."

"Okay," interjected [local pastor] Wilkins. "Now you have to stop for the summer, Big Cat. We're not asking for a two-year thing, or nothing like that. Just when the kids are outside."

"I guess I could work it on 59th, but that [business owner] keeps telling us he doesn't want us around, keeps calling the cops...."

"If I get him to leave you alone during the day, and you can hang out in the parking lot on the other side of the store, you'll leave the park for the summer."

"Yeah," Big Cat replied, dejected at the compromise. "Okay, we'll be gone."

—sociologist Sudhir Alladi Venkatesh (2006: 294–295) describing a conversation among three people he interviewed during his research into gangs and Chicago's underground economy

© A. Ramey/PhotoEdit

▲ **Members of the California group known as the Culver City Boyz typify how gang members use items of clothing and gang signs made with their hands to assert their membership in the group and solidarity with one another. Some people might view this conduct as deviant behavior, whereas many gang members view it as an act of conformity.**

Chapter Focus Question:

What do studies of adolescent peer cliques and youth gangs tell us about deviance?

Sociologists and criminologists typically define a *gang* as an ongoing group of people, often young, who band together for purposes generally considered to be deviant or criminal by the larger society. For more than a century, gang behavior has been of special interest to sociologists (see Puffer, 1912), who generally agree that youth gangs can be found in many settings and among all racial and ethnic categories. The U.S. Department of Justice estimates that in 2008 (the latest year for which figures are available) there were 27,900 gangs with about 774,000 members of all ages in the United States. Nearly one-third of all cities, suburban areas, towns, and rural areas experience gang problems each year; however, most gang-related problems occur in cities with populations of more than 250,000 (Egley, Howell, and Moore, 2010).

Today, some gang members are committing sophisticated crimes that are different from the drive-by shootings and narcotics distribution of other gangs. Some twenty-first-century gang members use the Internet to carry out crimes such as identity theft, bank fraud, check kiting, criminal impersonation, and other technology-related crimes that were largely unheard of in the past (Ferrell, 2008).

As unusual as it may sound, important similarities exist between youth gangs and adolescent peer cliques, which are typically viewed as conforming to most social norms. At the most basic level, *cliques* are friendship circles, whose members identify one another as mutually connected (Adler and Adler, 1998). Members of contemporary cliques not only meet in person but often communicate with each other by cell phone, Facebook, Twitter, and other social media. Unlike informal friendship networks, cliques have a distinctive organizational structure. According to the sociologists Patricia A. Adler and Peter Adler (1998: 56), cliques "have a hierarchical structure, being dominated by leaders, and are exclusive in nature, so that not all individuals who desire membership are accepted." Moreover, sociologists have found that cliques function as "bodies of power" in schools by "incorporating the most popular individuals, offering the most exciting social lives, and commanding the most interest and attention from classmates" (Adler and Adler, 1998: 56).

Although cliques may have some similarities with gangs, there are also significant differences: Gangs play a large role in the economy of many low-income urban neighborhoods, where residents often believe that they must do whatever is necessary to survive. Some activities in the underground economy include the performance of unregulated, unreported, and untaxed work, whereas others involve more widely recognized criminal activities such as the sale of drugs by gang members. According to Venkatesh (2006), one remarkable thing about studying deviance and crime in settings such as "Maquis Park" (a pseudonym for a real Southside Chicago neighborhood) was learning that residents and gang members sometimes forge temporary alliances and engage in self-initiated policing so that neighborhood children may play safely at the park or enjoy other everyday activities without fear of harm. Venkatesh's study reveals people's efforts to survive with the resources they amass in the underground economy, as well as residents' willingness to negotiate with gang members if it will help restore a sense of order to their neighborhood. This unique form of community policing often takes place without the assistance of law enforcement officials.

Highlighted Learning Objectives

- Define deviant behavior and explain when deviance is considered a crime.
- Identify the major theoretical perspectives on deviance.
- Describe how crimes are classified.
- Explain how the criminal justice system deals with crime.

In this chapter we look at the relationship among conformity, deviance, and crime; even in times of economic hardships and other national crises, "everyday" deviance and crime occur as usual. People do not stop activities that might be viewed by others—or by law enforcement officials—as violating social norms. An example is gang behavior, which is used in this chapter as an example of deviant behavior. For individuals who find a source of identity, self-worth, and a feeling of protection by virtue of gang membership, no radical change occurs in daily life even as events around them may change. Youth gangs have been present in the United States for many years because they meet the perceived needs of members. Some gangs may be thought of as being very similar to youth cliques, whereas other gangs engage in activities that constitute crime. Before reading on, take the Sociology and Everyday Life quiz on peer cliques, youth gangs, and deviance.

What Is Deviance?

***Deviance* is any behavior, belief, or condition that violates significant social norms in the society or group in which it occurs.** We are most familiar with *behavioral* deviance, based on a person's intentional or inadvertent actions. For example, a person may engage in intentional deviance by drinking too much or robbing a bank, or in inadvertent deviance by losing money in a Las Vegas casino or laughing at a funeral.

© Everett Collection

▲ Although most people think of a high school clique as being far different from a gang, patterns of inclusion and exclusion operate similarly in both groups. In this still from the movie *Mean Girls*, note the three young women on the right-hand side. In what ways are they excluding the young woman (played by Lindsey Lohan) on the left?

Although we usually think of deviance as a type of behavior, people may be regarded as deviant if they express a radical or unusual *belief system*. Members of cults (such as Moonies and satanists) or of far-right-wing or far-left-wing political groups may be considered deviant when their religious or political beliefs become known to people with more-conventional cultural beliefs. However, individuals who are considered to be "deviant" by one category of people may be seen as conformists by another group. For example, adolescents in some peer cliques and youth gangs may shun mainstream cultural beliefs and values but routinely conform to subcultural codes of dress, attitude (such as defiant individualism), and behavior (Jankowski, 1991). Those who think of themselves as "Goths" may wear black trench coats, paint their fingernails black, and listen to countercultural music.

In addition to their behavior and beliefs, individuals may also be regarded as deviant because they possess a specific *condition* or *characteristic*. A wide range of conditions have been identified as "deviant," including being obese (Degher and Hughes, 1991; Goode, 1996) and having AIDS (Weitz, 2004). For example, research by the sociologist Rose Weitz (2004) has shown that persons with AIDS live with a stigma that affects their relationships with other people, including family members, friends, lovers, colleagues, and health care workers. Chapter 5 defines a *stigma* as any physical or social attribute or sign that so devalues a person's social identity that it disqualifies the person from full social acceptance (Goffman, 1963b). Based on this definition, the stigmatized person has a "spoiled identity" as a result of being negatively evaluated by others (Goffman, 1963b). To avoid or reduce stigma, many people seek to conceal the characteristic or condition that might lead to stigmatization.

Who Defines Deviance?

Are some behaviors, beliefs, and conditions inherently deviant? In commonsense thinking, deviance is often viewed as inherent in certain kinds of behavior or people. For sociologists, however, deviance is a formal property of social situations and social structure. As the sociologist Kai T. Erikson (1964: 11) explains,

> Deviance is not a property inherent in certain forms of behavior; it is a property conferred upon these forms by the audiences which directly or indirectly witness them. The critical variable in the study of deviance, then, is the social audience

sociology and everyday life

How Much Do You Know About Peer Cliques, Youth Gangs, and Deviance?

True	False	
T	F	1. According to some sociologists, deviance may serve a useful purpose in society.
T	F	2. Peer cliques on high school campuses have few similarities to youth gangs.
T	F	3. Most people join gangs to escape from broken homes caused by divorce or the death of a parent.
T	F	4. Juvenile gangs are an urban problem; few rural areas have problems with gangs.
T	F	5. Street crime has a much higher economic cost to society than crimes committed in executive suites or by government officials.
T	F	6. Rising crime rates are accurately reflected by the extensive crime coverage in the media and the growing number of crime dramas (such as *CSI* and *Law & Order* and their spin-offs) on television.
T	F	7. Studies have shown that peer cliques have become increasingly important to adolescents over the past two decades.
T	F	8. Gangs are an international problem.

Answers on page 182.

rather than the individual actor, since it is the audience which eventually determines whether or not any episode of behavior or any class of episodes is labeled deviant.

Based on this statement, we can conclude that deviance is *relative*—that is, an act becomes deviant when it is socially defined as such. Definitions of deviance vary widely from place to place, from time to time, and from group to group (see "Sociology Works!"). Today, for example, some women wear blue jeans and very short hair to college classes; some men wear an earring and long hair. In the past, such looks violated established dress codes in many schools, and administrators probably would have asked these students to change their appearance or leave school.

Deviant behavior also varies in its *degree of seriousness*, ranging from mild transgressions of folkways, to more serious infringements of mores, to quite serious violations of the law. Have you kept a library book past its due date or cut classes? If so, you have violated folkways. Others probably view your infraction as relatively minor; at most, you might have to pay a fine or receive a lower grade. Violations of mores—such as falsifying a college application or cheating on an examination—are viewed as more serious infractions and are punishable by stronger sanctions, such as academic probation or expulsion. Some forms of deviant behavior violate the criminal law, which defines the behaviors that society labels as criminal. A ***crime*** **is a behavior that violates criminal law and is punishable with fines, jail terms, and/or other negative sanctions.** Crimes range from minor offenses (such as traffic violations) to major offenses (such as murder). A subcategory, ***juvenile delinquency,*** **refers to a violation of law or the commission of a status offense by young people.** Note that the legal concept of juvenile delinquency includes not only crimes but also status offenses, which are illegal only when committed by younger people (such as cutting school or running away from home).

What Is Social Control?

Societies not only have norms and laws that govern acceptable behavior; they also have various mechanisms to control people's behavior. ***Social control*** **refers to the systematic practices that social groups develop in order to encourage conformity**

deviance any behavior, belief, or condition that violates significant social norms in the society or group in which it occurs.

crime a behavior that violates criminal law and is punishable with fines, jail terms, and/or other negative sanctions.

juvenile delinquency a violation of law or the commission of a status offense by young people.

social control systematic practices developed by social groups to encourage conformity to norms, rules, and laws and to discourage deviance.

sociology and everyday life

ANSWERS to the Sociology Quiz on Peer Cliques, Youth Gangs, and Deviance

1. **True.** From Durkheim to contemporary functionalists, theorists have regarded some degree of deviance as functional for societies.
2. **False.** Many social scientists believe that there are striking similarities between adolescent cliques and youth gangs, including the demands that are placed on members in each category to conform to group norms pertaining to behavior, appearance, and the people with whom one is allowed to associate.
3. **False.** Recent studies have found that people join gangs for a variety of reasons, including the desire to gain access to money, recreation, and protection.
4. **False.** Gangs are frequently thought of as an urban problem because central-city gangs organized around drug dealing have become prominent in recent years; however, gangs are found in suburban counties, smaller cities, Native American reservations, and rural counties throughout the country as well.
5. **False.** Although street crime—such as assault and robbery—often has a greater psychological cost, crimes committed by persons in top positions in business (such as accounting and tax fraud) or government (including the Pentagon) have a far greater economic cost, especially for U.S. taxpayers.
6. **False.** Despite extensive news media coverage and popular culture representations of a rapidly growing crime rate in the United states, crime rates overall are declining.
7. **True.** As more youths grow up in single-parent households or in households where both parents are employed, many adolescents have turned to members of their peer cliques to satisfy their emotional needs and to gain information.
8. **True.** Gangs are found in nations around the world. In countries such as Japan, youth gangs are often points of entry into adult crime organizations.

Sources: Based on Adler and Adler, 2003; and FBI, 2010.

to norms, rules, and laws and to discourage deviance. Social control mechanisms may be either internal or external. Internal social control takes place through the socialization process: Individuals *internalize* societal norms and values that prescribe how people should behave and then follow those norms and values in their everyday lives. By contrast, external social control involves the use of negative sanctions that proscribe certain behaviors and set forth the punishments for rule breakers and nonconformists. In contemporary societies the criminal justice system, which includes the police, the courts, and the prisons, is the primary mechanism of external social control.

If most actions deemed deviant do little or no direct harm to society or its members, why is social control so important to groups and societies? Why is the same belief or action punished in one group or society and not in another? These questions pose interesting theoretical concerns and research topics for sociologists and criminologists who examine issues pertaining to law, social control, and the criminal justice system. ***Criminology* is the systematic study of crime and the criminal justice system, including the police, courts, and prisons.**

The primary interest of sociologists and criminologists is not questions of how crime and criminals can best be controlled but rather social control as a social product. Sociologists do not judge certain kinds of behavior or people as being "good" or "bad." Instead, they attempt to determine what types of behavior are defined as deviant, who does the defining, how and why people become deviants, and how society deals with deviants. Although sociologists have developed a number of theories to explain deviance and crime, no one perspective is a comprehensive explanation of all deviance. Each theory provides a different lens through which we can examine aspects of deviant behavior.

Functionalist Perspectives on Deviance

As we have seen in previous chapters, functionalists focus on societal stability and the ways in which various parts of society contribute to the whole. According to functionalists, a certain amount of deviance contributes to the smooth functioning of society.

sociology works!

Social Definitions of Deviance: Have You Seen Bigfoot or a UFO Lately?

It was December 2006, and [sociologist] Carson Mencken sat shivering and perched in a tree in the Sam Houston National Forest. It was 1 A.M. The temperature was 19 degrees. Mencken had accompanied a group of Bigfoot hunters as they set out to do "call blasting" in their quest for Bigfoot, of which some 800 to 1,000 specimens are thought to roam about East Texas. Giant tape recorders and speakers were set up in the trees to assist in luring the legendary creature.

"The idea is to play Bigfoot sounds and, if there is a Bigfoot in the forest, to get that Bigfoot to respond to the sounds you are hearing," explained one of the hunters. . . . Suddenly, the Bigfoot hunter in the tree with Mencken said, "Alright, here we go!"

"He pushed this button, and the most horrifying scream went off from these big, ol' speakers," Mencken said. "As soon as that howl went off, every animal in those woods freaked out, as you can well imagine. . . . And I started to think, I really hope that there's not a Bigfoot, because if that's what it sounds like, we're in a lot of trouble." (qtd. in Aydelotte, 2010)

My mind's open to anything. After all, they just found another planet. So, who knows? Anything's possible.

—Jim Maier, a resident of Seneca, Illinois, explaining the possible existence of Bigfoot, an eight-foot-tall "wild man" who is allegedly covered in hair, has a strong odor, and walks on much larger feet than those of a typical human being (qtd. in Wischnowsky, 2005)

Bigfoot is one of those things that people like to believe in. . . . Regardless of whether there are such things as Bigfoot, people like that thrill of uncertainty, that sense of danger. It's exciting to try and discover the unknown. And it's a lot more fun to have that little bit of doubt when you're sitting out in the woods.

—sociologist Christopher Bader describing why tales of the improbable are believable to some individuals, whereas others think that people who spend countless hours waiting to catch sight of Bigfoot are engaged in deviant behavior (qtd. in Wischnowsky, 2005)

Sociology contributes to our thinking about conformity and deviance by making us aware that the people we are around help us define what we think of as "normal" beliefs and actions. If we are surrounded by individuals who believe that a Bigfoot or UFO (unidentified flying object) sighting is just around the corner, we may think of such beliefs as normal and gain a personal sense of belonging when we go out and wait with these individuals for Bigfoot or a flying saucer to show up. For this reason, some people join groups such as the Bigfoot Field Researchers Organization (**www.bfro.net**) so that they can share their outings, compare field notes on recent sightings, and feel that they are part of an important group or a clique. Among other Bigfoot believers, followers are treated with respect when they record sightings rather than receiving blank stares or comments like "You've got to be kidding?" all the while they are being labeled as "weird" or "deviant" by outsiders who are nonbelievers (Wischnowsky, 2005).

Looking at the seemingly deviant behavior of going out on Bigfoot or UFO sightings from a sociological perspective, researchers such as Christopher Bader, F. Carson Mencken, and Joseph Baker (2010) place these actions within a larger social context. One context they use for studying people's fascination with Bigfoot sightings and other paranormal occurrences is the sociology of religion. According to Bader, many people who believe in Bigfoot or UFOs "believe without the kinds of evidence that would convince outsiders—it's a matter of faith" (qtd. in Weiss, 2004). This faith may be intensified by use of the Internet, where true believers may easily report their sightings without fear of ridicule or being identified as deviant by outsiders.

reflect & analyze

At your college or university, what beliefs and actions of individuals and groups might be classified as conformity by some people but identified as deviance by others? For example, do some students and/or professors believe that certain buildings are haunted and stay away from those areas?

What Causes Deviance, and Why Is It Functional for Society?

Sociologist Emile Durkheim believed that deviance is rooted in societal factors such as rapid social change and lack of social integration among people. As you will recall, Durkheim attributed the social upheaval he saw at the end of the nineteenth century to the shift from mechanical to organic solidarity, which was brought about by rapid industrialization and urbanization. Although many people continued to follow the dominant morals (norms, values, and laws) as best they could, rapid social change contributed to *anomie*—a social condition in which people experience a sense of futility because social norms are weak, absent, or conflicting. According

criminology the systematic study of crime and the criminal justice system, including the police, courts, and prisons.

© AP Photo/U.S. Marshall

▲ Shown here are eight of the nine members arrested in 2010 from a Michigan group called Hutaree, which federal authorities described as consisting of domestic terrorists who planned to wage war against the government. How is such a group an example of deviance?

to Durkheim, as social integration (bonding and community involvement) decreased, deviance and crime increased. However, from his perspective this was not altogether bad because he believed that deviance has positive social functions in terms of its consequences. For Durkheim (1964a/1895), deviance is a natural and inevitable part of all societies. Likewise, contemporary functionalist theorists suggest that deviance is universal because it serves three important functions:

1. *Deviance clarifies rules.* By punishing deviant behavior, society reaffirms its commitment to the rules and clarifies their meaning.
2. *Deviance unites a group.* When deviant behavior is seen as a threat to group solidarity and people unite in opposition to that behavior, their loyalties to society are reinforced.
3. *Deviance promotes social change.* Deviants may violate norms in order to get them changed. For example, acts of *civil disobedience*—including lunch counter sit-ins and bus boycotts—were used to protest and eventually correct injustices such as segregated buses and lunch counters in the South.

Functionalists acknowledge that deviance may also be dysfunctional for society. If too many people violate the norms, everyday existence may become unpredictable, chaotic, and even violent. If even a few people commit acts that are so violent that they threaten the survival of a society, then deviant acts move into the realm of the criminal and even the unthinkable. Of course, the example that stands out in everyone's mind is terrorist attacks around the world and the fear that remains constantly present as a result.

Although there is a wide array of contemporary functionalist theories regarding deviance and crime, many of these theories focus on social structure. For this reason, the first theory we will discuss is referred to as a structural functionalist approach. It describes the relationship between the society's economic structure and why people might engage in various forms of deviant behavior.

Strain Theory: Goals and Means to Achieve Them

Modifying Durkheim's (1964a/1895) concept of *anomie*, the sociologist Robert Merton (1938, 1968) developed strain theory. According to ***strain theory,*** **people feel strain when they are exposed to cultural goals that they are unable to obtain because they do not have access to culturally approved means of achieving those goals.** The goals may be material possessions and money; the approved means may include an education and jobs. When denied legitimate access to these goals, some people seek access through deviant means.

Merton identified five ways in which people adapt to cultural goals and approved ways of achieving them: conformity, innovation, ritualism, retreatism, and rebellion (see ■ Table 7.1). According to Merton, *conformity* occurs when people accept culturally approved goals and pursue them through approved means. Persons who want to achieve success through conformity work hard, save their money, and so on. Even people who find that they are blocked from achieving a high level of education or a lucrative career may take a lower-paying job and attend school part time, join the

table 7.1

Merton's Strain Theory of Deviance

Mode of Adaptation	Method of Adaptation	Seeks Culture's Goals	Follows Culture's Approved Ways
Conformity	Accepts culturally approved goals; pursues them through culturally approved means	Yes	Yes
Innovation	Accepts culturally approved goals; adopts disapproved means of achieving them	Yes	No
Ritualism	Abandons society's goals but continues to conform to approved means	No	Yes
Retreatism	Abandons both approved goals and the approved means to achieve them	No	No
Rebellion	Challenges both the approved goals and the approved means to achieve them	No—seeks to replace	No—seeks to replace

military, or seek alternative (but legal) avenues, such as playing the lottery, to "strike it rich."

Conformity is also crucial for members of middle- and upper-class teen cliques, who often gather in small groups to share activities and confidences. Some youths are members of a variety of cliques, and peer approval is of crucial significance to them—being one of the "in" crowd, not a "loner," is a significant goal for many teenagers. In the aftermath of the recent school shootings, for example, journalists trekked to school campuses to report that athletes ("jocks"), cheerleaders, and other "popular" students enforce the social code at high schools (Adler, 1999; Cohen, 1999).

▲ The sociologist Robert Merton identified five ways in which people adapt to cultural goals and approved ways of achieving them. Consider the young woman shown here. Which of Merton's modes of adaptation might best explain her views on social life?

Merton classified the remaining four types of adaptation as deviance:

- *Innovation* occurs when people accept society's goals but adopt disapproved means for achieving them. Innovations for acquiring material possessions or money cover a wide variety of illegal activities, including theft and drug dealing.
- *Ritualism* occurs when people give up on societal goals but still adhere to the socially approved means for achieving them. Ritualism is the opposite of innovation; persons who cannot obtain expensive material possessions or wealth may nevertheless seek to maintain the respect of others by being a "hard worker" or "good citizen."
- *Retreatism* occurs when people abandon both the approved goals and the approved means of achieving them. Merton included persons such as skid-row alcoholics and drug addicts in this category; however, not all retreatists are destitute. Some may be middle- or upper-income individuals who see themselves as rejecting the conventional trappings of success or the means necessary to acquire them.
- *Rebellion* occurs when people challenge both the approved goals and the approved means for achieving them and advocate an alternative set of goals or means. To achieve their alternative goals, rebels may use violence (such as rioting) or may register their displeasure with society through acts of vandalism or graffiti (as further discussed in "You Can Make a Difference").

strain theory the proposition that people feel strain when they are exposed to cultural goals that they are unable to obtain because they do not have access to culturally approved means of achieving those goals.

you can make a difference

Graffiti Hurts®, and You Can Help!

Graffiti is always a big concern. It has a lot of implications once it takes over an area. It quickly becomes a public safety issue, an economic development issue and a crime issue.

—Dewey Bartlett, Tulsa mayor, introducing a program in which students at Spartan College of Aeronautics and Technology serve as volunteers to help rid this Oklahoma city of graffiti (qtd. in Barber, 2010)

Did you know that:

- Graffiti costs U.S. communities more than $8 billion per year?
- In southern Nevada alone, taxpayers pay more than $30 million a year to clean up graffiti vandalism?
- Durham, North Carolina, has special training programs such as "the Graffiti Deciphering, Interdiction and Investigation Course" that are designed for police officers, probation and correction officers, federal agents, and other law enforcement officials so that they can learn about gang identification and deciphering graffiti?
- Graffiti Hurts® is a program of Keep America Beautiful (see **www.kab.org**) that provides people with information on how to rid their community of graffiti?

Frankly, most of us don't pay attention to graffiti: It has become a part of the urban or rural landscape we see so often as we rush past. But in many communities, graffiti is a major concern not only because of its appearance and the cost of cleaning it up, but also because graffiti provides gangs with an illegal form of communication.

Although not all graffiti is done by street gangs, some gang members use graffiti to increase their visibility, mark their territory, threaten rival gangs, and intimidate local residents (Salt Lake City Sheriff's Department, 2007). According to law enforcement officials, taggers are usually less violent than are members of traditional street gangs, and their "art" is usually more "artistic" and less threatening than street-gang graffiti. However, the work of both taggers and street-gang members defaces walls, buses, subways, and other public areas.

Is there anything we can do when we see graffiti to get it removed and to improve the appearance of our community? How might we lessen the opportunities for gang members to use graffiti to communicate with one another and to threaten outsiders? Here are some suggestions from law enforcement officials:

- Report graffiti immediately.
- If you see graffiti in progress, report it: Do not confront or challenge a person who is tagging a wall or writing graffiti on a public space.
- Make sure that owners of private property or public officials are notified about the graffiti: It is important that the graffiti be painted over immediately. Studies show that if graffiti is left up, it becomes a status symbol, and the area is likely to be hit again and again.
- Look for adopt-a-wall programs or other groups in which volunteers assist in cleaning off or painting over graffiti.
- Find out if your city has a hotline where you can report graffiti. Many cities have instituted these hotlines so that graffiti can be quickly removed from both public and private property. (National Crime Prevention Council, 2011; New York State Troopers, 2011)

Although graffiti may appear to be a small issue, this kind of behavior is one telling sign that law enforcement officials use when identifying possible criminal trends. For example, a dramatic increase in the amount of graffiti in a community may be a sign that gang membership is growing and that gang activities are becoming more confrontational toward rival gangs and toward society as a whole.

Opportunity Theory: Access to Illegitimate Opportunities

Expanding on Merton's strain theory, sociologists Richard Cloward and Lloyd Ohlin (1960) suggested that for deviance to occur, people must have access to ***illegitimate opportunity structures*—circumstances that provide an opportunity for people to acquire through illegitimate activities what they cannot achieve through legitimate channels.** For example, gang members may have insufficient legitimate means to achieve conventional goals of status and wealth but have illegitimate opportunity structures—such as theft, drug dealing, or robbery—through which they can achieve these goals. In his study of the "Diamonds," a Chicago street gang whose members are second-generation Puerto Rican youths, sociologist Felix M. Padilla (1993) found that gang membership was linked to the members' belief that they might reach their aspirations by transforming the gang into a business enterprise. Coco, one of the Diamonds, explains the importance of sticking together in the gang's income-generating business organization:

> We are a group, a community, a family—we have to learn to live together. If we separate, we will never have a chance. We need each other even to make sure that we have a spot for selling our supply [of drugs]. You know, there is people around

▲ Conflict theorists suggest that criminal law is unequally enforced along class lines. Consider this setting, in which low-income defendants are arraigned by a judge who sees them only on a television monitor. Do you think, as a rule, that these defendants will be as well represented by attorneys as a wealthier defendant might be?

> here, like some opposition, that want to take over your *negocio* [business]. And they think that they can do this very easy. So we stick together, and that makes other people think twice about trying to take over what is yours. In our case, the opposition has never tried messing with our hood, and that's because they know it's protected real good by us fellas. (qtd. in Padilla, 1993: 104)

Based on their research, Cloward and Ohlin (1960) identified three basic gang types—criminal, conflict, and retreatist—which emerge on the basis of what type of illegitimate opportunity structure is available in a specific area. The *criminal gang* is devoted to theft, extortion, and other illegal means of securing an income. For young men who grow up in a criminal gang, running drug houses and selling drugs on street corners make it possible for them to support themselves and their families as well as purchase material possessions to impress others. By contrast, *conflict gangs* emerge in communities that do not provide either legitimate or illegitimate opportunities. Members of conflict gangs seek to acquire a "rep" (reputation) by fighting over "turf" (territory) and adopting a value system of toughness, courage, and similar qualities. On some Native American reservations, for example, homegrown gangs routinely fight their rivals, often over a minor incident or slight, and engage in thefts, assaults, and property crimes in some of the nation's poorest, most neglected places, including the Pine Ridge Indian Reservation (Eckholm, 2009). Unlike criminal and conflict gangs, members of *retreatist gangs* are unable to gain success through legitimate means and are unwilling to do so through illegal ones. As a result, the consumption of drugs is stressed, and addiction is prevalent.

Sociologist Lewis Yablonsky (1997) has updated Cloward and Ohlin's findings on delinquent gangs. According to Yablonsky, today's gangs are more likely to use and sell drugs, and to carry more lethal weapons, than gang members in the past. Today's gangs have become more varied in their activities and are more likely to engage in intraracial conflicts, with "black on black and Chicano on Chicano violence," whereas minority gangs in the past tended to band together to defend their turf from gangs of different racial and ethnic backgrounds (Yablonsky, 1997: 3).

How useful are social structural approaches such as opportunity theory and strain theory in explaining deviant behavior? Although there are weaknesses in these approaches, they focus our attention on one crucial issue: the close association between certain forms of deviance and social class position. According to criminologist Anne Campbell (1984: 267), gangs are a "microcosm of American society, a mirror image in which power, possession, rank, and role . . . are found within a subcultural life of poverty and crime."

Conflict Perspectives on Deviance

Who determines what kinds of behavior are deviant or criminal? Different branches of conflict theory offer somewhat divergent answers to this question. One branch emphasizes power as the central factor in defining deviance and crime: People in positions of power maintain their advantage by using the law to protect their interests. Another branch emphasizes the relationship between deviance and capitalism, whereas a third focuses on feminist perspectives and the confluence of race, class, and gender issues in regard to deviance and crime.

Deviance and Power Relations

Conflict theorists who focus on power relations in society suggest that the lifestyles considered deviant

illegitimate opportunity structures circumstances that provide an opportunity for people to acquire through illegitimate activities what they cannot achieve through legitimate channels.

by political and economic elites are often defined as illegal. According to this approach, norms and laws are established for the benefit of those in power and do not reflect any absolute standard of right and wrong (Turk, 1969, 1977). As a result, the activities of poor and lower-income individuals are more likely to be defined as criminal than those of persons from middle- and upper-income backgrounds. Moreover, the criminal justice system is more focused on, and is less forgiving of, deviant and criminal behavior engaged in by people in specific categories. For example, research shows that young, single, urban males are more likely to be perceived as members of the *dangerous classes* and receive stricter sentences in criminal courts (Miethe and Moore, 1987). Power differentials are also evident in how victims of crime are treated. When the victims are wealthy, white, and male, law enforcement officials are more likely to put forth more extensive efforts to apprehend the perpetrator as contrasted with cases in which the victims are poor, black, and female (Smith, Visher, and Davidson, 1984). More-recent research generally supports this assertion (Wonders, 1996).

© Nick Koudis/Photodisc/Getty Images

▲ According to Karl Marx, capitalism produces haves and have-nots, and each group engages in different types of crime. Statistically, the man being arrested here is much more likely to be suspected of a financial crime than a violent crime.

Deviance and Capitalism

A second branch of conflict theory—Marxist/critical theory—views deviance and crime as a function of the capitalist economic system. Although the early economist and social thinker Karl Marx wrote very little about deviance and crime, many of his ideas are found in a critical approach that has emerged from earlier Marxist and radical perspectives on criminology. The critical approach is based on the assumption that the laws and the criminal justice system protect the power and privilege of the capitalist class. As you may recall from Chapter 1, Marx based his critique of capitalism on the inherent conflict that he believed existed between the capitalists (bourgeoisie) and the working class (proletariat). In a capitalist society, social institutions (such as law, politics, and education, which make up the superstructure) legitimize existing class inequalities and maintain the capitalists' superior position in the class structure. According to Marx, capitalism produces haves and have-nots, who engage in different forms of deviance and crime.

According to the sociologist Richard Quinney (2001/1974), people with economic and political power define as criminal any behavior that threatens their own interests. The powerful use law to control those who are without power. For example, drug laws enacted early in the twentieth century were actively enforced in an effort to control immigrant workers, especially the Chinese, who were being exploited by the railroads and other industries (Tracy, 1980). By contrast, antitrust legislation passed at about the same time was seldom enforced against large corporations owned by prominent families such as the Rockefellers, Carnegies, and Mellons. Having antitrust laws on the books merely shored up the government's legitimacy by making it appear responsive to public concerns about big business (Barnett, 1979).

In sum, the Marxist/critical approach argues that criminal law protects the interests of the affluent and powerful. The way that laws are written and enforced benefits the capitalist class by ensuring that individuals at the bottom of the social class structure do not infringe on the property or threaten the safety of those at the top (Reiman, 1998). However,

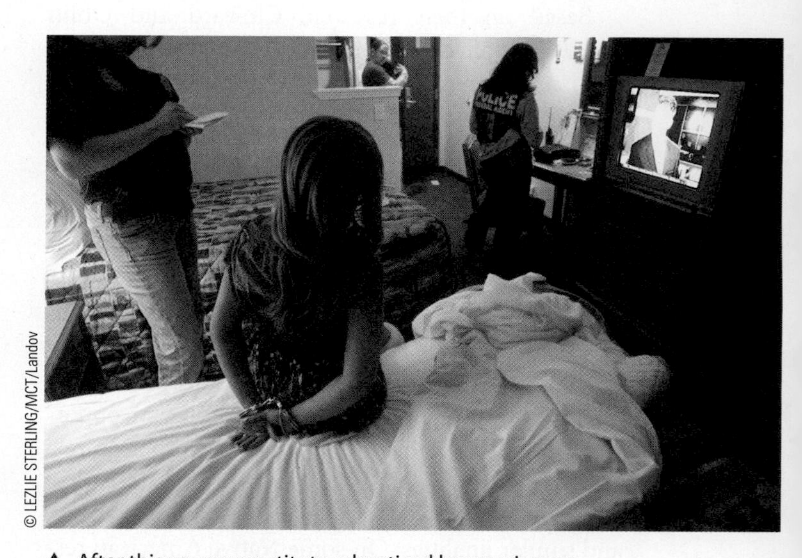

© LEZLIE STERLING/MCT/Landov

▲ After this young prostitute advertised her services on Craigslist, the Sacramento vice squad and the FBI arranged a meeting that led to her arrest. Which of the feminist theories of women's crime best explains this young woman's offense?

others assert that critical theorists have not shown that powerful economic and political elites actually manipulate lawmaking and law enforcement for their own benefit. Rather, people of all classes share a consensus about the criminality of certain acts. For example, laws that prohibit murder, rape, and armed robbery protect not only middle- and upper-income people but also low-income people, who are frequently the victims of such violent crimes.

Feminist Approaches

Can theories developed to explain male behavior be used to understand female deviance and crime? According to feminist scholars, the answer is no. A new interest in women and deviance developed in 1975 when two books—Freda Adler's *Sisters in Crime* and Rita James Simons's *Women and Crime*—declared that women's crime rates were going to increase significantly as a result of the women's liberation movement. Although this so-called *emancipation theory* of female crime has been refuted by subsequent analysts, Adler's and Simons's works encouraged feminist scholars (both women and men) to examine more closely the relationship among gender, deviance, and crime. Other feminist scholars such as Kathleen Daly and Meda Chesney-Lind (1988) developed theories and conducted research to fill the void in our knowledge about gender and crime. For example, in a study of the female offender, Chesney-Lind (1997) examined the cultural factors in women's lives that may contribute to their involvement in criminal behavior. Although there is no single feminist perspective on deviance and crime, three schools of thought have emerged.

Why do women engage in deviant behavior and commit crimes? According to the *liberal feminist approach*, women's deviance and crime are a rational response to the gender discrimination that women experience in families and the workplace. From this view, lower-income and minority women typically have fewer opportunities not only for education and good jobs but also for "high-end" criminal endeavors. As some feminist theorists have noted, a woman is no more likely to be a big-time drug dealer or an organized crime boss than she is to be a corporate director (Daly and Chesney-Lind, 1988; Simpson, 1989).

By contrast, the *radical feminist approach* views the cause of women's crime as originating in patriarchy (male domination over women). This approach focuses on social forces that shape women's lives and experiences and shows how exploitation may trigger deviant behavior and criminal activities. From this view, arrests and prosecution for crimes such as prostitution reflect our society's sexual double standard whereby it is acceptable for a man to pay for sex but unacceptable for a woman to accept money for such services. Although state laws usually view both the female prostitute and the male customer as violating the law, in most states the woman is far more likely than the man to be arrested, brought to trial, convicted, and sentenced.

The third school of feminist thought, the *Marxist (socialist) feminist approach*, is based on the assumption that women are exploited by both capitalism and patriarchy. Because most females have relatively low-wage jobs (if any) and few economic resources, crimes such as prostitution and shoplifting become a means to earn money or acquire consumer goods. However, instead of freeing women from their problems, prostitution institutionalizes women's dependence on men and results in a form of female sexual slavery (Vito and Holmes, 1994). Lower-income women are further victimized by the fact that they are often the targets of violent acts by lower-class males, who perceive themselves as being powerless in the capitalist economic system.

Some feminist scholars have noted that these approaches to explaining deviance and crime neglect the centrality of race and ethnicity and focus on the problems and perspectives of women who are white, middle and upper income, and heterosexual without taking into account the views of women of color, lesbians, and women with disabilities (Martin and Jurik, 1996).

Approaches Focusing on the Interaction of Race, Class, and Gender

Some studies have focused on the simultaneous effects of race, class, and gender on deviant behavior. In one study the sociologist Regina Arnold (1990) examined the relationship between women's earlier victimization in their family and their subsequent involvement in the criminal justice system. Arnold interviewed African American women serving criminal sentences and found that adolescent females are often "labeled and processed as deviants—and subsequently as criminals—for refusing to accept or participate in their own victimization." Arnold attributes many of the women's offenses to living in families in which sexual abuse, incest, and other violence left them few choices except to engage in deviance. Economic marginality and racism also contributed to their victimization: "To be young, Black, poor, and female is to be in a high-risk category for victimization and stigmatization on many levels" (Arnold, 1990: 156). According to Arnold, the criminal behavior of the women in her study was linked to class, gender, and racial oppression, which they experienced daily in their families and at school and work.

Symbolic Interactionist Perspectives on Deviance

Symbolic interactionists focus on *social processes*, such as how people develop a self-concept and learn conforming behavior through socialization. According to this approach, deviance is learned in the same way as conformity—through interaction with others. Although there are a number of symbolic interactionist perspectives on deviance, we will examine three major approaches—differential association and differential reinforcement theories, control theory, and labeling theory.

Differential Association Theory and Differential Reinforcement Theory

How do people learn deviant behavior through their interactions with others? According to the sociologist Edwin Sutherland (1939), people learn the necessary techniques and the motives, drives, rationalizations, and attitudes of deviant behavior from people with whom they associate. ***Differential association theory* states that people have a greater tendency to deviate from societal norms when they frequently associate with individuals who are more favorable toward deviance than conformity.** From this approach, criminal behavior is learned within intimate personal groups, such as one's family and peer groups.

Differential association theory contributes to our knowledge of how deviant behavior reflects the individual's learned techniques, values, attitudes, motives, and rationalizations. It calls attention to the fact that criminal activity is more likely to occur when a person has frequent, intense, and long-lasting interactions with others who violate the law. However, it does not explain why many individuals who have been heavily exposed to people who violate the law still engage in conventional behavior most of the time. It should also be noted that some criminologists believe that differential association theory is more closely associated with a functionalist approach to deviance and crime than with a symbolic interactionist perspective.

Criminologist Ronald Akers (1998) has combined differential association theory with elements of psychological learning theory to create *differential reinforcement theory,* which suggests that both deviant behavior and conventional behavior are learned through the same social processes. Akers starts with the fact that people learn to evaluate their own behavior through interactions with significant others. If the persons and groups that a particular individual considers most significant in his or her life define deviant behavior as being "right," the individual is more likely to engage in deviant behavior; likewise, if the person's most significant friends and groups define deviant behavior as "wrong," the person is less likely to engage in that behavior. This approach helps explain not only juvenile gang behavior but also how peer cliques on high school campuses have such a powerful influence on people's behavior. For example, when clique members at Glenbrook, a suburban Chicago high school, jealously "guarded" their favorite locations at the school, one student described her response to the powerful pressures to conform as follows:

> As an experiment . . . Lauren Barry, a pink-haired trophy-case kid at Glenbrook, switched identities with a well-dressed girl from "the wall." Barry walked around all day in the girl's expensive jeans and Doc Martens, carrying a shopping bag from Abercrombie & Fitch. "People kept saying, 'Oh, you look so pretty,'" she recalls. "I felt really uncomfortable." It was interesting, but the next day, and ever since, she's been back in her regular clothes. (Adler, 1999: 58)

Another such approach to studying deviance is rational choice theory, which suggests that people weigh the rewards and risks involved in certain types of behavior and then decide which course of action to follow.

Richard Wong/Alamy

▲ Is this example of graffiti likely to be the work of isolated artists or of gang members? In what ways do gangs reinforce such behavior?

Rational Choice Theory

Rational choice theory is based on the assumption that when people are faced with several courses of action, they will usually do what they believe is likely to have the best overall outcome (Elster, 1989). The

***rational choice theory of deviance* states that deviant behavior occurs when a person weighs the costs and benefits of nonconventional or criminal behavior and determines that the benefits will outweigh the risks involved in such actions.** Rational choice approaches suggest that most people who commit crimes do not engage in random acts of antisocial behavior. Instead, they make careful decisions based on weighing the available information regarding *situational factors,* such as the place of the crime, suitable targets, and the availability of people to deter the behavior, and *personal factors,* such as what rewards they may gain from their criminal behavior (Siegel, 2007).

How useful is rational choice theory in explaining deviance and crime? A major strength of this theory is that it explains why high-risk youths do not constantly engage in delinquent acts: They have learned to balance risk against the potential for criminal gain in each situation. Moreover, rational choice theory is not limited by the underlying assumption of most social structural theories, which is that the primary participants in deviant and criminal behaviors are people in the lower classes. Rational choice theory also has important policy implications regarding crime reduction or prevention, suggesting that people must be taught that the risks of engaging in criminal behavior far outweigh any benefits they may gain from their actions. Thus, people should be taught *not* to engage in crime.

Control Theory: Social Bonding

According to the sociologist Walter Reckless (1967), society produces pushes and pulls that move people toward criminal behavior; however, some people "insulate" themselves from such pressures by having positive self-esteem and good group cohesion. Reckless suggests that many people do not resort to deviance because of *inner containments*—such as self-control, a sense of responsibility, and resistance to diversions—and *outer containments*—such as supportive family and friends, reasonable social expectations, and supervision by others. Those with the strongest containment mechanisms are able to withstand external pressures that might cause them to participate in deviant behavior.

Extending Reckless's containment theory, sociologist Travis Hirschi's (1969) social control theory is based on the assumption that deviant behavior is minimized when people have strong bonds that bind them to families, schools, peers, churches, and other social institutions. ***Social bond theory* holds that the probability of deviant behavior increases when a person's ties to society are weakened or broken.** According to Hirschi, social bonding consists of (1) *attachment* to other people, (2) *commitment* to conformity, (3) *involvement* in conventional activities, and (4) *belief* in the legitimacy of conventional values and norms. Although Hirschi did not include females in his study, others who have replicated that study with both females and males have found that the theory appears to apply to each (see Naffine, 1987).

What does control theory have to say about delinquency and crime? Control theories suggest that the probability of delinquency increases when a person's social bonds are weak and when peers promote antisocial values and violent behavior. However, some critics assert that Hirschi was mistaken in his assumption that a weakened social bond leads to deviant behavior. The chain of events may be just the opposite: People who routinely engage in deviant behavior may find that their bonds to people who would be positive influences are weakened over time (Agnew, 1985; Siegel, 2007). Or, as labeling theory suggests, people may engage in deviant and criminal behavior because of destructive social interactions and encounters (Siegel, 2007).

Labeling Theory

***Labeling theory* states that deviance is a socially constructed process in which social control agencies designate certain people as deviants, and they, in turn, come to accept the label placed upon them and begin to act accordingly.** Based on the symbolic interaction theory of Charles H. Cooley and George H. Mead, labeling theory focuses on the variety of symbolic labels that people are given in their interactions with others.

How does the process of labeling occur? The act of fixing a person with a negative identity, such as

differential association theory the proposition that people have a greater tendency to deviate from societal norms when they frequently associate with individuals who are more favorable toward deviance than conformity.

rational choice theory of deviance the proposition that deviant behavior occurs when a person weighs the costs and benefits of nonconventional or criminal behavior and determines that the benefits will outweigh the risks involved in such actions.

social bond theory the proposition that the probability of deviant behavior increases when a person's ties to society are weakened or broken.

labeling theory the proposition that deviance is a socially constructed process in which social control agencies designate certain people as deviants, and they, in turn, come to accept the label placed upon them and begin to act accordingly.

Primary deviance		Secondary deviance		Tertiary deviance
Initial rule breaking	→	New identity accepted, deviance continues	→	Individual relabels behavior as nondeviant

▲ FIGURE 7.1 A CLOSER LOOK AT LABELING THEORY

"criminal" or "mentally ill," is directly related to the power and status of those persons who *do* the labeling and those who are *being labeled.* Behavior, then, is not deviant in and of itself; it is defined as such by a social audience (Erikson, 1962). According to the sociologist Howard Becker (1963), *moral entrepreneurs* are often the ones who create the rules about what constitutes deviant or conventional behavior. Becker believes that moral entrepreneurs use their own perspectives on "right" and "wrong" to establish the rules by which they expect other people to live. They also label others as deviant. Often these rules are enforced on persons with less power than the moral entrepreneurs. Becker (1963: 9) concludes that the deviant is "one to whom the label has successfully been applied; deviant behavior is behavior that people so label."

As the definition of labeling theory suggests, several stages may occur in the labeling process (see ▶ Figure 7.1). ***Primary deviance* refers to the initial act of rule breaking** (Lemert, 1951). However, if individuals accept the negative label that has been applied to them as a result of the primary deviance, they are more likely to continue to participate in the type of behavior that the label was initially meant to control. ***Secondary deviance* occurs when a person who has been labeled a deviant accepts that new identity and continues the deviant behavior.** For example, a person may shoplift an item of clothing from a department store but not be apprehended or labeled as a deviant. The person may subsequently decide to forgo such behavior in the future. However, if the person shoplifts the item, is apprehended, is labeled as a "thief," and subsequently accepts that label, then the person may shoplift items from stores on numerous occasions. A few people engage in ***tertiary deviance,* which occurs when a person who has been labeled a deviant seeks to normalize the behavior by relabeling it as nondeviant** (Kitsuse, 1980). An example would be drug users who believe that using marijuana or other illegal drugs is no more deviant than drinking alcoholic beverages and therefore should not be stigmatized.

Can labeling theory be applied to high school peer groups and gangs? In a classic study, the sociologist William Chambliss (1973) documented how the labeling process works in some high schools when he studied two groups of adolescent boys: the "Saints" and the "Roughnecks." Members of both groups were constantly involved in acts of truancy, drinking, wild parties, petty theft, and vandalism. Although the Saints committed more offenses than the Roughnecks, the Roughnecks were the ones who were labeled as "troublemakers" and arrested by law enforcement officials. By contrast, the Saints were described as being the "most likely to succeed," and none of the Saints were ever arrested. According to Chambliss (1973), the

▲ According to control theory, strong bonds—including close family ties—are a factor in explaining why many people do not engage in deviant behavior. Why do some sociologists believe that quality family time is more important in discouraging delinquent behavior than is time spent with other young people?

AP Photo/Brett Coomer

▲ Michel Foucault contended that new means of surveillance would make it possible for prison officials to use their knowledge of prisoners' activities as a form of power over the inmates. These guards are able to monitor the activities of many prisoners without ever leaving their station.

Roughnecks were more likely to be labeled as deviants because they came from lower-income families, did poorly in school, and were generally viewed negatively, whereas the Saints came from "good families," did well in school, and were generally viewed positively. Although both groups engaged in similar behavior, only the Roughnecks were stigmatized by a deviant label.

How successful is labeling theory in explaining deviance and social control? One contribution of labeling theory is that it calls attention to the way in which social control and personal identity are intertwined: Labeling may contribute to the acceptance of deviant roles and self-images. Critics argue that this does not explain what caused the original acts that constituted primary deviance, nor does it provide insight into why some people accept deviant labels and others do not (Cavender, 1995).

Postmodernist Perspectives on Deviance

Departing from other theoretical perspectives on deviance, some postmodern theorists emphasize that the study of deviance reveals how the powerful exert control over the powerless by taking away their free will to think and act as they might choose. From this approach, institutions such as schools, prisons, and mental hospitals use knowledge, norms, and values to categorize people into "deviant" subgroups such as slow learners, convicted felons, or criminally insane individuals, and then to control them through specific patterns of discipline.

An example of this idea is found in social theorist Michel Foucault's *Discipline and Punish* (1979), in which Foucault examines the intertwining nature of power, knowledge, and social control. In this study of prisons from the mid-1800s to the early 1900s, Foucault found that many penal institutions ceased torturing prisoners who disobeyed the rules and began using new surveillance techniques to maintain social control. Although the prisons appeared to be more humane in the post-torture era, Foucault contends that the new means of surveillance impinged more on prisoners and brought greater power to prison officials. To explain, he described the *Panoptican*—a structure that gives prison officials the possibility of complete observation of criminals at all times. Typically, the Panoptican was a tower located in the center of a circular prison from which guards could see all the cells. Although the prisoners knew they could be observed at any time, they did not actually know when their behavior was being scrutinized. As a result, prison officials were able to use their knowledge as a form of power over the inmates. Eventually, the guards did not even have to be present all the time because prisoners believed that they were under constant scrutiny by officials in the observation post. If we think of this in contemporary times, we can see how cameras, computers, and other devices have made continual surveillance quite easy in virtually all institutions. In such cases social control and discipline are based on the use of knowledge, power, and technology.

Foucault's view on deviance and social control has influenced other social analysts, including Shoshana Zuboff (1988), who views the computer as a modern Panoptican that gives workplace supervisors virtually unlimited capabilities for surveillance over subordinates. Today, cell phones and the Internet provide additional opportunities for surveillance by government officials and others who are not visible to the individuals who are being watched.

We have examined functionalist, conflict, interactionist, and postmodernist perspectives on social control, deviance, and crime (see the Concept Quick Review) All of these explanations contribute to our understanding of the causes and consequences of deviant behavior; however, we now turn to the subject of crime itself.

primary deviance the initial act of rule-breaking.

secondary deviance the process that occurs when a person who has been labeled a deviant accepts that new identity and continues the deviant behavior.

tertiary deviance deviance that occurs when a person who has been labeled a deviant seeks to normalize the behavior by relabeling it as nondeviant.

[concept quick review]

Theoretical Perspectives on Deviance

	Theory	Key Elements
Functionalist Perspectives		
Robert Merton	Strain theory	Deviance occurs when access to the approved means of reaching culturally approved goals is blocked. Innovation, ritualism, retreatism, or rebellion may result.
Richard Cloward/ Lloyd Ohlin	Opportunity theory	Lower-class delinquents subscribe to middle-class values but cannot attain them. As a result, they form gangs to gain social status and may achieve their goals through illegitimate means.
Conflict Perspectives		
Karl Marx Richard Quinney	Critical approach	The powerful use law and the criminal justice system to protect their own class interests.
Kathleen Daly Meda Chesney-Lind	Feminist approach	Historically, women have been ignored in research on crime. Liberal feminism views women's deviance as arising from gender discrimination, radical feminism focuses on patriarchy, and socialist feminism emphasizes the effects of capitalism and patriarchy on women's deviance.
Symbolic Interactionist Perspectives		
Edwin Sutherland	Differential association	Deviant behavior is learned in interaction with others. A person becomes delinquent when exposure to lawbreaking attitudes is more extensive than exposure to law-abiding attitudes.
Travis Hirschi	Social control/social bonding	Social bonds keep people from becoming criminals. When ties to family, friends, and others become weak, an individual is most likely to engage in criminal behavior.
Howard Becker	Labeling theory	Acts are deviant or criminal because they have been labeled as such. Powerful groups often label less-powerful individuals.
Edwin Lemert	Primary/secondary deviance	Primary deviance is the initial act. Secondary deviance occurs when a person accepts the label of "deviant" and continues to engage in the behavior that initially produced the label.
Postmodernist Perspective		
Michel Foucault	Knowledge as power	Power, knowledge, and social control are intertwined. In prisons, for example, new means of surveillance that make prisoners think they are being watched all the time give officials knowledge that inmates do not have. Thus, the officials have a form of power over the inmates.

Crime Classifications and Statistics

Crime in the United States can be divided into different categories. We will look first at the legal classifications of crime and then at categories typically used by sociologists and criminologists.

How the Law Classifies Crime

Crimes are divided into felonies and misdemeanors. The distinction between the two is based on the seriousness of the crime. A *felony* is a serious crime such as rape, homicide, or aggravated assault, for which punishment typically ranges from more than a year's imprisonment to death. A *misdemeanor* is a minor crime that is typically punished by less than one year in jail. In either event a fine may be part of the sanction as well. Actions that constitute felonies and misdemeanors are determined by the legislatures in the various states; thus, their definitions vary from jurisdiction to jurisdiction.

Other Crime Categories

The *Uniform Crime Report* (UCR) is the major source of information on crimes reported in the United States. The UCR has been compiled since 1930 by the

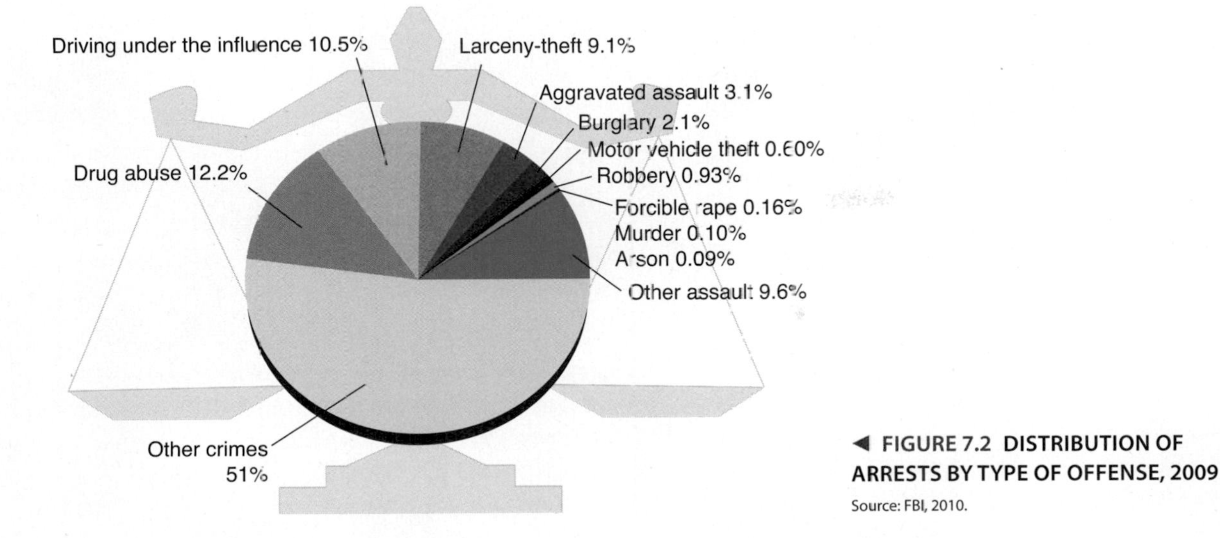

◀ **FIGURE 7.2 DISTRIBUTION OF ARRESTS BY TYPE OF OFFENSE, 2009**

Source: FBI, 2010.

Federal Bureau of Investigation based on information filed by law enforcement agencies throughout the country. When we read that the rate of certain types of crimes has increased or decreased when compared with prior years, for example, this information is usually based on UCR data. The UCR focuses on violent crime and property crime (which, prior to 2004, were jointly referred to in that report as "index crimes"), but it also contains data on other types of crime (see ▶ Figure 7.2). In 2009 about 13.7 million arrests were made in the United States for all criminal infractions (excluding traffic violations). Although the UCR gives some indication of crime in the United States, the figures do not reflect the actual number and kinds of crimes, as will be discussed later.

Violent Crime *Violent crime* **consists of actions—murder, forcible rape, robbery, and aggravated assault—involving force or the threat of force against others.** Although only 4.3 percent of all arrests in the United States in 2009 were for violent crimes, this category is probably the most anxiety-provoking of all criminal behavior: Most of us know someone who has been a victim of violent crime, or we have been so ourselves. Victims are often physically injured or even lose their lives; the psychological trauma may last for years after the event. Violent crime receives the most sustained attention from law enforcement officials and the media (see "Framing Violent Crime in the Media").

Property Crime *Property crimes* **include burglary (breaking into private property to commit a serious crime), motor vehicle theft, larceny-theft (theft of property worth $50 or more), and arson.** Some offenses, such as robbery, are both violent crimes and property crimes. In the United States a property crime occurs, on average, once every 3.4 seconds; a violent crime occurs, on average, once every 23.9 seconds (see ▶ Figure 7.3). In most property crimes the primary motive is to obtain money or some other desired valuable.

Public Order Crime *Public order crimes* involve an illegal action voluntarily engaged in by the participants, such as prostitution, illegal gambling, the private use of illegal drugs, and illegal pornography. Many people assert that such conduct should not be labeled as a crime; these offenses are often referred to as ***victimless crimes*** **because they involve a willing exchange of illegal goods or services among adults.** However, morals crimes can include children and adolescents as well as adults. Young children and adolescents may unwillingly become child pornography "stars" or prostitutes.

Occupational and Corporate Crime Although the sociologist Edwin Sutherland (1949) developed the theory of white-collar crime more than sixty years ago, it was not until the 1980s that the public became fully aware of its nature. ***Occupational (white-collar) crime*** **comprises illegal activities committed by people in the course of their employment or financial affairs.**

violent crime actions—murder, forcible rape, robbery, and aggravated assault—involving force or the threat of force against others.

property crimes burglary (breaking into private property to commit a serious crime), motor vehicle theft, larceny-theft (theft of property worth $50 or more), and arson.

victimless crimes crimes involving a willing exchange of illegal goods or services among adults.

occupational (white-collar) crime illegal activities committed by people in the course of their employment or financial affairs.

framing violent crime in the media

Fear and Disbelief in the Face of Evil

They were a model family living in an affluent suburb. William Petit was a prominent doctor. His daughter was on the way to Dartmouth, hoping to follow in his footsteps. His wife had multiple sclerosis and the family was active in efforts to raise money to fight the disease. But a chance encounter with a career criminal at a supermarket in July 2007 destroyed the family, authorities say. Joshua Komisarjevsky spotted Jennifer Hawke-Petit and her two daughters at the store and followed them home, then returned later with his friend Steven Hayes and together they severely beat Petit and killed his wife and daughters, authorities say. (Christoffersen, 2010)

This all-too-familiar crime story involves a family that was destroyed by a brutally violent home invasion. Although millions of people will never experience such horror, continual reports from various media sources provide us with endless details of crime and the fear it produces in victims, families, neighbors, and other community members. In the Petit case, for example, media accounts provide extensive details about the violent actions of two paroled burglars who tormented a family of four for seven hours before strangling the wife and tying up the two daughters, leaving them to die in a gasoline-fueled house fire. Only the husband escaped with his life.

How does media reporting of violent crime compare with the reporting of other types of crime? The media are much more likely to report on violent crime than property crime. Several factors are important in determining how much coverage a specific violent crime will receive:

1. How brutal was the crime?
2. How much did the victims suffer?
3. Will media audiences feel threatened and believe that something similar could happen to them?

In the case of extremely violent crime, media framing typically uses a fear-based approach to describe what happened. Fear framing uses a crime event to create conflicting emotions of disbelief and concern in media audiences and to suggest that ordinary citizens are in danger of encountering evil in ordinary locations such as the grocery store parking lot (as in the Petit crime).

Many newspaper, TV, and Internet accounts of crime emphasize human emotions of fear, disbelief, and distrust. Without a systematic framework for thinking about serious violent crimes, media audiences are left with representations of crime as the product of "bad people" and evil in the world—ideas which divert attention from important linkages that these crimes have to the larger social structure or culture (Reiner, 2007). For people who are already fearful of violent crime, media representations confirm these concerns and make individuals more afraid to leave home and interact with others. The sheer normalcy of family life, such as in the Petit family, that is shattered by one night of violence leaves millions worrying about their own safety even if the statistical odds of a similar event occurring at their own home are astronomically small. Still, the overall media impact remains strong, according to Rich Hanley, Quinnipiac University journalism director, who stated that the Petit murder left the state shocked and people feeling vulnerable in the sense that it happened in a town where violence rarely occurs and it happened in a way that shook civilization (qtd. in Christoffersen, 2010).

reflect & analyze

Do the media place too much emphasis on violent crime, as this discussion suggests? Do journalists provide too many graphic details about gruesome events, or does the public have a right to know all of the details?

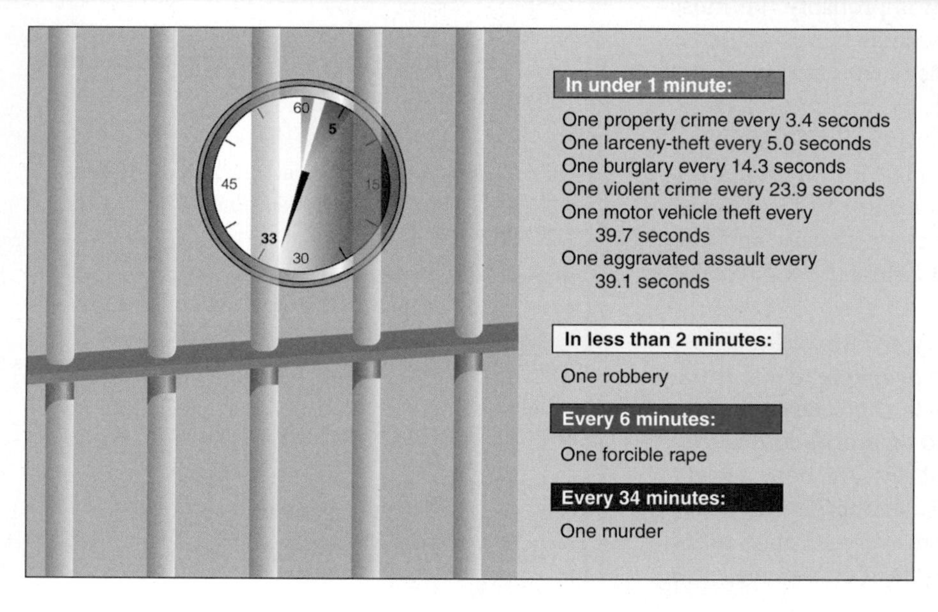

◀ FIGURE 7.3 THE FBI CRIME CLOCK

Source: FBI, 2010.

In addition to acting for their own financial benefit, some white-collar offenders become involved in criminal conspiracies designed to improve the market share or profitability of their companies. This is known as ***corporate crime*****—illegal acts committed by corporate employees on behalf of the corporation and with its support.** Examples include antitrust violations; tax evasion; misrepresentations in advertising; infringements on patents, copyrights, and trademarks; price fixing; and financial fraud. These crimes are a result of deliberate decisions made by corporate personnel to enhance resources or profits at the expense of competitors, consumers, and the general public.

Although people who commit occupational and corporate crimes can be arrested, fined, and sent to prison, many people often have not regarded such behavior as "criminal." People who tend to condemn street crime are less sure of how their own (or their friends') financial and corporate behavior should be judged. At most, punishment for such offenses has usually been a fine or a relatively brief prison sentence.

Until recently, public concern and media attention focused primarily on the street crimes disproportionately committed by persons who were poor, powerless, and nonwhite. Today, however, part of our focus has shifted to crimes committed in corporate suites, such as fraud, tax evasion, and insider trading by executives at large and well-known corporations. Bernard Madoff, the former chairperson of NASDAQ, admitted to defrauding his clients of up to $50 billion in a massive scheme that took place over a number of years. Madoff used his social connections to raise large sums of money for a fund that he used for his own gain. Clients invested in the fund in hopes that Madoff would manage their money wisely and that they would earn large returns on their investments. Instead, he lived lavishly and used new money that came in from investors to pay off existing clients who wanted to cash out of his fund rather than using the new money for the purpose intended. However, Madoff is not an isolated example of such criminal endeavors. Over the past decade, numerous occupational and corporate criminals, including Dennis Kozlowski, the former chief executive of Tyco International, who was convicted of looting more than $600 million from his company, and former Enron executives Ken Lay and Jeff Skilling, who were convicted of corporate conspiracy and fraud in connection with the collapse of one-time energy giant Enron, have all engaged in activities that have cost other people billions of dollars and, in some cases, their life savings.

Corporate crimes are often more costly in terms of money and lives lost than street crimes. Thousands of jobs and billions of dollars have been lost annually as a result of corporate crime. Deaths resulting from corporate crimes such as polluting the air and water, manufacturing defective products, and selling unsafe foods and drugs far exceed the number of deaths caused by homicides each year. Other costs include the effect on the moral climate of society (Clinard and Yeager, 1980; Simon, 2008). Throughout the United States the confidence of everyday people in the nation's economy has been shaken badly by the greedy and illegal behavior of corporate insiders.

Organized Crime ***Organized crime*** **is a business operation that supplies illegal goods and services for profit.** Premeditated, continuous illegal activities of organized crime include murder, extortion, narcotics trafficking, prostitution, loan-sharking, money laundering, and large-scale theft such as truck hijackings. No single organization controls all organized crime; rather, many groups operate at all levels of society. Some people believe that organized crime is becoming a thing of the past; however, recent arrests of high-profile members of the Gambino and Colombo crime families in New York on charges ranging from classic mob hits that eliminated perceived rivals to allegations of corruption among dockworkers suggest otherwise (Hays, 2011). Organized crime has also become transnational in nature. The globalization of the economy and the introduction of better communications technology have made it possible for groups around the world to operate in the United States and other nations (see "Transnational Crime and the Global Criminal Economy" on page 211). The Federal Bureau of Investigation has identified several major categories of organized crime groups, including (1) Italian organized crime and racketeering, (2) Eurasian/Middle Eastern organized crime, and (3) Asian and African criminal enterprises, as shown in ▶ Figure 7.4.

Organized crime thrives because there is great demand for illegal goods and services. Criminal organizations initially gain control of illegal activities by combining threats and promises. For example, small-time operators running drug or prostitution rings may be threatened with violence if they compete with organized crime or fail to make required payoffs (Cressey, 1969).

Apart from their illegal enterprises, organized crime groups have infiltrated the world of legitimate

corporate crime illegal acts committed by corporate employees on behalf of the corporation and with its support.

organized crime a business operation that supplies illegal goods and services for profit.

▲ **FIGURE 7.4 ORGANIZED CRIME THREATS IN THE UNITED STATES**

Source: http://www.fbi.gov/hq/cid/orgcrime/ocshome.htm.

business. Known linkages between legitimate businesses and organized crime exist in banking, hotels and motels, real estate, garbage collection, vending machines, construction, delivery and long-distance hauling, garment manufacture, insurance, stocks and bonds, vacation resorts, and funeral homes (National Council on Crime and Delinquency, 1969). In addition, some law enforcement and government officials are corrupted through bribery, campaign contributions, and favors intended to buy them off.

▲ Over the years, there have been many notorious leaders of organized crime syndicates. Show here is John A. Gotti, who assumed control of the Gambino family after his father, John Gotti, was sentenced to life in prison for murder and racketeering. The younger Gotti also pled guilty to racketeering, in 1999, and spent six years in prison.

Based on current economic problems in the United States, some criminologists believe that organized crime will have an even greater effect on our nation in the future. According to these analysts, organized crime may further weaken the U.S. economy because illegal activities such as tax-evasion scams and cigarette trafficking bring about losses in tax revenue for state and federal governments. Organized crime groups that are involved in areas such as commodities, credit, insurance, stocks, securities, and investments will also have the ability to further weaken the already-troubled financial and housing markets (see Finklea, 2009).

Political Crime The term ***political crime*** **refers to illegal or unethical acts involving the usurpation of power by government officials, or illegal/unethical acts perpetrated against the government by outsiders seeking to make a political statement, undermine the government, or overthrow it.** Government officials may use their authority unethically or illegally for the purpose of material gain or political power (Simon, 2008). They may engage in graft (taking advantage of their political position to gain money or property) through bribery, kickbacks, or "insider" deals that financially benefit them. For example, in the late 1980s several top Pentagon officials were found guilty of receiving bribes for passing classified information on to major defense contractors that had garnered many lucrative contracts from the government (Simon, 2008).

Other types of corruption have been costly for taxpayers, including dubious use of public funds and public property, corruption in the regulation of commercial activities (such as food inspection), graft in zoning and land-use decisions, and campaign contributions and other favors to legislators that corrupt the legislative process. Whereas some political crimes are for personal material gain, others (such as illegal wiretapping and political "dirty tricks") are aimed at gaining or maintaining political office or influence.

Some acts committed by agents of the government against persons and groups believed to be threats to national security are also classified as political crimes. Four types of political deviance have been attributed to some officials: (1) secrecy and deception designed to manipulate public opinion, (2) abuse of power, (3) prosecution of individuals because of their political activities, and (4) official violence, such as police brutality against people of color or the use of citizens as unwilling guinea pigs in scientific research (Simon, 2008).

Political crimes also include illegal or unethical acts perpetrated against the government by outsiders seeking to make a political statement or to undermine or overthrow the government. Examples include treason, acts of political sabotage, and terrorist attacks on public buildings.

AP Photo/Stephen Chernin

▲ Political crime frequently involves the use of an office for personal material gain. Here, former New York State Comptroller Alan Hevesi awaits sentencing after being convicted of influence peddling involving the state's huge public pension fund.

Crime Statistics

How useful are crime statistics as a source of information about crime? As mentioned previously, official crime statistics provide important information on crime; however, the data reflect only those crimes that have been reported to the police.

Why are some crimes not reported? People are more likely to report crime when they believe that something can be done about it (apprehension of the perpetrator or retrieval of their property, for example). About half of all assault and robbery victims do not report the crime because they may be embarrassed or fear reprisal by the perpetrator. Thus, the number of crimes reported to police represents only the proverbial "tip of the iceberg" when compared with all offenses actually committed. Official statistics are problematic in social science research because of these limitations.

The National Crime Victimization Survey was developed by the Bureau of Justice Statistics as an alternative means of collecting crime statistics. In this annual survey the members of 100,000 randomly selected households are interviewed to determine whether they have been the victims of crime, even if the crime was not reported to the police. The most recent victimization survey suggests that 50 percent of all violent crimes and 61 percent of all property crimes are not reported to the police and are thus not reflected in the UCR (U.S. Bureau of Justice Statistics, 2008). If these percentages are accurate, then reported crime is indeed the tip of the iceberg of all violent and property crimes committed in this country.

Studies based on anonymous self-reports of criminal behavior also reveal much higher rates of crime than those found in official statistics. For example, self-reports tend to indicate that adolescents of all classes violate criminal laws. However, official statistics show that those who are arrested and placed in juvenile facilities typically have limited financial resources, have repeatedly committed serious offenses, or both (Steffensmeier and Allan, 2000). Data collected for the Juvenile Court Statistics program also reflect class and racial bias in criminal justice enforcement. Not all children who commit juvenile offenses are apprehended and referred to court. Children from white, affluent families are more likely to have their cases handled outside the juvenile justice system (for example, a youth may be sent to a private school or hospital rather than to a juvenile correctional facility).

Many crimes committed by persons of higher socioeconomic status in the course of business are handled by an administrative or quasi-judicial body, such as the Securities and Exchange Commission or the Federal Trade Commission, or by civil courts. As a result, many elite crimes are never classified as "crimes," nor are the businesspeople who commit them labeled as "criminals."

Qi Heng/Xinhua/Newscom

▲ Sometimes, Americans' fears about violent crime are justified in extreme ways. On January 8, 2011, Jared Loughner opened fire in a Tucson, Arizona, grocery store, killing six people and wounding fourteen others, including U.S. Congresswoman Gabrielle Giffords. Loughner has since been found mentally incompetent to stand trial.

Terrorism and Crime

In the twenty-first century the United States and other nations are confronted with a difficult prospect: how

political crime illegal or unethical acts involving the usurpation of power by government officials, or illegal/unethical acts perpetrated against the government by outsiders seeking to make a political statement, undermine the government, or overthrow it.

to deal with terrorism. ***Terrorism* is the calculated, unlawful use of physical force or threats of violence against persons or property in order to intimidate or coerce a government, organization, or individual for the purpose of gaining some political, religious, economic, or social objective.** A frequently asked question today is this: What is the difference between terrorism and organized crime? According to authorities, the principal distinction between organized crime groups and terrorist groups is motivation: "Money motivates organized crime, and ideology motivates terrorism" (Finklea, 2009: 23). However, money is still the linking element between organized crime and terrorism because terrorist organizations typically obtain money for their activities from criminal acts such as money laundering and drug trafficking.

How are sociologists and criminologists to explain world terrorism, which may have its origins in more than one nation and include diverse "cells" of terrorists who operate in a somewhat gang-like manner but are believed to be following directives from leaders elsewhere? In order to deal with the aftermath of terrorist attacks, government officials typically focus on "known enemies" such as Osama bin Laden. The nebulous nature of the "enemy" and the problems faced by any one government trying to identify and apprehend the perpetrators of acts of terrorism have resulted in a global "war on terror." Social scientists who use a rational choice approach suggest that terrorists are rational actors who constantly calculate the gains and losses of participation in violent—and sometimes suicidal—acts against others.

▲ Most of the crimes that women commit are nonviolent ones. Nevertheless, many women are incarcerated. What effects might a mother's imprisonment have on the lives of her children?

Street Crimes and Criminals

Given the limitations of official statistics, is it possible to determine who commits crimes? We have much more information available about conventional (street) crime than elite crime; therefore, statistics concerning street crime do not show who commits all types of crime. Gender, age, class, and race are important factors in official statistics pertaining to street crime.

Gender and Crime It goes without saying that there is a gender gap in crime statistics: Males are arrested for significantly more crimes than females. In 2009 nearly 75 percent of all persons arrested nationwide were male. Males made up about 81 percent of persons arrested for violent crime and almost 63 percent of persons arrested for property crimes (FBI, 2010). Females have higher arrests rates than males for prostitution, embezzlement, and runaways, but in all other categories males have higher rates.

Before further consideration of differences in crime rates by males and females, three similarities should be noted. First, the three most common arrest categories for both men and women are driving under the influence of alcohol or drugs (DUI), larceny, and minor or criminal mischief types of offenses. Second, liquor law violations (such as underage drinking), simple assault, and disorderly conduct are middle-range offenses for both men and women. Third, the rate of arrests for murder, arson, and embezzlement is relatively low for both men and women.

The most important gender differences in arrest rates are reflected in the proportionately greater involvement of men in major property crimes (such as robbery and larceny-theft) and violent crime, as shown in ▶ Figure 7.5. In 2009 men accounted for more than 88 percent of murders and robberies and about 56 percent of all larceny-thefts in the United States. Property crimes for which women are most frequently arrested are nonviolent in nature, including shoplifting, theft of services, passing bad checks, credit card fraud, and employee pilferage. Often when women are arrested for serious violent and property crimes, they are seen as accomplices of the men who planned the crime and instigated its

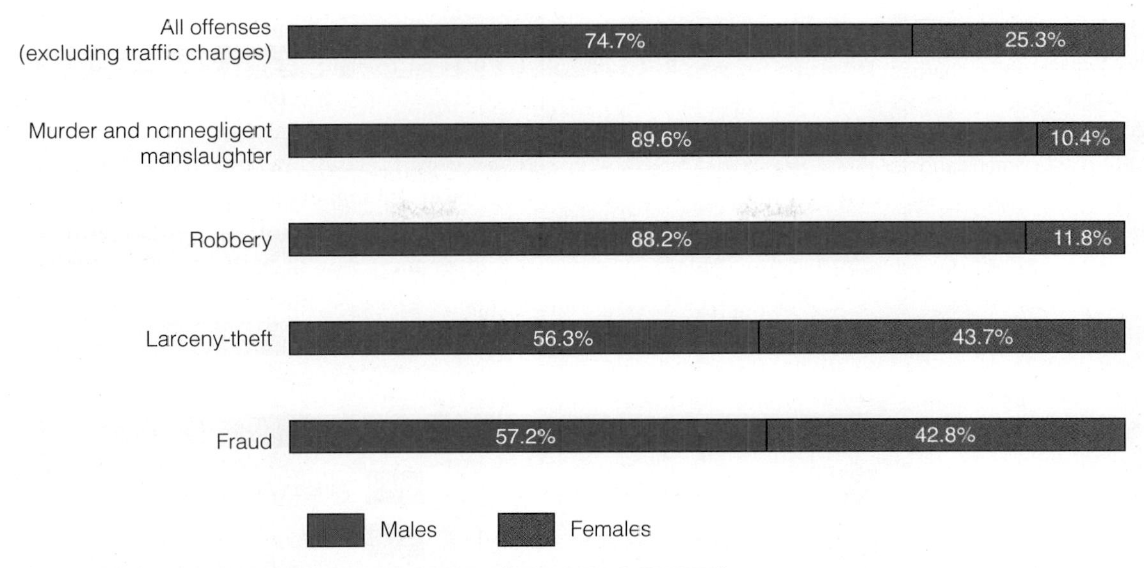

▲ **FIGURE 7.5 ARREST RATES BY GENDER, 2009 (SELECTED OFFENSES)**

Source: FBI, 2010.

commission; however, this assumption frequently does not prove true today. Some studies have found that some women play an active role in planning and carrying out robberies and other major crimes.

Age and Crime Of all factors associated with crime, the age of the offender is one of the most significant. Arrest rates for violent crime and property crime are highest for people between the ages of 13 and 25, with the peak being between ages 16 and 17. In 2009 persons under age 25 accounted for about 45 percent of all arrests for violent crime and almost 56 percent of property crime arrests (FBI, 2010). Individuals under age 18 accounted for 24 percent of all arrests for robbery, burglary, and larceny-theft.

Scholars do not agree on the reasons for this age distribution. In one study the sociologist Mark Warr (1993) found that peer influences (defined as exposure to delinquent peers, time spent with peers, and loyalty to peers) tend to be more significant in explaining delinquent behavior than age itself. More-recent studies have tended to confirm this finding. However, others argue that people simply "age out" of committing crimes, particularly those offenses that require physical strength or speed to get away from victims or law enforcement personnel.

The median age of those arrested for aggravated assault and homicide is somewhat older, generally in the late twenties. Typically, white-collar criminals are even older because it takes time to acquire both a high-ranking position and the skills needed to commit this particular type of crime.

Rates of arrest remain higher for males than females at every age and for nearly all offenses. This female-to-male ratio remains fairly constant across all age categories. The most significant gender difference in the age curve is for prostitution (a nonviolent crime). In 2009, 60 percent of all women arrested for prostitution were under age 35. For individuals over age 45, many more men than women are arrested for sex-related offenses (including procuring the services of a prostitute). This difference has been attributed to a more stringent enforcement of prostitution statutes when young females are involved (Chesney-Lind, 1997). It has also been suggested that opportunities for prostitution are greater for younger women. This age difference may not have the same impact on males, who continue to purchase sexual services from young females or males (Steffensmeier and Allan, 2000).

Social Class and Crime Individuals from all classes commit crimes; they simply commit different kinds of crimes. Persons from lower socioeconomic backgrounds are more likely to be arrested for violent and property crimes. By contrast, persons from the upper part of the class structure generally commit white-collar or elite crimes, although only a very small proportion of these individuals will ever be arrested or convicted of a crime.

What about social class and violence by youths? Between 1992 and 2008, there were 683 violent deaths in U.S. schools (U.S. Department of Education, 2009). Most of these deaths were not attributed to lower-income, inner-city youths, as popular stereotypes might suggest. Instead, these acts of violence largely

terrorism the calculated, unlawful use of physical force or threats of violence against persons or property in order to intimidate or coerce a government, organization, or individual for the purpose of gaining some political, religious, economic, or social objective.

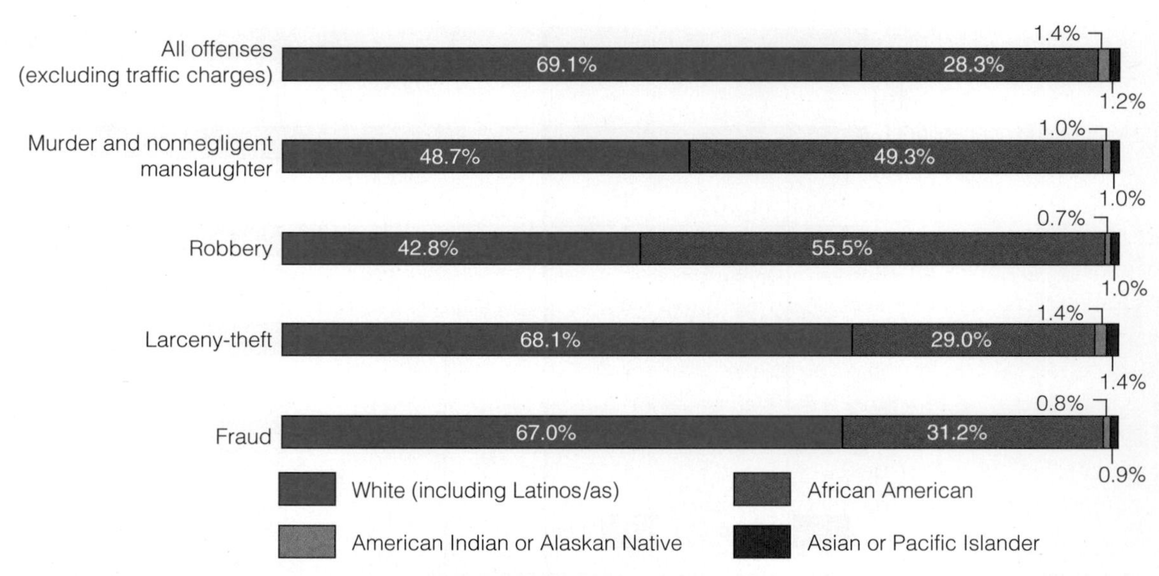

▲ **FIGURE 7.6 ARREST BY RACE, 2009 (SELECTED OFFENSES)**

Note: Classifications as used in the Uniform Crime Report.

Source: FBI, 2010.

were perpetrated by more-affluent young people who had no apparent financial hardship.

Similarly, membership in today's youth gangs cannot be identified with just one class. Across class lines, the percentage of students reporting the presence of gangs at their school increased from 21 to 23 percent between 2003 and 2008. Students between the ages of 12 and 18 were asked to indicate if gangs were present at their school, regardless of whether or not gang members were involved in any violent or illegal activity. Twenty-five percent of students at public schools reported gang activity at their school, but only four percent of private school students reported that they had knowledge of gang members at their school. This is not surprising, given the fact that the U.S. Department of Education estimates that 50 percent of gang members are part of the nation's underclass—the class comprising families whose members are poor, seldom employed, and caught in patterns of long-term deprivation. According to studies from the Department of Education, however, about 35 percent of gang members are working class, whereas 15 percent are middle or upper-middle class. Today, females from working-class and lower-income families are more visible in female gangs and in formerly all-male gangs.

In any case, official statistics are not an accurate reflection of the relationship between class and crime. Self-report data from offenders themselves may be used to gain information on family income, years of education, and occupational status; however, such reports rely on respondents to report information accurately and truthfully.

Race and Crime Who is most likely to be arrested for committing a crime? In 2009 whites (including Hispanics or Latinos/as) accounted for about 69 percent of all arrests, as shown in ▶ Figure 7.6. Compared with African Americans, arrest rates for whites were higher for nonviolent property crimes such as fraud and larceny-theft but were lower for violent crimes such as robbery. In 2009 whites accounted for slightly more than 67 percent of all arrests for property crimes and about 58 percent of arrests for violent crimes. African Americans, who account for about 13 percent of the U.S. population, made up 39 percent of arrests for violent crimes and 30 percent of arrests for property crimes (FBI, 2010).

Although official arrest records reveal certain trends, these data tell us very little about the actual dynamics of crime by racial–ethnic category. According to official statistics, African Americans are overrepresented in arrest data. In 2009 African Americans made up about 13 percent of the U.S. population but accounted for about 28 percent of all arrests. More specifically, African Americans made up 75 percent of arrests for gambling, 55 percent of arrests for robbery, 51 percent of arrests for suspicion, and 49 percent of arrests for murder. Likewise, African Americans are more likely than people in other racial or ethnic classifications to become the victims of crime (FBI, 2010).

It is now impossible to separate out arrest rates for white (non-Hispanic) offenders and Hispanic (Latino/a) offenders because the Uniform Crime Report no longer reports these data separately. In 2007, the last year for which data were reported separately, Latinos/as made up about 13 percent of the U.S. population and accounted for about 13 percent of all arrests. Latinos have higher arrest rates for nonviolent drug offenses and immigration violations. In 2008

Latinos/as made up almost one-third (33 percent) of all inmates incarcerated in the federal prison system.

In 2008 Native Americans (designated in the UCR as "American Indian" or "Alaskan Native") accounted for 1.3 percent of all arrests; however, the majority of these arrests were for larceny-theft, assaults, vandalism, and alcohol- and drug-related violations (FBI, 2009). In that same year, 1.1 percent of all arrests were of Asian Americans, Native Hawaiians, or Pacific Islanders. Among the higher percentages of arrests for members of this category were prostitution, commercialized vice, and runaways under 18 years of age (FBI, 2009).

Are arrest statistics a true reflection of crimes committed? The Federal Bureau of Investigation indicates that they are not, as did the late criminologist Coramae Richey Mann (1993), who argued that arrest statistics are skewed because reporting practices differ in accordance with race and class. According to Mann, arrest statistics reflect the UCR's focus on violent and property crimes, especially property crimes, which are committed primarily by low-income people. This emphasis draws attention away from the white-collar and elite crimes committed by middle- and upper-income people. Police may also demonstrate bias and racism in their decisions regarding whom to question, detain, or arrest under certain circumstances (Mann, 1993). Some law enforcement officials believe that problems such as these primarily occurred in the past; however, issues still arise in the twenty-first century about police brutality against persons of color and about unequal treatment of individuals who reside in racially segregated, low-income areas of urban centers and rural communities.

Another reason that statistics may show a disproportionate number of people of color being arrested is the focus of law enforcement on certain types of crime and certain neighborhoods in which crime is considered more prevalent. As discussed previously, many poor, young, central-city males turn to forms of criminal activity because of their belief that no opportunities exist for them to earn a living wage through legitimate employment. Because of the trend of law enforcement efforts to focus on drug-related offenses, arrest rates for young people of color have risen rapidly. These young people are also more likely to live in central-city areas, where there are more police patrols to make arrests.

Immigration crackdowns in various cities, such as Los Angeles, San Diego, and San Antonio, have produced higher rates of arrests for Latinos/as who are in the United States legally, as well as those who are undocumented workers seeking to provide a living for family members "back home" in Mexico, South America, or other areas of the world. Individuals who are in possession of drugs or have outstanding warrants for other violations may be arrested by federal authorities conducting immigration stings even when the primary focus of such raids is to identify individuals who are illegally residing in the United States.

Finally, arrests should not be equated with guilt: Being arrested does not mean that a person is guilty of the crime with which he or she has been charged. In the United States, individuals accused of crimes are, at least theoretically, "innocent until proven guilty" (Mann, 1993).

Crime Victims

How can we learn more about crime victims? The National Crime Victimization Survey (NCVS), compiled by the U.S. Census Bureau for the Bureau of Justice Statistics, provides annual data about crimes (reported and not reported to the police) from which we can find out more about who is actually victimized by crime. Based on data from the 2008 NCVS, violent and property crime rates are at or near their lowest levels since the 1980s. The violent crime rate in 2008 was 19.3 victimizations per 1,000 people age 12 or over; the property crime rate was 135 victimizations per 1,000 households (Rand, 2009).

Victimization surveys indicate that men are the most frequent victims of most crimes of violence and theft. Among males who are now 12 years old, an estimated 89 percent will be the victim of a violent crime at least once during their lifetime, as compared with 73 percent of females. The elderly also tend to be more fearful of crime but are the least likely to be victimized. Young men of color between the ages of 12 and 24 have the highest criminal victimization rates. African American males are more likely to be victimized than African American females, younger African Americans are more likely to be the victims of violent crime than older African Americans, African Americans with lower annual incomes are at greater risk of violence than those in households with higher annual incomes, and African Americans living in urban areas are more likely than those in suburban or rural areas to be victims of crime (Harrell, 2007).

A study by the Justice Department found that Native Americans are more likely to be victims of violent crimes than are members of any other racial category and that the rate of violent crimes against Native American women was about 50 percent higher than that for African American men (Perry, 2004). During the period covered in the study (1992–2002), Native Americans were the victims of violent crimes at a rate more than twice the national average. They were also more likely to be the victims of violent crimes committed by members of a race other than their own (Perry, 2004). According to the survey, the average annual rate at which Native Americans were victims of violent crime—101 crimes per 1,000 people, ages 12 or older—is about two-and-a-half times the national average of 41 crimes per 1,000 people who are above the age of 12. By comparison, the average annual rate for whites was 41 crimes per 1,000 people; for African Americans, 50 per 1,000; and for Asian Americans, 22 per 1,000 (Perry, 2004).

The burden of robbery victimization falls more heavily on some categories of people than others. NCVS data indicate that males are robbed at almost twice the rate of females. African Americans are more than twice as likely to be robbed as whites. Young people have a much greater likelihood of being robbed than middle-aged and older persons. Persons from lower-income families are more likely to be robbed than people from higher-income families (Harrell, 2007).

Victimization studies show that Asian Americans, Native Hawaiians, and other Pacific Islanders have the lowest rates of both violent and property crime victimizations. These population groups make up about 4 percent of the U.S. population but are victims in only 2 percent of nonfatal violent crimes and 3 percent of property crimes per year (Harrell, 2009). Asian Americans are less likely to be murdered by someone of their same race than are white victims (78 percent) or black victims (92 percent), and most nonfatal crimes against Asians are committed by strangers (Harrell, 2009). Shifting to property crimes, Asian households have lower rates of property crime than non-Asian households at nearly every income level and in every U.S. region.

Across racial/ethnic categories, households in the lowest income group (less than $7,500 per year) experienced property crime rates that were about 1.5 times higher than the rates for households earning $75,000 or more per year. Property crime rates for burglary were more than three times higher in the lowest income group as compared with highest income category. Motor vehicle theft is the primary exception: No significant difference exists in the rates of motor vehicle theft across household income levels.

What can we learn from victimization studies? Because all crimes are not reported, we can find out more about the nature and extent to which people in specific regions of the country, income categories, racial or ethnic groupings, ages, and other demographic characteristics are victimized by violent and property crimes in the United States. These data can be compared with official crime statistics to see if arrest rates and convictions are an accurate reflection of crime in the United States. The NCVS particularly helps us to learn about the types and number of offenses that go unreported to police or other law enforcement officials who are part of the criminal justice system.

© AP Photo/Bebeto Matthews

▲ The problems that some police departments have with racial issues can generate tragic consequences. Omar Edwards, an off-duty African American New York City police officer, was running after a man who had tried to break into Edwards's car. Three Caucasian plainclothes officers noticed the altercation and ordered the men to stop. When Edwards turned to them with his gun in his hand, one of the officers shot him fatally.

The Criminal Justice System

Of all of the agencies of social control (including families, schools, and churches) in contemporary societies, only the criminal justice system has the power to control crime and punish those who are convicted of criminal conduct. The ***criminal justice system*** **refers to the more than 55,000 local, state, and federal agencies that enforce laws, adjudicate crimes, and treat and rehabilitate criminals.** The system includes the police, the courts, and correctional facilities, and it employs more than 2 million people in 17,000 police agencies, nearly 17,000 courts, more than 8,000 prosecutorial agencies,

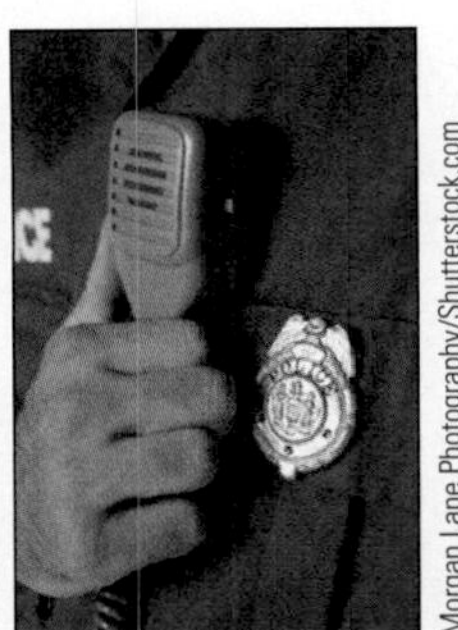
Morgan Lane Photography/Shutterstock.com

Police

- Enforce specific laws
- Investigate specific crimes
- Search people, vicinities, buildings
- Arrest or detain people

Fuse/Image100/Jupiter Images

Prosecutors

- File charges or petitions for judicial decision
- Seek indictments
- Drop cases
- Reduce charges
- Recommend sentences

Image Source/Jupiter Images

Judges or Magistrates

- Set bail or conditions for release
- Accept pleas
- Determine delinquency
- Dismiss charges
- Impose sentences
- Revoke probation

▲ FIGURE 7.7 DISCRETIONARY POWERS IN LAW ENFORCEMENT

about 6,000 correctional institutions, and more than 3,500 probation and parole departments. More than $150 billion is spent annually for civil and criminal justice, which amounts to more than $500 for every person living in the United States (Siegel, 2006).

The term *criminal justice system* is somewhat misleading because it implies that law enforcement agencies, courts, and correctional facilities constitute one large, integrated system, when, in reality, the criminal justice system is made up of many bureaucracies that have considerable discretion in how decisions are made. *Discretion* refers to the use of personal judgment by police officers, prosecutors, judges, and other criminal justice system officials regarding whether and how to proceed in a given situation (see ▶ Figure 7.7). The police are a prime example of discretionary processes because they have the power to selectively enforce the law and have on many occasions been accused of being too harsh or too lenient on alleged offenders.

The Police

The role of the police in the criminal justice system continues to expand. The police are responsible for crime control and maintenance of order, but local police departments now serve numerous other human-service functions, including improving community relations, resolving family disputes, and helping people during emergencies. It should be remembered that not all "police officers" are employed by local police departments; they are employed in more than 25,000 governmental agencies ranging from local jurisdictions to federal levels. However, we will focus primarily on metropolitan police departments because they constitute the vast majority of the law enforcement community.

Metropolitan police departments are made up of a chain of command (similar to the military), with ranks such as officer, sergeant, lieutenant, and captain, and each rank must follow specific rules and procedures. However, individual officers maintain a degree of discretion in the decisions they make as they respond to calls and try to apprehend fleeing or violent offenders. The problem of police discretion is most acute when decisions are made to use force (such as grabbing, pushing, or hitting a suspect) or deadly force (shooting and killing a suspect). Generally, deadly force is allowed only in situations in which a suspect is engaged in a felony, is fleeing the scene of a felony, or is resisting arrest and has endangered someone's life.

Although many police departments have worked to improve their public image in recent years, the practice of *racial profiling*—the use of ethnic or racial background as a means of identifying criminal suspects—remains a highly charged issue. Officers in some police departments have singled out for discriminatory treatment African Americans, Latinos/as, and other people of color, treating them more harshly than white (Euro-American) individuals. However, police department officials typically contend that race is only one factor in determining why individuals are questioned or detained as they go about everyday activities such as driving a car or walking down the street. By contrast, equal-justice advocacy groups argue that differential treatment of minority-group members amounts to a race-based double standard, which they believe exists not only in police work but throughout the criminal justice system (see Cole, 2000).

The belief that differential treatment takes place on the basis of race contributes to a negative image of

criminal justice system the more than 55,000 local, state, and federal agencies that enforce laws, adjudicate crimes, and treat and rehabilitate criminals.

sociology and social policy

Juvenile Offenders and "Equal Justice Under Law"

When you walk into the U.S. Supreme Court building in Washington, D.C., it is impossible to miss the engraved statement overhead: "Equal Justice Under Law." Do young people, regardless of race, class, or gender, receive the same treatment under the law?

In courtrooms throughout the nation, judges have a wide range of discretion in their decisions regarding juveniles who are alleged to have committed some criminal or status offense. Whereas judges in television courtroom dramas are often African Americans, women, or members of other subordinate groups, "real-life" judges typically come from capitalist or managerial and professional backgrounds. Because more than 90 percent are white and most are male, their decisions may reflect a built-in class, racial, and gender bias.

Juvenile courts were established under a different premise than courts for adults. Under the doctrine of *parens patriae* (the state as parent), the official purpose of juvenile courts has been to care for, rather than punish, youthful offenders. In theory, less weight is given to offenses and more weight to the youth's physical, mental, or social condition. The juvenile court seeks to change or resocialize offenders through treatment or therapy, not to punish them. Consequently, judges in juvenile courts are given relatively wide latitude, or discretion, in the decisions they mete out regarding young offenders.

Unlike adult offenders, juveniles are not always represented by legal counsel. A juvenile hearing is not a trial but rather an informal private hearing before a judge or probation officer, with only the young person and a parent or guardian present. No jury is convened, and the juvenile offender does not cross-examine her or his accusers. In addition, the offender is not "sentenced"; rather, the case is "adjudicated" or "disposed of." Finally, the offender is not "punished" but instead may be "remanded to the custody" of a youth authority in order to receive training, treatment, or care.

Because of judicial discretion, courts may treat juveniles differently based on gender. Considerable disparity exists in the disposition of juvenile cases, with much of the variation thought to result from judges' beliefs rather than objective facts in the case. Female offenders are more likely than males to be institutionalized for committing status offenses such as truancy, running

police among many people of color who believe that they have been hassled by police officers, and this assumption is intensified by the fact that police departments have typically been made up of white male personnel at all levels. In recent years this situation has slowly begun to change. Currently, about 22 percent of all *sworn officers*—those who have taken an oath and been given the powers to make arrests and use necessary force in accordance with their duties—are women and minorities (Cole and Smith, 2004). The largest percentage of minority and women police officers are located in cities with a population of 250,000 or more. African Americans make up a larger percentage of the police department in cities with a larger proportion of African American residents (such as Detroit), but Latinos/as constitute a larger percentage in cities such as San Antonio and El Paso, Texas, where Latinos/as make up a larger proportion of the population. Women officers of all races are more likely to be employed in departments in cities of more than 250,000 as compared with smaller communities (cities of less than 50,000), where women officers constitute a small percentage of the force.

Police departments now place greater emphasis on *community-oriented policing*—an approach to law enforcement in which officers maintain a presence in the community, walking up and down the streets or

▲ After Phillip Alpert, 18, of Orlando, Florida, had an argument with his 16-year-old girlfriend, he used the Internet to send a nude photo of her to her friends and family, a practice called "sexting." Much to Alpert's dismay, he was arrested on a felony charge of distributing child pornography, was convicted and given five years' probation, and was placed on Florida's registered sex offenders list. His attorney—and others—see the last part of the punishment as an example of cultural lag, wherein the law has not yet appropriately addressed the cybertechnology, and the attorney is trying to get Albert's name off the list.

Juvenile detention centers are not luxury facilities, as a view of this room in a Texas institution makes clear.

away from home, and other offenses that serve as "buffer charges" for suspected sexual misconduct (Chesney-Lind, 1989).

Disparity also exists on the basis of race and class. Judges tend to see youths from white, middle- or upper-class families as being very much like their own children and to believe that the families will take care of the problem on their own. They may view juveniles from lower-income families or other racial–ethnic groups as delinquents in need of attention from authorities. Furthermore, some judges view gang members from impoverished central cities as "guilty by association" because of their companions.

The political climate may have an effect on how judges dispose of juvenile cases. In the process of dealing with the public perception that the juvenile justice system is too lenient, some judges may have inadvertently contributed to other problems. Many more youths have been remanded to overcrowded juvenile detention facilities that are unable to provide necessary educational, health, and social services. Based on a judge's discretion, many juvenile offenders are incarcerated under indeterminate sentences and placed in a detention facility that may serve merely as a school for adult criminality.

reflect & analyze

Do you believe that the approach to juvenile justice that is used in many U.S. states is fair? Why or why not? Have you or someone you know had a direct experience with the juvenile courts? If so, what was the effect of that experience?

Sources: Based on Barlow and Kauzlarich, 2002; Chesney-Lind, 1989; and Inciardi, Horowitz, and Pottieger, 1993.

riding bicycles, getting to know people, and holding public service meetings at schools, churches, and other neighborhood settings. Community-oriented policing is often limited by budget constraints and the lack of available personnel to conduct this type of "hands-on" community involvement. In many jurisdictions, police officers believe that they have only enough time to keep up with reports of serious crime and life-threatening occurrences and that the level of available personnel and resources does not allow officers to take on a greatly expanded role in the community.

The Courts

Criminal courts determine the guilt or innocence of those persons accused of committing a crime. In theory, justice is determined in an adversarial process in which the prosecutor (an attorney who represents the state) argues that the accused is guilty, and the defense attorney asserts that the accused is innocent. In reality, judges wield a great deal of discretion. Working with prosecutors, they decide whom to release and whom to hold for further hearings, and what sentences to impose on those persons who are convicted.

Prosecuting attorneys also have considerable leeway in deciding which cases to prosecute and when to negotiate a plea bargain with a defense attorney. As cases are sorted through the legal machinery, a steady attrition occurs. At each stage, various officials determine what alternatives will be available for those cases still remaining in the system. These discretionary decisions often have a disproportionate impact on youthful offenders who are poor (see "Sociology and Social Policy").

About 90 percent of criminal cases are never tried in court; instead, they are resolved by plea bargaining, a process in which the prosecution negotiates a reduced sentence for the accused in exchange for a guilty plea (Senna and Siegel, 2002). Defendants (especially those who are poor and cannot afford to pay an attorney) may be urged to plead guilty to a lesser crime in return for not being tried for the more serious crime for which they were arrested. Prison sentences given in plea bargains vary widely from one region to another and even from judge to judge within one state.

Those who advocate the practice of plea bargaining believe that it allows for individualized justice for alleged offenders because judges, prosecutors, and defense attorneys can agree to a plea and to a punishment that best fits the offense and the offender. They also believe that this process helps reduce the backlog of criminal cases in the court system as well as the lengthy process often involved in a criminal trial. However, those who seek to abolish plea bargaining believe that this practice leads to innocent people pleading guilty to crimes they have not committed or pleading guilty to a crime other than the one they actually committed because they are offered a lesser sentence. More-serious crimes, such as murder, felonious assault, and rape, are more likely to proceed to trial than other forms of criminal conduct; however, many of these cases do not reach the trial stage.

One of the most important activities of the court system is establishing the sentence of the accused after he or she has been found guilty or has pleaded guilty. Typically, sentencing involves the following kinds of sentences or dispositions: fines, probation, alternative or intermediate sanctions (such as house arrest or electronic monitoring), incarceration, and capital punishment (Siegel, 2006). However, adult courts operate differently from those established for juvenile offenders.

Punishment and Corrections

Although the United States makes up less than 5 percent of the world's population, our nation accounts for almost 25 percent of the world's prison population. Some analysts suggest that our laws prescribe greater punishment for some offenses than those in other nations, resulting in Americans being locked up for crimes that rarely would result in prison sentences in other countries (Liptak, 2008). About 2.3 million people in the United States are being "punished" at any given time through jail terms and prison sentences. It should be noted that jails differ from prisons. Most jails are run by a local government or a sheriff's department. They are designed to hold people before they make bail, when they are awaiting trial, or when they are serving short sentences for committing a misdemeanor. By contrast, prisons are operated by state governments and the Federal Bureau of Prisons, and are designed to hold individuals convicted of felonies. Some prisons are operated by private contractors that build and control the facilities while receiving public monies for their operation. Both jails and prisons are based on the assumption that punishment and/or corrections are necessary to protect the public good and to effectively deal with those who violate laws.

Punishment **is any action designed to deprive a person of things of value (including liberty) because of some offense the person is thought to have committed** (Barlow and Kauzlarich, 2002). Historically, punishment has had four major goals:

1. *Retribution* is punishment that a person receives for infringing on the rights of others (Cole and Smith, 2004). Retribution imposes a penalty on the offender and is based on the premise that the punishment should fit the crime: The greater the degree of social harm, the more the offender should be punished. For example, an individual who murders should be punished more severely than one who shoplifts.
2. *General deterrence* seeks to reduce criminal activity by instilling a fear of punishment in the general public. However, we most often focus on *specific deterrence*, which inflicts punishment on specific criminals to discourage them from committing future crimes. Recently, criminologists have debated whether imprisonment has a deterrent effect, given the fact that high rates (between 30 and 50 percent) of those who are released from prison become recidivists (previous offenders who commit new crimes).
3. *Incapacitation* is based on the assumption that offenders who are detained in prison or are executed will be unable to commit additional crimes. This approach is often expressed as "Lock 'em up and throw away the key!" In recent years more emphasis has been placed on *selective incapacitation*, which means that offenders who repeat certain kinds of crimes are sentenced to long prison terms (Cole and Smith, 2004).
4. *Rehabilitation* seeks to return offenders to the community as law-abiding citizens by providing therapy or vocational or educational training. Based on this approach, offenders are treated, not punished, so that they will not continue their criminal activity. However, many correctional facilities are seriously understaffed and underfunded in the rehabilitation programs that exist. The job skills (such as agricultural work) that many offenders learn in prison do not transfer to the outside world, nor are offenders given any assistance in finding work that fits their skills once they are released.

Recently, newer approaches have been advocated for dealing with criminal behavior. Key among these is the idea of *restoration,* which is designed to repair the damage done to the victim and the community by an offender's criminal act (Cole and Smith, 2004). This approach is based on the *restorative justice perspective,* which states that the criminal justice system should promote a peaceful and just society; therefore, the system should focus on peacemaking rather than on punishing offenders. Advocates of this approach believe that punishment of offenders actually encourages crime rather than deterring it and are in favor of approaches such as probation with treatment. Opponents of this approach suggest that increased punishment of offenders leads to lower crime rates and that the restorative justice approach amounts to "coddling criminals." However, numerous restorative justice programs are now in operation, and many are associated with community policing programs as they seek to help offenders realize the damage that they have done to their victims and the community and to be reintegrated into society (Senna and Siegel, 2002).

Instead of the term *punishment*, the term *corrections* is often used. Criminologists George F. Cole and Christopher E. Smith (2004: 409) explain corrections as follows:

> Corrections refers to the great number of programs, services, facilities, and organizations responsible for the management of people accused or convicted of criminal offenses. In addition to prisons and jails, corrections includes probation, halfway houses, education and work release programs, parole supervision, counseling, and community service. Correctional programs operate in Salvation Army hostels, forest camps, medical clinics, and urban storefronts.

As Cole and Smith (2004) explain, corrections is a major activity in the United States today. Consider the fact that about 6.5 million adults (more than one out of every twenty men and one out of every hundred women) are under some form of correctional control. The rate of African American males under some form of correctional supervision is even greater (one out of every six African American adult men and one out of every three African American men in their twenties). Some analysts believe that these figures are a reflection of centuries of underlying racial, ethnic, and class-based inequalities in the United States as well as sentencing disparities that reflect race-based differences in the criminal justice system. However, others argue that newer practices such as determinate or mandatory sentences may help to reduce such disparities over time. A *determinate sentence* sets the term of imprisonment at a fixed period of time (such as three years) for a specific offense. *Mandatory sentencing guidelines* are established by law and require that a person convicted of a specific offense or series of offenses be given a penalty within a fixed range. Although these practices limit judicial discretion in sentencing, many critics are concerned about the effects of these sentencing approaches. Another area of great discord within and outside the criminal justice system is the issue of the death penalty.

The Death Penalty

Historically, removal from the group has been considered one of the ultimate forms of punishment. For many years capital punishment, or the death penalty, has been used in the United States as an appropriate and justifiable response to very serious crimes. In 2009, 52 inmates were executed (as contrasted with an all-time high of 98 in 1999), and about 3,300 people awaited execution, having received the death penalty under federal law or the laws of one of the states that have the death penalty (Death Penalty Information Center, 2010). By far, the largest number of people on death row are in states such as California, Florida, Texas, Pennsylvania, and Alabama (see ▶ Figure 7.8).

Because of the finality of the death penalty, it has been a subject of much controversy and numerous Supreme Court debates about the decision-making process involved in capital cases. In 1972 the U.S. Supreme Court ruled (in *Furman v. Georgia*) that *arbitrary* application of the death penalty violates the Eighth Amendment to the Constitution but that the death penalty itself is not unconstitutional. In other words, capital punishment is legal if it is fairly imposed. Although there have been a number of cases involving death penalty issues before the Supreme Court since that time, the court typically has upheld the constitutionality of this practice. Yet the fact remains that racial disparities are highly evident in the death row census. African Americans make up about 42 percent of the death row population but less than 13 percent of the U.S. population. In 96 percent of the states where there have been reviews of race and the death penalty, a distinct pattern emerged in regard to either race-of-victim or race-of-defendant discrimination: 243 black defendants were executed when the victim was white, while only 15 persons were executed in white defendant/black victim cases (Death Penalty Information Center, 2010).

People who have lost relatives and friends as a result of criminal activity often see the death penalty as justified. However, capital punishment raises many doubts for those who fear that innocent individuals may be executed for crimes they did not commit. For still others, the problem of racial discrimination in the sentencing process poses troubling questions. Other questions that remain today involve the execution of those who are believed to be insane and of those defendants who did not have effective legal counsel during their trial. In 2002, for example, the Supreme Court ruled (in *Atkins v. Virginia*) that executing the mentally retarded is unconstitutional. In another landmark case (*Ring v. Arizona*), the Court ruled that juries, not judges, must decide whether a convicted murderer should receive the death penalty (Cole and Smith, 2004).

Executions resumed in 2008 after a *de facto* moratorium was lifted by the Supreme Court when the justices decided to uphold lethal injection. However, only southern states returned to regular executions, and executions in the South accounted for over 80 percent of all executions in 2009. With 24 executions, Texas accounted for about 50 percent of all executions in

punishment any action designed to deprive a person of things of value (including liberty) because of some offense the person is thought to have committed.

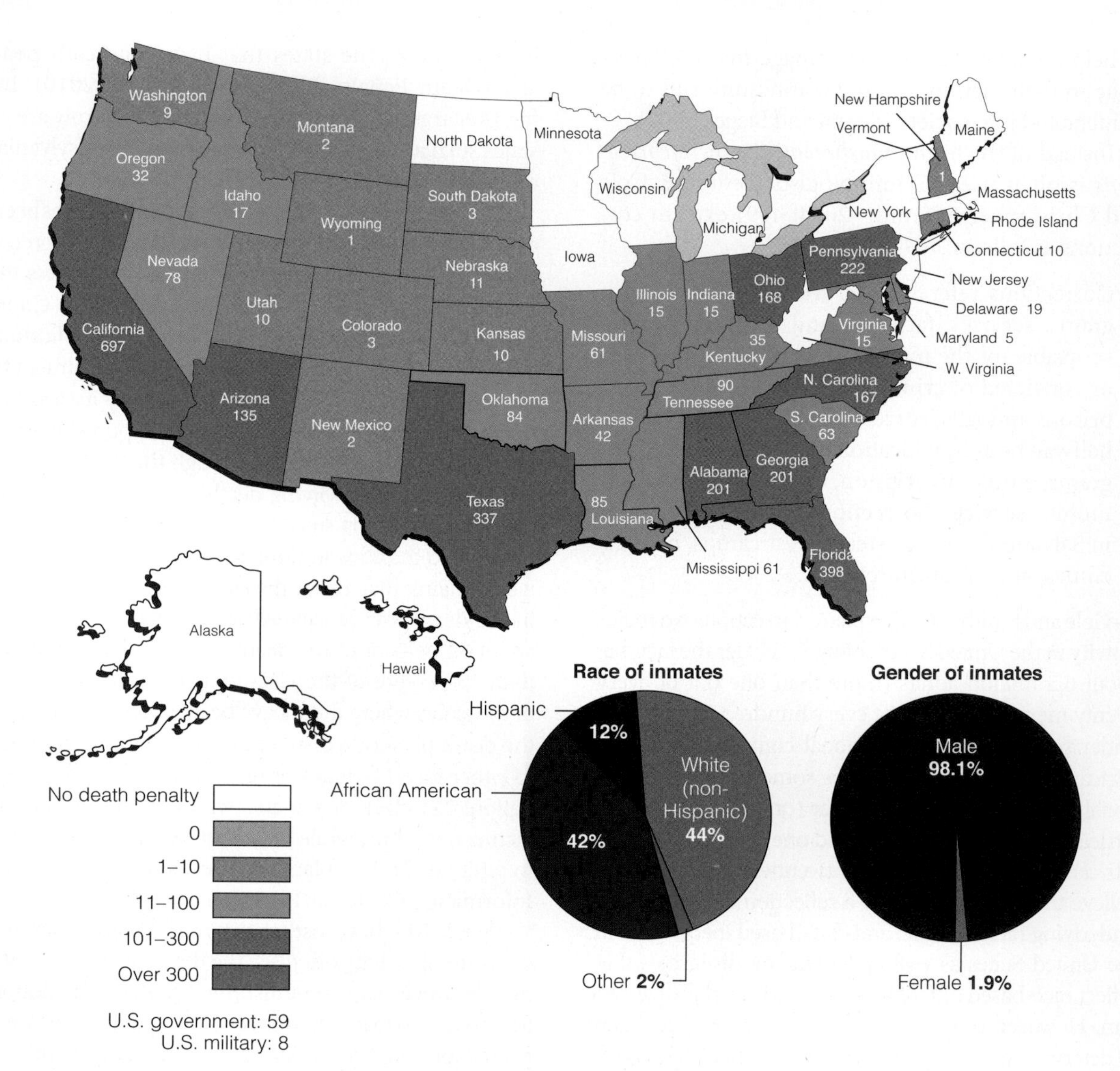

▲ **FIGURE 7.8 DEATH ROW CENSUS, JANUARY 1, 2010**

Death row inmates are heavily concentrated in certain states. African Americans, who make up less than 13 percent of the U.S. population, account for approximately 42 percent of inmates on death row. In addition to those persons held by state governments, 67 inmates were held on death row by the federal government or the U.S. military.

Source: Death Penalty Information Center, 2010.

that year (Death Penalty Information Center, 2010). Although only 106 new death sentences were handed down in 2009 (down from a high of 328 in 1994), the issue of the death penalty is far from resolved; the debate, which has taken place for more than two hundred years, no doubt will continue into the second and third decades of the twenty-first century.

U.S. Deviance and Crime in the Future

Two pressing questions pertaining to deviance and crime in the United States will face us in the future: Is the solution to our "crime problem" more law and order? Is equal justice under the law possible?

Although many people in the United States agree that crime is one of the most important problems in this country, they are divided over what to do about it. Some of the frustration about crime might be based on unfounded fears; studies show that the overall crime rate has been decreasing slightly in recent years.

One thing is clear: The existing criminal justice system cannot solve the "crime problem." If roughly 20 percent of all crimes result in arrest, only half of those lead to a conviction in serious cases, and less than 5 percent of those result in a jail term, the "lock 'em up and throw the key away" approach has little chance of succeeding. Nor does the high rate of recidivism among those who have been incarcerated speak well for the rehabilitative efforts of our existing correctional facilities. Reducing street crime may hinge on finding ways to short-circuit criminal behavior.

One of the greatest challenges is juvenile offenders, who may become the adult criminals of tomorrow.

© Robin Nelson/PhotoEdit

▲ In recent years, military-style boot camps such as this one have been used as an alternative to prison and long jail terms for nonviolent offenders under age 30. Critics argue that structural solutions—not stopgap measures such as these camps—are needed to reduce crime.

However, instead of military-style boot camps or other stopgap measures, *structural solutions*—such as more and better education and jobs, affordable housing, more equality and less discrimination, and socially productive activities—are needed to reduce street crime. In the past, structural solutions such as these have made it possible for immigrants who initially committed street crimes to leave the streets, get jobs, and lead productive lives. Ultimately, the best approach for reducing delinquency and crime would be prevention: to work with young people *before* they become juvenile offenders to help them establish family relationships, build self-esteem, choose a career, and get an education that will help them pursue that career. Sociologist Elliott Currie (1998) has proposed that an initial goal in working to prevent delinquency and crime is to pinpoint specifically what kinds of preventive programs work and to establish priorities that make prevention possible. Among these priorities are preventing child abuse and neglect, enhancing children's intellectual and social development, providing support and guidance to vulnerable adolescents, and working intensively with juvenile offenders (Currie, 1998).

Is equal justice under the law possible? As long as racism, sexism, classism, and ageism exist in our society, people will see deviant and criminal behavior through a selective lens. To solve the problems addressed in this chapter, we must ask ourselves what we can do to ensure the rights of everyone, including the poor, people of color, and women and men alike. Many of us can counter classism, racism, sexism, and ageism where they occur. Perhaps the only way that the United States can have equal justice under the law (and, perhaps, less crime as a result) in the future is to promote social justice for individuals regardless of their race, class, gender, or age.

Transnational Crime and the Global Criminal Economy

Transnational crime occurs across multiple national borders. This type of crime not only involves crossing borders between countries but also crimes in which crossing national borders is *essential* to the criminal activity. Much of transnational crime is conducted by organized criminal groups that use systematic violence and corruption to achieve their goals. Often these criminal networks are able to prey on less-powerful governments that do not have the resources to oppose them. According to the National Institute of Justice (2011), transnational crime should be of concern to people living in the United States because this type of criminal activity has a detrimental effect on everyone, not just the nations that are destabilized by these activities, often through the use of bribery, violence, or terror.

Transnational crime is fueled by globalization, which has brought increased travel, expanded international trade, and advances in telecommunications and computer technology. This type of criminal activity cannot be controlled by one nation alone. In 2000 the United Nations Convention against Transnational Organized Crime was signed by member nations in an effort to counter this type of crime. However, terrorist activities in the United States in 2001, followed by various acts of terrorism in other nations, led some analysts to believe that terrorism and international organized crime intersect and overlap each other. Other analysts disagree to the extent that this occurs. They argue that transnational criminal organizations and terrorist groups often adopt similar methods but strive for divergent ends: Crime is economically driven; terrorism is rooted in political goals.

How much money and other resources change hands in the global criminal economy? Although the exact amount of profits and financial flows originating in the global criminal economy is impossible to determine, the United Nations Conference on Global Organized Crime estimated that more than $600 billion (in U.S. currency) per year is accrued in the global trade in drugs alone. Today, profits from all kinds of global criminal activities are estimated to range from $750 billion to more than $2 trillion per year (United Nations Development Programme, 2009). Some analysts believe that even these figures underestimate the true nature and extent of the global criminal economy (Castells, 1998). The highest-income-producing activities of global criminal organizations include trafficking in drugs, weapons, and nuclear material; smuggling of things and people (including many migrants);

trafficking in women and children for the sex industry; and trafficking in body parts such as corneas and major organs for the medical industry. Undergirding the entire criminal system are money laundering and various complex financial schemes and international trade networks that make it possible for people to use the resources they obtain through illegal activity for the purposes of consumption and investment in the ("legitimate") formal economy.

Can anything be done about transnational crime? Recent studies have concluded that reducing global crime will require a global response, including the cooperation of law enforcement agencies, prosecutors, and intelligence services across geopolitical boundaries. However, this approach is problematic because countries such as the United States often have difficulty getting the various law enforcement agencies to cooperate within their own nation. Similarly, law enforcement agencies in high-income nations such as the United States and Canada are often suspicious of law enforcement agencies in low-income countries, believing that these law enforcement officers are corrupt. Regulation by the international community (for example, through the United Nations) would also be necessary to control global criminal activities such as international money laundering and trafficking in people and controlled substances such as drugs and weapons. However, development and enforcement of international agreements on activities such as the smuggling of migrants or the trafficking of women and children for the sex industry have been extremely limited thus far. Many analysts acknowledge that economic globalization has provided great opportunities for wealth through global organized crime (Castells, 1998; United Nations Development Programme, 2009).

chapter review Q & A

Use these questions and answers to check how well you've achieved the learning objectives set out at the beginning of this chapter.

- **How do sociologists view deviance?**

Sociologists are interested in what types of behavior are defined by societies as "deviant," who does that defining, how individuals become deviant, and how those individuals are dealt with by society.

- **What are the main functionalist theories for explaining deviance?**

Functionalist perspectives on deviance include strain theory and opportunity theory. Strain theory focuses on the idea that when people are denied legitimate access to cultural goals, such as a good job or a nice home, they may engage in illegal behavior to obtain them. Opportunity theory suggests that for deviance to occur, people must have access to illegitimate means to acquire what they want but cannot obtain through legitimate means.

- **How do conflict and feminist perspectives explain deviance?**

Conflict perspectives on deviance focus on inequalities in society. Marxist conflict theorists link deviance and crime to the capitalist society, which divides people into haves and have-nots, leaving crime as the only source of support for those at the bottom of the economic ladder. Feminist approaches to deviance focus on the relationship between gender and deviance.

- **How do symbolic interactionists view deviance?**

According to symbolic interactionists, deviance is learned through interaction with others. Differential association theory states that individuals have a greater tendency to deviate from societal norms when they frequently associate with persons who tend toward deviance instead of conformity. According to social control theories, everyone is capable of committing crimes, but social bonding (attachments to family and to other social institutions) keeps many from doing so. According to labeling theory, deviant behavior is that which is labeled deviant by those in powerful positions.

- **What is the postmodernist view on deviance?**

Postmodernist views on deviance focus on how the powerful control others through discipline and surveillance. This control may be maintained through largely invisible forces such as the Panoptican, as described by Michel Foucault, or by newer technologies that place everyone—not just "deviants"—under constant surveillance by authorities, who use their knowledge as power over others.

- **How do sociologists classify crime?**

Sociologists identify six main categories of crime: violent crime (murder, forcible rape, robbery, and aggravated assault), property crime (burglary, motor vehicle theft, larceny-theft, and arson), public order crimes (sometimes referred to as "morals" crimes), occupational (white-collar) crime, organized crime, and political crime.

- **What are the main sources of crime statistics?**

Official crime statistics are taken from the Uniform Crime Report, which lists crimes reported to the police, and the National Crime Victimization Survey, which interviews households to determine the incidence of crimes, including those not reported to police. Studies show that many more crimes are committed than are officially reported.

- **How are age and class related to crime statistics?**

Age is the key factor in crime statistics. Persons under age 25 account for almost one-half of all arrests for violent crime and almost 55 percent of all arrests for property crime. Persons from lower socioeconomic backgrounds are more likely to be arrested for violent and property crimes; white-collar crime is more likely to occur among the upper socioeconomic classes.

- **Who are the most frequent victims of crime?**

Young males of color between ages 12 and 24 have the highest criminal victimization rates. The elderly tend to be fearful of crime but are the least likely to be victimized.

- **How is discretion used in the criminal justice system?**

The criminal justice system, including the police, the courts, and prisons, often has considerable discretion in dealing with offenders. The police use discretion in deciding whether to act on a situation. Prosecutors and judges use discretion in deciding which cases to pursue and how to handle them.

key terms

corporate crime 197
crime 181
criminal justice system 204
criminology 182
deviance 180
differential association theory 190
illegitimate opportunity structures 186
juvenile delinquency 181
labeling theory 191
occupational (white-collar) crime 195
organized crime 197
political crime 198
primary deviance 192
property crimes 195
punishment 208
rational choice theory of deviance 191
secondary deviance 192
social bond theory 191
social control 181
strain theory 184
terrorism 200
tertiary deviance 192
victimless crimes 195
violent crime 195

questions for critical thinking

1. Does public toleration of deviance lead to increased crime rates? If people were forced to conform to stricter standards of behavior, would there be less crime in the United States?
2. Should so-called victimless crimes, such as prostitution and recreational drug use, be decriminalized? Do these crimes harm society?
3. As a sociologist armed with a sociological imagination, how would you propose to deal with the problem of crime in the United States and around the world?

turning to video

Watch the BBC video *Obedience in North Korea* (running time 4:24), available through **CengageBrain.com**. This video gives you some glimpses of what life is like in North Korea, a society where everything is highly regimented and questioning authority is a crime. As you watch the video, think about ways in which you conform to the social norms of your culture. What are some of the ways that you conform to social norms? Why do you try to meet these expectations? What do you think would happen if you deviated from these norms?

online study resources

Go to CENGAGE brain.com to access online study resources, including the Sociology CourseMate for this text as well as special features such as video, an interactive sociology time line and interactive maps, General Social Survey (GSS) data, and U.S. Census 2010 data.

CourseMate brings course concepts to life with interactive learning, study, and exam-preparation tools that support the printed textbook. A textbook-specific website, **Sociology CourseMate** includes an integrated interactive eBook and other interactive learning tools, including quizzes, flash cards, and videos.

Visit **www.cengagebrain.com** to access your account and purchase materials.

8 Class and Stratification in the United States

I was born in Mexico [in 1991]. When I was three, we crossed into the United States illegally. I had such a typical American experience. It was like honors and AP classes and pressure for National Honor Society and cross-country captain. I graduated in the top 5 percent of my class. Like really just your run-of-the-mill American experience when it comes to high school. And I never took into consideration that being undocumented was going to be a huge hurdle to cross. It means that I am going to graduate in a year and a half with two degrees in political science and English literature, and I am not going to be able to put those to good use. It just sort of hit me in the face recently that I had a year and a half left and I've worked so hard, my parents worked so hard, to get me through college. They did all of this so that I could get a job in which hopefully I could help other people but like also myself, and I'm just going to have to keep being a waitress and getting paid [so little]. That's horrible! It's a horrible feeling to have. To know that you worked so hard to get that little degree, diploma, and that in the end it's just going to be a piece of paper in your living room. If your name is out there [as being undocumented], then there is always this fear of being deported to a country you have never lived in or could not be able to live in because

AP Photo/John Smierciak

▲ In 2010 the U.S. Senate considered the DREAM Act, legislation that would have granted amnesty to some children of undocumented residents. Here, supporters of the bill greet Senator Robert Menendez and Senator Bill Nelson as they approach the Senate. The legislation failed to pass.

you don't know the inner workings of their society. And there's this fear of like "I don't want to leave my home" and this—right here—this is my home, and I would not want to have to go and live somewhere else. And there is that fear every day.

—Maricela Aguilar, a college student, expressing her concern that she has publicly identified herself as "undocumented" because she made it known that she was in the country illegally when she stood on the Milwaukee federal courthouse steps and openly supported a proposed congressional bill known as the "Dream Act" that would have benefited undocumented college students (qtd. in nytimes.com, 2011)

Chapter Focus Question:

How is the American Dream influenced by social stratification?

Many people living in the United States want to achieve the American Dream. Throughout the history of this country, people have aspired to have more than their parents and grandparents. Immigrant college students without documentation—including Maricela Aguilar (whom we met above)—are no exception. Recently, many of these students had their hopes of attaining U.S. citizenship and gaining access to the American Dream heightened when members of Congress considered a bill that would create a path to citizenship for certain young illegal immigrants who came to the U.S. as children, completed at least two years of college or military service, passed a criminal background check, and met other requirements. However, the hopes of students such as Maricela Aguilar were dashed when the U.S. Senate blocked the "Dream Act," leaving an estimated 1.2 million illegal immigrant college students and recent graduates in a state of legal limbo, with only limited economic opportunity. Many of the students have lived in this country for their entire life or for most of it. Should these students be granted citizenship and an opportunity to achieve the American Dream? Political leaders and other policy makers continue to debate the pros and cons of this question.

From a sociological perspective, this issue brings us to an important question in studying class and stratification in the United States: What is the American Dream? Simply stated, the American Dream is the belief that if people work hard and play by the rules, they will have a chance to get ahead. Moreover, each generation will be able to have a higher standard of living than that of its parents. The American Dream is based on the assumption that people in the United States have equality of opportunity regardless of their race, creed, color, national origin, gender, or religion.

For middle- and upper-income people, the American Dream typically means that each subsequent generation will be able to acquire more material possessions and wealth than people in the preceding generation. To some people, achieving the American Dream means having a secure job, owning a home, and getting a good education for their children. To others, it is the promise that anyone may rise from poverty to wealth (from "rags to riches") if he or she works hard enough.

Highlighted Learning Objectives

- Describe how prestige, power, and wealth determine social class.
- Explain why ownership of resources is a key factor in a conflict perspective on class.
- Identify ways in which social stratification and poverty are linked in a nation.

When we talk about the American Dream, it is important to realize that not all people will achieve success. The way a society is stratified has a major influence on a person's position in the class structure. In this chapter we examine systems of social stratification and describe how the U.S. class system may make it easier for some individuals to attain (or maintain) top positions in society while others have great difficulty moving up from poverty or low-income origins. Before we explore class and stratification, test your knowledge of wealth, poverty, and the American Dream by taking the Sociology and Everyday Life quiz.

What Is Social Stratification?

***Social stratification* is the hierarchical arrangement of large social groups based on their control over basic resources** (Feagin and Feagin, 2008). Stratification involves patterns of structural inequality that are associated with membership in each of these groups, as well as the ideologies that support inequality. Sociologists examine the social groups that make up the hierarchy in a society and seek to determine how inequalities are structured and persist over time.

Max Weber's term ***life chances* refers to the extent to which individuals have access to important societal resources such as food, clothing, shelter, education, and health care.** According to sociologists, more-affluent people typically have better life chances than the less affluent because they have greater access to quality education, safe neighborhoods, high-quality nutrition and health care, police and private security protection, and an extensive array of other goods and services. In contrast, persons with low- and poverty-level incomes tend to have limited access to these resources. *Resources* are anything valued in a society, ranging from money and property to medical care and education; they are considered to be scarce because of their unequal distribution among social categories. If we think about the valued resources available in the United States, for example, the differences in life chances are readily apparent. As one analyst suggested, "Poverty narrows and closes life chances. The victims of poverty experience a kind of arteriosclerosis of opportunity. Being poor not only means economic insecurity, it also wreaks havoc on one's mental and physical health" (Ropers, 1991: 25). Our life chances are intertwined with our class, race, gender, and age.

All societies distinguish among people by age. Young children typically have less authority and responsibility than older persons. Older persons, especially those without wealth or power, may find themselves at the bottom of the social hierarchy. Similarly, all societies differentiate between females and males: Women are often treated as subordinate to men. From society to society, people are treated differently as a result of their religion, race/ethnicity, appearance, physical strength, disabilities, or other distinguishing characteristics. All of these differentiations result in inequality. However, systems of stratification are also linked to the specific economic and social structure of a society and to a nation's position in the system of global stratification, which is so significant for understanding social inequality that we will devote the next chapter to this topic (Chapter 9).

Systems of Stratification

Around the globe, one of the most important characteristics of systems of stratification is their degree of flexibility. Sociologists distinguish among such systems based on the extent to which they are open or closed. In an *open system* the boundaries between levels in the hierarchies are more flexible and may be influenced (positively or negatively) by people's achieved statuses. Open systems are assumed to have some degree of social mobility. ***Social mobility* is the movement of individuals or groups from one level in a stratification system to another.** This movement can be either upward or downward. ***Intergenerational mobility* is the social movement experienced by family members from one generation to the next.** For example, Sarah's father is a carpenter who makes good wages in good economic times but is often unemployed when the construction industry slows to a standstill. Sarah becomes a neurologist, earning $350,000 a year, and moves from the working class to the upper-middle class. Between her father's generation and her own, Sarah has experienced upward social mobility.

By contrast, ***intragenerational mobility* is the social movement of individuals within their own lifetime.** Consider, for example, RaShandra, who began her career as a high-tech factory worker and through increased experience and taking specialized courses in her field became an entrepreneur, starting her own highly successful Internet-based business. RaShandra's advancement is an example of upward intragenerational social mobility. Unlike RaShandra, who sought upward mobility through her work, many others seek upward mobility through luck or talent, such as playing the lottery or competing on a television show in hopes of winning the "big" prize (see "Framing Class in the Media" on page 218). In a *closed system* the boundaries between levels in the hierarchies of social stratification are rigid, and people's positions are set by ascribed

sociology and everyday life

How Much Do You Know About Wealth, Poverty, and the American Dream?

True	False	
T	F	1. People no longer believe in the American Dream.
T	F	2. Individuals over age 65 have the highest rate of poverty.
T	F	3. Men account for two out of three impoverished adults in the United States.
T	F	4. About 5 percent of U.S. residents live in households whose members sometimes do not get enough to eat.
T	F	5. Income is more unevenly distributed than wealth.
T	F	6. People who are poor usually have personal attributes that contribute to their impoverishment.
T	F	7. A number of people living below the official poverty line have full-time jobs.
T	F	8. One in three U.S. children will be poor at some point in their childhood.

Answers on page 219.

status. Open and closed systems are ideal-type constructs; no actual stratification system is completely open or closed. The systems of stratification that we will examine—slavery, caste, and class—are characterized by different hierarchical structures and varying degrees of mobility. Let's examine these three systems of stratification to determine how people acquire their positions in each and what potential for social movement they have.

Slavery

***Slavery* is an extreme form of stratification in which some people are owned by others.** It is a closed system in which people designated as "slaves" are treated as property and have little or no control over their lives. According to some social analysts, throughout recorded history only five societies have been slave societies—those in which the social and economic impact of slavery was extensive: ancient Greece, the Roman Empire, the United States, the Caribbean, and Brazil (Finley, 1980). Others suggest that slavery also existed in the Americas prior to European settlement, and throughout Africa and Asia (Engerman, 1995).

Those of us living in the United States are most aware of the legacy of slavery in our own country. Beginning in the 1600s, slaves were forcibly imported to the United States as a source of cheap labor. Slavery was defined in law and custom by the 1750s, making it possible for one person to own another person. In fact, early U.S. presidents including George Washington, James Madison, and Thomas Jefferson owned slaves. As practiced

© Zubin Shroff/Stone/Getty Images

▲ Social mobility is the movement from one level in a stratification system to another. The background of the photo shows a traditional Indian marketplace; however, the man in the foreground shows signs of upward mobility.

social stratification the hierarchical arrangement of large social groups based on their control over basic resources.

life chances Max Weber's term for the extent to which individuals have access to important societal resources such as food, clothing, shelter, education, and health care.

social mobility the movement of individuals or groups from one level in a stratification system to another.

intergenerational mobility the social movement experienced by family members from one generation to the next.

intragenerational mobility the social movement of individuals within their own lifetime.

slavery an extreme form of stratification in which some people are owned by others.

framing class in the media

Free-Money Framing on TV Game Shows

If you and a friend were handed a million dollars in cash and had to answer only a few simple questions to keep it, could you do it? The Fox game show *Million Dollar Money Drop* gives a two-member team $1 million in cold, hard cash ($20 bills banded in 50 bundles of $20,000 each) to bet on the correct answers to seven multiple-choice questions such as this:

According to the Lenox Report, which of these is the most popular Thanksgiving side dish?

- Mashed potatoes
- Green bean casserole
- Cranberry sauce
- Candied yams

Contestants place bundles of money onto a row of trapdoors ("drops") that correspond to answers to each of the questions. When time is up and contestants have allocated their money on responses to each question, they watch the money drop(s) on which the cash is piled to find out if their choice was correct. If they are wrong, they see large stacks of money disappear through the money drop below. After seven questions, any money that is left belongs to the contestants.

On *Million Dollar Money Drop* and hundreds of other TV game shows, contestants complete for "free money," which supposedly is their own for the taking if they are able to correctly answer a few questions, endure an extreme obstacle course (ABC's *Wipeout*), select the briefcase with the most money in it (NBC's *Deal or No Deal*), or prove to be the best singer (Fox's *American Idol*) or most talented person (NBC's *America's Got Talent*). Shows such as these that give away large sums of money or other substantial prizes are extremely popular with audiences because these programs suggest that riches, fame, and happiness are just around the corner and that the American Dream can be achieved quickly and without much effort. This type of media framing would have us believe that it is fairly easy to find a shortcut to wealth and a higher position in the class structure through playing a simple game or winning a talent competition.

This sort of media framing depicts upward mobility as being similar to riding an express elevator to the top of a high-rise building: You can bypass many floors (or economic levels) in one swift move to the top. Game shows provide the illusion that a few lucky contestants can win instant riches without having to make the usual stops (such as acquiring a good education or accumulating years of work-related experience) that are typically necessary for an individual to experience upward social mobility. As defined in this chapter, *social mobility* is the movement of individuals or groups from one level in a stratification system to another. According to conflict theorists, upward social mobility has become

NBC/NBCU Photo Bank via AP Images

▲ Jessica Robinson reacts as *Deal or No Deal* host Howie Mandel opens "her" suitcase to reveal that she is a million-dollar winner. What messages do programs such as this send to viewers?

increasingly rare over the past century; however, this reality does not stop people from wishing and hoping that they will find the next express elevator to the top. The media (like state lotteries and other "giveaways") often foster the notion that upward mobility is more easily attained than is actually the case.

Shows such as *Million Dollar Money Drop* and *Deal or No Deal* are highly successful game shows because they are framed to suggest that the American Dream is just around the corner even in difficult economic times. Even if you have not seen the game shows discussed here, similar framing has been used on television entertainment shows for many years, and you can no doubt provide other examples of this approach. In each case the key message is that a quick and easy way exists to gain vast sums of money and, in some situations, to instantly achieve the American Dream if you are lucky, talented, or simply able to convince other people to vote for you.

reflect & analyze

Do the media help to *shape* our society and *create* cultural perceptions that we hold to be true about ourselves or individuals in the social world? Or are media representations merely a reflection of the nation and world in which we live?

sociology and everyday life

ANSWERS to the Sociology Quiz on Wealth, Poverty, and the American Dream

1. False. The American Dream appears to be alive and well. U.S. culture places a strong emphasis on the goal of monetary success, and many people use legal or illegal means to attempt to achieve that goal.

2. False. As a group, children have a higher rate of poverty than the elderly. Government programs such as Social Security have been indexed for inflation, whereas many of the programs for the young have been scaled back or eliminated. However, many elderly individuals still live in poverty.

3. False. Women, not men, account for two out of three impoverished adults in the United States. Reasons include the lack of job opportunities for women, lower pay than men for comparable jobs, lack of affordable day care for children, sexism in the workplace, and a number of other factors.

4. True. It is estimated that about 5 percent of the U.S. population (1 in 20 people) resides in household units where members do not get enough to eat.

5. False. Wealth is more unevenly distributed among the U.S. population than is income. However, both wealth and income are concentrated in very few hands compared with the size of the overall population.

6. False. According to one widely held stereotype, the poor are lazy and do not want to work. Rather than looking at the structural characteristics of society, people cite the alleged personal attributes of the poor as the reason for their plight.

7. True. Many of those who fall below the official poverty line are referred to as the "working poor" because they work full time but earn such low wages that they are still considered to be impoverished.

8. True. According to a study by the Children's Defense Fund, one in three U.S. children will live in a family that is below the official poverty line at some point in their childhood. For some of these children, poverty will be a persistent problem throughout their childhood and youth.

Sources: Based on Children's Defense Fund, 2001; Gilbert, 2010; and U.S. Census Bureau, American Community Survey, 2009.

in the United States, slavery had four primary characteristics: (1) it was for life and was inherited (children of slaves were considered to be slaves); (2) slaves were considered property, not human beings; (3) slaves were denied rights; and (4) coercion was used to keep slaves "in their place" (Noel, 1972). Although most slaves were powerless to bring about change, some were able to challenge slavery—or at least their position in the system—by engaging in activities such as sabotage, intentional carelessness, work slowdowns, or running away from owners and working for the abolition of slavery. Despite the fact that slavery in this country officially ended many years ago, sociologists such as Patricia Hill Collins (1990) believe that its legacy is deeply embedded in current patterns of prejudice and discrimination against African Americans.

Slavery is not simply an unfortunate historical legacy. Although legal slavery no longer exists, economist Stanley L. Engerman (1995: 175) believes that the world will not be completely free of slavery as long as there are "debt bondage, child labor, contract labor, and other varieties of coerced work for limited periods of time, with limited opportunities for mobility, and with limited political and economic power." See "Sociology in Global Perspective" on the next page for a discussion of global slavery in the twenty-first century.

The Caste System

Like slavery, caste is a closed system of social stratification. A ***caste system* is a system of social inequality in which people's status is permanently determined at birth based on their parents' ascribed characteristics.** Vestiges of caste systems exist in contemporary India and South Africa.

In India, caste is based in part on occupation; thus, families typically perform the same type of work from generation to generation. By contrast, the caste system of South Africa was based on racial classifications and the belief of white South Africans (Afrikaners) that they were morally superior to the black majority. Until the 1990s the Afrikaners controlled the government, the police, and the military by enforcing

caste system a system of social inequality in which people's status is permanently determined at birth based on their parents' ascribed characteristics.

sociology in global perspective

Slavery in the Twenty-First Century: A Global Problem

When it is mentioned we tend to think of people, almost always black people; degraded, abused and bound in chains, and we tend to think of such images, and the word *slavery* itself, as belonging to another era. We do not see slavery as belonging to our world, not as something which is still happening today.

—Rageh Omaar (2007), a BBC journalist, introducing his TV documentary on the world of modern child slavery

It is very uncomfortable for many of us to think about ways in which people traffic in human beings—particularly children in slavery. According to the United States Trafficking Victims Protection Act of 2000, trafficking is "the recruitment, harboring, transportation, provision, or obtaining of a person for labor or services, through the use of force, fraud, or coercion" (qtd. in Woods, 2010). Based on this definition, a person does not have to be physically transported from one location to another to be considered a trafficking victim. Estimates suggest that about 12.3 million people are trafficked or enslaved as forced laborers or sex workers. Some experts place the number much higher, as many as 27 million worldwide, depending on what definition of forced labor is used.

Although the legacy of slavery and forced labor remains a shameful part of the history of the United States, many people in this country prefer to think of slavery as something that occurred in the past but no longer exists. Unfortunately, social scientists, journalists, human rights activists, and other social analysts have systematically documented the presence of twenty-first-century slavery in the form of forced labor, child labor, commercial sexual exploitation, and other types of economic exploitation. In some cases, forced labor is imposed by the state, such as situations in which individuals are made to work by a government, a penal system, the military, or a rebel group. About 25 percent of forced labor is state imposed, while 75 percent is imposed by private operators who engage in prostitution, other commercial sexual activity, bonded labor, or forced domestic or agricultural labor (*BBC News*, 2005).

Globalization has further intensified the problem of slavery and trading in human beings. Tens of thousands of people from Asia, Latin America, Central Europe, and Eastern Europe have been trafficked into the United States or Canada, where they have been forced into prostitution, domestic work, agricultural labor, or factory work. In Latin America and the Caribbean, forced labor is found not only in urban areas but also in rural or indigenous populations in remote areas. Labor and sexual exploitation are two primary reasons why people are trafficked in Latin America, and children have been particularly vulnerable to domestic slavery (Skinner, 2008).

As with many other nations, child slavery has been a pressing problem in Africa and India. Chattel slavery has been found in some countries of Africa, where hundreds of thousands are born into slavery. Others are sold by their parents or abducted and required to work in various forms of labor. In India, E. Benjamin Skinner (2008), an international journalist, found that at least half a million children were in bondage, working in places ranging from private residences and tea stalls to carpet and sari factories around Delhi. Anti-Slavery International has documented the bondage of the children Skinner describes:

> Most slaves do not make products for export. Thousands of children work for no pay under threat of violence in begging stables around Mumbai or Diwali fireworks factories in Tamil Nadu. Across the country, perhaps 8 million toil in the oldest form of bondage, agricultural slavery. Some farmers enslave girls in cotton fields because, lore had it, the crops would not replenish if men reaped the harvest. In 2001, investigators found farm slaves literally in chains. . . . In southern India, tens of thousands of girls are *devadasi*—ritual sex slaves. (Skinner, 2008: 208–209)

Overall, Skinner's work documents trafficking networks and slave sales on five continents in the twenty-first century, so it comes as little surprise that he also found such practices in urban and suburban America. In describing the working conditions of "sex slaves" in spas and massage parlors in Houston, who were brought to the United States from Thailand, the journalist Mimi Swartz (2010: 107) stated that "The customers rarely seemed to grasp that the women were captives. They didn't see the other rooms. . . . These so-called spas were as tightly run as maximum-security prisons: Without permission, no one got in—or out."

From the works of journalists such as Skinner and Swartz, we learn that slavery has not disappeared; instead, it has become more global. However, it also happens in places very close to our own homes and our daily lives, whether or not we are observant of these harsh conditions of life.

reflect & analyze

Why is it important for us to be aware of problems such as human slavery even if we might believe they have no relevance to our life? Is there any relationship between the American Dream and the exploitation of people on a global basis?

apartheid—the separation of the races. Blacks were denied full citizenship and restricted to segregated hospitals, schools, residential neighborhoods, and other facilities. Whites held almost all of the desirable jobs; blacks worked as manual laborers and servants. In a caste system, marriage is endogamous, meaning that people are allowed to marry only within their own group. In India, parents traditionally have selected marriage partners for their children. In South Africa, interracial marriage was illegal until 1985.

◀ Systems of stratification include slavery, caste, and class. As shown in these photos, the life chances of people living in each of these systems differ widely.

© Hulton Archive/Getty Images

© Blend Images/John Lund/Drew Kelly/Collection Mix: Subjects/Getty Images

© Alan Sussman/The Image Works

Cultural beliefs and values sustain caste systems. Hinduism, the primary religion of India, reinforced the caste system by teaching that people should accept their fate in life and work hard as a moral duty. Caste systems grow weaker as societies industrialize; the values reinforcing the system break down, and people start to focus on the types of skills needed for industrialization.

As we have seen, in closed systems of stratification, group membership is hereditary, and it is almost impossible to move up within the structure. Custom and law frequently perpetuate privilege and ensure that higher-level positions are reserved for the children of the advantaged.

The Class System

The ***class system*** **is a type of stratification based on the ownership and control of resources and on the type of work people do.** At least theoretically, a class system is more open than a caste system because the boundaries between classes are less distinct than the boundaries between castes. In a class system, status comes at least partly through achievement rather than entirely by ascription.

In class systems, people may become members of a class other than that of their parents through both intergenerational and intragenerational mobility, either upward or downward. Horizontal mobility occurs when people experience a gain or loss in position and/or income that does not produce a change in their place in the class structure. For example, a person may get a pay increase and a more prestigious title but still not move from one class to another. By contrast, movement up or down the class structure is *vertical mobility*. Martin, a commercial artist who owns his own firm, is an example of vertical, intergenerational mobility:

> My family came out of a lot of poverty and were eager to escape it.... My [mother's parents] worked in a sweatshop. My grandfather to the day he died never earned more than $14 a week. My grandmother worked in knitting mills while she had five children.... My father quit school when he was in eighth grade and supported his mother and his two sisters when he was twelve years old. My grandfather died when my father was four and he basically raised his sisters. He got a man's job when he was twelve and took care of the three of them. (qtd. in Newman, 1993: 65)

Martin's situation reflects upward mobility; however, people may also experience downward mobility, caused by any number of reasons, including a lack of jobs, low wages and employment instability, marriage to someone with fewer resources and less power than oneself, and changing social conditions (Newman, 1988, 1993).

Classical Perspectives on Social Class

Early sociologists grappled with the definition of class and the criteria for determining people's location in the class structure. Both Karl Marx and Max Weber viewed class as an important determinant of social inequality and social change, and their works have had a profound influence on how we view the U.S. class system today.

Karl Marx: Relationship to the Means of Production

According to Karl Marx, class position and the extent of our income and wealth are determined by our work situation, or our relationship to the means of production. As we have previously seen, Marx stated that capitalistic societies consist of two classes—the capitalists and the workers. The ***capitalist class (bourgeoisie)*** **consists of those who own the means of production**—the land and capital necessary for factories and mines, for example. The ***working class (proletariat)*** **consists of those who must sell their labor to the owners in order to earn enough money to survive** (see ▶ Figure 8.1).

According to Marx, class relationships involve inequality and exploitation. The workers are exploited as capitalists maximize their profits by paying workers less than the resale value of what they produce but do not own. This exploitation results in workers' ***alienation*****—a feeling of powerlessness and estrangement from other people and from oneself.**

Scott Olson/Getty Images

▲ In 2010 Arizona passed the Support Our Law Enforcement and Safe Neighborhoods Act, a stringent law that gave police officers greatly enhanced powers to check people's immigration status. The marchers in this Phoenix demonstration took to the streets to make known their opposition to the new law.

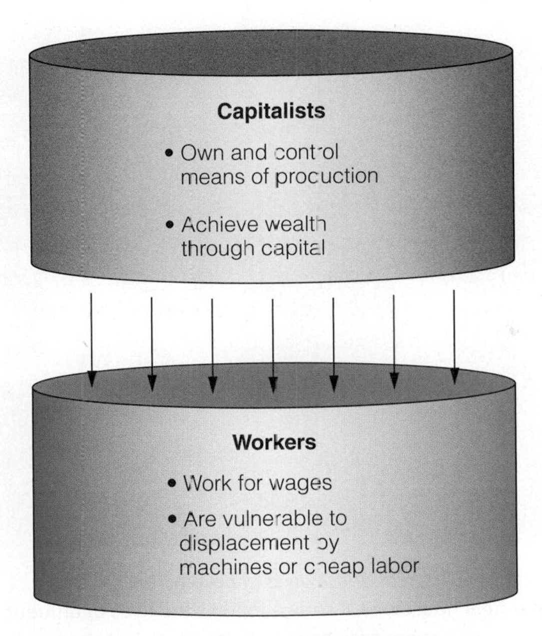

▲ **FIGURE 8.1 MARX'S VIEW OF STRATIFICATION**

In Marx's view alienation develops as workers manufacture goods that embody their creative talents but do not own the goods. Workers are also alienated from the work itself because they are forced to perform it in order to live. Because the workers' activities are not their own, they feel self-estrangement. Moreover, workers are separated from others in the factory because they individually sell their labor power to the capitalists as a commodity.

In Marx's view the capitalist class maintains its position at the top of the class structure by control of the society's *superstructure*, which is composed of the government, schools, churches, and other social institutions that produce and disseminate ideas perpetuating the existing system of exploitation. Marx predicted that the exploitation of workers by the capitalist class would ultimately lead to ***class conflict*—the struggle between the capitalist class and the working class.** According to Marx, when the workers realized that capitalists were the source of their oppression, they would overthrow the capitalists and their agents of social control, leading to the end of capitalism. The workers would then take over the government and create a more egalitarian society.

Why has no workers' revolution occurred? According to the sociologist Ralf Dahrendorf (1959), capitalism may have persisted because it has changed significantly since Marx's time. Individual capitalists no longer own and control factories and other means of production; today, ownership and control have largely been separated. For example, contemporary transnational corporations are owned by a multitude of stockholders but run by paid officers and managers. Similarly, many (but by no means all) workers have experienced a rising standard of living, which may have contributed to a feeling of complacency. During the twentieth century, workers pressed for salary increases and improvements in the workplace through their activism and labor union membership. They also gained more legal protection in the form of workers' rights and benefits such as workers' compensation insurance for job-related injuries and disabilities (Dahrendorf, 1959). For these reasons, and because of a myriad of other complex factors, the workers' revolution predicted by Marx never came to pass. However, the failure of his prediction does not mean that his analysis of capitalism and his theoretical contributions to sociology are without validity.

Marx had a number of important insights into capitalist societies. First, he recognized the economic basis of class systems (Gilbert, 2010). Second, he noted the relationship between people's

▲ Although unions have lost some of their importance over the last fifty years, this 2009 United Auto Workers rally in support of the Employee Free Choice Act—legislation that would make it easier for employees to join a union—shows that many workers still believe in a collective response to workplace issues.

class system a type of stratification based on the ownership and control of resources and on the type of work that people do.

capitalist class (or **bourgeoisie**) Karl Marx's term for the class that consists of those who own and control the means of production.

working class (or **proletariat**) those who must sell their labor to the owners in order to earn enough money to survive.

alienation a feeling of powerlessness and estrangement from other people and from oneself.

class conflict Karl Marx's term for the struggle between the capitalist class and the working class.

social location in the class structure and their values, beliefs, and behavior. Finally, he acknowledged that classes may have opposing (rather than complementary) interests. For example, capitalists' best interests are served by a decrease in labor costs and other expenses and a corresponding increase in profits; workers' best interests are served by well-paid jobs, safe working conditions, and job security.

Max Weber: Wealth, Prestige, and Power

Max Weber's analysis of class builds upon earlier theories of capitalism (particularly those by Marx) and of money (particularly those by Georg Simmel, as discussed in Chapter 1). Living in the late nineteenth and early twentieth centuries, Weber was in a unique position to see the transformation that occurred as individual, competitive, entrepreneurial capitalism went through the process of shifting to bureaucratic, industrial, corporate capitalism. As a result, Weber had more opportunity than Marx to see how capitalism changed over time.

Weber agreed with Marx's assertion that economic factors are important in understanding individual and group behavior. However, Weber emphasized that no single factor (such as economic divisions between capitalists and workers) was sufficient for defining the location of categories of people within the class structure. According to Weber, the access that people have to important societal resources (such as economic, social, and political power) is crucial in determining people's life chances. To highlight the importance of life chances for categories of people, Weber developed a multidimensional approach to social stratification that reflects the interplay among wealth, prestige, and power. In his analysis of these dimensions of class structure, Weber viewed the concept of "class" as an *ideal type* (that can be used to compare and contrast various societies) rather than as a specific social category of "real" people (Bourdieu, 1984).

***Wealth* is the value of all of a person's or family's economic assets, including income, personal property, and income-producing property.** Weber placed categories of people who have a similar level of wealth and income in the same class. For example, he identified a privileged commercial class of *entrepreneurs*—wealthy bankers, ship owners, professionals, and merchants who possess similar financial resources. He also described a class of *rentiers*—wealthy individuals who live off their investments and do not have to work. According to Weber, entrepreneurs and rentiers have much in common. Both are able to purchase expensive consumer goods, control other people's opportunities to acquire wealth and property, and monopolize costly status privileges (such as education) that provide contacts and skills for their children.

Isabella Vosmikova/© Bravo /Everett Collection

▲ Bravo's *The Real Housewives of Beverly Hills* (and its other "real housewives" series) follows the lives of affluent women who seemingly have nothing to do but shop, drink, and argue. Do you think that these women have a high level of prestige, as defined by Max Weber? Why or why not?

Weber divided those who work for wages into two classes: the middle class and the working class. The middle class consists of white-collar workers, public officials, managers, and professionals. The working class consists of skilled, semiskilled, and unskilled workers.

The second dimension of Weber's system of stratification is ***prestige*—the respect or regard with which a person or status position is regarded by others.** Fame, respect, honor, and esteem are the most common forms of prestige. A person who has a high level of prestige is assumed to receive deferential and respectful treatment from others. Weber suggested that individuals who share a common level of social prestige belong to the same status group regardless of their level of wealth. They tend to socialize with one another, marry within their own group of social equals, spend their leisure time together, and safeguard their status by restricting outsiders' opportunities to join their ranks.

The other dimension of Weber's system is ***power*—the ability of people or groups to achieve their goals despite opposition from others.** The powerful can shape society in accordance with their own interests and direct the actions of others. According to Weber, bureaucracies hold social power in modern societies; individual power depends on a person's position within the bureaucracy. Weber suggested that the power of modern bureaucracies was so strong that even a workers' revolution (as predicted by Marx) would not lessen social inequality.

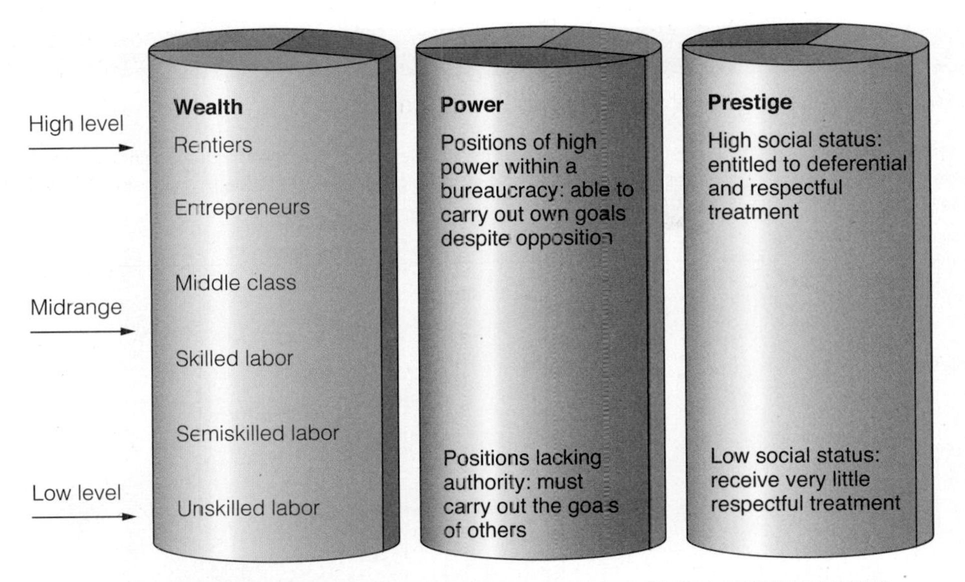

▲ **FIGURE 8.2 WEBER'S MULTIDIMENSIONAL APPROACH TO SOCIAL STRATIFICATION**
According to Max Weber, wealth, power, and prestige are separate continuums. Individuals may rank high in one dimension and low in another, or they may rank high or low in more than one dimension. Also, individuals may use their high rank in one dimension to achieve a comparable rank in another. How does Weber's model compare with Marx's approach as shown in Figure 8.1?

Weber stated that wealth, prestige, and power are separate continuums on which people can be ranked from high to low, as shown in ▶ Figure 8.2. Individuals may be high on one dimension while being low on another. For example, people may be very wealthy but have little political power (for example, a recluse who has inherited a large sum of money). They may also have prestige but not wealth (for instance, a college professor who receives teaching excellence awards but lives on a relatively low income). In Weber's multidimensional approach, people are ranked on all three dimensions. Sociologists often use the term ***socioeconomic status (SES)*** **to refer to a combined measure that, in order to determine class location, attempts to classify individuals, families, or households in terms of factors such as income, occupation, and education.**

What important contribution does Weber make to our understanding of social stratification and class? Weber's analysis of social stratification contributes to our understanding by emphasizing that people behave according to both their economic interests and their values. He also added to Marx's insights by developing a multidimensional explanation of the class structure and by identifying additional classes. Both Marx and Weber emphasized that capitalists and workers are the primary players in a class society, and both noted the importance of class to people's life chances. However, they saw different futures for capitalism and the social system. Marx saw these structures being overthrown; Weber saw the increasing bureaucratization of life even without capitalism.

A substantial advantage of Weber's theory is that it has made empirical investigation of the U.S. class structure possible. Through his distinctions among wealth, power, and prestige, Weber makes it possible for researchers to examine the different dimensions of social stratification. Weber's enlarged conceptual formulation of stratification is the theoretical foundation for mobility research by sociologists such as Peter Blau and Otis Duncan (1967), who measured the three dimensions from Weber's theory through a study of the occupational positions that individuals hold. According to Blau and Duncan, a person's occupational position is not identical to either economic class or prestige, but is closely related to both. As you might expect, different occupations have significantly different levels of status or prestige (see ■ Table 8.1). For many years, occupational ratings by prestige have remained remarkably consistent in the United States (Gilbert, 2010).

wealth the value of all of a person's or family's economic assets, including income, personal property, and income-producing property.

prestige the respect or regard with which a person or status position is regarded by others.

power the ability of people or groups to achieve their goals despite opposition from others.

socioeconomic status (SES) a combined measure that, in order to determine class location, attempts to classify individuals, families, or households in terms of factors such as income, occupation, and education.

table 8.1

Prestige Ratings for Selected Occupations in the United States

Respondents were asked to evaluate a list of occupations according to their prestige; the individual rankings were averaged and then converted into scores, with 1 the lowest possible score and 99 the highest possible score. How would you rate the prestige level of these various jobs?

Occupation	Score	Occupation	Score
Physician	86	Airplane pilot	61
Attorney	75	Police officer	60
College professor	74	Electrician	51
Architect	73	Funeral director	49
Aerospace engineer	72	Mail carrier	47
Dentist	72	Secretary	46
Clergy	69	Butcher	35
Pharmacist	68	Baker	35
Petroleum engineer	66	Garbage collector	28
Registered nurse	66	Bill collector	24
Accountant	65	Janitor	22
Grade school teacher	64	Maid	20

Source: Davis, Smith, and Marsden, 2007.

A significant limitation of occupational prestige rankings is that the level of prestige accorded to a position may not actually be based on the importance of the position to society. The highest ratings may be given to professionals—such as physicians and lawyers—because they have many years of training in their fields and some control their own work, not necessarily because these positions contribute the most to society.

Contemporary Sociological Models of the U.S. Class Structure

How many social classes exist in the United States? What criteria are used for determining class membership? No broad consensus exists about how to characterize the class structure in this country. In fact, many people deny that class distinctions exist. Most people like to think of themselves as middle class; it puts them in a comfortable middle position—neither rich nor poor. Sociologists have developed two models of the class structure: One is based on a Weberian approach, the other on a Marxian approach. We will examine both models briefly.

The Weberian Model of the U.S. Class Structure

Expanding on Weber's analysis of class structure, sociologist Dennis Gilbert (2010) uses a model of social classes based on three elements: (1) education, (2) occupation of family head, and (3) family income (see ▶ Figure 8.3).

The Upper (Capitalist) Class The upper class is the wealthiest and most powerful class in the United States. About 1 percent of the population is included in this class, whose members own substantial income-producing assets and operate on both

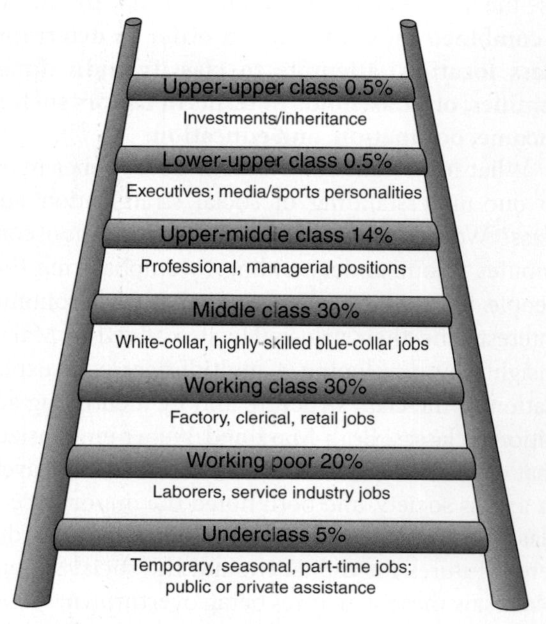

▲ FIGURE 8.3 STRATIFICATION BASED ON EDUCATION, OCCUPATION, AND INCOME

the national and international levels. According to Gilbert (2010), people in this class have an influence on the economy and society far beyond their numbers.

Some models further divide the upper class into upper-upper ("old money") and lower-upper ("new money") categories (Warner and Lunt, 1941; Coleman and Rainwater, 1978; Kendall, 2002). Members of the upper-upper class come from prominent families which possess great wealth that they have held for several generations. Family names—such as Rockefeller, Mellon, Du Pont, and Kennedy—are well-known and often held in high esteem. Persons in the upper-upper class tend to have strong feelings of in-group solidarity. They belong to the same exclusive clubs and support high culture (such as the opera, symphony orchestras, ballet, and art museums). Children are educated in prestigious private schools and Ivy League universities; many acquire strong feelings of privilege from birth, as upper-class author Lewis H. Lapham (1988: 14) states:

> Together with my classmates and peers, I was given to understand that it was sufficient accomplishment merely to have been born. Not that anybody ever said precisely that in so many words, but the assumption was plain enough, and I could confirm it by observing the mechanics of the local society. A man might become a drunkard, a concert pianist or an owner of companies, but none of these occupations would have an important bearing on his social rank.

Children of the upper class are socialized to view themselves as different from others; they also learn that they are expected to marry within their own class (Warner and Lunt, 1941; Mills, 1959a; Domhoff, 1983; Kendall, 2002).

Members of the lower-upper class may be extremely wealthy but not have attained as much prestige as members of the upper-upper class. The "new rich" have earned most of their money in their own lifetime as entrepreneurs, presidents of major corporations, sports or entertainment celebrities, or top-level professionals. For some members of the lower-upper class, the American Dream has become a reality. Others still desire the respect of members of the upper-upper class.

The Upper-Middle Class Persons in the upper-middle class are often highly educated professionals who have built careers as physicians, attorneys, stockbrokers, or corporate managers. Others derive their income from family-owned businesses. According to Gilbert (2010), about 14 percent of the U.S. population is in this category. A combination of three factors qualifies people for the upper-middle class: university degrees, authority and independence on the job, and high income. Of all the class categories, the upper-middle class is the one that is most shaped by formal education. Over the past fifty years, Asian Americans, Latinos/as, and African Americans have placed great importance on education as a means of attaining the American Dream. Many people of color have moved into the upper-middle class by acquiring higher levels of education.

The Middle Class In past decades a high school diploma was necessary to qualify for most middle-class jobs. Today, two-year or four-year college degrees have replaced the high school diploma as an entry-level requirement for employment in many middle-class occupations, including medical technicians, nurses, legal and medical assistants, lower-level managers, semiprofessionals, and nonretail salesworkers. An estimated 30 percent of the U.S. population is in the middle class even though most people in this country think of themselves as middle class. Nowhere is this myth of the vast middle class more prevalent than in television situation comedies, which for decades have focused on an idealized notion of the middle class or the debunking of that myth.

Traditionally, most middle-class occupations have been relatively secure and have provided more opportunities for advancement (especially with increasing levels of education and experience) than working-class positions. Recently, however, four factors have eroded the American Dream for this class: (1) escalating housing prices, (2) occupational insecurity, (3) blocked mobility on the job, and (4) the cost-of-living squeeze that has penalized younger workers, even when they have more education and better jobs than their parents (Pew Research Center, 2008).

The Working Class An estimated 30 percent of the U.S. population is in the working class. The core of this class is made up of semiskilled machine operators who work in factories and elsewhere. Members of the working class also include some workers in the service sector, as well as clerks and salespeople whose job responsibilities involve routine, mechanized tasks requiring little skill beyond basic literacy and a brief period of on-the-job training (Gilbert, 2010). Some people in the working class are employed in ***pink-collar occupations*—relatively low-paying, nonmanual, semiskilled positions primarily held by women,** such as day-care workers,

pink-collar occupations relatively low-paying, nonmanual, semiskilled positions primarily held by women.

checkout clerks, cashiers, and restaurant servers. Social scientists coined this term to describe the work of many women in the working class, as contrasted with the "blue-collar" occupations held by many working-class men.

How does life in the working-class family compare with that of individuals in middle-class families? According to sociologists, working-class families not only earn less than middle-class families, but they also have less financial security, particularly because of high rates of layoffs and plant closings in some regions of the country. Few people in the working class have more than a high school diploma, and many have less, which makes job opportunities scarce for them in a "high-tech" society (Gilbert, 2010). Others find themselves in low-paying jobs in the service sector of the economy, particularly fast-food restaurants, a condition that often places them among the working poor.

The Working Poor The working poor account for about 20 percent of the U.S. population. Members of the working-poor class live from just above to just below the poverty line; they typically hold unskilled jobs, seasonal migrant jobs in agriculture, lower-paid factory jobs, and service jobs (such as counter help at restaurants). Employed single mothers often belong to this class; consequently, children are overrepresented in this category. African Americans and other people of color are also overrepresented among the working poor. To cite only one example, in the United States today there are two white hospital orderlies to every one white physician, whereas there are twenty-five African American orderlies to every one African American physician (Gilbert, 2010). For the working poor, living from paycheck to paycheck makes it impossible to save money for emergencies such as periodic or seasonal unemployment, which is a constant threat to any economic stability they may have.

Social critic and journalist Barbara Ehrenreich (2001) left her upper-middle-class lifestyle for a period of time to see if it was possible for the working poor to live on the wages that they were being paid as restaurant servers, salesclerks at discount department stores, aides at nursing homes, housecleaners for franchise maid services, or similar jobs. She conducted her research by actually holding those jobs for periods of time and seeing if she could live on the wages that she received. Through her research Ehrenreich persuasively demonstrated that people who work full time, year-round, for poverty-level wages must develop survival strategies that include such things as help from relatives or constantly moving from one residence to another in order to have a place to live. Like many other researchers, Ehrenreich found that minimum-wage jobs cannot cover the full cost of living, such as rent, food, and the rest of an adult's monthly needs, even without taking into consideration the needs of children or other family members.

At some point in our lives, most of us have held a job paying the minimum wage, and we know the limitations of trying to survive on such low earnings. The federal *minimum wage* is the hourly rate that (with certain exceptions) is the lowest amount an employer can legally pay its employees (each state may adopt a higher minimum wage, but not a lower one). In 2009 the federal minimum wage was raised to $7.25 per hour. A person earning minimum wage and working forty hours every week, fifty-two weeks per year (in other words, no time off, no vacation), would still earn an amount slightly above the *official poverty line*

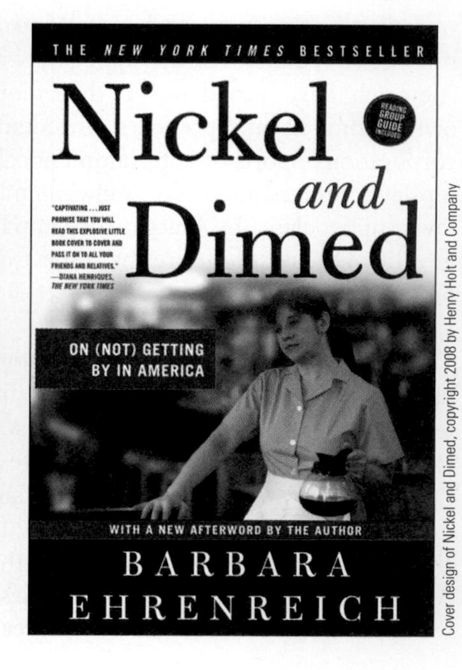

▲ In *Nickel and Dimed: On (Not) Getting by in America*, Barbara Ehrenreich (left) recounts her attempt to replicate the lives of the working poor by working a series of low-paying jobs and trying to survive on her wages.

(and slightly below that line for a person with two children). In 2011 seven states had a minimum wage above the federal requirement: Washington state's minimum wage was the highest, at $8.67 per hour. Social analysts believe that increasing social and economic inequality in the United States can be partly attributed to the vast divide between the low wages paid to workers, based on a low federal minimum wage, and the astronomically high salaries and compensation packages given to major corporate CEOs.

The Underclass According to Gilbert (2010), people in the ***underclass*** **are poor, seldom employed, and caught in long-term deprivation that results from low levels of education and income and high rates of unemployment.** Some are unable to work because of age or disability; others experience discrimination based on race/ethnicity. Single mothers are overrepresented in this class because of the lack of jobs, the lack of affordable child care, and many other impediments to a mother's future and that of her children. People without a "living wage" often must rely on public or private assistance programs for their survival. About 3 to 5 percent of the U.S. population is in this category, and the chances of their children moving out of poverty are about fifty-fifty (Gilbert, 2010).

Studies by various social scientists have found that meaningful employment opportunities are the critical missing link for people on the lowest rungs of the class ladder. According to these analysts, job creation is essential in order for people to have the opportunity to earn a decent wage; have medical coverage; live meaningful, productive lives; and raise their children in a safe environment (see Fine and Weis, 1998; Nelson and Smith, 1999; Newman, 1999; W. Wilson, 1996). These issues are closely tied to the American Dream we have been discussing in this chapter.

© Billy Hustace/Stone/Getty Images

▲ In which segment of the class structure would sociologists place clerical workers such as those in this office mail room? What are the key elements of that social class?

The Marxian Model of the U.S. Class Structure

The earliest Marxian model of class structure identified ownership or nonownership of the means of production as the distinguishing feature of classes. From this perspective, classes are social groups organized around property ownership, and social stratification is created and maintained by one group in order to protect and enhance its own economic interests. Moreover, societies are organized around classes in conflict over scarce resources. Inequality results from the more powerful exploiting the less powerful.

Contemporary Marxian (or conflict) models examine class in terms of people's relationship to others in the production process. For example, conflict theorists attempt to determine the degree of control that workers have over the decision-making process and the extent to which they are able to plan and implement their own work. They also analyze the type of supervisory authority, if any, that a worker has over other workers. According to this approach, most employees are a part of the working class because they do not control either their own labor or that of others.

Erik Olin Wright (1979, 1985, 1997), one of the leading stratification theorists to examine social class from a Marxian perspective, has concluded that Marx's definition of "workers" does not fit the occupations found in advanced capitalist societies. For example, many top executives, managers, and supervisors who do not own the means of production (and thus would be "workers" in Marx's model) act like capitalists in their zeal to control workers and maximize profits. Likewise, some experts hold positions in which they have control over money and the use of their own time even though they are not owners. Wright views Marx's category of "capitalist" as being too broad as well. For instance, small-business owners might be viewed as capitalists because they own their own tools and have a few people working for them, but they have little in common with large-scale capitalists and do not share the interests of factory workers. ▶ Figure 8.4 compares Marx's model and Wright's model.

underclass those who are poor, seldom employed, and caught in long-term deprivation that results from low levels of education and income and high rates of unemployment.

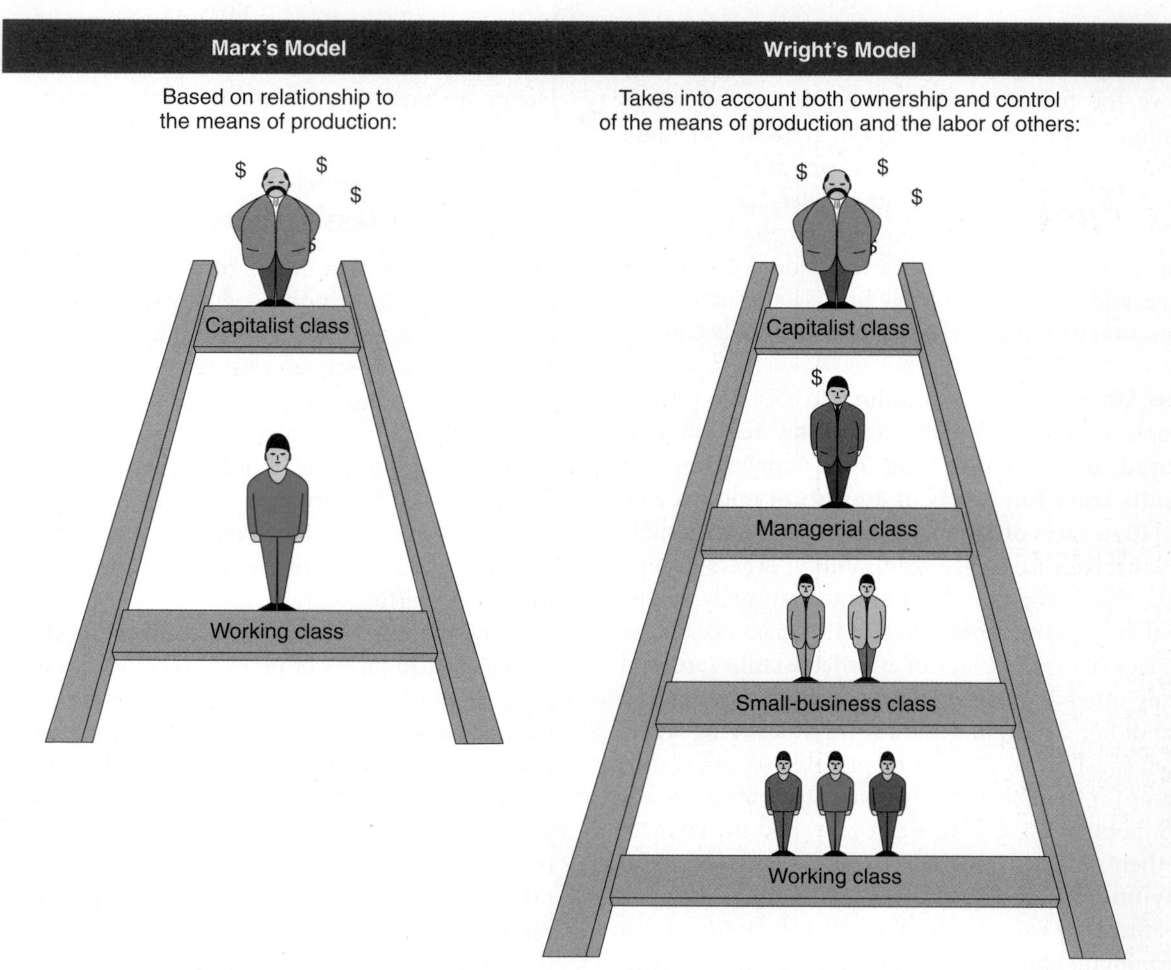

▲ **FIGURE 8.4 COMPARISON OF MARX'S AND WRIGHT'S MODELS OF CLASS STRUCTURE**

Wright (1979) also argues that classes in modern capitalism cannot be defined simply in terms of different levels of wealth, power, and prestige, as in the Weberian model. Consequently, he outlines four criteria for placement in the class structure: (1) ownership of the means of production, (2) purchase of the labor of others (employing others), (3) control of the labor of others (supervising others on the job), and (4) sale of one's own labor (being employed by someone else). Wright (1978) assumes that these criteria can be used to determine the class placement of all workers, regardless of race/ethnicity, in a capitalist society. Let's take a brief look at Wright's (1979, 1985) four classes—(1) the capitalist class, (2) the managerial class, (3) the small-business class, and (4) the working class—so that you can compare them to those found in the Weberian model.

The Capitalist Class According to Wright, this class holds most of the wealth and power in society through ownership of capital—for example, banks, corporations, factories, mines, news and entertainment industries, and agribusiness firms. The "ruling elites," or "ruling class," within the capitalist class hold political power and are often elected or appointed to influential political and regulatory positions.

This class is composed of individuals who have inherited fortunes, own major corporations, or are top corporate executives with extensive stock holdings or control of company investments. Even though many top executives have only limited *legal ownership* of their corporations, they have substantial economic ownership and exert extensive control over investments, distribution of profits, and management of resources. The major sources of income for the capitalist class are profits, interest, and very high salaries. Members of this class make important decisions about the workplace, including which products and services to make available to consumers and how many workers to hire or fire.

According to *Forbes* magazine's 2010 list of the richest people in the world, Bill Gates (cofounder of Microsoft Corporation, the world's largest microcomputer software company) was the second wealthiest capitalist, with a net worth of $53 billion, down from his $63 billion figure in 2000 (*Forbes*, 2010). Investor Warren Buffet came in third, with $47 billion. The title for wealthiest person in the world went to Carlos Slim Helu, the Mexican

© AP Photo/Adrian Wyld/CP

© ERIC DRAPER/WHITE HOUSE/UPI/Landov

▲ After founding Microsoft and becoming one of the world's richest people, Bill Gates (top) has devoted recent years to the foundation that he began with his wife, Melinda, to combat poverty and disease in Africa. His career has followed a traditional version of the American Dream: the entrepreneur who becomes very wealthy and then philanthropic. The extended family of George and Barbara Bush (bottom) is a current example of another type of success story, following Erik Olin Wright's theory of the interplay of money and political power: Three generations of the Bush family have held federal office.

telecom entrepreneur, whose net worth was $53.5 billion. Although men who made the *Forbes* list of the wealthiest people gained their fortunes through entrepreneurship or being CEOs of large corporations, many women who made the list acquired their wealth typically through inheritance, marriage, or both. In 2010 only fifteen women were heads of Fortune 500 companies. Women celebrities and television entrepreneurs such as Oprah Winfrey ($2.3 billion net worth) typically have far less wealth, relatively speaking, than the top male billionaires.

The Managerial Class People in the managerial class have substantial control over the means of production and over workers. However, these upper-level managers, supervisors, and professionals usually do not participate in key corporate decisions such as how to invest profits. Lower-level managers may have some control over employment practices, including the hiring and firing of some workers.

Top professionals such as physicians, attorneys, accountants, and engineers may control the structure of their own work; however, they typically do not own the means of production and may not have supervisory authority over more than a few people. Even so, they may influence the organization of work and the treatment of other workers. Members of the capitalist class often depend on these professionals for their specialized knowledge.

The Small-Business Class This class consists of small-business owners and crafts-people who may hire a small number of employees but largely do their own work. Some members own businesses such as "mom-and-pop" grocery stores, retail clothing stores, and jewelry stores. Others are doctors and lawyers who receive relatively high incomes from selling their own services. Some of these professionals now share attributes with members of the capitalist class because they have formed corporations that hire and control the employees who produce profits for the professionals.

It is in the small-business class that we find many people's hopes of achieving the American Dream. Recent economic trends, including corporate downsizing, telecommuting, and the movement of jobs to other countries, have encouraged more people to think about starting their own business. As a result, more people today are self-employed or own a small business than at any time in the past, and more of the owners are members of racial/ethnic minority groups. Between 2002 and 2007 (the latest year for which data are available), the number of black-owned businesses increased at triple the national rate (census.gov, 2011), and the number of Hispanic-owned businesses increased at more than double the national rate (census.gov, 2010a). More women of all races are in the small-business class than was true previously. According to Census Bureau reports, businesses where women were owners or half-owners represented almost 46 percent of all firms in the United States. However, women alone accounted for about 29 percent of all businesses nationwide, and 88 percent of these women-owned businesses had no paid employees and were in fields such as health care and social assistance, repair and maintenance, and personal and laundry services,

© PhotoEdit

© Michael Newman/PhotoEdit

▲ Many immigrants believe that they can achieve the American Dream by starting a small business, such as the stores shown here in San Francisco's Chinatown. Skilled laborers and tradespeople sometimes take another approach to small business by offering services such as landscaping, plumbing, carpentry, and, as seen here, mobile dog grooming.

meaning that there was a substantial gap in revenues between businesses owned exclusively by women and those owned by men (census.gov, 2010b). Recent trends indicate that small-business ownership may have stalled during the recession beginning in 2008 and that there will be a corresponding increase in bankruptcies and failures of minority-owned businesses (Morello, 2010).

Throughout U.S. history, immigrants and people of color have owned small businesses, viewing such enterprises as a way to achieve the American Dream (see this chapter's Photo Essay).

The Working Class The working class is made up of a number of subgroups, one of which is blue-collar workers, some of whom are highly skilled and well paid and others of whom are unskilled and poorly paid. Skilled blue-collar workers include electricians, plumbers, and carpenters; unskilled blue-collar workers include janitors and gardeners.

White-collar workers are another subgroup of the working class. Referred to by some as a "new middle class," these workers are actually members of the working class because they do not own the means of production, do not control the work of others, and are relatively powerless in the workplace. Secretaries, other clerical workers, and salesworkers are members of the white-collar faction of the working class. They take orders from others and tend to work under constant supervision. Thus, these workers are at the bottom of the class structure in terms of domination and control in the workplace. The working class contains about half of all employees in the United States.

Although Marxian and Weberian models of the U.S. class structure show differences in people's occupations and access to valued resources, neither fully reflects the nature and extent of inequality in the United States. In the next section we will take a closer look at the unequal distribution of income and wealth in the United States and the effects of inequality on people's opportunities and life chances.

Inequality in the United States

Throughout human history, people have argued about the distribution of scarce resources in society. Disagreements often center on whether the share we get is a fair reward for our effort and hard work. Recently, social analysts have pointed out that (except during temporary economic downturns) the old maxim "the rich get richer" continues to be valid in the United States. To understand how this happens, we must take a closer look at the distribution of income and wealth in this country.

Distribution of Income and Wealth

Money is essential for acquiring goods and services. People without money cannot purchase food, shelter, clothing, medical care, legal aid, education, and the other things they need or desire. Money—in the form of both income and wealth—is very unevenly distributed in the United States. Median household income varies widely from one state to another, for example (see ▶ Map 8.1). For the entire nation the real median household income (adjusted for inflation) was

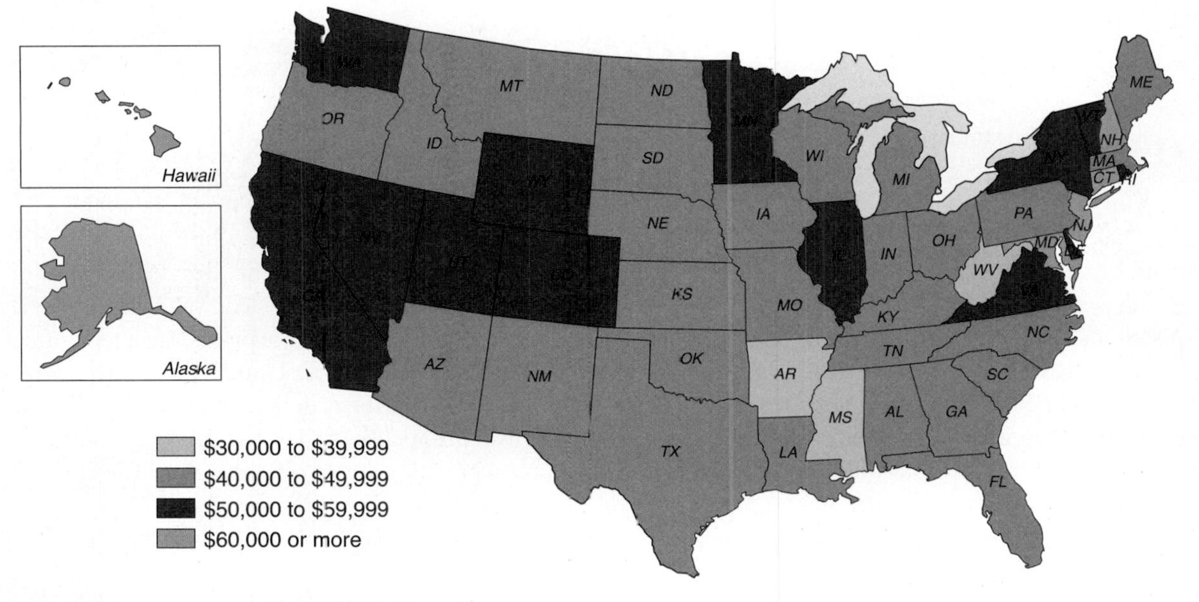

▲ **MAP 8.1 MEDIAN INCOME BY STATE: 2009**

What factors contribute to the uneven distribution of income in the United States?

Source: census.gov, 2010.

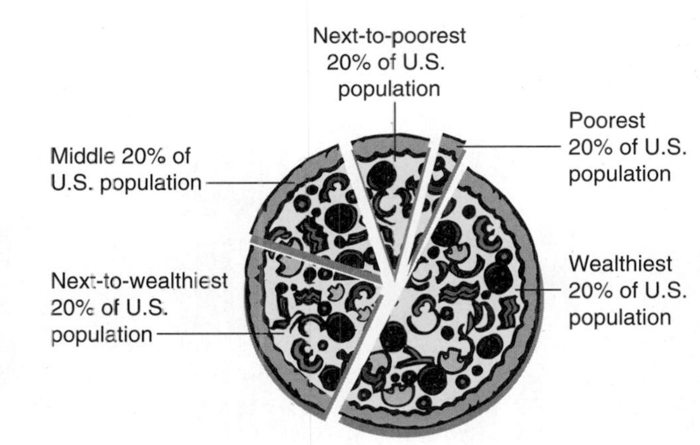

▶ **FIGURE 8.5 DISTRIBUTION OF PRETAX INCOME IN THE UNITED STATES**

Thinking of personal income in the United States (before taxes) as a large pizza helps us to see which segments of the population receive the largest and smallest portions.

Source: U.S. Census Bureau, 2010b.

$49,777 in 2009, down from $51,965 in 2007. Real median household income has fallen in recent years, and the decline has been widespread since the recession that started in 2007–2008 (DeNavas-Walt, Proctor, and Smith, 2010). Some estimates of household income are adjusted for inflation while others are not, so these figures can be tricky to interpret. However, one fact remains across decades of reporting of median household income: Among high-income nations, the United States remains number one in inequality of income distribution.

Income Inequality ***Income* is the economic gain derived from wages, salaries, income transfers (governmental aid), and ownership of property.** Or, to put it another way, "income refers to money, wages, and payments that periodically are received as returns for an occupation or investment" (Kerbo, 2000: 19). Data from the U.S. Census Bureau typically provide income estimates that are based solely on money income before taxes and do not include the value of noncash benefits such as health care coverage or retirement benefits.

Sociologist Dennis Gilbert (2010) compares the distribution of income to a national pie that has been sliced into portions, ranging in size from stingy to generous, for distribution among segments of the population. As shown in ▶ Figure 8.5, in 2008 the wealthiest 20 percent of households received about 50 percent of the total income "pie" while the poorest 20 percent of households received less than 3 percent of all income. If the distribution of income were equal, each quintile (one-fifth) would receive exactly 20 percent (Gilbert, 2010).

Since the 1980s the gap between the rich and the poor in the United States has continued to widen

income the economic gain derived from wages, salaries, income transfers (governmental aid), and ownership of property.

photo essay The New Realities of the American Dream

According to some social analysts, many Americans made poor choices in the years leading up to the Great Recession, which began in 2007–2008. However, many other problems occurred because of events that individuals could not control. A *cascade* effect plagued a great number of households and individuals.

Here is an example. The Smiths are a family of four: mother, father, daughter, son. Prior to 2008, Mrs. Smith worked in a travel agency. Mr. Smith worked for a wholesale electronics distributor. They had bought a new house in 2006 that cost three times as much as their previous house. In 2008 their mortgage payments went up, but Mrs. Smith had lost her job at the travel agency. (Americans were cutting back on their vacations, and many of those who still planned vacations preferred to use online services instead of travel agencies.) Mr. Smith had his hours reduced at work, and his company then required him to pay more for his and his family's health insurance. After that, the company transferred him to another location 150 miles from his home. Besides the commute of more than two hours each way, Mr. Smith also faced swiftly rising gasoline prices. He and his family needed to sell their house and move to the new location, but the mortgage was underwater (the house was worth much less than the mortgage), and there was no practical way to sell the house. The Smiths are barely getting by.

As you look at the photographs on these pages, think of people you know (or members of your family) who are facing these and similar problems in today's distressed economy. Will the American Dream survive?

Kirby Hamilton/iStockphoto.com

◀ **A DREAM DENIED**

A cornerstone of the American Dream is home ownership. During the late 1990s and into the middle of the next decade, more and more Americans were able to buy homes. However, many of them bought homes that they couldn't really afford, and—because of shady lending practices—many Americans signed their names to mortgages that they had no hope of paying off. When the real estate market imploded later in that decade, a recession ensued, and many people living "on the edge" had no other option but to default on their mortgage and lose their home. Not since the Great Depression (1929–1941) has the residential real estate situation been so shaky.

JustASC/Shutterstock.com

◀ **TIME ON THEIR HANDS**

One of the most discouraging features of the Great Recession has been the persistent unemployment that has affected the entire country. At the time of this writing, the national unemployment rate is about 9 percent; however, that number does not include people who have given up trying to find a new job and people who are working one or more jobs part time. The real unemployment rate is 16 percent—one in six adults of working age.

Harper Smith/NBCU Photo Bank via Getty Images

▲ **JOB FLIGHT**

Many Americans are out of work because jobs that they might have filled have been moved to other countries where labor costs are lower. NBC's *Outsourced* was a comedy about a young American, played by Ben Rappaport (left), who moves to India to manage the call center for a U.S. novelties company. The series explored many of the ramifications of such an arrangement—not all of them humorous.

John Woodworth/Alamy

▲ **TAKIN' IT TO THE STREETS**

Unemployment is especially hard on workers who are middle aged or older. Many remain unemployed for years, and if they do find work, it is frequently not the same level of position as they previously occupied, and it often pays substantially less. The man shown here has obviously given up on the traditional job-search process and has adopted an interesting method of self-marketing.

reflect & analyze

1. **Since the U.S. economic recession started in 2007, the number of undocumented workers in this country has shrunk dramatically. Do you think that the workers have given up their hope of achieving the American Dream, or have they deferred it?**
2. **Does the U.S. government have the obligation to keep the American Dream alive for people of lower incomes? Are there alternatives to government-funded student aid programs, for example?**
3. **Do you know anyone whose life was changed as a result of playing a state lottery? Was the change positive or detrimental to the person's life? What conclusions can you draw from this person's experience?**

David Grossman/Alamy

▲ **UNANTICIPATED CONSEQUENCES**

Adding insult to injury in the Great Recession, gasoline prices have increased sharply in the past five years. As well, people have been spending less money in general, distrustful of their employment situation and of the national economy as a whole. The result has been increased use of public transportation, nonmotorized vehicles such as bicycles, and simply walking. Additionally, people are being more careful with how they spend their money, often choosing to delay major and minor purchases and getting the most of the items that they already own. The result has been less waste and less of a collective carbon footprint.

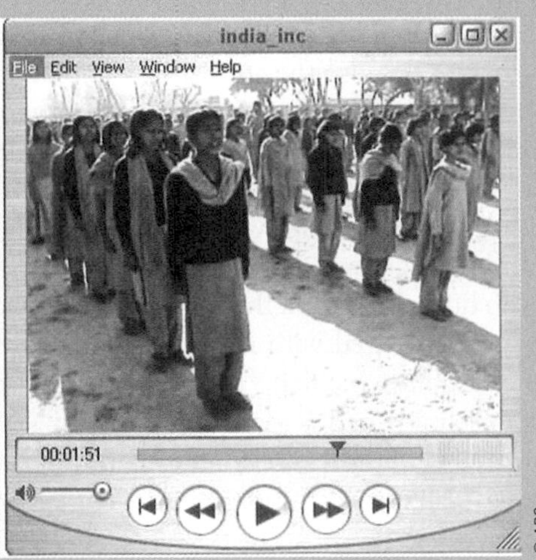

© ABC

▶ turning to video

Watch the ABC video *India Inc: Economic Explosion* (running time 2:38), available through **CengageBrain .com**. Because of the country's growing economy (it's the second-fastest-growing economy in the world), there are more middle-class Indians with buying power than the entire U.S. population. Many American jobs are being outsourced to India, and India has even begun hiring out-of-work Americans. As you watch this news report, think about the photographs, commentary, and questions you encountered in this photo essay. After you've watched the video, consider two more questions: How does another country's growth affect the American Dream, and to what degree is it possible that India will replace the United States as the land of opportunity?

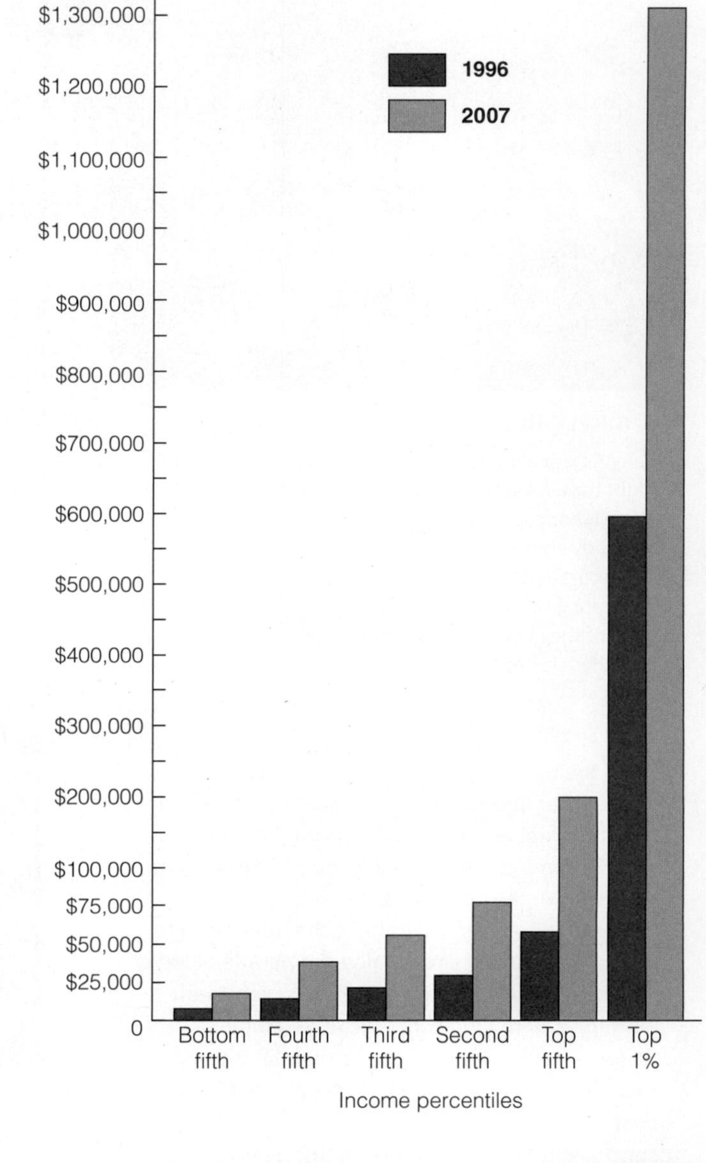

▶ **FIGURE 8.6 AVERAGE FAMILY INCOME IN THE UNITED STATES**

This chart shows the distribution of after-tax household income in the United States. Notice the dramatic increase in income for the top 1 percent of U.S. families. During the past decade the difference in income between the richest and poorest has become even more pronounced.

Source: Tax Policy Center, 2009.

until it has become a chasm. The poor are more likely to remain poor, and the affluent have been more likely to stay affluent. While the top 1 percent of households in the United States have an after-tax income of more than $1.3 million per year, the bottom 20 percent earn an average of $17,700 (see ▶ Figure 8.6). Although the Great Recession may have reduced the annual income of some rich Americans, this does not mean that they have become poor. Their income may have dropped by several million dollars, causing them to slightly modify their affluent lifestyle, but their overall standard of living is protected by other wealth they possess. For example, the net worth of top U.S. billionaires such as Bill Gates and Warren E. Buffett may have dropped several billion dollars, but they are far from living in poverty. However, the picture is different for people situated in the middle and bottom sectors of the income pie who are faced with high mortgage rates or rent payments and increasing costs for food, fuel, transportation, and other necessities.

Income distribution varies by race/ethnicity as well as class. ▶ Figure 8.7 compares median household income by race/ethnicity, showing not only the disparity among groups but also the consistency of that disparity. Although households across all racial/ethnic categories have experienced some decline in real annual median income, the income gap between African American households and white (non-Hispanic) and Asian and Pacific Islanders is striking. In 2009 African American households had the lowest median income, $32,584, as compared with Asian households, which had the highest median, $65,469. Non-Hispanic white households had a median income of $54,461, as compared with the $38,039 of Hispanic households (DeNavas-Walt, Proctor, and Smith, 2010).

Wealth Inequality Income is only one aspect of wealth. Wealth includes property such as buildings, land, farms, houses, factories, and cars, as well as other assets such as bank accounts, corporate stocks, bonds, and insurance policies. Wealth is computed

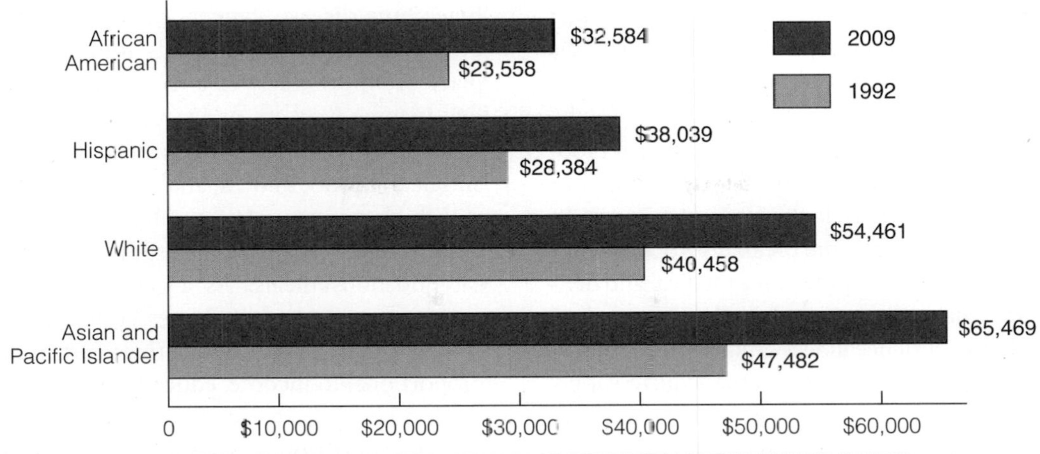

▲ **FIGURE 8.7 MEDIAN HOUSEHOLD INCOME BY RACE/ETHNICITY IN THE UNITED STATES**

Amounts shown in constant dollars.

Sources: DeNavas-Walt, Cleveland, and Webster, 2003; DeNavas-Walt, Proctor, and Smith, 2010.

by subtracting all debt obligations and converting the remaining assets into cash (U.S. Congress, 1986). For most people in the United States, wealth is invested primarily in property that generates no income, such as a home or car. By contrast, the wealth of a small number of elites is often in the form of income-producing property.

To see how wealth inequality has increased in recent decades, let's compare two studies. An earlier study by the Joint Economic Committee of Congress divided the population into four categories: (1) the super-rich (0.5 percent of households), who own 35 percent of the nation's wealth, with net assets averaging almost $9 million; (2) the very rich (the next 0.5 percent of households), who own about 7 percent of the nation's wealth, with net assets ranging from $1.4 million to $2.5 million; (3) the rich (9 percent of households), who own 30 percent of the wealth, with net assets of a little over $400,000; and (4) everybody else (the bottom 90 percent), who own about 28 percent of the nation's wealth. However, by 1995 another study indicated that the holdings of super-rich households had risen from 35 percent to almost 40 percent of all assets in the nation (stocks, bonds, cash, life insurance policies, paintings, jewelry, and other tangible assets) (Rothchild, 1995).

In the 2000s wealth continues to be highly concentrated in the hands of a very few. In 2007, for example, the wealthiest 1 percent of families owned slightly more than one-third (34.6 percent) of all privately held wealth in the United States. The next wealthiest 19 percent (the managerial, professional, and small-business category) owned about 50.5 percent, leaving the bottom 80 percent (made up of wage and salary workers) with only 15 percent of all privately held wealth (Domhoff, 2011).

For the upper class, wealth often comes from interest, dividends, and inheritance. One study looking at the *Forbes* 400 list of the wealthiest U.S. citizens found that nearly half of the people on that list had inherited sufficient wealth to put them on the list (Gilbert, 2010). Inheritors are often three or four generations removed from the individuals who amassed the original wealth (Odendahl, 1990). After inheriting a fortune, John D. Rockefeller, Jr., stated that "I was born into [wealth] and there was nothing I could do about it. It was there, like air or food or any other element. The only question with wealth is what to do with it" (qtd. in Glastris, 1990: 26). On the other hand, wealth sometimes dissipates across generations as it is passed on to increasing numbers of family members, who divide into smaller segments what was formerly a large fortune.

Consequences of Inequality

Income and wealth are not simply statistics; they are intricately related to the American Dream and our individual life chances. Persons with a high income or substantial wealth have more control over their own lives. They have greater access to goods and services; they can afford better housing, more education, and a wider range of medical services. Persons with less income, especially those living in poverty, must spend their limited resources to acquire the basic necessities of life.

Physical Health, Mental Health, and Nutrition Recent reports from the National Center for Health Statistics (2010) emphasize that the socioeconomic circumstances of persons, as well as the places where they live and work, strongly influence their health. People who are wealthy and well educated and who have high-paying jobs are much more likely to be healthy than are poor people. As people's economic status increases, so does their health status. The poor have shorter life expectancies and are at greater risk for chronic illnesses such as diabetes, heart disease, and cancer, as well as infectious diseases such as tuberculosis.

Children born into poor families are at much greater risk of dying during their first year of life. Some die from disease, accidents, or violence. Others are unable to survive because they are born with low birth weight, a condition linked to birth defects and increased probability of infant mortality. Low birth weight in infants is attributed, at least in part, to the inadequate nutrition received by many low-income pregnant women. Most of the poor do not receive preventive medical and dental checkups; many do not receive adequate medical care after they experience illness or injury.

Many high-poverty areas lack an adequate supply of doctors and medical facilities. Even in areas where such services are available, the inability to pay often prevents people from seeking medical care when it is needed. Some "charity" clinics and hospitals may provide indigent patients (those who cannot pay) with minimal emergency care but make them feel stigmatized in the process. For many of the working poor, health insurance is out of the question. The Census Bureau classifies health insurance coverage as private coverage or government coverage. Private health insurance is a plan provided through an employer or a union, or is purchased by an individual from a private company. By contrast, government health insurance includes such programs as Medicare, Medicaid, military health care, the Children's Health Insurance Program (CHIP), and individual state health plans.

About 50.7 million people in the United States were without health insurance coverage in 2009—an increase of more than 3 million from 46.3 million the preceding year (DeNavas-Walt, Proctor, and Smith, 2010). It is too soon to know what, if any, impact that the 2010 Affordable Care Plan, passed by Congress and signed into law by President Barack Obama, will have on the uninsured. The law is supposed to make health insurance coverage more widely available to all Americans, but it is being gradually implemented, and some provisions are being hotly contested by a number of states (see Chapter 18).

Where do people currently get their health insurance coverage? Many people rely on their employers for health coverage; however, some employers are cutting back on health coverage, particularly for employees' family members. Between 2008 and 2009 the percentage of people covered by private health insurance decreased from 66.7 percent to 63.9 percent. Among those covered by employment-based health insurance, the percentage decreased from 58.5 percent in 2008 to 55.8 percent in 2009. Of all age groups, persons between the ages of 18 and 34 are the most likely to be uninsured; Medicare and other benefit programs provide medical care to most persons 65 and over (DeNavas-Walt, Proctor, and Smith, 2010).

The places where people work also affect their health. Many lower-paying jobs are often the most dangerous and have the greatest health hazards. Black lung disease, cancers caused by asbestos, and other environmental hazards found in the workplace are more likely to affect manual laborers and low-income workers, as are job-related accidents.

Although the precise relationship between class and health is not known, analysts suggest that people with higher income and wealth tend to smoke less, exercise more, maintain a healthy body weight, and eat nutritious meals. As a category, more-affluent persons tend to be less depressed and face less psychological stress, conditions that tend to be directly proportional to income, education, and job status.

Good health is basic to good life chances; in turn, adequate nutrition is essential for good health. Hunger is related to class position and income inequality. Surveys estimate that 13 percent of children under age 12 are hungry or at risk of being hungry. Among the working poor, almost 75 percent of the children are thought to be in this category. After spending 60 percent of their income on housing, low-income families are unable to provide adequate food for their children. Between one-third and one-half of all children living in poverty consume significantly less than the federally recommended guidelines for caloric and nutritional intake (Children's Defense Fund, 2008). Lack of adequate nutrition has been linked to children's problems in school.

Studying the problem of hunger has become more complex in recent years because the U.S. Department of Agriculture stopped using the word *hunger* in its reports in 2006. *Food insecure* is now used to identify people in various categories, including those who are unable to afford the basics, those who are unable to get to the grocery store, and those who are unable to find fresh, nutritious produce and other foods to eat because they are surrounded with fast-food stores that do not provide foods with proper nutrition (Dolnick, 2010).

Between 2007 and 2009 the number of people who received food stamps in the United States increased by nearly a third. By 2010, the program was feeding more than 36 million people, or one in eight Americans and one in four children (DeParle and Gebeloff, 2010). Almost 90 percent of people using food stamps live near or below the federal poverty line (about $22,000 a year for a family of four). Some of them are newly jobless, whereas others are chronically unemployed. Government officials now refer to food stamps as "nutritional aid" in an effort to reduce the stigma attached to using the stamps as being a form of "public assistance" or "welfare" (DeParle and Gebeloff, 2009, 2010).

Housing As discussed in Chapter 5, homelessness is a major problem in the United States. The lack of affordable housing is a pressing concern for many low-income individuals and families. With the economic prosperity of the 1990s, low-cost housing

▶ Conflict theorists see schools as agents of the capitalist class system that perpetuate social inequality: Upper-class students are educated in well-appointed environments such as the one shown here, whereas children of the poor tend to go to antiquated schools with limited funds.

© Image Source/Getty Images

units in many cities were replaced with expensive condominiums and luxury single-family residences for affluent people. Since then, housing costs have remained high compared to many families' ability to pay for food, shelter, clothing, and other necessities. The economic crisis of the 2000s contributed to increased numbers of foreclosures that have placed more families in danger of becoming homeless.

Lack of *affordable* housing is one central problem brought about by economic inequality. Another concern is *substandard* housing, which refers to facilities that have inadequate heating, air conditioning, plumbing, electricity, or structural durability. Structural problems—attributable to faulty construction or lack of adequate maintenance—exacerbate the potential for other problems such as damage from fire, falling objects, or floors and stairways collapsing.

Education Educational opportunities and life chances are directly linked. Some functionalist theorists view education as the "elevator" to social mobility. Improvements in the educational achievement levels (measured in number of years of schooling completed) of the poor, people of color, and white women have been cited as evidence that students' abilities are now more important than their class, race, or gender. From this perspective, inequality in education is declining, and students have an opportunity to achieve upward mobility through achievements at school. Functionalists generally see the education system as flexible, allowing most students the opportunity to attend college if they apply themselves (Ballantine and Hammack, 2009).

In contrast, most conflict theorists stress that schools are agencies for reproducing the capitalist class system and perpetuating inequality in society. From this perspective, lack of education perpetuates poverty. More than half of the children in young families (parent or parents younger than 30) live in poverty if the head of the household is not a high school graduate. Even if the parent has completed some college (but not earned a bachelor's degree), about 25 percent of children live in poverty. Parents with low educational attainment and limited income are often not able to provide the same opportunities for their children as families where at least one parent has more years of formal education (see ▶ Figure 8.8).

Today, great disparities exist in the distribution of educational resources. Because of massive shortfalls

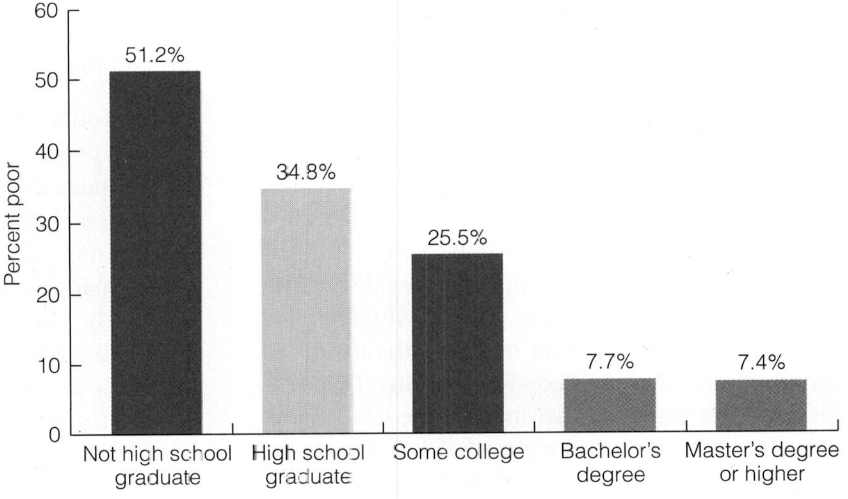

▶ **FIGURE 8.8 HOW EDUCATION IS RELATED TO POVERTY AMONG CHILDREN**

Source: Children's Defense Fund, 2010.

in state and federal budgets, funding for education is being cut drastically at all levels. Because funding for education comes primarily from local property taxes, school districts in wealthy suburban areas generally pay higher teachers' salaries, have newer buildings, and provide state-of-the-art equipment. By contrast, schools in poorer areas have a limited funding base. Students in central-city schools and poverty-stricken rural areas often attend dilapidated schools that lack essential equipment and teaching materials. Author Jonathan Kozol (1991, qtd. in Feagin and Feagin, 1994: 191) documented the effect of a two-tiered system on students:

> Kindergartners are so full of hope, cheerfulness, high expectations. By the time they get into fourth grade, many begin to lose heart. They see the score, understanding they're not getting what others are getting.... They see suburban schools on television.... They begin to get the point that they are not valued much in our society. By the time they are in junior high, they understand it. "We have eyes and we can see; we have hearts and we can feel.... We know the difference."

Poverty extracts such a toll that many young people will not have the opportunity to finish high school, much less enter college.

Poverty in the United States

When many people think about poverty, they think of people who are unemployed or on welfare. However, many hardworking people with full-time jobs live in poverty. The U.S. Social Security Administration has established an ***official poverty line,* which is based on what is considered to be the minimum amount of money required for living at a subsistence level.** The poverty level is computed by determining the cost of a minimally nutritious diet (a low-cost food budget on which a family could survive nutritionally on a short-term, emergency basis) and multiplying this figure by three to allow for nonfood costs. In 2009, 43.6 million people lived below the official government poverty level of $22,050 for a family of four. In 2011, the official government poverty level for the same family was $22,350, only a slight increase. The official poverty rate in 2009 was 14.3 percent of the U.S. population, up from 13.2 percent in 2008 (DeNavas-Walt, Proctor, and Smith, 2010).

When sociologists define poverty, they distinguish between absolute and relative poverty. ***Absolute poverty* exists when people do not have the means to secure the most basic necessities of life.** This definition comes closest to that used by the federal government. Absolute poverty often has life-threatening consequences, such as when a homeless person freezes to death on a park bench. By comparison, ***relative poverty* exists when people may be able to afford basic necessities but are still unable to maintain an average standard of living.** A family must have income substantially above the official poverty line in order to afford the basic necessities, even when these are purchased at the lowest possible cost. At about 155 percent of the official poverty line, families could live on an economy budget. What is it like to live on the economy budget? John Schwarz and Thomas Volgy (1992: 43) offer the following distressing description:

> Members of families existing on the economy budget never go out to eat, for it is not included in the food budget; they never go out to a movie, concert, or ball game or indeed to any public or private establishment that charges admission, for there is no entertainment budget; they have no cable television, for the same reason; they never purchase alcohol or cigarettes; never take a vacation or holiday that involves any motel or hotel or, again, any meals out; never hire a baby-sitter or have any other paid child care; never give an allowance or other spending money to the children; never purchase any lessons or home-learning tools for the children; never buy books or records for the adults or children, or any toys, except in the small amounts available for birthday or Christmas presents ($50 per person over the year); never pay for a haircut; never buy a magazine; have no money for the feeding or veterinary care of any pets; and, never spend any money for preschool for the children, or educational trips for them away from home, or any summer camp or other activity with a fee.

Concerns such as these are shared across the United States. However, poverty rates vary widely across the nation. The percentage of people living below the poverty line is much higher in some states and regions of the country than in others (see ▶ Map 8.2).

Who Are the Poor?

Poverty in the United States is not randomly distributed, but rather is highly concentrated according to age and race/ethnicity.

Age Today, children are at a much greater risk of living in poverty than are older persons. Social scientists estimate that a large number of children—about one in five—live in poverty today. In the past, persons over age 65 were at the greatest risk of being poor, and indeed many are still poor. Older women are twice as likely to be poor as older men; older African Americans and Latinos/as are much more likely to live below the poverty line than are non-Latino/a whites. However, government programs such as Social Security and pension plans provide for something closer to an adequate standard of living for older adults than do social welfare programs for children and their parents (see ▶ Figure 8.9).

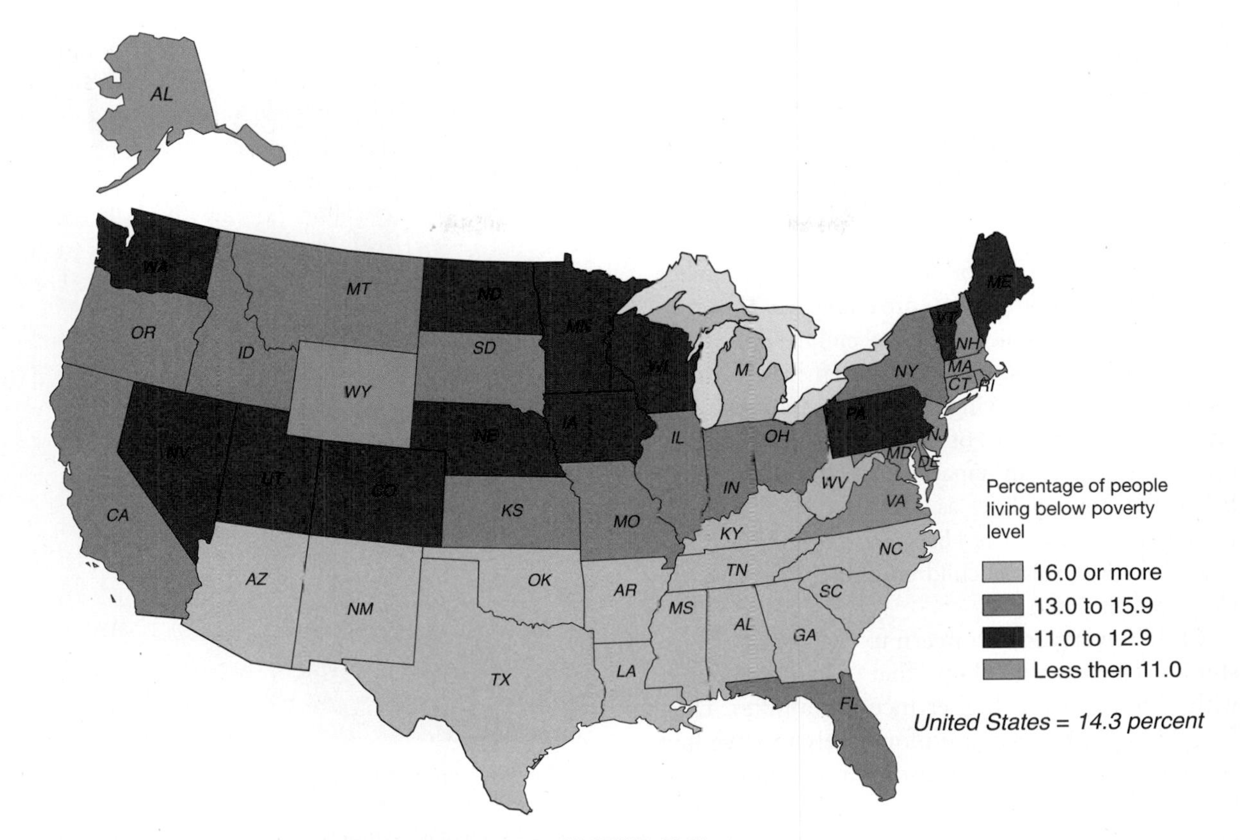

▲ **MAP 8.2 PERCENTAGE OF PEOPLE IN POVERTY BY STATE: 2009**

Source: Bishaw and Renwick, 2009.

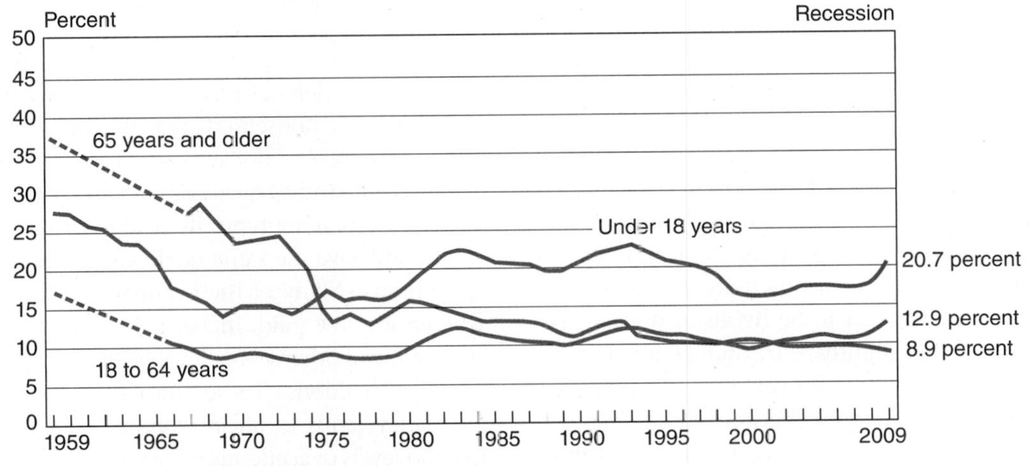

▲ **FIGURE 8.9 POVERTY RATES BY AGE: 1959 TO 2009**

Note: The data points are placed at the midpoints of the respective years. Data for people aged 18 to 64 and 65 and older are not available from 1960 to 1965.

Source: U.S. Census Bureau. Current Population Survey, 1960 to 2010 Annual Social and Economic Supplements.

As shown in Figure 8.9, children are the most vulnerable to poverty today. In 2009 both the poverty rate and the number in poverty increased for children under 18 years old, rising to almost 21 percent from 19 percent in 2008. This is a continuing upward trend that dates back to 2000. The number of children in poverty increased to 15.5 million in 2009, as compared with 14.1 million in 2008 and 13.3 million in 2007. The precarious position of African American and Latino/a children is even more striking in that about 34 percent of all African Americans under age 18 live in poverty; 28 percent of Latino/a children

official poverty line the federal income standard that is based on what is considered to be the minimum amount of money required for living at a subsistence level.

absolute poverty a level of economic deprivation that exists when people do not have the means to secure the most basic necessities of life.

relative poverty a condition that exists when people may be able to afford basic necessities but are still unable to maintain an average standard of living.

are also poor, as compared with almost 14 percent of non-Latino/a white children and 12 percent of Asian and Pacific Islanders (U.S. Census Bureau, 2010b).

What do such statistics indicate about the future of our society? Children as a group are poorer now than they were at the beginning of the 1980s, whether they live in one- or two-parent families. The majority of children classified as "poor" live in two-parent families in which one or both parents are employed. However, in all racial and ethnic categories, children in single-parent households headed by women have a much greater likelihood of living in poverty. Approximately one-third (32.3 percent) of white (non-Hispanic) children under age 18 in female-headed households live below the poverty line, as sharply contrasted with more than 50 percent of Hispanic (Latino/a) and African American (black) children in the same category (Moore et al., 2009).

Child poverty is of concern to everyone because studies consistently show that when compared with children from higher-income families, children living in poverty are more likely to have low academic achievement, to drop out of school, and to have physical and mental health problems. Nor does the future look bright: Many governmental programs established to alleviate childhood poverty and malnutrition have been seriously cut back or eliminated altogether. In 2010 almost 6.5 million children lived in extreme poverty—defined as income that is 150 percent or more below the official poverty line (Children's Defense Fund, 2011).

Gender Women in the United States are more likely to be poor than men. About two-thirds of all adults living in poverty are women. African American and Hispanic (Latina) women are at least twice as likely as white women to be living in poverty. It should be noted that slightly more than 50 percent of all adult women (age 18 and over) living in poverty are single with no dependent children. About 25 percent of poor adult women are single with dependent children. In 2009 single-parent families headed by women had a 29.9 percent poverty rate, as compared with a 16.9 percent rate for male-householder-with-no-wife-present families and a 5.8 percent rate for two-parent families. Elderly women are far more likely to be poor than elderly men. Approximately 15 percent of women over 75 years of age are poor, as compared to 6 percent of men.

Why are more women living in poverty? One of the earliest explanations came from sociologist Diana Pearce (1978), who coined a term to describe this problem: The ***feminization of poverty*** **refers to the trend in which women are disproportionately represented among individuals living in poverty.** According to Pearce (1978), women have a higher risk of

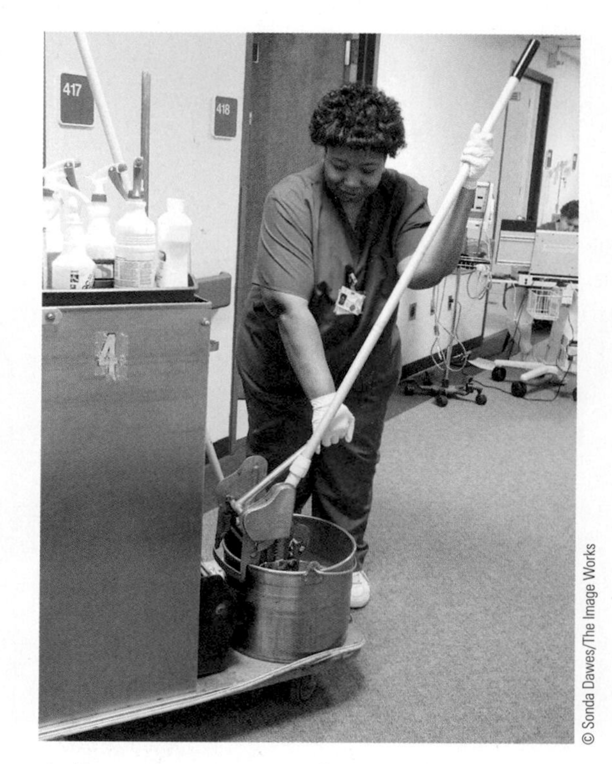

▲ Many women are among the "working poor," who, although employed full time, have jobs in service occupations that are typically lower paying and less secure than jobs in other sectors of the labor market. Does the nature of women's work contribute to the feminization of poverty in the United States?

being poor because they bear the major economic and emotional burdens of raising children when they are single heads of households but earn on average about 77 cents for every dollar that a male worker earns. More-recent studies have confirmed her theory, finding that women are typically paid less than men even when they have the same qualifications and work the same hours. Many of the occupations dominated by women are low paid, including teaching, child care, cleaning, and other lower-level health care and service industry positions. These studies also confirm that pregnancy tends to affect women's work opportunities and levels of academic achievement. Coupled with the fact that women are more likely than men to bear the cost of raising children and to spend more time providing unpaid caregiving for family members of all ages, many women find themselves in a situation where it is difficult to improve their standard of living (Cawthorne, 2008). However, we should note that all women are not equally vulnerable to poverty. Many in the upper and upper-middle classes have the financial resources, education, and skills to support themselves regardless of the presence of a man in the household. But this should not diminish the fact that poverty is everyone's problem: When women are impoverished, so are their children. And many of the poor are men, especially those who are chronically unemployed, older, homeless, and dealing with disabilities.

Race/Ethnicity According to some stereotypes, most of the poor and virtually all welfare recipients are people of color. However, this stereotype is false; white Americans (non-Latinos/as) account for approximately two-thirds of those below the official poverty line. However, such stereotypes are perpetuated because a disproportionate percentage of the impoverished in the United States is made up of African Americans, Latinos/as, and Native Americans. Approximately 26 percent of African Americans and 25 percent of Hispanics (Latinas/os) were among the officially poor in 2009, as compared with 9.4 percent of non-Latino/a whites and 12.5 percent of Asian Americans (DeNavas-Walt, Proctor, and Smith, 2010). Some analysts estimate Native American poverty rates to be as high as 33 percent, but this information is not compiled by the U.S. Census Bureau.

Economic and Structural Sources of Poverty

Social inequality and poverty have both economic and structural sources. Unemployment is a major cause of contemporary poverty. Tough economic times provide fewer opportunities for individuals to get an entry-level position that may help them to gain a toehold in U.S. society. Massive plant closings and layoffs in various employment sectors contribute to a trickle-down effect that causes workers in many fields throughout the nation to lose their jobs or take severe pay cuts.

© AP Photo/Rich Pedroncelli

▲ This California electronic benefit transfer (EBT) card represents a modern approach to helping people of limited income purchase groceries. Data-encoded cards such as this one were developed to prevent the trading or selling of traditional food stamps. However, one drawback of this technology is that many of California's popular farmers' markets are not able to process EBT cards.

Low wages paid for many jobs is another major cause: Half of all families living in poverty are headed by someone who is employed, and one-third of those family heads work full time. A person with full-time employment in a minimum-wage job cannot keep a family of four from sinking below the official poverty line.

Structural problems contribute to both unemployment and underemployment. These structural problems include transformations from employment in the manufacturing sector to service-sector and information-processing employment. In the process, major corporations have displaced millions of workers through downsizing, layoffs, and moving jobs to other nations where "cheap labor" is assumed to exist because people initially will work for lower wages. The proliferation of computers, the Internet, and other technology has produced ***job deskilling*****—a reduction in the proficiency needed to perform a specific job that may lead to a corresponding reduction in the wages for that job.** The shift from manufacturing to service occupations has resulted in the loss of higher-paying positions and their replacement with lower-paying and less-secure positions that do not offer the wages, job stability, or advancement potential of the disappearing manufacturing jobs. Many of the new jobs are located in the suburbs, thus making them inaccessible to central-city residents.

The problems of unemployment, underemployment, and poverty-level wages are even greater for people of color and young people in large urban centers, where the unemployment rate for African Americans is almost double that for whites (U.S. Bureau of Labor Statistics, 2007).

The Great Recession of the 2000s has contributed to job instability, high rates of unemployment, housing issues with high rents and massive numbers of foreclosures, and concern about the overall strength of the economy. Although these factors particularly affect those who are better off in a society, these problems also diminish the chances for people living in poverty to gain more financial security for themselves and their children.

Solving the Poverty Problem

The United States has attempted to solve the poverty problem in several ways. One of the most enduring is referred to as social welfare. When most people

feminization of poverty the trend in which women are disproportionately represented among individuals living in poverty.

job deskilling a reduction in the proficiency needed to perform a specific job that may lead to a corresponding reduction in the wages for that job.

sociology works!

Reducing Structural Barriers to Achieving the American Dream

Our society recognizes a moral obligation to provide a helping hand to those in need, but those in poverty have been getting only the back of the hand. They receive little or no public assistance. Instead, they are scolded and told that they have caused their own misfortunes. This is our "compassion gap"—a deep divide between our moral commitments and how we actually treat those in poverty.

In this statement the sociologists Fred Block, Anna C. Korteweg, and Kerry Woodward (2008: 166) describe the contradiction between our nation's alleged moral commitment to alleviating poverty and how we actually treat people who live in poverty. Children, single mothers with children, and people of color (particularly African Americans and Latinos/as) make up a disproportionate segment of the nation's poorest groups, and individuals in these categories are the persons most disadvantaged by arguments asserting that the poor have no one but themselves to blame for their poverty.

Numerous sociological studies regarding wealth and poverty demonstrate how structural factors contribute to the ability of some individuals to achieve the American Dream whereas others are hampered in achieving that goal by factors that are beyond their control. Yet, according to the Economic Mobility Project, policy makers do little to alleviate poverty in the United States because of the widely held belief that the American Dream should provide everyone with equality of *opportunity* but not necessarily equality of *outcome*: "The belief in America as a land of opportunity may also explain why rising inequality in the United States has yielded so little in terms of responsiveness from policy makers: if the American Dream is alive and well, then there is no need for government intervention to smooth the rough edges of capitalism. Diligence and skill, the argument goes, will yield a fair distribution of rewards" (Pew Charitable Trusts, 2007). However, research continues to reveal that structural factors beyond the control of individuals are important in determining where a person's place will be in the U.S. class system.

Can we keep the American Dream alive for all? According to Block, Korteweg, and Woodward (2008), we must take a number of specific steps to revitalize the American Dream for more people in this country and to reverse the compassion gap. We must make people aware of how far social reality has departed from the ideals of the American Dream. As a nation we must also take action to deal with the costs of four critical services that have risen much more rapidly than people's wages and the rate of inflation. These four critical services are *health care, higher education, high-quality child care*, and *housing*. As sociologists and other social analysts have suggested, if we as a nation are to claim that we have a commitment to compassion, we must make it our collective responsibility to help remove the structural barriers that currently reduce opportunities, mobility, and a chance for a better way of life for millions of Americans: "True compassion requires that we build a society in which every person has a first chance, a second chance, and, if needed, a third and fourth chance, to achieve the American Dream. We . . . need to use every instrument we have—faith groups, unions, community groups, and most of all government programs—to address the structural problems that reproduce poverty in our affluent society" (Block, Korteweg, and Woodward, 2008: 175).

reflect & analyze

Consider the area where you live. Do you know of people there whose situations could be improved if they were given a greater opportunity to take charge of their lives? How could such a change happen?

think of "welfare," they think of food stamps and programs such as Temporary Assistance for Needy Families (TANF) or the earlier program it replaced, Aid to Families with Dependent Children (AFDC). However, the primary beneficiaries of social welfare programs are not poor. Some analysts estimate that approximately 80 percent of all social welfare benefits are paid to people who do not qualify as "poor." For example, many recipients of Social Security are older people in middle- and upper-income categories.

When older persons, including members of Congress, accept Social Security payments, they are not stigmatized, nor should they be. Similarly, veterans who receive benefits from the Veterans Benefits Administration are not viewed as "slackers," and this is entirely appropriate. Unemployed workers who receive unemployment compensation are viewed with sympathy because they cannot find work and must take care of their families. By contrast, poor women and children who receive minimal benefits from welfare programs tend to be stigmatized and sometimes humiliated, even when our nation describes itself as having compassion for the less fortunate (see "Sociology Works!").

Are there solutions to the poverty problem? Often, how a person answers this question is related to which of the sociological explanations of social inequality that individual embraces. Some people believe that government and other public programs are the answer; others believe that only individual initiative is the answer to solving the poverty problem in the United States and other nations.

Sociological Explanations of Social Inequality in the United States

Obviously, some people are disadvantaged as a result of social inequality. Therefore, is inequality always harmful to society?

Functionalist Perspectives

According to the sociologists Kingsley Davis and Wilbert Moore (1945), inequality is not only inevitable but also necessary for the smooth functioning of society. The Davis–Moore thesis, which has become the definitive functionalist explanation for social inequality, can be summarized as follows:

1. All societies have important tasks that must be accomplished and certain positions that must be filled.
2. Some positions are more important for the survival of society than others.
3. The most qualified people must fill the most important positions.
4. The positions that are the most important for society and that require scarce talent, extensive training, or both must be the most highly rewarded.
5. The most highly rewarded positions should be those that are functionally unique (no other position can perform the same function) and on which other positions rely for expertise, direction, or financing.

Davis and Moore use the physician as an example of a functionally unique position. Doctors are very important to society and require extensive training, but individuals would not be motivated to go through years of costly and stressful medical training without incentives to do so. The Davis–Moore thesis assumes that social stratification results in ***meritocracy*****—a hierarchy in which all positions are rewarded based on people's ability and credentials.**

Critics have suggested that the Davis–Moore thesis ignores inequalities based on inherited wealth and intergenerational family status (Rossides, 1986). The thesis assumes that economic rewards and prestige are the only effective motivators for people and fails to take into account other intrinsic aspects of work, such as self-fulfillment (Tumin, 1953). It also does not adequately explain how such a reward system guarantees that the most qualified people will gain access to the most highly rewarded positions.

© Stevin Rubin/The Image Works

▲ According to a functional perspective, people such as these Harvard Law School graduates attain high positions in society because they are the most qualified and they work the hardest. Is our society a meritocracy? How would conflict theorists answer this question?

Conflict Perspectives

From a conflict perspective, people with economic and political power are able to shape and distribute the rewards, resources, privileges, and opportunities in society for their own benefit. Conflict theorists do not believe that inequality serves as a motivating force for people; they argue that powerful individuals and groups use ideology to maintain their favored positions at the expense of others. Core values in the United States emphasize the importance of material possessions, hard work, individual initiative to get ahead, and behavior that supports the existing social structure. These same values support the prevailing resource-distribution system and contribute to social inequality.

Are wealthy people smarter than others? According to conflict theorists, certain stereotypes suggest that this is the case; however, the wealthy may actually be "smarter" than others only in the sense of having "chosen" to be born to wealthy parents from whom they could inherit assets. Conflict theorists also note that laws and informal social norms support inequality in the United States. For the first half of the twentieth century, both legalized and institutionalized segregation and discrimination reinforced employment discrimination and produced higher levels of economic inequality. Although laws have been passed to make these overt acts of discrimination illegal, many forms of discrimination still exist in educational and employment opportunities.

meritocracy a hierarchy in which all positions are rewarded based on people's ability and credentials.

[concept quick review]

Sociological Explanations of Social Inequality in the United States	
Functionalist Perspectives	Some degree of social inequality is necessary for the smooth functioning of society (in order to fill the most important positions) and thus is inevitable.
Conflict Perspectives	Powerful individuals and groups use ideology to maintain their favored positions in society at the expense of others, and wealth is not necessary in order to motivate people.
Symbolic Interactionist Perspectives	The beliefs and actions of people reflect their class location in society.

Symbolic Interactionist Perspectives

Symbolic interactionists focus on microlevel concerns and usually do not analyze larger structural factors that contribute to inequality and poverty. However, many significant insights into the effects of wealth and poverty on people's lives and social interactions can be derived from applying a symbolic interactionist approach. Using qualitative research methods and influenced by a symbolic interactionist approach, researchers have collected the personal narratives of people across all social classes, ranging from the wealthiest to the poorest people in the United States.

A few studies provide rare insights into the social interactions between people from vastly divergent class locations. Sociologist Judith Rollins's (1985) study of the relationship between household workers and their employers is one example. Based on in-depth interviews and participant observation, Rollins examined rituals of deference that were often demanded by elite white women of their domestic workers, who were frequently women of color. According to the sociologist Erving Goffman (1967), *deference* is a type of ceremonial activity that functions as a symbolic means whereby appreciation is regularly conveyed to a recipient. In fact, deferential behavior between nonequals (such as employers and employees) confirms the inequality of the relationship and each party's position in the relationship relative to the other. Rollins identified three types of linguistic deference between domestic workers and their employers: use of the first names of the workers, contrasted with titles and last names (Mrs. Adams, for example) of the employers; use of the term *girls* to refer to female household workers regardless of their age; and deferential references to employers, such as "Yes, ma'am." Spatial demeanor, including touching and how close one person stands to another, is an additional factor in deference rituals across class lines. Rollins (1985: 232) concludes that

> The employer, in her more powerful position, sets the essential tone of the relationship; and that tone . . . is one that functions to reinforce the inequality of the relationship, to strengthen the employer's belief in the rightness of her advantaged class and racial position, and to provide her with justification for the inegalitarian social system.

Many concepts introduced by the sociologist Erving Goffman (1959, 1967) could be used as springboards for examining microlevel relationships between inequality and people's everyday interactions. What could you learn about class-based inequality in the United States by using a symbolic interactionist approach to examine a setting with which you are familiar?

The Concept Quick Review summarizes the three major perspectives on social inequality in the United States.

U.S. Stratification in the Future

The United States is facing one of the greatest economic challenges it has experienced since the Great Depression, in the 1930s. Although we have strong hopes that the American Dream will remain alive and well, many people are concerned that national and international economic problems will create a lack of upward mobility for more Americans. The nationwide slump in housing and jobs has distressed people across income levels, and high rates of unemployment and a fluctuating stock market produce weekly predictions that the economy is either getting better or worse.

So this brings us to an important final question in this chapter: Will social inequality in the United States increase, decrease, or remain the same in the future? Many social scientists believe that existing trends point to an increase. First, the purchasing power of the dollar has stagnated or declined since the early 1970s. As families started to lose ground financially, more family members (especially women) entered the labor force in an attempt to support themselves and their families (Gilbert, 2010). Economist and former Secretary of Labor Robert Reich (1993) has noted that in recent years the employed have been traveling on two escalators—one going up and the other going down. The gap between the

you can make a difference

Students Helping Others Through Campus Kitchen

My life has gotten to the point where if I'm not in class I'm sleeping or doing something for Campus Kitchen.... So, I'll let y'll in on a little secret: Campus Kitchen is worth being excited about. For the uninitiated, Campus Kitchen is a student-led organization. We rescue and cook food to distribute to those in need. Five days a week students go to the dining halls and pick up pans of food and take them to the Salvation Army. On Tuesday afternoons, students are busy in the Family and Consumer Sciences kitchen creating healthy snacks to be given to children at local schools. On Thursdays, the kitchen crew cooks a meal that is delivered to the women and children at the Family Abuse Center.

Excited yet?

—Amy Heard (2011), a Baylor University student, describing how much she enjoys volunteering for Campus Kitchen

What is Campus Kitchen? Currently, more than 25 colleges are involved in Campus Kitchen, an on-campus student service program that is part of the Campus Kitchen Project, begun by Robert Egger, director of the nonprofit D.C. Central Kitchen in Washington, D.C. As Amy Heard explains, students go to dining halls, cafeterias, and local food banks or restaurants to pick up unserved, usable food and to make sure that meals gets to a local organization that feeds persons in need. Each college provide on-campus space for the "Campus Kitchen," such as a dining hall at off-hours or a classroom/kitchen, where students can prepare meals and snacks using the donated food. Students then deliver the meals to individuals and families in need of food assistance and to organizations such as homeless shelters and soup kitchens in their community.

In addition to helping start Campus Kitchen, Robert Egger has been an inspiration for many others who make a difference by helping feed the hungry. Egger is one of the people responsible for an innovative chef-training program that feeds hope as well as hunger. At the Central Kitchen, located in the nation's capital, staff and guest chefs annually train around 48 homeless persons in three-month-long kitchen-arts courses. While the trainees are learning about food preparation, which will help them get starting jobs in the restaurant industry, they are also helping feed about 3,000 homeless persons each day. Much of the food is prepared using donated goods such as turkeys that people have received as gifts at office parties and given to the kitchen, and leftover food from grocery stores, restaurants, hotels, and college cafeterias. Central Kitchen got its start using leftovers from President George H. W. Bush's inaugural banquet in the late 1980s (Clines, 1996). Donated food for use at both Central Kitchen and Campus Kitchen has gotten a boost from the Bill Emerson Good Samaritan Food Donation Act, which exempts nonprofit organizations and *gleaners*—volunteers who collect what is left in the field after harvesting—from liability for problems with food that they contribute in good faith (see Burros, 1996).

Can you think of ways that leftover food could be recovered from your college or university or other places where you eat so the food could be redistributed to persons in need?

- Campus Kitchen Project
  www.campuskitchens.org/national
- D.C. Central Kitchen
  www.dccentralkitchen.org

AP Photo/Amy Sinisterra

Here is an example of Campus Kitchen at work. These Gonzaga University undergraduates are using leftovers from the dining hall to put together meals for the needy. Does your college have a similar program?

earnings of workers and the income of managers and top executives has widened even in an era when large salaries and even larger bonuses for CEOs have been frowned on by political leaders and everyday people.

Second, wealth continues to become more concentrated at the top of the U.S. class structure. As the rich have grown richer, more people have found themselves among the ranks of the poor. Third, federal tax laws in recent years have benefited corporations and wealthy families at the expense of middle- and lower-income families, and even if changes are made in the tax code, wealthier individuals and corporations typically find new ways to shelter their incomes. Finally, as previously mentioned, structural sources of upward mobility are shrinking, whereas the rate of downward mobility has increased.

Are we sabotaging our future if we do not work constructively to eliminate poverty? It has been said that a chain is no stronger than its weakest link. If we apply this idea to the problem of poverty, then it is to our advantage to see that those who cannot find work or do not have a job that provides a living wage receive adequate training and employment. Innovative programs can combine job training with producing something useful to meet the immediate needs of people living in poverty. Children of today—the adults of tomorrow—need nutrition, education, health care, and safety as they grow up (see "You Can Make a Difference").

Some social analysts believe that the United States will become a better nation if it attempts to regain the American Dream by attacking poverty. According to the sociologist Michael Harrington (1985: 13), if we join in solidarity with the poor, we will "rediscover our own best selves . . . we will regain the vision of America."

chapter review Q & A

Use these questions and answers to check how well you've achieved the learning objectives set out at the beginning of this chapter.

- **What is social stratification, and how does it affect our daily life?**

Social stratification is the hierarchical arrangement of large social groups based on their control over basic resources. People are treated differently based on where they are positioned within the social hierarchies of class, race, gender, and age.

- **What are the major systems of stratification?**

Stratification systems include slavery, caste, and class. Slavery, an extreme form of stratification in which people are owned by others, is a closed system. The caste system is also a closed one in which people's status is determined at birth based on their parents' position in society. The class system, which exists in the United States, is a type of stratification based on ownership of resources and on the type of work that people do.

- **How did classical sociologists such as Karl Marx and Max Weber view social class?**

Karl Marx and Max Weber acknowledged social class as a key determinant of social inequality and social change. For Marx, people's relationship to the means of production determines their class position. Weber developed a multidimensional concept of stratification that focuses on the interplay of wealth, prestige, and power.

- **What are some of the consequences of inequality in the United States?**

The stratification of society into different social groups results in wide discrepancies in income and wealth and in variable access to available goods and services. People with high income or wealth have greater opportunity to control their own lives. People with less income have fewer life chances and must spend their limited resources to acquire basic necessities.

- **How do sociologists view poverty?**

Sociologists distinguish between absolute poverty and relative poverty. Absolute poverty exists when people do not have the means to secure the basic necessities of life. Relative poverty exists when people may be able to afford basic necessities but are still unable to maintain an average standard of living.

- **Who are the poor?**

Age, gender, and race tend to be factors in poverty. Children have a greater risk of being poor than do the elderly, and women have a higher rate of poverty than do men. Although whites account for approximately two-thirds of those below the poverty line, people of color account for a disproportionate share of the impoverished in the United States.

- **What is the functionalist view on class?**

Functionalist perspectives view classes as broad groupings of people who share similar levels of privilege on the basis of their roles in the occupational structure. According to the Davis–Moore thesis, stratification exists in all societies, and some inequality is not only inevitable but also necessary for the ongoing functioning of society. The positions that are most important within society and that require the most talent and training must be highly rewarded.

- **What is the conflict view on class?**

Conflict perspectives on class are based on the assumption that social stratification is created and maintained by one group (typically the capitalist class) in order to enhance and protect its own economic interests. Conflict theorists measure class according to people's relationships with others in the production process.

- **What is the symbolic interactionist view on class?**

Unlike functionalist and conflict perspectives that focus on macrolevel inequalities in societies, symbolic interactionist views focus on microlevel inequalities such as how class location may positively or negatively influence one's identity and everyday social interactions. Symbolic interactionists use terms such as *social cohesion* and *deference* to explain how class binds some individuals together while categorically separating out others.

key terms

absolute poverty 240
alienation 222
capitalist class (bourgeoisie) 222
caste system 219
class conflict 223
class system 222
feminization of poverty 242
income 233
intergenerational mobility 216
intragenerational mobility 216
job deskilling 243
life chances 216
meritocracy 245
official poverty line 240
pink-collar occupations 227
power 224
prestige 224
relative poverty 240
slavery 217
social mobility 216
social stratification 216
socioeconomic status (SES) 225
underclass 229
wealth 224
working class (proletariat) 222

questions for critical thinking

1. Based on the Weberian and Marxian models of class structure, what is the class location of each of your ten closest friends or acquaintances? What is their location in relationship to yours? To one another's? What does their location tell you about friendship and social class?
2. Should employment be based on meritocracy, need or affirmative action policies?
3. What might happen in the United States if the gap between rich and poor continues to widen?

turning to video

Watch the ABC video *Growing Up Fast: Children of Poverty* (running time 9:22), available through **CengageBrain.com.** According to the video, 13 million children in the United States live below the poverty line, and the obstacles for them to reach the American Dream are tremendous. As you watch the video, think about your own obstacles and opportunities in graduating from high school and attending college. After you've watched the video, consider these questions: Why do we have such high child poverty rates in the United States, and what are the consequences? What kinds of things can be done to alleviate child poverty?

online study resources

Go to CENGAGE brain.com to access online study resources, including the Sociology CourseMate for this text as well as special features such as video, an interactive sociology time line and interactive maps, General Social Survey (GSS) data, and U.S. Census 2010 data.

CourseMate brings course concepts to life with interactive learning, study, and exam-preparation tools that support the printed textbook. A textbook-specific website, **Sociology CourseMate** includes an integrated interactive eBook and other interactive learning tools, including quizzes, flash cards, and videos.

Visit **www.cengagebrain.com** to access your account and purchase materials.

9 Global Stratification

National Public Radio, *Marketplace,* February 9, 2011:
Journalist Kai Ryssdal: Tens of thousands of Egyptians took to the streets once again today.... Across the region—from Tunisia to Egypt and over to Jordan—people are demanding change. They want political change, obviously, but they also want economic change. They want a chance at prosperity and an end to the wealthy elites who they believe are sucking their countries dry.

National Public Radio, *Marketplace,* February 10, 2011:
Journalist Kai Ryssdal: We're going back to the Middle East now and the regional part of this larger story. Behind the political protests of the past few weeks are some economic realities. High unemployment. Low wages. And widespread poverty.... From Jordan today, *Marketplace*'s Alisa Roth [tells us] what happens to the very poor.

Alisa Roth: Siham Taher is showing me the long-handled squeegee she keeps in her kitchen. She needs it because the kitchen floods every time it rains. She and her family of seven live in this two-room concrete house in the outskirts of Zarqa, which is an industrial city about an hour north of Amman. The rain comes through

UPI/Landov

▲ Tired of long-term unemployment and poverty, Egyptians demanded that their government be replaced by one that was more responsive to their needs. The massive street demonstration shown here was just one of many that shook Egypt and several other countries in the Middle East in 2011.

the space where the corrugated metal roof and the wall don't meet. The bathroom is a hole in the floor. Here in the bedroom she points out the crack in the ceiling. She's afraid it'll collapse on her children's heads some night.

Here in Jordan, as in Egypt and many other countries in the Middle East, there is a large gap between rich and poor. Here families living below the poverty line, like Taher's, make up more than 13 percent of the population. And there's not much chance for them to escape.

—U.S. media reports on National Public Radio (NPR) describing how the riots and other turmoil in a number of Middle Eastern countries occurred against a backdrop of poverty where wealthy elites spend $15 million on a birthday party but poor people cannot put food on the table for their children (Beard, 2011)

Chapter Focus Question:

How do global stratification and economic inequality affect the life chances of people around the world?

In the 2010s numerous protests and riots occurred in nations where people responded to economic crises brought about by food shortages, the rising costs of goods and services, and a lack of jobs and income. Global poverty affects people in a variety of ways: For some, it means absolute poverty because they do not have even the basic necessities to survive. For others, poverty is relative, which means that their standard of living remains below that of other people in their nation and around the world.

Regardless of where people live, social and economic inequalities are pressing daily concerns. Poverty and inequality know no political boundaries or national borders. In this chapter we examine global stratification and inequality, and then discuss perspectives that have been developed to explain the nature and extent of this problem. Before reading on, test your knowledge of global wealth and poverty by taking the Sociology and Everyday Life quiz.

Highlighted Learning Objectives

- Define global stratification and explain how it contributes to economic inequality within and between nations.
- Identify how global poverty and human development are related.
- Discuss modernization theory and list its stages.
- Describe how conflict theorists explain patterns of global stratification.

sociology and everyday life

How Much Do You Know About Global Wealth and Poverty?

True	False	
T	F	1. Poverty has been increasing in the United States but decreasing in other nations because of globalization.
T	F	2. The assets of the 200 richest people are more than the combined income of over 40 percent of the world's population.
T	F	3. More than one billion people worldwide live below the international poverty line, earning less than $1.25 each day.
T	F	4. Although poverty is a problem in most areas of the world, relatively few people die of causes arising from poverty.
T	F	5. In low-income countries the problem of poverty is unequally shared between men and women.
T	F	6. The majority of people with incomes below the poverty line live in urban areas of the world.
T	F	7. Two-thirds of adults (15 years and older) worldwide who are not able to read and write are men.
T	F	8. Poor people in low-income countries meet most of their energy needs by burning wood, dung, and agricultural wastes, which increases health hazards and environmental degradation.

Answers on page 254.

Wealth and Poverty in Global Perspective

What do we mean by global stratification? ***Global stratification*** **refers to the unequal distribution of wealth, power, and prestige on a global basis, resulting in people having vastly different lifestyles and life chances both within and among the nations of the world.** Just as the United States is divided into classes, the world is divided into unequal segments characterized by extreme differences in wealth and poverty. For example, the income gap between the richest and the poorest 20 percent of the world population continues to widen (see ▶ Figure 9.1).

As previously defined, *high-income countries* are nations characterized by highly industrialized economies; technologically advanced industrial, administrative, and service occupations; and relatively high levels of national and per capita (per person) income. In contrast, *middle-income countries* are nations with industrializing economies, particularly

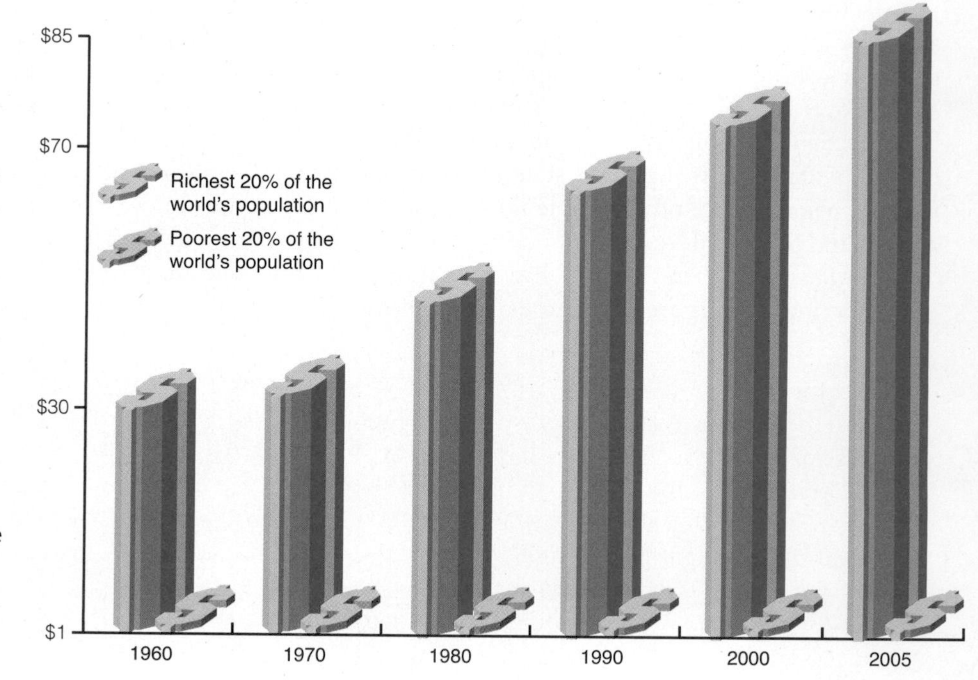

▶ FIGURE 9.1 INCOME GAP BETWEEN THE WORLD'S RICHEST AND POOREST PEOPLE

The income gap between the richest and poorest people in the world continued to grow between 1960 and 2005. As this figure shows, in 1960 the highest-income 20 percent of the world's population received $30 for each dollar received by the lowest-income 20 percent. By 2005, the disparity had increased: $85 to $1.

Source: World Bank, 2010.

in urban areas, and moderate levels of national and personal income. Middle-income countries are subdivided into lower-middle and upper-middle income. *Low-income countries* are primarily agrarian nations with little industrialization and low levels of national and personal income.

Just as the differences between the richest and poorest people in the United States have increased, the gap in global income differences between rich and poor countries has continued to widen over the past 50 years. In 1960 the wealthiest 20 percent of the world population had more than 30 times the income of the poorest 20 percent. By 2005, the wealthiest 20 percent of the world population had 85 times the income of the poorest 20 percent (World Bank, 2010). Income disparities *within* countries are often even more pronounced.

Although some progress has been made in reducing extreme poverty and child mortality rates while improving health and literacy rates in some lower-income countries, the overall picture remains bleak. The 2008 U.S. financial crisis led to a global economic recession, which spread from high-income economies to lower-income economies. The economic slowdown particularly harmed countries located in sub-Saharan Africa, home of thirty of the forty-three low-income economies (World Bank, 2010).

Many people have sought to address the issue of world poverty and to determine ways in which resources can be used to meet the urgent challenge of poverty. However, not much progress has been made on this front despite a great deal of talk and billions of dollars in "foreign aid" flowing from high-income nations to low-income nations. The idea of "development" has become the primary means used in attempts to reduce social and economic inequalities and alleviate the worst effects of poverty in the less industrialized nations of the world. Often, the nations that have not been able to reduce or eliminate poverty are chastised for not making the necessary social and economic reforms to make change possible. As one social analyst has suggested,

> The *problem* of inequality lies not in poverty, but in excess. "The problem of the world's poor," defined more accurately, turns out to be "the problem of the world's rich." This means that the solution to the problem is not a massive change in the culture of poverty so as to place it on the path of development, but a massive change in the culture of superfluity in order to place it on the path of counter development. It does not call for a new value system forcing the world's majority to feel shame at their traditionally moderate consumption habits, but for a new value system forcing the world's rich to see the shame and vulgarity of their over consumption habits, and the double vulgarity of standing on other people's shoulders to achieve those consumption habits. (Lummis, 1992: 50)

As this statement suggests, the increasing interdependence of all the world's nations was largely overlooked or ignored until increasing emphasis was placed on the global marketplace and the global economy. In addition, there are a number of problems inherent in studying global stratification, one of which is what terminology should be used to describe various nations.

Problems in Studying Global Inequality

One of the primary problems encountered by social scientists studying global stratification and social and economic inequality is what terminology should be used to refer to the distribution of resources in various nations. During the past sixty years, major changes have occurred in the way that inequality is addressed by organizations such as the United Nations and the World Bank. Most definitions of inequality are based on comparisons of levels of income or economic development, whereby countries are identified in terms of the "three worlds" or upon their levels of economic development.

The "Three Worlds" Approach

After World War II, the terms "First World," "Second World," and "Third World" were introduced by social analysts to distinguish among nations on the basis of their levels of economic development and the standard of living of their citizens. *First World* nations consisted of the rich, industrialized nations that primarily had capitalist economic systems and democratic political systems. The most frequently noted First World nations were the United States, Canada, Japan, Great Britain, Australia, and New Zealand *Second World* nations were countries with at least a moderate level of economic development and a moderate standard of living. These nations included China, North Korea, Vietnam, Cuba, and portions of the former Soviet Union. According to social analysts, although the quality of life in Second World nations was not comparable to that of

global stratification the unequal distribution of wealth, power, and prestige on a global basis, resulting in people having vastly different lifestyles and life chances both within and among the nations of the world.

sociology and everyday life

ANSWERS to the Sociology Quiz on Global Wealth and Poverty

1. False. In the twenty-first century, the income gap between the richest and poorest has increased on a global basis as well as in the United States.

2. True. The assets of the 200 richest people are more than the combined income of 50 percent of the world's population (United Nations Development Programme, 2010).

3. True. The World Bank estimates that more than 1.4 billion people worldwide live below the international poverty line, which is defined as earning less than $1.25 each day (World Bank, 2009).

4. False. One of the consequences of extreme poverty is hunger, and millions of people—including 6 million children under the age of 5 years—die of hunger-related diseases or chronic malnutrition each year (PBS, 2008).

5. True. In almost all low-income countries (as well as middle- and high-income countries), poverty is a more chronic problem for women because of sexual discrimination, resulting in a lack of educational and employment opportunities.

6. False. Although the number of poor people residing in urban areas is growing rapidly, the majority of people with incomes below the poverty line live in rural areas of the world.

7. False. Two-thirds of adults worldwide who are unable to read and write are women. Gender disparity is greatest in southern Asia, where 73 percent of all men but only 51 percent of women are able to read and write.

8. True. Although these fuels are inefficient and harmful to health, many low-income people cannot afford appliances, connection charges, and so forth. In some areas, electric hookups are not available.

© Viviane Moos/Corbis News/Corbis

▲ Vast inequalities in income and lifestyle are evident in this photo of slums and nearby higher-priced housing in Mumbai, India. What visible patterns of economic inequality exist in the United States?

life in the First World, it was far greater than that of people living in the *Third World*—the poorest countries, with little or no industrialization and the lowest standards of living, shortest life expectancies, and highest rates of mortality.

The Levels of Development Approach

Among the most controversial terminology used for describing world poverty and global stratification has been the language of development. Terminology based on levels of development includes concepts such as developed nations, developing nations, less-developed nations, and underdevelopment. Let's look first at the contemporary origins of the idea of "underdevelopment" and "underdeveloped nations."

Following World War II, the concepts of *underdevelopment* and *underdeveloped nations* emerged out of the Marshall Plan (named after U.S. Secretary of State George C. Marshall), which provided massive sums of money in direct aid and loans to rebuild the European economic base destroyed during World War II. Given the Marshall Plan's success in rebuilding much of Europe, U.S. political leaders decided that the Southern Hemisphere nations that had recently been released from European colonialism could also benefit from a massive financial infusion and rapid economic development. Leaders of the developed nations argued that urgent problems such as poverty, disease, and famine could be reduced through the transfer of finance, technology, and experience from the developed nations to lesser-developed countries. From this viewpoint, economic development is the primary way to solve the poverty problem: Hadn't economic growth brought the developed nations to their own high standard of living?

Ideas regarding *underdevelopment* were popularized by President Harry S Truman in his 1949 inaugural address. According to Truman, the nations in the Southern Hemisphere were "underdeveloped areas" because of their low gross national product, which today is referred to as *gross national income* (GNI)—a term that refers to all the goods and services produced in a country in a given year, plus the net income earned outside the country by individuals or corporations. If nations could increase their GNI, then social and economic inequality among the citizens within the country could also be reduced. Accordingly, Truman believed that it was necessary to assist the people of economically underdeveloped areas to raise their *standard of living*, by which he meant material well-being that can be measured by the quality of goods and services that may be purchased by the per capita national income. Thus, an increase in the standard of living meant that a nation was moving toward economic development, which typically included the improved exploitation of natural resources by industrial development.

What has happened to the issue of development since the post–World War II era? After several decades of economic development fostered by organizations such as the United Nations and the World Bank, it became apparent by the 1970s that improving a country's GNI did not tend to reduce the poverty of the poorest people in that country. In fact, global poverty and inequality were increasing, and the initial optimism of a speedy end to underdevelopment faded.

Why did inequality increase even with greater economic development? Some analysts in the developed nations began to link growing social and economic inequality on a global basis to relatively high rates of population growth taking place in the underdeveloped nations. Organizations such as the United Nations and the World Health Organization stepped up their efforts to provide family planning services to the populations so that they could control their own fertility. More recently, however, population researchers have become aware that issues such as population growth, economic development, and environmental problems must be seen as interdependent concerns. This changing perception culminated in the U.N. Conference on Environment and Development in Rio de Janeiro, Brazil (the "Earth Summit"), in 1992; as a result, terms such as *underdevelopment* have largely been dropped in favor of measurements such as sustainable development, and economies are now classified by their levels of income or their ranking on the Human Development Index.

Classification of Economies by Income

The World Bank classifies nations into four economic categories and establishes the upper and lower limits for the gross national income (GNI) in each category. *Low-income economies* had a GNI per capita of $995 or less in 2009; *lower-middle-income economies* had a GNI per capita of $996 to $3,945; *upper-middle-income economies* had a GNI per capita between $3,946 and $12,195; and *high-income economies* had a GNI per capita of $12,196 or more.

Low-Income Economies

About half the world's population lives in the thirty-nine low-income economies, where most people engage in agricultural pursuits, reside in nonurban areas, and are impoverished (World Bank, 2010). As shown in ▶ Map 9.1, low-income economies are primarily found in countries in Asia and Africa, where half of the world's population resides.

Among those most affected by poverty in low-income economies are women and children. Mayra Buvinić, former head of the women in development program for the Inter-American Development Bank, describes the plight of one Nigerian woman as an example:

> On the outskirts of Ibadan, Nigeria, Ade cultivates a small, sparsely planted plot with a baby on her back and other visibly undernourished children nearby. Her efforts to grow an improved soybean variety, which could have improved her children's diet, failed because she lacked the extra time to tend the new crop, did not have a spouse who would help her, and could not afford hired labor. (Buvinić, 1997: 38)

Robert Harding Picture Library Ltd/Alamy

▲ Global inequality is most striking in nations that are sometimes referred to as "Third World" or "underdeveloped." In this Dhaka, Bangladesh, garment factory, female workers produce clothing only for export to other countries.

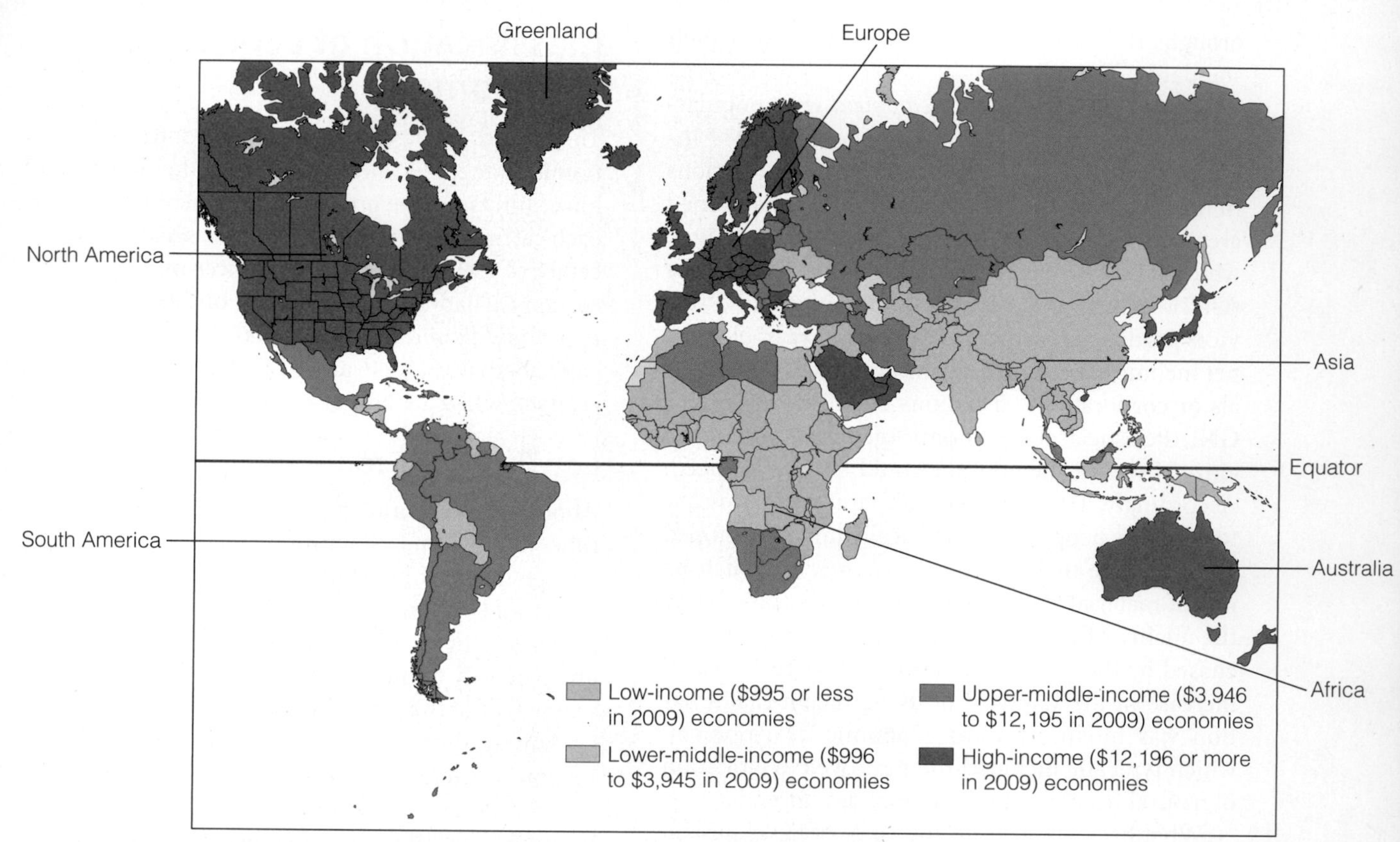

▲ **MAP 9.1 HIGH-, MIDDLE-, AND LOW-INCOME ECONOMIES IN GLOBAL PERSPECTIVE**
How does the United States compare with other nations with regard to income?
Source: World Bank, 2010.

According to Buvinić, Ade's life is typical of many women worldwide who face obstacles to increasing their economic power because they do not have the time to invest in the additional work that could bring in more income. Also, many poor women worldwide do not have access to commercial credit and have been trained only in traditionally female skills that produce low wages. Despite the fact that women have made some gains in terms of well-being, the income gap between men and women remains wide in low-income, developing nations.

Middle-Income Economies

About one-third of the world's population resides in a middle-income economy. The World Bank subdivides the middle-income economies into two categories. Fifty-six economies were included in the lower-middle-income category (a GNI per capita of $996 to $3,945), including Angola, Bolivia, Cameroon, China, Ecuador, Egypt, India, Indonesia, Iraq, Pakistan, Thailand, and Vietnam. China is an important example of a lower-middle-income economy because that nation's rural–urban divide has brought increasing income and wealth to urban areas through world trade and the presence of multinational corporations, as contrasted with rural areas that remain dependent on agriculture, particularly rice, for economic survival. Although some people have grown wealthy in lower-middle-income nations, many others continue to live in poverty, defined as below $1.25 per day in purchasing power (World Bank, 2010). In recent years, millions of people have migrated from lower- and lower-middle-income countries to upper-middle-income nations in hopes of finding better economic conditions, but they have still remained impoverished (see "Sociology in Global Perspective").

As compared with lower-middle-income economies, people in the forty-eight nations classified as upper-middle-income economies ($3,946 to $12,195 per capita in 2009) typically have a somewhat higher standard of living. Nations with upper-middle-income economies include Argentina, Chile, Mexico, Peru, Russian Federation, and South Africa. Some of these nations export a diverse variety of goods and services, ranging from manufactured goods to raw materials and fuels. For example, Kazakhstan, in Central Asia, is well known for tobacco growing and harvesting; however, this country has been accused of hazardous child labor practices (see "Sociology and Social Policy"). Some nations in the upper-middle-income category have extremely high levels of indebtedness and few resources for fighting poverty.

sociology in global perspective

Marginal Migration: Moving to a Less-Poor Nation

> We are forced to come back here—not because we like it, but because we are poor. When we cross the border, we are a little better off. We are able to buy shoes and maybe a chicken.
>
> —Anes Moises explaining why he and many other Haitian migrants live and work in the Dominican Republic (qtd. in DeParle, 2007: A1)

Even before a devastating earthquake hit in 2010, Haiti was the poorest country in the Western Hemisphere, with about 80 percent of the population living below the poverty line. As a result, many Haitians, including Anes Moises, migrated to countries such as the Dominican Republic in hopes of finding work. Anes found work in the banana fields, where he earned six times as much as he would in his homeland of Haiti. For Anes and the millions of "south-to-south" migrants—people who have moved from a low-income country to one that is slightly better off—even marginal gains in money and quality of life are important. As one journalist described the living conditions of Haitian migrants in the Dominican Republic, "The scrap-wood shanties on a muddy hillside are a poor man's promised land" (DeParle, 2007: A1). Although we hear much more about the more than 80 million migrants who have moved "south to north"—from a lower-income nation to a country with a high-income economy (such as from Mexico to the United States)—the lived experiences of "south-to-south" migrants are also important in understanding global wealth and poverty in the twenty-first century.

What are the typical characteristics of people who move from one poor nation to a somewhat-less-poor one? Do their efforts make a difference in their economic status? Recent studies have found that south-to-south migrants are often poorer than individuals who migrate from lower-income to high-income nations. South-to-south migrants are also more likely to travel without proper documentation and be more vulnerable to unscrupulous people and to apprehension by law enforcement officials than are south-to-north migrants (Ratha and Shaw, 2007). Many south-to-south migrants send money back home to extremely poor family members who reside in remote rural areas. Some analysts estimate that the money sent by these migrants has a significant economic impact on the lives of people in the poorest nations of the world. For example, Haitians residing in the Dominican Republic typically send about $135 million a year to relatives back home (DeParle, 2007: A16). Ironically, numerous jobs are available for Haitians in the Dominican Republic because many Dominicans have migrated to the United States (south-to-north migration) in hopes of finding better jobs and higher wages. We can identify similar patterns that exist across many nations as some individuals migrate to high-income economies while others move from a poor country to a slightly less poor one:

> Nicaraguans build Costa Rican buildings. Paraguayans pick Argentine crops. Nepalis dig Indian mines. Indonesians clean Malaysian homes. Farm hands from Burkina Faso tend the fields in Ivory Coast. Some save for the more expensive journeys north, while others find the move from one poor land to another all they will ever afford. With rich countries tightening their borders, migration within the developing world is likely to grow. (DeParle, 2007: A16).

Natural disasters such as earthquakes only make matters worse: The already poor have even less in the aftermath of such crises, and they have little hope that anything will change for the better in their own country.

reflect & analyze

How do global stratification and inequality affect people's decisions about migrating to find work? What are the costs and benefits of migrating to another country to find work?

© Dell Inc./Business Wire/Getty Images Publicity/Getty Images

◀ Where are these Dell Computer service technicians working? Dell's headquarters is located in Round Rock, Texas (near Austin), so you might assume that these employees are working there or somewhere else in the United States. However, part of the global workforce in high-income nations, the people you see here are employed at the Dell Enterprise Command Center in Limerick, Ireland.

sociology and social policy

Should We Be Concerned About Child Labor in the Global Economy?

News Article 1:

> One woman said children as young as 10 working in the fields developed red rashes on their stomachs and necks as they harvested tobacco for use in cigarettes made by Philip Morris . . . . Human Rights Watch plans to release a report on Wednesday showing that child and forced labor is widespread on farms that supply a cigarette factory owned by Philip Morris International in Kazakhstan, in Central Asia. (Kramer, 2010)

News Article 2:

> Philip Morris International Inc. said it would change its policies after a report said it bought tobacco from Eurasian farms that used forced and child labor. (UPI.com, 2010)

International child labor is more common than we might think. An estimated 158 million children ages five to fourteen are engaged in child labor—one in six children in the world—and many of these children are engaged in hazardous situations or conditions (unicef.org, 2011). Tobacco growing and harvesting are particularly dangerous because of factors such as the physical difficulty of the work; long hours; exposure to high heat, the sun, pesticides, and fumes from tobacco plants; and other health risks associated with the handling of tobacco plants. According to Philip Morris spokespersons, the company was unaware that child labor was involved in the growing and harvesting of the tobacco. Furthermore, no tobacco raised on the farms that employ child labor went into cigarettes sold outside of the former Soviet countries (UPI.com, 2010). Because many of the children's parents were migrant laborers from neighboring Central Asian countries, particularly impoverished Kyrgyzstan (officially the Kyrgyz Republic), and the cigarettes were not sold in the United States, some analysts have stated that the United States should not get involved with child labor practices in other nations.

However, in the United States, as in most other high-income countries, most people believe that young children should not be toiling in agricultural fields and factories—whether in this country or in some low-income nation—and that they certainly should not be working long hours, be exposed to high levels of nicotine, or drink from irrigation channels contaminated with pesticides, as was reported in the Philip Morris case. Rather, we believe that children should be in school, obtaining an education that will allow them to better themselves, and that they should have plenty of spare time to be, well, just to be *kids*.

Our nation's laws—at both federal and state levels—restrict the number of hours that people under age sixteen can work and also list certain occupations that are deemed to be too hazardous for young workers to perform. But what about other countries? Should we have any say about their social policies in regard to child labor practices? Are we engaging in ethnocentrism if we try to impose our laws on people in other nations? In some nations, child labor in agricultural fields or factories is viewed as a necessity for the family's and the country's economic survival. In the tobacco fields, for example, many migrant workers are based on a piecework basis, by the ton of harvested tobacco. Parents can earn more money if they bring their children to help with the harvest, and even then, many families make only a few hundred dollars, after paying for travel and lodging, for a hard half-year's work. Do we really know what is best for children in other nations? And even if we *would* impose our beliefs about child labor on people elsewhere, what can we do about child labor in those countries, anyway?

The United Nations has attempted to do something about exploitative or dangerous child labor. Article 32 of the Convention on the Rights of the Child recognizes the right of all children to be protected from any work that threatens their health, education, or development and requires nations to set age minimums for employment and to regulate working conditions—and to enforce those requirements with appropriate penalties. The problem is that each nation sets its own minimum age and its own definition of exploitative working conditions, and each nation acts on its own in deciding how to enforce those laws. In many nations, the minimum age for full-time work is lower than the required age for compulsory education, producing an inherent conflict with the goal of protecting children from work that interferes with their education, and if a child's parents don't see as much value in an education as in the child producing income for the family, they may prefer for the child to earn even a small income that the family needs in order to survive.

One way that people in many nations, including the United States, have attempted to reduce or eliminate exploitative or dangerous child labor is by boycotting—refusing to buy—products from companies and nations that permit such practices. Some analysts believe that the best way to end abusive child labor is public awareness of the problem, putting pressure on political officials and business elites around the globe, and demanding better educational opportunities for children in low-income countries. To accomplish such a goal, however, we must be more proactive about global abuses of child labor, and we would have to be willing to pay more for the goods we buy that are manufactured or grown in low-income nations so that the parents could earn enough not to need their children's wages for survival. Are we willing to pay the price?

reflect & analyze

Do you think we should be concerned about child labor practices in other nations? Can you think of examples that show how our purchasing habits affect people worldwide?

[concept quick review]

Classification of Economies by Income

	Low-Income Economies	Middle-Income Economies	High-Income Economies
Previous Categorization	Third World, underdeveloped	Second World, developing	First World, developed
2009 per Capita Income (GNI)	$995 or less	Lower-middle: $996–$3,945 Upper-middle: $3,946–$12,195	$12,196 or more
Type of Economy	Largely agricultural	Diverse, from agricultural to manufacturing	Information-based and postindustrial

High-Income Economies

High-income economies are found in sixty-nine nations, including the United States, Canada, Japan, Australia, Portugal, Ireland, Israel, Italy, Norway, and Germany. Nations recently moving into the high-income category include Latvia, Slovenia, and Poland. High-income economies are defined as having a gross national income per capita of $12,196 or more. According to the World Bank, people in high-income economies typically have a higher standard of living than those in low- and middle-income economies. Nations with high-income economies continue to dominate the world economy, despite the fact that shifts in the global marketplace have affected some workers who have found themselves without work because of *capital flight*—the movement of jobs and economic resources from one nation to another. This flight often occurred because transnational corporations became aware of pools of cheap labor in lower-income nations where workers did not have legal protection or unions to help boost their wages and improve their work conditions. Some of the nations in the high-income category do not have as high a rate of annual economic growth as newly industrializing nations, particularly in East Asia and Latin America, that are still in the process of development. (The Concept Quick Review describes economies classified by income.)

Measuring Global Wealth and Poverty

On a global basis, measuring wealth and poverty is a difficult task because of problems in acquiring comparable data from various nations. As well, over time, some indicators, such as the literacy rate, become less useful in helping analysts determine what progress is being made in reducing poverty.

Absolute, Relative, and Subjective Poverty

How is poverty defined on a global basis? Isn't it more a matter of comparison than an absolute standard? According to social scientists, defining poverty involves more than comparisons of personal or household income; it also involves social judgments made by researchers. From this point of view, *absolute poverty*—previously defined as a condition in which people do not have the means to secure the most basic necessities of life—would be measured by comparing personal or household income or expenses with the cost of buying a given quantity of goods and services. The World Bank has defined absolute poverty as living on less than $1.25 a day. Similarly, *relative poverty*—which exists when people may be able to afford basic necessities but are still unable to maintain an average standard of living—would be measured by comparing one person's income with the incomes of others. Finally, *subjective poverty* would be measured by comparing the actual income against the income

▲ Malnutrition is a widespread health problem in many low-income nations. What kind of health problems are closely linked to middle-income and high-income countries?

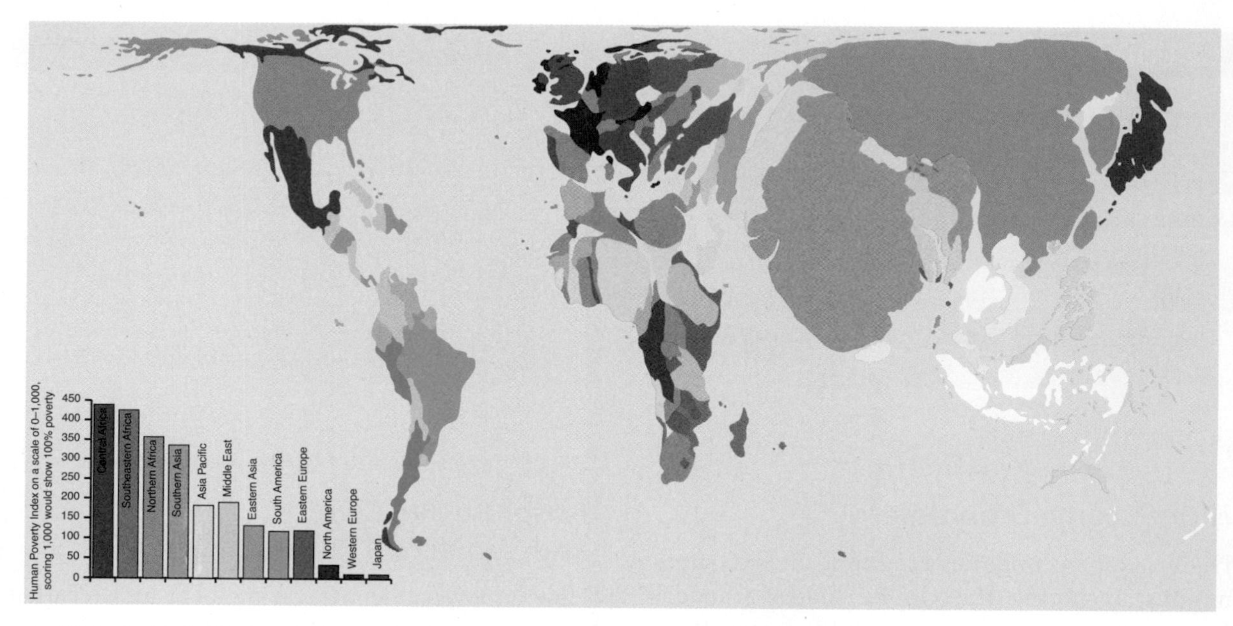

▲ **MAP 9.2 PROPORTION OF WORLD'S POPULATION LIVING IN POVERTY (BY REGION)**

Source: Copyright 2006 SASI Group (University of Sheffield) and Mark Newman (University of Virginia). Map courtesy of worldmapper.org.

earner's expectations and perceptions. However, for low-income nations in a state of economic transition, data on income and levels of consumption are typically difficult to obtain and are often ambiguous when they are available. Defining levels of poverty involves several dimensions: (1) how many people are poor, (2) how far below the poverty line people's incomes fall, and (3) how long they have been poor (is the poverty temporary or long term?). ▶ Map 9.2 provides a unique portrayal of human poverty in which the territory size shows the proportion of the world population living in poverty in each region.

The Gini Coefficient and Global Quality-of-Life Issues

One measure of income inequality is the *Gini coefficient*, which measures the degree of inequality in the distribution of family income in a country. The lower a country's score on the Gini coefficient, the more equal the income distribution. The index ranges from zero (meaning that everyone has the same income) to 100 (one person receives all the income). Based on this measure, income inequality is highest in countries such as Namibia (70.7), Cambodia (66.0), and South Africa (65.0) and lowest in nations such as Norway (25.0) and Sweden (23.0) (CIA, 2010). The United States is ranked 43rd in the world, with a Gini coefficient of 45. On this measure the United States fares better than a number of countries in the Middle East, including Tunisia (62nd), Jordan, (63rd), and Egypt (91st), locations where protesters and rioters demanded greater economic equality and political change in the 2010s, but worse than a number of European nations. As compared to the Middle Eastern countries, the United States has lower poverty rates, better educational and job opportunities, and a better quality of life as measured by people's perceptions of their health and well-being, their community, crime and safety, education and work, and the environment.

Global Poverty and Human Development Issues

Income disparities are not the only factor that defines poverty and its effect on people. Although the average income per person in lower-income countries has doubled in the past thirty years and for many years economic growth has been seen as the primary way to achieve development in low-income economies, the United Nations since the 1970s has more actively focused on human development as a crucial factor in fighting poverty. In 1990 the United Nations Development Programme introduced the Human Development Index (HDI), establishing three new criteria for measuring the level of development in a country: life expectancy, education, and living standards. According to the United Nations, human development is the process of "expanding choices that people have in life, to lead a life to its full potential and in dignity, through expanding capabilities and through people taking action themselves to improve their lives" (United Nations Development Programme, 2009). ▶ Figure 9.2 compares life expectancy and per capita gross national income in regions around the world.

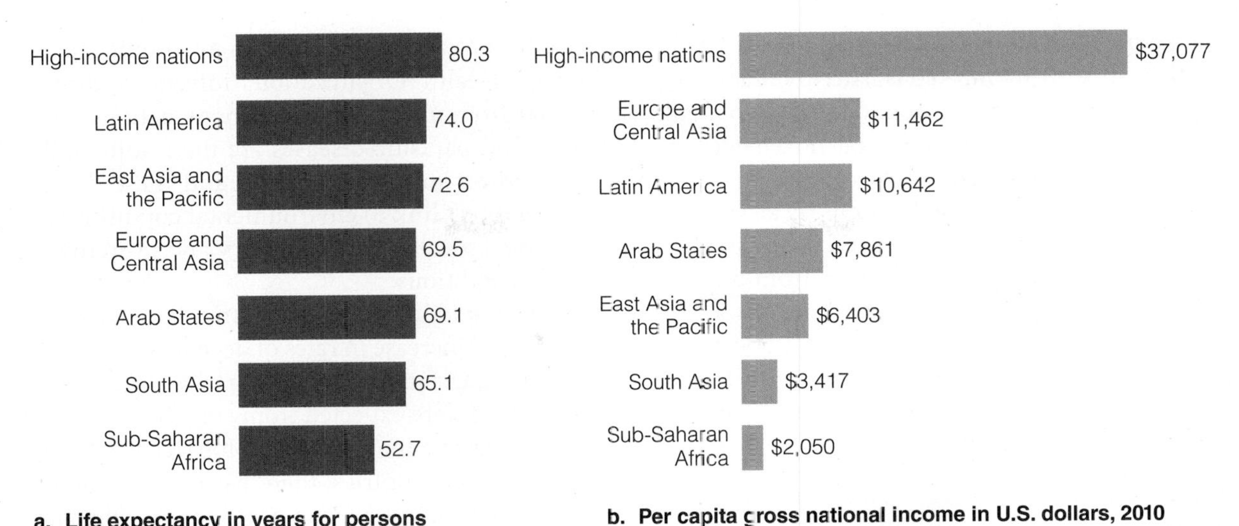

▲ **FIGURE 9.2 INDICATORS OF HUMAN DEVELOPMENT**

Source: United Nations Development Programme, 2010.

Beginning in 2010, the *Human Development Report* included a new top category of nations, "Very High Human Development." According to the report, people who live in countries in the highest human development categories can expect to be better educated, to live longer, and to earn more. For all categories of development, three dimensions are included in the HDI: (1) life expectancy at birth, (2) mean years of schooling and expected years of schooling, and (3) gross national income (GNI) per capita. The top four countries identified as having "Very High Human Development" are Norway, Australia, New Zealand, and the United States. By contrast, the bottom three countries in the "Low Human Development" category are Niger, Democratic Republic of the Congo, and Zimbabwe (United Nations Development Programme, 2010). To show how these categories translate into real life, consider the fact that a child born in Norway in 2010 has a life expectancy of 81 years, as contrasted with a life expectancy of 47 for a child born in the same year in Zimbabwe. Or compare the gross national income per capita of $176 (U.S. currency) in Zimbabwe to the GNI of $58,810 in Norway. How does the United States compare with these? The 2010 life expectancy rate for the United States was 79.6 years, and the GNI was $47,094.

Life Expectancy

As the example above shows, although some advances have been made in middle- and low-income countries regarding life expectancy, major problems still exist. The average life expectancy at birth of people in middle-income countries remains about 12 years less than that of people in high-income countries. Moreover, the life expectancy of people in low-income nations is as much as 30 years less than that of people in high-income nations. A child born in a low-HDI country has a life expectancy of about 56 years, which is 13 years less than in medium-HDI countries and 24 years less than in very-high-HDI countries (United Nations Development Programme, 2010). Despite these gains, life expectancy fell below 1970 levels in six sub-Saharan African countries and three countries in the former Soviet Union. The decline in life expectancy in some sub-Saharan African countries can be linked to the HIV epidemic; declines in the former Soviet Union are concentrated among men, perhaps linked to excessive alcohol consumption and stress brought about by a transition to a market economy that produced high rates of unemployment, inflation, and financial insecurity (United Nations Development Programme, 2010).

One major cause of shorter life expectancy in low-income nations is the high rate of infant mortality. The infant mortality rate (deaths per thousand live births) is more than eight times higher in low-income countries than in high-income countries (World Bank, 2009). Low-income countries typically have higher rates of illness and disease, and they do not have adequate health care facilities. Malnutrition is a common problem among children, many of whom are underweight, stunted, and anemic—a nutritional deficiency with serious consequences for child mortality. Consider this journalist's description of a child she saw in Haiti (prior to the deadly 2010 earthquake):

Like any baby, Wisly Dorvil is easy to love. Unlike others, this 13-month-old is hard to hold.

That's because his 10-pound frame is so fragile that even the most minimal of movements can dislocate his shoulders.

As lifeless as a rag doll, Dorvil is starving. He has large, brown eyes and a feeble smile, but a stomach so tender that he suffers from ongoing bouts of vomiting and diarrhea.

Fortunately, though, Dorvil recently came to the attention of U.S. aid workers. With round-the-clock feeding, he is expected to survive.

Others are not so lucky. (Emling, 1997a: A17)

Among adults and children alike, life expectancies are strongly affected by hunger and malnutrition. It is estimated that people in the United States spend more than $60 billion each year on weight loss products (Williams, 2011), whereas the world's poorest people suffer from chronic malnutrition, and many die each year from hunger-related diseases. Inadequate nutrition affects people's ability to work and to earn the income necessary for a minimum standard of living. Although some gains have been made in reducing the rate of malnourishment in some lower-income nations, about one billion people around the world are malnourished, and 63 percent of these are in Asian and the Pacific, 26 percent in Sub-Saharan Africa, and 1 percent in developed countries (United Nations Development Programme, 2010).

On the plus side of the life-expectancy problem, some nations have made positive gains, seeing average life expectancy increase in the past four decades. For example, life expectancy has risen by more than 18 years in the Arab states and by more than 8 years in some nations of sub-Saharan Africa (United Nations Development Programme, 2010). Of course, problems of illness and mortality remain grave in regions such as sub-Saharan Africa, where overall longevity and quality of life are highly problematic.

Health

Health is defined in the constitution of the World Health Organization as "a state of complete physical, mental and social well-being and not merely the absence of disease or infirmity." Many people in low-income nations are far from having physical, mental, and social well-being. In fact, more than 25 million people die each year from AIDS, malaria, tuberculosis, pneumonia, diarrheal diseases, measles, and other infectious and parasitic illnesses. As of 2009, about 33.3 million adults and children were living with HIV/AIDS (World Health Organization, 2009). According to the World Health Organization, infectious diseases are far from under control in many nations: Infectious and parasitic diseases are the leading killers of children and young adults, and these diseases have a direct link to environmental conditions and poverty, especially to unsanitary and overcrowded living conditions.

Some middle-income countries are experiencing a rapid increase in rates of degenerative diseases such as cancer and coronary heart disease, and many more deaths are expected from smoking-related illnesses. Despite the decrease in tobacco smoking in high-income countries, there has been an increase in per capita consumption of tobacco in some low- and middle-income countries.

Health is also affected by war and conflict. Countries such as Afghanistan, Cambodia, Mozambique, and Iraq have suffered from widespread conflict, and other nations join their ranks regularly. These crises not only contribute to the number of injuries and deaths but also seriously weaken the ability of a nation's health-related infrastructure to deliver medicine and immunizations. As a result, people are more vulnerable to disease and mortality.

Education and Literacy

According to the *Human Development Report* (United Nations Development Programme, 2010: 36), education is an important component in improving life chances: "Educated people are more aware of how to avoid health risks and to live longer and more comfortable lives. They also tend to earn higher wages and have better jobs." Progress in education has been made in many nations, and people around the world have higher levels of education than in the past. For this reason the *Human Development Report* uses "mean years of schooling"—completed years of educational attainment—and "expected years of schooling"—the years of schooling that a child can expect to receive given current enrollment rates—to measure progress in education.

Educational attainment has increased for both men and women, but a large gap remains in many developing countries. Although gender gaps are relatively small for young children in developing countries, they are more pronounced for older children in rural areas. In Bolivia, for example, 35 percent of rural girls and 71 percent of urban boys are enrolled in school. Gains in higher education are most evident in countries with higher scores on the Human Development Index, and women's enrollment in higher education has risen in many parts of the world. In the Arab States, for example, an average of 132 women are now enrolled for every 100 men, but in countries such as Guinea and Niger, three men are enrolled for every woman. Despite higher spending and larger enrollments, some analysts emphasize that this does not necessarily mean that students are receiving better schooling. Socioeconomic factors play a significant role in the kind of education that children receive: "Children from well-off families are likely to be better nourished and healthier and have more access to materials than poorer children, and their parents can do more to help them" (United Nations Development Programme, 2010: 40).

What is the relationship between education and literacy? Literacy was previously used by the United Nations as a measure to determine education in relation to levels of human development. However, as improvements occurred in global literacy rates, researchers found that other measures of educational attainment were more useful in assessing the knowledge dimension of human development. Examples of progress in literacy include the youth literacy rate, which exceeds 95 percent in 63 of the 104 countries with available data. According to the *Human Development Report* (United Nations Development Programme, 2010), fewer people worldwide now lack basic reading and writing skills. However, the illiteracy rate of older adults remains high: In developing countries almost 36 percent of people ages 65–74 have never attended school and are not functionally literate, as compared to only 7 percent of those ages 15–24. Lack of literacy remains a greater problem for women because many nations do not value the potential contributions of women and provide only limited opportunities to learn how to read and write. Women in the poorest nations have a literacy rate of 45.9 percent, as compared to 64.5 percent for men. In sharp contrast, women in high-income nations have a 98.7 percent literacy rate, as compared to 99.3 percent for men (United Nations Development Programme, 2010). Literacy is crucial for everyone, but it is particularly important for women because it is closely linked to decreases in fertility, improved child health, and increased earnings potential.

Persistent Gaps in Human Development

Some middle- and lower-income countries have made progress in certain indicators of human development. The gap between some richer and middle- or lower-income nations has narrowed significantly for life expectancy, educational attainment, and daily calorie supply; however, the overall picture for the world's poorest people remains dismal. The gap between the richest and poorest countries has widened to a gulf. According to the United Nations, the richest country today (Liechtenstein) is three times richer than the richest country in 1970, and the poorest country (Zimbabwe) is about 25 percent poorer than the poorest country in 1970 (also Zimbabwe). Moreover, despite enormous increases in material well-being in high-income nations, the real average income of people in 13 countries (in the bottom 25 percent of world income distribution) is lower than it was in 1970 (United Nations Development Programme, 2010).

Poverty, food shortages, hunger, and rapidly growing populations are pressing problems for at least two billion people, most of them women and children living in a state of absolute poverty. Although more women around the globe have paid employment than in the past, more and more women are still finding themselves in poverty because of increases in single-person and single-parent households headed by women and the fact that low-wage work is often the only source of livelihood available to them.

Human development research has reached a surprising conclusion: Economic growth and higher incomes in low- and medium-development nations are not always necessary to bring about improvements in health and education. According to the *Human Development Report*, technological

Borderlands/Alamy

▲ In an effort to reduce poverty, some nations have developed adult literacy programs so that people can gain an education that will help lift them out of poverty. In regions such as South Sudan, women's literacy is a particularly crucial issue.

improvements and changes in societal structure allow even poorer countries to bring about significant changes in health and education even without significant gains in income.

A Multidimensional Measure of Poverty

Looking at human development requires that we also examine deprivation. According to the United Nations Development Programme (2010: 94), "The dimensions of poverty go far beyond inadequate income—to poor health and nutrition, low education and skills, inadequate livelihoods, bad housing conditions, social exclusion and lack of participation." The Multidimensional Poverty Index (MPI) was introduced by the United Nations to help identify overlapping deprivations that are suffered by households in health, education, and living standards. About 1.75 billion people in the 104 countries covered by the MPI experience multidimensional poverty. (This is a larger number than the estimated 1.44 billion people who are defined as poor based on income, because they live on $1.25 a day or less, but it is lower than the estimated 2.6 billion people living on less than $2 a day.) Using the MPI, sub-Saharan Africa has the highest incident of multidimensional poverty (for example, 93 percent in Niger); however, more than half (844 million) of the world's multidimensionally poor live in South Asia. The three dimensions of the MPI—health, education, and living standards—are subdivided into ten indicators:

- Health—*nutrition* and *child mortality*
- Education—*years of schooling* (deprived if no household member has completed five years of school) and *children enrolled*
- Living standards—*cooking fuel, toilet, water, electricity, floor* (deprived if the household has dirt, sand, or dung floor), and *assets* (deprived if the household does not own more than one of the following: radio, television, telephone, bike, or motorbike)

To be considered multidimensionally poor, households must be deprived in at least six standard-of-living indicators or in three standard-of-living indicators and one health or education indicator.

A major contribution of the Multidimensional Poverty Index is that it focuses on many aspects of poverty and calls our attention to the idea that human development involves much more than money: It includes life chances and opportunities that contribute to human well-being. Overall, countries with less human development have more multidimensional inequality and poverty. As a result, the MPI is most useful in analyzing poverty in the less-developed countries of South Asia and Sub-Saharan Africa and in the poorest Latin American countries.

Theories of Global Inequality

Why is the majority of the world's population growing richer while the poorest 20 percent—more than one billion people—are so poor that they are effectively excluded from even a moderate standard of living? Social scientists have developed a variety of theories that view the causes and consequences of global inequality somewhat differently. We will examine the development approach and modernization theory, dependency theory, world systems theory, and the new international division of labor theory.

Development Approach and Modernization Theory

According to some social scientists, global wealth and poverty are linked to the level of industrialization and economic development in a given society. Although the process by which a nation industrializes may vary somewhat, industrialization almost inevitably brings with it a higher standard of living and some degree of social mobility for individual participants in the society. Specifically, the traditional caste system becomes obsolete as industrialization progresses. Family status, race/ethnicity, and gender are said to become less significant in industrialized nations than in agrarian-based societies. As societies industrialize, they also urbanize as workers locate their residences near factories, offices, and other places of work. Consequently, urban values and folkways overshadow the beliefs and practices of the rural areas. Analysts using a development framework typically view industrialization and economic development as essential steps that nations must go through in order to reduce poverty and increase life chances for their citizens.

Earlier in the chapter we discussed the post–World War II Marshall Plan, under which massive financial aid was provided to the European nations to help rebuild infrastructure lost in the war. Based on the success of this infusion of cash in bringing about modernization, President Truman and many other politicians and leaders in the business community believed that it should be possible to help so-called underdeveloped nations modernize in the same manner.

The most widely known development theory is ***modernization theory*—a perspective that links global inequality to different levels of economic development and suggests that low-income economies can move to middle- and high-income economies by achieving self-sustained economic growth.** According to modernization theory, the low-income, less-developed nations can improve their standard of living only with a period of intensive economic growth and accompanying changes in people's beliefs, values, and attitudes toward work. As a result of modernization, the values of people in developing countries supposedly become more similar to those of people in high-income nations. The number of hours that people work at their jobs each week is one measure of the extent to which individuals subscribe to the *work ethic*, a core value widely believed to be of great significance in the modernization process.

Perhaps the best-known modernization theory is that of Walt W. Rostow (1971, 1978), who, as an economic advisor to U.S. President John F. Kennedy, was highly instrumental in shaping foreign policy toward Latin America in the 1960s. Rostow suggested that all countries go through four stages of economic development, with identical content, regardless of when these nations started the process of industrialization. He compared the stages of economic development to an airplane ride. The first stage is the *traditional stage*, in which very little social change takes place, and people do not think much about changing their current circumstances. According to Rostow, societies in this stage are slow to change because the people hold a fatalistic value system, do not subscribe to the work ethic, and save very little money. The second stage is the *take-off stage*—a period of economic growth accompanied by a growing belief in individualism, competition, and achievement. During this stage, people start to look toward the future, to save and invest money, and to discard traditional values. According to Rostow's modernization theory, the development of capitalism is essential for the transformation from a traditional, simple society to a modern, complex one. With the financial help and advice of the high-income countries, low-income countries will eventually be able to "fly" and enter the third stage of economic development. In the third stage the country moves toward *technological maturity*. At this point the country will improve its technology, reinvest in new industries, and embrace the beliefs, values, and social institutions of the high-income, developed nations. In the fourth and final stage the country reaches the phase of *high mass consumption* and a correspondingly high standard of living.

Modernization theory has had both its advocates and its critics. According to proponents of this approach, studies have supported the assertion that economic development occurs more rapidly in a capitalist economy. In fact, the countries that have been most successful in moving from low- to middle-income status typically have been those that are most centrally involved in the global capitalist economy. For example, the nations of East Asia have successfully made the transition from low-income to higher-income economies through factors such as a high rate of savings, an aggressive work ethic among employers and employees, and the fostering of a market economy.

Critics of modernization theory point out that it tends to be Eurocentric in its analysis of low-income countries, which it implicitly labels as backward (see Evans and Stephens, 1988). In particular, modernization theory does not take into account the possibility that all nations do not

sevenke/Shutterstock.com

▲ Poverty is persistent in many low-income nations, where the poor are caught in situations that they cannot control. When Chinese garlic farmers grow large quantities of the spice, the market price and their corresponding profits are very low. When the farmers limit garlic production, the price soars, but most of the profits are taken by garlic brokers, not the farmers.

modernization theory a perspective that links global inequality to different levels of economic development and suggests that low-income economies can move to middle- and high-income economies by achieving self-sustained economic growth.

industrialize in the same manner. In contrast, some analysts have suggested that modernization of low-income nations today will require novel policies, sequences, and ideologies that are not accounted for by Rostow's approach.

Which sociological perspective is most closely associated with the development approach? Modernization theory is based on a market-oriented perspective which assumes that "pure" capitalism is good and that the best economic outcomes occur when governments follow the policy of laissez-faire (or hands-off) business, giving capitalists the opportunity to make the "best" economic decisions, unfettered by government restraints or cumbersome rules and regulations. In today's global economy, however, many analysts believe that national governments are no longer central corporate decision makers and that transnational corporations determine global economic expansion and contraction. Therefore, corporate decisions to relocate manufacturing processes around the world make the rules and regulations of any one nation irrelevant and national boundaries obsolete (Gereffi, 1994). Just as modernization theory most closely approximates a functionalist approach to explaining inequality, dependency theory, world systems theory, and the new international division of labor theory are perspectives rooted in the conflict approach. All four of these approaches are depicted in ▶ Figure 9.3.

Dependency Theory

***Dependency theory* states that global poverty can at least partially be attributed to the fact that the high-income countries have exploited the low-income countries.** Analyzing events as part of a particular historical process—the expansion of global capitalism—dependency

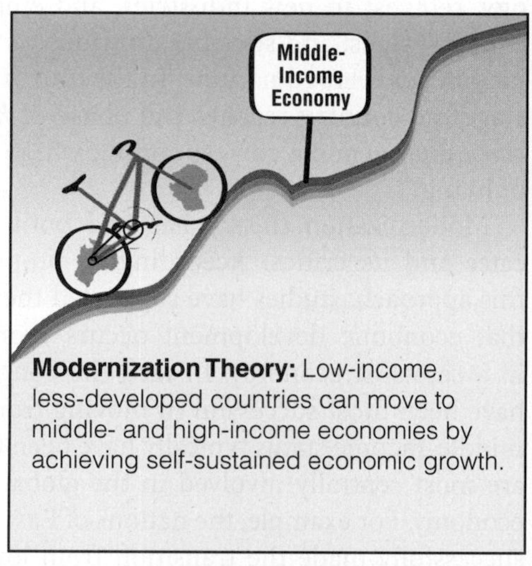

Modernization Theory: Low-income, less-developed countries can move to middle- and high-income economies by achieving self-sustained economic growth.

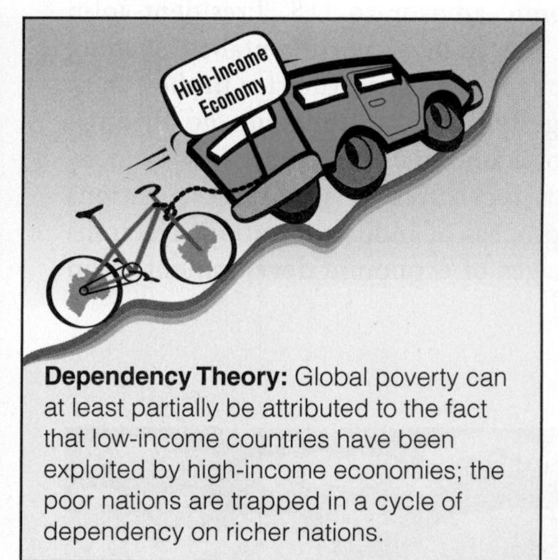

Dependency Theory: Global poverty can at least partially be attributed to the fact that low-income countries have been exploited by high-income economies; the poor nations are trapped in a cycle of dependency on richer nations.

World Systems Theory: How a country is incorporated into the global capitalist economy (e.g., a core, semiperipheral, or peripheral nation) is the key feature in determining how economic development takes place in that nation.

The New International Division of Labor Theory: Commodity production is split into fragments, each of which can be moved (e.g., by a transnational corporation) to whichever part of the world can provide the best combination of capital and labor.

▲ **FIGURE 9.3 APPROACHES TO STUDYING GLOBAL INEQUALITY**
What causes global inequality? Social scientists have developed a variety of explanations, including the four theories shown here.

theorists see the greed of the rich countries as a source of increasing impoverishment of the poorer nations and their people. Dependency theory disputes the notion of the development approach, and modernization theory specifically, that economic growth is the key to meeting important human needs in societies. In contrast, the poorer nations are trapped in a cycle of structural dependency on the richer nations because of their need for infusions of foreign capital and external markets for their raw materials, making it impossible for the poorer nations to pursue their own economic and human development agendas. For this reason, dependency theorists believe that countries such as Brazil, Nigeria, India, and Kenya cannot reach the sustained economic growth patterns of the more-advanced capitalist economies.

Dependency theory has been most often applied to the newly industrializing countries (NICs) of Latin America, whereas scholars examining the NICs of East Asia found that dependency theory had little or no relevance to economic growth and development in that part of the world. Therefore, dependency theory had to be expanded to encompass transnational economic linkages that affect developing countries, including foreign aid, foreign trade, foreign direct investment, and foreign loans. On the one hand, in Latin America and sub-Saharan Africa, transnational linkages such as foreign aid, investments by transnational corporations, foreign debt, and export trade have been significant impediments to development within a country. On the other hand, East Asian countries such as Taiwan, South Korea, and Singapore have historically also had high rates of dependency on foreign aid, foreign trade, and interdependence with transnational corporations but have still experienced high rates of economic growth despite dependency. According to the sociologist Gary Gereffi (1994), differences in outcome are probably associated with differences in the timing and sequencing of a nation's relationship with external entities such as foreign governments and transnational corporations.

Dependency theory makes a positive contribution to our understanding of global poverty by noting that "underdevelopment" is not necessarily the cause of inequality. Rather, it points out that exploitation not only of one country by another but of countries by transnational corporations may limit or retard economic growth and human development in some nations. However, what remains unexplained is how East Asia and India had successful "dependency management" whereas many Latin American countries did not (Gereffi, 1994).

Hobert Harding Picture Library Ltd/Alamy

▲ A variety of factors, including foreign investment and the presence of multinational corporations, have contributed to the economic growth of nations such as China.

World Systems Theory

World systems theory suggests that what exists under capitalism is a truly global system that is held together by economic ties. From this approach, global inequality does not emerge solely as a result of the exploitation of one country by another. Instead, economic domination involves a complex world system in which the industrialized, high-income nations benefit from other nations and exploit their citizens. This theory is most closely associated with the sociologist Immanuel Wallerstein (1979, 1984), who believed that a country's mode of incorporation into the capitalist work economy is the key feature in determining how economic development takes place in that nation. According to *world systems theory*, the capitalist world economy is a global system divided into a hierarchy of three major types of nations—core, semiperipheral, and peripheral—in which upward or downward mobility is conditioned by the resources and obstacles that characterize the international system.

dependency theory the belief that global poverty can at least partially be attributed to the fact that the high-income countries have exploited the low-income countries.

***Core nations* are dominant capitalist centers characterized by high levels of industrialization and urbanization.** Core nations such as the United States, Japan, and Germany possess most of the world's capital and technology. Even more importantly for their position of domination, they exert massive control over world trade and economic agreements across national boundaries. Some cities in core nations are referred to as *global cities* because they serve as international centers for political, economic, and cultural concerns. New York, Tokyo, and London are the largest global cities, and they are often referred to as the "command posts" of the world economy.

***Semiperipheral nations* are more developed than peripheral nations but less developed than core nations.** Nations in this category typically provide labor and raw materials to core nations within the world system. These nations constitute a midpoint between the core and peripheral nations that promotes the stability and legitimacy of the three-tiered world economy. These nations include South Korea and Taiwan in East Asia, Mexico and Brazil in Latin America, India in South Asia, and Nigeria and South Africa in Africa. Only two global cities are located in semiperipheral nations: São Paulo, Brazil, which is the center of the Brazilian economy, and Singapore, which is the economic center of a multicountry region in Southeast Asia. According to Wallerstein, semiperipheral nations exploit peripheral nations, just as the core nations exploit both the semiperipheral and the peripheral nations.

Most low-income countries in Africa, South America, and the Caribbean are ***peripheral nations*—nations that are dependent on core nations for capital, have little or no industrialization (other than what may be brought in by core nations), and have uneven patterns of urbanization.** According to Wallerstein (1979, 1984), the wealthy in peripheral nations benefit from the labor of poor workers and from their own economic relations with core-nation capitalists, whom they uphold in order to maintain their own wealth and position. At a global level, uneven economic growth results from capital investment by core nations; disparity between the rich and the poor within the major cities in these nations is increased in the process. The U.S./Mexican border is an example of disparity and urban growth: Transnational corporations have built *maquiladora* plants so that goods can be assembled by low-wage workers to keep production costs down. ▶ Figure 9.4 describes this process. In 2011 there were more than 3,000 *maquiladora* plants operating in Mexico despite some job losses because of the economic recession. These plants produce a wide array of products, including apparel, computers, televisions, small

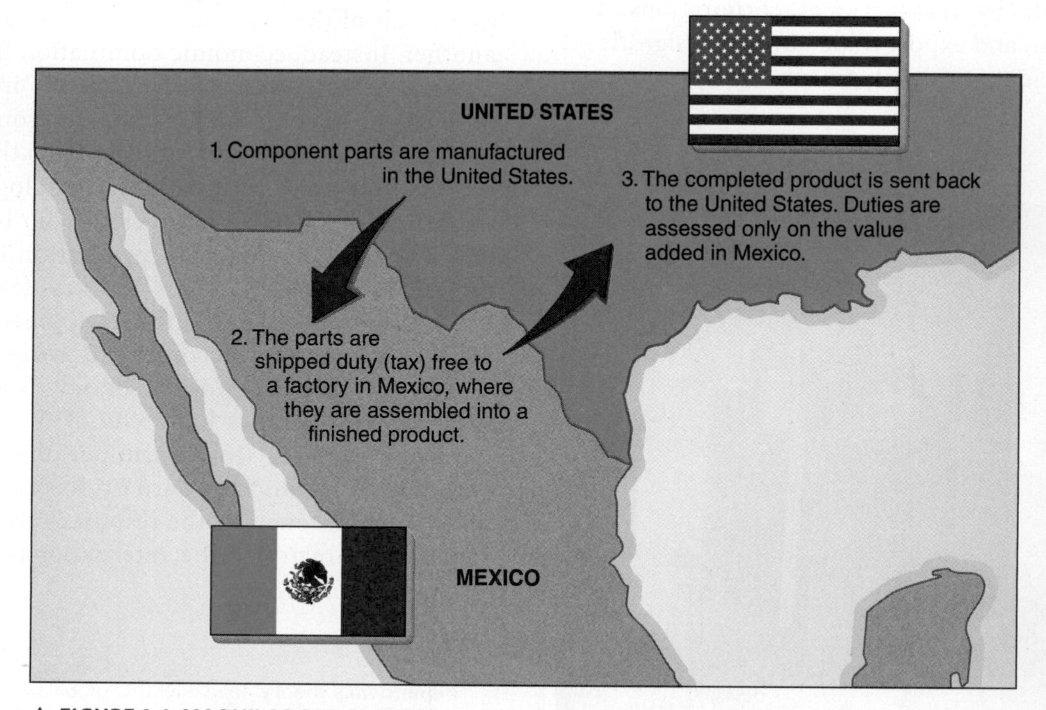

▲ **FIGURE 9.4 MAQUILADORA PLANTS**
Here is the process by which transnational corporations establish plants in Mexico so that profits can be increased by using low-wage workers there to assemble products that are then brought into the United States for sale.

appliances, and motor vehicles and parts. Apparently, the so-called drug wars along the Mexican border have not seriously affected *maquiladora* plants: Political leaders in border cities such as Harlingen and El Paso state that they strongly favor these plants because they provide job opportunities and enhance economic development on both sides of the border.

As Wallerstein's world system theory would suggest, one of the threats to the Mexican *maquiladora* plants is the movement of work by some transnational corporations to China and countries in Central America.

Not all social analysts agree with Wallerstein's perspective on the hierarchical position of nations in the global economy. However, most scholars acknowledge that nations throughout the world are influenced by a relatively small number of cities and transnational corporations that have prompted a shift from an international to a more global economy (see Knox and Taylor, 1995; Wilson, 1997). Wallerstein (1991) acknowledges that world systems theory is an "incomplete, unfinished critique" of long-term, large-scale social change that influences global inequality, and he continues to develop his perspective.

The New International Division of Labor Theory

Although the term *world trade* has long implied that there is a division of labor between societies, the nature and extent of this division have recently been reassessed based on the changing nature of the world economy. According to the *new international division of labor theory*, commodity production is being split into fragments that can be assigned to whichever part of the world can provide the most profitable combination of capital and labor. Consequently, the new international division of labor has changed the pattern of geographic specialization between countries, whereby high-income countries have now become dependent on low-income countries for labor. The low-income countries provide transnational corporations with a situation in which they can pay lower wages and taxes and face fewer regulations regarding workplace conditions and environmental protection (Waters, 1995). Overall, a global manufacturing system has emerged in which transnational corporations establish labor-intensive, assembly-oriented export production, ranging from textiles and clothing to technologically sophisticated exports such as computers, in middle- and lower-income nations (Gereffi, 1994). At the same time, manufacturing technologies are shifting from the large-scale, mass-production assembly lines of the past toward a more flexible production process involving microelectronic technologies. Even service industries—such as processing insurance claims forms—that were formerly thought to be less mobile have become exportable through electronic transmission and the Internet. The global nature of these activities has been referred to as *global commodity chains*, a complex pattern of international labor and production processes that results in a finished commodity ready for sale in the marketplace.

Some commodity chains are producer-driven, whereas others are buyer-driven. *Producer-driven commodity chains* is the term used to describe industries in which transnational corporations play a central part in controlling the production process. Industries that produce automobiles, computers, and other capital- and technology-intensive products are typically producer-driven. In contrast, *buyer-driven commodity chains* is the term used to refer to industries in which large retailers, brand-name merchandisers, and trading companies set up decentralized production networks in various middle- and low-income countries. This type of chain is most common in labor-intensive, consumer-goods industries such as toys, garments, and footwear. Athletic footwear companies such as Nike and Reebok are examples of the buyer-driven model. Because these products tend to be labor intensive at the manufacturing stage, the typical factory system is very competitive and globally decentralized. Workers in buyer-driven commodity chains are sometimes exploited by low wages, long hours, and poor working conditions.

In view of the global economic crisis of 2008–2009, scholars have recently reexamined the effects of global value chains (GVCs) that

core nations according to world systems theory, dominant capitalist centers characterized by high levels of industrialization and urbanization.

semiperipheral nations according to world systems theory, nations that are more developed than peripheral nations but less developed than core nations.

peripheral nations according to world systems theory, nations that are dependent on core nations for capital, have little or no industrialization (other than what may be brought in by core nations), and have uneven patterns of urbanization.

"encompass the full range of activities that are required to bring a good or service from conception through the different phases of production—provision of raw materials; the input of various components, subassemblies, and producer services; the assembly of finished goods—to delivery to final consumers, as well as disposal after use" (Cattaneo, Gereffi, and Staritz, 2010). These researchers have studied how the global spread of industries affects both corporations and countries, as well as how quickly a financial crisis in one part of the world affects the entire world economy. Consider these examples: When U.S. consumers put off buying a new car, it affects not only the U.S. auto industry but also the Liberian rubber sector that produces the material for the tires, as well as every other sector that is involved in the global automotive supply chain. Likewise, when U.S. consumers slowed down on buying new computers and cell phones, this drop in sales affected not only cell phone assembly plants in China but also U.S. factories that export semiconductors and components to China, where they are installed in cell phones that are then exported back to the United States.

Although most discussions of the international division of labor focus on the effects on people residing in industrialized urban areas of developing nations, millions of people continue to live in grinding poverty in rural regions of these countries (see "Sociology Works!").

Global Inequality in the Future

Social inequality is vast both within and among the countries of the world. Even in high-income nations where wealth is highly concentrated, many poor people coexist with the affluent. In middle- and low-income countries, there are small pockets of wealth in the midst of poverty and despair. Although some political and business elites in local economies benefit greatly from partnerships with transnational corporations, everyday people residing in these nations have often continued to be exploited in both industrial and agricultural work. In China, for example, some people have accumulated vast wealth in urban areas while poverty has increased dramatically in some regions and particularly in rural areas.

What are the future prospects for greater equality across and within nations? Not all social scientists agree on the answer to this question. Depending on the theoretical framework that they apply in studying global inequality, social analysts may describe either an optimistic or a pessimistic scenario for the future. Moreover, some analysts highlight the human rights issues embedded in global inequality, whereas others focus primarily on an economic framework.

In some regions, persistent and growing poverty continues to undermine human development and future possibilities for socioeconomic change. Gross inequality has high financial and quality-of-life costs to people, even among those who are not the poorest of the poor. In the future, continued population growth, urbanization, environmental degradation, and violent conflict threaten even the meager living conditions of those residing in low-income nations. From this approach, the future looks dim not only for people in low-income and middle-income countries but also for those in high-income countries, who will see their quality of life diminish as natural resources are depleted, the environment is polluted, and high rates of immigration and global political unrest threaten the high standard of living that many people have previously enjoyed. According to some social analysts, transnational corporations and financial institutions such as the World Bank and the International Monetary Fund will further solidify and control a globalized economy, which will transfer the power to make significant choices to these organizations and away from the people and their governments. Further loss of resources and means of livelihood will affect people and countries around the globe.

As a result of global corporate domination, there could be a leveling out of average income around the world, with wages falling in high-income countries and wages increasing significantly in low- and middle-income countries. If this pessimistic scenario occurs, there is likely to be greater polarization of the rich and the poor and more potential for ethnic and national conflicts over such issues as worsening environmental degradation and who has the right to natural resources. For example, pulp-and-paper companies in Indonesia, along with palm-oil plantation owners, have continued clearing land for crops by burning off vast tracts of jungle, producing high levels of smog and pollution across seven Southeast Asian nations and creating havoc for millions of people.

On the other hand, a more optimistic scenario is also possible. With modern technology and worldwide economic growth, it might be possible to reduce absolute poverty and to increase people's opportunities. Among the trends cited by the *Human Development Report* (United Nations Development Programme, 2008) that have the potential to bring about more sustainable patterns of development are

sociology works!

Why *Place* Matters in Global Poverty

> We're deadly poor. We grow just enough food for ourselves to eat, with no surplus grain. We don't have to pay the grain tax anymore, but our lives aren't much better.
>
> —Zhou Zhiwen, a woman who lives in Yangmiao, China, describing what rural poverty is like in her village (qtd. in French, 2008: YT4)

Zhou Zhiwen lives in an area of China that has largely been untouched by the economic boom in her country. Even with the recent abolition of agricultural taxes for people who are impoverished, local villagers such as Zhou continue to live in abject poverty. More people have risen out of poverty in China's urban centers in recent decades, but poverty in the rural areas, mountainous regions, and deserts remains severe. According to some villagers, the central government is "out of touch with rural realities in places like this," and officials have made little effort to take care of the rural poor (French, 2008). China's poverty is widespread, and the income gap between rural and urban residents has widened over the past three decades (IFAD, 2002). Many people live close to, or below, the minimum standard for poverty: Approximately 350 million people in China live below the international poverty line of $1.25 per day (World Bank, 2007).

For many years, sociologists studying poverty have focused on differences in rural and urban poverty. Throughout the world, they have found that *place* does matter when it comes to finding the deepest pockets of poverty (Rural Policy Research Institute, 2004). Where people live strongly influences how much money they will make, and income inequalities are important indicators of the life chances of entire families. In some developing countries the rural poor rely primarily on agriculture, fishing, forestry, and sometimes small-scale industries and services for their livelihood (Khan, 2001). When they are unable to derive sufficient economic resources from these endeavors, little else is available for them. Some migrate to urban centers in hopes of finding new opportunities, but many remain behind, living in grinding poverty. When sociologists speak of "place," they are referring to such things as an area's natural environment, which includes its climate, natural resources, and degree of isolation (Rural Policy Research Institute, 2004). Place also involves the economic structure in the area, such as the extent to which adequate amounts of food can be raised to meet people's needs, or whether an individual can earn sufficient money to purchase food.

Can rural poverty be reduced? According to some social policy analysts, broad economic stability, competitive markets, and public investment in *physical* and *social* infrastructure are important prerequisites for a reduction in rural poverty in developing nations (Khan, 2001). From this perspective, a major reduction in rural poverty in China will occur only if people have access to land and credit, education, health care, support services, and food through well-designed public works programs and other transfer mechanisms.

reflect & analyze

Whether changes that reduce poverty will occur in China's future remains to be seen, but the sociological premise that place matters in regard to poverty remains a valid assumption in helping us explain global poverty and inequality. Can you apply this idea to rural and urban areas with which you are familiar in the United States? Why are issues such as this important to each of us even if we do not live in a rural area and have no personal experience with poverty?

the socioeconomic progress made in many low- and middle-income countries over the past thirty years as technological, social, and environmental improvements have occurred. For example, technological innovation continues to improve living standards for some people. Fertility rates are declining in some regions (but remain high in others, where there remains grave cause for concern about the availability of adequate natural resources for the future). Finally, health and education may continue to improve in lower-income countries. According to the *Human Development Report* (United Nations Development Programme, 2008), healthy, educated populations are crucial for the future in order to reduce global poverty. The education of women is of primary importance in the future if global inequality is to be reduced. As one analyst stated, "If you educate a boy, you educate a human being. If you educate a girl, you educate generations" (Buvinić, 1997: 49). All aspects of schooling and training are crucial for the future, including agricultural extension services in rural areas to help women farmers in regions such as western Kenya produce more crops to feed their families. As we saw earlier in the chapter, easier access to water can make a crucial difference in people's lives. Mayra Buvinić, of the Inter-American Development Bank, puts global poverty in perspective for people living in high-income countries by pointing out that their problems are our problems. She provides the following example:

you can make a difference

Global Networking to Reduce World Hunger and Poverty

We, the people of the world, will mobilize the forces of transnational civil society behind a widely shared agenda that binds our many social movements in pursuit of just, sustainable, and participatory human societies. In so doing we are forging our own instruments and processes for redefining the nature and meaning of human progress and for transforming those institutions that no longer respond to our needs. We welcome to our cause all people who share our commitment to peaceful and democratic change in the interest of our living planet and the human societies it sustains.

—International NGO Forum, United Nations Conference on Environment and Development, Rio de Janeiro, Brazil, June 12, 1992 (qtd. in Korten, 1996: 333)

If everyone lit just one little candle, what a bright world this would be.

—line from the 1950s theme song for Bishop Fulton J. Sheen's television series *Life Is Worth Living* (Sheen, 1995: 245)

When many of us think about problems such as world poverty, we tend to see ourselves as powerless to bring about change in so vast an issue. However, a recurring message from social activists and religious leaders is that each person can contribute something to the betterment of other people and sometimes the entire world.

An initial way for each of us to be involved is to become more informed about global issues and to learn how we can contribute time and resources to organizations seeking to address social issues such as illiteracy and hunger. We can also find out about meetings and activities of organizations and participate in online discussion forums where we can express our opinions, ask questions, share information, and interact with other people interested in topics such as international relief and development. At first, it may not feel like you are doing much to address global problems; however, information and education are the first steps in promoting greater understanding of social problems and of the world's people, whether they reside in high-, middle-, or low-income countries and regardless of their individual socioeconomic position. Likewise, it is important to help our own nation's children understand that they can make a difference in ending hunger in the United States and other nations.

Would you like to function as a catalyst for change? You can learn how to proceed by gathering information from organizations that seek to reduce problems such as poverty and to provide forums for interacting with other people. Here are a few starting points for your search:

- CARE International is a confederation of 10 national members in North America, Europe, Japan, and Australia. CARE assists the world's poor in their efforts to achieve social and economic well-being. Its work reaches 25 million people in 53 nations in Africa, Asia, Latin America, and Eastern Europe. Programs include emergency relief, education, health and population, children's health, reproductive health, water and sanitation, small economic activity development, agriculture, community development, and environment: **www.care.org**

Other organizations fighting world hunger and health problems include the following:

- WhyHunger: **www.whyhunger.org**
- "Kids Can Make a Difference," an innovative program developed by the International Education and Resource Network: **www.kidscanmakeadifference.org**
- World Health Organization: **www.who.int**

AP Photo/Jennifer Graylock

Actress Drew Barrymore has been designated a "Friend of the UN" because of her support for the United Nations' humanitarian efforts.

© Scott Olson/Getty Images News/Getty Images

▲ The current trend of using grain-based ethanol to power cars and trucks has proved to be a financial boon to some U.S. farmers but has had another effect as well: The price of grain has increased worldwide, making it even harder for people in developing countries to procure enough food to eat.

> Reina is a former guerilla fighter in El Salvador who is being taught how to bake bread under a post-civil war reconstruction program. But as she says, "the only thing I have is this training and I don't want to be a baker. I have other dreams for my life."
>
> Once upon a time, women like Reina . . . only migrated [to the United States] to follow or find a husband. This is no longer the case. It is likely that Reina, with few opportunities in her *own* country, will sooner or later join the rising number of female migrants who leave families and children behind to seek better paying work in the United States and other industrial countries. Wisely spent foreign aid can give Reina the chance to realize her dreams in her *own* country. (Buvinić, 1997: 38, 52)

From this viewpoint, we can enjoy prosperity only by ensuring that other people have the opportunity to survive and thrive in their own surroundings (see "You Can Make a Difference"). The problems associated with global poverty are therefore of interest to a wide-ranging set of countries and people.

chapter review Q & A

Use these questions and answers to check how well you've achieved the learning objectives set out at the beginning of this chapter.

- **What is global stratification, and how does it contribute to economic inequality?**

Global stratification refers to the unequal distribution of wealth, power, and prestige on a global basis, which results in people having vastly different lifestyles and life chances both within and among the nations of the world. Today, the income gap between the richest and the poorest 20 percent of the world population continues to widen, and within some nations the poorest one-fifth of the population has an income that is only a slight fraction of the overall average per capita income for that country.

- **How are global poverty and human development related?**

Income disparities are not the only factor that defines poverty and its effect on people. The United Nations' Human Development Index measures the level of development in a country through indicators such as life expectancy, infant mortality rate, proportion of underweight children under age five (a measure of nourishment and health), and adult literacy rate for low-income, middle-income, and high-income countries.

- **What is modernization theory?**

Modernization theory is a perspective that links global inequality to different levels of economic development and suggests that low-income economies can move to middle- and high-income economies by achieving self-sustained economic growth.

- **How does dependency theory differ from modernization theory?**

Dependency theory states that global poverty can at least partially be attributed to the fact that the high-income countries have exploited the low-income countries. Whereas modernization theory focuses on how societies can reduce inequality through industrialization and economic development, dependency theorists see the greed of the rich countries as a source of increasing impoverishment of the poorer nations and their people.

- **What is world systems theory, and how does it view the global economy?**

According to world systems theory, the capitalist world economy is a global system divided into a hierarchy of three major types of nations: Core nations are dominant capitalist centers characterized by high levels of industrialization and urbanization, semiperipheral nations are more developed than peripheral nations but less developed than core nations, and peripheral nations are those countries that are dependent on core nations for capital, have little or no industrialization (other than what may be brought in by core nations), and have uneven patterns of urbanization.

- **What is the new international division of labor theory?**

The new international division of labor theory is based on the assumption that commodity production is split into fragments that can be assigned to whichever part of the world can provide the most profitable combination of capital and labor. This division of labor has changed the pattern of geographic specialization between countries, whereby high-income countries have become dependent on low-income countries for labor. The low-income countries provide transnational corporations with a situation in which they can pay lower wages and taxes, and face fewer regulations regarding workplace conditions and environmental protection.

key terms

core nations 268
dependency theory 266
global stratification 252
modernization theory 265
peripheral nations 268
semiperipheral nations 268

questions for critical thinking

1. You have decided to study global wealth and poverty. How would you approach your study? What research methods would provide the best data for analysis? What might you find if you compared your research data with popular presentations—such as films and advertising—of everyday life in low- and middle-income countries?
2. How would you compare the lives of poor people living in the low-income nations of the world with those in central cities and rural areas of the United States? In what ways are their lives similar? In what ways are they different?
3. Should U.S. foreign policy include provisions for reducing poverty in other nations of the world? Should U.S. domestic policy include provisions for reducing poverty in the United States? How are these issues similar? How are they different?
4. Using the theories discussed in this chapter, devise a plan to alleviate global poverty. Assume that you have the necessary wealth, political power, and other resources necessary to reduce the problem. Share your plan with others in your class, and create a consolidated plan that represents the best ideas and suggestions presented.

turning to video

Watch the BBC video *Haiti Rebuilding* (running time 2:56), available through **CengageBrain.com**. A month after the 2010 earthquake, more than a million people in the country still needed basic help. This video covers the rebuilding efforts in one community and the difficulties that still lay ahead at the time—and may not yet have been overcome. As you watch the video, think about the reasons a major natural disaster in a less wealthy nation may be so much more devastating than in a wealthier nation. After you watch the video, consider this question: Is international aid in a time of crisis the best a country such as Haiti can hope for from the global community? Why or why not? What other options exist?

online study resources

Go to CENGAGE brain.com to access online study resources, including the Sociology CourseMate for this text as well as special features such as video, an interactive sociology time line and interactive maps, General Social Survey (GSS) data, and U.S. Census 2010 data.

CourseMate brings course concepts to life with interactive learning, study, and exam-preparation tools that support the printed textbook. A textbook-specific website, **Sociology CourseMate** includes an integrated interactive eBook and other interactive learning tools, including quizzes, flash cards, and videos.

Visit **www.cengagebrain.com** to access your account and purchase materials.

10 Race and Ethnicity

My whole life, up until I was eight years old, I thought that I was white. Looking back, I thought how did I think that, because all my brothers and sisters have blond hair and blue eyes, and I obviously don't. I don't consider myself to be half anything unless someone asks what I'm really composed of because I don't see myself as relating sometimes to this and sometimes to that—I'm just me!

—Laura Wood, a University of Maryland student, explaining how she perceives her mixed-race identity as the daughter of a black father and a white mother (qtd. in Orr and Saulny, 2011)

The federal Department of Education would categorize Michelle López-Mullins—a university student who is of Peruvian, Chinese, Irish, Shawnee and Cherokee descent—as "Hispanic." But the National Center for Health Statistics, the government agency that tracks data on births and deaths, would pronounce her "Asian" racially and Hispanic ethnically. And what does Ms. López-Mullins' birth certificate from the State of Maryland say? It doesn't mention her race.

Ms. López-Mullins, 20, usually marks "other" on surveys these days, but when she filled out a census form last year, she chose Asian, Hispanic, Native American and white.

—Susan Saulny (2011b: A1), a *New York Times* journalist, describing the numerous ways that persons such as Michelle López-Mullins are classified in the United States, a rapidly changing, multiracial nation

▲ **In the United States, having more than one racial background, formerly a rare situation, is becoming more and more common.**

Chapter Focus Question:

How have racial and ethnic identities changed over time in the United States?

Look around you! Today's college campus is a reflection of important changes that are occurring in the racial and ethnic composition of the United States, including a dramatic increase in the number of individuals who identify themselves as "mixed race." According to Susan Saulny (2011a: A1), "The crop of students moving through college right now includes the largest group of mixed-race people ever to come of age in the United States, and they are only the vanguard: the country is in the midst of a demographic shift driven by immigration and intermarriage."

Similar to the changing face of many college campuses, college and professional sports has a growing number of athletes who self-identify as "mixed race." Professional golfer Tiger Woods, who has received extensive publicity about his professional and private life, is one of the best-known "mixed-race" athletes because of media coverage regarding the fact that he is one-half African American, one-fourth Chinese, and one-fourth Native American (Mixedfolks.com, 2011).

For more than a century, most people in the United States thought of themselves in singular racial terms, and on official documents they were identified as being of one racial or ethnic category. However, this country is in the middle of a major demographic shift because multiracial and multiethnic Americans are rapidly growing in number, and the U.S. Census Bureau and other governmental agencies are increasingly grouping them together as "mixed race."

How has this happened? Today, one in seven new marriages is between spouses of different races or ethnicities, and immigration in the past two decades has brought large numbers of new residents, particularly from Asia and Mexico, to this country. In the late 1990s the U.S. Census Bureau abandoned the idea of allowing individuals to classify themselves simply as "multiracial" and instead gave people the option to check more than one racial or ethnic category on census and other government forms. As a result, many young people today no longer adhere to "old" color lines that have defined this country for generations: They favor a much more fluid sense of identity, as indicated by the situations of Laura Wood and Michelle López-Mullins.

In this chapter we examine prejudice, discrimination, sociological perspectives on race and ethnicity, and commonalities and differences in the experiences of racial and ethnic groups in the United States. In the process, sports is used as an example of the effects of race and ethnicity on people's lives because this area of social life shows us how, for more than a century, people who have been singled out for negative treatment on the basis of their perceived race or ethnicity have sought to overcome prejudice and discrimination through their determination in endeavors that can be judged based on objective standards ("What's the final score?") rather than subjective standards ("Do I like this person based on characteristics such as race or ethnicity?"). Before reading on, test your knowledge about race, ethnicity, and sports by taking the Sociology and Everyday Life quiz.

Highlighted Learning Objectives

- Distinguish between race and ethnicity.
- Describe what this statement means: "Race is a social construct."
- Distinguish between prejudice and discrimination.
- Identify the major sociological perspectives on race and ethnic relations.
- Discuss the unique historical experiences of the racial and ethnic groups in the United States.

sociology and everyday life

How Much Do You Know About Race, Ethnicity, and Sports?

True	False	
T	F	1. Some deeply held ideas and race and racial differences are often expressed in our beliefs about sports and athletic ability.
T	F	2. Most elite sprinters are of West African descent because they possess a greater percentage of fast-twitch muscles that can be attributed to a specific racial gene.
T	F	3. The chances (statistical odds) of becoming a professional athlete are better than the chance of getting struck by lightning or writing a *New York Times* best seller.
T	F	4. In the past decade the number of African Americans in football and basketball head coaching positions at colleges and universities has increased significantly.
T	F	5. In the 2010s all university presidents at the 120 FBS (NCAA Division I) colleges are white.
T	F	6. The number of African American coaches in charge of NFL teams has remained relatively stagnant in the 2000s.
T	F	7. The number of Latinos and Asian Americans in college head coaching positions has increased significantly in recent years.
T	F	8. All of the conference commissioners of the Football Bowl Subdivision (FBS), formerly known as NCAA Division I-A, are white men.

Answers on page 280.

Race and Ethnicity

What is race? Some people think it refers to skin color (the Caucasian "race"); others use it to refer to a religion (the Jewish "race"), nationality (the British "race"), or the entire human species (the human "race") (Marger, 2009). Popular usages of race have been based on the assumption that a race is a grouping or classification based on *genetic* variations in physical appearance, particularly skin color. However, social scientists and biologists dispute the idea that biological race is a meaningful concept. In fact, the idea of race has little meaning in a biological sense because of the enormous amount of interbreeding that has taken place within the human population. For these reasons, sociologists sometimes place "race" in quotation marks to show that categorizing individuals and population groups on biological characteristics is neither accurate nor based on valid distinctions between the genetic makeup of differently identified "races."

Race is a *socially constructed reality*, not a biological one. Understanding what we mean when we say that race is a social construct is important to our understanding of how race affects all aspects of social life and society. Race as a *social construct* means that races as such do not actually exist, but some groups are still racially defined because the *idea* persists in many people's minds that races are distinct biological categories with physically distinguishable characteristics and a shared common cultural heritage. However, research on the human genome has been unable to identify any racially based genetic differences in human beings, and fossil and DNA evidence also points to humans all being of one race. However, race continues to be an important concern in the twenty-first century, not because it is a biological reality, but because it takes on a life of its own when it is socially defined and shapes how we see others and ourselves. Race also has significant social consequences such as which individuals experience prejudice and discrimination and which have the best life chances and opportunities. When we look at race in this way, the *social significance* that people accord to race is more important than any biological differences that might exist among people who are placed in arbitrary racial categories (Frankenberg, 1993).

A ***race*** **is a category of people who have been singled out as inferior or superior, often on the basis of real or alleged physical characteristics such as skin color, hair texture, eye shape, or other subjectively selected attributes** (Feagin and Feagin, 2011). Categories of people frequently thought of as racial groups include Native Americans, Mexican Americans, African Americans, and Asian Americans.

As compared with race, *ethnicity* defines individuals who are believed to share common characteristics that differentiate them from the other collectivities in a society. An ***ethnic group*** **is a collection of people distinguished, by others or by themselves, primarily on the basis of cultural or nationality characteristics** (Feagin and Feagin, 2011). Characteristics often included in identifying various ethnic groups include national origin, language, religion, and culture. Examples of ethnic groups include Jewish Americans, Irish Americans, and Italian Americans. Ethnic groups share five main characteristics: (1) *unique cultural traits*, such as language, clothing, holidays, or religious practices; (2) *a sense of community*; (3) *a feeling of ethnocentrism*; (4) *ascribed membership from birth*; and (5) *territoriality*, or the tendency to occupy a distinct geographic area (such as Little Italy, Little Havana, or Little Russia) by choice and/or for self-protection. Although some people do not identify with any ethnic group, others participate in social interaction with individuals in their ethnic group and feel a sense of common identity based on cultural characteristics such as language, religion, or politics. However, ethnic groups are not only influenced by their own past history but also by patterns of ethnic domination and subordination in societies. It is important to note that terminology pertaining to racial–ethnic groups is continually in flux, and people within the category as well as outsiders often contest these changes. Examples include the use of *African American*, as compared to *black*, and *Hispanic*, as compared to *Latino/a*.

The Social Significance of Race and Ethnicity

Race and ethnicity take on great social significance because how people act in regard to these terms drastically affects other people's lives, including what opportunities they have, how they are treated, and even how long they live. According to the sociologists Michael Omi and Howard Winant (1994: 158), race "permeates every institution, every relationship, and every individual" in the United States:

> As we … compare real estate prices in different neighborhoods, select a radio channel to enjoy while we drive to work, size up a potential client, customer, neighbor, or teacher, stand in line at the unemployment office, or carry out a thousand other normal tasks, we are compelled to think racially, to use the racial categories and meaning systems into which we have been socialized. (Omi and Winant, 1994: 158)

Historically, stratification based on race and ethnicity has pervaded all aspects of political, economic, and social life. Consider sports as an example. Throughout the early history of the game of baseball, many African Americans had outstanding skills as players but were categorically excluded from Major League teams because of their skin color. Even after Jackie Robinson broke the "color line" to become the first African American in the Major Leagues in 1947, his experience was marred by racial slurs, hate letters, death threats against his infant son, and assaults on his wife (Ashe, 1988; Peterson, 1992/1970). With some professional athletes from diverse racial–ethnic categories having multimillion-dollar contracts and lucrative endorsement deals, it is easy to assume that racism in sports—as well as in the larger society—is a thing of the past. However, this *commercialization* of sports does not mean that racial prejudice and discrimination no longer exist.

Racial Classifications and the Meaning of Race

If we examine racial classifications throughout history, we find that in ancient Greece and Rome a person's race was the group to which she or he belonged, associated with an ancestral place and culture. From the Middle Ages until about the eighteenth century, a person's race was based on family and ancestral ties, in the sense of a *line*, or ties to a national group. During the eighteenth century, physical differences such as the darker skin hues of Africans became associated with race, but racial divisions were typically based on differences in religion and cultural tradition rather than on human biology. With the intense (though misguided) efforts that surrounded the attempt to justify black slavery and white dominance in all areas of life during the second half of the nineteenth century, *races* came to be defined as distinct biological categories of people who were not all members of the same family but who shared inherited physical and cultural traits that were alleged to be different from those traits shared by people in other races. Hierarchies of races were established, placing the "white race" at the top, the "black race" at the bottom, and others in between.

However, racial classifications in the United States have changed over the past century. If we look at U.S. Census Bureau classifications, for example, we can see how the meaning of race continues to change. First, race is defined by perceived skin color: white or nonwhite. Whereas one category exists for "whites" (who vary considerably in actual skin color and physical appearance), all of the remaining categories are considered "nonwhite."

Second, racial purity is assumed to exist. Prior to the 2000 census, for example, the true diversity of the U.S. population was not revealed in census data

race a category of people who have been singled out as inferior or superior, often on the basis of real or alleged physical characteristics such as skin color, hair texture, eye shape, or other subjectively selected attributes.

ethnic group a collection of people distinguished, by others or by themselves, primarily on the basis of cultural or nationality characteristics.

sociology and everyday life

ANSWERS to the Sociology Quiz on Race, Ethnicity, and Sports

1. True. According to the American Anthropological Association, many people continue to believe that differences in athletic abilities can largely be attributed to racial physiology (such as in the term "natural athlete") rather than personal attributes such as "hard worker" or "good decision maker."

2. False. This is an example of the "White Men Can't Jump" assumption that links racial differences to athletic ability; however, no "racial" gene has been found to account for differences in muscle fiber in black runners and white runners. (Human muscles contain a genetically determined mixture of both slow and fast fiber types. Fast-twitch fibers produce force at a higher rate for short bursts of speed and can be an asset to a short-distance sprinter. Slow-twitch fibers produce a lower rate of force that lasts longer and can be an asset to a distance runner.)

3. False. A person has a better chance of getting struck by lightning or writing a *New York Times* best seller (or marrying a millionaire) than becoming a professional athlete, where the current odds are about 24,550 to one.

4. True. Research has found improvements in the number of African Americans in football head coaching positions. As of 2011, there were eleven African American coaches in the College Football Bowl Subdivision (FBS), but the largest increases have come in men's basketball head coaching positions, where 21 percent of all head coaches were African American (down from an all-time high of 25.2 percent in the 2005–2006 season).

5. False. As of 2011, there were five African American, two Asian American, and two Latino presidents of FBS (Division I) universities; however, there were no Native American presidents. This constitutes a positive gain by African Americans in the role of university president at athletic powerhouse schools.

6. True. In 2011 there were seven African American head coaches in the NFL. This number has remained constant since 2008.

7. False. The number of Latinos and Asian Americans in college coaching positions has not grown significantly in recent years.

8. True. All 30 (100 percent) of the FBS (Division I) conference commissioners, who make major decisions about how their conferences are run, are white men, despite the growing number of coaches and athletes who are persons of color and/or women.

Sources: Based on American Anthropological Association, 2007; and Lapchick, 2010.

because multiracial individuals were forced to either select a single race as being their "race" or to select the vague category of "other." Census 2000 made it possible—for the first time—for individuals to classify themselves as being of more than one race (see "Census Profiles: Percentage Distribution of Persons Reporting Two or More Races"). In 2007 updates of the 2000 census, more than one in fifty people self-identified as "multiracial." The pattern of race reporting for foreign-born residents differed from that of native-born Americans: The foreign born most often listed their nation of origin when they were asked to identify their race or ethnicity (Roberts, 2010a).

Third, categories of official racial classifications may (over time) create a sense of group membership or "consciousness of kind" for people within a somewhat arbitrary classification. When people of European descent were classified as "white," some began to see themselves as different from "nonwhite." Consequently, Jewish, Italian, and Irish immigrants may have felt more a part of the Northern European white mainstream in the late eighteenth and early nineteenth centuries. Whether

▲ Miami's Little Havana is an ethnic enclave where people participate in social interaction with other individuals in their ethnic group and feel a sense of shared identity. Ethnic enclaves provide economic and psychological support for recent immigrants as well as for those who were born in the United States.

Percentage Distribution of Persons Reporting Two or More Races

Of the more than 6.8 million people (2.4 percent of the U.S. population) who reported being of two or more races in Census 2000, slightly over 400,000 of them reported being of three or more races. However, the largest combinations are those shown here:

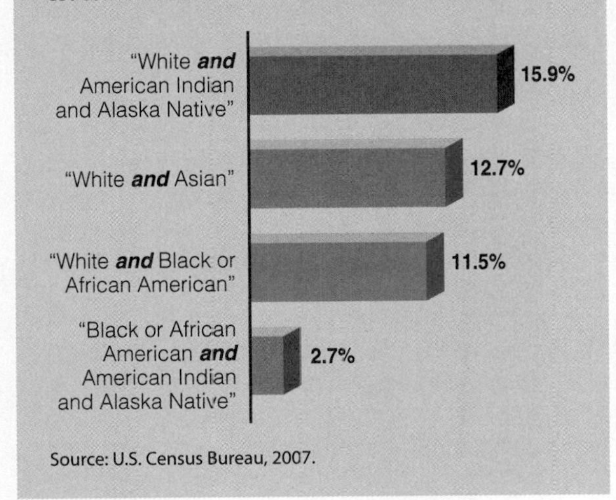

Source: U.S. Census Bureau, 2007.

Chinese Americans, Japanese Americans, Korean Americans, and Filipino Americans come to think of themselves collectively as "Asian Americans" because of official classifications remains to be seen.

AP Photo/Bill Hudson

▲ Contemporary prejudice and discrimination cannot be understood without taking into account the historical background. For African Americans, gains in civil rights were accomplished despite substantial resistance from whites. Today, integration in education, housing, and many other areas of social life remains a pressing social issue.

In the future, increasing numbers of children in the United States will be likely to have a mixed racial or ethnic heritage such as this student describes:

> I am part French, part Cherokee Indian, part Filipino, and part black. Our family taught us to be aware of all these groups, and just to be ourselves. But I have never known what I am. People have asked if I am a Gypsy, or a Portuguese, or a Mexican, or lots of other things. It seems to make people curious, uneasy, and sometimes belligerent. Students I don't even know stop me on campus and ask, "What are you anyway?" (qtd. in Davis, 1991: 133)

The way people are classified remains important because such classifications affect their access to employment, housing, social services, federal aid, and many other "publicly or privately valued goods" (Omi and Winant, 1994: 3).

Dominant and Subordinate Groups

The terms *majority group* and *minority group* are widely used, but their meanings are less clear as the composition of the U.S. population continues to change. Accordingly, many sociologists prefer the terms *dominant* and *subordinate* to identify power relationships that are based on perceived racial, ethnic, or other attributes and identities. To sociologists, a ***dominant group* is one that is advantaged and has superior resources and rights in a society** (Feagin and Feagin, 2011). In the United States, whites with Northern European ancestry (often referred to as Euro-Americans, white Anglo-Saxon Protestants, or WASPs) have been considered to be the dominant group for many years. A ***subordinate group* is one whose members, because of physical or cultural characteristics, are disadvantaged and subjected to unequal treatment by the dominant group and who regard themselves as objects of collective discrimination.** Historically, African Americans and other persons of color have been considered to be subordinate-group members, particularly when they are from lower-income categories.

It is important to note that, in the sociological sense, the word *group* as used in these two terms is misleading because people who merely share ascribed racial or ethnic characteristics do not

dominant group a group that is advantaged and has superior resources and rights in a society.

subordinate group a group whose members, because of physical or cultural characteristics, are disadvantaged and subjected to unequal treatment by the dominant group and who regard themselves as objects of collective discrimination.

sociology in global perspective

Twenty-First-Century Racism in European Football

I was their target. Some of them hit me and I retaliated to defend myself. It's a disgrace. I'm not going to let things lie.... This season the atmosphere in the stands has got worse. Twice already, I have been the victim of monkey chants, of racist insults. It was the same thing again on Saturday night. They managed to get inflatable bananas into the stadium that they were shaking around every time I touched the ball.

—Djibril Cisse, a French professional soccer player born in Africa's Ivory Coast, describing the racism and abuse that he experienced at the Athens Derby game between archrivals Olympiakos and Panathinaikos of the Greek Football Association (qtd. in farenet.org, 2011)

I just wanted to make it public. People have got to be aware that certain things in sport are not acceptable.... I just think there are certain boundaries that shouldn't be crossed. Race is not a topic to make fun of. From what I gathered, [the offending player on the other team] was just trying to provoke me. There's other means of provoking a player without crossing that threshold.

—Oguchi Onyewu, a professional soccer player in Europe who traces his origins to South Africa, has been subjected to monkey noises and racial remarks by fans in the past but recently has also experienced such harsh actions by another player. (qtd. in Longman, 2009: Y1)

For a number of years the problem of racism has been a real issue for the European football family and for the image of football in general. Black footballers and fans have been abused, harassed, and attacked. However, organizations such as the UEFA (Union of European Football Association) and FARE (Football Against Racism in Europe) continue to fight against all forms of discrimination in sports and elsewhere.

Soccer is Europe's equivalent of football in the United States, and European fans are at least as passionate about sports as are fans in the United States. However, racism is an issue in European soccer just as it is an issue in U.S. sports. UEFA, which is European soccer's ruling body, has received numerous reports of racist behavior at soccer matches throughout Europe, and officials have decided that UEFA must work to reduce racism. Accusations of racist behavior toward players of other nationalities or ethnic groups have been lodged against sports fans and some team members at various athletic events throughout Europe. At some soccer matches, fans shout racist comments during the game; throw bottles, cans, and other missiles at players; and pass out derogatory literature about other ethnic groups or nationalities outside the stadium.

In an effort to eliminate such behavior, UEFA and FARE have developed educational programs for schools and are attempting to educate the general public about the destructive nature of racism. UEFA has developed a specific plan that calls for these actions (among others) on the part of soccer clubs:

- Statements must be made in printed programs and on the grounds where the competition takes place that racism will not be tolerated and specific actions will be taken against those who engage in racist chanting or similar behavior.
- Regular public address announcements should be made condemning racist chanting at matches.
- Make it a condition for season-ticket holders that they cannot take part in any racist abuse.
- Take disciplinary action against any players who engage in racial abuse.
- Remove all racist graffiti from the grounds immediately. (uefa.com, 2011)

Will these steps help to reduce racist behavior at European sporting events? Although these actions are a step in the right direction, over the past decade such behavior has continued at some games. But it is important that UEFA officials and others have acknowledged that a problem does exist and that the problem is not limited to a small handful of fans and players. In the final analysis, whether in Europe or other regions of the world, it will require the consistent efforts of everyone to reduce racist behavior at football games and throughout everyday life.

As Djibril Cisse (shown here) points out, sporting events, such as soccer in Europe and football in the United States, are sometimes tainted by the racist behavior of players or by the actions of some fans, who demonstrate racist behavior by their comments about or actions toward players, officials, or other fans. What can be done to reduce such harmful behavior?

reflect & analyze

Have you witnessed truly obnoxious verbal behavior by other spectators at a sporting event? How did you react? Also, some people believe that such behavior is constitutionally protected—a matter of "free speech." Do you agree? Why or why not?

constitute a group. However, the terms *dominant group* and *subordinate group* do give us a way to describe relationships of advantage/disadvantage and power/exploitation that exist in contemporary nations.

Prejudice

Although there are various meanings of the word *prejudice*, sociologists define ***prejudice* as a negative attitude based on faulty generalizations about members of specific racial, ethnic, or other groups.** The term *prejudice* is from the Latin words *prae* ("before") and *judicium* ("judgment"), which means that people may be biased either for or against members of other groups even before they have had any contact with them. Although prejudice can be either *positive* (bias in favor of a group—often our own) or *negative* (bias against a group—one we deem less worthy than our own), it most often refers to the negative attitudes that people may have about members of other racial or ethnic groups.

Stereotypes

Prejudice is rooted in ethnocentrism and stereotypes. When used in the context of racial and ethnic relations, *ethnocentrism* refers to the tendency to regard one's own culture and group as the standard—and thus superior—whereas all other groups are seen as inferior. Ethnocentrism is maintained and perpetuated by ***stereotypes*—overgeneralizations about the appearance, behavior, or other characteristics of members of particular categories.** Although stereotypes can be either positive or negative, examples of negative stereotyping abound in sports. Think about the Native American names, images, and mascots used by sports teams such as the Atlanta Braves, Cleveland Indians, and Washington Redskins. Members of Native American groups have been actively working to eliminate the use of stereotypic mascots (with feathers, buckskins, beads, spears, and "warpaint"), "Indian chants," and gestures (such as the "tomahawk chop"), which they claim trivializes and exploits Native American culture. College and university sports teams with Native American names and logos also remain the subject of controversy in the twenty-first century. According to sociologist Jay Coakley (2004), the use of stereotypes and words such as *redskin* symbolizes a lack of understanding of the culture and heritage of native peoples and is offensive to many Native Americans. Although some people see these names and activities as "innocent fun," others view them as a form of racism.

Racism

What is racism? ***Racism* is a set of attitudes, beliefs, and practices that is used to justify the superior treatment of one racial or ethnic group and the inferior treatment of another racial or ethnic group.** The world has seen a long history of racism: It can be traced from the earliest civilizations. At various times throughout U.S. history, various categories of people, including the Irish Americans, Italian Americans, Jewish Americans, African Americans, and Latinos/as, have been the objects of racist ideology.

Racism may be overt or subtle. Overt racism is more blatant and may take the form of public statements about the "inferiority" of members of a racial or ethnic group. In sports, for example, calling a player of color a derogatory name, participating in racist chanting during a sporting event, and writing racist graffiti in a team's locker room are all forms of overt racism. These racist actions are blatant, but subtle forms of racism are often hidden from sight and more difficult to prove. Examples of subtle racism in sports include those descriptions of African American athletes which suggest that they have "natural" abilities and are better suited for team positions requiring speed and agility. By contrast, whites are described as having the intelligence, dependability, and leadership and decision-making skills needed in positions requiring higher levels of responsibility and control.

Racism tends to intensify in times of economic uncertainty and high rates of immigration. Recently, relatively high rates of immigration in nations such as the United States, Canada, England, France, and Germany have been accompanied by an upsurge in racism and racial conflict (see "Sociology in Global Perspective"). Sometimes, intergroup racism and conflicts further exacerbate strained relationships between dominant-group members and subordinate racial and ethnic group members. For example, when animosities have run very high among African American and Salvadoran groups in Long Island, New York, some white Americans have pointed to those hostilities as evidence that both groups are inferior and not deserving of assistance from the U.S. government or from charitable organizations such as the Catholic church (Mahler, 1995).

prejudice a negative attitude based on faulty generalizations about members of specific racial, ethnic, or other groups.

stereotypes overgeneralizations about the appearance, behavior, or other characteristics of members of particular categories.

racism a set of attitudes, beliefs, and practices that is used to justify the superior treatment of one racial or ethnic group and the inferior treatment of another racial or ethnic group.

Theories of Prejudice

Are some people more prejudiced than others? To answer this question, some theories focus on how individuals may transfer their internal psychological problem onto an external object or person. Others look at factors such as social learning and personality types.

The *frustration–aggression hypothesis* states that people who are frustrated in their efforts to achieve a highly desired goal will respond with a pattern of aggression toward others (Dollard et al., 1939). The object of their aggression becomes the ***scapegoat*—a person or group that is incapable of offering resistance to the hostility or aggression of others** (Marger, 2009). Scapegoats are often used as substitutes for the actual source of the frustration. For example, members of subordinate racial and ethnic groups are often blamed for societal problems (such as unemployment or an economic recession) over which they have no control.

According to some symbolic interactionists, prejudice results from social learning; in other words, it is learned from observing and imitating significant others, such as parents and peers. Initially, children do not have a frame of reference from which to question the prejudices of their relatives and friends. When they are rewarded with smiles or laughs for telling derogatory jokes or making negative comments about outgroup members, children's prejudiced attitudes may be reinforced.

Psychologist Theodor W. Adorno and his colleagues (1950) concluded that highly prejudiced individuals tend to have an ***authoritarian personality*, which is characterized by excessive conformity, submissiveness to authority, intolerance, insecurity, a high level of superstition, and rigid, stereotypic thinking** (Adorno et al., 1950). This type of personality is most likely to develop in a family environment in which dominating parents who are anxious about status use physical discipline but show very little love in raising their children (Adorno et al., 1950). Other scholars have linked prejudiced attitudes to traits such as submissiveness to authority, extreme anger toward outgroups, and conservative religious and political beliefs (Altemeyer, 1981, 1988; Weigel and Howes, 1985).

© Michael Greenlar/The Image Works

▲ According to the frustration–aggression hypothesis, members of white supremacy groups such as the Ku Klux Klan often use members of subordinate racial and ethnic groups as scapegoats for societal problems over which they have no control.

Discrimination

Whereas prejudice is an attitude, ***discrimination* involves actions or practices of dominant-group members (or their representatives) that have a harmful impact on members of a subordinate group.** Prejudiced attitudes do not always lead to discriminatory behavior. As shown in ▶ Figure 10.1, the sociologist Robert Merton (1949) identified four combinations of attitudes and responses. Unprejudiced nondiscriminators are not personally prejudiced and do not discriminate against others. For example, two players on a professional sports team may be best friends although they are of different races. Unprejudiced discriminators may have no personal prejudice but still engage in discriminatory behavior because of peer-group pressure or economic, political, or social interests. For example, in some sports a coach might feel no prejudice toward African American players but believe that white fans will accept only a certain percentage of people of color on the team. Prejudiced nondiscriminators hold personal prejudices but do not discriminate because of peer pressure, legal demands, or a desire for profits. For example, a coach with prejudiced beliefs may hire an African American player to enhance the team's ability to win (Coakley, 2004). Finally, prejudiced discriminators hold personal prejudices and actively discriminate against others. For example, a baseball umpire who is personally prejudiced against African Americans may intentionally call a play incorrectly based on that prejudice.

Discriminatory actions vary in severity from the use of derogatory labels to violence against individuals and groups. The ultimate form of discrimination occurs when people are considered to be unworthy to live because of their race or ethnicity. ***Genocide* is the deliberate, systematic killing of an entire people or nation.** Examples of genocide include the killing of thousands of Native Americans by white settlers in North America and the extermination of six million European Jews by Nazi Germany. More recently, the term *ethnic cleansing* has been used to

	Prejudiced attitude?	Discriminatory behavior?
Unprejudiced nondiscriminator	No	No
Unprejudiced discriminator	No	Yes
Prejudiced nondiscriminator	Yes	No
Prejudiced discriminator	Yes	Yes

▲ **FIGURE 10.1 MERTON'S TYPOLOGY OF PREJUDICE AND DISCRIMINATION**

Merton's typology shows that some people may be prejudiced but not discriminate against others. Do you think that it is possible for a person to discriminate against some people without holding a prejudiced attitude toward them? Why or why not?

define a policy of "cleansing" geographic areas by forcing persons of other races or religions to flee—or die.

Discrimination also varies in how it is carried out. Individuals may act on their own, or they may operate within the context of large-scale organizations and institutions, such as schools, churches, corporations, and governmental agencies. How does individual discrimination differ from institutional discrimination? ***Individual discrimination* consists of one-on-one acts by members of the dominant group that harm members of the subordinate group or their property.** For example, a person may decide not to rent an apartment to someone of a different race. By contrast, ***institutional discrimination* consists of the day-to-day practices of organizations and institutions that have a harmful impact on members of subordinate groups.** For example, a bank might consistently deny loans to people of a certain race. The individuals who implement the policies and procedures of organizations carry out institutional discrimination.

Sociologist Joe R. Feagin (Feagin and Feagin, 2011) has identified four major types of discrimination:

1. *Isolate discrimination* is harmful action intentionally taken by a dominant-group member against a member of a subordinate group. This type of discrimination occurs without the support of other members of the dominant group in the immediate social or community context. For example, a prejudiced judge may give harsher sentences to all African American defendants but may not be supported by the judicial system in that action.
2. *Small-group discrimination* is harmful action intentionally taken by a limited number of dominant-group members against members of subordinate groups. Existing norms or other dominant-group members in the immediate social or community context do not support this type of discrimination. For example, a small group of white students may deface a professor's office with racist epithets without the support of other students or faculty members.
3. *Direct institutionalized discrimination* is organizationally prescribed or community-prescribed action that intentionally has a differential and negative impact on members of subordinate groups. These actions are routinely carried out by a number of dominant-group members based on the norms of the immediate organization or community. Intentional exclusion of people of color from public accommodations is an example of this type of discrimination.
4. *Indirect institutionalized discrimination* refers to practices that have a harmful effect on subordinate-group members even though the organizationally or community-prescribed norms or regulations guiding these actions were initially established with no intent to harm. For example, special education classes were originally intended to provide extra educational opportunities for children with various types of disabilities. However, critics claim that these programs have amounted to racial segregation in many school districts.

scapegoat a person or group that is incapable of offering resistance to the hostility or aggression of others.

authoritarian personality a personality type characterized by excessive conformity, submissiveness to authority, intolerance, insecurity, a high level of superstition, and rigid, stereotypic thinking.

discrimination actions or practices of dominant-group members (or their representatives) that have a harmful effect on members of a subordinate group.

genocide the deliberate, systematic killing of an entire people or nation.

individual discrimination behavior consisting of one-on-one acts by members of the dominant group that harm members of the subordinate group or their property.

institutional discrimination the day-to-day practices of organizations and institutions that have a harmful impact on members of subordinate groups.

sociology works!

Attacking Discrimination to Reduce Prejudice?

Question: Do you think it is possible to reduce racial and ethnic prejudice in the United States by attacking discrimination at the societal level?

Answer: Well, I think since it's individuals who hate each other and do mean stuff, they should work it out for themselves. We don't need the government telling us what to do or how to behave. Like, my dad runs a company, and he says that the government should "butt out" of our business.

—"Brian," an introductory sociology student, stating why he believes that the government has "no business" trying to reduce discrimination (author's notes)

Brian's comment is typical of how many people feel about court rulings and government initiatives over the past sixty years that have sought to reduce the corrosive effects of racial and ethnic discrimination in American life. Based on the widely held axiom that "prejudice causes discrimination," many people argue that discrimination can be alleviated only through changing the attitudes of individuals. From this perspective, discrimination will go away over time if people are encouraged to shed their negative attitudes about members of other racial or ethnic groups. Discarding negative stereotypes about other groups and bringing to light the truth about popular myths regarding the superiority of one's own race, ethnic group, or nationality are widely seen as the best ways of bringing about positive social change. Diversity training sessions held in schools and at the workplace are a classic example of this approach.

Since the civil rights era in the 1960s, however, sociologists have demonstrated that discrimination can be tackled up front and now—rather than waiting for prejudice to diminish—so that people in subordinate racial/ethnic categories can gain a greater measure of human dignity and a variety of opportunities that they otherwise would not have. For example, establishing social policies and laws to eliminate specific practices of segregation and discrimination in education, employment, housing, health care, law enforcement, and other areas of public life has served as a significant starting point for social change that has positively affected generations of people of color, women of all racial and ethnic categories, religious minorities, and many others who have lived outside the mainstream of social life. By enacting legislation that prohibits discrimination based on race, ethnic origin, and color, our nation has sought to provide people with greater equality before the law and access to crucial opportunities and social resources. These changes have, over time, reduced the prejudiced attitudes of many people on a wide variety of issues.

Although progress has been made in reducing some aspects of overt prejudice and institutional discrimination, racism is clearly not a thing of the past. However, if previous sociological research tells us anything about the future, it is that we must continue to tackle not only individual prejudices and discriminatory conduct but also the larger, societal patterns of discrimination—embedded in the organizations and institutions of which we are a part—that restrict freedom, opportunities, and quality of life for all people.

reflect & analyze

Using sports as an example, let's think about these questions: How might college sporting events serve to reduce prejudice and discrimination on campus and beyond? How might these same events serve to perpetuate negative stereotypes and popular myths about racial and ethnic "differences"? What do you think? (See the "You Can Make a Difference" box [page 306] for additional discussion on this topic.)

Various types of racial and ethnic discrimination call for divergent remedies if we are to reduce discriminatory actions and practices in contemporary social life. Since the 1950s and 1960s, many U.S. sociologists have analyzed the complex relationship between prejudice and discrimination. Some have reached the conclusion that prejudice is difficult, if not seemingly impossible, to eradicate because of the deeply held racist beliefs and attitudes that are often passed on from person to person and from one generation to the next. However, the persistence of prejudicial attitudes and beliefs does not mean that racial and ethnic discrimination should be allowed to flourish until such a time as prejudice is effectively eliminated. From this approach, discrimination must be tackled aggressively through demands for change and through policies that specifically target patterns of discrimination (see "Sociology Works!").

Sociological Perspectives on Race and Ethnic Relations

Symbolic interactionist, functionalist, and conflict analysts examine race and ethnic relations in different ways. Functionalists focus on the macrolevel intergroup processes that occur between members of dominant and subordinate groups in society. Conflict theorists analyze power and economic differentials between the dominant group and subordinate groups. Symbolic interactionists examine how microlevel contacts between people may produce either greater racial tolerance or increased levels of hostility.

© AP Photo/Sergey Ponomarev

▲ Following the March 2010 suicide bombings in Moscow's subway, the media reported fears of violence against individuals thought to be from Russia's North Caucasus region, which had been home to the attackers. How do symbolic interactionist perspectives help us to understand such fears?

Symbolic Interactionist Perspectives

What happens when people from different racial and ethnic groups come into contact with one another? In the *contact hypothesis*, symbolic interactionists point out that contact between people from divergent groups should lead to favorable attitudes and behavior when certain factors are present. Members of each group must (1) have equal status, (2) pursue the same goals, (3) cooperate with one another to achieve their goals, and (4) receive positive feedback when they interact with one another in positive, nondiscriminatory ways (Allport, 1958; Coakley, 2004).

What happens when individuals meet someone who does not conform to their existing stereotype? Frequently, they ignore anything that contradicts the stereotype, or they interpret the situation to support their prejudices (Coakley, 2004). For example, a person who does not fit the stereotype may be seen as an exception—"You're not like other [persons of a particular race]."

When a person is seen as conforming to a stereotype, he or she may be treated simply as one of "you people." Former Los Angeles Lakers basketball star Earvin "Magic" Johnson (1992: 31–32) described how he was categorized along with all other African Americans when he was bused to a predominantly white school:

> On the first day of [basketball] practice, my teammates froze me out. Time after time I was wide open, but nobody threw me the ball. At first I thought they just didn't see me. But I woke up after a kid named Danny Parks looked right at me and then took a long jumper. Which he missed.
>
> I was furious, but I didn't say a word. Shortly after that, I grabbed a defensive rebound and took the ball all the way down for a basket. I did it again and a third time, too.
>
> Finally Parks got angry and said, "Hey, pass the [bleeping] ball."
>
> That did it. I slammed down the ball and glared at him. Then I exploded. "I *knew* this would happen!" I said. "That's why I didn't want to come to this [bleeping] school in the first place!"
>
> "Oh, yeah? Well, you people are all the same," he said. "You think you're gonna come in here and do whatever you want? Look, hotshot, your job is to get the rebound. Let us do the shooting."

The interaction between Johnson and Parks demonstrates that when people from different racial and ethnic groups come into contact with one another, they may treat one another as stereotypes, not as individuals. Eventually, Johnson and Parks were able to work out most of their differences. "There's nothing like winning to help people get along," Johnson explained (1992: 32). Although we might hope that nothing like this happens today, there is much evidence that covert discrimination occurs in many sports and social settings.

Symbolic interactionist perspectives make us aware of the importance of intergroup contact and the fact that it may either intensify or reduce racial and ethnic stereotyping and prejudice.

Functionalist Perspectives

How do members of subordinate racial and ethnic groups become a part of the dominant group?

To answer this question, early functionalists studied immigration and patterns of dominant- and subordinate-group interactions.

Assimilation ***Assimilation*** **is a process by which members of subordinate racial and ethnic groups become absorbed into the dominant culture.** To some analysts, assimilation is functional because it contributes to the stability of society by minimizing group differences that might otherwise result in hostility and violence.

Assimilation occurs at several distinct levels, including the cultural, structural, biological, and psychological stages. *Cultural assimilation*, or *acculturation*, occurs when members of an ethnic group adopt dominant-group traits, such as language, dress, values, religion, and food preferences. Cultural assimilation in this country initially followed an "Anglo conformity" model: Members of subordinate ethnic groups were expected to conform to the culture of the dominant white Anglo-Saxon population (Gordon, 1964). However, members of some groups refused to be assimilated and sought to maintain their unique cultural identity.

Structural assimilation, or *integration*, occurs when members of subordinate racial or ethnic groups gain acceptance in everyday social interaction with members of the dominant group. This type of assimilation typically starts in large, impersonal settings such as schools and workplaces, and only later (if at all) results in close friendships and intermarriage. *Biological assimilation*, or *amalgamation*, occurs when members of one group marry those of other social or ethnic groups. Biological assimilation has been more complete in some other countries, such as Mexico and Brazil, than in the United States.

Psychological assimilation involves a change in racial or ethnic self-identification on the part of an individual. Rejection by the dominant group may prevent psychological assimilation by members of some subordinate racial and ethnic groups, especially those with visible characteristics such as skin color or facial features that differ from those of the dominant group.

Ethnic Pluralism Instead of complete assimilation, many groups share elements of the mainstream culture while remaining culturally distinct from both the dominant group and other social and ethnic groups. ***Ethnic pluralism*** **is the coexistence of a variety of distinct racial and ethnic groups within one society.**

Equalitarian pluralism, or *accommodation*, is a situation in which ethnic groups coexist in equality with one another. Switzerland has been described as a model of equalitarian pluralism; more than six million people with French, German, and Italian cultural heritages peacefully coexist there. *Inequalitarian pluralism*, or *segregation*, exists when specific ethnic groups are set apart from the dominant group and have unequal access to power and privilege.

© Bettmann/CORBIS

© TIM JOHNSON/Reuters/Landov Media

▲ Are sports a source of upward mobility for recent immigrants and ethnic minorities, as was true for some in previous generations? Early-twentieth-century Jewish American and Italian American boxers, for example, not only produced intragroup ethnic pride but also earned a livelihood through boxing matches. And, as this recent NBA matchup shows, U.S. sports now attract immigrants from all over the world.

***Segregation* is the spatial and social separation of categories of people by race, ethnicity, class, gender, and/or religion.** Law may enforce segregation. *De jure segregation* refers to laws that systematically enforced the physical and social separation of African Americans in all areas of public life. An example of de jure segregation was the Jim Crow laws, which legalized the separation of the races in public accommodations (such as hotels, restaurants, transportation, hospitals, jails, schools, churches, and cemeteries) in the southern United States after the Civil War (Feagin and Feagin, 2011).

Segregation may also be enforced by custom. *De facto segregation*—racial separation and inequality enforced by custom—is more difficult to document than de jure segregation. For example, residential segregation is still prevalent in many U.S. cities; owners, landlords, real estate agents, and apartment managers often use informal mechanisms to maintain their properties for "whites only." Even middle-class people of color find that racial polarization is fundamental to the residential layout of many cities.

Although functionalist explanations provide a description of how some early white ethnic immigrants assimilated into the cultural mainstream, they do not adequately account for the persistent racial segregation and economic inequality experienced by people of color.

Conflict Perspectives

Conflict theorists focus on economic stratification and access to power in their analyses of race and ethnic relations. Some emphasize the caste-like nature of racial stratification, others analyze class-based discrimination, and still others examine internal colonialism and gendered racism.

The Caste Perspective The caste perspective views racial and ethnic inequality as a permanent feature of U.S. society. According to this approach, the African American experience must be viewed as different from that of other racial or ethnic groups. African Americans were the only group to be subjected to slavery; when slavery was abolished, a caste system was instituted to maintain economic and social inequality between whites and African Americans (Feagin and Feagin, 2011).

The caste system was strengthened by *antimiscegenation laws*, which prohibited sexual intercourse or marriage between persons of different races. Most states had such laws, which were later expanded to include relationships between whites and Chinese, Japanese, and Filipinos. These laws were not declared unconstitutional until 1967 (Frankenberg, 1993).

Although the caste perspective points out that racial stratification may be permanent because of structural elements such as the law, it has been criticized for not examining the role of class in perpetuating racial inequality.

Class Perspectives Class perspectives emphasize the role of the capitalist class in racial exploitation. Based on early theories of race relations by the African American scholar W. E. B. Du Bois, the sociologist Oliver C. Cox (1948) suggested that African Americans were enslaved because they were the cheapest and best workers the owners could find for heavy labor in the mines and on plantations. Thus, the profit motive of capitalists, not skin color or racial prejudice, accounts for slavery.

More recently, sociologists have debated the relative importance of class and race in explaining the unequal life chances of African Americans. Sociologist William Julius Wilson (1996) has suggested that race, cultural factors, social psychological variables, and social class must all be taken into account in examining the life chances of "inner-city residents." His analysis focuses on how class-based economic determinants of social inequality, such as deindustrialization and the decline of the central (inner) city, have affected many African Americans, especially in the Northeast. African Americans were among the most severely affected by the loss of factory jobs because work in the manufacturing sector had previously made upward mobility possible. Wilson (1996) is not suggesting that prejudice and discrimination have been eradicated; rather, he is arguing that they may be less important than class in explaining the current status of African Americans.

How do conflict theorists view the relationship among race, class, and sports? Simply stated, sports reflects the interests of the wealthy and powerful. At all levels, sports exploits athletes (even highly paid ones) in order to gain high levels of profit and prestige for coaches, managers, and owners. African American athletes and central-city youths in particular are exploited by the message of rampant consumerism. Many are given the unrealistic expectation that sports can be a ticket out of the ghetto or barrio. If they try hard enough (and wear the right athletic gear), they too can become wealthy and famous.

assimilation a process by which members of subordinate racial and ethnic groups become absorbed into the dominant culture.

ethnic pluralism the coexistence of a variety of distinct racial and ethnic groups within one society.

segregation the spatial and social separation of categories of people by race, ethnicity, class, gender, and/or religion.

Internal Colonialism Why do some racial and ethnic groups continue to experience subjugation after many years? According to the sociologist Robert Blauner (1972), groups that have been subjected to internal colonialism remain in subordinate positions longer than groups that voluntarily migrated to the United States. ***Internal colonialism* occurs when members of a racial or ethnic group are conquered or colonized and forcibly placed under the economic and political control of the dominant group.**

In the United States, indigenous populations (including groups known today as Native Americans and Mexican Americans) were colonized by Euro-Americans and others who invaded their lands and conquered them. In the process, indigenous groups lost property, political rights, aspects of their culture, and often their lives. The capitalist class acquired cheap labor and land through this government-sanctioned racial exploitation (Blauner, 1972). The effects of past internal colonialism are reflected today in the number of Native Americans who live on government reservations and in the poverty of Mexican Americans who lost their land and had no right to vote.

The internal colonialism perspective is rooted in the historical foundations of racial and ethnic inequality in the United States. However, it tends to view all voluntary immigrants as having many more opportunities than do members of colonized groups. Thus, this model does not explain the continued exploitation of some immigrant groups, such as the Chinese, Filipinos, Cubans, Vietnamese, and Haitians, and the greater acceptance of others, primarily those from Northern Europe (Cashmore, 1996).

© Shelly Katz/Getty Images

▲ Grinding poverty is a pressing problem for families living along the border between the United States and Mexico. Economic development has been limited in areas where *colonias* such as this one are located, and the wealthy have derived far more benefit than others from recent changes in the global economy.

The Split-Labor-Market Theory Who benefits from the exploitation of people of color? Dual or split-labor-market theory states that white workers and members of the capitalist class both benefit from the exploitation of people of color. ***Split labor market* refers to the division of the economy into two areas of employment: a primary sector or upper tier, composed of higher-paid (usually dominant-group) workers in more-secure jobs, and a secondary sector or lower tier, composed of lower-paid (often subordinate-group) workers**

© AP Photo/Eric Risberg

▲ In 2009, controversial radio-talk-show host Rush Limbaugh was part of a group that attempted an ultimately unsuccessful bid to buy the NFL's St. Louis Rams. Many professional franchises are seen as investments, owned by people with little previous connection to sports.

in jobs with little security and hazardous working conditions (Bonacich, 1972, 1976). According to this perspective, white workers in the upper tier may use racial discrimination against nonwhites to protect their positions. These actions most often occur when upper-tier workers feel threatened by lower-tier workers hired by capitalists to reduce labor costs and maximize corporate profits. In the past, immigrants were a source of cheap labor that employers could use to break strikes and keep wages down. Throughout U.S. history, higher-paid workers have responded with racial hostility and joined movements to curtail immigration and thus do away with the source of cheap labor (Marger, 2009).

Proponents of the split-labor-market theory suggest that white workers benefit from racial and ethnic antagonisms. However, these analysts typically do not examine the interactive effects of race, class, and gender in the workplace.

Perspectives on Race and Gender The term ***gendered racism*** **refers to the interactive effect of racism and sexism on the exploitation of women of color.** According to the social psychologist Philomena Essed (1991), women's particular position must be explored within each racial or ethnic group because their experiences will not have been the same as men's in each grouping.

Capitalists do not equally exploit all workers. Gender and race or ethnicity are important in this exploitation. Historically, white men have monopolized the high-paying primary labor market. Many people of color and white women hold lower-tier jobs. Below that tier is the underground sector of the economy, characterized by illegal or quasi-legal activities such as drug trafficking, prostitution, and working in sweatshops that do not meet minimum wage and safety standards. Many undocumented workers and some white women and people of color attempt to earn a living in this sector (Amott and Matthaei, 1996).

The ***theory of racial formation*** **states that actions of the government substantially define racial and ethnic relations in the United States**. Government actions range from race-related legislation to imprisonment of members of groups believed to be a threat to society. Sociologists Michael Omi and Howard Winant (1994) suggest that the U.S. government has shaped the politics of race through actions and policies that cause people to be treated differently because of their race. For example, immigration legislation reflects racial biases. The Naturalization Law of 1790 permitted only white immigrants to qualify for naturalization; the Immigration Act of 1924 favored northern Europeans and excluded Asians and southern and eastern Europeans.

Social protest movements of various racial and ethnic groups periodically challenge the government's definition of racial realities. When this social rearticulation occurs, people's understanding about race may be restructured somewhat. For example, the African American protest movements of the 1950s and 1960s helped redefine the rights of people of color in the United States.

An Alternative Perspective: Critical Race Theory

Emerging out of scholarly law studies on racial and ethnic inequality, *critical race theory* derives its foundation from the U.S. civil rights tradition. Critical race theory has several major premises, including the belief that racism is such an ingrained feature of U.S. society that it appears to be ordinary and natural to many people (Delgado, 1995). As a result, civil rights legislation and affirmative action laws (formal equality) may remedy some of the more overt, blatant forms of racial injustice but have little effect on subtle, business-as-usual forms of racism that people of color experience as they go about their everyday lives. Although many more minority-group members participate in collegiate and professional sports, studies of sports and media show that overt and covert forms of racism persist in the twenty-first century (see "Framing Race in the Media").

According to this approach, the best way to document racism and ongoing inequality in society is to listen to the lived experiences of people who have experienced such discrimination. In this way we can learn what actually happens in regard to racial oppression and the many effects it has on people, including

internal colonialism according to conflict theorists, a practice that occurs when members of a racial or ethnic group are conquered or colonized and forcibly placed under the economic and political control of the dominant group.

split labor market a term used to describe the division of the economy into two areas of employment: a primary sector or upper tier, composed of higher-paid (usually dominant-group) workers in more-secure jobs, and a secondary sector or lower tier, composed of lower-paid (often subordinate-group) workers in jobs with little security and hazardous working conditions.

gendered racism the interactive effect of racism and sexism on the exploitation of women of color.

theory of racial formation the idea that actions of the government substantially define racial and ethnic relations in the United States.

framing race in the media

"Tellin' It Like It Is" in Sports Reporting?

The most popular sports in the United States are the NFL, MLB, and NBA. The majority of these players are minorities (NFL & NBA black and MLB foreign born) and just like in society when minorities are dominating something there will always be a racial element to it.... I don't believe the media are racists. What I do believe is the media have a bias against athletes who don't fit into their image of a *"classic athlete."* I call it *"Michael Jordan Syndrome."* Has nothing to do with race, but more to do with image.... The media expects every athlete to act, speak and be like Mike and if they aren't they are treated and covered differently. This isn't racism, it is prejudice, but it feeds into the racial overtones fans already feel about certain athletes.

—Robert Littal (2011), the African American creator and CEO of BlackSportsOnline, discussing his ideas about race and sports reporting, in an era when middle-age, white men have dominated sports journalism. Reprinted by permission of Robert Littal. www.blacksportsonline.com

Although African Americans and other persons of color have made large strides as college and professional athletes, far fewer minority-group members have entered the upper echelons of sports reporting. For example, white Americans account for about 94 percent of sports editors, 89 percent of assistant sports editors, 88 percent of columnists, and 87 percent of reporters. Compare these figures to the fact that African Americans account for more than 80 percent of NBA players and 70 percent of NFL players. However, regardless of their racial or ethnic identity, sports journalists face questions about how to cover issues of race and racism in sports (Lapchick, 2010).

How do sports journalists typically frame sports stories that involve racially related issues? Although some sports journalists make a conscientious effort to discuss relevant race-related issues as they occur and to offer balanced coverage of all participants' viewpoints, two major negative framing devices are prevalent among today's sports journalists. The first is the *color-blind approach*, which denies the racial underpinnings of an incident where discrimination was evident, such as when a coach, player, fan, or sports reporter refers to another participant by an offensive, racially derogatory name or description. Color-blind framing that downplays racial slurs focuses on *individuals* but denies a larger environment of *institutional racism*, where groups and organizations have not fully confronted and condemned long-held patterns of prejudice and discrimination. From this perspective, discriminatory language and actions are not about racism but about isolated individuals who exhibit bad sports behavior and should receive the equivalent of a "foul" call in a game.

In contrast to the color-blind approach, the *new racism approach* demonizes black athletes and demands that their attitudes, appearance, and actions should be patterned after those of white athletes or respected African American sports figures such as Michael Jordan. According to researchers at Penn State University's John Curley Center for Sports Journalism (2007), sports reporting often emphasizes individualism while ignoring societal problems: "We see it in sports with the general demonizing of black athletes coupled with a failure to recognize institutional racism in sports." The irony of framing white athletes as role models for African American athletes is that some white athletes are able to maintain their popularity and "good-guy" reputations despite their well-documented off-the-field bad behavior, even as some black athletes with similar behavior become synonymous with "bad boys" and all that's wrong with college or professional sports (Rogers, 2011).

Is there a way to improve reporting so that it is a more accurate reflection of what is going on in the world of sports,

alienation, depression, and certain physical illnesses. Central to this argument is the belief that *interest convergence* is a crucial factor in bringing about social change. According to the legal scholar Derrick Bell, white elites tolerate or encourage racial advances for people of color *only* if the dominant-group members believe that their own self-interest will be served in so doing (cited in Delgado, 1995). From this approach, civil rights laws have typically benefited white Americans as much (or more) as people of color because these laws have been used as mechanisms to ensure that "racial progress occurs at just the right pace: change that is too rapid would be unsettling to society at large; change that is too slow could prove destabilizing" (Delgado, 1995: xiv). The Concept Quick Review on page 294 outlines the key aspects of each sociological perspective on race and ethnic relations.

Racial and Ethnic Groups in the United States

How do racial and ethnic groups come into contact with one another? How do they adjust to one another and to the dominant group over time? Sociologists have explored these questions extensively; however, a detailed historical account of the unique experiences of each group is beyond the scope of this chapter. Instead, we will look briefly at intergroup contacts. In the process, sports will be used as an example of how members of some groups have attempted to gain upward mobility and become integrated into society.

Native Americans

Native Americans are believed to have migrated to North America from Asia thousands of years ago, as

▲ Happy after his team's victory over the Denver Broncos, cornerback Antoine Cason of the San Diego Chargers takes part in a postgame interview. Is it fair that Cason, or any other African American athlete, should be judged against the example set by Michael Jordan?

particularly situations involving race and possible racism? Richard Lapchick (2010), Director of the Institute for Diversity and Ethics in Sports, believes that achieving greater diversity in the media will enhance the overall quality of sports reporting, both in how journalists get information correct and in how they find different angles in coverage that do not always focus on sensationalized, often trivial, negative behavior. In the words of sports journalist Thabiti Lewis (2010: xiv), "As a writer, I deem it my duty to utilize sport as a vehicle for discussing and correcting the subtle nuances of the complex images and realities of race. It is essential if we desire a clearer understanding of the impact of sport in America. . . . But the bitter truth is that in American sports culture—like American society—the treatment and depiction of athletes of color is plagued by race, which influences White self-perception, while attacking Black self-reliance, self-culture, and individual expression."

reflect & analyze

Have popular culture images and ads showing African American male athletes selling sneakers, hot dogs, batteries, underwear, and other products lulled us into believing that race is not an issue in sports and that racism is dead in the United States? How do you think sports reporting might be improved to better reflect the accomplishments and concerns of athletes across lines of race and gender?

shown on the time line in ▶ Figure 10.2. One of the most widely accepted beliefs about this migration is that the first groups of Mongolians made their way across a natural bridge of land called Beringia into present-day Alaska. From there, they moved to what is now Canada and the northern United States, eventually making their way as far south as the tip of South America (Cashmore, 1996).

As schoolchildren are taught, the explorer Christopher Columbus first encountered the native inhabitants in 1492 and referred to them as "Indians." When European settlers (or invaders) arrived on this continent, the native inhabitants' way of life was changed forever. Experts estimate that approximately two million native inhabitants lived in North America at that time (Cashmore, 1996); however, their numbers had been reduced to fewer than 240,000 by 1900.

Genocide, Forced Migration, and Forced Assimilation Native Americans have been the victims of genocide and forced migration. Although the United States never had an official policy that set in motion a pattern of deliberate extermination, many Native Americans were either massacred or died from European diseases (such as typhoid, smallpox, and measles) and starvation (Wagner and Stearn, 1945; Cook, 1973). In battle, Native Americans were often no match for the Europeans, who had "modern" weaponry (Amott and Matthaei, 1996). Europeans justified their aggression by stereotyping the Native Americans as "savages" and "heathens" (Takaki, 1993).

After the Revolutionary War, the federal government offered treaties to the Native Americans so that more of their land could be acquired for the

[concept quick review]

Sociological Perspectives on Race and Ethnic Relations

	Focus	Theory/Hypothesis
Symbolic Interactionist	Microlevel contacts between individuals	Contact hypothesis
Functionalist	Macrolevel intergroup processes	1. Assimilation a. cultural b. biological c. structural d. psychological 2. Ethnic pluralism a. equalitarian pluralism b. inequalitarian pluralism (segregation)
Conflict	Power/economic differentials between dominant and subordinate groups	1. Caste perspective 2. Class perspective 3. Internal colonialism 4. Split labor market 5. Gendered racism 6. Racial formation
Critical Race Theory	Racism is an ingrained feature of society that affects everyone's daily life.	Laws may remedy overt discrimination but have little effect on subtle racism. Interest convergence is required for social change.

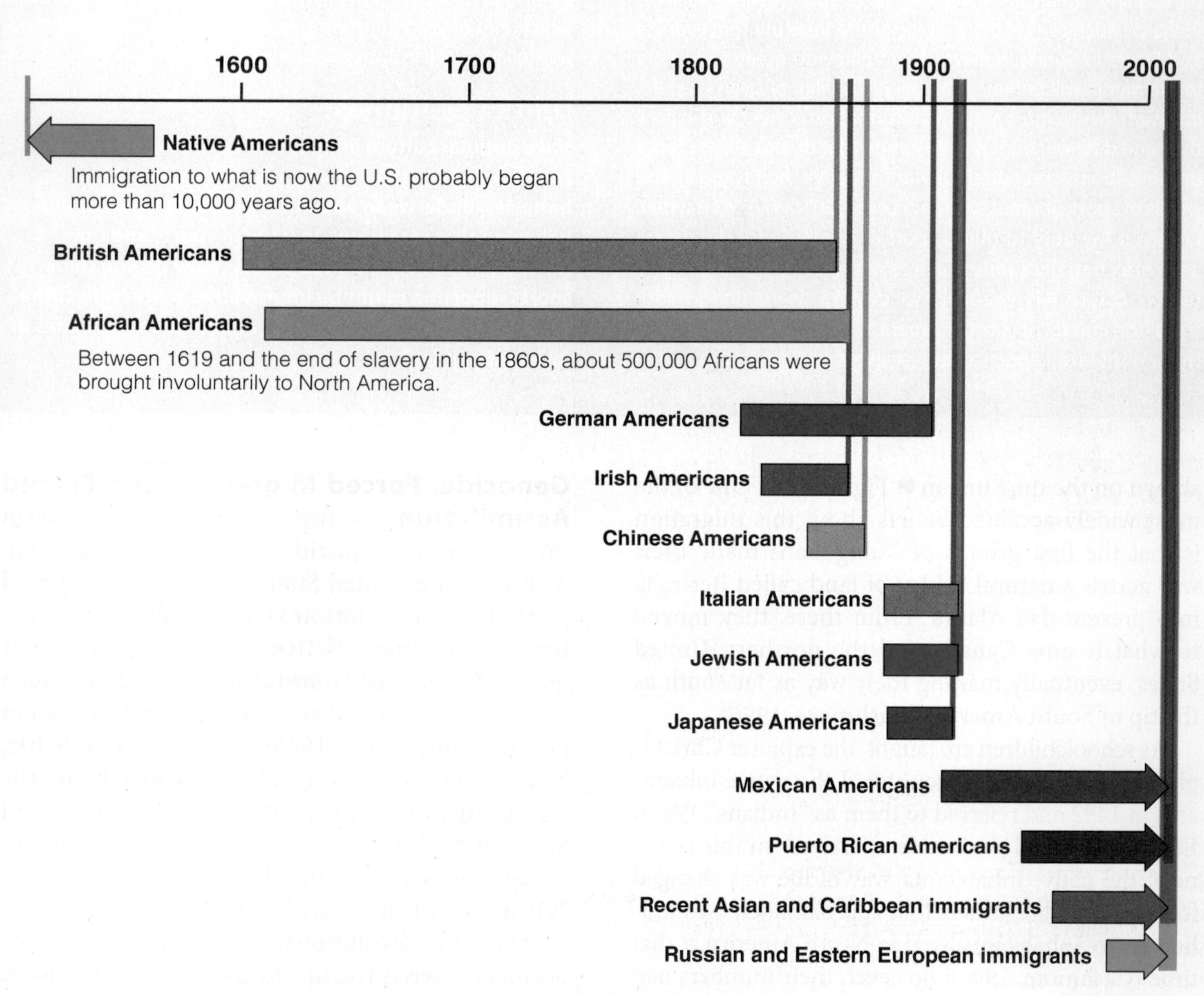

▲ FIGURE 10.2 TIME LINE OF RACIAL AND ETHNIC GROUPS IN THE UNITED STATES

growing white population. Scholars note that the government broke treaty after treaty as it engaged in a policy of wholesale removal of indigenous nations in order to clear the land for settlement by Anglo-Saxon "pioneers" (Green, 1977). Entire nations were forced to move in order to accommodate the white settlers. The "Trail of Tears" was one of the most disastrous of the forced migrations. In the coldest part of the winter of 1832, over half of the Cherokee Nation died during or as a result of their forced relocation from the southeastern United States to the Indian Territory in Oklahoma (Thornton, 1984).

Native Americans were subjected to forced assimilation on the reservations after 1871 (Takaki, 1993). Native American children were placed in boarding schools operated by the Bureau of Indian Affairs to hasten their assimilation into the dominant culture. About 98 percent of native lands had been expropriated by 1920 (see McDonnell, 1991). This process was aided by the Dawes Act (1887), which allowed the federal government to usurp Native American lands for the benefit of corporations and other non-native settlers who sought to turn a profit from oil and gas exploration and grazing.

Native Americans Today Currently, about five million Native Americans and Alaska Natives, including those of more than one race, live in the United States. They make up 1.6 percent of the total population and include a wide variety of groups, such as the Aleuts, Inuit (Eskimos), Cherokee, Navajo, Choctaw, Chippewa, Sioux, and more than 500 other nations of varying sizes and different locales. There is a wide diversity among the people in this category: Each nation has its own culture, history, and unique identity, and more than 250 Native American languages are spoken today. Slightly more than 20 percent of American Indians and Alaska Natives age five and older reported that they spoke a language other than English at home. Although Native Americans live in a number of states, they are concentrated in specific regions of the country. Less than one-third of Native Americans live on one of the 310 federal Indian reservations in this country.

Data continue to show that Native Americans are the most disadvantaged racial or ethnic group in the United States in terms of income, employment, housing, nutrition, and health. In 2009 nearly one-fourth (23.6 percent) of American Indians and Alaska Natives were in poverty. The life chances of Native Americans who live on reservations are especially limited. They have the highest rates of infant mortality and death by exposure and malnutrition. They also have high rates of suicide, substance abuse, and school violence (CDC Morbidity and Mortality Weekly Report, 2011).

Historically, Native Americans have had very limited educational opportunities and a very high rate of unemployment. Since the introduction of six tribally controlled community colleges in the 1970s, a growing network of 37 tribal colleges and universities now serves over 30,000 students from more than 250 tribal nations (see ▶ Map 10.1).

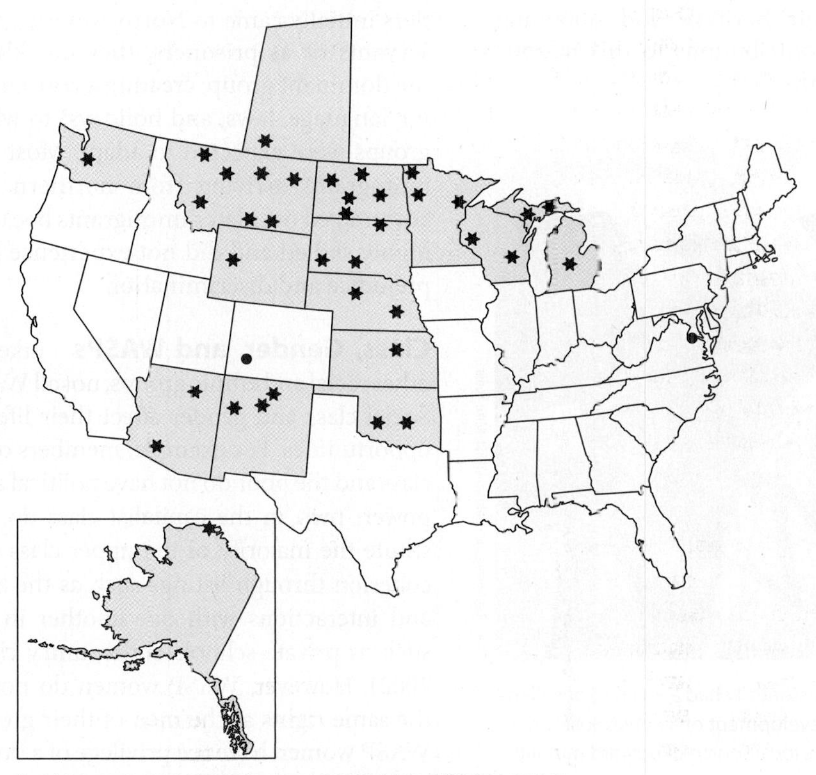

▲ **MAP 10.1 U.S. TRIBAL COLLEGES AND UNIVERSITIES**

Source: American Indian College Fund, 2011.

This network has been successful in providing some Native Americans with the education they need to move into the ranks of the skilled working class and beyond. Across the nation, Native Americans own and operate many types of enterprises, such as construction companies, computer graphic design firms, grocery stores, and management consulting businesses. Casino gambling operations and cigarette shops on Native American reservations—resulting from a reinterpretation of federal law in the 1990s—have brought more income to some of the nations, but this change has not been without its critics, many of whom believe that these businesses result in new problems for Native Americans.

In 2009 Native Americans received a $3.4 billion settlement from the federal government after the conclusion of *Cobell v. Salazar*, a thirteen-year-old lawsuit that accused the government of mishandling revenues generated by the extraction of natural resources from American Indian land trusts as a result of the Dawes Act. Although the federal government was responsible for leasing tribal lands for use by mining, lumber, oil, and gas industries and passing on royalty payments to the Native Americans to whom the lands belonged, Native Americans derived little benefit because of the government's massive abuse of the trust funds.

Native Americans are currently in a transition from a history marked by prejudice and discrimination to a contemporary life in which they may find new opportunities. Many see the challenge for Native Americans today as erasing negative stereotypes while maintaining their heritage and obtaining recognition for their contributions to this nation's development and growth.

▲ Native Americans have historically had a low rate of college attendance. However, the development of a network of tribal colleges has provided them a local source of upward mobility.

Native Americans and Sports Early in the twentieth century, Native Americans such as Jim Thorpe gained national visibility as athletes in football, baseball, and track and field. Teams at boarding schools such as the Carlisle Indian Industrial School in Pennsylvania and the Haskell Institute in Kansas were well-known. However, after the first three decades of the twentieth century, Native Americans became much less prominent in sports. Although some Navajo athletes have been very successful in basketball and some Choctaws have excelled in baseball, Native Americans have seldom been able to compete at the college, professional, or Olympic level. Native American scholar Joseph B. Oxendine (2003) attributes the lack of athletic participation to these factors: (1) a reduction in opportunities for developing sports skills, (2) restricted opportunities for participation, and (3) a lessening of Native Americans' interest in competing with and against non-Native Americans.

However, in the twenty-first century, Native Americans slowly began making their mark in a few professional sports, such as golf (Notah Begay III), lacrosse (Brett Bucktooth), bowling (Mike Edwards), rodeo (Clint Harry), and baseball (Kyle Lohse).

White Anglo-Saxon Protestants (British Americans)

Whereas Native Americans have been among the most disadvantaged peoples, white Anglo-Saxon Protestants (WASPs) have been the most privileged group in this country. Although many English settlers initially came to North America as indentured servants or as prisoners, they quickly emerged as the dominant group, creating a core culture (including language, laws, and holidays) to which all other groups were expected to adapt. Most of the WASP immigrants arriving from northern Europe were advantaged over later immigrants because they were highly skilled and did not experience high levels of prejudice and discrimination.

Class, Gender, and WASPs Like members of other racial and ethnic groups, not all WASPs are alike. Social class and gender affect their life chances and opportunities. For example, members of the working class and the poor do not have political and economic power; men in the capitalist class do. WASPs constitute the majority of the upper class and maintain cohesion through listings such as the *Social Register* and interactions with one another in elite settings such as private schools and country clubs (Kendall, 2002). However, WASP women do not always have the same rights as the men of their group. Although WASP women have the privilege of a dominant racial position, they do not have the gender-related privileges of men (Amott and Matthaei, 1996).

WASPs and Sports Family background, social class, and gender play an important role in the sports participation of WASPs. Contemporary North American football was invented at the Ivy League colleges and was dominated by young, affluent WASPs who had the time and money to attend college and participate in sports activities. Today, whites are more likely than any other racial or ethnic group to become professional athletes in all sports except football and basketball. Although current data are not available to document differences among racial and ethnic categories by types of sports, we know that the probability of competing in athletics beyond the high school interscholastic level is extremely low. For example, only 0.03 percent of high school men's basketball players will become professional athletes, as will only 0.02 percent of women's basketball players. For football, the percentage of high school players who will become professional athletes is 0.09 percent; for baseball, 0.5 percent; for men's ice hockey, 0.4 percent; and for men's soccer, 0.08 percent. Even the odds of advancing from high school athletics to NCAA college sports remain low: 2.9 percent for men's basketball, 3.1 percent for women's basketball, 5.8 percent for football, 5.6 percent for baseball, 12.9 percent for men's ice hockey, and 5.7 percent for men's soccer (coasports.org, 2009).

Affluent WASP women participated in intercollegiate women's basketball in the late 1800s, and various other sporting events were used as a means to break free of restrictive codes of femininity. Until recently, however, most women have had little chance for any involvement in college and professional sports.

African Americans

The African American (black) experience has been one uniquely marked by slavery, segregation, and persistent discrimination. There is a lack of consensus about whether *African American* or *black* is the most appropriate term to refer to the 41.8 million black residents of the United States (including those who report more than one race). Those who prefer the term *black* point out that it incorporates many African-descent groups living in this country that do not use *African American* as a racial or ethnic self-description. For example, people who trace their origins to Haiti, Puerto Rico, or Jamaica typically identify themselves as "black" but not as "African American" (Cashmore, 1996). Although African Americans reside throughout the United States, eighteen states have a estimated black population of at least one million. About 3.5 million African Americans lived in New York State in 2009, but Mississippi had the largest share of blacks (38 percent) in its total population (see ▶ Map 10.2).

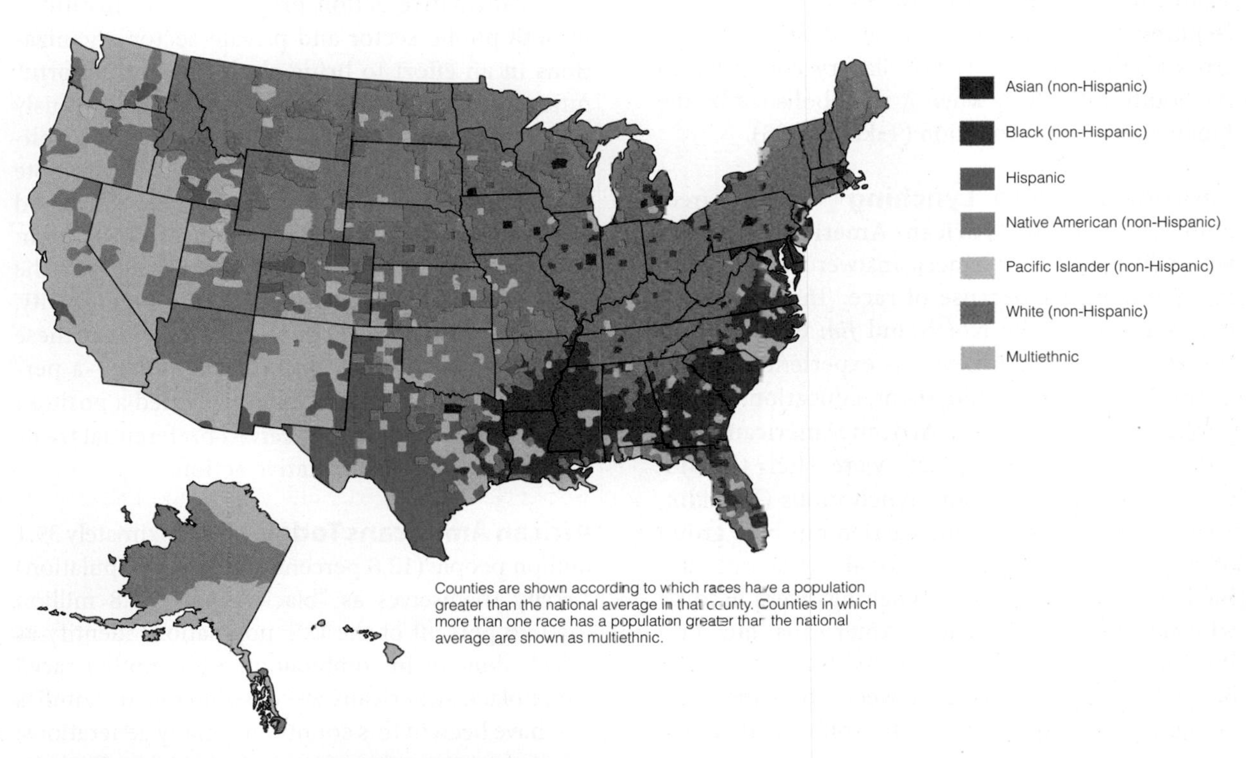

▲ **MAP 10.2 U.S. RACIAL AND ETHNIC DISTRIBUTION**
Although minority populations do continue to grow, regional differences in racial makeup are still quite pronounced, as this map shows.
Source: U.S. Census Bureau, 2007.

Although the earliest African Americans probably arrived in North America with the Spanish conquerors in the fifteenth century, most historians trace their arrival to about 1619, when the first groups of indentured servants were brought to the colony of Virginia. However, by the 1660s indentured servanthood had turned into full-fledged slavery because of the enactment of laws that sanctioned the enslavement of African Americans. Although the initial status of persons of African descent in this country may not have been too different from that of the English indentured servants, all of that changed with the passage of laws turning human beings into property and making slavery a status from which neither individuals nor their children could escape (Franklin, 1980).

Between 1619 and the 1860s about 500,000 Africans were forcibly brought to North America, primarily to work on southern plantations, and these actions were justified by the devaluation and stereotyping of African Americans. Some analysts believe that the central factor associated with the development of slavery in this country was the plantation system, which was heavily dependent on cheap and dependable manual labor. Slavery was primarily beneficial to the wealthy southern plantation owners, but many of the stereotypes used to justify slavery were eventually institutionalized in southern custom and practice (Wilson, 1978). However, some slaves and whites engaged in active resistance against slavery and its barbaric practices, eventually resulting in slavery being outlawed in the northern states by the late 1700s. Slavery continued in the South until 1863, when it was abolished by the Emancipation Proclamation (Takaki, 1993).

Segregation and Lynching Gaining freedom did not give African Americans equality with whites. African Americans were subjected to many indignities because of race. Through informal practices in the North and *Jim Crow laws* in the South, African Americans experienced segregation in housing, employment, education, and all public accommodations. African Americans who did not stay in their "place" were often the victims of violent attacks and lynch mobs (Franklin, 1980). *Lynching* is a killing carried out by a group of vigilantes seeking revenge for an actual or imagined crime by the victim. Lynchings were used by whites to intimidate African Americans into staying "in their place." It is estimated that as many as 6,000 lynchings occurred between 1892 and 1921 (Feagin and Feagin, 2011). In spite of all odds, many African American women and men resisted oppression and did not give up in their struggle for equality (Amott and Matthaei, 1996).

Discrimination In the twentieth century the lives of many African Americans were changed by industrialization and two world wars. When factories were built in the northern United States, many African American families left the rural South in hopes of finding jobs and a better life.

During World Wars I and II, African Americans were a vital source of labor in war production industries; however, racial discrimination continued both on and off the job. In World War II, many African Americans fought for their country in segregated units in the military; after the war, they sought—and were denied—equal opportunities in the country for which they had risked their lives.

African Americans began to demand sweeping societal changes in the 1950s. Initially, the Reverend Dr. Martin Luther King, Jr., and the civil rights movement used *civil disobedience*—nonviolent action seeking to change a policy or law by refusing to comply with it—to call attention to racial inequality and to demand greater inclusion of African Americans in all areas of public life. Subsequently, leaders of the Black Power movement, including Malcolm X and Marcus Garvey, advocated black pride and racial awareness among African Americans. Gradually, the courts and the federal government outlawed racial segregation. For example, the Civil Rights Acts of 1964 and 1965 sought to do away with discrimination in education, housing, employment, and health care. Affirmative action programs were instituted in both public-sector and private-sector organizations in an effort to bring about greater opportunities for African Americans and other previously excluded groups. *Affirmative action* refers to policies or procedures that are intended to promote equal opportunity for categories of people deemed to have been previously excluded from equality in education, employment, and other fields on the basis of characteristics such as race or ethnicity. Critics of affirmative action often assert that these policies amount to *reverse discrimination*—a person who is better qualified being denied a position because another person received preferential treatment as a result of affirmative action.

African Americans Today Approximately 39.1 million people (12.8 percent of the U.S. population) classify themselves as "black alone"; 41.8 million (or 13.6 percent of the U.S. population) identify as "black alone or in combination with another race." Some black Americans are descendants of families that have been in this country for many generations; others are recent immigrants from Africa and the Caribbean. Black Haitians make up the largest group of recent Caribbean immigrants; others come from

Jamaica and Trinidad and Tobago. Recent African immigrants are primarily from Nigeria, Ethiopia, Ghana, and Kenya. They have been simultaneously "pushed" out of their countries of origin by severe economic and political turmoil and "pulled" by perceived opportunities for a better life in the United States. Recent immigrants are often victimized by the same racism that has plagued African Americans as a people for centuries.

Since the 1960s many African Americans have made significant gains in politics, education, employment, and income. Between 1964 and 1999 the number of African Americans elected to political office increased from about 100 to almost 9,000 nationwide (Joint Center for Political and Economic Studies, 2003). African Americans have won mayoral elections in many major cities that have large African American populations, such as Atlanta, Houston, New Orleans, Philadelphia, and Washington, D.C. Despite these political gains, African Americans still represent less than 3 percent of all elected officials in the United States.

Some African Americans have made impressive occupational gains and joined the ranks of professionals in the upper middle class. Others have achieved great wealth and fame as entertainers, professional athletes, and entrepreneurs. However, even those who make millions of dollars a year and live in affluent neighborhoods are not always exempt from racial prejudice and discrimination. And although some African Americans have made substantial occupational and educational gains, many more have not. The African American unemployment rate remains twice as high as that of whites.

African Americans and Sports In recent decades many African Americans have seen sports as a possible source of upward mobility because other means have been unavailable. However, their achievements in sports have often been attributed to "natural ability" and not determination and hard work. Sociologists have rejected such biological explanations for African Americans' success in sports and have focused instead on explanations rooted in the structure of society.

During the slavery era a few African Americans gained better treatment and, occasionally, freedom by winning boxing matches on which their owners had bet large sums of money (McPherson, Curtis, and Loy, 1989). After emancipation, some African Americans found jobs in horse racing and baseball. For example, fourteen of the fifteen jockeys in the first Kentucky Derby (in 1875) were African Americans. A number of African Americans played on baseball teams; a few played in the Major Leagues until the Jim Crow laws forced them out. Then they formed their own "Negro" baseball and basketball leagues (Peterson, 1992/1970).

Since Jackie Robinson broke baseball's "color line," in 1947, many African American athletes have played collegiate and professional sports. Even now, however, persistent class inequalities between whites and African Americans are reflected in the fact that, until recently, African Americans have primarily excelled in sports (such as basketball or football) that do not require much expensive equipment and specialized facilities in order to develop athletic skills (Coakley, 2004). According to one sports analyst,

▲ In August 2008, Barack Obama made history by becoming the first African American to receive the presidential nomination of a major political party, and on Election Day he was voted in as the first African American president of the United States.

> African Americans typically participate in certain sports and not others because of the *sports opportunity structure*—the availability of facilities, coaching, and competition in the schools and community recreation programs in their area. (Phillips, 1993)

Regardless of the sport in which they participate, African Americans continue to experience inequalities in coaching opportunities. As of 2011, only 7 of the 32 National Football League head coaches and only 11 (about 10 percent) of the 120 Division I-A (now known as "Football Bowl Subdivision") head coaches in college football were African Americans. Although few African American coaches head up teams in the Football Bowl Subdivision, almost 30 percent of the assistant coaches and more than 50 percent of the players are African American. Today, African Americans remain significantly underrepresented in other sports, including hockey, skiing, figure skating, golf, volleyball, softball, swimming, gymnastics, sailing, soccer, bowling, cycling, and tennis.

White Ethnic Americans

The American Dream initially brought many white ethnics to the United States. The term *white ethnic Americans* is applied to a wide diversity of immigrants who trace their origins to Ireland and to eastern and southern European countries such as Poland, Italy, Greece, Germany, Yugoslavia, and Russia and other former Soviet republics. Unlike the WASPs, who immigrated primarily from northern Europe and assumed a dominant cultural position in society, white ethnic Americans arrived late in the nineteenth century and early in the twentieth century to find relatively high levels of prejudice and discrimination directed at them by nativist organizations that hoped to curb the entry of non-WASP European immigrants. Because many of the people in white ethnic American categories were not Protestant, they experienced discrimination because they were Catholic, Jewish, or members of other religious bodies, such as the Eastern Orthodox churches (Farley, 2000).

Discrimination Against White Ethnics

Many white ethnic immigrants entered the United States between 1830 and 1924. Irish Catholics were among the first to arrive, with more than four million Irish fleeing the potato famine and economic crisis in Ireland and seeking jobs in the United States (Feagin and Feagin, 2011). When they arrived, they found that British Americans controlled the major institutions of society. The next arrivals were Italians who had been recruited for low-wage industrial and construction jobs. British Americans viewed Irish and Italian immigrants as "foreigners": The Irish were stereotyped as apelike, filthy, bad-tempered, and heavy drinkers; the Italians were depicted as lawless, knife-wielding thugs looking for a fight, "dagos," and "wops" (short for "without papers") (Feagin and Feagin, 2011).

Both Irish Americans and Italian Americans were subjected to institutionalized discrimination in employment. Employment ads read "Help Wanted—No Irish Need Apply" and listed daily wages at $1.30–$1.50 for "whites" and $1.15–$1.25 for "Italians" (Gambino, 1975: 77). In spite of discrimination, white ethnics worked hard to establish

◀ After coming to the United States in the nineteenth century, many white ethnic immigrants faced severe poverty, an issue explored by director Martin Scorsese in his 2002 film *Gangs of New York*.

themselves in the United States, often founding mutual self-help organizations and becoming politically active (Mangione and Morreale, 1992).

Between 1880 and 1920 more than two million Jewish immigrants arrived in the United States and settled in the Northeast. Jewish Americans differ from other white ethnic groups in that some focus their identity primarily on their religion whereas others define their Jewishness in terms of ethnic group membership (Feagin and Feagin, 2011). In any case, Jews continued to be the victims of *anti-Semitism*—prejudice, hostile attitudes, and discriminatory behavior targeted at Jews. For example, signs in hotels read "No Jews Allowed," and some "help wanted" ads stated "Christians Only" (Levine, 1992: 55). In spite of persistent discrimination, Jewish Americans achieved substantial success in many areas, including business, education, the arts and sciences, law, and medicine.

White Ethnics Today Although the Census Bureau does not keep records of all white ethnic groups, it maintains records on Irish Americans for Irish-American Heritage Month (March) and St. Patrick's Day. Today, 36.9 million U.S. residents claim Irish ancestry, a number that is more than eight times the population of Ireland itself (4.5 million). Nearly one-fourth of the population of Massachusetts is composed of residents of Irish ancestry, as compared with 12 percent for the nation as a whole. Households headed by an Irish American have a $56,383 median income, as compared to $50,221 for all households. About 10 percent of people of Irish ancestry were in poverty in 2009, as compared to 14 percent of all U.S. households. Like other white ethnic groups, Irish Americans have higher median household incomes and lower rates of poverty than a number of other racial or ethnic minorities in the United States. Many analysts attribute the success of Irish Americans to their earlier arrival in the United States and higher rates of educational attainment and assimilation than some other racial and ethnic groups.

White Ethnics and Sports Sports provided a pathway to assimilation for many white ethnics. The earliest collegiate football players who were not white Anglo-Saxon Protestants were of Irish, Italian, and Jewish ancestry. Sports participation provided educational opportunities that some white ethnics would not have had otherwise.

Boxing became a way to make a living for white ethnics who did not participate in collegiate sports. Boxing promoters encouraged ethnic rivalries to increase their profits, pitting Italians against Irish or Jews, and whites against African Americans (Levine, 1992; Mangione and Morreale, 1992). Eventually, Italian Americans graduated from boxing into baseball and football. Jewish Americans found that sports lessened the shock of assimilation and gave them an opportunity to refute stereotypes about their physical weaknesses and counter anti-Semitic charges that they were "unfit to become Americans" (Levine, 1992: 272).

Today, assimilation is so complete that little attention is paid to the origins of white ethnic athletes. As former Pittsburgh Steeler running back Franco Harris stated, "I didn't know I was part Italian until I became famous" (qtd. in Mangione and Morreale, 1992: 384).

Asian Americans

The U.S. Census Bureau uses the term *Asian Americans* to designate the many diverse groups with roots in Asia. Chinese and Japanese immigrants were among the earliest Asian Americans. Many Filipinos, Asian Indians, Koreans, Vietnamese, Cambodians, Pakistani, and Indonesians have arrived more recently. Today, Asian Americans belong to the fastest-growing ethnic minority group in the United States and constitute about 5 percent of the nation's population. In 2009 about 16 million people in the United States identified themselves as residents of Asian descent or Asian in combination with one or more other races.

Chinese Americans The initial wave of Chinese immigration occurred between 1850 and 1880, when more than 200,000 Chinese men were "pushed" from China by harsh economic conditions and "pulled" to the United States by the promise of gold in California and employment opportunities in the construction of transcontinental railroads. Far fewer Chinese women immigrated; however, many of them were brought to the United States against their will and forced into prostitution, where they were treated like slaves (Takaki, 1993).

Chinese Americans were subjected to extreme prejudice and stereotyped as "coolies," "heathens," and "Chinks." Some Asian immigrants were attacked and even lynched by working-class whites who feared that they were losing their jobs to the immigrants. Passage of the Chinese Exclusion Act of 1882 brought Chinese immigration to a halt. The Exclusion Act was not repealed until World War II, when Chinese Americans who were contributing to the war effort by working in defense plants pushed for its repeal (Takaki, 1993). After immigration laws were further relaxed in the 1960s, the second and largest wave of Chinese immigration occurred, with immigrants coming primarily from Hong Kong and

Taiwan. These recent immigrants have had more education and workplace skills than earlier arrivals, and they brought families and capital with them to pursue the American Dream.

Today, about 3.8 million Asians of Chinese descent live in the United States. Many reside in large urban enclaves in California, Texas, New York, and Hawaii. As a group, Asian Americans have enjoyed considerable upward mobility: The median household income for single-race Asian Americans (Chinese and others) is $68,780. Some Chinese Americans have become highly successful professionals, business entrepreneurs, and technology experts. However, many Chinese Americans remain in the lower tier of the working class—providing low-wage labor in personal services, repair, and maintenance. In 2009, 2.6 million people age five and older in the United States reported that they spoke Chinese at home.

Japanese Americans Most of the early Japanese immigrants were men who worked on sugar plantations in the Hawaiian Islands in the 1860s. Like Chinese immigrants, the Japanese American workers were viewed as a threat by white workers, and immigration of Japanese men was curbed in 1908. However, Japanese women were permitted to enter the United States for several years thereafter because of the shortage of women on the West Coast. Although some Japanese women married white men, laws prohibiting interracial marriage stopped this practice.

▲ During World War II, nearly 120,000 Japanese Americans—many of whom are still alive today—were interned in camps such as the Manzanar Relocation Center in California, where the statue memorializes their ordeal.

With the exception of the enslavement of African Americans, Japanese Americans experienced one of the most vicious forms of discrimination ever sanctioned by U.S. laws. During World War II, when the United States was at war with Japan, nearly 120,000 Japanese Americans were placed in internment camps, where they remained for more than two years despite the total lack of evidence that they posed a security threat to this country (Takaki, 1993). This action was a direct violation of the citizenship rights of many *Nisei* (second-generation Japanese Americans), who were born in the United States (see Daniels, 1993). Ironically, only Japanese Americans were singled out for such harsh treatment; German Americans avoided this fate even though the United States was also at war with Germany. Four decades later the U.S. government issued an apology for its actions and eventually paid $20,000 each to some of those who had been placed in internment camps (Daniels, 1993; Takaki, 1993).

Since World War II, many Japanese Americans have been very successful. The median income of Japanese Americans is more than 30 percent above the national average. However, most Japanese Americans (and other Asian Americans) live in states that not only have higher incomes but also higher costs of living than the national average. In addition, many Asian American families have more persons in the paid labor force than do other families (Takaki, 1993).

Korean Americans The first wave of Korean immigrants were male workers who arrived in Hawaii between 1903 and 1910. The second wave came to the U.S. mainland following the Korean War in 1954 and was made up primarily of the wives of servicemen and Korean children who had lost their parents in the war. The third wave arrived after the Immigration Act of 1965 permitted well-educated professionals to migrate to the United States. Korean Americans have helped one another open small businesses by pooling money through the *kye*—an association that grants members money on a rotating basis to gain access to more capital.

Today, many Korean Americans live in California and New York, where there is a concentration of Korean-owned grocery stores, businesses, and churches. Unlike earlier Korean immigrants, more-recent arrivals have come as settlers and have brought their families with them. More than one million people age five and older speak Korean at home.

Filipino Americans Today, Filipino Americans constitute the second largest category of Asian Americans, with about 3.1 million U.S. residents reporting that they are Filipino alone or in combination with one or more categories. To understand the

Alan C. Heison/Shutterstock.com

▲ Increasing numbers of Asian Americans are distinguishing themselves in college and professional athletics. Ichiro Suzuki, of the Seattle Mariners, is so famous that he usually is called by his first name only.

status of Filipino Americans, it is important to look at the complex relationship between the Philippine Islands and the U.S. government. After Spain lost the Spanish-American War, the United States established colonial rule over the islands, a rule that lasted from 1898 to 1946. Despite control by the United States, Filipinos were not granted U.S. citizenship, but male Filipinos were allowed to migrate to Hawaii and the U.S. mainland to work in agriculture and in fish canneries in Seattle and Alaska. Like other Asian Americans, Filipino Americans were accused of taking jobs away from white workers and suppressing wages, and Congress restricted Filipino immigration to fifty people per year between the Great Depression and the aftermath of World War II.

The second wave of Filipino immigrants came following the Immigration Act of 1965, when large numbers of physicians, nurses, technical workers, and other professionals moved to the U.S. mainland. Most Filipinos have not had the start-up capital necessary to open their own businesses, and many have been employed in the low-wage sector of the service economy. However, the average household income of Filipino American families is relatively high because about 75 percent of Filipino American women are employed, and nearly half have a four-year college degree (Espiritu, 1995).

Today, about 25 percent of all Filipino Americans live in Southern California, particularly the greater Los Angeles area, where many own businesses such as restaurants or work in medical, dental, or optical careers.

Indochinese Americans Indochinese Americans include people from Vietnam, Cambodia, Thailand, and Laos, most of whom have come to the United States in the past three decades. Vietnamese refugees who had the resources to flee at the beginning of the Vietnam War were the first to arrive. Next came Cambodians and lowland Laotians, referred to as "boat people" by the media. Many who tried to immigrate did not survive at sea; others were turned back when they reached this country or were kept in refugee camps for long periods of time. When they arrived in the United States, inflation was high, the country was in a recession, and many native-born citizens feared that they would lose their jobs to these new refugees, who were willing to work very hard for low wages.

Today, many Indochinese Americans are first- or second-generation residents of the United States; about half live in the western states, especially California. Even though most Indochinese Americans speak Vietnamese or other languages at home, their children speak English fluently and have done very well in school, where some of them have been stereotyped as "brains."

Asian Americans and Sports Until recently, Asian Americans received little recognition in sports. However, Apolo Ohno (speed skating), Julie Chu (ice hockey), Jeremy Lin (basketball), Ed Wang (football), and Tim Lincecum (baseball) have been recognized as top athletes, as have past and present winners in women's ice skating and gymnastics. As one sport analyst stated about the importance of having outstanding Asian American athletes, "[These athletes] are of Asian descent, but more importantly they are Asian Americans whose actions reflect upon the United States. . . . As role models, particularly for the Asian-American community, they exemplify success, integrity, discipline and a dedicated work ethic" (Shum, 1997).

Latinos/as (Hispanic Americans)

The terms *Latino* (for males), *Latina* (for females), and *Hispanic* are used interchangeably to refer to people who trace their origins to Spanish-speaking Latin America and the Iberian peninsula. However, as racial–ethnic scholars have pointed out, the label *Hispanic* was first used by the U.S. government to designate people of Latin American and Spanish descent living in the United States, and it has not been fully accepted as a source of identity by some of the more than 48.8 million Latinos/as who live in the United States today and constitute 16 percent of the nation's total population. Instead, many of the people who trace their roots to Spanish-speaking countries think of themselves as Mexican Americans, Chicanos/as, Puerto Ricans, Cuban Americans, Salvadorans, Guatemalans, Nicaraguans, Costa Ricans,

Argentines, Hondurans, Dominicans, or members of other categories. Many also think of themselves as having a combination of Spanish, African, and Native American ancestry.

Mexican Americans or Chicanos/as Mexican Americans—including both native- and foreign-born people of Mexican origin—are the largest segment (approximately two-thirds) of the Latino/a population in the United States. Most Mexican Americans live in the southwestern region of the United States, although more have moved throughout the United States in recent years.

Immigration from Mexico is the primary vehicle by which the Mexican American population grew in this country. Initially, Mexican-origin workers came to work in agriculture, where they were viewed as a readily available cheap and seasonal labor force. Many initially entered the United States as undocumented workers ("illegal aliens"); however, they were more vulnerable to deportation than other illegal immigrants because of their visibility and the proximity of their country of origin. For more than a century, there has been a "revolving door" between the United States and Mexico that has been open when workers were needed and closed during periods of economic recession and high rates of U.S. unemployment.

Mexican Americans have long been seen as a source of cheap labor, while—ironically—at the same time, they have been stereotyped as lazy and unwilling to work. As has been true of other groups, when white workers viewed Mexican Americans as a threat to their jobs, they demanded that the "illegal aliens" be sent back to Mexico. Consequently, U.S. citizens who happen to be Mexican American have been asked for proof of their citizenship, especially when anti-immigration sentiments are running high. Many Mexican American families have lived in the United States for five or six generations—they have fought in wars, made educational and political gains, and consider themselves to be solid U.S. citizens. Thus, it is a source of great frustration for them to be viewed as illegal immigrants or to be asked "How long have you been in this country?"

AP Photo/John Smierciak

▲ Hispanics play a very large role in Major League Baseball. Here, Brian Fuentes is pitching the tenth inning for the Oakland Athletics in a win against the Chicago White Sox.

Puerto Ricans Today, Puerto Rican Americans make up 9 percent of Hispanic-origin people in the United States. When Puerto Rico became a territory of the United States, in 1917, Puerto Ricans acquired U.S. citizenship and the right to move freely to and from the mainland. In the 1950s many migrated to the mainland when the Puerto Rican sugar industry collapsed, settling primarily in New York and New Jersey. Although living conditions have improved substantially for some Puerto Ricans, life has been difficult for the many living in poverty in Spanish Harlem and other barrios. Nevertheless, in recent years Puerto Ricans have made dramatic advances in education, the arts, and politics. Increasing numbers have become lawyers, physicians, and college professors (see Rodriguez, 1989).

Cuban Americans Cuban Americans live primarily in the Southeast, especially Florida. As a group, they have fared somewhat better than other Latinos/as because many Cuban immigrants were affluent professionals and businesspeople who fled Cuba after Fidel Castro's 1959 Marxist revolution. This early wave of Cuban immigrants has median incomes well above those of other Latinos/as; however, this group is still below the national average. The second wave of Cuban Americans, arriving in the 1970s, has fared worse. Many had been released from prisons and mental hospitals in Cuba, and their arrival fueled an upsurge in prejudice against all Cuban Americans. The more recent arrivals have developed their own ethnic and economic enclaves in Miami's Little Havana, and many of the earlier immigrants have become mainstream professionals and entrepreneurs.

Latinos/as and Sports For most of the twentieth century, Latinos have played Major League Baseball. Originally, Cubans, Puerto Ricans, and Venezuelans were selected for their light skin as well as for their skill as players (Hoose, 1989). Today, Latinos represent more than 20 percent of all major leaguers. If not for a 1974 U.S. Labor Department quota limiting how many foreign-born players can play professional baseball, this number might be even larger (Hoose, 1989).

Recently, Latinos in sports have gained more recognition as books and websites have been created to describe their accomplishments. For example, the website Latino Legends in Sports was created in 1999 to inform people about the contributions of Latino and Latina athletes (see **www.latinosportslegends.com**).

Education is a crucial issue for Latinos/as. Because of past discrimination and unequal educational opportunities, many Latinos/as currently have low levels of educational attainment. Many are unable to attend college or participate in collegiate sports, which is essential for being drafted in professional sports other than baseball. Consequently, the overall number of Latinas/os in college and professional sports is low compared to the rest of the U.S. population who are in this age bracket.

Middle Eastern Americans

Since 1970, many immigrants have arrived in the United States from countries located in the "Middle East," which is the geographic region from Afghanistan to Libya and including Arabia, Cyprus, and Asiatic Turkey. Placing people in the "Middle Eastern" American category is somewhat like placing wide diversities of people in the categories of Asian American or Latino/a; some U.S. residents trace their origins to countries such as Bahrain, Egypt, Iran, Iraq, Kuwait, Lebanon, Oman, Qatar, Saudi Arabia, Syria, UAE (United Arab Emirates), and Yemen. Middle Eastern Americans speak a variety of languages and have diverse religious backgrounds: Some are Muslim, some are Coptic Christian, and others are Melkite Catholic. Although some are from working-class families, Lebanese Americans, Syrian Americans, Iranian Americans, and Kuwaiti Americans primarily come from middle- and upper-income family backgrounds. For example, numerous Iranian Americans are scientists, professionals, and entrepreneurs.

In cities across the United States, Muslims have established social, economic, and ethnic enclaves. On the Internet, they have created websites that provide information about Islamic centers, schools, and lists of businesses and services run by those who adhere to Islam, one of the fastest-growing religions in this country. In cities such as Seattle, incorporation into the economic mainstream has been relatively easy for Palestinian immigrants who left their homeland in the 1980s. Some have found well-paid employment with corporations such as Microsoft because they bring educational skills and talents to the information-based economy, including the ability to translate software into Arabic for Middle Eastern markets (M. Ramirez, 1999). In the United States, Islamic schools and centers often bring together people from a diversity of countries such as Egypt and Pakistan. Many Muslim leaders and parents focus on how to raise children to be good Muslims and good U.S. citizens. However, recent immigrants continue to be torn between establishing roots in the United States and the continuing divisions and strife that exist in their homelands. Some Middle Eastern Americans experience prejudice and discrimination based on their speech patterns, appearance (such as the *hijabs*, or "head-to-toe covering" that leaves only the face exposed, which many girls and women wear), or the assumption that "all Middle Easterners" are somehow associated with terrorism.

Following the September 11, 2001, attacks on the United States by terrorists whose origins were traced to the Middle East, hate crimes and other forms of discrimination against people who were assumed to be Arabs, Arab Americans, or Muslims escalated in this country. With the passage of the U.S. Patriot Act—a law giving the federal government greater authority to engage in searches and surveillance with less judicial review than previously—in the aftermath of the terrorist attacks, many Arab Americans have expressed concern that this new law could be used to target people who appear to be of Middle Eastern origins.

▲ Muslims in the United States who wear traditional attire may face prejudice and/or discrimination as they go about their daily lives.

you can make a difference

Working for Racial Harmony

Suppose that you are talking with several friends about a series of racist incidents at your college. Having studied the sociological imagination, you decide to launch an organization similar to No Time to Hate, which was started at Emory University several years ago to reduce racism on campus. In analyzing racism, your group identifies factors contributing to the problem: (1) divisiveness between different cultural and ethnic communities, (2) persistent lack of trust, (3) the fact that many people never really communicate with one another, (4) the need to bring different voices into the curriculum and college life generally, and (5) the need to learn respect for people from different backgrounds (Loeb, 1994). Your group also develops a set of questions to be answered regarding racism on campus:

- *Encouraging inclusion and acceptance.* Do members of our group reflect the college's racial and ethnic diversity? How much do I know about other people's history and culture? How can I become more tolerant—or accepting—of people who are different from me?
- *Raising consciousness.* What is racism? What causes it? Can people participate in racist language and behavior without realizing what they are doing? What is our college or university doing to reduce racism?
- *Becoming more self-aware.* How much do I know about my own family roots and ethnic background? How do the families and communities in which we grow up affect our perceptions of racial and ethnic relations?
- *Using available resources.* What resources are available for learning more about working to reduce racism? Here are some agencies to contact:
    - ACLU (American Civil Liberties Union), 125 Broad Street, 18th floor, New York, NY 10004. Online: **www.aclu.org**
    - ADL (Anti-Defamation League of B'nai B'rith), 823 United Nations Plaza, New York, NY 10017. Online: **www.adl.org**
    - NAACP (National Association for the Advancement of Colored People), 4805 Mt. Hope Drive, Baltimore, MD 21215. Online: **www.naacp.org**
    - National Council of La Raza, 1126 16th St., NW, Washington, DC 20036. Online: **www.nclr.org**

What additional items would you add to the list of problem areas on your campus? How might your group's objective be reached? Over time, students like you have changed many colleges and universities as a result of involvement!

Middle Eastern Americans and Sports

Although more Islamic schools are beginning to focus on sports, particularly for teenage boys, there has been limited emphasis on competitive athletics among many Middle Eastern Americans. Based on popular sporting events in their countries of origin, some Middle Eastern Americans play golf or soccer. As well, some Iranian Americans follow the soccer careers of professional players from Iran who now play for German, Austrian, Belgian, and Greek clubs. Keeping up with global sporting events is easy with all-sports television cable channels and websites that provide up-to-the-minute information about players and competitions. Over time, there will probably be greater participation by Middle Eastern American males in competitions such as soccer and golf; however, girls and women in Muslim families are typically not allowed to engage in athletic activities. Although little research has been done on this issue in the United States, one study of Islamic countries in the Middle East found that female athletes face strong cultural opposition to their sports participation (Dupre and Gains, 1997).

Global Racial and Ethnic Inequality in the Future

Throughout the world, many racial and ethnic groups seek *self-determination*—the right to choose their own way of life. As many nations are currently structured, however, self-determination is impossible.

Worldwide Racial and Ethnic Struggles

The cost of self-determination is the loss of life and property in ethnic warfare. In recent years the Cold War has given way to dozens of smaller wars over ethnic dominance. In Europe, for example, ethnic violence has persisted in Yugoslavia, Spain, Britain (between the Protestant majority and the Catholic minority in Northern Ireland), Romania, Russia, Moldova, and Georgia. Ethnic violence continues in the Middle East, Africa, Asia, and Latin America. Hundreds of thousands have died from warfare, disease, and refugee migration.

Ethnic wars have also have a high price for survivors, whose life chances can become bleaker even after the violence subsides. In the ethnic conflict between Abkhazians and Georgians in the former Soviet Union, for example, as many as two thousand people have been killed and more than eighty thousand displaced. Ethnic hatred also devastated the province of Kosovo, which is located in Serbia, and brought about the deaths of thousands of ethnic Albanians (Bennahum, 1999).

In the twenty-first century the struggle between the Israeli government and various Palestinian factions over the future and borders of Palestine continues to make headlines. Discord in this region has heightened tensions among people not only in Israel and Palestine but also in the United States and around the world as deadly clashes continue and political leaders are apparently unable to reach a lasting solution to the decades-long strife.

Growing Racial and Ethnic Diversity in the United States

Racial and ethnic diversity is increasing in the United States. African Americans, Latinos/as, Asian Americans, and Native Americans constitute one-fourth of the U.S. population, whereas whites are a shrinking percentage of the population. Today, white Americans make up 70 percent of the population, in contrast to 80 percent in 1980. It is predicted that by 2056, the roots of the average U.S. resident will be in Africa, Asia, Hispanic countries, the Pacific islands, and the Middle East—not white Europe.

What effect will these changes have on racial and ethnic relations? Several possibilities exist. On the one hand, conflicts may become more overt and confrontational as people continue to use *sincere fictions*—personal beliefs that reflect larger societal mythologies, such as "I am not a racist" or "I have never discriminated against anyone"—even when these are inaccurate perceptions (Feagin and Vera, 1995). Interethnic tensions may increase as competition for education, jobs, and other resources continues to grow.

On the other hand, there is reason for cautious optimism. Throughout U.S. history, members of diverse racial and ethnic groups have struggled to gain the freedom and rights that were previously withheld from them. Today, minority grassroots organizations are pressing for affordable housing, job training, and educational opportunities. As discussed in "You Can Make a Difference," movements composed of both whites and people of color continue to oppose racism in everyday life, to seek to heal divisions among racial groups, and to teach children about racial tolerance. Many groups hope not only to affect their own microcosm but also to contribute to worldwide efforts to end racism.

To eliminate racial discrimination, it will be necessary to equalize opportunities in schools and workplaces. As Michael Omi and Howard Winant (1994: 158) have emphasized,

> Today more than ever, opposing racism requires that we notice race, not ignore it, that we afford it the recognition it deserves and the subtlety it embodies. By noticing race we can begin to challenge racism, with its ever-more-absurd reduction of human experience to an essence attributed to all without regard for historical or social context.... By noticing race we can develop the political insight and mobilization necessary to make the U.S. a more racially just and egalitarian society.

chapter review Q & A

Use these questions and answers to check how well you've achieved the learning objectives set out at the beginning of this chapter.

- **How do race and ethnicity differ?**

A race is a category of people who have been singled out as inferior or superior, often on the basis of physical characteristics such as skin color, hair texture, or eye shape. An ethnic group is a collection of people distinguished primarily by cultural or national characteristics, including unique cultural traits, a sense of community, a feeling of ethnocentrism, ascribed membership, and territoriality.

- **What are dominant groups and subordinate groups?**

A dominant group is an advantaged group that has superior resources and rights in society. A subordinate group is a disadvantaged group whose members are subjected to unequal treatment by the dominant group. Use of the terms *dominant* and *subordinate* reflects the importance of power in relationships.

• How is prejudice related to discrimination?

Prejudice is a negative attitude often based on stereotypes, which are overgeneralizations about the appearance, behavior, or other characteristics of all members of a group. Discrimination involves actions or practices of dominant-group members that have a harmful effect on members of a subordinate group.

• What are the major psychological explanations of prejudice?

According to the frustration–aggression hypothesis of prejudice, people frustrated in their efforts to achieve a highly desired goal may respond with aggression toward others, who then become scapegoats. Another theory of prejudice focuses on the authoritarian personality, marked by excessive conformity, submissiveness to authority, intolerance, insecurity, superstition, and rigid thinking.

• How do individual discrimination and institutional discrimination differ?

Individual discrimination involves actions by individual members of the dominant group that harm members of subordinate groups or their property. Institutional discrimination involves day-to-day practices of organizations and institutions that have a harmful effect on members of subordinate groups.

• How do sociologists view racial and ethnic group relations?

Symbolic interactionists suggest that increased contact between people from divergent groups should lead to favorable attitudes and behavior when members of each group (1) have equal status, (2) pursue the same goals, (3) cooperate with one another to achieve goals, and (4) receive positive feedback when they interact with one another. Functionalists stress that members of subordinate groups become a part of the mainstream through assimilation, the process by which members of subordinate groups become absorbed into the dominant culture. Conflict theorists focus on economic stratification and access to power in race and ethnic relations. The caste perspective views inequality as a permanent feature of society, whereas class perspectives focus on the link between capitalism and racial exploitation. According to racial formation theory, the actions of the U.S. government substantially define racial and ethnic relations.

• How have the experiences of various racial–ethnic groups differed in the United States?

Native Americans suffered greatly from the actions of European settlers, who seized their lands and made them victims of forced migration and genocide. Today, they lead lives characterized by poverty and lack of opportunity. White Anglo-Saxon Protestants are the most privileged group in the United States, although social class and gender affect their life chances. White ethnic Americans, whose ancestors migrated from southern and eastern European countries, have gradually made their way into the mainstream of U.S. society. Following the abolishment of slavery in 1863, African Americans were still subjected to segregation, discrimination, and lynchings. Despite civil rights legislation and economic and political gains by many African Americans, racial prejudice and discrimination continue to exist. Asian American immigrants as a group have enjoyed considerable upward mobility in U.S. society in recent decades, but many Asian Americans still struggle to survive by working at low-paying jobs and living in urban ethnic enclaves. Although some Latinos/as have made substantial political, economic, and professional gains in U.S. society, as a group they are nevertheless subjected to anti-immigration sentiments. Middle Eastern immigrants to the United States speak a variety of languages and have diverse religious backgrounds. Because they generally come from middle-class backgrounds, they have made inroads into mainstream U.S. society.

key terms

assimilation 288
authoritarian personality 284
discrimination 284
dominant group 281
ethnic group 278
ethnic pluralism 288
gendered racism 291
genocide 284
individual discrimination 285
institutional discrimination 285
internal colonialism 290
prejudice 283
race 278
racism 283
scapegoat 284
segregation 289
split labor market 290
stereotypes 283
subordinate group 281
theory of racial formation 291

questions for critical thinking

1. Do you consider yourself defined more strongly by your race or by your ethnicity? How so?
2. Given that subordinate groups have some common experiences, why is there such deep conflict between some of these groups?
3. What would need to happen in the United States, both individually and institutionally, for a positive form of ethnic pluralism to flourish in the twenty-first century?

turning to video

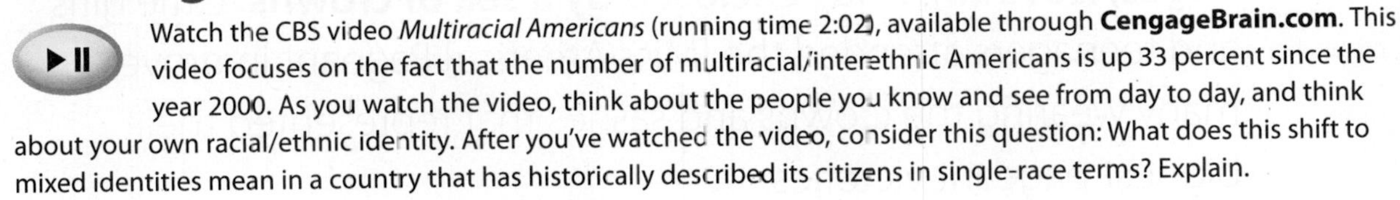

Watch the CBS video *Multiracial Americans* (running time 2:02), available through **CengageBrain.com**. This video focuses on the fact that the number of multiracial/interethnic Americans is up 33 percent since the year 2000. As you watch the video, think about the people you know and see from day to day, and think about your own racial/ethnic identity. After you've watched the video, consider this question: What does this shift to mixed identities mean in a country that has historically described its citizens in single-race terms? Explain.

online study resources

Go to CENGAGE brain.com to access online study resources, including the Sociology CourseMate for this text as well as special features such as video, an interactive sociology time line and interactive maps, General Social Survey (GSS) data, and U.S. Census 2010 data.

CourseMate brings course concepts to life with interactive learning, study, and exam-preparation tools that support the printed textbook. A textbook-specific website, **Sociology CourseMate** includes an integrated interactive eBook and other interactive learning tools, including quizzes, flash cards, and videos.

Visit **www.cengagebrain.com** to access your account and purchase materials.

11 Sex and Gender

As I sat in the theater at the Aladdin Hotel on the Strip [in Las Vegas, Nevada], I was enclosed by a sea of crowns. Little girls and teenagers attended the [Miss America] Pageant in droves, many wearing the crowns and sashes that represented their biggest pageant victories.

Sitting in the middle of the cheering section for Miss Kansas, surrounded by cardboard daisies on wooden sticks with Miss Kansas's face in the center and shouts of "You go girl!" I felt as if I were at a political convention or a religious revival. Also marooned in the Miss Kansas section, without any daisies, were a little girl, her mother (who had twice competed in the Miss Nevada state pageant), and her grandmother, who sat in front of me. The twelve-year-old watched the Pageant with wide eyes the whole night. She wore no crown or sash, but she looked smart in a black velvet dress.

© Tim Shaffer/Reuters/Landov

▲ **Jennifer Berry, Miss Oklahoma, accepts her crown as the 2006 Miss America at the Aladdin Casino in Las Vegas. What is your opinion of pageants such as this one?**

After the talent segment, the girl turned to me and asked, "When I'm in the Miss America Pageant, I want to play the piano and the saxophone for my talent. I can switch back and forth. Do you think that would work?"

"Well, it would certainly be different," I replied.

"Good, then that would help me win."

"So, you really want to be Miss America someday?"

The girl nodded her head, her face solemn. Before replying, I paused. "Well, you can do that. But, you know, there are so many other things to do besides being a beauty queen."

The little girl did not hear me. She was rapturously watching as Miss Oklahoma was crowned Miss America 2006. All the other girls in the audience, those with crowns and those without, stood together, mouthing the words to the famous theme song as the new Miss America was serenaded by the voice of the great Pageant emcee, Bert Parks, who died in 1992: *"There she is, Miss America, there she is, your ideal. . . ."*

—Hilary Levey (2007: 72), then a graduate student in sociology at Princeton University, describes her thoughts upon attending a Miss America Pageant. Although Ms. Levey never competed in a beauty pageant, her mother was named Miss America (from Michigan) in 1970.

Chapter Focus Question:

How do expectations about female and male appearance, especially weight, reflect gender inequality?

Many little girls are similar to the one whom Hilary Levey encountered at the Miss America Pageant: They have their hearts set on being chosen as the winner of a beauty and/or talent competition such as Miss America or Miss USA. Tens of thousands of beauty pageants—ranging from local beach bikini pageants to international scholarship competitions—are held annually. Two competitions—the Miss America Scholarship program and the Miss USA Pageant—involve more than 7,500 local and regional pageants across the country each year (Banet-Weiser, 1999).

Of course, all pageants are not identical. For example, organizers of the Miss America Pageant claim that their competition focuses on both talent and beauty, while Miss USA originated as a "bathing beauty" competition that was sponsored by a swimwear company. Today, Miss USA and its younger counterpart, Miss Teen USA, continue to look for female models who look outstanding in swimsuits and evening gowns, and who can promote products ranging from suntan lotion to flashy diamonds (Angelotti, 2006). Regardless of somewhat different stated goals, these talent and beauty competitions are really about physical beauty and appearance.

For this reason, competitions such as Miss America, Miss USA, Miss Universe, and Miss Teen USA have been the subject of both praise and criticism for the ways in which they portray girls and young women. Some individuals believe that beauty pageants are good for women because they encourage individual achievement and promote self-confidence. Pageant winners are often praised for being positive role models for young women, particularly if the titleholder remains scandal-free during the year of her reign. However, some critics of beauty pageants claim that these events promote an unrealistic beauty ideal that is not attainable for most people and that is not necessarily desirable in the real world (see Banet-Weiser, 1999). Other critics believe that pageants are degrading to women because the contestants are ranked "like prize horses" and given a sash to put around their neck (Corsbie-Massay, 2005: 1). Some feminist analysts argue that beauty pageants objectify women (Watson and Martin, 2004).

Highlighted Learning Objectives

- Explain how society's resources and economic structure influence gender stratification.
- Describe the primary agents of gender socialization.
- Identify ways in which the contemporary workplace reflects gender stratification.
- Compare and contrast functionalist, conflict, and feminist perspectives on gender stratification.

What is objectification? *Objectification* is the process whereby some people treat other individuals as if they were objects or things, not human beings. For example, we objectify women—or men—when we judge them strictly on the basis of their physical appearance rather than on their individual qualities, attributes, or actions (Schur, 1983). Although men may be objectified in some societies, the objectification of girls and women is widespread and particularly common in the United States and many other nations (see ■ Table 11.1). In regard to beauty pageants, organizers seek to deflect this criticism by providing contestants with an opportunity to talk about themselves and their interests or to answer questions that supposedly will show that they are intelligent and knowledgeable about current events. At the end of each pageant, however, the winner's physical attractiveness is most often highlighted rather than the true substance of her life (Angelotti, 2006). Although some people think of the Miss America Pageant and similar competitions as a vestige of the past, many women and men in the twenty-first century are strongly influenced by the images that these competitions project regarding beauty, body image, race/ethnicity, identity, and consumerism (Watson and Martin, 2004).

Some differences between men and women are biological in nature; however, many differences between the sexes are socially constructed. Studying sociology makes us aware of differences that relate to gender (a social concept) as well as differences that are based on a person's biological makeup, or sex. In this chapter we examine the issue of gender: what it is and how it affects us. Before reading on, test your knowledge about body image and gender by taking the Sociology and Everyday Life quiz.

Sex: The Biological Dimension

Whereas the word *gender* is often used to refer to the distinctive qualities of men and women (masculinity and femininity) that are culturally created, ***sex* refers to the biological and anatomical differences between females and males**. At the core of these differences is the chromosomal information transmitted at the moment a child is conceived. The mother contributes an X chromosome and the father either an X (which produces a female embryo) or a Y (which produces a male embryo). At birth, male and female infants are distinguished by ***primary sex characteristics:* the genitalia used in the reproductive process**. At puberty, an increased production of hormones results in the development of ***secondary sex characteristics:* the physical traits (other than reproductive organs) that identify an individual's sex**. For women, these include larger breasts, wider hips, and narrower shoulders; a layer of fatty tissue throughout the body; and menstruation. For men, they include development of enlarged genitals, a deeper voice, greater height, a more muscular build, and more body and facial hair.

Intersexed and Transgendered Persons

Sex is not always clear-cut. An ***intersexed person* is an individual who is born with a reproductive or sexual anatomy that does not correspond to typical**

table 11.1
The Objectification of Women

General Aspects of Objectification	Objectification Based on Cultural Preoccupation with "Looks"
Women are responded to primarily as "females," whereas their personal qualities and accomplishments are of secondary importance.	Women are often seen as the objects of sexual attraction, not full human beings—for example, when they are stared at.
Women are seen as "all alike."	Women are seen by some as depersonalized body parts—for example, "a piece of ass."
Women are seen as being subordinate and passive, so things can easily be "done to a woman"—for example, discrimination, harassment, and violence.	Depersonalized female sexuality is used for cultural and economic purposes—such as in the media, advertising, the fashion and cosmetics industries, and pornography.
Women are seen as easily ignored or trivialized.	Women are seen as being "decorative" and status-conferring objects to be bought (sometimes collected) and displayed by men and sometimes by other women.
	Women are evaluated according to prevailing, narrow "beauty" standards and often feel pressure to conform to appearance norms.

Source: Schur, 1983.

sociology and everyday life

How Much Do You Know About Body Image and Gender?

True	False	
T	F	1. Most people have an accurate perception of their physical appearance.
T	F	2. Recent studies show that up to 95 percent of men express dissatisfaction with some aspect of their bodies.
T	F	3. Many young girls and women believe that being even slightly "overweight" makes them less "feminine."
T	F	4. Physical attractiveness is a more central part of self-concept for women than for men.
T	F	5. Contestants in beauty pageants such as Miss America have remained about the same in body size throughout the history of these competitions.
T	F	6. Thinness has always been the "ideal" body image for women.
T	F	7. Women bodybuilders have gained full acceptance in society.
T	F	8. The media play a significant role in shaping societal perceptions about the ideal female body.

Answers on page 314.

definitions of male or female; in other words, the person's sexual differentiation is ambiguous. Formerly referred to as *hermaphrodites* by some in the medical community, intersexed persons may appear to be female on the outside at birth but have mostly male-type anatomy on the inside, or they may be born with genitals that appear to be in between the usual male and female types. For example, a chromosomally normal (XY) male may be born with a penis just one centimeter long and a urinary opening similar to that of a female. However, although intersexuality is considered to be an inborn condition, intersexed anatomy is not always known or visible at birth. In fact, intersexual anatomy sometimes does not become apparent until puberty, or when an adult is found to be infertile, or when an autopsy is performed at death. It is possible for some intersexual people to live and die with intersexed anatomy but never know that the condition exists. According to the Intersex Society of North America (2011),

> Intersex is a socially constructed category that reflects real biological variation. Nature presents us with sex anatomy spectrums [, but] nature doesn't decide where the category of "male" ends and the category of "intersex" begins, or where the category of "intersex" ends and the category of "female" begins. *Humans decide*. Humans (today, typically doctors) decide how small a penis has to be, or how unusual a combination of parts has to be, before it counts as intersex. Humans decide whether a person with XXY chromosomes and XY chromosomes and androgen insensitivity will count as intersex.

Some people may be genetically of one sex but have a gender identity of the other. That is true for a ***transgendered person*—an individual whose gender identity (self-identification as woman, man, neither, or both) does not match the person's assigned sex (identification by others as male, female, or intersex based on physical/genetic sex)**. Consequently, transgendered persons may believe that they have the opposite gender identity from that of their sex organs and may be aware of this conflict between gender identity and physical sex as early as the preschool years. Some transgendered individuals choose to take hormone treatments or have a sex-change operation to alter their genitalia so that they can have a body congruent with their sense of gender identity. Many then go on to lead lives that they view as being compatible with their

sex the biological and anatomical differences between females and males.

primary sex characteristics the genitalia used in the reproductive process.

secondary sex characteristics the physical traits (other than reproductive organs) that identify an individual's sex.

intersexed person an individual who is born with a reproductive or sexual anatomy that does not correspond to typical definitions of male or female; in other words, the person's sexual differentiation is ambiguous.

transgendered person an individual whose gender identity (self-identification as woman, man, neither, or both) does not match the person's assigned sex (identification by others as male, female, or intersex based on physical/genetic sex).

sociology and everyday life

ANSWERS to the Sociology Quiz on Body Image and Gender

1. False. Many people do not have a very accurate perception of their bodies. For example, many girls and women think of themselves as "fat" when they are not. Some boys and men believe that they need a well-developed chest and arm muscles, broad shoulders, and a narrow waist.

2. True. In recent studies, up to 95 percent of men believed they needed to improve some aspect of their bodies.

3. True. More than half of all adult women in the United States are currently dieting, and over three-fourths of normal-weight women think they are "too fat." Recently, very young girls have developed similar concerns.

4. True. Women have been socialized to believe that being physically attractive is very important. Studies have found that weight and body shape are the central determinants of women's perception of their physical attractiveness.

5. False. During the 1980s and 1990s, contestants in Miss America and other national beauty pageants decreased in body size, becoming much thinner and less curvaceous. Today, more emphasis is placed on being physically fit, but winning contestants typically have much lower body weight than the average woman of their height and age.

6. False. The "ideal" body image for women has changed a number of times. A positive view of body fat has prevailed for most of human history; however, in the twentieth century in the United States, this view gave way to "fat aversion."

7. False. Although bodybuilding among women has gained some degree of acceptance, women bodybuilders are still expected to be very "feminine" and not to overdevelop themselves.

8. True. Women in the United States are bombarded by advertising, television programs, and films containing images of women that typically represent an ideal which most real women cannot attain.

Sources: Based on Fallon, Katzman, and Wooley, 1994; Kilbourne, 1999; Seid, 1994; and Turner, 1997.

true gender identity. But the issue of hormonal and surgical sex reassignment remains highly politicized. The "Standards of Care," a set of guidelines set up by the Harry Benjamin International Gender Dysphoria Association, establishes standards by which transgendered persons may obtain hormonal and surgical sex reassignment to help ensure that people choosing such options are informed about what is involved in a gender transition.

Western societies acknowledge the existence of only two sexes; some other societies recognize three—men, women, and *berdaches* (or *hijras* or *xaniths*): biological males who behave, dress, work, and are treated in most respects as women. The closest approximation of a third sex in Western societies is a ***cross-dresser*** (formerly known as a *transvestite*), **a male who dresses as a woman or a female who dresses as a man but does not alter the genitalia**. Although cross-dressers are not treated as a third sex, they often "pass" for members of that sex because their appearance and mannerisms fall within the range of what is expected from members of the other sex. Most cross-dressers are heterosexual men, many of whom are married, but gay men, lesbians, and straight women may also be cross-dressers. Cross-dressing may occur in conjunction with homosexuality, but this is frequently not the case. Some researchers believe that both cross-dressing and homosexuality have a common prenatal cause such as a critically timed hormonal release due to stress in the mother or the presence of certain hormone-mimicking chemicals during critical steps of fetal development. Researchers continue to examine this issue and debate the origins of cross-dressing and homosexuality.

Sexual Orientation

Sexual orientation **refers to an individual's preference for emotional–sexual relationships with members of the opposite sex (heterosexuality), the same sex (homosexuality), or both** (bisexuality) (Lips, 2001). Some scholars believe that sexual orientation is rooted in biological factors that are present at birth; others believe that sexuality has both biological and social components and is not preordained at birth.

In referring to homosexuality, many people prefer to use the acronym *LGBT*. What does this term mean? In recent years, organizations representing lesbian, gay, bisexual, and transgendered persons

© AP Photo/Jeff Chiu

▲ In May 2010, Chastity Bono legally became Chaz Bono—and officially male. The only daughter of Sonny and Cher, Chastity had long known that she was a lesbian. But after a 2009 sex-change operation, the new Chaz is a heterosexual male.

have adopted this acronym. (Some groups use the acronym *GLBT*.) The term *gay* refers to males who prefer same-sex relationships; *lesbian* refers to females who prefer same-sex relationships. *Bisexual* is the term used to describe a person's physical or romantic attraction to both males and females.

What criteria have social scientists used to study sexual orientation? In a definitive study of sexuality conducted in the mid-1990s, researchers at the University of Chicago established three criteria for identifying people as homosexual or bisexual: (1) *sexual attraction* to persons of one's own gender, (2) *sexual involvement* with one or more persons of one's own gender, and (3) *self-identification* as a gay, lesbian, or bisexual (Michael et al., 1994). According to these criteria, then, having engaged in a homosexual act does not necessarily classify a person as homosexual. In fact, many respondents in the University of Chicago study indicated that although they had at least one homosexual encounter when they were younger, they were no longer involved in homosexual conduct and never identified themselves as gay, lesbian, or bisexual.

Studies have examined how sexual orientation is linked to identity. Sociologist Kristin G. Esterberg (1997) interviewed lesbian and bisexual women to determine how they "perform" lesbian or bisexual identity through daily activities such as choice of clothing and hairstyles, as well as how they use body language and talk. According to Esterberg (1997), some of the women viewed themselves as being "lesbian from birth," whereas others had experienced shifts in their identities, depending on social surroundings, age, and political conditions at specific periods in their lives. Another study looked at gay and bisexual men. Human development scholar Ritch C. Savin-Williams (2004) found that gay/bisexual youths often believe from an early age that they are different from other boys:

> The pattern that most characterized the youths' awareness, interpretation, and affective responses to childhood attractions consisted of an overwhelming desire to be in the company of men. They wanted to touch, smell, see, and hear masculinity. This awareness originated from earliest childhood memories; in this sense, they "always felt gay."

However, most of the boys and young men realized that these feelings were not typical of other males and were uncomfortable when others attempted to make them conform to the established cultural definitions of masculinity, such as showing a great interest in team sports, competition, and aggressive pursuits.

Discrimination Based on Sexual Orientation

The United States has numerous forms of discrimination based on sexual orientation. In most states, gay and lesbian couples cannot enter into legally recognized marital relationships because the states have passed constitutional amendments that limit marriage to a union between a man and a woman or because legislators have passed statutes with similar language. As well, lesbian and gay couples continue to fight for parental rights in a number of states because partners who want to adopt a child or are raising children together (typically from a previous heterosexual marriage) learn that only one partner is legally recognized as the child's parent or guardian.

Sexual orientation is sometimes a concern when gay individuals seek medical treatment because of the

cross-dresser a male who dresses like a woman or a woman who dresses like a man but does not alter the genitalia.

sexual orientation a person's preference for emotional–sexual relationships with members of the opposite sex (heterosexuality), the same sex (homosexuality), or both (bisexuality).

long-term stigma that associates homosexuality with HIV/AIDS. Some health care providers refuse to treat individuals whom they believe might be at high risk for HIV/AIDS, and others do not address the social and psychological needs of patients who are HIV-positive.

Occupational discrimination remains a pressing problem for people in the LGBT community. Despite laws in many states prohibiting discrimination in employment on the basis of sexual orientation, openly gay and lesbian people often experience bias in hiring, retention, and promotion in public-sector and private-sector employment. The Equal Employment Opportunity Commission, a federal agency, has documented the fact that heterosexuals frequently harass homosexuals in the workplace. The agency has also handled many cases of same-sex sexual harassment.

One of the most widely publicized forms of discrimination against gays and lesbians has been in the military, where a 1993 "Don't Ask, Don't Tell" policy was implemented during the Clinton administration. Under this policy, commanders were not allowed to ask about a serviceperson's sexual orientation, and gay men and lesbians were allowed to serve in the military only as long as they did not reveal their sexual orientation. Various studies showed that this policy led to differential treatment of many gay men and lesbians in the military, as well as causing extensive recruitment problems for various branches of the military service. This policy remained in effect until President Barack Obama signed into law the Don't Ask, Don't Tell Repeal Act of 2010, which allowed gay and lesbian Americans to serve their country openly.

© AP Photo/Rogelio V. Solis

▲ In April 2010, Constance McMillen, a high school student in Aberdeen, Mississippi, wanted to attend the senior prom with another young woman as her date. The county school board refused. After a court blocked the school board's decision, McMillen and her friend attended the prom to find only five other students there. The "real" prom was apparently held in a secret location.

Various organizations of gays, lesbians, and transgendered persons have been unified in their desire to reduce discrimination and other forms of ***homophobia*—extreme prejudice and sometimes discriminatory actions directed at gays, lesbians, bisexuals, transgendered persons, and others who are perceived as not being heterosexual**. Homophobia involves an aversion to LGBT people or their lifestyle or culture, and it sometimes includes behavior or an act, such as a hate crime, based on this aversion. Some analysts use the term *heterosexism* to describe an ideological system that denies, denigrates, and stigmatizes any nonheterosexual form of behavior, identity, relationship, or community. This term is used as a parallel to other forms of prejudice and discrimination, including racism, sexism, ageism, and anti-Semitism. Clearly, from this perspective, issues pertaining to homosexuality and heterosexism are not just biological issues but also social constructions that involve societal customs and institutions. Let's turn to the cultural dimension of gender to see how socially constructed differences between females and males are crucial in determining how we identify ourselves as girls or boys, women or men.

Gender: The Cultural Dimension

***Gender* refers to the culturally and socially constructed differences between females and males found in the meanings, beliefs, and practices associated with "femininity" and "masculinity."** Although biological differences between women and men are very important, in reality most "sex differences" are socially constructed "gender differences." According to sociologists, social and cultural processes, not biological "givens," are most important in defining what females and males are, what they should do, and what sorts of relations do or should exist between them. Sociologist Judith Lorber (1994: 6) summarizes the importance of gender:

> Gender is a human invention, like language, kinship, religion, and technology; like them, gender

© Alex Farnsworth/The Image Works

© Mohammed Ameen/Reuters/Landov

▲ In what ways do these two figures contradict traditional gender expectations? Do you see this trend as a healthy one?

organizes human social life in culturally patterned ways. Gender organizes social relations in everyday life as well as in the major social structures, such as social class and the hierarchies of bureaucratic organizations.

Virtually everything social in our lives is *gendered:* People continually distinguish between males and females and evaluate them differentially (see this chapter's Photo Essay). Gender is an integral part of the daily experiences of both women and men (Kimmel and Messner, 2004).

A microlevel analysis of gender focuses on how individuals learn gender roles and acquire a gender identity. ***Gender role* refers to the attitudes, behavior, and activities that are socially defined as appropriate for each sex and are learned through the socialization process** (Lips, 2001). For example, in U.S. society males are traditionally expected to demonstrate aggressiveness and toughness, whereas females are expected to be passive and nurturing. ***Gender identity* is a person's perception of the self as female or male**. Typically established between eighteen months and three years of age, gender identity is a powerful aspect of our self-concept. Although this identity is an individual perception, it is developed through interaction with others. As a result, most people form a gender identity that matches their biological sex: Most biological females think of themselves as female, and most biological males think of themselves as male. Body consciousness is a part of gender identity. ***Body consciousness* is how a person perceives and feels about his or her body**; it also includes an awareness of social conditions in society that contribute to this self-knowledge (Thompson, 1994). Consider, for example, these comments by Steve Michalik, a former Mr. Universe:

> I was small and weak, and my brother Anthony was big and graceful, and my old man made no bones about loving him and hating me.... The

homophobia extreme prejudice and sometimes discriminatory actions directed at gays, lesbians, bisexuals, and others who are perceived as not being heterosexual.

gender the culturally and socially constructed differences between females and males found in the meanings, beliefs, and practices associated with "femininity" and "masculinity."

gender role the attitudes, behavior, and activities that are socially defined as appropriate for each sex and are learned through the socialization process.

gender identity a person's perception of the self as female or male.

body consciousness a term that describes how a person perceives and feels about his or her body.

photo essay How Do We "Do Gender" in the Twenty-First Century?

What distinctive ways of acting and feeling are characteristic of women? Of men? For centuries, people have used a male/female dichotomy to answer these questions and, in the process, have identified women's and men's behaviors as opposites in many respects: Men are supposed to be "real men" and meet the normative conception of *masculinity* by being aggressive, independent, and powerful, whereas women are supposed to demonstrate *femininity* by being passive, dependent, and weak.

However, many theorists using a symbolic interactionist perspective suggest that gender is something that we *do* rather than being a set of masculine or feminine traits that resides within the individual. Sociologists Candace West and Don H. Zimmerman (1987) coined the term *doing gender* to refer to the process by which we *socially* create differences between males and females that are not based on natural, essential, or biological factors but instead are based on the things we do in our social interactions. According to West and Zimmerman (1987), *accountability* is involved in the process of doing gender: We know that our actions will be evaluated by others based on how well they think we meet the normative conceptions of appropriate attitudes and activities that are expected of people in our sex category (the socially required displays that identify a person as being either male or female).

What is the primary difference in these two approaches? These viewpoints are based on different assumptions. The male/female dichotomy is based on the assumption that women and men have inherently different traits, whereas the concept of *doing gender* is based on the assumption that, through our interactions with others, we produce, reproduce, and sustain the social meanings that are accorded to gender in any specific society at any specific point in time (in other words, those meanings may change from time to time and from place to place). By focusing on gender as an *accomplishment* (rather than something that is previously established), symbolic interactionists emphasize that some people's resistance to existing gendered norms is probable and that social change is possible. Symbolic interactionist theories also make us aware that change is less likely to take place when people feel constrained to be accountable to others for their behavior as women or men (Fenstermaker and West, 2002).

When you look at the pictures on these three pages, think about how the people in each setting are "doing gender" in everyday life. Are they doing gender based on what they perceive to be the normative expectations of others? Or are they doing gender as they see fit? What do you think?

▲ GENDER AND APPEARANCE

How do we do gender on a shopping trip? Consider, for example, the difference shown in this photo. For the man, a causal outing means wearing jeans, sandals, and a partially unbuttoned shirt. For a woman, a causal outing means a "pulled together" outfit with heels, a sexy dress, her hair in place, and carefully positioned sunglasses.

© Ariel Skelley/Getty Images

◀ GENDER AND HOME LIFE

Although in recent years more men have assumed greater responsibility at home with regard to household tasks and child rearing, women are more likely to be seen as the primary caregivers of children. Does our society still hold women, more than men, accountable for the well-being of children?

▶ **GENDER AND CAREERS**

In recent decades more women have become doctors and lawyers than in the past. How has this affected the way that people do gender in settings that reflect their profession? Do professional women look and act more like their male colleagues, or have men changed their appearance and activities at work as a result of having female colleagues?

© Tim Pannell/Corbis

© Ed Bock/CORBIS

◀ **GENDER AND SOCIAL INTERACTION**

How are the men shown here doing gender after their game of pick-up basketball? Consider, for example, what they are wearing, how they are sitting, and how they are communicating with one another. If three women were in this photo instead of three men, in what ways would their attire, actions, and expressions be different?

© moodboard/Alamy

▲ GENDER AND SUCCESS
When you hear someone referred to a "wealthy entrepreneur," do you think of a man or a woman? Are the typical trappings of success—such as luxury cars and expensive private airplanes—more associated with how men or how women do gender? Do you believe that this situation will change in the future?

reflect & analyze

1. If you went to a medical clinic for the first time and was told that your physician was "Dr. Smith," would you automatically assume that this person was a man? If so, why?
2. Is your social behavior dependent on gender context? In other words, do you behave differently when around members of your own sex than you do in mixed situations? Come up with a specific example of a situation where you felt the need to adapt your behavior because of social and gender expectations.
3. Do you think that wide differences in how two people look when they go out on a date can affect the outcome of their encounter? Why or why not?

▶ turning to video

Watch the ABC video *Are Men Smarter? Sexism in the Classroom* (running time 4:33), available through CengageBrain.com. This special report focuses on the experience of leading neurobiologist Ben Barres, who became a female-to-male transsexual at the age of forty. Having been active in his field as both Barbara and Ben, Dr. Barres has gained unique perspective on the biased perceptions of men and women in science. As you watch the video, think about the photographs, commentary, and questions that you encountered in this photo essay. After you've watched the video, consider this question: How do Ben Barres's life experiences and professional accomplishments reflect West and Zimmerman's concept of "doing gender"?

© ABC

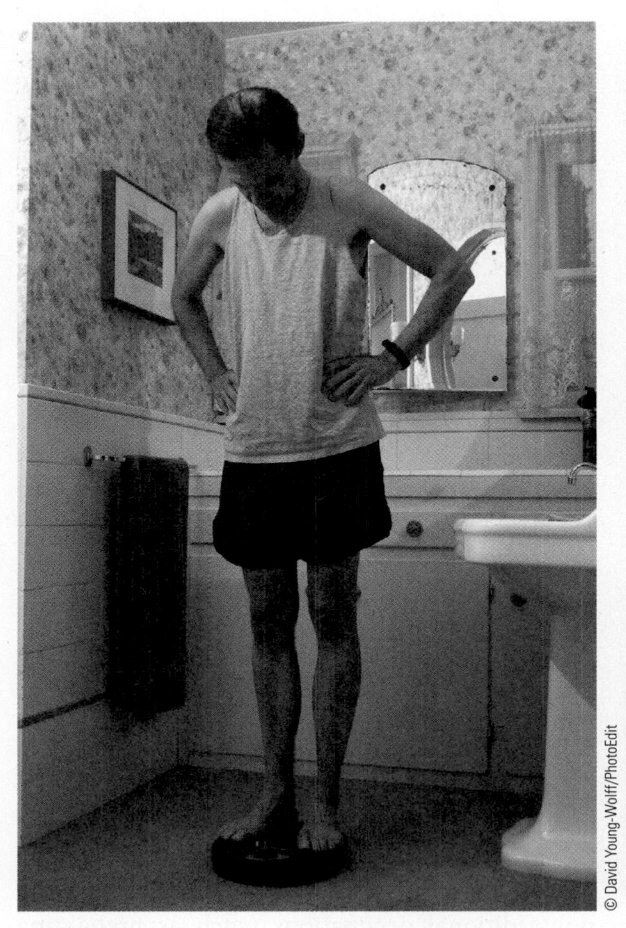
© David Young-Wolff/PhotoEdit

© A. Ramey/PhotoEdit

▲ Not all anorectics are women, and not all bodybuilders are men. However, Susan Bordo argues that these two issues are manifestations of the same desire: to avoid having soft, flabby flesh.

> minute I walked in from school, it was, "You worthless little s---, what are you doing home so early?" His favorite way to torture me was to tell me he was going to put me in a home. We'd be driving along in Brooklyn somewhere, and we'd pass a building with iron bars on the windows, and he'd stop the car and say to me, "Get out. This is the home we're putting you in." I'd be standing there sobbing on the curb—I was maybe eight or nine at the time. (qtd. in Klein, 1993: 273)

As we grow up, we become aware, as Michalik did, that the physical shape of our bodies subjects us to the approval or disapproval of others. Being small and weak may be considered positive attributes for women, but they are considered negative characteristics for "true men."

A macrolevel analysis of gender examines structural features, external to the individual, that perpetuate gender inequality. These structures have been referred to as *gendered institutions*, meaning that gender is one of the major ways by which social life is organized in all sectors of society. Gender is embedded in the images, ideas, and language of a society and is used as a means to divide up work, allocate resources, and distribute power. For example, every society uses gender to assign certain tasks—ranging from child rearing to warfare—to females and to males, and differentially rewards those who perform these duties.

These institutions are reinforced by a *gender belief system*, which includes all the ideas regarding masculine and feminine attributes that are held to be valid in a society. This belief system is legitimated by religion, science, law, and other societal values (Lorber, 2005). For example, gender belief systems may change over time as gender roles change. Many fathers take care of young children today, and there is a much greater acceptance of this change in roles. However, popular stereotypes about men and women, as well as cultural norms about gender-appropriate appearance and behavior, serve to reinforce gendered institutions in society.

The Social Significance of Gender

Gender is a social construction with important consequences in everyday life. Just as stereotypes regarding race/ethnicity have built-in notions of superiority and inferiority, gender stereotypes hold that men and women are inherently different in attributes, behavior, and aspirations. Stereotypes define men as strong, rational, dominant, independent, and less concerned with their appearance. Women are stereotyped as weak, emotional, nurturing, dependent, and anxious about their appearance.

The social significance of gender stereotypes is illustrated by eating problems. The three most common eating problems are anorexia, bulimia, and obesity. With *anorexia,* a person has lost at least 25 percent of body weight because of a compulsive fear of becoming fat (Lott, 1994). With *bulimia,* a person binges by consuming large quantities of food and then purges the food by induced vomiting, excessive exercise, taking laxatives, or fasting. With *obesity,* individuals are 20 percent or more above their desirable weight, as established by the medical profession. For a 5-foot-4-inch woman, that is about twenty-five pounds; for a 5-foot-10-inch man, it is about thirty pounds (Burros, 1994: 1).

Sociologist Becky W. Thompson argues that, based on stereotypes, the primary victims of eating problems are presumed to be white, middle-class, heterosexual women. However, such problems also exist among women of color, working-class women, lesbians, and some men. According to Thompson, explanations regarding the relationship between gender and eating problems must take into account a complex array of social factors, including gender socialization and women's responses to problems such as racism and emotional, physical, and sexual abuse (Thompson, 1994; see also Wooley, 1994).

Bodybuilding is another gendered experience. *Bodybuilding* is the process of deliberately cultivating an increase in the mass and strength of the skeletal muscles by means of lifting and pushing weights. In the past, bodybuilding was predominantly a male activity; musculature connoted power, domination, and virility. Today, however, an increasing number of women engage in this activity. As gendered experiences, eating problems and bodybuilding have more in common than we might think. Women's studies scholar Susan Bordo (2004) has noted that the anorexic body and the muscled body are not opposites, but instead are both united against the common enemy of soft, flabby flesh. In other words, the *body* may be objectified both through compulsive dieting and compulsive bodybuilding.

Sexism

***Sexism* is the subordination of one sex, usually female, based on the assumed superiority of the other sex.** Sexism directed at women has three components: (1) negative attitudes toward women; (2) stereotypical beliefs that reinforce, complement, or justify the prejudice; and (3) discrimination—acts that exclude, distance, or keep women separate (Lott, 1994).

Can men be victims of sexism? Although women are more often the target of sexist remarks and practices, men can be victims of sexist assumptions. Examples of sexism directed against men are the assumption that men should not be employed in certain female-dominated occupations, such as nurse or elementary school teacher, and the belief that it is somehow more harmful for families when female soldiers are killed in battle than male soldiers.

Like racism, sexism is used to justify discriminatory treatment. Obvious manifestations of sexism are found in the undervaluing of women's work and in hiring and promotion practices that effectively exclude women from an organization or confine them to the bottom of the organizational hierarchy. Even today, some women who enter nontraditional occupations (such as firefighting and welding) or professions (such as dentistry, architecture, or investment banking) encounter hurdles that men do not face (see "Sociology Works!").

Sexism is interwoven with ***patriarchy*—a hierarchical system of social organization in which men control cultural, political, and economic structures.** By contrast, ***matriarchy* is a hierarchical system of social organization in which women control cultural, political, and economic structures;** however, few (if any) societies have been organized in this manner. Patriarchy is reflected in the way that men may think of their position as men as a given, whereas women may deliberate on what their position in society should be (see "Sociology in Global Perspective" for an example). As the sociologist Virginia Cyrus (1993: 6) explains, "Under patriarchy, men are seen as 'natural' heads of households, Presidential candidates, corporate executives, college presidents, etc. Women, on the other hand, are men's subordinates, playing such supportive roles as housewife, mother, nurse, and secretary." Gender inequality and a division of labor based on male dominance are nearly universal, as we will see in the following discussion on the origins of gender-based stratification.

sociology works!

Institutional Discrimination: Women in a Locker-Room Culture

News Item: "Goldman Sachs Sued for Sexual Harassment." Three former employees filed a gender bias lawsuit against the firm on behalf of a number of women employees. The suit claims that Goldman pays women less than men, denies them promotions, gives them unfairly harsh feedback, blocks business opportunities, and refuses to adequately train them. (Alden, 2010)

News Item: "Bank of America and Merrill Lynch Sex Discrimination Lawsuit." Three female financial advisers filed a sexual discrimination suit against Bank of America and Merrill Lynch alleging the companies gave their male counterparts bigger bonuses and better opportunities. The women also claimed that they were punished when they complained about the inequities. (Egan, 2010)

For decades, sociologists have called attention to the fact that both individual discrimination and institutional discrimination—based on sex, race/ethnicity, and other devalued characteristics and attributes of subordinate-group members—are widespread. Previous sex-discrimination lawsuits typically involved details about the crude behavior of male employees toward women, or pornography in the workplace, or male bosses' demands that women employees accompany them to strip clubs and other objectionable locations as part of their work-related duties. However, cases such as Goldman Sachs and Bank of American/Merrill Lynch involve none of these issues and instead focus on women's opportunities for training, gaining new clients, promotion, and pay and bonus equity—all factors related to institutional discrimination, previously defined (in Chapter 10) as the day-to-day practices of organizations and institutions that have a harmful impact on members of subordinate groups. According to many sex-discrimination lawsuits, a "locker-room culture" or an "outdated corporate culture" prevails in many prestigious organizations where elite men make everyday decisions with little or no consideration for women who hold positions as managing directors, vice presidents, and associate employees. For example, the Goldman Sachs lawsuit alleges that women constitute only 14 percent of all partners, 17 percent of managing directors, and 29 percent of vice presidents at Goldman.

Sociological theorizing and research have increased public awareness that the day-to-day practices of organizations and institutions may have a negative and differential effect on individuals who have historically been excluded from workplace settings such as prestigious banks and Wall Street firms, which traditionally have been dominated by white males. Although many gains have been made through legislation and litigation to reduce institutional discrimination, recent lawsuits demonstrate that much remains to be done before women truly have equal opportunities in the workplace.

reflect & analyze

Why is it important for all employees to feel that they are being treated fairly at work? Are some employment settings more resistant to change than others? What do you think?

Gender Stratification in Historical and Contemporary Perspective

How do tasks in a society come to be defined as "men's work" or "women's work"? Three factors are important in determining the gendered division of labor in a society: (1) the type of subsistence base, (2) the supply of and demand for labor, and (3) the extent to which women's child-rearing activities are compatible with certain types of work. As defined in Chapter 5, *subsistence* refers to the means by which a society gains the basic necessities of life, including food, shelter, and clothing. You may recall that societies are classified based on subsistence, as hunting and gathering societies, horticultural and pastoral societies, agrarian societies, industrial societies, and postindustrial societies. The first three of these categories are all *preindustrial* societies in which gender stratification is different from the type found in industrial and postindustrial societies.

sexism the subordination of one sex, usually female, based on the assumed superiority of the other sex.

patriarchy a hierarchical system of social organization in which men control cultural, political, and economic structures.

matriarchy a hierarchical system of social organization in which women control cultural, political, and economic structures.

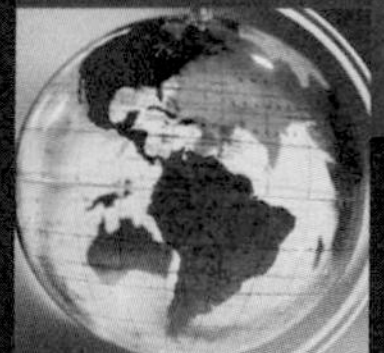

sociology in global perspective

The Rise of Islamic Feminism in the Middle East?

> I would like for all of the young Muslim girls to be able to relate to Iman, whether they wear the hijab [head scarf] or not. Boys will also enjoy Iman's adventures because she is one tough, smart girl! Iman gets her super powers from having very strong faith in Allah, or God. She solves many of the problems by explaining certain parts of the Koran that relate to the story.
>
> —Rima Khoreibi, an author from Dubai (United Arab Emirates), explaining that she has written a book about an Islamic superhero who is female because she would like to dispel a widely held belief that sexism in her culture is deeply rooted in Islam (see theadventuresof iman.com, 2007; Kristof, 2006)

Although Rima Khoreibi and many others who have written fictional and nonfictional accounts of girls and women living in the Middle East typically do not deny that sexism exists in their region or that sexism is deeply interwoven with patriarchy around the world, they dispute the perception that Islam is inherently misogynistic (possessing hatred or strong prejudice toward women). As defined in this chapter, *patriarchy* is a hierarchical system of social organization in which cultural, political, and economic structures are controlled by men. The influence of religion on patriarchy is a topic of great interest to contemporary scholars, particularly those applying a feminist approach to their explanations of why social inequalities persist between women and men and how these inequalities are greater in some regions of the world than in others.

According to some gender-studies specialists, a newer form of feminist thinking is emerging among Muslim women. Often referred to as "feminist Islam" or "Islamic feminism," this approach is based on the belief that greater gender equality may be possible in the Muslim world if the teachings of Islam, as set forth in the Koran (or Qur'an)—the Islamic holy book—are followed more closely. Islamic feminism is based on the principle that Muslim women should retain their allegiance to Islam as an essential part of their self-determination and identity but that they should also work to change patriarchal control over the basic Islamic worldview (Wadud, 2002). According to the journalist Nicholas D. Kristof (2006), both Islam and evangelical Christianity have been on the rise in recent years because both religions provide "a firm moral code, spiritual reassurance and orderliness to people vexed by chaos and immorality around them, and . . . dignity to the poor."

Islamic feminists believe that the rise of Islam might contribute to greater, rather than less, equality for women. From this perspective, stories about characters such as Iman may help girls and young women realize that they can maintain their deep religious convictions and their head scarf (*hijab*) while, at the same time, working for greater equality for women and more opportunities for themselves. In *The Adventures of Iman*, the female hero always wears a pink scarf around her neck, and she uses the scarf to cover her hair when she is praying to Allah. Iman quotes the Qur'an when she is explaining to others that Muslims are expected to be tolerant, kind, and righteous. For Iman, religion is a form of empowerment, not an extension of patriarchy.

Islamic feminism is quite different from what most people think of as Western feminism (particularly in regard to issues such as the wearing of the *hijab* or the fact that in Saudi Arabia, a woman may own a motor vehicle but may not legally drive it). However, change is clearly under way in many regions of the Middle East and other areas of the world as rapid economic development and urbanization quickly change the lives of many people.

reflect & analyze

Why is women's inequality a complex issue to study across nations? What part does culture play in defining the roles of women and men in various societies? How do religious beliefs influence what we perceive as "appropriate" or "inappropriate" behaviors for men, women, and children? What do you think?

Courtesy of Rima Khoreibi

The home page of *The Adventures of Iman*.

Preindustrial Societies

The earliest known division of labor between women and men is in hunting and gathering societies. While the men hunt for wild game, women gather roots and berries. A relatively equitable relationship exists because neither sex has the ability to provide all the food necessary for survival. When wild game is nearby, both men and women may hunt. When it is far away, hunting becomes incompatible with child rearing (which women tend to do because they breast-feed their young), and women are placed at a disadvantage in terms of contributing to the food supply (Lorber, 1994). In most hunting and gathering societies, women are full economic partners with men; relations between them tend to be cooperative and relatively egalitarian (Chafetz, 1984; Bonvillain, 2001). Little social stratification of any kind is found because people do not acquire a food surplus.

In horticultural societies, which first developed ten to twelve thousand years ago, a steady source of food becomes available. People are able to grow their own food because of hand tools, such as the digging stick and the hoe. Women make an important contribution to food production because hoe cultivation is compatible with child care. A fairly high degree of gender equality exists because neither sex controls the food supply.

When inadequate moisture in an area makes planting crops impossible, *pastoralism*—the domestication of large animals to provide food—develops. Men primarily do herding, and women contribute relatively little to subsistence production in such societies. In some herding societies, women have relatively low status; their primary value is their ability to produce male offspring so that the family lineage can be preserved and enough males will exist to protect the group against attack (Nielsen, 1990). Even so, the relationship between men and women is more equitable than it is in agrarian societies, which first developed about eight thousand to ten thousand years ago.

In agrarian societies, gender inequality and male dominance become institutionalized. Agrarian societies rely on agriculture—farming done by animal-drawn or mechanically powered plows and equipment. Because agrarian tasks require more labor and greater physical strength than horticultural ones, men become more involved in food production. It has been suggested that women are excluded from these tasks because they are viewed as too weak for the work and because child-care responsibilities are considered incompatible with the full-time labor that the tasks require (Nielsen, 1990). Most of the world's population currently lives in agrarian societies in various stages of industrialization.

Why does gender inequality increase in agrarian societies? Scholars cannot agree on an answer; however, some suggest that it results from private ownership of property. When people no longer have to move continually in search of food, they can acquire a surplus. Men gain control over the disposition of the surplus and the kinship system, and this control serves men's interests (Lorber, 1994). The importance of producing "legitimate" heirs to inherit the surplus increases significantly, and women's lives become more secluded and restricted as men attempt to ensure the legitimacy of their children. Premarital virginity and marital fidelity are required; indiscretions are punished (Nielsen, 1990). However, some scholars argue that male dominance existed before the private ownership of property (Firestone, 1970; Lerner, 1986).

Industrial Societies

An *industrial society* is one in which factory or mechanized production has replaced agriculture as the major form of economic activity. As societies industrialize, the status of women tends to decline further. Industrialization in the United States created a gap between the nonpaid work performed by women at home and the paid work that increasingly was performed by men and unmarried girls (Amott and Matthaei, 1996). When families needed extra money, their daughters worked in the textile mills until they married. In 1900, for example, 22 percent of single white women who had been born in the United States were in the paid labor force. Because factory work was not compatible with child-care responsibilities, only 3 percent of married women were so employed, and only when they were extremely poor (Amott and Matthaei, 1996). As it became more difficult to make a living by farming, many men found work in the factories, where their primary responsibility was often supervising the work of women and children. Men began to press for a clear division between "men's work" and "women's work," as well as corresponding pay differentials (higher for men, lower for women).

In the United States the division of labor between men and women in the middle and upper classes became much more distinct with industrialization. The men were responsible for being "breadwinners;" the women were seen as "homemakers." In this new "cult of domesticity" (also referred to as the "cult of true womanhood"), the home became a private, personal sphere in which women created a haven for the family. Those who supported the cult of domesticity argued that women were the natural keepers of the domestic sphere and that children were the mother's responsibility. Meanwhile, the "breadwinner" role placed enormous pressures on men to support their families—being a good provider was considered to be a sign of manhood (Amott and

▲ In contemporary societies, women do a wide variety of work and are responsible for many diverse tasks. The women shown here are employed in the agricultural and the postindustrial sectors of the U.S. economy. How might issues of gender inequality differ for these two women? What issues might be the same for both of them?

Matthaei, 1996). However, this gendered division of labor increased the economic and political subordination of women. As a result, many white women focused their efforts on acquiring a husband who was capable of bringing home a good wage. Single women and widows and their children tended to live a bleak existence, crowded into run-down areas of cities, where they were often unable to support themselves on their meager wages.

The cult of true womanhood not only increased white women's dependence on men but also became a source of discrimination against women of color, based on both their race and the fact that many of them had to work in order to survive. Employed, working-class white women were similarly stereotyped at the same time that they became more economically dependent on their husbands because their wages were so much lower.

As people moved from a rural, agricultural lifestyle to an urban existence, body consciousness increased. People who worked in offices often became sedentary and exhibited physical deterioration from their lack of activity. As gymnasiums were built to fight this lack of physical fitness, a new image of masculinity developed. Whereas the "burly farmer" or "robust workman" had previously been the idealized image of masculinity, now the middle-class man who exercised and lifted weights came to embody this ideal (Klein, 1993).

In the late nineteenth century, middle-class women started to become preoccupied with body fitness. As industrialization progressed and food became more plentiful, the social symbolism of body weight and size changed. Previously, it had been considered a sign of high status to be somewhat overweight, but now a slender body reflected an enhanced social status. To the status-seeking middle-class man, a slender wife became a symbol of the husband's success. At the same time, excess body weight was seen as a reflection of moral or personal inadequacy, or a lack of willpower (Bordo, 2004).

Postindustrial Societies

Chapter 5 defines *postindustrial societies* as ones in which technology supports a service- and information-based economy. In such societies the division of labor in paid employment is increasingly based on whether people provide or apply information or are employed in service jobs such as fast-food restaurant counter help or health care workers. For both women and men in the labor force, formal education is increasingly crucial for economic and social success. However, as some women have moved into entrepreneurial, managerial, and professional occupations, many others have remained in the low-paying service sector, which affords few opportunities for upward advancement.

Will technology change the gendered division of labor in postindustrial societies? Scholars do not agree on the effects of computers, the Internet, cellular phones, and many newer forms of communications technology on the role of women in society. For example, some feminist writers had a pessimistic view of the impact of computers and monitors on women's health and safety, predicting that women in secretarial and administrative roles would experience an increase in eyestrain, headaches, and problems such as carpal tunnel syndrome. However, some medical experts now believe that such problems extend to both men and women, as computers have become omnipresent in more people's lives. The term "24/7" has come to mean that a person is available "twenty-four hours a day, seven days a week" via cell phones, pocket pagers, fax machines, e-mail, and other means of communication, whether the individual is at the office or four thousand miles away on "vacation."

How do new technologies influence gender relations in the workplace? Although some analysts presumed that technological developments would reduce the boundaries between women's and men's work, researchers have found that the gender stereotyping associated with specific jobs has remained remarkably stable even when the nature of work and the skills required to perform it have been radically transformed. Today, men and women continue to be segregated into different occupations, and this segregation is particularly visible within individual workplaces (as discussed later in the chapter).

How does the division of labor change in families in postindustrial societies? For a variety of reasons, women head more households with no adult male present. As shown in the Census Profiles feature, the percentage of U.S. households headed by a single mother with children under eighteen has increased. Chapter 15 ("Families and Intimate Relationships") discusses a number of reasons that the current division of labor in household chores in some families is between a woman and her children rather than between women and men. Consider, for example, that almost one-fourth (23 percent) of all U.S. children live with their mother only (as contrasted with just 5 percent who reside with their father only); among African American children, 48 percent live with their mother only (U.S. Census Bureau, 2007). This means that women in these households truly have a double burden, both from family responsibilities and from the necessity of holding gainful employment in the labor force.

Single Mothers with Children Under 18

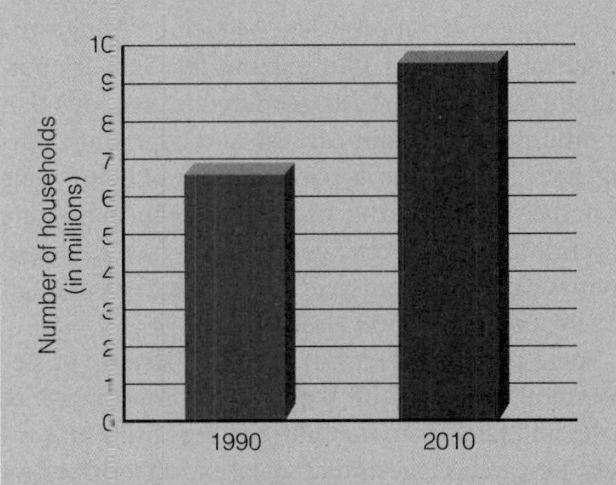

As shown above, the number of U.S. families headed by single mothers increased from almost 6,600,000 in 1990 to an estimated 9,924,000 in 2010. (For census purposes, a single mother is identified as a woman who is widowed, divorced, separated, or never married and who has children under 18 living at home.) In contrast, it is estimated that only 1,762,000 single male householders had children under age 18 living at home. In your opinion, do these figures have implications for gender roles and family relationships in the future?

Source: U.S. Census Bureau, 2010b.

Even in single-person or two-parent households, programming "labor-saving" devices (if they can be afforded) often means that a person must have some leisure time to learn how to do the programming. According to analysts, leisure is deeply divided along gender lines, and women have less time to "play in the house" than do men and boys. Some websites seek to appeal to women who have economic resources but are short on time, making it possible for them to shop, gather information, "telebank," and communicate with others at all hours of the day and night.

In postindustrial societies such as the United States, more than 60 percent of adult women are in the labor force, meaning that finding time to care for children, help aging parents, and meet the demands of the workplace will continue to place a heavy burden on women, despite living in an information- and service-oriented economy.

How people accept new technologies and the effect these technologies have on gender stratification are related to how people are socialized into gender roles. However, gender-based stratification remains rooted in the larger social structures of society, which individuals have little ability to control.

Gender and Socialization

We learn gender-appropriate behavior through the socialization process. Our parents, teachers, friends, and the media all serve as gendered institutions that communicate to us our earliest, and often most lasting, beliefs about the social meanings of being male or female and thinking and behaving in masculine or feminine ways. Some gender roles have changed dramatically in recent years; others have remained largely unchanged over time.

Some parents prefer boys to girls because of stereotypical ideas about the relative importance of males and females to the future of the family and society. Research suggests that social expectations play a major role in this preference. We are socialized to believe that it is important to have a son, especially for a first or only child. For many years it was assumed that only a male child could support his parents in their later years and carry on the family name.

Across cultures, boys are preferred to girls, especially when law or economic conditions limit the number of children that parents can have. For example, in China, which strictly regulates the allowable number of children to one per family, a disproportionate number of female fetuses are aborted, resulting in a shortage of women. However, in the aftermath of the catastrophic earthquakes of early 2008, the government is revising this policy to some degree (see Jacobs, 2008).

Parents and Gender Socialization

From birth, parents act toward children on the basis of the child's sex. Baby boys are perceived to be less fragile than girls and tend to be treated more roughly by their parents. Girl babies are thought to be "cute, sweet, and cuddly" and receive more gentle treatment. Parents strongly influence the gender-role development of children by passing on—both overtly and covertly—their own beliefs about gender. When girl babies cry, parents respond to them more quickly, and parents are more prone to talk and sing to girl babies (Wharton, 2004).

Children's toys reflect their parents' gender expectations (Thorne, 1993). Gender-appropriate toys for boys include computer games, trucks and other vehicles, sports equipment, and war toys such as guns and soldiers. Girls' toys include "Barbie" dolls, play makeup, and homemaking items. Parents' choices of toys for their children are not likely to change in the near future. A group of college students in one study were shown slides of toys and asked to decide which ones they would buy for girls and boys. Most said they would buy guns, soldiers, jeeps, carpenter tools, and red bicycles for boys; girls would get baby

◀ Are children's toys a reflection of their own preferences and choices? How do toys reflect gender socialization by parents and other adults?

dolls, dishes, sewing kits, jewelry boxes, and pink bicycles (Fisher-Thompson, 1990).

When children are old enough to help with household chores, they are often assigned different tasks. Maintenance chores (such as mowing the lawn) are assigned to boys, whereas domestic chores (such as shopping, cooking, and clearing the table) are assigned to girls. Chores may also become linked with future occupational choices and personal characteristics. Girls who are responsible for domestic chores such as caring for younger brothers and sisters may learn nurturing behaviors that later translate into employment as a nurse or schoolteacher. Boys may learn about computers and other types of technology that lead to different career options.

In the past, most studies of gender socialization focused on white, middle-class families and paid little attention to ethnic differences (Raffaelli and Ontai, 2004). According to earlier studies, children from middle- and upper-income families are less likely to be assigned gender-linked chores than children from lower-income backgrounds. In addition, gender-linked chore assignments occur less frequently in African American families, where both sons and daughters tend to be socialized toward independence, employment, and child care (Bardwell, Cochran, and Walker, 1986; Hale-Benson, 1986). Sociologist Patricia Hill Collins (1991) suggests that African American mothers are less likely to socialize their daughters into roles as subordinates; instead, they are likely to teach them a critical posture that allows them to cope with contradictions.

© Barbara Campbell

▲ Parents, peers, and the larger society all influence our perceptions about gender-appropriate behavior.

In contrast, a recent study of gender socialization in U.S. Latino/a families suggests that adolescent females of Mexican, Puerto Rican, Cuban, or other Central or South American descent receive different gender socialization by their parents than do their male siblings (Raffaelli and Ontai, 2004). Latinas are given more stringent curfews and are allowed less interaction with members of the opposite sex than are the adolescent males in their families. Rules for dating, school activities, and part-time jobs are more stringent for the girls because many parents want to protect their daughters and keep them closer to home.

Across classes and racial/ethnic categories, mothers typically play a stronger role in gender socialization of daughters, whereas fathers do more to socialize sons than daughters (McHale, Crouter, and Tucker, 1999). However, many parents are aware of the effect that gender socialization has on their children and make a conscientious effort to provide gender-neutral experiences for them.

Peers and Gender Socialization

Peers help children learn prevailing gender-role stereotypes, as well as gender-appropriate and gender-inappropriate behavior (Hibbard and Buhrmester, 1998). During the preschool years, same-sex peers have a powerful effect on how children see their gender roles (Maccoby and Jacklin, 1987); children are more socially acceptable to their peers when they conform to implicit societal norms governing the "appropriate" ways that girls and boys should act in social situations and what prohibitions exist in such cases (Martin, 1989).

Male peer groups place more pressure on boys to do "masculine" things than female peer groups place on girls to do "feminine" things. For example, girls wear jeans and other "boy" clothes, play soccer and softball, and engage in other activities traditionally associated with males. By contrast, if a boy wears a dress, plays hopscotch with girls, and engages in other activities associated with being female, his peers will ridicule him. This distinction between the relative value of boys' and girls' behaviors strengthens the cultural message that masculine activities and behavior are more important and more acceptable (Wood, 1999).

During adolescence, peers are often stronger and more effective agents of gender socialization than adults are (Hibbard and Buhrmester, 1998). Peers are thought to be especially important in boys' development of gender identity. Male bonding that occurs during adolescence is believed to reinforce masculine identity (Gaylin, 1992) and to encourage gender-stereotypical attitudes and behavior (Huston, 1985; Martin, 1989). For example, male peers have a tendency to ridicule and bully others about their appearance, size, and weight. Aleta Walker painfully recalls walking down the halls at school when boys would flatten themselves against the lockers and cry, "Wide load!" At lunchtime, the boys made a production of watching her eat lunch and frequently made sound like grunts or moos (Kolata, 1993). Because peer acceptance is so important for both males and females during their first two decades, such actions can have very harmful consequences for the victims.

As young adults, men and women still receive many gender-related messages from peers. Among college students, for example, peer groups are organized largely around gender relations and play an important role in career choices and the establishment of long-term, intimate relationships. In a study of women college students at two universities (one primarily white, the other predominantly African American), anthropologists Dorothy C. Holland and Margaret A. Eisenhart (1990) found that the peer system propelled women into a world of romance in which their attractiveness to men counted most. Although peers initially did not influence the women's choices of majors and careers, they did influence whether the women continued to pursue their original goals, changed their course of action, or were "derailed."

Teachers, Schools, and Gender Socialization

From kindergarten through college, schools operate as a gendered institution. Teachers provide important messages about gender through both the formal content of classroom assignments and informal interactions with students. Sometimes, gender-related messages from teachers and other students reinforce gender roles that have been taught at home; however, teachers may also contradict parental socialization. During the early years of a child's schooling, teachers' influence is very powerful; many children spend more hours per day with their teachers than they do with their own parents.

According to some researchers, the quantity and quality of teacher–student interactions often vary between the education of girls and that of boys (Wellhousen and Yin, 1997). One of the messages that teachers may communicate to students is that boys are more important than girls. Research spanning the past thirty years shows that unintentional gender bias occurs in virtually all educational settings. ***Gender bias* consists of showing favoritism toward one gender over the other.** Researchers consistently find that teachers devote more time, effort, and attention to boys than to girls (Sadker and Sadker, 1994). Males receive more praise for their contributions and are called on more frequently in class, even when they do not volunteer.

Teacher–student interactions influence not only students' learning but also their self-esteem (Sadker and Sadker, 1985, 1986, 1994). A comprehensive study of gender bias in schools suggested that girls' self-esteem is undermined in school through such experiences as (1) a relative lack of attention from teachers; (2) sexual harassment by male peers; (3) the stereotyping and invisibility of females in textbooks, especially in science and math texts; and (4) test bias based on assumptions about the relative importance of quantitative and visual–spatial ability, as compared with verbal ability, which restricts some girls' chances of being admitted to the most prestigious colleges and being awarded scholarships.

Teachers also influence how students treat one another during school hours. Many teachers use sex segregation as a way to organize students, resulting in unnecessary competition between females and males. In

◀ Teachers often use competition between boys and girls because they hope to make a learning activity more interesting. Here, a middle-school girl leads other girls against boys in a Spanish translation contest. What are the advantages and disadvantages of gender-based competition in classroom settings?

addition, teachers may take a "boys will be boys" attitude when girls complain of sexual harassment. Even though law prohibits sexual harassment, and teachers and administrators are obligated to investigate such incidents, the complaints may be dealt with superficially. If that happens, the school setting can become a hostile environment rather than a site for learning.

Sports and Gender Socialization

Children spend more than half of their nonschool time in play and games, but the type of games played differs with the child's sex. Studies indicate that boys are socialized to participate in highly competitive, rule-oriented games with a larger number of participants than games played by girls. Girls have been socialized to play exclusively with others of their own age, in groups of two or three, in activities such as hopscotch and jump rope that involve a minimum of competitiveness. Other research shows that boys express much more favorable attitudes toward physical exertion and exercise than girls do. Some analysts believe this difference in attitude is linked to ideas about what is gender-appropriate behavior for boys and girls (Brustad, 1996). For males, competitive sports becomes a means of "constructing a masculine identity, a legitimated outlet for violence and aggression, and an avenue for upward mobility" (Lorber, 1994: 43). Recently, more girls have started to play soccer and softball and to participate in sports formerly regarded as exclusively "male" activities. Girls who go against the grain and participate in masculine play as children are more likely to participate in sports as young women and adults (Giuliano, Popp, and Knight, 2000; Greendorfer, 1993).

Jamie Sabau/Stringer/Getty Images

▲ In recent years, women have expanded their involvement in college and professional sports; however, most sports remain rigidly divided into female events and male events. Do you think that media coverage of women's and men's college and professional sporting events differs?

Many women athletes believe that they have to manage the contradictory statuses of being both "women" and "athletes." One study found that women college basketball players dealt with this contradiction by dividing their lives into segments. On the basketball court, the women "did athlete": They pushed, shoved, fouled, ran hard, sweated, and cursed. Off the court, they "did woman": After the game, they showered, dressed, applied makeup, and styled their hair, even if they were only getting in a van for a long ride home (Watson, 1987).

Most sports are rigidly divided into female and male events. Assumptions about male and female physiology and athletic capabilities influence the types of sports in which members of each sex are encouraged to participate. For example, women who engage in activities that are assumed to be "masculine" (such as bodybuilding) may either ignore their critics or attempt to redefine the activity or its result as "feminine" or womanly (Klein, 1993).

Mass Media and Gender Socialization

The media, including newspapers, magazines, television, and movies, are powerful sources of gender stereotyping. Although some critics argue that the media simply reflect existing gender roles in society, others point out that the media have a unique ability to shape ideas. Think of the impact that television might have on children if they spend one-third of their waking time watching it, as has been estimated. From children's cartoons to adult shows, television programs are sex-typed, and many are male oriented. More male than female roles are shown, and male characters act strikingly different from female ones. Typically, males are more aggressive, constructive, and direct, and are rewarded for their actions. By contrast, females are depicted as acting deferential toward other people or as manipulating them through helplessness or seductiveness to get their way.

In prime-time television, a number of significant changes in the past three decades have reduced gender stereotyping; however, men still outnumber women as leading characters, and they are often "in charge" in any setting where both men's and women's roles are portrayed. In the popular ABC series *Grey's Anatomy,* for example, the number of women's and men's roles is evenly balanced, but the male

gender bias behavior that shows favoritism toward one gender over the other.

framing gender in the media

Celebrity Weight Battles Versus "Thinspiration" in Media Representations

When Kirstie Alley recently stepped onto the scale for the first time in 15 months, she screamed. When Oprah Winfrey realized she was the "dreaded 2-0-0," she wrote . . . "I'm mad at myself. I'm embarrassed. In the last year, so many celebrities have shared their body battles with us. . . . Up, down. Up down. Up."

—*New York Times* columnist Jan Hoffman (2009) discussing how media audiences "binge" on stories about celebrity weight battles

When you have famous people turning their weight tribulations into mass-media extravaganzas, they're contributing to a culture where passing comments on strangers' bodies is considered O.K.

—Lesley Kinzel, who writes the blog *Fatshionista* and weighs about 300 pounds, pointing out how shows such as *The Biggest Loser* and *Bulging Brides* encourage people to see that it is acceptable to make fun of people and to see other individuals' attempts at weight loss as TV entertainment (qtd. in Hoffman, 2009)

More and more celebrities are losing weight very quickly. Kate Bosworth, Ellen Pompeo, Keira Knightly. The list goes on. Stars are just getting smaller and smaller.

—*US Weekly* executive editor Caroline Schaefer (qtd. in ABCNews.com, 2006) explaining why celebrity-oriented magazines such as hers focus on shrinking celebrities and extreme diets

It matters not what female celebrity you are talking about or which year it is; media framing of stories about today's female Hollywood stars and other women celebrities remains the same: "How do they look?" "Are they overweight?" "Underweight?" "Anorexic?" Typically, celebrities who maintain an average body weight receive little attention about this issue, but those who are judged as overweight or underweight receive unprecedented coverage about their

characters are typically the top surgeons at the hospital, whereas the female characters are residents, interns, or nurses. In shows with predominantly female characters, such as ABC's *Desperate Housewives,* the women are typically very attractive, thin, and ultimately either hysterical or compliant when dealing with male characters (Stanley, 2004).

Advertising—whether on television and billboards or in magazines and newspapers—can be very persuasive. The intended message is clear to many people: If they embrace traditional notions of masculinity and femininity, their personal and social success is assured; if they purchase the right products and services, they can enhance their appearance and gain power over other people. In commercials, men's roles are typically portrayed differently from women's roles: Men are more likely to be shown working or playing outside the house rather than inside, whereas women are more likely to be doing domestic tasks such as cooking, cleaning, shopping, or taking care of the children. As such, television commercials may act as agents of socialization, showing children and others what women's and men's designated activities are (Kaufman, 1999).

A study by the sociologist Anthony J. Cortese (2004) found that women—regardless of what they were doing in a particular ad—were frequently shown in advertising as being young, beautiful, and seductive. Although such depictions may sell products, they may also have the effect of influencing how we perceive others and ourselves with regard to issues of power and subordination.

Studies have found that mass media have a particularly strong influence on how people of all ages think about weight and body image. However, young girls and women are especially affected by the cult of thinness, as well as the media representations of celebrity weight gain or loss that are so often shown on television and sensationalized on websites (see "Framing Gender in the Media").

▲ Cosmetic surgery can be dangerous. Dr. Donda West, mother of rapper Kanye West, died in 2007 after an elective procedure.

physical appearance. Negative media framing of stories about women's weight is further encouraged by celebrities who willingly make derogatory remarks about their appearance: "Hideous!" (Kirstie Alley) or "Embarrassing" (Oprah Winfrey). By contrast, journalists who refer to female celebrities who are thin or underweight as "dangerously skinny or anorexic" put them on the defensive.

If the media are setting the standards for beauty and appropriate weight, does this harm anyone? Yes, it may have a negative influence on younger and older persons alike! Research has shown that 80 percent of ten-year-olds fear that they are fat, and many of these perceptions are based on media representations. More than half of teenage girls and nearly one-third of teenage boys engage in unhealthy weight-control behaviors, including skipping meals, fasting, smoking cigarettes, vomiting, and taking laxatives, some of which they also learned from media coverage (ABCnews.com, 2006). Moreover, "fat bashing" and "thinspiration" in the media may affect people of all ages. In the words of one journalist, media accounts of celebrity weight battles and their "mortification-of-the-flesh" narratives may be "toxic, undermining [heavy women's] hard-won self-esteem and exacerbating the derision they face" (Hoffman, 2009). Similarly, media framing of stories about "thinspiration" and "rapidly shrinking" stars may influence people to believe that you can never be too thin (or too rich). To counter this perception, celebrities such as Kate Bosworth and Kate Winslet have gone public with their concern that the wrong idea is being conveyed to young girls about eating and weight. These stars emphasize that they do not have an eating disorder but instead spend large amounts of time exercising and eating nutritious food. Some even go so far as to appear on TV talk shows describing how they eat calorie-laden foods such as macaroni and cheese and have lavish desserts while on vacation. Even this "I'm just like everyone" approach may discourage overweight women, who wonder, "Why can't I eat like that without gaining ten pounds?"

reflect & analyze

Are media stories about celebrity weight gain or loss just entertainment for audiences? Should we be concerned that media representations about weight and appearance may influence how people, particularly girls and women, think about eating and weight? What do you think?

Adult Gender Socialization

Gender socialization continues as women and men complete their training or education and join the workforce. Men and women are taught the "appropriate" type of conduct for persons of their sex in a particular job or occupation—both by their employers and by coworkers. However, men's socialization usually does not include a measure of whether their work can be successfully combined with having a family; it is often assumed that men can and will do both. Even today, the reason given for women not entering some careers and professions is that this kind of work is not suitable for women because of their physical capabilities or assumed child-care responsibilities.

Different gender socialization may occur as people reach their forties and enter "middle age." A double standard of aging exists that affects women more than men. Often, men are considered to be at the height of their success as their hair turns gray and their face gains a few wrinkles. By contrast, not only do other people in society make middle-aged women feel as if they are "over the hill," but multimillion-dollar advertising campaigns also continually call attention to women's every weakness, every pound gained, and every bit of flabby flesh, wrinkle, or gray hair. Increasingly, both women and men have turned to "miracle" products, and sometimes to cosmetic surgery, to reduce the visible signs of aging.

Knowledge of how we develop a gender-related self-concept and learn to feel, think, and act in feminine or masculine ways is important for an understanding of ourselves. Examining gender socialization makes us aware of the impact of our

▲ Singer and actor Sheryl Crow displays her highly toned body on the cover of a fitness and image magazine. Can such an emphasis on sculpted bodies be harmful to women who are not celebrities?

parents, siblings, teachers, friends, and the media on our perspectives about gender. However, the gender socialization perspective has been criticized on several accounts. Childhood gender-role socialization may not affect people as much as some analysts have suggested. For example, the types of jobs that people take as adults may have less to do with how they were socialized in childhood than with how they are treated in the workplace. From this perspective, women and men will act in ways that bring them the most rewards and produce the fewest punishments. Also, gender socialization theories can be used to blame women for their own subordination by not taking into account structural barriers that perpetuate gender inequality. We will now examine a few of those structural forces.

Contemporary Gender Inequality

According to feminist scholars, women experience gender inequality as a result of past and present economic, political, and educational discrimination. Women's position in the U.S. workforce reflects the years of subordination that they have experienced in society.

Gendered Division of Paid Work in the United States

Where people are located in the occupational structure of the labor market has a major impact on their earnings. The workplace is another example of a gendered institution. In industrialized countries, most jobs are segregated by gender and by race/ethnicity. Sociologist Judith Lorber (1994: 194) gives this example:

> In a workplace in New York City—for instance, a handbag factory—a walk through the various departments might reveal that the owners and managers are white men; their secretaries and bookkeepers are white and Asian women; the order takers and data processors are African American women; the factory hands are [Latinos] cutting pieces and [Latinas] sewing them together; African American men are packing and loading the finished product; and non-English-speaking Eastern European women are cleaning up after everyone. The workplace as a whole seems integrated by race, ethnic group, and gender, but the individual jobs are markedly segregated according to social characteristics.

Lorber notes that in most workplaces, employees are either gender segregated or all of the same gender. *Gender-segregated work* refers to the concentration of women and men in different occupations, jobs, and places of work. In 2010, for example, 97 percent of all secretaries in the United States were women; 87 percent of all engineers were men (U.S. Census Bureau, 2010b). To eliminate gender-segregated jobs in the United States, more than half of all men or all women workers would have to change occupations. Moreover, women are severely underrepresented at the top of U.S. corporations. Women hold less than 20 percent of the executive jobs at Fortune 500 companies, and only 15 women are the CEOs of such a company. Today, women also hold slightly more than 15 percent of all seats on the boards of

© Will Hart/PhotoEdit

© Spencer Grant/PhotoEdit

▲ What stereotypes are associated with men in female-oriented positions? What about with women in male-oriented occupations? Do you think that such stereotypes will change in the near future?

Fortune 500 companies, and this percentage has remained about the same since 2003.

Although the degree of gender segregation in the professional labor market (including physicians, dentists, lawyers, accountants, and managers) has declined since the 1970s, racial–ethnic segregation has remained deeply embedded in the social structure. As the sociologist Elizabeth Higginbotham (1994) points out, for many years African American professional women found themselves limited to employment in certain sectors of the labor market. Although some change has occurred in recent years, women of color are more likely than their white counterparts to be concentrated in public-sector employment (as public schoolteachers, welfare workers, librarians, public defenders, and faculty members at public colleges, for example) rather than in the private sector (for example, in large corporations, major law firms, and private educational institutions). Across all categories of occupations, white women and all people of color are not evenly represented, as shown in ■ Table 11.2.

Labor market segmentation—the division of jobs into categories with distinct working conditions—results in women having separate and unequal jobs (Amott and Matthaei, 1996; Lorber, 2005). The pay gap between men and women is the best-documented consequence of gender-segregated work. Most women work in lower-paying, less-prestigious jobs, with little opportunity for advancement. Because many employers assume that men are the breadwinners, men are expected to make more money than women in order to support their families. For many years, women have been viewed as supplemental wage earners in a male-headed household, regardless of the women's marital status. Consequently, women have not been seen as legitimate workers but mainly as wives and mothers (Lorber, 2005). Such thinking has especially harmful consequences for the many women who are the only breadwinner in their family because of divorce, widowhood, unmarried status, or other reasons.

Gender-segregated work affects both men and women. Men are often kept out of certain types of jobs. Those who enter female-dominated occupations often have to justify themselves and prove that they are "real men." They have to fight stereotypes (gay, "wimpy," and passive) about why they are interested in such work (Williams, 2004). Even if these assumptions do not push men out of female-dominated occupations, they affect how the men manage their gender identity at work. For example, men in occupations such as nursing tend to emphasize their masculinity, attempt to distance themselves from female colleagues, and try to move quickly into management and supervisory positions (Williams, 2004).

Occupational gender segregation contributes to stratification in society. Job segregation is structural; it does not occur simply because individual workers have different abilities, motivations, and material needs. As a result of gender and racial segregation, employers are able to pay many men of color and

table 11.2

Percentage of the Workforce Represented by Women, African Americans, Hispanics, and Asian Americans in Selected Occupations

The U.S. Census Bureau accumulates data that show what percentage of the total workforce is made up of women, African Americans, and Hispanics. As used in this table, *women* refers to females in all racial–ethnic categories, whereas *African American, Hispanic,* and *Asian Americans* refer to both women and men.

	Women	African Americans	Hispanics	Asian Americans
All occupations	47.3	10.7	14.0	4.7
Managerial, professional, and related occupations	51.4	8.4	7.3	6.2
Management occupations	37.4	6.2	7.5	4.4
Professional and related occupations	52.5	9.4	7.1	7.1
Sales and office occupations	63.0	11.2	12.4	4.2
Sales occupations	49.6	9.6	11.9	4.8
Office and administrative support	74.5	12.7	12.7	3.6
Service occupations (all)	57.2	15.4	20.6	4.6
Food preparation and serving	55.7	11.4	21.6	3.6
Cleaning and building service workers	40.6	13.8	34.1	3.6
Health care service support occupations	89.4	25.3	13.8	3.9
Grounds maintenance workers	5.3	7.0	39.6	1.5

Source: U.S. Census Bureau, 2010b.

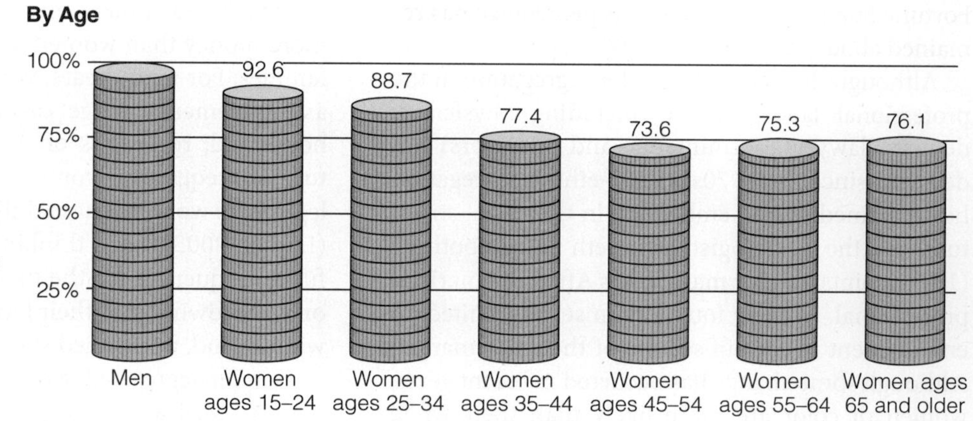

▶ **FIGURE 11.1 THE WAGE GAP**

Although men's average wages vary depending upon age, women's average wages are always lower than those of men in the same age group, and the older women get, the greater the gap.

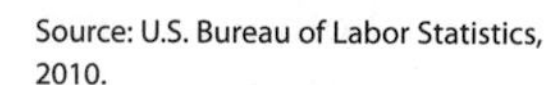

Source: U.S. Bureau of Labor Statistics, 2010.

all women less money, promote them less often, and provide fewer benefits.

Pay Equity (Comparable Worth)

Occupational segregation contributes to a ***pay gap*—the disparity between women's and men's earnings**. It is calculated by dividing women's earnings by men's to yield a percentage, also known as the earnings ratio. As ▶ Figure 11.1 shows, women in all age categories receive less pay than men. Overall, women who were full-time wage and salary workers earned about 79 cents for every dollar earned by men in 2009. The recession beginning in 2008 affected the wage gap, as 4.5 million men and 1.3 million women have been forced into part-time or part-year employment. The financial crisis has dragged down the median annual earnings of all men by 4.1 percent since 2007 and dropped women's annual earnings by 2.8 percent over the same time period. Earnings differences between women and men were widest for white Americans and Asian Americans. White women's earnings were 79 percent of their white male counterparts in 2009, while Asian American women earned 82 percent as much. By comparison, Hispanic women (Latinas) earned about 90 percent of their Hispanic male counterparts, and African American women earned 94 percent as much as African American men (U.S. Bureau of Labor Statistics, 2010). In each of the 50 states and the District of Columbia, women's median earnings were less than men's median earnings (see ▶ Map 11.1).

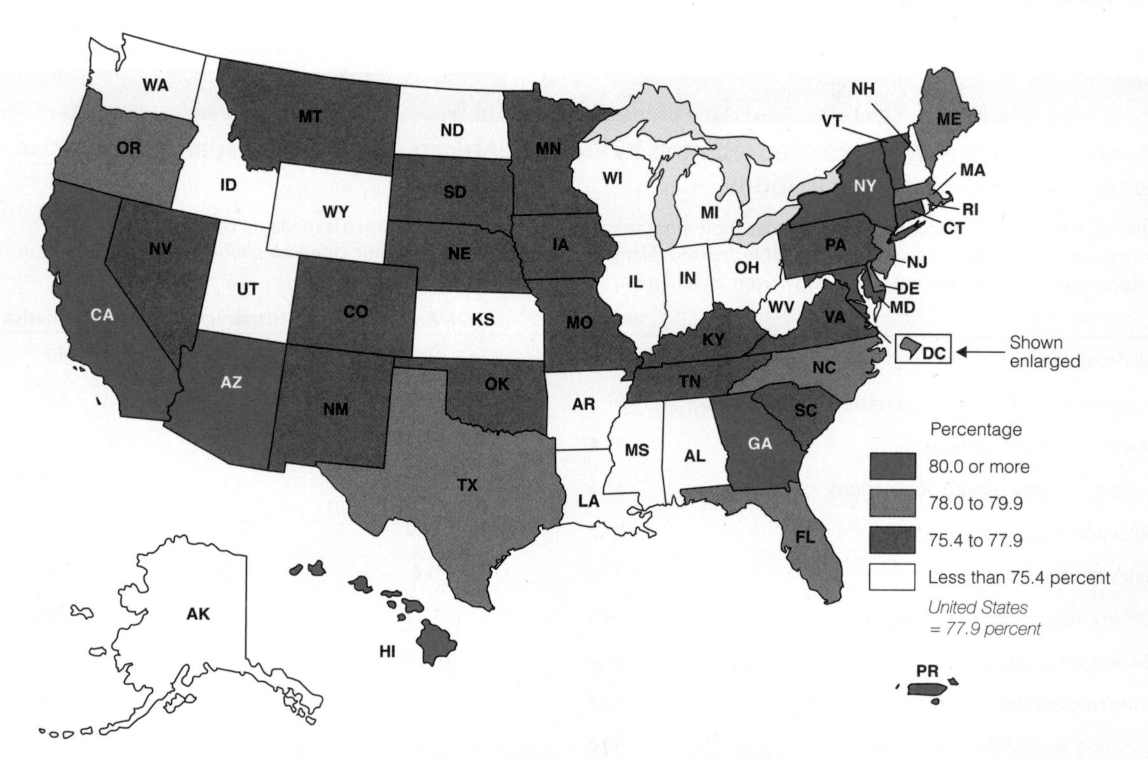

▲ **MAP 11.1 WOMEN'S EARNINGS AS A PERCENTAGE OF MEN'S EARNINGS BY STATE AND PUERTO RICO: 2008**

Sources: U.S. Census Bureau, American Community Survey, 2008; and Puerto Rico Community Survey, 2008.

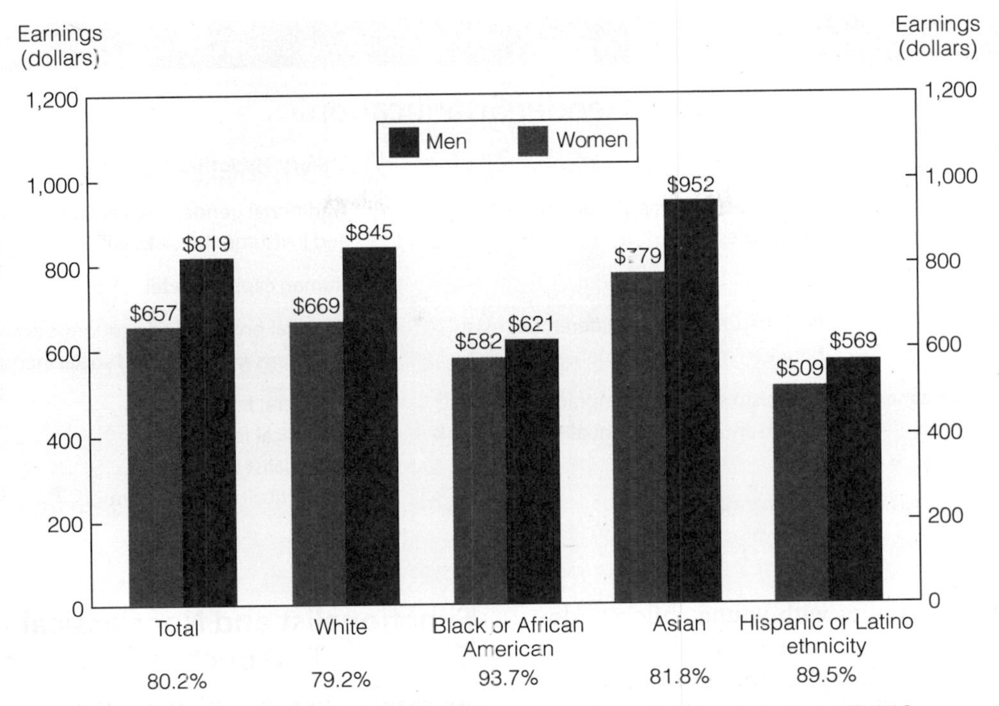

▲ **FIGURE 11.2 WOMEN'S WAGES AS A PERCENTAGE OF MEN'S IN EACH RACIAL/ETHNIC CATEGORY**

Source: U.S. Bureau of Labor Statistics, 2010.

Pay equity or ***comparable worth*** **is the belief that wages ought to reflect the worth of a job, not the gender or race of the worker.** How can the comparable worth of different kinds of jobs be determined? One way is to compare the actual work of women's and men's jobs and see if there is a disparity in the salaries paid for each. To do this, analysts break a job into components—such as the education, training, and skills required, the extent of responsibility for others' work, and the working conditions—and then allocate points for each (Lorber, 2005). For pay equity to exist, men and women in occupations that receive the same number of points should be paid the same. However, pay equity exists for very few jobs.

Paid Work and Family Work

As previously discussed, the first big change in the relationship between family and work occurred with the Industrial Revolution and the rise of capitalism. The cult of domesticity kept many middle- and upper-class women out of the workforce during this period. Primarily, working-class and poor women were the ones who had to deal with the work/family conflict. Today, however, the issue spans the entire economic spectrum (see ▶ Figure 11.2). The typical married woman in the United States combines paid work in the labor force and family work as a homemaker. Although this change has occurred at the societal level, individual women bear the brunt of the problem.

Even with dramatic changes in women's workforce participation, the sexual division of labor in the family remains essentially unchanged. Most married women now share responsibility for the breadwinner role, yet many men do not equally share domestic responsibilities. Consequently, many women have a "double day" or "second shift," because of their dual responsibilities for paid and unpaid work (Hochschild, 1989, 2003). Working women have less time to spend on housework; if husbands do not participate in routine domestic chores, some chores simply do not get done or get done less often. Although the income that many women earn is essential to the economic survival of their families, they still must spend part of their earnings on family maintenance, such as day-care centers, fast-food restaurants, and laundries, in an attempt to keep up with their obligations.

pay gap a term used to describe the disparity between women's and men's earnings.

comparable worth (or **pay equity**) the belief that wages ought to reflect the worth of a job, not the gender or race of the worker.

[concept quick review]

Sociological Perspectives on Gender Stratification

Perspective	Focus	Theory/Hypothesis
Functionalist	Macrolevel analysis of women's and men's roles	Traditional gender roles ensure that expressive and instrumental tasks will be performed. Human capital model
Conflict	Power and economic differentials exist between men and women.	Unequal political and economic power heightens gender-based social inequalities.
Feminist Approaches	Feminism should be embraced to reduce sexism and gender inequality.	1. Liberal feminism 2. Radical feminism 3. Socialist feminism 4. Multicultural feminism

Especially in families with young children, domestic responsibilities consume a great deal of time and energy. Although some kinds of housework can be put off, the needs of children often cannot be ignored or delayed. When children are ill or school events cannot be scheduled around work, parents (especially mothers) may experience stressful role conflicts ("Shall I be a good employee or a good mother?"). Many working women care not only for themselves, their husbands, and their children but also for elderly parents or in-laws. Some analysts refer to these women as "the sandwich generation"—caught between the needs of their young children and of their elderly relatives. Many women try to solve their time crunch by forgoing leisure time and sleep. When Arlie Hochschild interviewed working mothers, she found that they talked about sleep "the way a hungry person talks about food" (1989: 9). Perhaps this is one reason that, in later research, Hochschild (1997) learned that some married women with children found more fulfillment at work and that they worked longer hours because they liked work better than facing the pressures of home.

Perspectives on Gender Stratification

Sociological perspectives on gender stratification vary in their approach to examining gender roles and power relationships in society. Some focus on the roles of women and men in the domestic sphere; others note the inequalities arising from a gendered division of labor in the workplace. Still others attempt to integrate both the public and private spheres into their analyses. The Concept Quick Review outlines the key aspects of each sociological perspective on gender socialization.

Functionalist and Neoclassical Economic Perspectives

As seen earlier, functionalist theory views men and women as having distinct roles that are important for the survival of the family and society. The most basic division of labor is biological: Men are physically stronger, and women are the only ones able to bear and nurse children. Gendered belief systems foster assumptions about appropriate behavior for men and women and may have an impact on the types of work that those women and men perform.

The Importance of Traditional Gender Roles According to functional analysts such as Talcott Parsons (1955), women's roles as nurturers and caregivers are even more pronounced in contemporary industrialized societies. While the husband performs the *instrumental* tasks of providing economic support and making decisions, the wife assumes the *expressive* tasks of providing affection and emotional support for the family. This division of family labor ensures that important societal tasks will be fulfilled; it also provides stability for family members.

This view has been adopted by a number of politically conservative analysts who assert that relationships between men and women are damaged when changes in gender roles occur, and family life suffers as a consequence. From this perspective, the traditional division of labor between men and women is the natural order of the universe.

The Human Capital Model Functionalist explanations of occupational gender segregation are similar to neoclassical economic perspectives, such as the human capital model (Horan, 1978; Kemp, 1994).

According to this model, individuals vary widely in the amount of human capital they bring to the labor market. *Human capital* is acquired by education and job training; it is the source of a person's productivity and can be measured in terms of the return on the investment (wages) and the cost (schooling or training) (Stevenson, 1988; Kemp, 1994).

From this perspective, what individuals earn is the result of their own choices (the kinds of training, education, and experience they accumulate, for example) and of the labor-market need (demand) for and availability (supply) of certain kinds of workers at specific points in time. For example, human capital analysts argue that women diminish their human capital when they leave the labor force to engage in childbearing and child-care activities. While women are out of the labor force, their human capital deteriorates from nonuse. When they return to work, women earn lower wages than men because they have fewer years of work experience and have "atrophied human capital" because their education and training may have become obsolete (Kemp, 1994: 70).

Evaluation of Functionalist and Neoclassical Economic Perspectives Although Parsons and other functionalists did not specifically endorse the gendered division of labor, their analysis suggests that it is natural and perhaps inevitable. However, critics argue that problems inherent in traditional gender roles, including the personal role strains of men and women and the social costs to society, are minimized by this approach. For example, men are assumed to be "money machines" for their families when they might prefer to spend more time in child-rearing activities. Also, the woman's place is assumed to be in the home, an assumption that ignores the fact that many women hold jobs because of economic necessity.

In addition, the functionalist approach does not take a critical look at the structure of society (especially the economic inequalities) that makes educational and occupational opportunities more available to some than to others. Furthermore, it fails to examine the underlying power relations between men and women or to consider the fact that the tasks assigned to women and to men are unequally valued by society (Kemp, 1994). Similarly, the human capital model is rooted in the premise that individuals are evaluated based on their human capital in an open, competitive market where education, training, and other job-enhancing characteristics are taken into account. From this perspective, those who make less money (often men of color and all women) have no one to blame but themselves.

Critics note that instead of blaming people for their choices, we must acknowledge other realities. Wage discrimination occurs in two ways: (1) the wages are higher in male-dominated jobs, occupations, and segments of the labor market, regardless of whether women take time for family duties, and (2) in any job, women and people of color will be paid less (Lorber, 1994).

© Peter Dazeley/The Image Bank/Getty Images

▲ According to the human capital model, women may earn less in the labor market because of their child-rearing responsibilities. What other sociological explanations are offered for the lower wages that women receive?

Conflict Perspectives

According to many conflict analysts, the gendered division of labor within families and in the workplace results from male control of and dominance over women and resources. Differentials between men and women may exist in terms of economic, political, physical, and/or interpersonal power. The importance of a male monopoly in any of these arenas depends on the significance of that type of power in a society (Richardson, 1993). In hunting and gathering and horticultural societies, male dominance over women is limited because all members of the society must work in order to survive (Collins, 1971; Nielsen, 1990). In agrarian societies, however, male sexual dominance is at its peak. Male heads of household gain a monopoly not only on physical power but also on economic power, and women become sexual property.

Although men's ability to use physical power to control women diminishes in industrial societies, men still remain the head of household and control

you can make a difference

"Love Your Body": Women's Activism on Campus and in the Community

Do You Love What You See When You Look in the Mirror? Hollywood and the fashion, cosmetics and diet industries work hard to make each of us believe that our bodies are unacceptable and need constant improvement.

—promotion for "Love Your Body Day," sponsored by nowfoundation.org (2011)

Although this message appears to be for girls and women only, many boys and men are also concerned about their physical appearance, as well as how girls and women are represented in the media. Both men and women can make a difference by becoming involved in a campus or community organization that helps people gain a better understanding of body-image issues. Some work can be done within the groups of which you are already a member:

- Participate in the national Love Your Body Day, which is a day of action to speak out against ads and images of women that are offensive, dangerous, and disrespectful.
- Discourage sexist ads and media reporting about women (for example, a focus on weight, hair, clothing, or other physical attributes rather than on their accomplishments) by submitting letters to the newspaper or otherwise encouraging journalists to rethink how they frame stories about girls and women.
- Think of on-campus traditions or events that promote negative body-image stereotypes. Examples are parties or mixers where students are encouraged to wear scant clothing that makes some women uncomfortable, particularly those who are already struggling with body-image concerns. Actively encourage the organizers of such events to rethink "theme party" clothing or other kinds of dress that contribute to body-image problems.
- Host a forum on women's health issues with doctors in your community discussing what good health means for women's bodies and why some diet and exercise regimes may be harmful.
- Promote positive body image on campus by encouraging your club or Greek organization to host a "Friends Don't Let Friends Fat Talk" day in the Student Union or a dining hall. Have students write down on an index card their negative body-image thoughts such as "I hate my thighs." Then ask students to wad up the cards and throw those thoughts into trash cans. Or students can be encouraged to write

the property. In addition, men gain more power through their predominance in the most highly paid and prestigious occupations and the highest elected offices. By contrast, women have the ability in the marriage market to trade their sexual resources, companionship, and emotional support for men's financial support and social status. As a result, women as a group remain subordinate to men (Collins, 1971; Nielsen, 1990).

All men are not equally privileged; some analysts argue that women and men in the upper classes are more privileged, because of their economic power, than men in lower-class positions and all people of color (Lorber, 1994). In industrialized societies, persons who occupy elite positions in corporations, universities, the mass media, and government or who have great wealth have the most power (Richardson, 1993). Most of these are men, however.

Conflict theorists in the Marxist tradition assert that gender stratification results from private ownership of the means of production; some men not only gain control over property and the distribution of goods but also gain power over women. According

Scott J. Ferrell/Contributor/Congressional Quarterly/Getty Images

▲ Although the demographic makeup of the U.S. Senate has been gradually changing in recent decades, men still dominate it, a fact that the conflict perspective attributes to a very old pattern in human societies.

positive body-image statements and post them on a paper banner in the Student Union or dining area.

Other opportunities for involvement exist through local, state, and national organizations if you are interested. Here are three places to start:

- Reflections Body Image Program by Delta Delta Delta **www.bodyimageprogram.org**
- The Center for Living, Learning and Leading. Phone: (817) 633-8001. E-mail: reflections@trideltaeo.org.
- The National Organization for Women (NOW), 1100 H St. NW, 3rd floor, Washington, DC 20005. (202) 628-8669. NOW works to end gender bias and seeks greater representation of women in all areas of public life. On the Internet, NOW provides links to other feminist resources: **www.now.org**
- The National Organization for Men Against Sexism (NOMAS), P.O. Box 455, Louisville, CO 80027. (303) 666-7043. NOMAS has a *profeminist stance* that seeks to end sexism and an *affirmative stance* on the rights of gay men and lesbians: **www.nomas.org**

© David J. Green/Alamy

People today, especially women, are conditioned to believe that their bodies are unacceptable if they don't match an "ideal" body type. Is this woman's self-perception likely to be accurate?

to Friedrich Engels and Karl Marx, marriage serves to enforce male dominance. Men of the capitalist class instituted monogamous marriage (a gendered institution) so that they could be certain of the paternity of their offspring, especially sons, whom they wanted to inherit their wealth. Feminist analysts have examined this theory, among others, as they have sought to explain male domination and gender stratification.

Feminist Perspectives

***Feminism*—the belief that women and men are equal and should be valued equally and have equal rights**—is embraced by many men as well as women. It holds in common with men's studies the view that gender is a socially constructed concept that has important consequences in the lives of all people (Craig, 1992). According to the sociologist Ben Agger (1993), men can be feminists and propose feminist theories; both women and men have much in common as they seek to gain a better understanding of the causes and consequences of gender inequality. Over the past three decades many different organizations have been formed to advocate causes uniquely affecting women or men and to help people gain a better understanding of gender inequality (see "You Can Make a Difference").

Feminist theory seeks to identify ways in which norms, roles, institutions, and internalized expectations limit women's behavior. It also seeks to demonstrate how women's personal control operates even within the constraints of relative lack of power (Stewart, 1994).

Liberal Feminism In liberal feminism, gender equality is equated with equality of opportunity. The roots of women's oppression lie in women's lack of equal civil rights and educational opportunities.

feminism the belief that women and men are equal and should be valued equally and have equal rights.

Only when these constraints on women's participation are removed will women have the same chance for success as men. This approach notes the importance of gender-role socialization and suggests that changes need to be made in what children learn from their families, teachers, and the media about appropriate masculine and feminine attitudes and behavior. Liberal feminists fight for better child-care options, a woman's right to choose an abortion, and the elimination of sex discrimination in the workplace.

Radical Feminism According to radical feminists, male domination causes all forms of human oppression, including racism and classism (Tong, 1989). Radical feminists often trace the roots of patriarchy to women's childbearing and child-rearing responsibilities, which make them dependent on men (Firestone, 1970; Chafetz, 1984). In the radical feminist view, men's oppression of women is deliberate, and ideological justification for this subordination is provided by other institutions such as the media and religion. For women's condition to improve, radical feminists claim, patriarchy must be abolished. If institutions are currently gendered, alternative institutions—such as women's organizations seeking better health care, day care, and shelters for victims of domestic violence and rape—should be developed to meet women's needs.

Socialist Feminism Socialist feminists suggest that women's oppression results from their dual roles as paid *and* unpaid workers in a capitalist economy. In the workplace, women are exploited by capitalism; at home, they are exploited by patriarchy (Kemp, 1994). Women are easily exploited in both sectors; they are paid low wages and have few economic resources. Gendered job segregation is "the primary mechanism in capitalist society that maintains the superiority of men over women, because it enforces lower wages for women in the labor market" (Hartmann, 1976: 139). As a result, women must do domestic labor either to gain a better-paid man's economic support or to stretch their own wages (Lorber, 1994). According to socialist feminists, the only way to achieve gender equality is to eliminate capitalism and develop a socialist economy that would bring equal pay and rights to women.

Multicultural Feminism Recently, academics and activists have been rethinking the experiences of women of color from a feminist perspective. The experiences of African American women and Latinas/Chicanas have been of particular interest to some social analysts. Building on the civil rights and feminist movements of the late 1960s and early 1970s, some contemporary black feminists have focused on the cultural experiences of African American women. A central assumption of this analysis is that race, class, and gender are forces that simultaneously oppress African American women (Hull, Bell-Scott, and Smith, 1982). The effects of these three statuses cannot be adequately explained as "double" or "triple" jeopardy (race + class + gender = a poor African American woman) because these ascribed characteristics are not simply added to one another. Instead, they are multiplicative in nature (race × class × gender); different characteristics may be more significant in one situation than another. For example, a well-to-do white woman (class) may be in a position of privilege when compared to people of color (race) and men from lower socioeconomic positions (class), yet be in a subordinate position as compared with a white man (gender) from the capitalist class (Andersen and Collins, 1998). In order to analyze the complex relationship among these characteristics, the lived experiences of African American women and other previously "silenced people" must be heard and examined within the context of particular historical and social conditions.

Another example of multicultural feminist studies is the work of the psychologist Aida Hurtado (1996), who explored the cultural identification of Latinas/Chicanas. According to Hurtado, distinct differences exist between the worldviews of the white (non-Latina) women who participate in the women's movement and those of many Chicanas, who have a strong sense of identity with their own communities. Hurtado (1996) suggests that women of color do not possess the "relational privilege" that white women have because of their proximity to white patriarchy through husbands, fathers, sons, and others. Like other multicultural feminists, Hurtado calls for a "politics of inclusion," creating social structures that lead to positive behavior and bring more people into a dialogue about how to improve social life and reduce inequalities.

Evaluation of Conflict and Feminist Perspectives Conflict and feminist perspectives provide insights into the structural aspects of gender inequality in society. These approaches emphasize factors external to individuals that contribute to the oppression of white women and people of color; however, they have been criticized for emphasizing the differences between men and women without taking into account the commonalities they share. Feminist approaches have also been criticized for their emphasis on male dominance without a corresponding analysis of the ways in

© Larry Kolvoord/The Image Works

▲ Latinas have become increasingly involved in social activism for causes that they believe are important. This woman is showing her support for amnesty for undocumented workers in the United States.

which some men may also be oppressed by patriarchy and capitalism.

Gender Issues in the Future

Over the past century, women made significant progress in the labor force. Laws were passed to prohibit sexual discrimination in the workplace and school. Affirmative action programs helped make women more visible in education, government, and the professional world. More women entered the political arena as candidates instead of as volunteers in the campaign offices of male candidates (Lott, 1994).

Many men joined movements to raise their consciousness, realizing that what is harmful to women may also be harmful to men. For example, women's lower wages in the labor force suppress men's wages as well; in a two-paycheck family, women who are paid less contribute less to the family's finances, thus placing a greater burden on men to earn more money. In the midst of these changes, however, many gender issues remain unresolved in the twenty-first century.

In the labor force, gender segregation and the wage gap are still problems. As the United States attempts to climb out of the worst economic recession in decades, job loss has affected both women and men. However, data from the Bureau of Labor Statistics show that the wages of the typical woman who has a job have risen slightly faster than those of the typical man. Rather than this being considered a gain for women, some analysts suggest that it is a situation where everyone is losing but that men are simply losing more because of job insecurity or loss, declining real wages, and loss of benefits such as health insurance and pension funds.

In the United States and other nations of the world, gender equity, political opportunities, education, and health care remain pressing problems for women. Gender issues and imbalances can contribute not only to individual problems but also to societal problems, such as the destabilization of nations in the global economy. Gender inequality is also an international problem because it is related to violence against women, trafficking, and other crimes against girls and women. To bring about social change for women, it is important for them to be equal players in the economy and the political process (International Foundation for Electoral Systems, 2011). According to Hillary Rodham Clinton (2011), U.S. Secretary of State,

> Governments and business leaders worldwide should view investing in women as a strategy for job creation and economic growth. And while many are doing so, the pool of talented women remains underutilized, underpaid and underrepresented over all in business and society. Worldwide, women do two-thirds of the work, yet they earn just one-third of the income and own less than 2 percent of the land. . . . If we invest in women's education and give them the opportunity to access credit or start a small business, we add fuel to a powerful engine for progress for women, their families, their communities and their countries.

As Clinton suggests, an investment in girls and women, whether in the United States or in other nations of the world, will strengthen other efforts to deal with social problems such as violence against women, inequality, and poverty. However, we must ask this: How will economic problems around the world affect gender inequality in the twenty-first century? Sadly, fallout from the economic crisis that began in 2008 and still remains will probably be most strongly felt by people who are already in the greatest social and financial peril. What do you think might be done to provide more equal opportunities for girls and women even in difficult economic times?

chapter review Q&A

Use these questions and answers to check how well you've achieved the learning objectives set out at the beginning of this chapter.

- **How do sex and gender differ?**

Sex refers to the biological categories and manifestations of femaleness and maleness; gender refers to the socially constructed differences between females and males. In short, sex is what we (generally) are born with; gender is what we acquire through socialization.

- **How do gender roles and gender identity differ from gendered institutions?**

Gender role encompasses the attitudes, behaviors, and activities that are socially assigned to each sex and that are learned through socialization. Gender identity is an individual's perception of self as either female or male. By contrast, gendered institutions are those structural features that perpetuate gender inequality.

- **How does the nature of work affect gender equity in societies?**

In most hunting and gathering societies, fairly equitable relationships exist between women and men because neither sex has the ability to provide all of the food necessary for survival. In horticultural societies, a fair degree of gender equality exists because neither sex controls the food supply. In agrarian societies, male dominance is overt; agrarian tasks require more labor and physical strength, and females are often excluded from these tasks because they are viewed as too weak or too tied to child-rearing activities. In industrialized societies, a gap exists between nonpaid work performed by women at home and paid work performed by men and women. A wage gap also exists between women and men in the marketplace.

- **What are the key agents of gender socialization?**

Parents, peers, teachers and schools, sports, and the media are agents of socialization that tend to reinforce stereotypes of appropriate gender behavior.

- **What causes gender inequality in the United States?**

Gender inequality results from economic, political, and educational discrimination against women. In most workplaces, jobs are either gender segregated or the majority of employees are of the same gender. Although the degree of gender segregation in the professional workplace has declined since the 1970s, racial and ethnic segregation remains deeply embedded.

- **How is occupational segregation related to the pay gap?**

Many women work in lower-paying, less-prestigious jobs than men. This occupational segregation leads to a disparity, or pay gap, between women's and men's earnings. Even when women are employed in the same job as men, on average they do not receive the same, or comparable, pay.

- **How do functionalists and conflict theorists differ in their view of division of labor by gender?**

According to functionalist analysts, women's roles as caregivers in contemporary industrialized societies are crucial in ensuring that key societal tasks are fulfilled. While the husband performs the instrumental tasks of economic support and decision making, the wife assumes the expressive tasks of providing affection and emotional support for the family. According to conflict analysts, the gendered division of labor within families and the workplace—particularly in agrarian and industrial societies—results from male control and dominance over women and resources.

key terms

questions for critical thinking

1. Do the media reflect societal attitudes on gender, or do the media determine and teach gender behavior? (As a related activity, watch television for several hours, and list the roles for women and men depicted in programs and those represented in advertising.)
2. Examine the various academic departments at your college. What is the gender breakdown of the faculty in selected departments? What is the gender breakdown of undergraduates and graduate students in those departments? Are there major differences among various academic areas of teaching and study? What hypothesis can you come up with to explain your observations?

turning to video

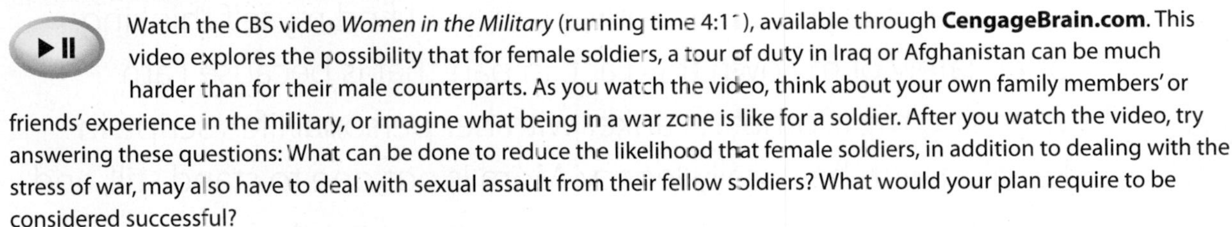

Watch the CBS video *Women in the Military* (running time 4:1[illegible]), available through **CengageBrain.com**. This video explores the possibility that for female soldiers, a tour of duty in Iraq or Afghanistan can be much harder than for their male counterparts. As you watch the video, think about your own family members' or friends' experience in the military, or imagine what being in a war zone is like for a soldier. After you watch the video, try answering these questions: What can be done to reduce the likelihood that female soldiers, in addition to dealing with the stress of war, may also have to deal with sexual assault from their fellow soldiers? What would your plan require to be considered successful?

online study resources

Go to CENGAGE brain.com to access online study resources, including the Sociology CourseMate for this text as well as special features such as video, an interactive sociology time line and interactive maps, General Social Survey (GSS) data, and U.S. Census 2010 data.

CourseMate brings course concepts to life with interactive learning, study, and exam-preparation tools that support the printed textbook. A textbook-specific website, **Sociology CourseMate** includes an integrated interactive eBook and other interactive learning tools, including quizzes, flash cards, and videos.

Visit **www.cengagebrain.com** to access your account and purchase materials.

12 Aging and Inequality Based on Age

The women in my family, at least on my mother's side, seem to live long and well. My grandmother Pearl's "third act" was one of worldwide travel and voracious learning. . . . I am a good 20-something years from retirement, yet I find myself thinking often lately of my own third act. In part that is because I am watching my mother . . . tinker with her personal life's script. In keeping with her family legacy, Mom is not one to stand still, and over the years has accumulated a couple of master's degrees, a Ph.D. and a law degree, along with a kaleidoscope of work experience. After my father died two years ago, it seemed only logical that my mother would mourn, then take a few exotic trips and find a more challenging job.

But just because Mom was ready for her third act didn't mean the working world was ready for Mom. Unlike so many career shifts she had made over the years, this one did not go smoothly. Her calls to prospective employers often went unanswered. Her résumé did not always open doors. She looks as if she's in her 50's, but her résumé makes it clear that she's in her 60's, and suddenly the

Joe McBride/Getty Images

▲ Many older individuals actively challenge the stereotypes and assumptions of others about what it means to be a "senior citizen."

years of experience that have been her greatest strength somehow disqualified her.

Mom did not take this quietly. "I know more than I did 20 years ago, and my brain works as well as it did 30 years ago," she said. The problem, she points out, is not with her generation, but with mine. "Employers—who are your age—have dismissed people who are my age," she said.

—Journalist Lisa Belkin (2006) describes the problem that many older workers, including her mother, encounter in finding new jobs. However, for Belkin's mother the story had a positive outcome: She eventually found the fulfilling job that she was seeking.

Chapter Focus Question:

Given the fact that aging is an inevitable consequence of living—unless a person dies young—why do many people in the United States devalue older persons?

If we apply our sociological imagination to the issue of aging and work as described by Lisa Belkin, we see that views on age and the problems that people experience in growing older are not just personal problems but also public issues of concern to everyone. During the U.S. recession, many older workers have either continued working (6.6 million people over age 65 were in the workforce in 2009) or they have looked for work, like Lisa Belkin's mother. Eventually, all of us will be affected by aging and possible problems such as finding work when we are considered to be "too old." ***Aging* is the physical, psychological, and social processes associated with growing older.** In the United States and some other high-income countries, older people are the targets of prejudice and discrimination based on myths about aging. For example, older persons may be viewed as incompetent solely on the basis of their age. Although some older people may need assistance from others and support from society, many others are physically, socially, and financially independent.

Today, people age 65 and older make up about 13 percent of the total population in the United States, and it is estimated that people in this age group will constitute 20 percent of the total population by 2050. Almost 90 percent of the people living in the high-income countries today are expected to survive to age 65. However, by sharp contrast, in low-income countries such as Zambia, Uganda, and Rwanda, nearly half of the total population in each nation is not expected to survive to *age 40* (United Nations Development Programme, 2010). In this chapter we examine the sociological aspects of aging. We will also examine how older people seek dignity, autonomy, and empowerment in societies such as the United States that may devalue those who do not fit the ideal norms of youth, beauty, physical fitness, and self-sufficiency. Before reading on, test your knowledge about aging and age-based discrimination in the United States by taking the Sociology and Everyday Life quiz.

Highlighted Learning Objectives

- Explain how functional age differs from chronological age.
- Describe how age helps determine a person's roles and statuses in society.
- Discuss the negative effects of ageism on older persons.
- Identify specific actions that might be taken to bring about a more equitable society for older people.

aging the physical, psychological, and social processes associated with growing older.

sociology and everyday life

How Much Do You Know About Aging and Age-Based Discrimination?

True	False	
T	F	1. U.S. Supreme Court rulings have made it easier for individuals to show that they have been discriminated against based on their age.
T	F	2. Women in the United States have a longer life expectancy than do men.
T	F	3. Scientific studies have documented the fact that women age faster than men do.
T	F	4. Most older persons are economically secure today as a result of Social Security, Medicare, and retirement plans.
T	F	5. Studies show that advertising no longer stereotypes older persons.
T	F	6. People over age 85 make up one of the fastest-growing segments of the U.S. population.
T	F	7. Organizations representing older individuals have demanded the same rights and privileges as those accorded to younger persons.
T	F	8. The rate of elder abuse in the United States has been greatly exaggerated by the media.

Answers on page 350.

The Social Significance of Age

"How old are you?" This is one of the most frequently asked questions in the United States. Beyond indicating how old or young a person is, age is socially significant because it defines what is appropriate for or expected of people at various stages. For example, child development specialists have identified stages of cognitive development based on children's ages:

> [W]e do not expect our preschool children, much less our infants, to have adult-like memories or to be completely logical. We are seldom surprised when a 4-year-old is misled by appearances; we express little dismay when our 2½ year old calls a duck a chicken.... But we would be surprised if our 7-year-olds continued to think segmented routes were shorter than other identical routes or if they continued to insist on calling all reasonably shaggy-looking pigs "doggy." We expect some intellectual (or cognitive) differences between preschoolers and older children. (Lefrançois, 1996: 196)

At the other end of the age continuum, a 75-year-old grandmother who travels through her neighborhood on in-line skates will probably raise eyebrows and perhaps garner media coverage about her actions because she is defying norms regarding age-appropriate behavior.

When people say "Act your age," they are referring to ***chronological age*—a person's age based on date of birth**. However, most of us actually estimate a person's age on the basis of ***functional age*—observable individual attributes such as physical appearance, mobility, strength, coordination, and mental capacity that are used to assign people to age categories.** Because we typically do not have access to other people's birth certificates to learn their chronological age, visible characteristics—such as youthful appearance or gray hair and wrinkled skin—may become our criteria for determining whether someone is "young" or "old." According to the historian Lois W. Banner (1993: 15), "Appearance, more than any other factor, has occasioned the objectification of aging. We define someone as old because he or she looks old." In fact, feminist scholars believe that functional age is so subjective that it is evaluated differently for women and men—as men age, they

Monkey Business Images/Shutterstock.com

▲ What can people across generations learn by spending time with each other? Can the learning process flow in both directions?

a. U.S. Population Growth, 1980–2000
The percentage of persons 65 years of age and above increased dramatically between 1980 and 2000.

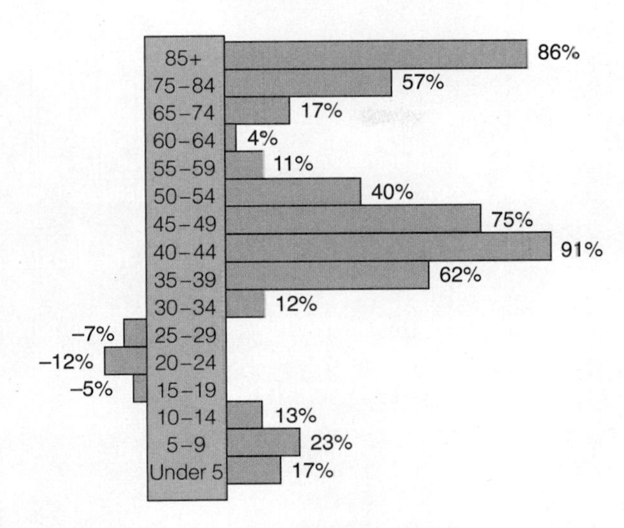

b. Selected Life Expectancies by Race, Ethnicity, and Sex, 2000
There are significant racial–ethnic and sex differences in life expectancy.

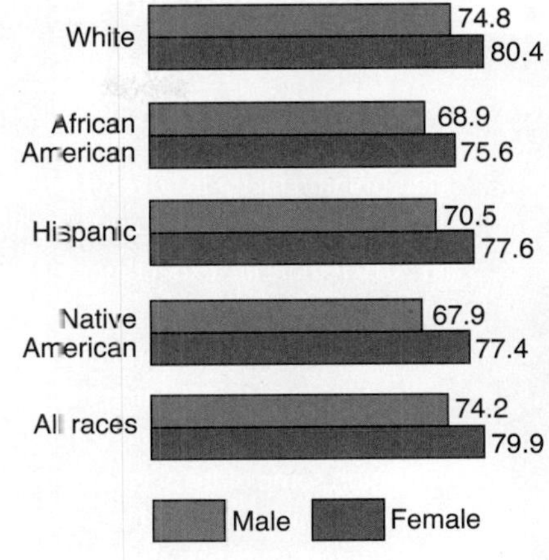

c. Percentage Distribution of U.S. Population by Age, 2000–2050 (projected)
Projections indicate that an increasing percentage of the U.S. population will be over age 65; one of the fastest-growing categories is persons age 85 and over.

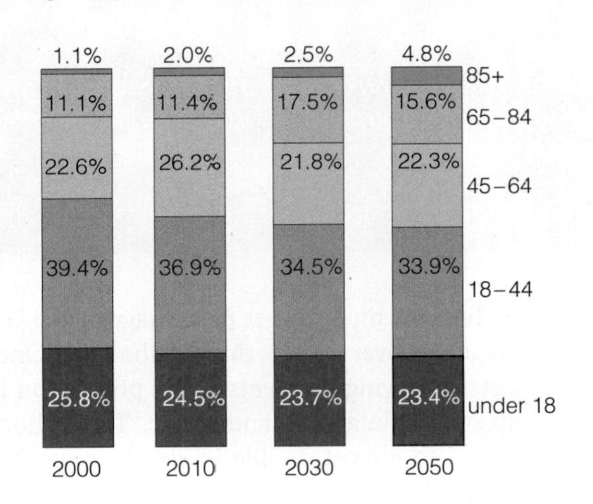

d. U.S. Age Pyramid by Age and Sex, 2000 (in millions)
Declining birthrates and increasing aging of the population are apparent in this age pyramid.

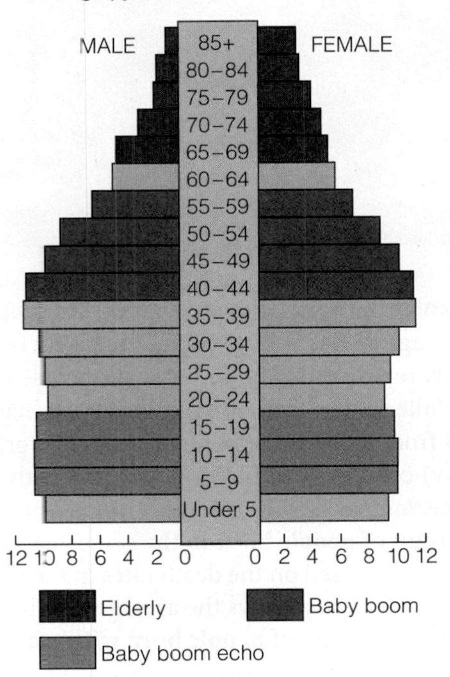

▶ **FIGURE 12.1 TRENDS IN AGING AND LIFE EXPECTANCY**

Source: U.S. Census Bureau, 2007.

are believed to become more distinguished or powerful, whereas when women grow older, they are thought to be "over-the-hill" or grandmotherly (Banner, 1993).

Trends in Aging

Over the past 30 years, the U.S. population has been aging. The median age (the age at which half the people are younger and half are older) has increased by slightly more than 6 years—from 30 in 1980 to 36.8 in 2010 (U.S. Census Bureau, 2010b). This change was partly a result of the Baby Boomers (people born between 1946 and 1964) moving into middle age and partly a result of more people living longer. As shown in ▶ Figure 12.1a, the number of older persons—age 65 and above—increased

chronological age a person's age based on date of birth.

functional age a term used to describe observable individual attributes such as physical appearance, mobility, strength, coordination, and mental capacity that are used to assign people to age categories.

sociology and everyday life

ANSWERS to the Sociology Quiz on Aging and Age-Based Discrimination

1. False. In 2009 the U.S. Supreme Court ruled that persons claiming age discrimination must prove that *age* was the determining factor in adverse job decisions, and few employers state "I won't hire you because of your age" or "I'm letting you go because of your age."

2. True. In 2010 female life expectancy (at birth) was 80.7 years, as compared with 75.7 for males. These figures vary by race and ethnicity.

3. False. No studies have documented that women actually age faster than men. However, some scholars have noted a "double standard" of aging that places older women at a disadvantage with respect to older men because women's worth in the United States is often defined in terms of physical appearance.

4. False. Although some older persons are economically secure, persons who rely solely on Social Security, Medicare, and/or pensions tend to live on low, fixed incomes that do not adequately meet their needs. According to government statistics, about 8.9 percent of people 65 and older live below the official poverty line; however, AARP studies conclude that about 18.7 percent of people over 65 are poor.

5. False. Studies have shown that advertisements frequently depict older persons negatively—for example, as chronically ill or absentminded.

6. True. The U.S. population is growing older, and persons age 85 and over constitute one of the fastest-growing segments of the population.

7. True. Organizations such as the Gray Panthers and AARP have been instrumental in the enactment of legislation beneficial to older persons.

8. False. Although cases of abuse and neglect of older persons are highly dramatized in the media, most coverage pertains to problems in hospitals, nursing homes, or other long-term care facilities. We know very little about the nature and extent of abuse that occurs in private homes.

Sources: U.S. Census Bureau, 2010b.

significantly between 1980 and 2000. The population over age 85 has been growing especially fast.

The increase in the number of older people living in the United States and other high-income nations resulted from an increase in life expectancy (greater longevity) combined with a stabilizing of birthrates. ***Life expectancy*—is the average number of years that a group of people born in the same year could expect to live.** Based on the death rates in the year of birth, life expectancy shows the average length of life of a ***cohort*—a group of people born within a specified period of time.** Cohorts may be established on the basis of one-, five-, or ten-year intervals; they may also be defined by events taking place at the time of their birth, such as Depression-era babies, Baby Boomers, or Millennials. For the cohort born in 2010, as an example, life expectancy at birth was 78.1 years for the overall U.S. population—75.7 for males and 80.7 for females. However, as Figure 12.1b shows, there are significant racial–ethnic and sex differences in life expectancy. Although the life expectancy of people of color has improved over the past 50 years, higher rates of illness and disability—attributed to poverty, inadequate health care, and greater exposure to environmental risk factors—still persist.

Today, a much larger percentage of the U.S. population is over age 65 than in the past. One of the fastest-growing segments of the population is made up of people age 85 and above. This cohort is expected to almost double in size between 2000 and 2025 (U.S. Census Bureau, 2009); by 2050, as Figure 12.1c shows, the Census Bureau predicts that the number of persons age 85 and over will have increased to about 20 million (almost 5 percent of the population). Even more astonishing is the fact that the number of centenarians (persons 100 years of age and above) in this country will increase more than 12 times, from 66,000 in 1999 to about 834,000 in 2050.

The distribution of the U.S. population is depicted in the "age pyramid" in Figure 12.1d. If, every year, the same number of people is born as in the previous year and a certain number die in each age group, the rendering of the population distribution should be pyramid shaped. As you will note, however, Figure 12.1d is not a perfect pyramid, but instead reflects declining birthrates among post–Baby Boomers.

As a result of changing population trends, research on aging has grown dramatically in the past 60 years. ***Gerontology* is the study of aging and older people.** A subfield of gerontology, *social* *gerontology*, is the study of the social (nonphysical) aspects of aging, including such topics as the societal consequences of an aging population and the personal experience of aging. According to gerontologists, age is viewed differently from society to society, and its perception changes over time.

Age in Historical and Contemporary Perspectives

People are assigned to different roles and positions based on the age structure and role structure in a particular society. *Age structure* is the number of people of each age level within the society; *role structure* is the number and type of positions available to them. Over the years the age continuum has been chopped up into finer and finer points. Two hundred years ago, people divided the age spectrum into "babyhood," a *very* short childhood, and then adulthood. What we would consider "childhood" today was quite different two hundred years ago, when agricultural societies needed a large number of strong arms and backs to work on the land to ensure survival. When 95 percent of the population had to be involved in food production, categories such as toddlers, preschoolers, preteens, teenagers, young adults, the middle-aged, and older persons did not exist.

If the physical labor of young people is necessary for society's survival, then young people are considered "little adults" and are expected to act like adults and do adult work. Older people are also expected to continue to be productive for the benefit of the society as long as they are physically able. In preindustrial societies, people of all ages help with the work, and little training is necessary for the roles that they fill. During the seventeenth and eighteenth centuries in the United States, for example, older individuals helped with the work and were respected because they were needed—and because few people lived that long.

Age in Global Perspective

Physical and sociocultural environments have different effects on how people experience aging and old age. In fact, concepts such as *young* and *old* may vary considerably from culture to culture. Unlike the sophisticated data-gathering techniques used to determine the number of older people in high-income and middle-income nations, we know less about the life expectancies and the aging populations in hunting and gathering, horticultural, pastoral, and agrarian societies. However, reaching the age of 30 or 40 is less likely for people in low-income (less-developed) nations than reaching the age of 70 or 80 in many high-income (developed) countries.

Preindustrial Societies

People in hunting and gathering societies are not able to accumulate a food surplus and must spend much of their time seeking food. They do not have permanent housing that protects them from the environment. In such societies, younger people may be viewed as a valuable asset in hunting and gathering food, whereas older people may be viewed as a liability because they typically move more slowly, are less agile, and may be perceived as being less productive.

Although more people reach older ages in horticultural, pastoral, and agrarian societies, life is still very hard for most people. It is possible to accumulate a surplus, so older individuals, particularly men, are often the most privileged in a society because they have the most wealth, power, and prestige. In agrarian societies, farming makes it possible for more people to live to adulthood and to more-advanced years. In these societies the proportion of older people living with other family members is extremely high, with few elderly living alone.

In recent years a growing number of people are reaching age 60 and above in some middle- and lower-income nations. Consider that India, for example, has about a billion people in its population. If only 5 percent of the population reaches age 60 or above, there will still be a significant increase in the number of older people in that country. Because so much of the world's population resides in India and other low-income nations, the proportion of older people in these countries will increase dramatically during the twenty-first century.

Industrial and Postindustrial Societies

In industrial societies, living standards improve and advances in medicine contribute to greater longevity for more people. Although it is often believed that less-industrialized countries accord greater honor, prestige, and respect to older people, some studies have found that the stereotypical belief that people in such nations will be taken care of by their relatives, particularly daughters and sons, is not necessarily true today (Martin and Kinsella, 1994).

In postindustrial societies, information technologies are extremely important, and a large proportion of the working population is employed

life expectancy the average number of years that a group of people born in the same year could expect to live.

cohort a group of people born within a specified period of time.

gerontology the study of aging and older people.

in service-sector occupations in the fields of education and health care, both of which may benefit older people. Some more-affluent older people may move away from family and friends upon retirement in pursuit of recreational facilities or a better climate (such as the popular move from the northeastern United States to the southern "sunbelt" states). Others may relocate to be closer to children or other relatives. The shift from a society that was primarily young to a society that is older will bring about major changes in societal patterns and in the needs of the population. Issues that must be addressed include the health care system, the Social Security system, transportation, housing, and recreation.

A Case Study: Aging in Japanese Society

The older population in Japan increased significantly over the past 40 years, and people age 65 and over now make up over one-fifth of the population. If this trend continues in Japan, about one-third of the total population will be age 65 or over by 2030. What caused this rapidly aging population? The answer lies in the declining birthrate: Fewer people are marrying, those who marry are waiting longer to have children, and married couples are having fewer children.

In the past it was widely assumed that older people in Japan were respected and revered; however, recent studies suggest that sociocultural changes and population shifts are producing a gradual change in the social importance of the elderly in that nation. When looking at living arrangements of individuals age 65 and over, for example, the number of three-generation households has declined from about 70 percent in 1980 to 44 percent in 2008. What does this mean? Although the number of older Japanese living with their children is still relatively high when compared to the United States or other Western countries, the number in Japan amounts to a remarkably steady decline. This might be attributed to changing demographic patterns, but it also may be related to younger and middle-aged couples working many hours a day and feeling that they have little time to take care of their parents.

▲ As the Japanese population rapidly ages, many people remain unemployed long past the time that their U.S. counterparts have retired.

Many older individuals in Japan are financially secure because they have public pensions and savings. About 70 percent of households with persons age 65 and over have pensions, and almost 30 percent of men age 65 and over are still in the labor force. Japanese women also lead the world in labor-force participation at age 65 and over, so many older households have income and savings to provide financial security. Despite the relative prosperity of many older Japanese residents, about 15 percent of the population age 65 and over report that they regularly have some financial problems or that they "really have problems" (International Longevity Center–Japan, 2010).

When the 2011 massive earthquake, tsunami, and nuclear peril devastated parts of Japan, tens of thousands of people were killed or seriously injured. Many older Japanese were among the missing or dead. Survivors' stories, including this one by Harumi Watanabe, often described unsuccessful efforts to save their elderly parents or other relatives:

> "There wasn't time to save them. They were old and too weak to walk so I couldn't get them to the car in time. [She held their hands when the tsunami's waves hit, but they were torn away from her by the force of the water, and she heard them yelling] "I can't breathe." Watanabe almost lost her life as well: "I stood on the furniture, but the water came up to my neck. There was only a narrow band of air below the ceiling. I thought I would die." (qtd. in msnbc.com, 2011)

Other stories of rescue among older Japanese were a source of encouragement to many who lost all material possessions. For example, one 63-year-old man was rescued from the roof of his house after it was swept out to sea. It remains to be seen what effects, if any, that these terrible natural disasters and their tragic aftermath will have on all the people of Japan, but particularly those who are among the nation's most vulnerable.

Age and the Life Course in Contemporary Society

During the twentieth century, life expectancy steadily increased as industrialized nations developed better water and sewage systems, improved

© George Shelley/Masterfile

▲ People in all cultures understand their responsibility to socialize the next generation. However, the physical arrangements associated with rearing children, including how they are moved from place to place, vary from culture to culture.

© Olivier Martel/Corbis

nutrition, and made tremendous advances in medical science. However, in the twenty-first century, children are often viewed as an economic liability; they cannot contribute to the family's financial well-being and must be supported. In industrialized and postindustrial societies, the skills necessary for many roles are more complex and the number of unskilled positions is more limited. Consequently, children are expected to attend school and learn the necessary skills for future employment rather than perform unskilled labor. Further, older people are typically expected to retire so that younger people can assume their economic, political, and social roles. However, when economic crises occur and many jobs are lost, age-based inequality tends to increase.

In the United States, age differentiation is typically based on categories such as infancy, childhood, adolescence, young adulthood, middle adulthood, and later adulthood. In Chapter 4 ("Socialization") we examined the socialization process that occurs when people are in various stages of the life course. However, these narrowly defined age categories have had a profound effect on our perceptions of people's capabilities, responsibilities, and entitlement. In this chapter we will look at what is considered appropriate for or expected of people at various ages. These expectations are somewhat arbitrarily determined and produce ***age stratification*—inequalities, differences, segregation, or conflict between age groups.** We will now examine some of those age groups and the unique problems associated with each one.

Infancy and Childhood

Infancy (birth to age 2) and childhood (ages 3 to 12) are typically thought of as carefree years; however, children are among the most powerless and vulnerable people in society. Historically, children were seen as the property of their parents, who could do with them as they pleased (Tower, 1996). In fact, whether an infant survives the first year of life depends on a wide variety of parental factors, as a community health scholar explains:

> All infants are not created equal. Those born to teenage mothers or to mothers who smoke cigarettes, drink alcohol, or take drugs are at higher risk for death in their first year. Those born in very rural areas or in inner cities are more likely to die as infants. Those born to black women are at twice the risk as those born to white women. Older mothers carry a high risk for conceiving an infant with Down's syndrome, and Native American women carry a high risk for having a baby with a serious birth defect. Add to the mix a mother's education; her economic, marital, and nutritional status; and whether she had adequate prenatal care, which all play into whether her infant will make it through the first year of life. But surviving the first year is only one piece of the equation. Quality of life is another. Infants who survive the first year can have lives so compromised that their future is seriously limited. . . . We cannot always predict which infants will survive, and we certainly cannot predict who will be happy. (Schneider, 1995: 26)

Moreover, early socialization plays a significant part in children's experiences and their quality of life. Many children are confronted with an array of problems in their families because of marital instability, an

age stratification inequalities, differences, segregation, or conflict between age groups.

increase in the number of single-parent households, and the percentage of families in which both parents are employed full time. These factors have heightened the need for high-quality, affordable child care for infants and young children. However, many parents have few options regarding who will take care of their children while they work. These statistics from the Children's Defense Fund (2010) point out potential problems of infancy and childhood:

> Each day in the United States, 4 children are killed by abuse or neglect; 5 children or teens commit suicide; 8 children or teens are killed by firearms; 2,421 children are confirmed as abused or neglected; 2,483 babies are born into poverty; and 3,477 children are arrested.

As these statistics show, childhood has many perils. In fact, two-thirds of all childhood deaths are caused by injuries. (Cancers, birth defects, heart disease, pneumonia, and HIV/AIDS cause the other third.) Although many previous childhood killers such as polio, measles, and diphtheria are now controlled through immunizations and antibiotics, motor vehicle accidents have become a major source of injury and death for infants and children. Despite laws and protective measures implemented to protect infants and children, far too many lose their lives at an early age because of the abuse, neglect, or negligence of adults.

Adolescence

In contemporary industrialized countries, adolescence roughly spans the teenage years between 13 and 19. Before the twentieth century, adolescence did not exist as an age category. Today, it is a period in which young people are expected to continue their education and perhaps hold a part-time job if they are able to find one. Most states have compulsory school-attendance laws requiring young people between certain ages, usually 6 to 16 or 18 years old, to attend school regularly; however, students who believe that they will receive little or no benefit from attending school or who believe that the money they make working is more important may find themselves labeled as "dropouts" or "juvenile offenders" for missing school. Juvenile laws define behavior such as truancy or running away from home as forms of delinquency—which would not be offenses if an adult committed them. Despite labor laws implemented to control working conditions for younger employees, many adolescents are employed in settings with hazardous working conditions, low wages, no benefits, and long work hours. Overall, the most significant concerns of teenagers from low-income and/or minority families are the lack of opportunity for education and future employment. Teens from low-income and minority families constitute the majority of young people in jails, prisons, and detention facilities—locations where some lawmakers and law enforcers believe that "out-of-control" adolescents will be less able to harm others.

William Perugin/Shutterstock.com

▲ The college years serve as an important bridge between adolescence and young adulthood for many young people.

Why do adolescents experience inequalities based on age in the United States? Adolescents are not granted full status as adults, but they are held more accountable than younger children. Early teens are considered too young to do "adult" things, such as stay out late at night, vote, drive a motor vehicle, use tobacco, or consume alcoholic beverages. In national surveys, teens consistently list the following as potential problems they face: drug and alcohol abuse; sexually transmitted diseases; teen pregnancy; eating disorders such as anorexia, bulimia, and binging; obesity; the presence of gangs in their school and neighborhood; feelings of tiredness and depression; fear of bullying; and excessive peer pressure. Of course, the United States is a diverse nation in terms of class, race/ethnicity, and other social attributes, making it difficult to make generalizations about the problems of all adolescents.

What are teenagers known for today? Teens are often known for their use of cell phones and social media. Rather than having face-to-face conversations with others, many teens prefer texting or talking on cell phones (Pew Internet & American Life Project, 2010). According to a recent study, one in three teens sends more than 100 text messages per day (3,000 texts per month); girls send and receive about 80 texts per day, but boys send and receive about 30 texts per day. Texting and extensive use of Facebook and Twitter may become areas of conflict between teens and parents. Texting while driving is of great concern, and more states are passing laws prohibiting this behavior. Extensive social media use has produced concerns about teen safety and the growing problem of bullying by friends and acquaintances. However, many parents want their children to have a cell phone because they believe that it makes them safer and because parents can more easily stay in touch with their children. Teens who are actively involved with their families and who give their time to volunteer work indicate that they are more satisfied with their lives and have higher self-esteem than those who are uninvolved. These positive attitudes tend to reduce risky behavior such as substance abuse, alcoholism, and unsafe sex, and to lead to a happier and more productive young adulthood (*Psychology Today,* 2007).

Young Adulthood

Young adulthood, which follows adolescence and lasts to about age 39, is socially significant. This is the age span in which many people are healthier and more energetic, have goals in life to which they aspire, and express enthusiasm for the future. During the earlier years of young adulthood, identity formation is important. Many young adults seek satisfactory sexual relations and hope to find a life partner. Self-growth is an important concern for many young adults today. In the past the standard expectation for these years was individuals would get married, have children, and hold down a meaningful job. People who did not fulfill these activities during young adulthood were viewed negatively. However, this is not necessarily true today. Problems in the larger society and global economy have made it more difficult for young adults to find permanent, well-paid employment and to have the necessary resources to get married, purchase a residence independent of one's parents, and have children. Although the middle to late thirties have been considered a time for "settling down," this is not universally true today, as a growing number of media reports tell of young adults moving back home with their parents or making arrangements to "double up" with other relatives or friends because of high rates of job loss, mortgage foreclosures, and other economic crises reflecting national and international patterns.

High levels of debt and financial stress are common among young adults. Those who do not attend or graduate from college are faced with limited employment opportunities. Some who complete college find that they owe large student loans but have limited financial resources to pay them off. Lack of affordable housing and expensive food and transportation costs also contribute to financial stress among young adults. For some, finding a job is more difficult than for others. Race/ethnicity, class, and gender influence people's career opportunities. People who are unable to earn income have both present and future problems: They have problems living in the here and now, but they also cannot save money or pay into retirement plans and Social Security, which leaves them further disadvantaged as they enter middle and late adulthood.

Middle Adulthood

Prior to the twentieth century, life expectancy in the United States was only about 47 years, so the concept of middle adulthood—people between the ages of 40 and 65—did not exist until fairly recently. Middle adulthood for some people represents the time during which (1) they have the highest levels of income and prestige, (2) they leave the problems of child rearing behind them and are content with their spouse of many years, and (3) they may have grandchildren, who give them another tie to the future. Job satisfaction and commitment to work are relatively high, and individuals enjoy what they are doing. Some people have adjusted their aspirations to goals that are attainable; others may have a feeling of sadness and frustration over unaccomplished goals. For many in middle adulthood, social stability is based on enduring relationships with spouses, friends, and workplace colleagues.

A major characteristic of middle adulthood is the physical and biological changes that occur in the body. Normal changes in appearance occur during these years, and although these changes have little relationship to a person's health or physical functioning, they are socially significant to many people. As people progress through middle adulthood, they experience *senescence* (primary aging) in the form of molecular and cellular changes in the body. Wrinkles and gray hair are visible signs of senescence. Less-visible signs include arthritis and a gradual dulling of the senses of taste, smell, touch, and vision. Typically, reflexes begin to slow down, but the actual nature and extent of change vary widely from person to person. And stereotypes and self-stereotypes about aging may often be important in determining changes in a person's work ability (see "Sociology Works!").

People also experience a change of life in this stage. Women undergo *menopause*—the cessation of the menstrual cycle caused by a gradual decline in the body's production of the "female" hormones estrogen and progesterone. Menopause typically occurs between the midforties and the early fifties and signals the end of a woman's childbearing capabilities. Some women may experience irregular menstrual cycles for several years, followed by hot flashes, retention of body fluids, swollen breasts, and other aches and pains. Other women may have

sociology works!

"I Think I Can't, I Think I Can't!": The Self-Fulfilling Prophecy and Older People

Most of us are familiar with the children's story of *The Little Engine That Could,* in which the engine on the train said "I think I can, I think I can!" when he was trying to pull the train up a steep hill. Used for years as an example of the power of positive thinking, the reverse side of the story might be that if people think that they cannot do something, that task becomes more difficult, and perhaps even impossible, for them to accomplish. Sociologists have a similar corollary to this story of the train, and the idea has widely referred to as the "Thomas theorem" or self-fulfilling prophecy. As defined in Chapter 5, a *self-fulfilling prophecy* is a false belief or prediction that produces behavior that makes the originally false belief come true. The original statement of this prophecy, named after its possible originator, W. I. Thomas, is as follows: "If [people] define situations as real, they are real in their consequences."

How does this statement apply to the lives of older people? Researchers have found a number of ways in which the self-fulfilling prophecy is applicable to how people perceive the aging process and how they view themselves in nations that value young people more than older people. Consider this question, for example: Is work ability strictly related to a person's age? The Norwegian sociologist Per Erik Solem (2008) explored the influence of the work environment on how age-related subjective changes occurred in people's work ability. He found that although age and physical health are obviously associated with decline in the work ability of older individuals, stereotypes and self-stereotypes (negative assumptions about themselves that are embraced by individuals who are being stereotyped) about aging are also important factors in producing a decline in a person's work ability. According to Solem (2008: 44), objective capacities such as physical strength and endurance may be important in performing certain tasks, such as lifting heavy equipment or nursing bedridden patients; however, "what workers believe they are able to do, influences to what extent they use their potential of objective abilities." In other words, if older individuals *subjectively* define themselves as less able than they actually are, given their *objective* potential, then they may be less likely to perform a task that might be quite within their capabilities to do.

Although physical changes associated with aging will always be a factor in some types of job performance, particularly those requiring quick reaction times or heavy physical labor, many occupations could benefit from older workers' experience and expertise. For this reason, Solem (2008) suggests that older workers should be provided with new learning opportunities and a chance to maintain their subjective work ability throughout their careers.

Studying and applying sociology not only helps us to see that cultural changes are needed if we are to rid ourselves and our society of age-based negative stereotypes about older individuals, but it also reminds us that the self-fulfilling prophecy is applicable at any stage of life. "I think I can" is a positive psychosocial force, whereas "I think I can't" is a negative force that works to our detriment.

reflect & analyze

Can you think of periods of time in your life when a negative stereotype and/or self-stereotype may have contributed to your not doing something that you had the objective ability to do? In addition to issues pertaining to age, do you think the Thomas theorem is also applicable to problems associated with race, class, and/or gender?

few or no noticeable physical symptoms. The psychological aspects of menopause are often as important as any physical effects. In one study, Anne Fausto-Sterling (1985) concluded that many women respond negatively to menopause because of negative stereotypes associated with menopausal and postmenopausal women. These stereotypes make the natural process of aging in women appear abnormal when compared with the aging process of men. Actually, many women experience a new interest in sexual activity because they no longer have to worry about the possibility of becoming pregnant. On the other hand, a few women have recently chosen to produce children using new medical technologies long after they have undergone menopause.

Men undergo a *climacteric,* in which the production of the "male" hormone testosterone decreases. Some have argued that this change in hormone levels produces nervousness and depression in men; however, it is not clear whether these emotional changes are attributable to biological changes or to a more general "midlife crisis," in which men assess what they have accomplished. Ironically, even as such biological changes may have a liberating effect on some people, they may also reinforce societal stereotypes of older people, especially women, as "sexless." Recently, intensive marketing of products such as Viagra for erectile dysfunction has made people of all ages more aware not only of the potential sexual problems associated with aging but also with the possibility of reducing or solving these problems with the use of prescription drugs.

Along with primary aging, people in middle adulthood also experience *secondary aging,* which occurs as a result of environmental factors and lifestyle choices. For example, smoking, drinking heavily, and engaging in little or no physical activity are factors that affect the aging process. People who live in regions with high levels of environmental degradation and other forms of pollution are also at greater risk of aging more rapidly and having chronic illnesses and diseases associated with these external factors.

In middle adulthood, many people face family-related issues such as the empty nest syndrome, divorce, death of a spouse, remarriage, or learning that they are part of the "sandwich generation," which means that they not only may have responsibility for their children but also for taking care of their aging parents. More recently, the term "club sandwich generation" has been coined to describe people who are sandwiched between their aging parents, their children, and their grandchildren (Abaya, 2011). The burden of these responsibilities disproportionately falls on women: "It's not like your husband is going to wash his own mother," a female forty-two-year-old former lawyer noted. "There's just no amount of feminism that is going to change that" (Vincent, 2004).

Middle adulthood has both high and low points. Given society's current structure, many people know that their status may begin to change significantly when they reach the end of this period of their lives. Those who had few opportunities available earlier in life tend to become increasingly disadvantaged as they grow older.

Late Adulthood

For many years, those who lived to late adulthood, which begins at about age 65 or 70, were thought of as a crucial link that tied multiple generations of family members together. Many older adults lived with their children and grandchildren and served as storytellers and caregivers for young children and as a valuable source of knowledge and experience for everyone. Today, many people are living longer, remaining productively employed, and residing in their own home. Improvements in health care, nutrition, exercise, and general living and working conditions have greatly increased the chances of people to live into late adulthood.

In the past, age 65 was referred to as the "normal" retirement age; however, with changes in Social Security regulations that provide for full retirement benefits to be paid only after a person reaches 66 or 67 years of age (based upon the individual's year of birth), many older persons have chosen to retire after the traditional age of 65. *Retirement* is the institutionalized separation of an individual from an occupational position, with continuation of income through a retirement pension based on prior years of service. Retirement means the end of a status that has long been a source of income and a means of personal identity. Perhaps the loss of a valued status explains why many retired persons introduce themselves by saying "I'm retired now, but I was a (banker, lawyer, plumber, supervisor, and so on) for forty years." As shown by ► Map 12.1, the percentage of the population age 65 and above varies from state to state, with Florida, Maine, West Virginia, and Pennsylvania having the highest proportion of people age 65 and over.

Some gerontologists subdivide late adulthood into three categories: (1) the "young-old" (ages 65 to 74),

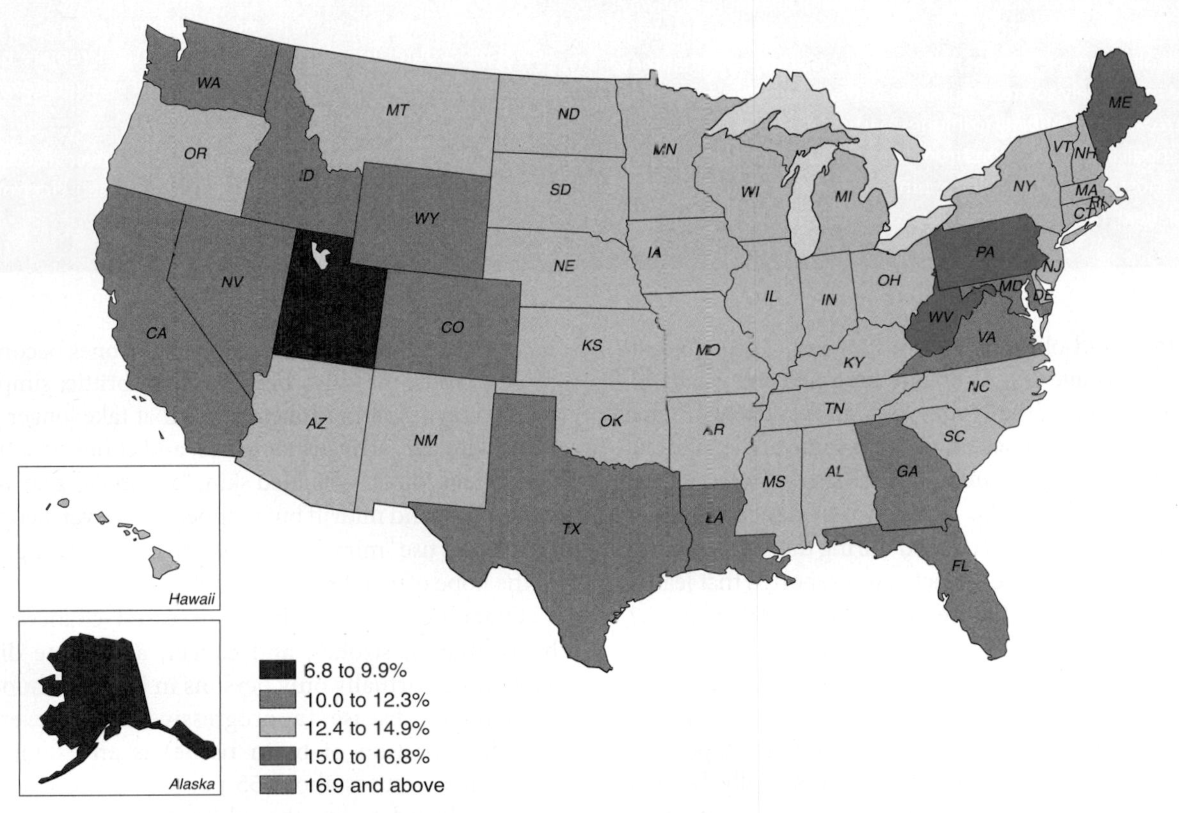

▲ MAP 12.1 PERCENTAGE OF RESIDENT POPULATION AGE 65+ PER STATE

Source: U.S. Census Bureau Population Estimates, 2010.

sociology and social policy

Driving While Elderly: An Update on Policies Pertaining to Mature Drivers

You are 84 years old and yesterday you killed my son. I don't know if you're even sound enough to understand what you've done but I have no sympathy for you or any of your family. And I pray EVERY day that you remain on this planet you see my son when you close your eyes. . . . We very much support a mandatory limit on the driving age for seniors.

—Amanda Wesling, the mother of an eight-year-old boy killed when an elderly driver drove her car into the boy's elementary school, expressing her anger and frustration that the woman was still allowed to drive (qtd. in Suhr, 2007)

Ironically, the driver involved in this deadly accident was on her way to a driving class at a senior citizens' center because it was time to renew her driver's license. When an accident such as this occurs, journalists and social media bloggers often focus on the age of the driver and make generalizations about how older drivers are unsafe on the roads. It is true that drivers over age 70 are keeping their licenses longer and driving more miles than earlier generations. This trend has led to dire predictions that aging Baby Boomers will have higher rates of motor vehicle crashes. However, research from the Insurance Institute for Highway Safety shows that fatal car accidents involving older drivers have actually declined in the past decade. Carefully consider these statistics: Compared to drivers aged 55 to 64, drivers over the age of 65 are almost twice as likely to *die* in car crashes. The likelihood increases to two-and-a-half times for those over 75, and four times for those 85 and older (Governors Highway Safety Association, 2011). However, what is missing is the fact that older people are not necessarily worse drivers or that they are more likely to crash but instead that older drivers have an *increased susceptibility* to injuries, particularly chest injuries, and to medical complications from being involved in a crash.

Most of us can recall the time when we anxiously awaited our next birthday so that we would be old enough to get our first driver's license. All states in the United States have a minimum age for getting a learner's permit or a first driver's license, but far fewer states have had policies regarding drivers over the age of 65. However, this is changing with the aging of the U.S. population. By 2020, nearly 50 million Americans over age 65 will be eligible for driver's licenses, and half of them will be age 75 or older. Mature drivers sometimes face impairments in three functions that may affect their driving abilities: vision, cognition, and motor function:

1. *Vision*. Adequate visual acuity and field of vision are necessary for safe driving but tend to decline with age. Glare, impaired contrast sensitivity, and increased time needed to adjust to changes in light levels are problems commonly experienced by mature drivers.
2. *Cognition*. Driving requires a variety of high-level cognitive skills, including memory, visual processing, and attention. Medical conditions (such as Alzheimer's and dementia) and medications that are often prescribed for the older population may affect cognitive levels.
3. *Motor function*. Motor abilities such as muscle strength, endurance, and flexibility are necessary for operating vehicle controls and turning. Changes related to aging such as arthritis can decrease an individual's ability to drive safely and comfortably. (Governors Highway Safety Association, 2011)

Although licensing and renewal policies vary throughout the United States, more states are implementing special renewal procedures for older drivers. Some states require one or more of the following for license renewal if a driver is over a specific age (usually 65 or 70):

- Shorter renewal intervals will be required for driver's licenses.
- Renewals must be done in person rather than electronically or by mail.

(2) the "old-old" (ages 75 to 85), and (3) the "oldest-old" (over age 85). Although these are somewhat arbitrary divisions, the "young-old" are less likely to suffer from disabling illnesses, whereas some of the "old-old" are more likely to suffer such illnesses. However, one study found that the prevalence of disability among those 85 and over decreased during the 1980s because of better health care. In fact, it was reported that Jeanne Calment of Paris, France, who died in 1997 at age 122, rode a bicycle until she was 100 (Whitney, 1997).

The rate of biological and psychological changes in older persons may be as important as their chronological age in determining how they and others perceive them. As adults grow older, they actually become shorter, partly because bones that have become more porous with age develop curvature. A loss of three inches in height is not uncommon. As bones become more porous, they also become more brittle; simply falling may result in broken bones that take longer to heal. With age, arthritis increases, and connective tissue stiffens joints. Wrinkled skin, "age spots," gray (or white) hair, and midriff bulge appear; however, people sometimes use "miracle" creams and cosmetic surgery in the hope of avoiding looking older.

Older persons also have increased chances of heart attacks, strokes, and cancer, and some diseases affect virtually only persons in late adulthood. Alzheimer's disease (a progressive and irreversible deterioration of brain tissue) is an example; Alzheimer's causes about 55 percent of all organic mental disorders in the older population. Persons with this disease have an impaired ability to

Photo by Melvin Levine/Time & Life Pictures/Getty Images

Although the vast majority of motor vehicle accidents are caused by younger drivers, when drivers in their seventies or eighties are involved in major accidents, widespread media coverage typically focuses on the driver's age. What effect, if any, will the aging baby boomer population have on social policies relating to driving? Keep in mind that many baby boomers are already in their sixties.

- Vision, reaction, and/or road tests (not routinely required of younger drivers) may be required for renewal.

Clearly, many issues are involved in social policies pertaining to older drivers. On the one hand, older drivers point out that most cities lack adequate public transportation and that not being able to drive makes them a burden on other people, if other people are even available to drive them around to doctors' appointments, for groceries and other errands, and for recreational activities. On the other hand, pedestrians, other drivers, and the general public have a vested interest in not having individuals of *any* age drive who are unsafe and who might, for whatever reason, constitute a threat to others.

As we think about social policy pertaining to older people, particularly in regard to the right to drive, it is important to have empathy for people who will be affected by rules requiring them to prove their competence, but it is also important to remember that if we are fortunate to live long enough, each of us will also face such questions as we reach later stages in our life.

reflect & analyze

What safety issues are similar for young and older drivers? How can a nation balance the rights of individuals to engage in activities such as driving a motor vehicle with the need to protect the safety of others? What part do social rules play in dealing with issues such as these?

function in everyday social roles; eventually, they cease to be able to recognize people they have always known and lose all sense of their own identity. Finally, they may revert to a speechless, infantile state such that others must feed them, dress them, sit them on the toilet, and lead them around. The disease can last up to 20 years; currently, there is no cure. An estimated 5.2 million people age 65 and over have Alzheimer's disease. Nearly half of all people age 85 and older (43 percent) are reported to have the disease. Women account for almost two-thirds of all people living with Alzheimer's in the United States (Alzheimer's Association, 2011). Fortunately, most older individuals do not suffer from Alzheimer's and are not incapacitated by their physical condition.

Although older people experience some decline in strength, flexibility, stamina, and other physical capabilities, much of that decline does not result simply from the aging process and is avoidable; with proper exercise, some of it is even reversible. With the physical changes come changes in the roles that older adults are expected (or even allowed) to perform. For example, people may lose some of the abilities necessary for driving a car safely, such as vision or reflexes. Although it is not true of all older persons, the average individual over age 65 does not react as rapidly as the average person who is younger than 65. The issue of elderly drivers has been widely debated in the media and political arenas after accidents were caused by older drivers (see "Sociology and Social Policy").

The physical and psychological changes that come with increasing age can cause stress. According to Erik Erikson (1963), older people must resolve a tension of "integrity versus despair." They must accept that the life cycle is inevitable, that their lives eventually will end, and that achieving inner harmony requires accepting both one's past accomplishments and past disappointments. And this is what the later years mean for many older adults, a chance to enjoy life and participate in leisure activities.

Will the life stages as we currently understand them accurately reflect aging in the future? Research continues to show that there are limited commonalities between those who are age 65 and those who are centenarians; however, many people tend to place everyone from 65 upward in categories such as "old," "elderly," or "senior citizen." In the future we will probably see such categorizations revised as growing numbers of older people reject such labels as forms of "ageism." Some analysts believe that the existing life-course and life-stages models will be modified to reflect a sense of "old age" beginning at age 75 or 80 while new stages will be added for those who reach age 90 and 100 (Morgan and Kunkel, 1998).

Inequalities Related to Aging

In previous chapters we have seen how prejudice and discrimination may be directed toward individuals based on ascribed characteristics—such as race/ethnicity or gender—over which they have no control. The same holds true for age.

Ageism

Stereotypes regarding older persons reinforce *ageism,* defined in Chapter 4 as prejudice and discrimination against people on the basis of age, particularly against older persons. Ageism against older persons is rooted in the assumption that as people grow older, they become unattractive, unintelligent, asexual, unemployable, and mentally incompetent.

Ageism is reinforced by stereotypes, whereby people have narrow, fixed images of certain groups. One-sided and exaggerated images of older people are used repeatedly in everyday life. Older persons are often stereotyped as thinking and moving slowly; as being bound to themselves and their past, unable to change and grow; as being unable to move forward and often moving backward. They are viewed as cranky, sickly, and lacking in social value; as egocentric and demanding; as shallow, enfeebled, aimless, and absentminded.

As previously discussed, popular media and the marketing of products created for "senior citizens" contribute to negative images of older persons, many of whom are portrayed as being wrinkled, suffering from erectile dysfunction, needing better bladder control, and having chronic depression. Fortunately, some members of the media are making an effort to draw attention to the positive contributions, talents, and physical stamina of many older persons rather than focusing primarily on negative portrayals. For example, some media sources highlight artistic, literary, and athletic accomplishments of older people, such as individuals over age 80 who participate in marathons and other sporting events.

Despite some changes in media coverage of older people, many younger individuals still hold negative stereotypes of "the elderly." In a classic study, William C. Levin (1988) showed photographs of the same man (disguised to appear as ages 25, 52, and 73 in various photos) to a group of college students and asked them to evaluate these (apparently different) men for employment purposes. Based purely on the photographs, the "73-year-old" was viewed by many of the students as being less competent, less intelligent, and less reliable than the "25-year-old" and the "52-year-old." Although this study was conducted more than three decades ago, the findings remain consistent with contemporary studies of ageism and age-based stereotypes. Social psychologist and gerontologist Becca Levy's (2009) more-recent research also shows that ageist stereotypes persist, and a causal link may exist between such stereotypes and various outcomes for older people. A self-fulfilling prophecy may occur in regard to memory loss, cardiac reactivity to stress, and decreased longevity because older people have been bombarded by negative images of aging.

Many older people resist ageism by continuing to view themselves as being in middle adulthood long after their actual chronological age would suggest otherwise. In one study of people age 60 and over, 75 percent of the respondents stated that they thought of themselves as middle-aged and only 10 percent viewed themselves as being old. When the same people were interviewed again 10 years later, one-third still considered themselves to be middle-aged. Even at age 80, one out of four men and one out of five women said that the word *old* did not apply to them; this lack of willingness to acknowledge having reached an older age is a consequence of ageism in society.

Older people may also resist ageism by maintaining positive relationships with others and keeping up with newer technologies that help them communicate with children, grandchildren, and friends. Studies have found that social networking use among people age 50 and older has nearly doubled in recent years. Although social media use has grown dramatically across all age groups in the 2010s, older users are particularly enthusiastic about social media networking (see "Framing Aging in the Social Media").

Wealth, Poverty, and Aging

Many of the positive images of aging and suggestions on how to avoid the most negative aspects of ageism

framing aging in the media

Living (and Sometimes Dying) on Social Networking Sites

All I used to do was sit all day and fall asleep. I'd be miserable if I didn't [have Facebook and Twitter].
—Ivy Bean (qtd. in Phillips, 2009)

I wish I could make all my followers a cup of tea to say thank you for taking time out to send me lovely messages.
—Ivy Bean (qtd. in Phillips, 2010)

At 104 years of age, Ivy Bean was perhaps the oldest Facebook and Twitter user in the United Kingdom, if not the world. Bean first gained Internet access after moving into a residential care home in Bradford, England, in 2007. Ironically, Geek Squad, the home computer maintenance company, originally signed up Bean for online service as a publicity stunt because of her age, but little did it know that she would take such an avid interest in cybercommunication. In 2008 (at the age of 102), Bean first logged on to Facebook to communicate with her family and make new friends. After she reached the maximum 5,000 friends allowed by Facebook, Bean signed up for Twitter and began sharing daily insights with an estimated 56,000 followers. Upon her death in 2010, Bean's friends from Facebook and Twitter posted comments about how much they missed her comments. According to one journalist, Ivy "brought a life-affirming joy to many, and proved that old age need not necessarily mean fading quietly away" (Phillips, 2010).

Was Ivy Bean unique? Are older people interested in using social media? Although Bean was older than most social media users, she was one of a growing number of older individuals who are finding pleasure and hope on the Internet. According to a Pew Research Center study, adults age 50 and older are definitely interested in social networking. Social media use among Internet users age 65 and older grew 100 percent between 2009 and 2010 because older users have found that this form of communication fits well with their lifestyle and interests. The desire to connect with ever-more-distant children and grandchildren, as well as to keep up with their own parents and other relatives or friends, is a primary driver that gets older people started on social media. When they begin to share photos, videos, news, and status updates, they feel like they are more in touch with others even if they live alone (Madden, 2010).

Framing on social media sites is based on "ambient awareness"—a form of *social awareness* in which an individual becomes aware of other people's lives without having to be in the same place at the same time they are or having to carry on a face-to-face conversation. Ambient awareness means that through social networking one person is able to decode another person's thoughts and moods and to interpret how he or she is feeling even when they are not in the same physical location. For this reason, the joys of life and the sadness of illness and death are often shared on social networking sites. In the case of Ivy Bean, her friends and fan base found out about her illness from care-home staff members who used her Twitter account to report that she had been hospitalized. After her death, care-home staff and journalists informed Bean's "friends" of her passing.

From all accounts of her life, Ivy Bean was a person who lived to the fullest, and among the experiences she treasured most was the opportunity to use social media to reach out to friends, family, and other people who viewed her as a friend and a very special person.

reflect & analyze

Do you have relatively constant contact with others through social networking? Would you say that you have experienced ambient awareness of them based on what they write? Are you aware of older persons, such as Ivy Bean, who have found a new social world awaiting through social networking sites?

are based on an assumption of class privilege, meaning that people can afford plastic surgery, exercise classes, and social activities such as ballroom dancing or golf, and that they have available time and facilities to engage in pursuits that will "keep them young." However, many older people with meager incomes, little savings, and poor health, as well as those who are isolated in rural areas or high-crime sections of central cities, do not have the same opportunities to follow popular recommendations about successful aging. In fact, for many older people of color, aging is not so much a matter of seeking to defy one's age but rather of attempting to survive in a society that devalues both old age and minority status.

If we compare wealth (all economic resources of value, whether they produce cash or not) with income (available money or its equivalent in purchasing power), we find that older people tend to have more wealth but less income than younger people. Older people are more likely to own a home that has increased substantially in market value; however, some may not have the available cash to pay property taxes, to buy insurance, and to maintain the property. Among older persons, a wider range of assets and income is seen than in other age categories. For example, many of the wealthiest people on the *Forbes* list of the richest people in this country (see Chapter 8) are over 65 years of age. On the other hand, almost 10 percent of all people over 65 live in poverty, as shown in Figure 12.2.

Age, Gender, and Inequality

Age, gender, and poverty are intertwined. Men and women over the age of 50 have experienced problems in the recent economic crisis in the United States and other nations linked to the global economy. Many older employees have been lost jobs as businesses have closed or the number of employees has been reduced. Despite many years of experience, older workers find it more difficult to get new positions. It is often assumed that they are overqualified, that they are too expensive to hire, or that insurance and other benefits would be more costly for older workers.

While some women have made gains in recent decades, older women in the workplace tend to earn less than men their age, and they often work in gender-segregated jobs that provide less income and fewer opportunities for advancement (see Chapter 11). As a result, women do not garner economic security for their retirement years at the same rate that many men do. These factors have contributed to the economic marginality of the current cohort of older women. Many women who are over age 70 spent their early adult lives as financial dependents of husbands or as working nonmarried women trying to support themselves in a culture that did not see women as the heads of households or as sole providers of family income. Because they were not viewed as being responsible for a family's financial security, women were paid less; therefore, older women may have to rely on inadequate income-replacement programs originally designed to treat them as dependents. Women also have a greater risk of poverty in their later years; statistically, women tend to marry men who are older than themselves, and women live longer than men. As a result, about half of all women over age 65 are widowed and living on a fixed annual income.

As shown in ▶ Figure 12.2, the percentage of persons age 65 and older living below the poverty line decreased significantly between 1980 and 2008. This largely resulted from increasing benefit levels of ***entitlements*****—certain benefit payments made by the government,** including Social Security, Supplemental Social Income (SSI), Medicare, Medicaid, and civil service pensions, which are the primary source of income for many persons over age 65. Ninety percent of all people in the United States over age 65 draw Social Security benefits. Social Security is especially important for women, who constitute 56 percent of Social Security beneficiaries age 62 and older and 68 percent of beneficiaries age 85 and older (Center on Budget and Policy Priorities, 2010). However, Social Security benefits are not as high as many people believe: In 2010 the average Social Security retirement benefit was $1,170 a month, or about $14,000 a year. Benefits are often much lower for aged widows who receive their deceased husband's benefits and have no other source of income. Without Social Security benefits almost half of all elderly Americans (13 million people)

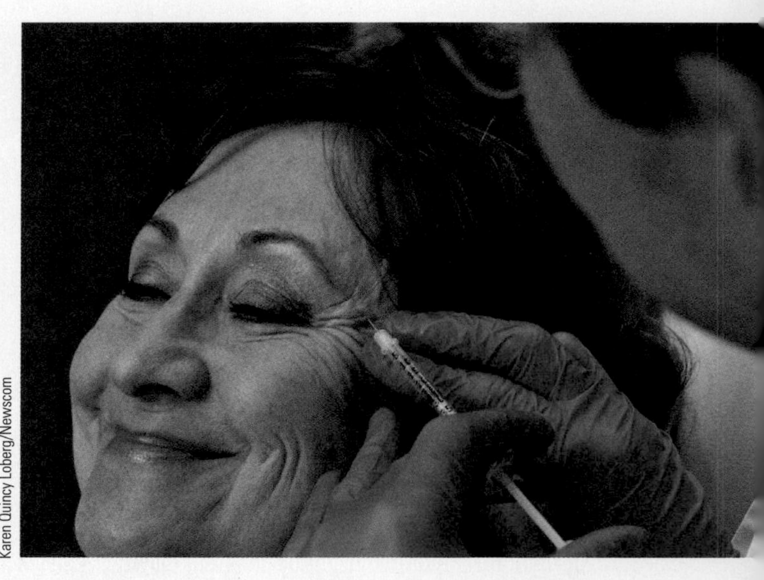
Karen Quincy Loberg/Newscom

▲ Not everyone has the health insurance or personal wealth needed for regular Botox injections, as this woman does. However, this wrinkle-removing procedure is not without risks: Patients have died afterward.

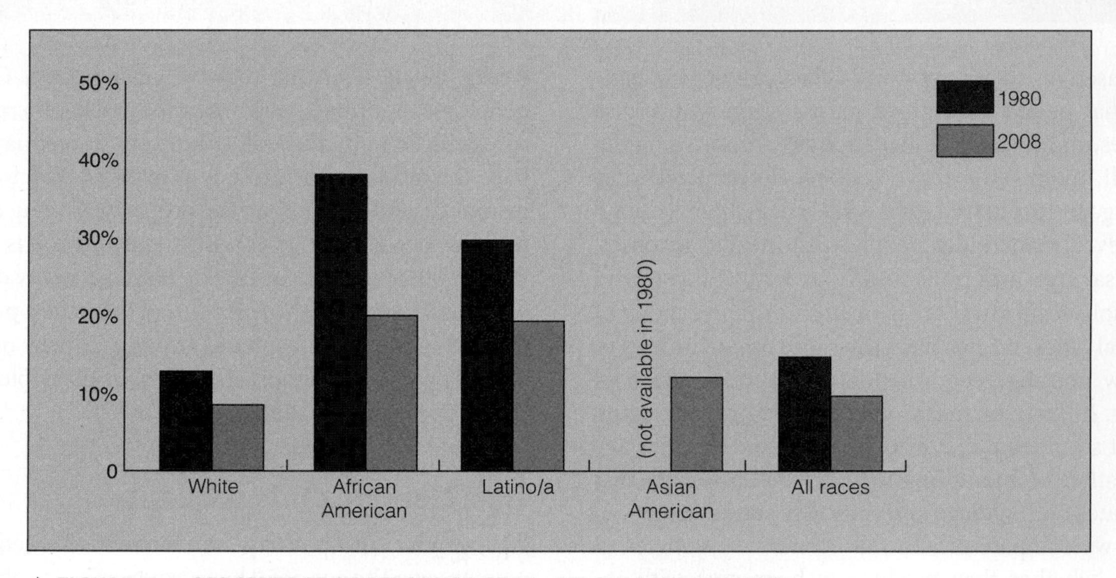

▲ **FIGURE 12.2 PERCENTAGE OF PERSONS AGE 65+ BELOW POVERTY LEVEL**

Source: U.S. Census Bureau, 2010b.

would fall below the official poverty line (Center on Budget and Policy Priorities, 2010).

Medicare, the other of the two largest entitlement programs, is a nationwide health care program for persons age 65 and older who are covered by Social Security or who are eligible to "buy into" the program by paying a monthly premium. Different parts of Medicare help cover specific health services. Part A is Hospital Insurance, which helps cover inpatient care in hospitals. It also helps cover skilled nursing facilities, hospice, and home health care. Part B is Medical Insurance, which helps cover doctors' services, hospital outpatient care, and home health care. Part D is Prescription Drug Coverage, which is an option run by Medicare-approved private insurance companies to help cover the cost of prescription drugs. Some sections of Medicare may be subject to change as the 2010 Affordable Health Care Law is gradually implemented.

In sum, age, gender, and inequality are issues that affect everyone, not just those who are older or who are women. The same may be said for age and race/ethnicity.

Age and Race/Ethnicity

Age, race/ethnicity, and economic inequality are closely intertwined. Inequalities that exist later in life originate in individuals' early participation in the labor force and are amplified in late adulthood. For example, older African Americans continue to feel the impact of segregated schools and overt patterns of job discrimination that were present during their early years. African Americans constitute about 9 percent of the population age 65 and over, but they account for almost 25 percent of older Americans in poverty. Latinos make up about 7 percent of the older population but account for 19 percent of older Americans in poverty. White Americans (non-Hispanics) account for 80 percent of the U.S. older population but make up only 7 percent of older Americans in poverty. Older white American (non-Hispanic) women and African American women have higher poverty rates than their male counterparts (Federal Interagency Forum on Aging Related Statistics, 2010).

As previously noted, the primary reason for the lower income status of many older African Americans can be traced to a pattern of limited employment opportunities and periods of unemployment throughout their lives, combined with their concentration in secondary-sector jobs, which pay lower wages, are sporadic, have few benefits, and were not covered by Social Security prior to the 1950s (Hooyman and Kiyak, 2011). Moreover, health problems may force some African Americans out of the labor force earlier than other workers because of a higher rate of chronic diseases such as hypertension, diabetes, and kidney failure (Hooyman and Kiyak, 2011).

Similarly, older Latinas/os have higher rates of poverty than whites (Anglos) because of lack of educational and employment opportunities. Some older Latinos/as entered the country illegally and have had limited opportunities for education and employment, leaving them with little or no Social Security or other benefits in their old age. High rates of poverty among older Latinas/os are associated with poor health conditions, lack of regular care by a physician, and fewer trips to the hospital for medical treatment of illness, disease, or injury.

Older Native Americans are among the most disadvantaged of all categories. Older Native Americans are more likely to live in high-poverty, rural areas than are other minority older populations. Some studies have found that about 50 percent of all older Native Americans live in poverty, having incomes that are between 40 and 60 percent less than those of older white Americans. In addition to experiencing educational and employment discrimination similar to that of African Americans and Latinos/as, older Native Americans were also the objects of historical oppression and federal policies toward the native nations that exacerbated patterns of economic impoverishment. Consequently, some older Native Americans have the worst living conditions and poorest health of all older people in this country. Research findings regarding the mental health of older Native Americans also indicate a high rate of depression, alcoholism and other drug abuse, and suicide (see Chapter 18, "Health, Health Care, and Disability").

Shifting our focus to Asian Americans who are now in their seventies and eighties, many who migrated to the United States prior to and during World War II (between 1939 and 1945) have fared less well in their old age than those who were either born in the United States or have arrived more recently. As discussed in Chapter 10, many older Asian Americans from Japan and China received less education and experienced more economic deprivation than did later cohorts of Japanese Americans and Chinese Americans. Today, many older Asian Americans remain in Chinatown, Japantown, or Koreatown, where others speak their language and provide goods and services that help them maintain their culture, and where mutual-aid and benevolent societies and recreational clubs provide them with social contacts and delivery of services. As a result, many older Asian Americans who qualify for

entitlements certain benefit payments made by the government.

various forms of entitlements, such as Supplemental Security Income, do not apply for it. Moreover, cultural values, including traditional healing practices, may help explain why many Asian Americans, particularly Chinese American elders, do not use physical and mental health services that are available to them. Overall, more research is needed on the unique needs of older people from diverse racial and ethnic categories.

Older People in Rural Areas

Rural counties in the United States have gotten older during the first decade of the twenty-first century as the percentage of the population under 25 has dropped in many rural counties. Although the proportion of people over 45 years of age has increased across the United States, the shift is more concentrated in rural counties. In 2009, 43.3 percent of people over age 45 lived in rural counties, as compared to 40.9 percent in exurban counties (outlying counties adjacent to urban areas) and 37.5 percent in urban counties. Rural counties in Nebraska, Arizona, Nevada, California, and Kentucky had the largest percentages of residents over age 65 (Gallardo, 2010).

The lives of many older people differ based on whether they reside in urban or rural areas. Some older individuals who reside in rural areas are middle- and upper-income people who have chosen to retire away from the noise and fast-paced lifestyle of urban areas. Others have lived in rural areas throughout their lives and have limited resources on which to survive. Despite the stereotypical image of the rural elderly living in a pleasant home located in an idyllic country setting, many rural elders, as compared with older urban residents, typically have lower incomes, are more likely to be poor, and have fewer years of schooling. The rural elderly also tend to be in poorer health, and many of them are less likely to receive needed health care because many rural areas lack adequate medical and long-term care facilities (Coburn and Bolda, 1999). With the "graying of America," the population of older adults (age 65 and over) has continued to grow in rural areas; however, this growth varies from region to region.

Whether or not they are poor, the rural elderly receive a higher proportion of their income from Social Security payments than do the urban elderly. Housing also differs between rural and urban elderly, with older rural residents being more likely to own their own homes than older residents in urban settings. However, the homes of the elderly in rural areas are more likely to have a lower value in the real estate market and to be in greater need of repair than are those owned by urban elderly residents (Housing Assistance Council, 2003).

AP Photo/Charles Krupa

▲ Medicare is a hot-button political issue in the United States. Older Americans vote at a much higher rate than does any other age group, so elected politicians are very careful when dealing with elder benefits. AARP, a group supporting elder Americans, took what it saw as a reasonable position on a Medicare bill being debated in Congress. Obviously, these marchers thought differently.

Elder Abuse

Elder abuse **refers to physical abuse, psychological abuse, financial exploitation, and medical abuse or neglect of people age 65 or older** (Hooyman and Kiyak, 2011). Physical abuse involves the use of force to threaten or physically injure an older person. It may be identified by injuries such as bruises, welts, sprains, and dislocations, or problems such as malnutrition. In contrast, emotional or psychological abuse is made up of verbal attacks, threats, rejection, isolation, or belittling acts that cause or could cause mental anguish, pain, or distress to a senior. Financial exploitation typically involves theft, fraud, or misuse of the older person's money or property by another person. Medical abuse occurs when a person withholds, or improperly administers, medications or health aids such as dentures, glasses, or hearing aids. Neglect is a caregiver's failure or refusal to provide care sufficient for the older person to maintain physical and mental health (Hooyman and Kiyak, 2011). Other types of elder abuse include sexual abuse, including sexual contact that is forced, tricked, threatened, or otherwise coerced, and abandonment, which occurs when a person with a duty of care deserts a frail or vulnerable older person (National Center on Elder Abuse, 2010).

How prevalent is the problem of elder abuse? We do not know for certain because relatively few cases are identified. According to the National Center for Victims of Crime (2011), as many as 1.6 million older people in the United States are the victims of physical or mental abuse each year. Just as with violence

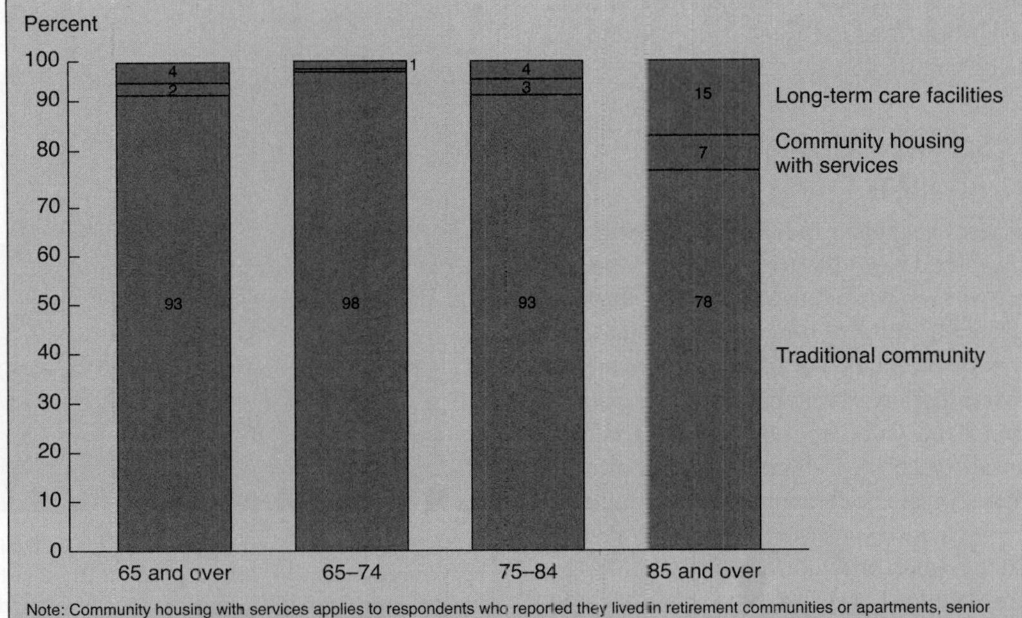

Note: Community housing with services applies to respondents who reported they lived in retirement communities or apartments, senior citizen housing, continuing care retirement facilities, assisted-living facilities, staged living communities, board-and-care facilities/homes, and other similar situations, and who reported they had access to one or more of the following services through their place of residence: meal preparation; cleaning or housekeeping services; laundry services; help with medications. Respondents were asked about access to these services, but not whether they actually used the services. A residence (or unit) is considered a long-term care facility if it is certified by Medicare or Medicaid; or has three or more beds and is licensed as a nursing home or other long-term care facility and provides at least one personal-care service; or provides 24-hour, seven-day-a-week supervision by a nonfamily, paid caregiver.
Reference population: These data refer to Medicare enrollees.

▲ **FIGURE 12.3 PERCENTAGE OF MEDICARE PARTICIPANTS AGE 65 1 IN SELECTED RESIDENTIAL SETTINGS, BY AGE GROUP, 2007**

Source: Federal Interagency Forum on Aging Related Statistics, 2010.

against children or women, it is difficult to determine how much abuse against older people occurs. Many victims are understandably reluctant to talk about it. Studies have shown that elder abuse may occur from age 50 on, but victims of elder abuse tend to be concentrated among those over age 75. Most of the victims are white, middle- to lower-middle-class Protestant women, age 75 to 85, who suffer some form of impairment. Sons, followed by daughters, are the most frequent abusers of older persons.

There has been a widespread belief that elder abuse occurs because the older person is *dependent on other people* (Brandl and Cook-Daniels, 2002). However, studies have found almost no evidence to support that conclusion. According to the sociologist Karl Pillemer (1985), no scientific support exists for the common assumption that dependency on the part of the older person leads to abuse. To the contrary, abusers are very likely to be *dependent on the older person* for housing and financial assistance (Brandl and Cook-Daniels, 2002).

Today, almost every state has enacted mandatory reporting laws regarding elder abuse or has provided some type of governmental protection for older persons. Cases of abuse and neglect of older people are highly dramatized in the media because they offend very central values in the United States—respect for and consideration of older persons.

Living Arrangements for Older Adults

Where older people live is linked to their income, health status, and the availability of caregivers. Among Medicare recipients, 98 percent of individuals between 65 and 74 reside in their own homes. Many live alone, and others reside with a spouse or relative. About 2 percent of Medicare recipients age 65 and over reside in community housing, where some services are provided (usually meal preparation, housekeeping, cleaning, laundry, and help with medications). As shown in ▶ Figure 12.3, about 15 percent of Medicare recipients age 85 and over are residents of long-term care facilities. Research suggests that availability of services in their own residence may help

elder abuse a term used to describe physical abuse, psychological abuse, financial exploitation, and medical abuse or neglect of people age 65 or older.

older persons maintain their independence and avoid institutionalization (Federal Interagency Forum on Aging Related Statistics, 2010).

Support Services, Homemaker Services, and Day Care

Support services help older individuals cope with the problems in their day-to-day lives. These services are often expensive even when they are provided through state or federally funded programs, hospitals, or community organizations. For older persons, homemaker services perform basic chores (such as light housecleaning and laundry); other services deliver meals to homes (such as Meals on Wheels) or provide transportation for medical appointments. Some programs provide balanced meals at set locations, such as churches, synagogues, or senior centers.

Day-care centers have also been developed to provide meals, services, and companionship to older people and in the process to help them maintain as much dignity and autonomy as possible. These centers typically provide transportation, activities, some medical personnel (such as a licensed practical nurse) on staff, and nutritious meals. The cost averages about $64 per day, adding up to $1,000 to $1,500 per month for a five-day week (Helpguide.org, 2011). Medicare does not cover the cost of adult day care; however, in some states, Medicaid covers the fees of older persons living below the poverty line.

Housing Alternatives

Some older adults remain in the residence where they have lived for many years—a process that gerontologists refer to as *aging in place*. Remaining in a person's customary residence is a symbol that he or she is able to maintain independence and preserve ties to his or her neighborhood and surrounding community. However, either by choice or necessity, some older adults move to smaller housing units or apartments.

This type of relocation frequently occurs when older people are living in a residence that does not meet their current needs. Moving to another location may be desirable or necessary when their family has grown smaller, the cost and time necessary to maintain the existing residence have become a strain on them, or individuals experience illness or disabilities that make it difficult for them to remain in the same location.

More housing alternatives are available to middle-income and upper-income older people than to low-income individuals. These include *retirement communities* such as Sun City, where residents must be age 55 or above. (Sun City is the name of several privately owned retirement communities located in such states as Arizona and Texas.) Residents of retirement communities purchase their housing units

Paul Doyle/Alamy

▲ Nursing homes are sometimes an inevitable step in the aging process. However, what distinguishes the British care home shown here from its American counterparts is that in the United States, people sometimes exhaust their insurance and end up spending their entire life savings for their necessary care.

and in some instances pay additional fees for the upkeep of shared areas and amenities such as a golf course, swimming pool, or other recreational facilities. Typically, residents of planned retirement communities are similar to one another with respect to race and ethnic background and social class.

People needing assistance with daily activities or desiring the regular companionship of other people may move to an *assisted-living facility*. Some of these facilities offer fully independent apartments and provide residents with support services such as bathing, help with dressing, food preparation, and taking medication. Some centers provide residents with transportation to medical appointments, beauty salons and barbershops, and social events in the community. However, cost is a compelling factor. Assisted-living centers with more-luxurious amenities are primarily available to retired professionals and other upper-middle- to upper-income people. A number of assisted-living facilities provide residents with the opportunity to age in place by also having long-term care facilities available if and when the need arises. Some of these facilities are well managed and carefully maintained; however, others are a major source of concern to residents, their families, and state regulators who are responsible for inspecting these facilities and responding to complaints filed against them. Some lower-income older people live in planned housing projects that are funded by federal, state, or local government agencies or by religious groups. Other low-income individuals receive care and assistance from relatives, neighbors, or members of religious congregations.

Nursing Homes

It sometimes becomes necessary for an older person to move into a long-term care facility or nursing home as a result of major physical and/or cognitive problems that prevent the person from living in any other setting. There are two types of care provided in nursing homes: skilled care and long-term care. Skilled care provides services that are rendered by a doctor, licensed nurse, physical therapist, occupational therapist, social worker, respiratory therapist, or other specialist. This type of care typically follows a hospital visit and involves short-term nursing care or therapy until the patient can return home. By contrast, long-term care is provided for individuals who can no longer take care of themselves and are likely to remain in the nursing home facility for the remainder of their lifetime (Medicare.gov, 2007).

About 16,000 nursing homes operate in the United States, with a total of more than 1.7 million beds and 1.4 million residents. At any given time about 83 percent of the beds are occupied. There has been some downward fluctuation in the total number of people living in nursing homes in recent years, largely because of the availability of home health care options and assisted-living facilities. As shown in Figure 12.3, most nursing home residents are 85 years of age or older, with relatively few being younger than age 65. Women make up about three-fourths of all residents and have relatively high rates of chronic illness or disability as they reach age 80 and over, making it necessary for them to receive assistance with various activities such as bathing, dressing, walking, and eating (AGS Foundation for Health in Aging, 2007).

Some nursing home residents have mental disturbances or have been diagnosed with dementia—a group of conditions that all gradually destroy brain cells and lead to a progressive decline in mental function. As previously discussed, Alzheimer's disease is the most common form of dementia, and a growing number of nursing home residents have this progressive brain disorder, which gradually destroys a person's memory and ability to learn new information, make judgments, and communicate with others. Some nursing home facilities are designed exclusively for residents with Alzheimer's disease because of the physical and behavioral changes (such as personality and behavioral disorders) that typically accompany this condition.

Nursing home care is paid for in a variety of ways, including Medicare, Medicaid, and long-term-care insurance. Medicare pays for skilled nursing or rehabilitative services but typically does not pay for long-term care. Medicaid, a joint state and federal program, pays most nursing home costs for people with limited income and assets if they are in a nursing home facility certified by the government. Some people have long-term-care insurance or managed-care plans through their health insurance policies; however, about half of all nursing home residents pay nursing home costs out of their own savings. This expense often has devastating consequences on their finances, and after these savings are depleted, they must turn to Medicaid for their long-term care (Medicare.gov, 2007).

How satisfactory are nursing homes in meeting the needs of older individuals? Although there have been frequent criticisms in the media and lawsuits brought against some nursing homes for problems such as neglect or mistreatment of patients, there has been a greater emphasis in both the public and private sector on quality care in nursing facilities. Regulatory changes in the late 1980s mandated that nursing homes must meet specific training guidelines and minimum staffing requirements. Many nursing homes have adopted codes that specify the rights of residents, including the right to be treated with dignity and respect; full disclosure about the services and costs associated with living at the nursing home; the right to manage one's own money or choose the person who will do this; the right to privacy and to keep personal belongings and property unless they constitute a health or safety hazard; the right to be informed about one's medical condition and medications and to see one's own doctor; and the right to refuse medications and treatments (Medicare.gov, 2007). Clearly, there is a difference in *providing care* and in *caring*, for the latter involves offering all residents the emotional support they need in a homelike setting where they can live out the remainder of their lives with a sense of well-being and dignity.

Sociological Perspectives on Aging

Sociologists and social gerontologists have developed a number of explanations regarding the social effects of aging. Some of the early theories were based on a microlevel analysis of how individuals adapt to changing social roles. More-recent theories have used a macrolevel approach to examine the inequalities produced by age stratification at the societal level.

Functionalist Perspectives on Aging

Functionalist explanations of aging focus on how older persons adjust to their changing roles in society. According to the sociologist Talcott Parsons (1960), the roles of older persons need to be redefined by society. He suggested that devaluing the contributions of older persons is dysfunctional for society; older persons often have knowledge and wisdom to share with younger people.

How does society cope with the disruptions resulting from its members growing older and dying?

According to ***disengagement theory,* older persons make a normal and healthy adjustment to aging when they detach themselves from their social roles and prepare for their eventual death** (Cumming and Henry, 1961). Gerontologists Elaine C. Cumming and William E. Henry (1961) noted that disengagement can be functional for both the individual and society. For example, the withdrawal of older persons from the workforce provides employment opportunities for younger people. Disengagement also aids a gradual and orderly transfer of statuses and roles from one generation to the next; an abrupt change would result in chaos. Retirement, then, can be thought of as recognition for years of service and the acknowledgment that the person no longer fits into the world of paid work. The younger workers who move into the vacated positions have received more up-to-date training—for example, the computer skills that are taught to most young people today.

Critics of this perspective object to the assumption that all older persons want to disengage while they are still productive and still gain satisfaction from their work. Disengagement may be functional for organizations but not for individuals. A corporation that encourages retirement by paying "retirement bonuses" or uses other means to encourage older workers to retire may be able to replace higher-paid, older workers with lower-paid, younger workers, but retirement is not always beneficial for older workers, particularly if they were not ready to retire. Contrary to disengagement theory, a number of studies have found that activity in society is *more* important with increasing age.

Symbolic Interactionist Perspectives on Aging

Symbolic interactionist perspectives examine the connection between personal satisfaction in a person's later years and a high level of activity. ***Activity theory* states that people tend to shift gears in late middle age and find substitutes for previous statuses, roles, and activities** (Havighurst, Neugarten, and Tobin, 1968). From this perspective, older people have the same social and psychological needs as middle-aged people and thus do not want to withdraw unless restricted by poor health or disability.

Whether retired persons invest their energies in grandchildren, traveling, hobbies, new work roles, or volunteering in the community, their social activity is directly related to longevity, happiness, health, and overall social well-being. Healthy people who remain active have a higher level of life satisfaction than do those who are inactive or in ill health (Havighurst, Neugarten, and Tobin, 1968). Among those whose mental capacities decline later in life, deterioration is most rapid in people who withdraw from social relationships and activities (see "You Can Make a Difference").

▲ According to the symbolic interactionist perspective, older people who invest their time and energy in volunteer work or in other enjoyable activities tend to be healthier, be happier, and live longer than those who disengage from society.

A variation on activity theory is the concept of *continuity*—that people are constantly attempting to maintain their self-esteem and lifelong principles and practices, and that they simply adjust to the feedback from and needs of others as they grow older (Williamson, Duffy Rinehart, and Blank, 1992). From this perspective, aging is a continuation of earlier life stages rather than a separate and unique period. Thus, values and behaviors that have previously been important to an individual will continue to be so as the person ages.

Another interactionist approach, the *social constructionist perspective on aging,* seeks to understand how individual processes of aging are influenced by social definitions and social structures. The social constructionist approach recognizes that individuals are active participants in their daily lives as they create and maintain social meanings for themselves and their acquaintances. In other words, through a person's relationships with others, the individual creates a "reality" that then structures his or her life. For example, one study examined how "frailty" is not a necessarily a specific physical condition as much as it is a social construct produced through the interactions of older people, their caretakers, and their health professionals. Through these interactions the subjective experience of frailty becomes interpreted and defined in a "medical/social idiom" that is framed in terms of surveillance ("keeping an eye on them") and independence (how much freedom they should have to move around on their own without falling, for example). Then rules (made by caregivers and/or medical professionals) for what the person should or should not do become facts and a social reality apart from the individual's actual

you can make a difference

Getting Behind the Wheel to Help Older People: Meals on Wheels

I don't have strength in my hands. I'll have something in my hands and all of a sudden it will fall. . . . [Meals on Wheels volunteers] always sort of bring you into their situation and make you feel like you are part of the outside world. For me, it's just hard to explain how much that means to me to have them come by here every day and have a few words of "How are you?" and back and forth. That they bring you something that is good for you—I just can't say enough about how much I appreciate that work on their part.

—a woman explaining how much Meals and Wheels of San Antonio, Texas, means to her (qtd. in MOWAA, 2011)

Like this grateful woman, thousands of women and men across the United States who are elderly, homebound, disabled, or frail have nutritious meals delivered to their residence by volunteers from Meals on Wheels, an organization that asks recipients to contribute what they can toward the cost of their meals but relies primarily on donations of time and money to cover the cost of the service that it provides. Although the majority of Meals on Wheels recipients are over age 60, some younger disabled or ill persons also benefit from this service.

What do "Wheels" volunteers do? On their designated service days, volunteers pick up prepared meals packed in insulated carriers at a neighborhood meal site and deliver the food to the people on their "route," which usually serves about ten people and takes no more than about an hour, even allowing for time to visit briefly with each food recipient. Along with delivering food to people's homes, Meals on Wheels volunteers also provide recipients with a warm smile and a caring attitude. Volunteers serve as a safety check to ensure that the older person is doing all right. In orientation sessions, volunteers learn what to do if they get to a residence and find that a person is in need of assistance (MOWAA, 2011).

What does it take to become a Meals on Wheels volunteer? "Wheels" volunteers must have a valid driver's license and proof of insurance. They must be 18 years of age or older, or if they are younger, they must be accompanied by a person over age 18 who meets the licensing and insurance requirements. Meals on Wheels organizations in various communities have different guidelines on how frequently they want volunteers to work, but a minimum of two days per month is a general requirement. Obviously, many local "Wheels" organizations would like for volunteers to work more often than that and attempt to assign volunteers to an area near their own home or workplace in order to make it easier to meet that goal (MOWAA, 2011).

What can you do to help provide food for elderly, homebound residents? Providing a little of our time and/or some financial support (even a small donation) greatly helps programs such as Meals on Wheels. Although the Meals on Wheels Association of America is a national organization that works with affiliate programs and member groups across the nation, Meals on Wheels is largely a grassroots operation that relies on local volunteers to provide time and money to help make a difference in the lives of the most vulnerable individuals in their community. A number of programs have been especially hard hit by a dramatic increase in the number of older residents in their area, the rising price of gasoline, and troubling economic times across the nation. If you are interested in helping ensure that older people in your community have nutritious food to eat, contact your local Meals on Wheels organization or visit the website of the Meals on Wheels Association of America (**www.mowaa.org**) for additional information.

Can you think of additional ways in which you might help older individuals in your community or in your own family receive food or other services they may need?

---

physical condition (Kaufman, 1994). Theories based on a symbolic interactionist tradition provide new understandings about the problems of aging and how interactions with others create social realities that become important in social–structural contexts (Bengtson, Burgess, and Parrott, 1997).

Conflict Perspectives on Aging

Conflict theorists view aging as especially problematic in contemporary capitalistic societies. As people grow older, their power tends to diminish unless they are able to maintain wealth. Consequently, those who have been disadvantaged in their younger years become even more so in late adulthood. Women age 75 and over are among the most disadvantaged because they often must rely solely on Social Security, having outlived their spouses and sometimes their children (Harrington Meyer, 1990).

disengagement theory the proposition that older persons make a normal and healthy adjustment to aging when they detach themselves from their social roles and prepare for their eventual death.

activity theory the proposition that people tend to shift gears in late middle age and find substitutes for previous statuses, roles, and activities.

© Sonda Dawes/The Image Works

▲ According to the conflict perspective, differences in class, gender, and race/ethnicity increasingly divide older people in the United States into the "haves" and the "have-nots." The woman pictured here, for instance, lives in a house without indoor plumbing. Older African Americans are especially affected by poverty because of overt discrimination in the past and more subtle forms of contemporary prejudice.

Underlying the capitalist system is an ideology which assumes that all people have equal access to the means of gaining wealth and that poverty results from individual weakness. When older people are in need, they may be viewed as not having worked hard enough or planned adequately for their retirement. The family and the private sector are seen as the "proper" agents to respond to their needs. To minimize the demand for governmental assistance, these services are made punitive and stigmatizing to those who need them. Class-based theories of inequality assert that government programs for older persons stratify society on the basis of class. Feminist approaches claim that these programs perpetuate inequalities on the basis of gender and race in addition to class. A political economy approach seeks to explain how economic and political forces come together to determine how resources are allocated in a society and why there is so often in high-income nations a variation in the treatment and status of older people. To better understand where these inequalities originated and how they are perpetuated, researchers using a political economy perspective examine public policies, economic trends, and social–structural factors. Examples include studies of the politicization of Alzheimer's disease (Robertson, 1990) and how "multiple jeopardies" of age, class, and race affect the lives of older, low-income women from subordinate racial or ethnic groups (Bengtson, Burgess, and Parrott, 1997).

Conflict analysis draws attention to the diversity in the older population. Differences in social class, gender, and race/ethnicity divide older people just as they do everyone else. Wealth cannot forestall aging indefinitely, but it can soften the economic hardships faced in later years. The conflict perspective adds to our understanding of aging by focusing on how capitalism devalues older people, especially women. However, critics assert that this approach ignores the fact that industrialization and capitalism have greatly enhanced the longevity and quality of life for many older persons.

The major sociological perspectives regarding aging are summarized in the Concept Quick Review.

[concept quick review]

Theoretical Perspectives on Aging

	Focus	Theory/Hypothesis
Functionalist Perspectives	How older people adjust to changing roles in society	Disengagement theory suggests that detachment and preparation for death are normal and healthy adjustments for older individuals.
Symbolic Interactionist Perspectives	Why microlevel contacts between individuals are particularly important for older people	Activity theory is based on the assumption that people are more satisfied in old age if they remain active and find new statuses, roles, and activities. Social constructionist approaches examine how individual processes of aging are influenced by social definitions and social structures.
Conflict Perspectives	How aging is difficult in a capitalist economy and why race, class, and gender are factors that make a difference in the well-being of older people	Inequality follows people across the life course, and poor and middle-income individuals often live on fixed incomes and must rely on Social Security and Medicare benefits.

Death and Dying

Historically, death was a common occurrence at all stages of the life course. Until the twentieth century, the chances that a newborn child would live to adulthood were very small. Poor nutrition, infectious diseases, accidents, and natural disasters took their toll on men and women of all ages. But in contemporary, industrial societies, death is looked on as unnatural because it has been largely removed from everyday life. Most deaths now occur among older persons, and the association of death with the aging process has contributed to ageism in our society; if people can deny aging, they feel that they can deny death.

In the past, explanations for death and dying were rooted in custom or religious beliefs. Today, people who have religious beliefs regarding living an afterlife typically have less anxiety about death, which they may view as the beginning of a better life (Feifel and Nagy, 1981). Research has shown that those who are most fearful of death are people who are confused or uncertain about their religious beliefs, not those who have confirmed their lack of religious belief (Hooyman and Kiyak, 2011).

How do people cope with dying? There are four widely known frameworks for explaining how people cope with the process of dying or with the loss of a loved one: the *stage-based approach,* the *trajectories of grief approach,* the *dying trajectory,* and the *task-based approach.*

The *stage-based approach* was popularized by psychiatrist Elisabeth Kübler-Ross (1969), who proposed five stages in the dying process: (1) denial and isolation ("Not me!"), (2) anger and resentment ("Why me?"), (3) bargaining and an attempt to postpone ("Yes me, but . . ."—negotiating for divine intervention), (4) depression and sense of loss, and (5) acceptance. She pointed out that these stages are not the same for all people; some of the stages may exist at the same time. Kübler-Ross (1969: 138) also stated that "the one thing that usually persists through all these stages is hope." Kübler-Ross's stages were attractive to the general public and the media because they provided common responses to a difficult situation. On the other hand, her stage-based model also generated a great deal of criticism. Some have pointed out that these stages have never been conclusively demonstrated or comprehensively explained.

The *trajectories of grief approach* was introduced by the clinical psychologist George Bonanno (2009, 2010), who claims that bereavement studies have disproved the stage-based approach and have helped researchers identify four common trajectories of grief:

- *Resilience*—the ability of people to maintain a relatively stable, healthy level of psychological and physical functioning while they are dealing with a highly disruptive event such as the death of a relative or a life-threatening situation.
- *Recovery*—a gradual return to previous levels of normal functioning after experiencing a period of psychological stress such as depression or post-traumatic stress disorder.
- *Chronic dysfunction*—lengthy suffering (sometimes several years or more) and the inability to function after experiencing grief.
- *Delayed grief or trauma*—experiencing what appears to be a normal adjustment after the loss of a loved one but then having an increase in distress and other symptoms months later. (based on Bonanno, 2010)

To explain his views on grief-related behavior, Bonanno coined the term "coping ugly" to describe how some seemingly inappropriate or counterintuitive behavior (such as telling jokes or laughing) may seem odd at the time but help a person move on after a loss. Although Bonanno's ideas about persistent resilience have been embraced by some gerontologists and bereavement specialists, others disagree and see this condition as a form of denial that might necessitate counseling or other treatment.

A third approach is referred to as the *dying trajectory,* which focuses on the perceived course of dying and the expected time of death. For example, a dying trajectory may be sudden, as in the case of a heart attack, or it may be slow, as in the case of lung cancer. According to the dying-trajectory approach, the process of dying involves three phases: the acute phase, characterized by the expression of maximum anxiety or fear; the chronic phase, characterized by a decline in anxiety as the person confronts reality; and the terminal phase, characterized by the dying person's withdrawal from others (Glaser and Strauss, 1968).

Finally, the *task-based approach* is based on the assumption that the dying person can and should go about daily activities and fulfill tasks that make the process of dying easier on family members and friends, as well as on the dying person. Physical tasks can be performed to satisfy bodily needs, whereas psychological tasks can be done to maximize psychological security, autonomy, and richness of experience. Social tasks sustain and enhance interpersonal attachments and address the social implications of dying. Spiritual tasks help people to identify, develop, or reaffirm sources of spiritual energy and to foster hope (Corr, Nabe, and Corr, 2003).

In the final analysis, however, how a person dies or experiences the loss of a loved one is shaped by many social and cultural factors. These endeavors are influenced by an individual's personality and philosophy of life, as well as the social context in which these events occur.

In recent years the process of dying has become an increasingly acceptable topic for public discussion. Such discussions helped further the hospice movement in the 1970s (Weitz, 2004). A ***hospice* is an organization that provides a homelike facility or home-based care (or both) for people who are terminally ill.** The hospice philosophy asserts that people should participate in their own care and have control over as many decisions pertaining to their life as possible. Pain and suffering should be minimized, but artificial measures should not be used to sustain life. This approach is family based and provides support for family members and friends, as well as for the person who is dying (see Corr, Nabe, and Corr, 2003). Although the hospice movement has been very successful (with some 1,800 facilities), critics claim that the movement has exchanged much of its initial philosophy and goals for social acceptance and financial support (Finn Paradis and Cummings, 1986; Weitz, 2004). Over time, hospice care has moved toward hospital standards because of the need for hospices to work with the federal government and the American Hospital Association to gain accreditation. Medicare funding and the resultant federal regulations have further changed hospices (see Weitz 2004).

Aging in the Future

The size of the older population in the United States will increase dramatically in the early decades of the twenty-first century. By the year 2050, there will be an estimated 88.5 million people age 65 and older (20 percent of the total population), as compared with 39.6 million (13 percent of the total population) in 2009. Thus, combined with decreasing birthrates, most of the population growth will occur in the older age cohorts during the next 50 years. More people will survive to age 85, and more will even reach the 95-and-over cohort. Among those 85 and older in 2050, about 33 percent are projected to be members of minority groups.

As the previous statement indicates, the demographics of the older population will undergo a dramatic transformation in the future. By 2050, the number of African Americans over age 65 will triple, moving them from 8 to 10 percent of all Americans age 65 and above. Older Latinos/as (Hispanics) will increase from fewer than 4 percent of all people over age 65 to nearly 16 percent of older adults, for an 11 percent gain. This demographic shift means that many older Americans will have increased needs for health and social services: African Americans and Latinos/as have special concerns regarding the onset of chronic illnesses at an earlier age than do white (non-Hispanic) Americans, higher incidences of obesity and late-onset diabetes, and social problems that contribute to mental health concerns such as higher rates of poverty, racially segregated communities, poor schools, high unemployment rates, and limited access to health care. In regard to access to health care for all older individuals, the effects of the 2010 Patient Protection and Affordable Care Act will become more apparent in the future as no-cost preventive medical services will become available and the closing of the income-based gap in Medicare prescription drug coverage, known as the "doughnut hole," is completed by 2020. It will be important to develop better and more comprehensive ways of assisting people from all racial/ethnic and nationality groupings so that they may be able to live full and productive lives.

In the future, who will assist older people with needs they cannot meet themselves? Family members in the future may be less willing or able to serve as caregivers. Women, the primary caregivers in the past, are faced with *triple* workdays if they attempt to combine working full time with caring for their children and assisting older relatives. More caregivers in the future will probably be paid employees who work in the individual's home or who are employed by independent or assisted-living facilities or nursing homes.

As biomedical research on aging continues, new discoveries in genetics may eliminate life-threatening diseases and make early identification of other diseases possible. Technological advances in the diagnosis, prevention, and treatment of Alzheimer's disease may revolutionize people's feelings about growing older. Advances in medical technology may lead to a more positive outlook on aging.

If these advances occur, will they help everyone or just some segments of the population? This is a very important question for the future. As we have seen, many of the benefits and opportunities of living in a highly technological, affluent society are not available to all people. Classism, racism, sexism, and ageism all serve to restrict individuals' access to education, medical care, housing, employment, and other valued goods and services in society.

For older persons, the issues discussed in this chapter are not merely sociological abstractions; they are an integral part of their everyday lives. Older people have resisted ageism through organizations such as the Gray Panthers, AARP, and the Older Women's League.

hospice an organization that provides a homelike facility or home-based care (or both) for people who are terminally ill.

chapter review Q&A

Use these questions and answers to check how well you've achieved the learning objectives set out at the beginning of this chapter.

- **What is aging, and what is the study of aging called?**

Aging refers to the physical, psychological, and social processes associated with growing older. Gerontology is the study of aging and older people. Social gerontology is the study of the social (nonphysical) aspects of aging, including the consequences of an aging population and the personal experience of aging.

- **How do views of aging differ in preindustrial and industrialized societies?**

In preindustrial societies, people of all ages are expected to share the work, and the contributions of older people are valued. In industrialized societies, however, older people are often expected to retire so that younger people may take their place.

- **What are ageism and elder abuse, and how are these perpetrated in the United States?**

Ageism is prejudice and discrimination against people on the basis of age, particularly against older persons. Ageism is reinforced by stereotypes of older people. Elder abuse includes physical abuse, psychological abuse, financial exploitation, and medical abuse or neglect of people age 65 or older. Passive neglect is the most common form of abuse.

- **How do functionalist and symbolic interactionist explanations of aging differ?**

Functionalist explanations of aging focus on how older persons adjust to their changing roles in society; the gradual transfer of statuses and roles from one generation to the next is necessary for the functioning of society. Activity theory, a part of the symbolic interactionist perspective, states that people change in late middle age and find substitutes for previous statuses, roles, and activities. This theory asserts that people do not want to withdraw unless restricted by poor health or disability.

- **What is the conflict perspective on aging and age-based inequality?**

Conflict theorists link the loss of status and power experienced by many older persons to their lack of ability to produce and maintain wealth in a capitalist economy.

- **What are the most common living arrangements for older people?**

Many older persons live alone or in an informal family setting. Support services and day care help older individuals who are frail or disabled cope with their daily needs, although many older people do not have the financial means to pay for these services. Nursing homes are the most restrictive environment for older persons. Many nursing home residents have major physical and/or cognitive problems that prevent them from living in any other setting, or they do not have available caregivers in their family.

- **How is death typically viewed in industrialized societies?**

In industrialized societies, death has been removed from everyday life and is often regarded as unnatural.

- **What stages in coping with dying were identified by Elisabeth Kübler-Ross?**

Kübler-Ross proposed five stages of coping with dying: denial, anger, bargaining, depression, and acceptance.

key terms

activity theory 368
age stratification 353
aging 347
chronological age 348
cohort 350
disengagement theory 368
elder abuse 364
entitlements 362
functional age 348
gerontology 350
hospice 372
life expectancy 350

questions for critical thinking

1. Why does activity theory contain more positive assumptions about older persons than disengagement theory does? Analyze your grandparents (or other older persons whom you know well or even yourself if you are older) in terms of disengagement theory and activity theory. Which theory seems to provide the most insights? Why?
2. How are race, class, gender, and aging related?
3. Is it necessary to have a mandatory retirement age? Why or why not?
4. How will the size of the older population in the United States affect society and programs such as Social Security in the future?

turning to video

Watch the ABC video *Sex in America: What Should the Age of Consent Be?* (running time 3:40), available through **CengageBrain.com**. This video discusses the age-of-consent laws and how they vary from state to state. As you watch the video, think about your own lifestyle choices and those of your peers in high school and college. After you've watched the video, consider these questions: In what ways do you think age-of-consent laws are harmful to teens? How are they helpful? Why do you think adolescents experience inequalities based on age?

online study resources

Go to CENGAGE brain.com to access online study resources, including the Sociology CourseMate for this text as well as special features such as video, an interactive sociology time line and interactive maps, General Social Survey (GSS) data, and U.S. Census 2010 data.

CourseMate brings course concepts to life with interactive learning, study, and exam-preparation tools that support the printed textbook. A textbook-specific website, **Sociology CourseMate** includes an integrated interactive eBook and other interactive learning tools, including quizzes, flash cards, and videos.

Visit **www.cengagebrain.com** to access your account and purchase materials.

13 The Economy and Work in Global Perspective

I didn't expect to be laid off. Not at all. I graduated from college in 2006 and quickly landed my first job as an IT specialist.... I knew the company wasn't doing as well as it had been, but I was still surprised.... I've cut back on everything. I eliminated going out for lunch and going out on the town on weekends. I go out about once every two months now. It's not the life I expected at 26. Every day, I search for new job postings. I've had some interviews but nothing solid has come from them. It's tough. I have ups and downs. Some days I feel hopeless, but I know I have to be optimistic.

—Jon Mikres describing how frustrating it is for him and many others like him to be laid off and unable to find another position (qtd. in CNNmoney.com, 2009)

It was always apparent to me that project [temporary] attorneys felt the passing of time differently compared to the law firm's permanent associates and partners. While we both shared the practice of billing our work by the hour, we nonetheless each experienced a unique temporal reality.... Our day-to-day tasks—reading and coding documents—were mind-numbingly repetitive. Our chances for extra rewards and advancement were negligible.... At the project's end, we did not

© Julian Winslow/Corbis

▲ Starting in 2008, many large U.S. corporations either closed their doors or severely curtailed their operations, cutting millions of jobs. Do you know anyone who suffered a job loss during this period?

experience the thrill of victory; we only felt that we had worked ourselves out of a job.

—Robert A. Brooks (2011), now a university professor, explaining what life was like when he worked as a temporary attorney who was hired by law firms to do endless document review for litigation or corporate transactions

Chapter Focus Question:

How are our beliefs about work influenced by changes in the economy and in the world in which we live?

As the comments of Jon and Robert suggest, problems of unemployment and underemployment are major concerns to millions of people in the United States. Although comments such as these about the U.S. economy and work represent only a few of the many narratives we might present to describe people's fears associated with losing a job or finding only temporary work when seeking full-time employment, these examples are typical of the significant changes that affect millions of people in the difficult economic times facing our nation and world.

A huge number of people have seen their jobs vanish as they joined the ranks of the unemployed. In March 2011, for example, the unemployment rate was 8.9 percent (13.7 million people), and the number of people employed part time for economic reasons (referred to as "involuntary part-time workers") such as Robert was 8.3 million. As a result of the Great Recession, which began in 2007–2008, some analysts estimate that more than 7.9 million jobs have been lost for good: Many of these positions will never return (Isidore, 2010). Among the hardest hit by this recession were younger people (ages 16 to 19 and 20 to 24). According to some analysts, this job gap might mean that people graduating from college today may earn approximately 17.5 percent less per year than comparable peers graduating in better labor markets, with this lower wage effect fading away only after 17 years of work. This problem was particularly intense for college graduating classes in 2008 through 2010 (Greenstone and Looney, 2010). Hopefully, this situation will improve in the next few years!

What is the connection between the work-related problems of unemployment and underemployment and the larger economic structure of our society and the global economy? In this chapter we will discuss the economy as a social institution and explain how our current problems are linked to the larger world of work—what people are paid for their labor, who gains and who loses in tough economic times, how people feel about their work, and what impact that recent changes in the economy may have on your future. Before reading on, test your knowledge about the economy and the world of work by taking the Sociology and Everyday Life quiz.

Highlighted Learning Objectives

- Compare the key characteristics of capitalism, socialism, and mixed economies.
- Identify factors that contribute to job satisfaction and to worker alienation.
- Describe the individual's role in the workforce.
- Discuss key reasons why the United States has a relatively high rate of unemployment.
- Describe ways in which workers attempt to gain control over their work situation.

sociology and everyday life

How Much Do You Know About the Economy and the World of Work?

True	False	
T	F	1. In today's economy, professions are largely indistinguishable from other occupations.
T	F	2. Workers' skills are usually upgraded when new technology is introduced in the workplace.
T	F	3. Many jobs in the service sector pay poorly and offer little job security.
T	F	4. In the United States and other nations, most assembly-line workers are white men.
T	F	5. Labor unions will probably cease to exist sometime during this century.
T	F	6. Few workers resist work conditions that they consider to be oppressive.
T	F	7. Around the world, positions with the most job security are located in large, transnational corporations.
T	F	8. Assembly lines are rapidly disappearing from all sectors of the U.S. economy.

Answers on page 380.

Comparing the Sociology of Economic Life with Economics

Perhaps you are wondering how a sociological perspective on the economy differs from the study of economics. Although aspects of the two disciplines overlap, each provides a unique perspective on economic institutions. Economists focus on the complex workings of economic systems (such as monetary policy, inflation, and the national debt), whereas sociologists focus on interconnections among the economy, other social institutions, and the social organization of work. At the macrolevel, sociologists may study the impact of transnational corporations on industrialized and low-income nations. At the microlevel, sociologists might study people's satisfaction with their jobs. To better understand these issues, we will examine how economic systems came into existence and how they have changed over time.

Economic Systems in Global Perspective

The ***economy*** **is the social institution that ensures the maintenance of society through the production, distribution, and consumption of goods and services.** *Goods* are tangible objects that are necessary (such as food, clothing, and shelter) or desired (such as iPads and the latest software). *Services* are intangible activities for which people are willing to pay (such as dry cleaning, a movie, or medical care). In high-income nations today, many of the goods and services that we consume are information goods. Examples include databases and surveys ("intermediate products") and the mass media, computer software, and the Internet ("information goods").

Some goods and services are produced by human labor (the plumber who unstops your sink, for example); others are primarily produced by capital (such as Internet and high-definition-TV access available through a service provider). *Labor* refers to the group of people who contribute their physical and intellectual services to the production process in return for wages that they are paid by firms (Boyes and Melvin, 2011). Capital is wealth (money or property) owned or used in business by a person or corporation. Obviously, money, or financial capital, is needed to invest in the physical capital (such as machinery, equipment, buildings, warehouses, and factories) used in production. For example, a person who owns a thousand shares of stock in a high-tech company owns financial capital, but these shares also represent an ownership interest in that corporation's physical capital.

To better understand the economy in the United States today, let's briefly look at three broad categories of economies: preindustrial, industrial, and postindustrial economies.

Preindustrial Economies

Preindustrial economies include hunting and gathering, horticultural and pastoral, and agrarian societies. Most workers engage in ***primary sector production*—the extraction of raw materials and natural resources from the environment**. These materials and resources are typically consumed or used without much processing. The

▲ Even as the United States and other high-income countries have increasingly relied on high-tech economies, other forms of work still exist in the agricultural and industrial sectors of the economy. Here, workers handpick a strawberry crop and work in a paper mill.

production units in hunting and gathering societies are small; family members produce most goods. The division of labor is by age and gender. The potential for producing surplus goods increases as people learn to domesticate animals and grow their own food.

In horticultural and pastoral societies, the economy becomes distinct from family life. The distribution process becomes more complex, with the accumulation of a *surplus* such that some people can engage in activities other than food production. In agrarian societies, production is primarily related to producing food. However, workers have a greater variety of specialized tasks, such as warlord or priest; for example, warriors are necessary to protect the surplus goods from plunder by outsiders. Once a surplus is accumulated, more people can also engage in trade. Initially, the surplus goods are distributed through a system of *barter*—the direct exchange of goods or services considered of equal value by the traders. However, bartering is limited as a method of distribution; equivalencies are difficult to determine (how many fish equal one rabbit?) because there is no way to assign a set value to the items being traded. As a result, *money*, a medium of exchange with a relatively fixed value, came into use in order to aid the distribution of goods and services in society.

What was the U.S. economy like in the preindustrial era? In the preindustrial economy of the colonial period (from the 1600s to the early 1700s), white men earned a livelihood through agricultural work or as small-business owners who ran establishments such as inns, taverns, and shops. During this period, white women worked primarily in their homes, doing such tasks as cooking, cleaning, and child care. Some also developed *cottage industries*—producing goods in their homes that could be sold to nonfamily members. However, a number of white women also worked outside their households as midwifes, physicians, nurses, teachers, innkeepers, and shopkeepers (Hesse-Biber and Carter, 2000). By contrast, the experiences of people of color were quite different in preindustrial America. According to the sociologists Sharlene Hesse-Biber and Gregg Lee Carter (2000), slavery, which came about largely as a result of the demand for cheap agricultural labor, was a major force in the exploitation of people of color, particularly women of color who suffered a double burden of both sexism and racism. By contrast, Native American women in some agricultural communities held greater power because they were able to maintain control over land, tools, and surplus food (Hesse-Biber and Carter, 2000).

Do preindustrial forms of work still exist in contemporary high-income nations? In short, yes. For example, portions of contemporary sub-Saharan Africa have a relatively high rate of exports in primary commodities, and foreign direct

economy the social institution that ensures the maintenance of society through the production, distribution, and consumption of goods and services.

primary sector production the sector of the economy that extracts raw materials and natural resources from the environment.

sociology and everyday life

ANSWERS to the Sociology Quiz on the Economy and the World of Work

1. **False.** Even in tough economic times, professions have five characteristics that distinguish them from other occupations: (1) abstract, specialized knowledge; (2) autonomy; (3) self-regulation; (4) authority over clients and subordinate occupational groups; and (5) a degree of altruism. As the example of the project lawyer shows, however, a profession may be degraded if the only work that many professionals in the field can find is temporary employment doing tedious, repetitive tasks that workers see as being beneath their education level and abilities.
2. **False.** Jobs are often deskilled when technology (such as bar-code scanners or computerized cash registers) is installed in the workplace. Some of the workers' skills are no longer needed because a "smart machine" now provides the answers.
3. **True.** Many jobs in the service sector, such as nurse's aide, child-care worker, hotel maid, and fast-food server, offer little job security and low pay.
4. **False.** Today, most assembly-line workers are young girls and women in developing nations, and women or men of color in the United States.
5. **False.** Sociologists who have examined organized labor generally predict that unions will continue to exist but in a much weakened state.
6. **False.** Many workers resist work conditions that they believe are unjust or oppressive. However, fewer workers are now involved in work-related sabotage or in other forms of labor activism.
7. **False.** Although it is difficult to determine which types of jobs are the most secure, many positions in transnational corporations have been lost through downsizing and plant relocations and closings.
8. **False.** According to some scholars, assembly-line type production will remain a fact of life for businesses ranging from fast-food restaurants to high-tech semiconductor plants. However, in certain sectors of the economy, such as the automobile industry, the number of jobs is rapidly declining.

Sources: Based on Leonhardt, 2011; and U.S. Bureau of Labor Statistics, 2011a, 2011c.

investment is concentrated in mineral extraction. Even in high-income nations such as the United States, entire families work in the agricultural sector of the economy, performing tasks such as picking ripened fruits and vegetables. Some parts of the agricultural sector coexist beside industrial and postindustrial sectors. For example, cherry pickers on the West Coast are employed in the same region as high-tech information workers who are employed by Microsoft and other software developers or computer manufacturers.

Industrial Economies

Industrial economies result from sweeping changes to the system of production and distribution of goods and services. Drawing on new forms of energy (such as steam, gasoline, and electricity) and machine technology, factories proliferate as the primary means of producing goods. Most workers engage in ***secondary sector production*—the processing of raw materials (from the primary sector) into finished goods**. For example, steelworkers process metal ore; autoworkers then convert the ore into automobiles, trucks, and buses. In industrial economies, work becomes specialized and repetitive, activities become bureaucratically organized, and workers primarily work with machines instead of with one another. With the emergence of mass production, larger surpluses are generated, typically benefiting some people and organizations but not others.

In sum, the typical characteristics of industrial economies include the following:

1. New forms of energy, mechanization, and the growth of the factory system.
2. Increased division of labor and specialization among workers.
3. Universal application of scientific methods to problem solving and profit making.
4. Introduction of wage labor, time discipline, and workers' deferred gratification, which means that employees should be diligent at work and pursue personal activities on their own time only.
5. Strengthening of bureaucratic organizational structure and enforcement of rules, policies, and procedures to make the workplace more efficient and profitable.

All these characteristics contribute to the development of industrial economies, greater productivity in the workplace, and a dramatic increase in consumption because many more goods are available at affordable prices.

Although industrialization brought about an increased standard of living for many people, the sociologist Thorstein Veblen (1857–1929) criticized industrialism in his famous book, *The Theory of the Leisure Class* (1967/1899). According to Veblen, the idle rich, who made vast sums of money through factory ownership and the hard work of their employees, represented a conspicuously consuming, parasitic leisure class. *Conspicuous consumption* is the ostentatious display of symbols of wealth, such as owning numerous mansions and expensive works of art, wearing extravagant jewelry and clothing, or otherwise flaunting the trappings of great wealth. By contrast, *conspicuous leisure* involves wasteful and highly visible leisure activities such as casino gambling or expensive sporting events that require costly gear or excessive travel expenses (such as going on a safari in Africa). If Veblen were alive today, do you think he would feel the same way about conspicuous consumption and perhaps incorporate the spending habits of Wall Street bankers, hedge fund managers, or billionaire entrepreneurs?

Postindustrial Economies

A postindustrial economy is based on ***tertiary sector production*—the provision of services rather than goods**—as a primary source of livelihood for workers and profit for owners and corporate shareholders. Tertiary sector production includes a wide range of activities, such as fast-food service, transportation, communication, education, real estate, advertising, sports, and entertainment. As shown in "Census Profiles," a majority of U.S. jobs are in tertiary sector employment, as contrasted with primary or secondary sector employment.

Civilian Occupations

One of the questions that the Census Bureau asks is about the occupations of people who are age 16 or over. The majority of people who are not full-time military employees responded in the most recent survey that they held jobs in the tertiary sector, including management, professional, service, and sales occupations, as contrasted with primary sector and secondary sector employment.

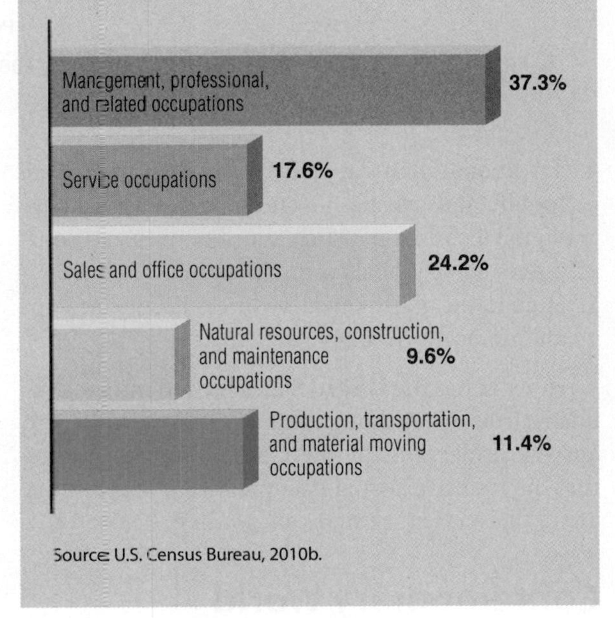

Source: U.S. Census Bureau, 2010b.

Graham Whitby Boot/Allstar/Sportsphoto Ltd./Allstar/Newscom

▲ Some of the greatest profits made in the postindustrial phase of the U.S. economy have come from computers, software, and information systems. Google Inc. is at the forefront of the information age.

Five characteristics are central to the postindustrial economy:

1. Service industries dominate over manufacturing.
2. Information and technological innovation displace property as the central preoccupation in the economy.
3. Professional and technical classes grow more predominant, and workplace culture shifts from factories to diversified work settings.

secondary sector production the sector of the economy that processes raw materials (from the primary sector) into finished goods.

tertiary sector production the sector of the economy that is involved in the provision of services rather than goods.

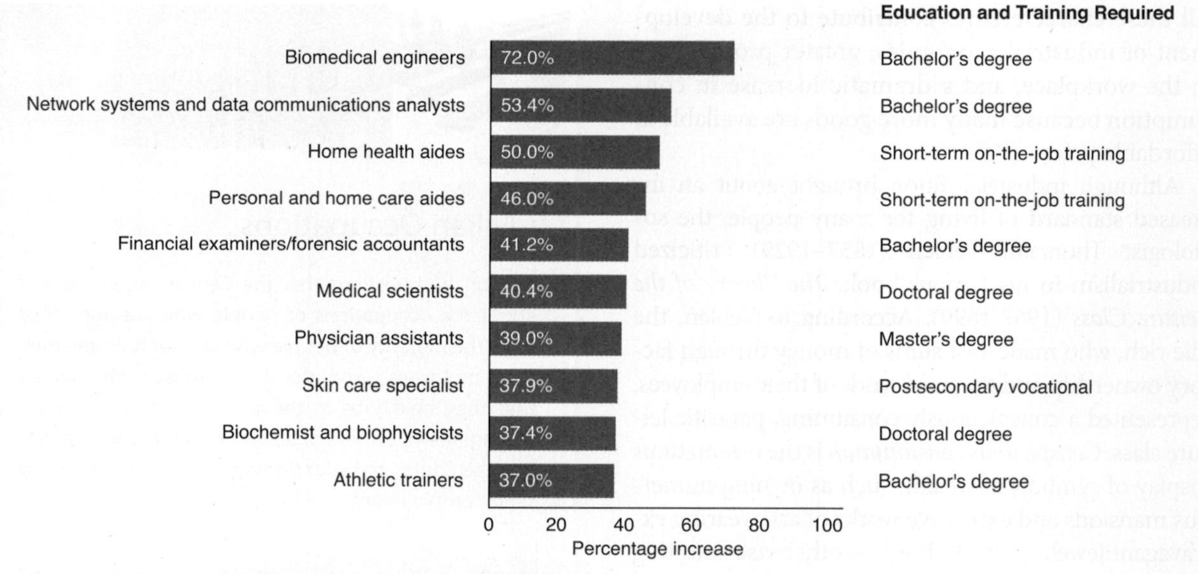

▲ **FIGURE 13.1 FASTEST-GROWING OCCUPATIONS, 2008–2018**
Source: U.S. Census Bureau, 2010b.

4. Traditional boundaries between work and home (public and private life) no longer exist because digital technologies such as cell phones and computers make global communication possible.
5. High levels of urbanization and a decline in population occur in many rural areas.

How far has the United States moved into postindustrialization? Although this question is difficult to answer, projections by the Census Bureau indicate that the fastest-growing occupations of this century are in the service sector, as shown in ▶ Figure 13.1.

Contemporary World Economic Systems

Capitalism and socialism are the principal economic models in industrialized countries. Sociologists use two criteria—property ownership and market control—to distinguish between types of economies.

Capitalism

***Capitalism* is an economic system characterized by private ownership of the means of production, from which personal profits can be derived through market competition and without government intervention.** Most of us think of ourselves as "owners" of private property because we own a car, a television, or other possessions. However, most of us are not capitalists; we *spend money* on the things we own rather than *make money* from them. Only a relatively few people own income-producing property from which a profit can be realized by producing and distributing goods and services. Everyone else is a consumer. "Ideal" capitalism has four distinctive features: (1) private ownership of the means of production, (2) pursuit of personal profit, (3) competition, and (4) lack of government intervention.

Private Ownership of the Means of Production Capitalist economies are based on the right of individuals to own income-producing property, such as land, water, mines, and factories, and the right to "buy" people's labor.

In the early stages of industrial capitalism (1850–1890), virtually all the capital for investment in the United States was individually owned—prior to the Civil War, an estimated 200 families controlled all major trade and financial organizations. By the 1890s, individual capitalists, including Andrew Carnegie, Cornelius Vanderbilt, and John D. Rockefeller, controlled most of the capital in commerce, agriculture, and industry (Feagin, Baker, and Feagin, 2006).

As workers grew tired of toiling for the benefit of capitalists instead of for themselves, some of them banded together to form the first national labor union, the Knights of Labor, in 1869. A ***labor union* is a group of employees who join together to bargain with an employer or a group of employers over wages, benefits, and working conditions.** The Knights of Labor included both skilled and unskilled laborers, but the American Federation of Labor (AFL), founded in 1886, targeted groups of skilled workers such as plumbers and carpenters; each of these craft unions maintained autonomy under the "umbrella" of the AFL.

Under early monopoly capitalism (1890–1940), most ownership rapidly shifted from individuals to huge ***corporations*****—large-scale organizations that have legal powers, such as the ability to enter into contracts and buy and sell property, separate from their individual owners.** During this period, major industries, including oil, sugar, and grain, came under the control of a few corporations owned by shareholders who hold or own shares of stock and cannot be personally blamed for the actions of the corporation. As automobile and steel plants shifted to mass production, the Congress of Industrial Organizations (CIO) was established in 1935 to represent both skilled and unskilled workers in industries such as automobile manufacturing. In 1937 GM workers held their first *sit-down strike* by refusing to work and paralyzing production, a move that came to dominate U.S. labor activism.

In advanced monopoly capitalism (1940–present), ownership and control of major industrial and business sectors have become increasingly concentrated, and many corporations have become more global in scope. ***Transnational corporations* are large corporations that are headquartered in one or a few countries but sell and produce goods and services in many countries.** These corporations play a major role in the economies and governments of many nations. The magnitude of these corporations is shown in ■ Table 13.1, which compares the revenues of the world's twenty largest transnational corporations with the gross domestic product of entire countries.

Transnational corporations are not dependent on the labor, capital, or technology of any one country and may move their operations to countries where wages and taxes are lower and the potential profits are higher. Corporate considerations of this kind help explain why many jobs formerly located in the United States have shifted to lower-income nations where few employment opportunities exist and workers can be paid significantly less than their U.S. counterparts. This appears to be a fact of life whether workers are reviewing legal documents, producing automobiles and computers, or cooking hamburgers in fast-food restaurants owned by transnational corporations. Some analysts believe that the so-called "global economy" under capitalism has produced a shrinking middle class in the United States, where almost 85 percent of all U.S. stocks are in the hands of 1 percent of the people, and 61 percent of people "always or usually" live from paycheck to paycheck (Snyder, 2010).

Pursuit of Personal Profit A tenet of capitalism is the belief that people are free to maximize their individual gain through personal profit; in the process, the entire society will benefit from their activities (Smith, 1976/1776). Economic development

table 13.1

Comparison of the Revenues of the World's 20 Largest Corporations with the GDP of Selected Countries (2010)

Company/Country	Revenues/GDP (in $ billions)
Saudi Arabia	**434**
Wal-Mart Stores	408
Venezuela	**285**
Royal Dutch Shell	285.1
Exxon Mobil	285
BP	246
Ireland	**204**
Toyota Motor	204
Japan Post Holdings	202
Philippines	**189**
Sinopec	188
State Grid	184
AXA	175
China Natl. Petroleum	165
Pakistan	**165**
Chevron	164
ING Group	163
General Electric	157
Total	156
Peru	**154**
Bank of America Corp.	150
Volkswagen	146
ConocoPhillips	140
New Zealand	**138**
BNP Paribas	131
Assicurazioni General	126
Allianz	126
Kuwait	**117**

Sources: CNNMoney.com, 2010; International Monetary Fund, 2010.

capitalism an economic system characterized by private ownership of the means of production, from which personal profits can be derived through market competition and without government intervention.

labor union a group of employees who join together to bargain with an employer or a group of employers over wages, benefits, and working conditions.

corporations large-scale organizations that have legal powers, such as the ability to enter into contracts and buy and sell property, separate from their individual owners.

transnational corporations large corporations that are headquartered in one or a few countries but sell and produce goods and services in many countries.

CHRISTOPH DERNBACH/epa/Corbis

▲ Apple Inc. is one of the most successful transnational corporations on the planet, making billions from its innovative computers and mobile communications technologies.

is assumed to benefit both capitalists and workers, and the general public also benefits from public expenditures (such as for roads, schools, and parks) made possible through an increase in business tax revenues.

During the period of industrial capitalism, however, specific individuals and families (not the general public) were the primary recipients of profits. For many generations, descendants of some of the early industrial capitalists have benefited from the economic deeds (and misdeeds) of their ancestors. In early monopoly capitalism, some stockholders derived massive profits from companies that held near-monopolies on specific goods and services.

In advanced (late) monopoly capitalism, profits have become even more concentrated. Although some people own *some* stock, they do not own *control;* in other words, they are unable to participate in establishing the policies that determine the size of the profit or the rate of return on investments (which affects the profits they derive).

Competition In theory, competition acts as a balance to excessive profits. When producers vie with one another for customers, they must be able to offer innovative goods and services at competitive prices. However, from the time of early industrial capitalism, the trend has been toward less, rather than more, competition among companies. In early monopoly capitalism, competition was diminished by increasing concentration *within* a particular industry, a classic case being the virtual monopoly on oil held by John D. Rockefeller's Standard Oil Company (Tarbell, 1925/1904; Lundberg, 1988). Today, Microsoft Corporation so dominates certain areas of the computer software industry that it has virtually no competitors in those areas.

In other situations, several companies may dominate certain industries. An ***oligopoly*** **exists when several companies overwhelmingly control an entire industry.** An example is the music industry, in which a few giant companies are behind many of the labels and artists (see ■ Table 13.2). More specifically, a ***shared monopoly*** **exists when four or fewer companies supply 50 percent or more of a particular market.** Examples include U.S. automobile manufacturers (referred to as the "Big Three") and cereal companies (three of which control the vast majority of the market). However, as we have seen in recent years, even some corporations, such as automobile manufacturers, that have been considered "too big to fail" have indeed had problems serious enough that government bailouts have been required to keep them in business and some of their workers employed.

In advanced monopoly capitalism, mergers also occur *across* industries: Corporations gain near-monopoly control over all aspects of the production and distribution of a product by acquiring both the companies that supply the raw materials and the companies that are the outlets for their products. For example, an oil company may hold leases on the land where the oil is pumped out of the ground, own the plants that convert the oil into gasoline, and own the individual gasoline stations that sell the product to the public.

Corporations with control both within and across industries are often formed by a series of mergers and acquisitions across industries. These corporations

Jason Kempin/Getty Images

▲ Capitalism is built on the pursuit of personal profit, and the opening-bell ceremony at the New York Stock Exchange is a symbol of that pursuit. Although corporations and shareholders may reap large economic gains from the stock market, most U.S. families do not own any type of stock, except that which may be part of their retirement accounts.

table 13.2
The Music Industry's Big Four

Company	Country	Leading Artists
Universal Music Group (MCA, Polygram)	France	Taylor Swift Lady Gaga Carrie Underwood Lil Wayne Eminem
Sony Music Entertainment (Columbia, RCA, Sony, Jive)	USA	Avril Lavigne Foo Fighters P!nk Kings of Leon
Warner Music Group (Atlantic, Elektra)	United States	Lupe Fiasco Death Cat for Cutie Rick Ross Linkin Park
EMI Group (Capitol, EMI, Virgin)	United Kingdom	Lady Antebellum Keith Urban Jay Z Alicia Keys

are referred to as ***conglomerates*—combinations of businesses in different commercial areas, all of which are owned by one holding company.** Media ownership is a case in point; companies such as Time Warner have extensive holdings in radio and television stations, cable television companies, book publishing firms, and film production and distribution companies, to name only a few.

Competition is reduced over the long run by ***interlocking corporate directorates*—members of the board of directors of one corporation who also sit on the board(s) of other corporations.** Although the Clayton Antitrust Act of 1914 made it illegal for a person to sit simultaneously on the boards of directors of two corporations that are in *direct* competition with each other, a person may serve simultaneously on the board of a financial institution (a bank, for example) and the board of a commercial corporation (a computer manufacturing company or a furniture store chain, for example) that borrows money from the bank. Directors of competing corporations may also serve together on the board of a third corporation that is not in direct competition with the other two. An example of interlocking directorates is depicted in ▶ Figure 13.2. Compensation for members of the boards of top corporations can be $1 million or more per person per year when stock, stock options and pensions are taken into account. To deflect public scrutiny, in recent years some corporate board members have "retired" from previous executive positions in high-powered corporations, banks, or law firms and thus have no visible conflict of interest.

Interlocking directorates diminish competition by producing interdependence. Individuals who serve on multiple boards are often able to forge cooperative arrangements that benefit their corporations but not necessarily the general public. When the same financial interests control several corporations, they are more likely to cooperate with one another than to compete.

Lack of Government Intervention Ideally, capitalism works best without government intervention in the marketplace. The policy of *laissez-faire* (les-ay-FARE, which means "leave alone") was advocated by economist Adam Smith in his 1776 treatise *An Inquiry into the Nature and Causes of the Wealth of Nations.* Smith argued that when people pursue their own selfish interests, they are guided "as if by an invisible hand" to promote the best interests of society (see Smith, 1976/1776). Today, terms such as *market economy* and *free enterprise* are often

oligopoly a condition that exists when several companies overwhelmingly control an entire industry.

shared monopoly a condition that exists when four or fewer companies supply 50 percent or more of a particular market.

conglomerate a combination of businesses in different commercial areas, all of which are owned by one holding company.

interlocking corporate directorates members of the board of directors of one corporation who also sit on the board(s) of other corporations.

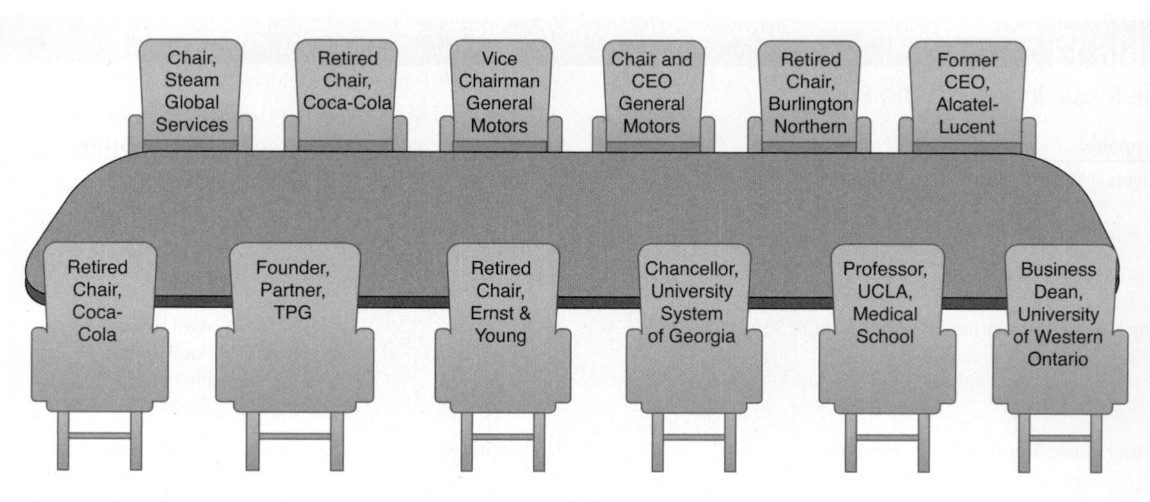

▲ **FIGURE 13.2** THE GENERAL MOTORS BOARD OF DIRECTORS
The 2010 General Motors Board of Directors shows the nature of interlocking directorates. On the chair representing each of the directors is the name of another entity each director is connected with, and his or her position with that entity.
Source: General Motors, 2010.

used, but the underlying assumption is the same: Free market competition, not the government, should regulate prices and wages. However, the "ideal" of unregulated markets benefiting all citizens has seldom been realized. Individuals and companies in pursuit of higher profits have run roughshod over weaker competitors, and small businesses have grown into large, monopolistic corporations. Accordingly, government regulations were implemented in an effort to curb the excesses of the marketplace brought about by laissez-faire policies.

However, much of what is referred to as government intervention has been in the form of aid to business. Between 1850 and 1900, corporations received government assistance in the form of public subsidies and protection from competition by tariffs, patents, and trademarks. The federal government also gave large tracts of land to the privately owned railroads to encourage their expansion across the nation. Antitrust laws originally intended to break up monopolies were used instead against labor unions that supported workers' interests (Parenti, 1996).

Overall, most corporations have gained much more than they have lost as a result of government involvement in the economy. The recent bailouts of various industries, including banks and automakers, are obvious examples of federal intervention to the tune of billions of dollars in taxpayers' monies and other funds.

Socialism

***Socialism* is an economic system characterized by public ownership of the means of production, the pursuit of collective goals, and centralized decision making.** Like "pure" capitalism, "pure" socialism does not exist. Karl Marx described socialism as a temporary stage en route to an ideal communist society. Although the terms *socialism* and *communism* are associated with Marx and are often used interchangeably, they are not identical. Marx defined communism as an economic system characterized by common ownership of all economic resources. In the *Communist Manifesto* and *Das Kapital,* he predicted that the working class would become increasingly impoverished and alienated under capitalism. As a result, the workers would become aware of their own class interests, revolt against the capitalists, and overthrow the

Brooks Kraft/Corbis

▲ For almost three decades, the federal government tended to reduce the amount of regulation that industries, banks, and investment firms faced. However, the 2008 collapse of the U.S. real estate market and the resulting failure (or near-failure) of huge investment firms have made more intense regulation a certainty. Here, Goldman Sachs executives were called before Congress in 2010 to account for their company's business practices in the years leading up to the meltdown.

Jeremy Nicholl/Alamy

▲ For many decades, Russia had a state-controlled economy. However, beginning in the 1990s, the government privatized many sectors of the economy, and new companies made vast profits in sectors such as oil exploration and similar endeavors. Here, workers in Siberia are setting pipe for the Yukos Oil Company.

entire system (see Turner, Beeghley, and Powers, 2007). After the revolution, private property would be abolished, and collectives of workers who would own the means of production would control capital. The government (previously used to further the interests of the capitalists) would no longer be necessary. People would contribute according to their abilities and receive according to their needs (Marx and Engels, 1967/1848; Marx, 1967/1867). Many of Marx's ideas have had a profound effect on how sociologists and other researchers view our contemporary economic and social problems (see "Sociology Works!"). "Ideal" socialism has three distinctive features: (1) public ownership of the means of production, (2) pursuit of collective goals, and (3) centralized decision making.

Public Ownership of the Means of Production In a truly socialist economy the means of production are owned and controlled by a collectivity or the state, not by private individuals or corporations. For example, prior to the early 1990s the state owned all the natural resources and almost all the capital in the Soviet Union. At least in theory, goods were produced to meet the needs of the people. Access to housing and medical care was considered to be a right.

Leaders of the Soviet Union and some Eastern European nations decided to abandon government ownership and control of the means of production because the system was unresponsive to the needs of the marketplace and offered no incentive for increased efficiency (Boyes and Melvin, 2011). Since the 1990s, Russia and other states in the former Soviet Union have attempted to privatize ownership of production. Economic reforms in the 1990s privatized most industries, with the exceptions of the energy and defense-related sectors. Today, the state-owned Russian oil company Rosneft makes billions of dollars annually from the sale of oil, particularly as concern has increased about the supply of Middle East oil (Kramer, 2011).

China—previously the world's other major communist economy—has privatized many state industries. In *privatization* resources are converted from state ownership to private ownership; the government takes an active role in developing, recognizing, and protecting private property rights (Boyes and Melvin, 2011). In the second decade of the twenty-first century, China has a hybrid political economy made up of both capitalism and an autocratic form of Communist Party governance. Double-digit economic growth has brought about an increase in annual urban income, life expectancy has increased by more than six years, and the rate of illiteracy has dropped significantly. With these improvements for middle-class Chinese, it is likely that the combination of communism and a modified form of capitalism will remain for the foreseeable future (Jacobs, 2011). Both Russia and China have undergone significant changes in recent years, moving from globally isolated, centrally controlled economies to more market-based and globally integrated economies.

Pursuit of Collective Goals Socialism is based on the pursuit of collective goals, rather than on personal profits. Equality in decision making replaces hierarchical relationships (such as between owners and workers or political leaders and citizens). Everyone shares in the goods and services of society, especially necessities such as food, clothing, shelter, and medical care, based on need, not on ability to pay. In reality, in nations such as China, members of the Communist Party are able to obtain low-interest loans from state-owned and state-operated banks as long as they play by party rules (Jacobs, 2011). In sum, even though pursuit of collective goals is one of the ideals of socialism, few societies can or do pursue purely collective goals.

Centralized Decision Making Another tenet of socialism is centralized decision making. In theory, economic decisions are based on the needs of society; the government is responsible for aiding the production and distribution of goods and services. Central planners set wages and prices to ensure that

socialism an economic system characterized by public ownership of the means of production, the pursuit of collective goals, and centralized decision making.

sociology works!

Marx: Not Completely Right, but Not Completely Wrong Either

The legacy of Karl Marx, the German economist and philosopher, has been a mixed one. Some people believe that his works should have no influence on contemporary thinking about the economy and political life because some of his predictions about capitalism proved to be incorrect; however, others emphasize that Marx was not completely wrong either, particularly when we look at his ideas in light of the current economic crisis. Slightly over 160 years ago, Karl Marx wrote *The Communist Manifesto,* and the journalist and social commentator Barbara Ehrenreich (2008) argues that, despite changing times, this document has retained relevance for today:

> The Manifesto makes for quaint reading today. All that talk about "production," for example: Did they actually make things in those days? Did the proletariat really slave away in factories instead of call centers? But on one point Marx and Engels proved right: Within capitalist societies, or at least the kind of wildly unregulated capitalism America has had, the rich got richer, the workers got poorer, and the erstwhile middle class has been sliding toward ruin. The last two outcomes are what Marx called "immiseration," which, in translation, is the process you're undergoing when you have cancer and no health insurance or a mortgage payment due and no paycheck coming in.

You may recall that Ehrenreich is a journalist and social activist who has written books such as *Nickel and Dimed* (previously mentioned in Chapter 8) and speaks on the behalf of the working class and the poor. After her time spent working as a manual laborer and hourly-wage earner at fast-food chains, hotels, and big-box stores such as Wal-Mart, Ehrenreich described how difficult it is for people earning the minimum wage or less to eke out a living as they become increasingly impoverished under capitalism. Some of her research and writing has been strongly influenced by conflict theorists such as Karl Marx.

Although the revolution of the workers and the fall of capitalism that Marx predicted (based on the workers becoming fed up with immiseration and "revolting, seizing the means of production, and insisting on running the show themselves") did not occur, Ehrenreich (2008) argues that this does not mean that Marx was incorrect in his assessment about the problems inherent in advanced capitalism:

> The revolution didn't happen, of course, at least not here. For the last several years, American workers have sweetly acquiesced to declining wages, rising prices, speed-ups at work, disappearing pensions, and increasingly threadbare health insurance. While CEO pay escalated to the 8-figure range and above, so-called ordinary Americans took on second jobs and crowded into multigenerational households with uncomfortable long waits for the bathroom. But all this immiseration—combined

the production process works. When problems such as shortages and unemployment arise, they can be dealt with quickly and effectively by the central government (Boyes and Melvin, 2011).

Mixed Economies

As we have seen, no economy is truly capitalist or socialist; most economies are mixtures of both. A ***mixed economy*** **combines elements of a market economy (capitalism) with elements of a command economy (socialism).** Sweden, Great Britain, and France have mixed economies, sometimes referred to as ***democratic socialism*****—an economic and political system that combines private ownership of some of the means of production, governmental distribution of some essential goods and services, and free elections.** For example, government ownership in Sweden is limited primarily to railroads, mineral resources, a public bank, and liquor and tobacco operations (Feagin, Baker, and Feagin, 2006). Compared with capitalist economies, however, the government in a mixed economy plays a larger role in setting rules, policies, and objectives.

The government is also heavily involved in providing services such as medical care, child care, and transportation. In Sweden, for example, all residents have health insurance, housing subsidies, child allowances, paid parental leave, and day-care subsidies. Recently, some analysts have suggested that the United States has assumed many of the characteristics of a ***welfare state,*** **a state in which there is extensive government action to provide support and services to the citizens,** as it has attempted to meet the basic needs of older persons, young children, unemployed people, and persons with a disability.

Perspectives on Economy and Work in the United States

Functionalists, conflict theorists, and symbolic interactionists view the economy and the nature of work from a variety of perspectives. We first examine functionalist and conflict views of the economy; then we focus on the symbolic interactionist perspective on job satisfaction and alienation.

© Alex Segre/Alamy

Karl Marx's beliefs about capitalism, socialism, and communism are widely known today, although it has been more than 100 years since his famous proclamation—as emblazoned on his tombstone outside London—"Workers of all lands unite."

with fabulous enrichment at the top—did end up destabilizing the capitalist system, if only because, in the last few years, America's substitute for decent wages has been easy credit. . . . Marx's argument was that the coexistence of great wealth for the few and growing poverty for the many is not only morally objectionable, it's also inherently unstable. [Marx] may have been wrong about the reasons for the instability, but no one can any longer deny it's there. When the greed of the rich collided with the needs of the poor—for a home, for example—the result was a global credit meltdown. Obviously, the way to address the crisis is to deal with the poverty and inequality that led to it. . . .

All sociologists do not agree with the ideas of Marx or with Ehrenreich's explanation of them, but the 160-year-old *Communist Manifesto* and the current writings of Ehrenreich do reflect the importance of taking into account the poor and acknowledging that everyone does not share equally in the hardships associated with tough economic times.

reflect & analyze

What alternative explanations do sociologists give for the current global economic crisis? What might we learn about the economy as a social institution by conducting research at locations frequented by the working class and the poor, such as food banks and homeless shelters? Is there a place for social activism in the study of sociology?

Functionalist Perspective

Functionalists view the economy as a vital social institution because it is the means by which needed goods and services are produced and distributed. When the economy runs smoothly, other parts of society function more effectively. However, if the system becomes unbalanced, such as when demand does not keep up with production, maladjustment occurs (in this case, a surplus). Some problems can be easily remedied in the marketplace (through "free enterprise") or through government intervention (such as paying farmers *not* to plant when there is an oversupply of a crop). However, other problems, such as periodic *peaks* (high points) and *troughs* (low points) in the business cycle, are more difficult to resolve. The *business cycle* is the rise and fall of economic activity relative to long-term growth in the economy.

From this perspective, peaks occur when "business" has confidence in the country's economic future. During a peak, or *expansion period,* the economy thrives, and upward social mobility for workers and their families becomes possible.

The American Dream of upward mobility is linked to peaks in the business cycle. Once the peak is reached, however, the economy turns down because too large a surplus of goods has been produced. In part, this is because of *inflation*—a sustained and continuous increase in prices. Inflation erodes the value of people's money, and they are no longer able to purchase as high a percentage of the goods that have been produced. Because of this lack of demand, fewer goods are produced, workers are laid off, credit becomes difficult to obtain, and people cut back on their

mixed economy an economic system that combines elements of a market economy (capitalism) with elements of a command economy (socialism).

democratic socialism an economic and political system that combines private ownership of some of the means of production, governmental distribution of some essential goods and services, and free elections.

welfare state a state in which there is extensive government action to provide support and services to the citizens.

purchases even more, fearing unemployment. Eventually, this produces a distrust of the economy, resulting in a *recession*—a decline in an economy's total production that lasts six months or longer. To combat a recession, the government lowers interest rates (to make borrowing easier and to get more money back into circulation) in an attempt to spur the beginning of the next expansion period.

Conflict Perspective

Conflict theorists have a different view of business cycles and the economic system. From a conflict perspective, business cycles are the result of capitalist greed. In order to maximize profits, capitalists suppress the wages of workers. As the prices of products increase, workers are not able to purchase them in the quantities that have been produced. The resulting surpluses cause capitalists to reduce production, close factories, and lay off workers, thus contributing to the growth of the reserve army of the unemployed, whose presence helps reduce the wages of the remaining workers. In some situations, workers are replaced with machines or nonunionized workers.

Karl Marx referred to the propensity of capitalists to maximize profits by reducing wages as the *falling rate of profit,* which he believed to be one of the inherent contradictions of capitalism that would produce its eventual downfall. According to the political sociologist Michael Parenti, business *is* the economic system. Parenti believes that political leaders treat the health of the capitalist economy as a necessary condition for the health of the nation and that the goals of big business (rapid growth, high profits, and secure markets at home and abroad) become the goals of government (Parenti, 1996). In sum, to some conflict theorists, capitalism is the problem; to some functionalist theorists, however, capitalism is the solution to society's problems.

Symbolic Interactionist Perspective

Sociologists who focus on microlevel analyses are interested in how the economic system and the social organization of work affect people's attitudes and behavior. Symbolic interactionists, in particular, have examined the factors that contribute to job satisfaction.

According to symbolic interactionists, work is an important source of self-identity for many people; it can help people feel positive about themselves, or it can cause them to feel alienated. *Job satisfaction* refers to people's attitudes toward their work, based on (1) their job responsibilities, (2) the organizational structure in which they work, and (3) their individual needs and values. Earlier studies have found that worker satisfaction is highest when employees have some degree of control over their work, when they are part of the decision-making process, when they are not too closely supervised, and when they feel that they play an important part in the outcome (Kohn et al., 1990). A more recent study by the Heldrich Center for Workplace Development at Rutgers University found that overall job satisfaction and job security declined significantly in the first decade of the twenty-first century. ▶ Figure 13.3 compares the percentages of people in 1999 and 2009 who stated that they were "very satisfied" with various aspects of their job. From this study, it appears that overall job satisfaction is down even when people state that the same key job attributes are central to their work life (Van Horn and Zukin, 2009).

Job satisfaction is often related to both intrinsic and extrinsic factors. Intrinsic factors pertain to the nature of the work itself, whereas extrinsic factors include such things as vacation and holiday policies, parking privileges, on-site day-care centers, and other amenities that contribute to workers' overall perception that their employer cares about them.

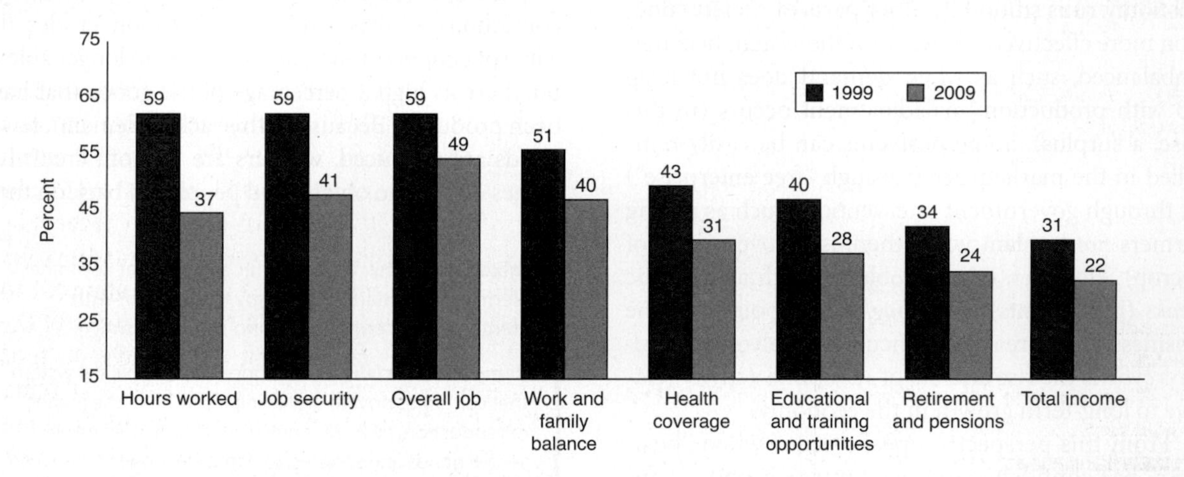

▲ **FIGURE 13.3 JOB SATISFACTION: 1999 AND 2009**

Percentages of workers reporting "very satisfied" on various aspects of their jobs.

Source: Van Horn and Zukin, 2009.

[concept quick review]

Perspectives on Economy and Work in the United States

Key Concept	
Functionalist Perspective	The economy is a vital social institution because it is the means by which needed goods and services are produced and distributed
Conflict Perspective	The capitalist economy is based on greed. In order to maximize profits, capitalists suppress the wages of workers, who, in turn, cannot purchase products, making it necessary for capitalists to reduce production, close factories, lay off workers, and adopt other remedies that are detrimental to workers and society.
Symbolic Interactionist Perspective	Many workers experience job satisfaction when they like their job responsibilities and the organizational structure in which they work and when their individual needs and values are met. Alienation occurs when workers do not gain a sense of self-identity from their jobs and when their work is done completely for material gain and not for personal satisfaction.

Alienation occurs when workers' needs for self-identity and meaning are not met and when work is done strictly for material gain, not a sense of personal satisfaction. According to Marx, workers are resistant to having very little power and no opportunities to make workplace decisions. This lack of control contributes to an ongoing struggle between workers and employers. Job segmentation, isolation of workers, and the discouragement of any type of pro-worker organizations (such as unions) further contribute to feelings of helplessness and frustration. Some occupations may be more closely associated with high levels of alienation than others.

The Concept Quick Review summarizes the three major sociological perspectives on the economy and work in the United States.

Zak Kendal Cultura/Newscom

▲ When workers' satisfaction with their jobs decreases, tensions can boil to the surface, leading to outbursts and behavior that is considered to be inappropriate in the workplace.

Work in the Contemporary United States

As we have seen, the kind of work that people perform has changed dramatically, and the gender distribution of persons in the labor force has also changed because of an increase in the number of women in the labor force. The term *labor force* refers to the number of people age 16 and over who are either employed or actively looking for work. It does not include active-duty military personnel or persons who are institutionalized, such as prison inmates.

Increase in Women's Paid Employment

The number of women, both single and married, in the labor force has increased since 1970, when about 53 percent of single women and 41 percent of married women were in the paid labor force, as compared to 64 percent of single women and 62 percent of married women in 2009 (U.S. Census Bureau, 2010b). By 2018, it is estimated that about 71 percent of men and 59 percent of women will hold paid employment; the recent recession and continuing high rates of job loss and unemployment may be factors that drive down labor-force participation rates in the future. In the 1950s men made up more than 80 percent of the labor force, as contrasted to less than 40 percent of women; today, those percentages are coming closer together.

The Kinds of Work We Do

The economy in the United States and other contemporary societies is partially based on the work (purposeful activity, labor, or toil) that people perform. However, work in high-income nations is highly differentiated and often fragmented because people have many kinds of occupations. Some occupations are referred to as professions.

Professions

Although sociologists do not always agree on exactly which occupations are professions, most of them agree that the term *professionals* includes most doctors, natural scientists, engineers, computer scientists, certified public accountants, economists, social scientists, psychotherapists, lawyers, policy experts of various sorts, professors, at least some journalists and editors, some clergy, and some artists and writers.

Characteristics of Professions

***Professions* are high-status, knowledge-based occupations that have five major characteristics** (Freidson, 1970, 1986; Larson, 1977):

1. *Abstract, specialized knowledge.* Professionals have abstract, specialized knowledge of their field based on formal education and interaction with colleagues.
2. *Autonomy.* Professionals are autonomous in that they can rely on their own judgment in selecting the relevant knowledge or the appropriate technique for dealing with a problem.
3. *Self-regulation.* In exchange for autonomy, professionals are theoretically self-regulating. All professions have licensing, accreditation, and regulatory associations that set professional standards and that require members to adhere to a code of ethics as a form of public accountability.
4. *Authority.* Because of their authority, professionals expect compliance with their directions and advice. Their authority is based on mastery of the body of specialized knowledge and on their profession's autonomy.
5. *Altruism.* Ideally, professionals have concern for others that goes beyond their self-interest or personal comfort so that they can help a patient or client.

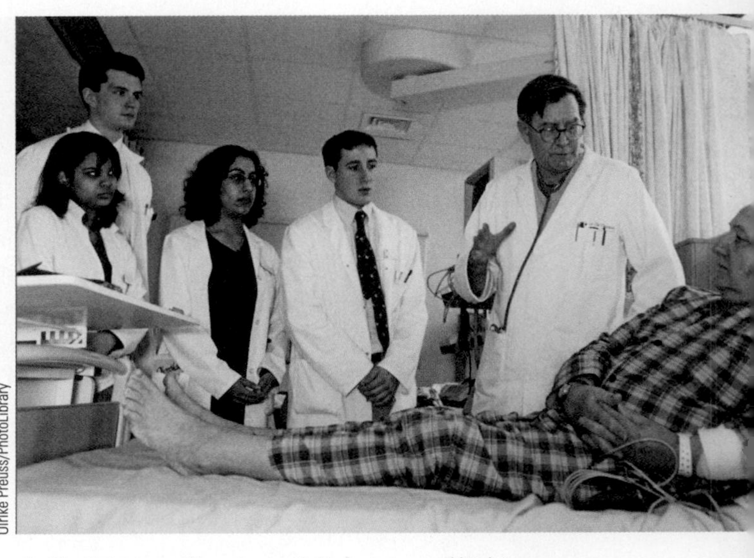

Ulrike Preuss/PhotoLibrary

▲ Physicians usually enjoy both high status and high income levels, making this profession a much-sought-after career choice. However, the process of becoming a licensed physician is long and very difficult.

Social Reproduction of Professionals

Although higher education is one of the primary qualifications for a profession, the emphasis on education gives children whose parents are professionals a disproportionate advantage early in life. There is a direct linkage between parental education/income and children's scores on college admissions tests such as the SAT, as shown in ▶ Figure 13.4. In turn, test scores are directly related to students' ability to gain admission to colleges and universities, which serve as springboards to most professions. Race and gender are also factors in access to the professions.

Deprofessionalization

Certain professions are undergoing a process of *deprofessionalization,* in which some of the characteristics of a profession are eliminated. Occupations such as pharmacist have already been *deskilled,* as Nino Guidici explains:

> In the old days [people] took druggists as doctors. . . . All we do [now] is count pills. Count out twelve on the counter, put 'em in here, count out twelve more. . . . Doctors used to write out their own formulas and we made most of these things. Most of the work is now done in the laboratory. The real druggist is found in the manufacturing firms. They're the factory workers and they're the pharmacists. We just get the name of the drugs and the number and the directions. It's a lot easier. (qtd. in Terkel, 1990/1972)

However, colleges of pharmacy in many universities have fought against deprofessionalization by upgrading the degrees awarded to pharmacy graduates from the traditional B.S. in pharmacy to a Ph.D. This upgrading of degrees has also occurred over the past two decades in law schools, where the Bachelor of Laws (LL.B.) has been changed to the Juris Doctor (J.D.) degree.

Other Occupations

***Occupations* are categories of jobs that involve similar activities at different work sites.** More than 600 different occupational categories and 35,000 occupation titles, ranging from motion picture cartoonist to drop-hammer operator, are currently listed by the U.S. Census Bureau. Historically, occupations have been classified as blue collar and white collar. Blue-collar workers were primarily factory and craft workers who did manual labor; white-collar workers were office workers and professionals. However, contemporary workers in the service

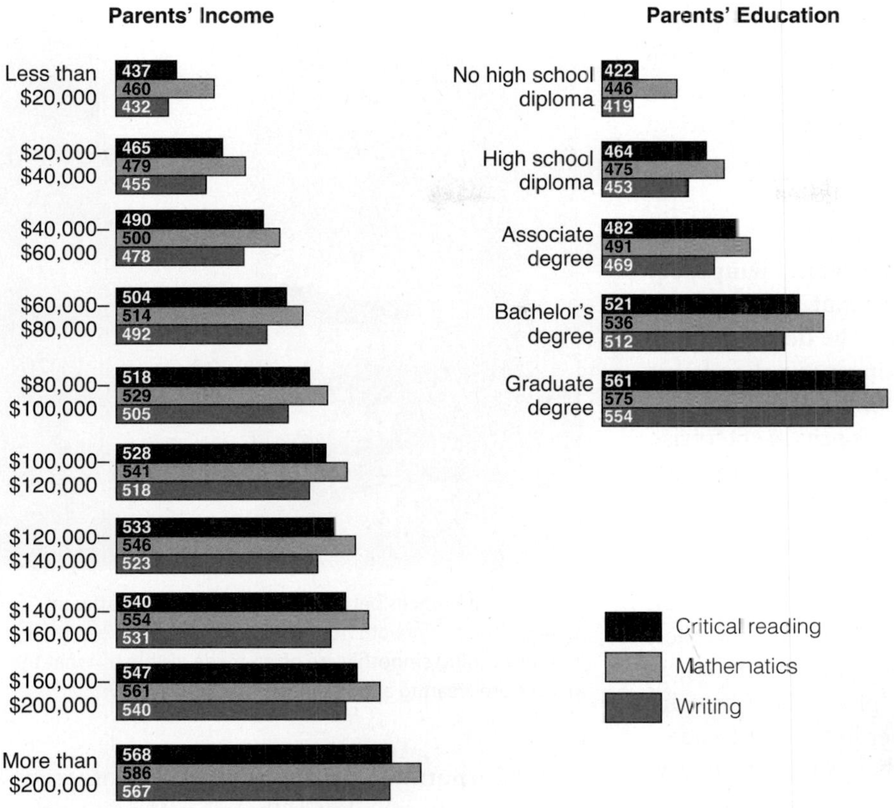

Parents' Income

Parents' Income	Critical reading	Mathematics	Writing
Less than $20,000	437	460	432
$20,000–$40,000	465	479	455
$40,000–$60,000	490	500	478
$60,000–$80,000	504	514	492
$80,000–$100,000	518	529	505
$100,000–$120,000	528	541	518
$120,000–$140,000	533	546	523
$140,000–$160,000	540	554	531
$160,000–$200,000	547	561	540
More than $200,000	568	586	567

Parents' Education

Parents' Education	Critical reading	Mathematics	Writing
No high school diploma	422	446	419
High school diploma	464	475	453
Associate degree	482	491	469
Bachelor's degree	521	536	512
Graduate degree	561	575	554

◀ **FIGURE 13.4 SAT SCORES BY PARENTS' INCOME AND EDUCATION, 2010**

Source: SAT, 2010.

sector do not easily fit into either of these categories; neither do the so-called pink-collar workers, primarily women, who are employed in occupations such as preschool teacher, dental assistant, secretary, and clerk. (The term refers to an era when some female restaurant employees were required to wear uniforms with a pink collar.)

Sociologists establish broad occupational categories by distinguishing between employment in the primary labor market and in the secondary labor market. The ***primary labor market* consists of high-paying jobs with good benefits that have some degree of security and the possibility of future advancement.** By contrast, the ***secondary labor market* consists of low-paying jobs with few benefits and very little job security or possibility for future advancement** (Bonacich, 1972).

Upper-Tier Jobs: Managers and Supervisors Managers are essential in contemporary bureaucracies, where work is highly specialized and authority structures are hierarchical. Workers at each level of the hierarchy take orders from their immediate superiors and perhaps give orders to a few subordinates. Upper-level managers are typically responsible for coordination of activities and control of workers.

Lower-Tier and Marginal Jobs Positions in the lower tier of the service sector are part of the secondary labor market, characterized by low wages, little job security, few chances for advancement, higher unemployment rates, and very limited (if any) unemployment benefits. Typical lower-tier positions include janitor, waitress, messenger, sales clerk, typist, file clerk, migrant laborer, and textile worker. Large numbers of young people, people of color, recent immigrants, and white women are employed in this sector.

***Marginal jobs* differ from the employment norms of the society in which they are located;** examples in the U.S. labor market include jobs in personal service industries such as eating and drinking places, hotels, and laundries, as well as private

professions high-status, knowledge-based occupations.

occupations categories of jobs that involve similar activities at different work sites.

primary labor market the sector of the labor market that consists of high-paying jobs with good benefits that have some degree of security and the possibility of future advancement.

secondary labor market the sector of the labor market that consists of low-paying jobs with few benefits and very little job security or possibility for future advancement.

marginal jobs jobs that differ from the employment norms of the society in which they are located.

household workers. Marginal jobs are frequently not covered by government work regulations—such as minimum standards of pay, working conditions, and safety standards—or do not offer sufficient hours of work each week to provide a living.

Contingent Work

***Contingent work* is part-time work, temporary work, or subcontracted work that offers advantages to employers but that can be detrimental to the welfare of workers.** Contingent work is found in every segment of the labor force. The federal government is part of this trend, as is private enterprise. In the health care field, physicians, nurses, and other workers are increasingly employed through temporary agencies. Employers benefit by hiring workers on a part-time or temporary basis; they are able to cut costs, maximize profits, and have workers available only when they need them. Temporary workers are the fastest-growing segment of the contingent workforce, and agencies that "place" them have increased dramatically in number in the last decade.

Subcontracted work is another form of contingent work that often cuts employers' costs at the expense of workers. Instead of employing a large workforce, many companies have significantly reduced the size of their payrolls and benefit plans by ***subcontracting*—an agreement in which a corporation contracts with other (usually smaller) firms to provide specialized components, products, or services to the larger corporation.** Hiring and paying workers become the responsibility of the subcontractor, not of the larger corporation.

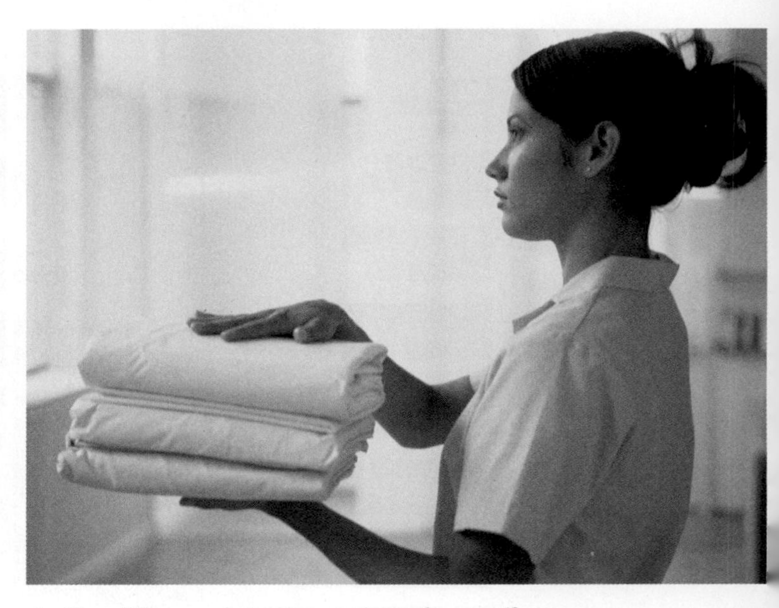

▲ Class differences between members of more-affluent families and the private household workers who keep their homes running smoothly are often made visible by what the people are wearing and what they are doing.

© Peter Hvizdak/The Image Works

▲ Occupational segregation by race and gender is clearly visible in personal service industries, such as restaurants and fast-food chains. Women and people of color are disproportionately represented in marginal jobs such as waitperson or fast-food server—jobs that typically do not meet societal norms for benefits and security.

The Underground (Informal) Economy

Some social analysts make a distinction between the legitimate and the underground (informal) economies in the United States. For the most part, the occupations previously described in this chapter operate within the *legitimate economy*: Taxes on income are paid by employers and employees, and individuals who hold jobs requiring a specialized license (such

© Bill Aron/PhotoEdit

▲ Hiring contingent workers can increase the profitability of many corporations. Other companies make their profits by furnishing these contingent workers to the corporations. What message does this ad convey to corporate employers?

© Kayte M. Deioma/PhotoEdit

▲ The underground (or shadow) economy involves activities such as cash payments to workers that are never reported as earnings and on which no taxes are paid. Is the underground economy simply a means of getting by, does it constitute a threat to the legitimate economy? What do you think?

as craftspeople or taxi drivers) possess the appropriate credentials for their work. By contrast, the *underground economy* is made up of a wide variety of activities through which people make money that they do not report to the government or through endeavors that may involve criminal behavior (Venkatesh, 2006). Sometimes referred to as the "informal" or "shadow economy," one segment of the underground economy is made up of workers who are paid "off the books," which means that they are paid in cash, their earnings are not reported, and no taxes are paid. Lawful jobs, such as nannies, construction workers, and landscape/yard workers, are often part of the shadow economy because workers and bosses make under-the-table deals so that both can gain through the transaction: Employers pay less for workers' services, and workers have more money to take home than if they paid taxes on their earnings.

The underground economy also involves trade in lawful goods that are sold "off the books" so that no taxes are paid and the sale of "designer alternative fashion" products that may be counterfeit ("knock-off") merchandise. Demand for such products is strong at all times but frequently increases in difficult economic times because many people retain a desire for luxury goods that are widely publicized by the media even when people have fewer resources to allocate to such purchases (see "Framing 'Luxury Consumption' in the Media").

According to one way of thinking, operating a business in the underground economy reveals capitalism at its best because it shows how the "free market" might work if there were no government intervention. However from another perspective, selling goods or services in the underground economy borders on—or moves into—criminal behavior. For some individuals the underground economy offers the only means for purchasing certain goods or for overcoming unemployment, particularly in low-income and poverty areas where people may feel alienated from the wider world and believe that they must use shady means to survive (see Venkatesh, 2006).

Unemployment

There are three major types of unemployment—cyclical, seasonal, and structural. *Cyclical unemployment* occurs as a result of lower rates of production during recessions in the business cycle; a recession is a decline in an economy's total production that lasts at least six months. Although massive layoffs initially occur, some of the workers will eventually be rehired, largely depending on the length and severity of the recession. *Seasonal unemployment* results from shifts in the demand for workers based on conditions such as the weather (in agriculture, the construction industry, and tourism) or the season (holidays and summer vacations). Both of these types of unemployment tend to be relatively temporary.

By contrast, structural unemployment may be permanent. *Structural unemployment* arises because the skills demanded by employers do not match the skills of the unemployed or because the unemployed do not live where the jobs are located. This type of unemployment often occurs when a number of plants in the same industry are closed or when new technology makes certain jobs obsolete. Structural unemployment often results from capital flight—the investment of capital in foreign facilities, as previously discussed. Today, many workers fear losing their jobs, exhausting their unemployment benefits (if any), and still not being able to find another job.

The ***unemployment rate* is the percentage of unemployed persons in the labor force actively seeking jobs.** The second decade of the twenty-first

contingent work part-time work, temporary work, or subcontracted work that offers advantages to employers but that can be detrimental to the welfare of workers.

subcontracting an agreement in which a corporation contracts with other (usually smaller) firms to provide specialized components, products, or services to the larger corporation.

unemployment rate the percentage of unemployed persons in the labor force actively seeking jobs.

framing "luxury consumption" in the media

From Glossy Ads to the Underground Economy

While some people love their [knockoff] Chanel bags (I spot those fakes all over campus!), others would rather rock Target than fake designer anything. . . . Personally, I'm not 100% against knockoffs—they make design affordable for everyone. . . . On the other hand, I'd never carry a faux Chanel bag—it would be sacrilegious! I don't think I could live with myself if I pretend a fake Chanel was real. Every time I wore it (which would be almost every day), I'd know I was lying to myself and the world. I'd rather save up and earn the real thing or not have one at all.

—Zephyr (2009), a student at the time she wrote this *College Fashion* blog, expressing the ambivalence many women feel about buying "knockoff" or "designer alternative" fashion handbags

From ads to media-produced "celebrity sightings" of stars and other high-profile women wearing luxury designer clothing or handbags by Chanel, Louis Vuitton, Prada, Gucci, and similar manufacturers, contemporary girls and women are bombarded by images of wealthy individuals carrying or wearing the latest designer products. Media framing of stories about celebrities often follows a "price-tag framing" approach. *Price-tag framing* describes the practice of making the cost of luxury items a key feature in media stories about the rich and famous (Kendall, 2011). Whereas straight news accounts typically provide information about the basic who, what, when, why, and how, price-tag framing focuses on *how much*. This type of framing encourages media audiences and consumers to focus on the cost of luxury items and on the celebrity status of those who own them.

What about people who desire to own luxury items but either cannot or will not spend the money necessary to make these purchases? This is where the underground economy and/or "designer alternative fashion" enters the picture. The underground economy for knockoff designer handbags runs from global tourist areas such as the Eiffel Tower in Paris and

century has seen a significant increase in unemployment. The U.S. unemployment rate in 2000 was 4.0 percent. By 2011, the overall rate hovered around 9 percent. (► Maps 13.1a and 13.1b show the detrimental effects of the Great Recession on unemployment in each state.) In 2011 the unemployment rate for adult men was 8.7 percent, as compared to 8.0 percent for adult women. However, the breakdown for unemployment across racial/ethnic and age categories tells a more complete story: Teenagers of all racial/ethnic categories had a 23.9 percent unemployment rate. African Americans of all ages had a 15.3 percent unemployment rate; Hispanics (Latinos/as) had an 11.6 percent rate, as compared

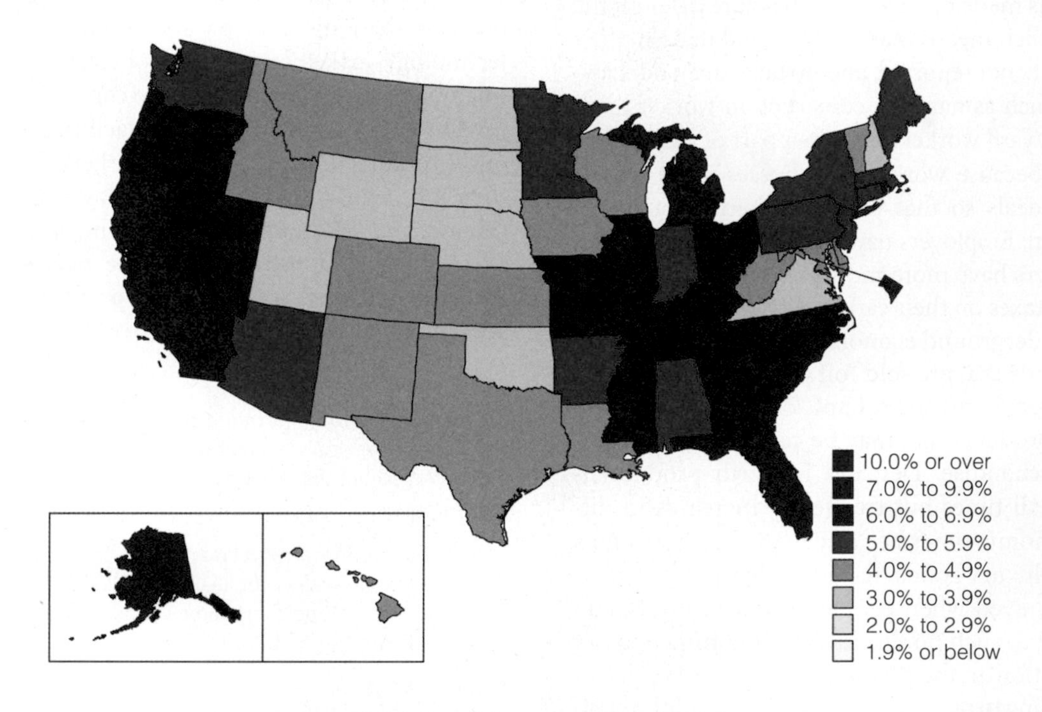

▲ **MAP 13.1A U.S. UNEMPLOYMENT RATES BY STATE BEFORE THE FALL OF 2008 (U.S. RATE: 5.8%)**

Source: Bureau of Labor Statistics Local Area Unemployment Statistics March 2010

streets bordering the Venice canals to "Counterfeit Triangle" along Canal Street in New York City and websites that use the names of designer labels to sell fake merchandise by falsely describing or representing goods that are offered for sale.

Although copying is rampant in the fashion industry, media representations that wearing a luxury designer product such as a handbag will make a person truly special may have contributed to the untold millions of dollars that are made worldwide in the underground economy because consumers, both men and women, want to feel special about themselves, and one way to do this is to assume that wearing high-priced goods makes them a better, more successful person. Publicizing the price and exclusivity of goods sets the wealthy apart from other people and in the process raises a barrier between the lifestyles of the rich and famous and everyone else. Some people seek to reduce that barrier by finding innovative ways to acquire alternative merchandise that is sold to them under the mistaken notion that you can outsmart others by paying much less for an item that is allegedly indistinguishable from a much more expensive product. From this viewpoint, equality does not exist in contemporary society, but media audiences are encouraged to view themselves as having a "right" to purchase items that will somehow make them equal to people above them in the social class hierarchy (Kendall, 2011).

If you are interested in doing further research on this topic, you will find many websites that offer fashion knockoffs for sale. Analyze some of these to see what claims are made and how these relate to "legitimate" media advertising and stories about luxury goods.

reflect & analyze

How do Veblen's ideas of conspicuous consumption (discussed elsewhere in this chapter) relate to the need of some people in other classes to emulate the excessive and extravagant purchases of the wealthy? Much has been written about U.S. people not being rebellious about growing wealth inequality: Do you believe that the ability of many people to purchase products that help them "live like the rich" could have anything to do with our perceived complacency over class divisions?

to a rate for whites (non-Hispanics) of 8.0 percent and for Asian Americans of 6.8 percent (U.S. Bureau of Labor Statistics, 2011a). However, like other types of "official" statistics, unemployment rates may be misleading. Individuals who become discouraged in their attempt to find work and no longer actively seek employment are not counted as unemployed. Some analysts believe that unemployment rates may drop for a period of time because people either do not seek work or they accept part-time or temporary jobs when they cannot find full-time employment. According to Bureau of

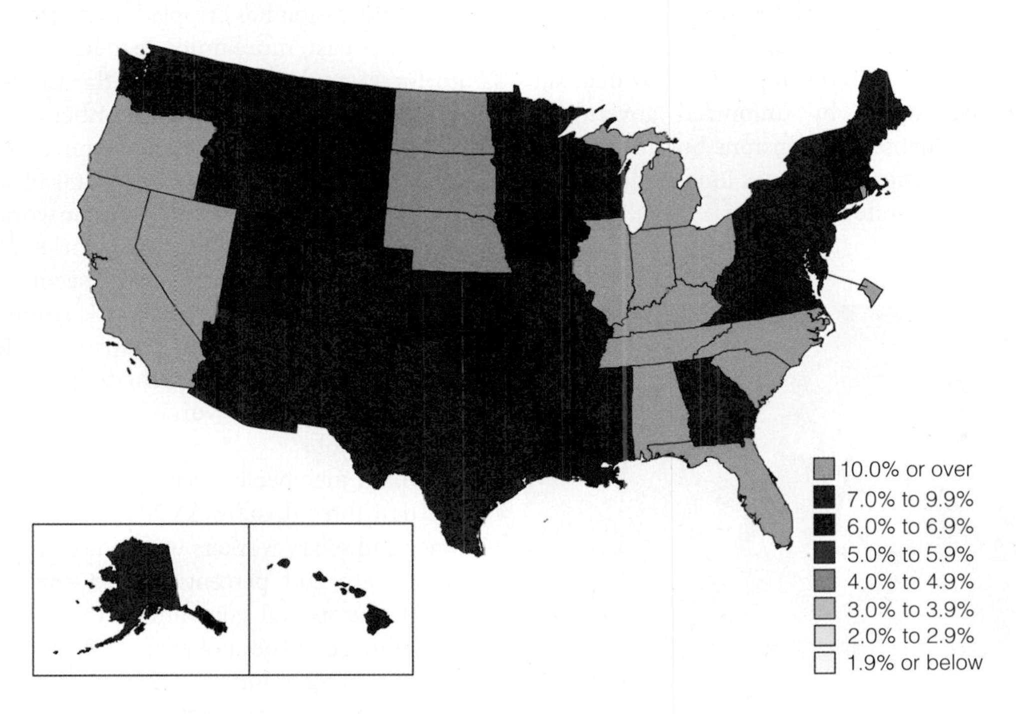

▲ **MAP 13.1B U.S. UNEMPLOYMENT RATES BY STATE AFTER THE GREAT RECESSION (U.S. RATE: 9.7%)**

Source: Bureau of Labor Statistics Local Area Unemployment Statistics March 2010

Labor Statistics (2011a) reports, 1 million "discouraged workers" are not currently looking for work because they believe that no jobs are available for them. Another 8.3 million people report that they are employed part time because they cannot find full-time employment.

Worker Resistance and Activism

In their individual and collective struggles to improve their work environment and gain some measure of control over their work-related activities, workers have used a number of methods to resist workplace alienation. Many have joined labor unions to gain strength through collective action.

Labor Unions

U.S. labor unions came into being in the mid-nineteenth century. Unions have been credited with gaining an eight-hour workday, a five-day workweek, health and retirement benefits, sick leave and unemployment insurance, and workplace health and safety standards for many employees. As one bumper sticker says, "Unions: The folks who brought you the weekend."

Most of these gains have occurred through *collective bargaining*—negotiations between employers and labor union leaders on behalf of workers. However, some states have passed laws making it harder for workers to organize or to engage in collective bargaining because of state and local budget shortfalls. For example, Wisconsin passed a law that bans collective bargaining by unionized government workers for benefits and pensions but allows them to bargain as a union for pay as long as their raises do not exceed the rate of inflation.

© AP Photo/Mark Lennihan

▲ During economic downturns and periods of high unemployment, good jobs become highly prized. These people are waiting for interviews at Con Edison, a large public utility in New York.

© PHOTOEDIT/PhotoEdit

▲ Seeking to improve economic and social opportunities for farmworkers, the late César Chávez held rallies and engaged in other protest activities in an effort to better the workers' lives.

There appears to be less public support for labor unions than in the past. A 2010 Pew Research poll found that just 41 percent of respondents stated that they had a favorable view of unions, which is the lowest level of support for unions in that poll's history (Surowiecki, 2011). Some analysts attribute this diminished support to the fact that many union members earn more money and have better pension plans than nonunionized workers. Also, in the past, more people may have had a better understanding of unions because more individuals belonged to unions or had family members who did, while unions have now largely disappeared from the private sector. In the final analysis, "The Great Depression invigorated the modern American labor movement. The Great Recession has crippled it" (Surowiecki, 2011).

In the past, more union leaders called for strikes to force employers to accept the union's position on wages and benefits. The number of workers involved in the actions declined from a peak of more than 2.5 million in 1971 to 45,000 in 2010, when there were only 11 major strikes or work stoppages that involved more than 1,000 workers (see ▶ Figure 13.5). This number was the second-lowest annual total since the major work stoppages began in 1947 (U.S. Bureau of Labor Statistics, 2011c). Strike activity has diminished significantly in recent years as more workers fear that they might lose their jobs.

Union membership has also been shrinking over the past three decades. In 2010 only 11.9 percent of wage and salary workers were union members, compared with 20.1 percent in 1983, the first year for which the federal government compiled such data (U.S. Bureau of Labor Statistics, 2011d). Union membership is higher for public sector workers (36.2 percent) than for private sector employees (6.9 percent). More men (12.6 percent) are union members than women (11.1 percent). In 2010, among major race and

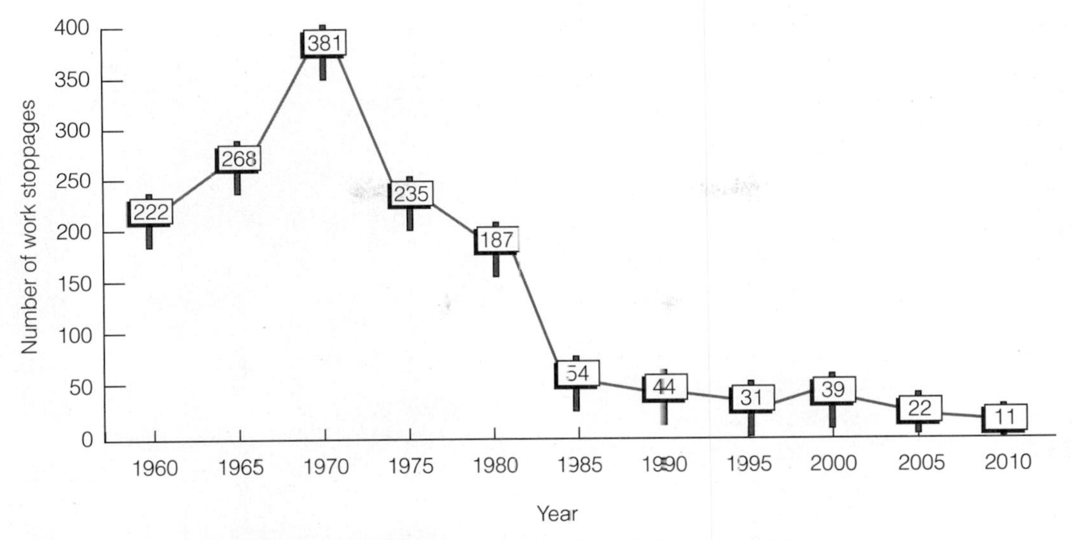

▲ **FIGURE 13.5 MAJOR WORK STOPPAGES IN THE UNITED STATES, 1960–2010**

Source: U.S. Census Bureau, 2010b.

ethnic groups, African American workers were more likely to be union members (13.4 percent) than workers who were white American (11.7 percent), Asian American (10.9 percent), or Hispanic (10.0 percent). The highest rate of union membership was among African American men (14.8 percent), whereas Asian American men had the lowest rate (9.4 percent).

If unions are in a period of decline, should we anticipate that this will have a detrimental effect on the hard-earned gains of workers in the United States? Clearly, union leaders have not given up on organized labor, as evidenced by a march by thousands of union leaders and workers in Los Angeles in 2011 that included nurses, electricians, teachers, and longshoremen in support of their own rights and those of Wisconsin union members to engage in collective bargaining. In the words of one union activist, "This is more than just about union-busting, this is about busting the middle class, and this is about future elections as well" (qtd. in Foxnews.com, 2011).

Employment Opportunities for Persons with a Disability

For many years, people with disabilities have been steered away from occupations by teachers, parents, prospective employers, and others who tended to focus more on what they could not do rather than what they are capable of doing. In 1990 the United States became the first nation to formally address the issue of equality for persons with a disability when Congress passed the Americans with Disabilities

◀ Workers with a disability are able to engage in a wide variety of occupations when they are offered the opportunity to do so.

© Michael Newman/PhotoEdit

you can make a difference

Focusing on What People with Disabilities *Can* Do, Not What They *Can't* Do!

I want to be allowed to compete for the same jobs as everyone else. I've worked hard to get good grades; I've never pitied myself; and I don't need other people to pity me. Just give me a chance to prove myself!

—Stephanie, a senior with a disability in one of the author's classes, stating how she hopes that she will be given the same chance as other students to use her college education in a tough job market

Sometimes employers hesitate to hire an adult with a disability because of a misguided sense that the person may not be able to handle the job when the going gets tough. I've used a wheelchair for most of my life, and I would argue that people with disabilities are in fact better equipped for acute problem-solving than their peers without disabilities. That's mainly because we're experts in finding creative ways to perform tasks that others may take for granted.

—Ralph Braun (2009), CEO of BraunAbility, explaining why he believes that employers' fears about hiring workers with a disability are unfounded

If you are a person with a disability or a person who plans to hire and supervise employees—some of whom may have a disability—you can make a difference in the workplace by learning more about disability employment and by promoting positive employment outcomes for people with disabilities. Here are some important job-search tips from the Massachusetts One-Stop Career Centers (Commonwealth of Massachusetts, 2011) for job seekers who have a disability:

- *Focus on your abilities and not disabilities.* While an employer cannot use the interview of the application process to inquire about a potential disability, it is helpful to bring up your disability if it relates to your ability to perform the job.
- *Practice interviewing before the interview.* Use mock interview sessions with trusted friends and relatives, be sure that clothing is clean and pressed, and make eye contact with the interviewer.
- *Be realistic about what types of positions will fit best for your disability.*
- *Read up on how to conduct a successful job search and interview.*
- *Learn as much as you can about the company or organization where you are interviewing.*
- *Become an in-house expert.* Add new job skills and become an expert in a particular area so that you will be more valuable to an employer.

It is important for workers with disabilities to overcome the steep occupational obstacles they face so that they will have

Act (ADA) to prohibit discrimination on the basis of disability. Combined with previous disability rights laws (such as those that provide for the elimination of architectural barriers from new, federally funded buildings and for the maximum integration of schoolchildren with disabilities), the ADA is a legal mandate for the full equality of people with disabilities. The federal law defines a person with a disability as an individual with at least one of the following conditions: He or she is deaf or has serious difficulty hearing; is blind or has serious difficulty seeing even when wearing glasses; has serious difficulty concentrating, remembering, or making decisions because of a physical, mental, or emotional condition; has serious difficulty walking or climbing stairs; has difficulty dressing or bathing; or has difficulty doing errands alone such as visiting a doctor's office or shopping because of a physical, mental, or emotional condition.

Despite the ADA and other laws regarding disability rights, many persons with a disability remain unemployed or have been tracked into disability-related service roles (such as helping other persons with a disability). The economic recession also hit workers with disabilities hard, particularly those individuals who have mobility impairments or difficulty performing routine daily activities. As of March 2011, persons age 16 years and over with one or more of these disabilities had an unemployment rate of 15.6 percent (accounting for 903,000 people). However, many other persons with disabilities were not in the labor force at all, nor were they actively seeking employment (U.S. Bureau of Labor Statistics, 2011b).

Many persons with a disability believe that they could work if they were offered the opportunity. However, even when persons with a disability are able to find jobs, they earn less than persons without a disability. On average, workers with a disability receive lower pay, less job security, and less access to health insurance, pension plans, and training than their nondisabled counterparts (Society for Human Resource Management, 2010). In the second decade of the twenty-first century, there is some cause for hope because increasing numbers of people are becoming involved in the campaign for disability employment (see "You Can Make a Difference").

an opportunity to achieve their ambitions and goals. It is also important for workers with disabilities to gain employment because they provide hope for others that they too can participate in the labor market and will not experience the same discrimination that some workers with a disability experienced in the past.

Although these are just a few suggestions for having a positive employment outcome when seeking a job, each of us can make a difference, regardless of our ability/disability status, if we recognize the value and talent that persons with disabilities bring to the workplace. According to the Campaign for Disability Employment (2011), "Whether good economic times or bad, it's the organizations that know how to identify and recognize talent that are most likely to succeed." If you are interested in becoming involved, search the Internet for sites such as the Campaign for Disability Employment (**www.whatcanyoudocampaign.org**), which explains how to promote positive employment outcomes for people with disabilities or how to find your own opportunities, if you are a person with a disability.

Finally, each of us can make a difference by actively working to dispel these commonly held myths about workers with disabilities (U.S. Department of Labor, 2008):

- Myth #1: Employees with disabilities have a higher absentee rate than employees without disabilities.
- Fact: Studies show that employees with disabilities are not absent any more than employees without disabilities.
- Myth #2: Persons with disabilities are unable to meet performance standards, thus making them a bad employment risk.
- Fact: Studies conducted by DuPont and others have shown that the vast majority of employees with disabilities rated average or better in job performance.
- Myth #3: Persons with disabilities have problems getting to work.
- Fact: Persons with disabilities are as successful at supplying their own transportation as are persons without a disability.
- Myth #4: Employees with disabilities are more likely to have accidents on the job than employees without disabilities.
- Fact: Safety records are virtually identical for workers with and without disabilities.

Lack of familiarity with disability issues and lack of involvement with individuals with a disability contribute to negative attitudes about employing persons with disabilities. You can make a difference by helping to dissipate myths and groundless fears that exclude many people from becoming fully productive members of the U.S. workforce.

These resources are available for further information:

- The Americans with Disabilities Act page provides information on the act and links to disability-related sites, organized by subject:
  www.ada.gov
- The International Center for Disability Resources on the Internet provides resources for persons with disabilities and includes links to other sites:
  www.icdri.org

The Global Economy in the Future

How will the U.S. economy look in the future? What about the global economy? Although sociologists do not have a crystal ball with which to predict the future, some general trends can be suggested.

The U.S. Economy

Many of the trends we examined in this chapter will produce dramatic changes in the organization of the economy and work in the twenty-first century. U.S. workers may find themselves fighting for a larger piece of an ever-shrinking economic pie. In April 2011 the U.S. national debt was at more than $14 trillion, making each citizen's share more than $46,000! To compare that figure to the national debt on the day you are reading this book, log on to **www.usdebtclock.org** and see how much it has increased.

Even as the gender and racial–ethnic composition of the labor force continues to diversify, many workers will remain in race- and/or gender-segregated occupations and industries. At the same time, workers may increasingly be fragmented into two major labor market divisions: (1) those who work in the innovative, primary sector and (2) those whose jobs are located in the growing secondary, marginal sector. In the innovative sector, increased productivity

▲ The next great wave of energy production is so-called "green" technology: solar- and wind-produced electricity, for example. However, in this area the United States lags behind other nations, including China, which was among the nations to aggressively subsidize alternative energy early on.

sociology and social policy

How Globalization Changes the Nature of Social Policy

In 1492 Christopher Columbus set sail for India, going west. He had the Niña, the Pinta and the Santa Maria. He never did find India, but he called the people he met "Indians" and came home and reported to his king and queen: "The world is round." I set off for India 512 years later. I knew just which direction I was going. I went east. I had Lufthansa business class, and I came home and reported only to my wife and only in a whisper: "The world is flat."

—author and newspaper columnist Thomas L. Friedman (2005) describing how the global economy is changing all areas of economic, political, and social life, including how we formulate social policy

What does Thomas Friedman mean when he states that "the world is flat"? As discussed in Friedman's (2007) best-selling book, *The World Is Flat 3.0: A Brief History of the Twenty-First Century*, *flat* means "level" or "connected" because (in his opinion) there is a more level global playing field in business (and almost any other endeavor) in the twenty-first century. According to Friedman, global telecommunications and the lowering of many trade and political barriers have brought about a new global era driven by *individuals*, not just by major corporations or giant trade organizations such as the World Bank. These individuals include entrepreneurs who create startup ventures around the world and computer freelancers whose work knows no boundaries (based on the idea of the older nation-state borders) when it comes to the transfer of information. Other factors that Friedman also believes have contributed to the "flattening" of the world include the streaming of the supply chain (Wal-Mart, for example), and the organization of information on the Internet by Google and Yahoo!, among others.

Worldwide, many freelancers and business entrepreneurs are not in the United States: They reside in nations such as India and China, where it is now possible to do more than merely compete in low-wage manufacturing and routine information labor (such as workers in call centers), but also in the top levels of research and design work and in professions such as law.

If Friedman's assertions are correct that the world has become flat and the United States is losing dominance in the global political and economic arena, where does this leave us in regard to social policy? Can we do something to stem the flow of jobs out of this country that has resulted from outsourcing and the shift of some information technologies from the United States to India, China, and other emerging nations?

Friedman suggests that if the United States is to remain competitive in the global economy, we must not continue to do things as they have previously been done. He believes we should have a thoughtful national discussion about what globalization means in all of our lives. Because there has been a shift from large-scale corporate players in the global economy to individual entrepreneurs and freelancers, we must look to each *individual* in our country to see how we can best play the economic game in the twenty-first century. In addition to nationwide policies, Friedman (2007) believes that our social policy regarding globalization must begin at home; children and young adults must be encouraged to rise to the economic challenge that faces them.

reflect & analyze

When social policy becomes personal (as Friedman believes), are we willing to engage in the changes it requires? Are Friedman's assumptions about the changing world order accurate? What do you think? What other arguments might be presented?

will be the watchword as corporations respond to heightened international competition. In the marginal sector, alienation will grow as temporary workers, sometimes professionals, look for avenues of upward mobility or at least a chance to make their work life more tolerable.

One of the greatest concerns for the future is the employment of U.S. workers. Well-paid jobs have been the backbone of the middle class in this country, as well as a major source of upward mobility for people throughout history, so there is cause for concern over the lack of availability of jobs for many people. There is also cause for concern because many jobs that are being created do not provide sufficient income for people to pay for fundamentals such as housing, utilities, food, health care, and transportation (Rich, 2011). In recent years, workers who have only a high school diploma have fared badly in the job market because of the recession and the slow recovery of the U.S. economy. By contrast, college-educated professionals have fared somewhat better, but those who have been best off during the tough economic times are persons who have seen their income and wealth diminish slightly, but overall they remain in the nation's top wealth categories. For example, in 2010 wealth was so concentrated in the United States that 25 people (all hedge fund chiefs) pocketed a total of $22.07 billion. By sharp contrast, it would take 441,400 people, each earning $50,000 per year, to pocket that same amount of money (Creswell, 2011). Despite issues such as these, some economists and political leaders suggest that we can be cautiously optimistic about the future of the U.S. economy, particularly if we understand its close ties to the global economy.

Global Economic Interdependence and Competition

The financial meltdown in the United States and around the world was a major concern in 2007 and 2008, during what was the largest economic slump since the Great Depression, in the 1930s. It was assumed that banks, regulators, and financial markets had built-in safeguards that would prevent economic disaster from occurring; however, many were shocked to learn how vulnerable that national and international financial markets actually are. The collapse of housing prices lead to an overall financial crisis in the United States that quickly spread to other nations, causing the world gross domestic product to fall 1.5 percent in 2009. Even economic downturns in the 1970s and 1980s had not produced such a major effect (Norris, 2011).

How have difficult economic times affected transnational corporations? Most social analysts predict that transnational corporations will become even more significant in the global economy. As they continue to compete for world market share, these corporations will become even less aligned with the values of any one nation. There was widespread shock, for example, when the *New York Times* reported that General Electric, our nation's largest corporation, generated worldwide profits of $14.2 billion, with $5.1 billion of the total coming from its U.S. operations, but paid nothing (zero) in taxes in 2010 to the U.S. government, claiming instead a tax benefit of $3.2 billion (Kocieniewski, 2011). The company paid no federal income taxes in 2008 or 2009 either.

According to one journalist, General Electric's "extraordinary success is based on an aggressive strategy that mixes fierce lobbying for tax breaks and innovative accounting that enables it to concentrate its profits offshore" (Kocieniewski, 2011: A1). It is estimated that since 2000, G.E. has spent $235.2 million in political money; in 2010 alone, the company spent $39.3 million and hired 195 lobbyists. Also, since 2002 G.E. has eliminated one-fifth of its U.S. workforce and increased employment in other countries (Kocieniewski, 2011).

We now live in a global community where transnational corporations such as G.E. are a way of life and in which telecommunications networks link workers around the world. Changes in the global economy require people in all nations—including the United States—to make changes in the way that things are done on the individual, regional, and national levels (see "Sociology and Social Policy"). Will we be able to make sufficient changes before our current economic problems become even worse?

chapter review Q & A

Use these questions and answers to check how well you've achieved the learning objectives set out at the beginning of this chapter.

- **What is the primary function of the economy?**

The economy is the social institution that ensures the maintenance of society through the production, distribution, and consumption of goods and services.

- **What are the three sectors of economic production?**

Economic production is divided into primary, secondary, and tertiary sectors. In primary sector production, workers extract raw materials and natural resources from the environment and use them without much processing. Industrial societies engage in secondary sector production, which is based on the processing of raw materials (from the primary sector) into finished goods. Postindustrial societies engage in tertiary sector production by providing services rather than goods.

- **How do the major contemporary economic systems differ?**

Capitalism is characterized by *private ownership* of the means of production, pursuit of personal profit, competition, and limited government intervention. By contrast, socialism is characterized by *public ownership* of the means of production, the pursuit of collective goals, and centralized decision making. In mixed economies, elements of a capitalist, market economy are combined with elements of a command, socialist economy. These mixed economies are often referred to as democratic socialism.

- **What are the functionalist, conflict, and symbolic interactionist perspectives on the economy and work?**

According to functionalists, the economy is a vital social institution because it is the means by which needed goods and services are produced and distributed. Conflict theorists suggest that the capitalist economy is based on greed. In order to maximize profits, capitalists suppress the wages of workers, who, in turn, cannot purchase products, making it necessary for capitalists to reduce production, close factories, lay off workers, and adopt other remedies that are detrimental to workers and society. Symbolic interactionists focus on the microlevel of the economic system, particularly on the social organization of work and its effects on workers' attitudes and behavior. Many workers experience job satisfaction when they like their job responsibilities and the organizational structure in which they work and when their individual needs and values are met. Alienation occurs when workers do not gain a sense of self-identity from their jobs and when their work is done completely for material gain and not for personal satisfaction.

- **What are occupations, and how do they differ in the primary and secondary labor markets?**

Occupations are categories of jobs that involve similar activities at different work sites. The primary labor market consists of well-paying jobs with good benefits that have some degree of security and the possibility for future advancement. The secondary labor market consists of low-paying jobs with few benefits and very little job security or possibility for future advancement.

- **What are the characteristics of professions?**

Professions are high-status, knowledge-based occupations characterized by abstract, specialized knowledge; autonomy; self-regulation; authority over clients and subordinate occupational groups; and a degree of altruism.

- **What are marginal jobs?**

Marginal jobs differ in some manner from mainstream employment norms: Jobs should be legal, be covered by government regulations, be relatively permanent, and provide adequate hours and pay in order to make a living. Marginal jobs fall below some or all of these norms.

- **What is contingent work?**

Contingent work is part-time work, temporary work, or subcontracted work that offers advantages to employers but may be detrimental to workers. Through the use of contingent workers, employers are able to cut costs and maximize profits, but workers have little or no job security.

key terms

questions for critical thinking

1. If you were the manager of a computer software division, how might you encourage innovation among your technical employees? How might you encourage efficiency? If you were the manager of a fast-food restaurant, how might you increase job satisfaction and decrease job alienation among your employees?
2. Using Chapter 2 as a guide, design a study to determine the degree of altruism in certain professions. What might be your hypothesis? What variables would you study? What research methods would provide the best data for analysis?
3. What types of occupations will have the highest prestige and income in 2030? The lowest prestige and income? What, if anything, does your answer reflect about the future of the U.S. economy?

turning to video

Watch the BBC video *Booming Thai Economy* (running time 2:36), available through **CengageBrain.com**. Many U.S. manufacturing jobs are being sent to places such as Thailand and India. As people in these countries move from being primarily producers of goods to also being consumers, what effect do you think this will have on the global economy? After you've watched the video consider this question: How does another country's growth affect the job market in America, and to what degree is it possible that India will replace the United States as the land of opportunity?

online study resources

Go to CENGAGE brain.com to access online study resources, including the Sociology CourseMate for this text as well as special features such as video, an interactive sociology time line and interactive maps, General Social Survey (GSS) data, and U.S. Census 2010 data.

CourseMate brings course concepts to life with interactive learning, study, and exam-preparation tools that support the printed textbook. A textbook-specific website, **Sociology CourseMate** includes an integrated interactive eBook and other interactive learning tools, including quizzes, flash cards, and videos.

Visit **www.cengagebrain.com** to access your account and purchase materials.

14 Politics and Government in Global Perspective

Do you think that mainstream TV news is boring? Does the idea of sitting down to watch *News Hour with Jim Lehrer* make you yawn . . . just thinking about it? If so, you're not alone—in fact, this is now the norm among the college-aged demographic. [One article states] that many young adults eschew traditional nightly news for [Jon Stewart's] *The Daily Show*, . . . which proudly bills itself as "the most trusted name in fake news." . . . Yes, it is largely satirical, but I can understand why many might watch the show as a main source of news. While it is based on real news, it is also written by comedy writers, and has ratings in mind—not necessarily the best interests of the American public or young people. . . . Yes, *The Daily Show* is funny. Yes, Jon Stewart is right on the nail with his ironic insight and sarcastic humor. However, it isn't a real news show. . . .

—Katie Stapleton-Paff (2007), a writer for the *Daily of the University of Washington*, discussing the fact that many people between the ages of eighteen and twenty-five use "comedy" programs as their prime source of news

© AP Photo/Jason DeCrow

▲ Jon Stewart, host of the extremely popular Comedy Central production *The Daily Show*, parodies mainstream news programming. Many people now watch Stewart's reports instead of watching network news programs. How might comedy news shows affect our perceptions of politics?

I love the mock interviews, but I never go into the show thinking I'm watching real news. What I see is what I get and in the case of *The Daily Show* I see a funny show that makes fun of the day's news, much like Jay Leno [of NBC's *Tonight Show*] does in his monologue every night. . . . Fair and balanced news is extremely hard to come by these days and if students can't get fair and balanced as well as entertaining, they'll just stick with what's entertaining.

—Allen, a blogger, responding to the article by Stapleton-Paff

There are days when I watch *The Daily Show*, and I kind of chuckle. There are days when I laugh out loud. There are days when I stand up and point to the TV and say, "You're damn right!" . . . The stock-in-trade of *The Daily Show* 's hypocrisy, exposing hypocrisy. And nobody else has the guts to do it. They really know how to crystallize an issue on all sides, see the silliness everywhere.

—Hub Brown, an associate dean at the S. I. Newhouse School of Public Communications and associate professor of journalism at Syracuse University, refers to himself as a "convert" after he began watching *The Daily Show* in response to students' comments that they preferred it to mainstream news programs (qtd. in Smolkin, 2007).

Chapter Focus Question:

What effect does the intertwining of politics and the media have on the United States and other nations?

At the same time that world problems continue to grow, ranging from national and global economic crises to unthinkable natural disasters in nations such as Japan, people are drawn to a much wider array of media sources for their news and information about current events. Not only are pseudo-news TV shows such as *The Daily Show* and *The Colbert Report* increasing in importance as a source of "news," particularly for younger viewers, but digital media are also causing people to spend more time each day with the news. According to one Pew Research Center survey (2011) that looks at where and how people get their news, people spend 57 minutes on average getting the news from TV, radio, or newspapers on a given day, but they also spend an additional 13 minutes getting news online, which increases their total news time to about 70 minutes. Of course, these data do not provide us with an indication about the quality of the content that people are receiving: Is it "real" news which provides facts that can be documented, or is it the "hot-topic" approach, which selectively sensationalizes some political or social event for the sake of humor or entertainment instead of conveying actual information?

For many people, satirical news programs on television and website blurbs have become major sources of

Highlighted Learning Objectives

- Compare the major political systems around the world.
- Explain how the center of power differs in the pluralist and the elite models of the U.S. power structure.
- Discuss how political parties and political attitudes shape the government.
- Identify key factors that contribute to the power of government bureaucracies.
- Discuss the place of democracy in the future.

information about what is going on in the world. The extent to which entertainment and "hard news" have become blurred is related to the extent to which our major social institution of politics and government is both reflective of and influenced by mass media that constitute another powerful social institution in our society and around the world.

In twenty-first-century America, the issue of politics and government in global perspective is a hot topic for concerned people because we live in an age of political and economy uncertainty and constant discord regarding many decisions made by our nation's political and business leaders. Sociologists are concerned about how the social institutions of politics and the economy operate and how the time that people spend with various forms of media influences their perspectives on life in the global community. In this chapter we discuss the intertwining nature of contemporary politics, government, and the media. Before reading on, test your knowledge of the media by taking the Sociology and Everyday Life quiz.

Politics, Power, and Authority

***Politics* is the social institution through which power is acquired and exercised by some people and groups.** In contemporary societies the government is the primary political system. ***Government* is the formal organization that has the legal and political authority to regulate the relationships among members of a society and between the society and those outside its borders.** Some social analysts refer to the government as the ***state*—the political entity that possesses a legitimate monopoly over the use of force within its territory to achieve its goals.**

Whereas political science focuses primarily on power and its distribution in different types of political systems, ***political sociology* is the area of sociology that examines the nature and consequences of power within or between societies, as well as the social and political conflicts that lead to changes in the allocation of power.** Political sociology primarily focuses on the *social circumstances* of politics and explores how the political arena and its actors are intertwined with social institutions such as the economy, religion, education, and the media.

What is the relationship between politics and media? Some sociologists suggest that the media often distort—either intentionally or unintentionally—the information they provide to citizens. According to the sociologist Michael Parenti (1998), the media have the power to influence public opinion in a way that favors management over labor, corporations over their critics, affluent whites over subordinate-group members, political officials over protestors, and free-market capitalism over public-sector development. Parenti's assertion raises an interesting issue about the distribution of power in the United States and other industrialized nations: Do the media distort information to suit their own interests?

© AP Photo/Will Burgess, Pool

© TG Stock/Contributor/Tim Graham/Getty Images

© Bloomberg/Contributor/Getty Images

▲ Max Weber's three types of global authority are shown here in global perspective. Pope Benedict XVI is an example of traditional authority sanctioned by custom. Mother Teresa exemplifies charismatic authority, for her leadership was based on personal qualities. The U.S. Supreme Court represents rational–legal authority, which depends upon established rules and procedures.

sociology and everyday life

How Much Do You Know About the Media?

True	False	
T	F	1. The Internet has displaced newspapers and radio as the primary source of news in the United States.
T	F	2. No media sources are publicly owned in the United States.
T	F	3. Each person in this country spends, on average, about 3,500 hours per year using media.
T	F	4. Websites are becoming increasingly profitable because visitors often click on ads and purchase products from website sponsors.
T	F	5. Thirty-minute nightly news programs on the major television networks (NBC, CBS, and ABC) typically contain slightly less than 24 minutes of news, and the rest is commercials, promotional announcements, and other non-news items.
T	F	6. Network and cable television channels have been very resistant to providing information through search engines such as Yahoo! News.
T	F	7. Newspapers have been losing subscribers and daily readers over the past decade.
T	F	8. Some analysts believe that "reality" TV shows blur the distinction between "news" and "entertainment."

Answers on page 410.

Power and Authority

***Power* is the ability of persons or groups to achieve their goals despite opposition from others** (Weber, 1968/1922). Through the use of persuasion, authority, or force, some people are able to get others to acquiesce to their demands. Consequently, power is a *social relationship* that involves both leaders and followers. Power is also a dimension in the structure of social stratification. Persons in positions of power control valuable resources of society—including wealth, status, comfort, and safety—and are able to direct the actions of others while protecting and enhancing the privileged social position of their class (Domhoff, 2002). For example, the sociologist G. William Domhoff (2002) argues that the media tend to reflect "the biases of those with access to them—corporate leaders, government officials, and policy experts." However, although Domhoff believes the media can amplify the message of powerful people and marginalize the concerns of others, he does not think the media are as important as government officials and corporate leaders are in the U.S. power equation.

What about power on a global basis? Although the most basic form of power is physical violence or force, most political leaders do not want to base their power on force alone. Instead, they seek to legitimize their power by turning it into ***authority*—power that people accept as legitimate rather than coercive.**

Ideal Types of Authority

Who is most likely to accept authority as legitimate and adhere to it? People have a greater tendency to accept authority as legitimate if they are economically or politically dependent on those who hold power. They may also accept authority more readily if it reflects their own beliefs and values (Turner, Beeghley, and Powers, 2007). Weber's outline of three *ideal types* of authority—traditional, charismatic, and rational–legal—shows how different bases of legitimacy are tied to a society's economy.

Traditional Authority According to Weber, ***traditional authority* is power that is legitimized**

politics the social institution through which power is acquired and exercised by some people and groups.

government the formal organization that has the legal and political authority to regulate the relationships among members of a society and between the society and those outside its borders.

state the political entity that possesses a legitimate monopoly over the use of force within its territory to achieve its goals.

political sociology the area of sociology that examines the nature and consequences of power within or between societies, as well as the social and political conflicts that lead to changes in the allocation of power.

power according to Max Weber, the ability of people or groups to achieve their goals despite opposition from others.

authority power that people accept as legitimate rather than coercive.

traditional authority power that is legitimized on the basis of long-standing custom.

sociology and everyday life

ANSWERS to the Sociology Quiz on the Media

1. True. According to recent surveys, more people in the United States get their news from the Internet than from newspapers or radio.

2. False. Although most media outlets are privately owned, Public Broadcasting Service (PBS) and National Public Radio (NPR) are funded by government support, grants from nonprofit foundations, and donations from viewers and listeners.

3. True. The hours per person spent using media increase annually, and current estimates suggest that an average of 3,500 hours per person per year will be spent using all forms of media, including television, radio, movies, print media, video games, and other types of entertainment.

4. False. Studies show that not as many website visitors are clicking on pop-up ads or purchasing products from website sponsors as originally had been projected by website producers who hoped to make a profit from providing information and entertainment online.

5. False. News programs such as *NBC Nightly News* typically have 19 minutes of news content and 11 minutes of commercials, promotional announcements, and other non-news items.

6. False. Although decision makers at network and cable television channels were initially resistant to providing free information through search engines such as Yahoo! News, their fear of losing viewers and sponsors led them to embrace the newer technologies in order to survive.

7. True. Subscription rates and daily paper purchases have decreased substantially over the past decade. Now some papers make their content available online for free, but more (including the *New York Times*) are charging for "premium content."

8. True. Because TV "reality shows" seek to imitate real life, some viewers have difficulty keeping firmly in mind what has actually happened and what is staged or faked for viewers' entertainment.

Sources: Based on Gilson, 2009; Pew Research Center, 2011; and Project for Excellence in Journalism, 2010.

on the basis of long-standing custom. In preindustrial societies, the authority of traditional leaders, such as kings, queens, pharaohs, emperors, and religious dignitaries, is usually grounded in religious beliefs and custom. For example, British kings and queens historically traced their authority from God. Members of subordinate classes obey a traditional leader's edicts out of economic and political dependency and sometimes personal loyalty. However, as societies industrialize, traditional authority is challenged by a more complex division of labor and by the wider diversity of people who now inhabit the area as a result of high immigration rates.

Gender, race, and class relations are closely intertwined with traditional authority. Political scientist Zillah R. Eisenstein (1994) suggests that *racialized patriarchy*—the continual interplay of race and gender—reinforces traditional structures of power in contemporary societies. According to Eisenstein (1994: 2), "Patriarchy differentiates women from men while privileging men. Racism simultaneously differentiates people of color from whites and privileges whiteness. These processes are distinct but intertwined."

Charismatic Authority ***Charismatic authority* is power legitimized on the basis of a leader's exceptional personal qualities or the demonstration of extraordinary insight and accomplishment that inspire loyalty and obedience from followers.** Charismatic leaders may be politicians, soldiers, and entertainers, among others.

Charismatic authority tends to be temporary and relatively unstable; it derives primarily from individual leaders (who may change their minds, leave, or die) and from an administrative structure usually limited to a small number of faithful followers. For this reason, charismatic authority often becomes routinized. The ***routinization of charisma* occurs when charismatic authority is succeeded by a bureaucracy controlled by a rationally established authority or by a combination of traditional and bureaucratic authority** (Turner, Beeghley, and Powers, 2007). According to Weber (1968/1922: 1148), "It is the fate of charisma to recede . . . after it has entered the permanent structures of social action."

Rational–Legal Authority According to Weber, ***rational–legal authority* is power legitimized by law or written rules and regulations.** Rational–legal

[concept quick review]

Weber's Three Types of Authority

Type	Description	Examples
Traditional	Legitimized by long-standing custom	Patrimony (authority resides in traditional leader supported by larger social structures, as in old British monarchy)
	Subject to erosion as traditions weaken	Patriarchy (rule by men occupying traditional positions of authority, as in the family)
Charismatic	Based on leader's personal qualities	Napoleon Adolf Hitler
	Temporary and unstable	Martin Luther King, Jr. César Chávez Mother Teresa
Rational–Legal	Legitimized by rationally established rules and procedures	Modern British Parliament
	Authority resides in the office, not the person	U.S. presidency, Congress, federal bureaucracy

authority—also known as *bureaucratic authority*—is based on an organizational structure that includes a clearly defined division of labor, hierarchy of authority, formal rules, and impersonality. Power is legitimized by procedures; if leaders obtain their positions in a procedurally correct manner (such as by election or appointment), they have the right to act.

Rational–legal authority is held by elected or appointed government officials and by officers in a formal organization. However, authority is invested in the *office*, not in the *person* who holds the office. For example, although the U.S. Constitution grants rational–legal authority to the office of the presidency, a president who fails to uphold the public trust may be removed from office. In contemporary society the media may play an important role in bringing to light allegations about presidents or other elected officials. Examples include the media blitzes surrounding the 1970s Watergate investigation that led to the resignation of President Richard M. Nixon, the late-1990s political firestorm over campaign fund-raising and the sex scandal involving President Bill Clinton, and more-recent ethics violations and convictions on charges of money laundering involving members of Congress during the Obama administration.

In a rational–legal system the governmental bureaucracy is the apparatus responsible for creating and enforcing rules in the public interest. Weber believed that rational–legal authority was the only means to attain efficient, flexible, and competent regulation under a rule of law (Turner, Beeghley, and Powers, 2007). Weber's three types of authority are summarized in the Concept Quick Review.

Political Systems in Global Perspective

Political systems as we know them today have evolved slowly. In the earliest societies, politics was not an entity separate from other aspects of life. Political institutions first emerged in agrarian societies as they acquired surpluses and developed greater social inequality. Elites took control of politics and used custom or traditional authority to justify their position. When cities developed circa 3500–3000 B.C.E., the *city-state*—a city whose power extended to adjacent areas—became the center of political power.

charismatic authority power legitimized on the basis of a leader's exceptional personal qualities or the demonstration of extraordinary insight and accomplishment that inspire loyalty and obedience from followers.

routinization of charisma the process by which charismatic authority is succeeded by a bureaucracy controlled by a rationally established authority or by a combination of traditional and bureaucratic authority.

rational–legal authority power legitimized by law or written rules and procedures. Also referred to as *bureaucratic authority*.

Nation-states as we know them began to develop in Europe between the twelfth and fifteenth centuries (see Tilly, 1975). A *nation-state* is a unit of political organization that has recognizable national boundaries and whose citizens possess specific legal rights and obligations. Nation-states emerge as countries develop specific geographic territories and acquire greater ability to defend their borders. Improvements in communication and transportation make it possible for people in a larger geographic area to share a common language and culture. As charismatic and traditional authority are superseded by rational–legal authority, legal standards come to prevail in all areas of life, and the nation-state claims a monopoly over the legitimate use of force.

About 193 nation-states currently exist throughout the world; today, everyone is born, lives, and dies under the auspices of a nation-state. Four main types of political systems are found in nation-states: monarchy, authoritarianism, totalitarianism, and democracy.

KIRSTY WIGGLESWORTH/AFP/Getty Images

▲ Through its many ups and downs, the British royal family has remained a symbol of Great Britain's monarchy, today headed by Queen Elizabeth. Monarchies typically pass power from generation to generation, and Queen Elizabeth's grandsons, Prince Harry and Prince William (shown here, after his wedding), represent the future of the royal family's rule.

Monarchy

***Monarchy* is a political system in which power resides in one person or family and is passed from generation to generation through lines of inheritance.** Monarchies are most common in agrarian societies and are associated with traditional authority patterns. However, the relative power of monarchs has varied across nations, depending on religious, political, and economic conditions.

Absolute monarchs claim a hereditary right to rule (based on membership in a noble family) or a divine right to rule (a God-given right to rule that legitimizes the exercise of power). In limited monarchies, rulers depend on powerful members of the nobility to retain their thrones. Unlike absolute monarchs, *limited monarchs* are not considered to be above the law. In *constitutional monarchies*, the royalty serve as symbolic rulers or heads of state while actual authority is held by elected officials in national parliaments. In present-day monarchies such as the United Kingdom, Sweden, Spain, and the Netherlands, members of royal families primarily perform ceremonial functions. In the United Kingdom, for example, the media often focus large amounts of time and attention on the royal family but concentrate on the personal lives of its members.

Authoritarianism

***Authoritarianism* is a political system controlled by rulers who deny popular participation in government.** A few authoritarian regimes have been absolute monarchies whose rulers claimed a hereditary right to their position. Today, Saudi Arabia and Kuwait are examples of authoritarian absolute monarchies. In *dictatorships* power is gained and held by a single individual. Pure dictatorships are rare; all rulers need the support of the military and the backing of business elites to maintain their position. *Military juntas* result when military officers seize power from the government, as has happened in recent decades in Argentina, Chile, and Haiti. Today, authoritarian regimes exist in Fidel Castro's Cuba and in the People's Republic of China. Authoritarian regimes seek to control the media and to suppress coverage of any topics or information that does not reflect upon the regime in a favorable light.

Totalitarianism

***Totalitarianism* is a political system in which the state seeks to regulate all aspects of people's public and private lives.** Totalitarianism relies on modern technology to monitor and control people; mass propaganda and electronic surveillance are widely used to influence people's thinking and control their actions. One example of a totalitarian regime was the National Socialist (Nazi) Party in Germany during World War II; military leaders there sought to control all aspects of national life, not just government operations. Other examples include the former Soviet Union and contemporary Iraq prior to the end of Saddam Hussein's regime.

To keep people from rebelling, totalitarian governments enforce conformity: People are denied the right to assemble for political purposes, access to information is strictly controlled, and secret police enforce compliance, creating an environment of constant fear and suspicion.

Many nations do not recognize totalitarian regimes as being the legitimate government of a particular country. Afghanistan in the year 2001 was an example. As the war on terrorism began in the aftermath of the September 11 terrorist attacks on the United States, many people developed a heightened awareness of the Taliban regime, which ruled most of Afghanistan and was engaged in fierce fighting to capture the rest of the country. The Taliban regime maintained absolute control over the Afghan people in most of that country, including requiring that all Muslims take part in prayer five times each day and that women wear the *hijab* (veil). Since U.S. military action commenced in Afghanistan, most of what U.S. residents know about the Taliban and about the war has been based on media reports and "expert opinions" expressed on television and the Internet.

Democracy

***Democracy* is a political system in which the people hold the ruling power either directly or through elected representatives.** The literal meaning of *democracy* is "rule by the people" (from the Greek words *demos*, meaning "the people," and *kratein*, meaning "to rule"). In an ideal-type democracy, people would actively and directly rule themselves. *Direct participatory democracy* requires that citizens be able to meet together regularly to debate and decide the issues of the day. Because there are about 312 million people in the United States today, it would be impossible for everyone to come together in one place for a meeting because an area of more than eighty square miles would be necessary for the gathering.

In countries such as the United States, Canada, Australia, and the United Kingdom, people have a voice in the government through ***representative democracy*, whereby citizens elect representatives to serve as bridges between themselves and the government.** The U.S. Constitution requires that each state have two senators and a minimum of one member in the House of Representatives. The number of voting representatives in the House (435) has not changed since the apportionment following the 1910 census; however, those 435 seats are reapportioned based on an increase or decrease in a state's population as shown in census data gathered every ten years.

In a representative democracy, elected representatives are supposed to convey the concerns and interests of those they represent, and the government is expected to be responsive to the wishes of the people. Elected officials are held accountable to the people through elections. However, representative democracy is not always equally accessible to all people in a nation. Throughout U.S. history, members of subordinate racial–ethnic groups have been denied full participation in the democratic process. Gender and social class have also limited some people's democratic participation.

Even representative democracies are not all alike. As compared to the winner-takes-all elections in the United States, which are usually decided by the candidate who wins the most votes, the majority of European elections are based on a system of proportional representation, meaning that each party is represented in the national legislature according to the proportion of votes that party received. For example, a party that won 40 percent of the vote would receive 40 seats in a 100-seat legislative body, and a party receiving 20 percent of the votes would receive 20 seats.

Perspectives on Power and Political Systems

Is political power in the United States concentrated in the hands of the few or distributed among the many? Sociologists and political scientists have suggested many different answers to this question; however, two prevalent models of power have emerged: pluralist and elite.

Functionalist Perspectives: The Pluralist Model

The pluralist model is rooted in a functionalist perspective, which assumes that people share a consensus on central concerns, such as freedom and protection from harm, and that the government serves important functions that no other institution can

monarchy a political system in which power resides in one person or family and is passed from generation to generation through lines of inheritance.

authoritarianism a political system controlled by rulers who deny popular participation in government.

totalitarianism a political system in which the state seeks to regulate all aspects of people's public and private lives.

democracy a political system in which the people hold the ruling power either directly or through elected representatives.

representative democracy a form of democracy whereby citizens elect representatives to serve as bridges between themselves and the government.

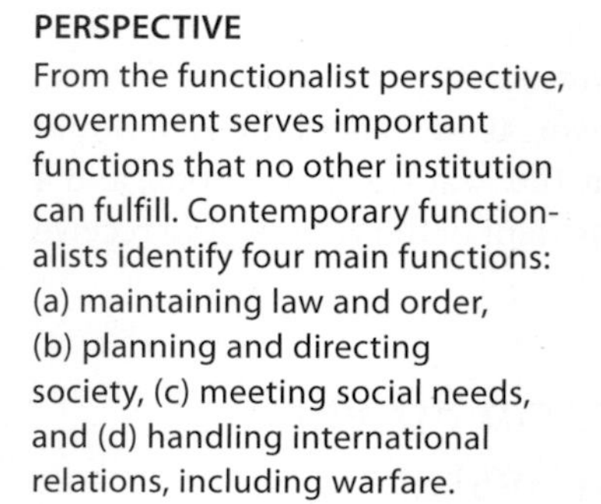

▶ **FIGURE 14.1 GOVERNMENT FROM A FUNCTIONALIST PERSPECTIVE**
From the functionalist perspective, government serves important functions that no other institution can fulfill. Contemporary functionalists identify four main functions: (a) maintaining law and order, (b) planning and directing society, (c) meeting social needs, and (d) handling international relations, including warfare.

fulfill. According to Emile Durkheim (1933/1893), the purpose of government is to socialize people to be good citizens, to regulate the economy so that it operates effectively, and to provide necessary services for citizens. Contemporary functionalists state the four main functions as follows: (1) maintaining law and order, (2) planning and directing society, (3) meeting social needs, and (4) handling international relations, including warfare (see ▶ Figure 14.1).

But what happens when people do not agree on specific issues or concerns? Functionalists suggest that divergent viewpoints lead to a system of political pluralism in which the government functions as an arbiter between competing interests and viewpoints. According to the ***pluralist model*****, power in political systems is widely dispersed throughout many competing interest groups** (Dahl, 1961).

In the pluralist model the diverse needs of women and men, people of all religions and racial–ethnic backgrounds, and the wealthy, middle class, and poor are met by political leaders who engage in a process of bargaining, accommodation, and compromise.

Competition among leadership groups in government, business, labor, education, law, medicine, and consumer organizations, among others, helps prevent abuse of power by any one group. Everyday people can influence public policy by voting in elections, participating in existing special interest groups, or forming new ones to gain access to the political system. In sum, power is widely dispersed, and leadership groups that wield influence on some decisions are not the same groups that may be influential in other decisions (Dye, Zeigler, and Schubert, 2012)

Special Interest Groups ***Special interest groups*** **are political coalitions made up of individuals or groups that share a specific interest they wish to protect or advance with the help of the political system.** Examples of special interest groups include the AFL-CIO (representing the majority of labor unions) and public interest or citizens' groups such as the American Conservative Union and Zero Population Growth.

What purpose do special interest groups serve in the political process? According to some analysts, special interest groups help people advocate their own interests and further their causes. Broad categories of special interest groups include banking, business, education, energy, the environment, health, labor, persons with a disability, religious groups, retired persons, women, and those espousing a specific ideological viewpoint; obviously, many groups overlap in interests and membership. Special interest groups are also referred to as *pressure groups* (because they put pressure on political leaders) or *lobbies*. Lobbies are often referred to in terms of the organization they represent or the single issue on which they focus—for example, the "gun lobby" and the "dairy lobby." The people who are paid to influence legislation on behalf of specific clients are referred to as *lobbyists*.

Over the past four decades, special interest groups have become more involved in "single-issue politics," in which political candidates are often supported or rejected solely on the basis of their views on a specific issue—such as abortion, gun control, gay and lesbian rights, or the environment. Single-issue groups derive their strength from the intensity of their beliefs; leaders have little room to compromise on issues.

Political Action Committees For many years, funding of lobbying efforts has been a hotly debated issue. Numerous attempts have been made to limit campaign contributions and expenditures to ensure that wealthy and influential individuals and organizations are not able to silence the voices of people who do not have equal resources. Reforms in campaign finance laws in the 1970s set limits on direct contributions to political candidates and led to the creation of ***political action committees* (PACs)—organizations of special interest groups that solicit contributions from donors and fund campaigns to help elect (or defeat) candidates based on their stances on specific issues.** As the cost of running for political office has skyrocketed, candidates have relied more on PACs for financial assistance. Advertising, staff, direct-mail operations, telephone banks, computers, consultants, travel expenses, office rentals, and other expenses incurred in political campaigns make PAC money vital to candidates.

Some PACs represent the "public interest" and ideological interest groups such as gay rights or the National Rifle Association. Other PACs represent the capitalistic interests of large corporations. Realistically, PACs do not represent members of the least-privileged sectors of society. As one senator pointed out, "There aren't any Poor PACs or Food Stamp PACs or Nutrition PACs or Medicare PACs" (qtd. in Greenberg and Page, 1993: 240). Critics of pluralism argue that "Big Business" wields such disproportionate power in U.S. politics that it undermines the democratic process (see Lindblom, 1977; Domhoff, 1978).

As an outgrowth of record-setting campaign spending in the 1996 national election, campaign financing abuses were alleged by both Republicans and Democrats in Washington. At the center of the controversy was the issue of "soft money" contributions, which are made outside the limits imposed by federal election laws. In 2002 Congress passed the McCain–Feingold campaign finance law, prohibiting soft money contributions in federal elections, and the U.S. Supreme Court upheld the soft money provisions of that law in 2003. However, the McCain–Feingold law applies only to federal elections and does not bar soft money contributions in state and local elections.

▲ Special interest groups help people advocate their interests and further their causes. Advocates may run for public office and gain a wider voice in the political process. An example is Ben Nighthorse Campbell of Colorado. Campbell, who was a U.S. senator from 1993 to 2005, is also a Cheyenne chief. He is shown here walking to the Senate floor after participating in a ground-breaking ceremony for the National Museum of the American Indian in Washington, D.C.

In 2010 the U.S. Supreme Court ruled in *Citizens United v. Federal Election Commission* that corporations and other organizations could bypass existing spending limits by giving unlimited

pluralist model an analysis of political systems that views power as widely dispersed throughout many competing interest groups.

special interest groups political coalitions composed of individuals or groups that share a specific interest that they wish to protect or advance with the help of the political system.

political action committees (PACs) organizations of special interest groups that solicit contributions from donors and fund campaigns to help elect (or defeat) candidates based on their stances on specific issues.

sociology works!

C. Wright Mills's Ideas About the Media: Ahead of His Time?

> In a mass society the dominant type of communication is the formal media, and the [various audiences] become mere *media markets:* all those exposed to the contents of a given mass media.
>
> —In *The Power Elite*, the influential sociologist C. Wright Mills (1959a: 299) provided this early critique of the U.S. mass media and their effect on the general public

During the mid-twentieth century, when Mills wrote these words, the press in the United States was often described as a "watchdog for the public." It was widely believed that journalists would reliably report what the public needed to know to stay informed, including any government and/or corporate dishonesty or ineptitude (Headley, 1985: 333). However, Mills rejected this notion because he believed the media industry lulled individuals into complacency and persuaded them to accept the status quo rather than encouraging the audience member to become a "public" person who formulates his or her own opinion on a given topic. As one contemporary scholar explained, "To recover Mills's 'public man' who is capable of formulating opinions, we would have to salvage whatever is left of him after he is stripped of MTV, CNN, and http://www., whose Italian translation, *ragnatele mondiale*, is useful in conveying the sense 'spider-web' and therefore the suspicion that we may be flies" (Cinquemani, 1997: 89).

Do the contemporary media inform people or trap them in a "spiderweb"? Mills hoped that individuals would develop the capacity for critical judgment based on information they received from the media industry; however, he was doubtful that this would occur for two important sociological reasons: (1) media communication involves a limited number of people who communicate *to* a great number of others ("the masses"), and (2) audiences have no effective way of answering back, making mass communication a one-way process.

As part of his legacy for making sociology work in the real world, Mills made an important prediction. According to Mills (1956: 303–304), critical thinking is more likely to occur among individuals under the following circumstances: "If public communications are so organized that there is a chance immediately and effectively to answer back any opinion expressed in public. Opinion formed by such discussion readily finds an outlet in effective action, even against—if necessary—the prevailing system of authority." If media audiences are able to respond immediately and provide their own opinions, media communications may become a two-way street. Technologies and methods of communication—including cell phones, text messaging, e-mail, and websites—now provide us with opportunities for media interactivity through which we can respond to media discourse and make our thoughts known.

amounts to "independent groups" that support candidates (but not the candidates themselves). In this decision the Court struck down a provision of the McCain–Feingold Act that prohibited both for-profit and not-for-profit corporations and unions from broadcasting "electioneering communications," which was defined as a broadcast, cable, or satellite communication that mentioned a candidate within sixty days of a general election or thirty days of a primary. The Court's decision was based on protection of First Amendment rights based on the assumption that a decision to spend money in support of a political cause or candidate was similar to giving a speech or carrying a campaign sign and thus protected by the First Amendment (Toobin, 2011). This controversial decision was criticized by many people, including President Barack Obama, who argued that special interests and their lobbyists had been granted even more power than they already held in Washington, while average Americans who make small contributions had been further downgraded in their efforts to support a political candidate. No doubt, this issue will continue well into the future and generate much conflict.

Conflict Perspectives: Elite Models

Although conflict theorists acknowledge that the government serves a number of useful purposes in society, they assert that government exists for the benefit of wealthy or politically powerful elites who use the government to impose their will on the masses. According to the ***elite model*****, power in political systems is concentrated in the hands of a small group of elites, and the masses are relatively powerless.** The pluralist model and the elite model are compared in ▶ Figure 14.2.

Contemporary elite models are based on the assumption that decisions are made by the elites, who agree on the basic values and goals of society. However, the needs and concerns of the masses are not often given consideration by those in the elite. According to this approach, power is highly

(c) Petrified Collection/Getty Images

C. Wright Mills believed that mass communication was a one-way street: The audience had no way to react to it and tended to become passive recipients of its content. However, today's digital technology allows people to respond quickly and, through developments such as blogs, to provide news content and opinions of their own.

© HANS DERYK/Reuters/Landov

Sociology works today in Mills's ideas because he encourages us to think for ourselves and to express our ideas and opinions rather than being passive recipients of information from the media. Mills challenges us to think about the extent to which we rely on *mediated* experiences of others, such as the ideas presented by television commentators or online bloggers, as compared with becoming our own "public person" and responding with thoughts and judgments of our own.

reflect & analyze

How might we relate Mills's ideas to the contemporary role of the media industries in shaping political and economic "realities" around the world?

concentrated at the top of a pyramid-shaped social hierarchy, and public policy reflects the values and preferences of the elite, not the preferences of the people (Dye, Zeigler, and Schubert, 2012).

C. Wright Mills and the Power Elite Who makes up the U.S. power elite? According to the sociologist C. Wright Mills (1959a), the ***power elite* is made up of leaders at the top of business, the executive branch of the federal government, and the military.** Of these three, Mills speculated that the "corporate rich" (the highest-paid officers of the biggest corporations) were the most powerful because of their unique ability to parlay the vast economic resources at their disposal into political power (see "Sociology Works!"). At the middle level of the pyramid, Mills placed the legislative branch of government, special interest groups, and local opinion leaders. The bottom (and widest layer) of the pyramid is occupied by the unorganized masses, who are relatively powerless and are vulnerable to economic and political exploitation.

G. William Domhoff and the Ruling Class Sociologist G. William Domhoff (2002) asserts that this nation in fact has a *ruling class*—the corporate rich, who constitute less than 1 percent of the U.S. population. Domhoff uses the term *ruling class* to signify a relatively fixed group of privileged people who wield power sufficient to constrain political processes and serve underlying capitalist interests. Although the power elite controls the everyday operation of the political system, who *governs* is less important than who *rules*.

According to Domhoff (2002), the corporate rich influence the political process in three ways. First, they affect the candidate selection process by helping

elite model a view of society that sees power in political systems as being concentrated in the hands of a small group of elites whereas the masses are relatively powerless.

power elite C. Wright Mills's term for the group made up of leaders at the top of business, the executive branch of the federal government, and the military.

PLURALIST MODEL

ELITE MODEL

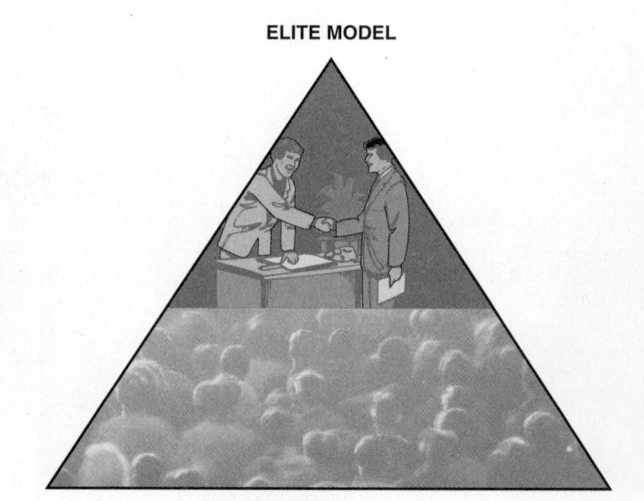

- Decisions are made on behalf of the people by leaders who engage in bargaining, accommodation, and compromise.
- Competition among leadership groups makes abuse of power by any one group difficult.
- Power is widely dispersed, and people can influence public policy by voting.
- Public policy reflects a balance among competing interest groups.

- Decisions are made by a small group of elite people.
- Consensus exists among the elite on the basic values and goals of society.
- Power is highly concentrated at the top of a pyramid-shaped social hierachy.
- Public policy reflects the values and preferences of the elite.

▲ FIGURE 14.2 PLURALIST AND ELITE MODELS

to finance campaigns and providing favors to political candidates. Second, through participation in the special interest process, the corporate rich are able to obtain favors, tax breaks, and favorable regulatory rulings. Finally, the corporate rich gain access to the policy-making process through their appointments to governmental advisory committees, presidential commissions, and other governmental positions.

Power elite models call our attention to a central concern in contemporary U.S. society: the ability of democracy and its ideals to survive in the context of the increasingly concentrated power held by capitalist oligarchies such as the media giants that we discuss in this chapter.

The U.S. Political System

The U.S. political system is made up of formal elements, such as the legislative process and the duties of the president, and informal elements, such as the role of political parties in the election process. We now turn to an examination of these informal elements, including political parties, political socialization, and voter participation.

Political Parties and Elections

A ***political party*** **is an organization whose purpose is to gain and hold legitimate control of government;** it is usually composed of people with similar attitudes, interests, and socioeconomic status. A political party (1) develops and articulates policy positions, (2) educates voters about issues and simplifies the choices for them, and (3) recruits candidates who agree with those policies, helps those candidates win office, and holds the candidates responsible for implementing the party's policy positions. In carrying out these functions, a party may try to modify the demands of special interests, build a consensus that could win majority support, and provide simple and identifiable choices for the voters on election day. Political parties create a *platform*, a formal statement of the party's political positions on various social and economic issues.

Since the Civil War, the Democratic and Republican parties have dominated the U.S. political system. Although one party may control the presidency for several terms, at some point the voters elect the other party's nominee, and control shifts. See ▶ Figure 14.3 for a look at the major political parties in U.S. history.

How well do the parties measure up to the ideal-type characteristics of a political party? Although both parties have been successful in getting their candidates elected at various times, they generally do not meet the ideal characteristics for a political party because they do not offer voters clear policy

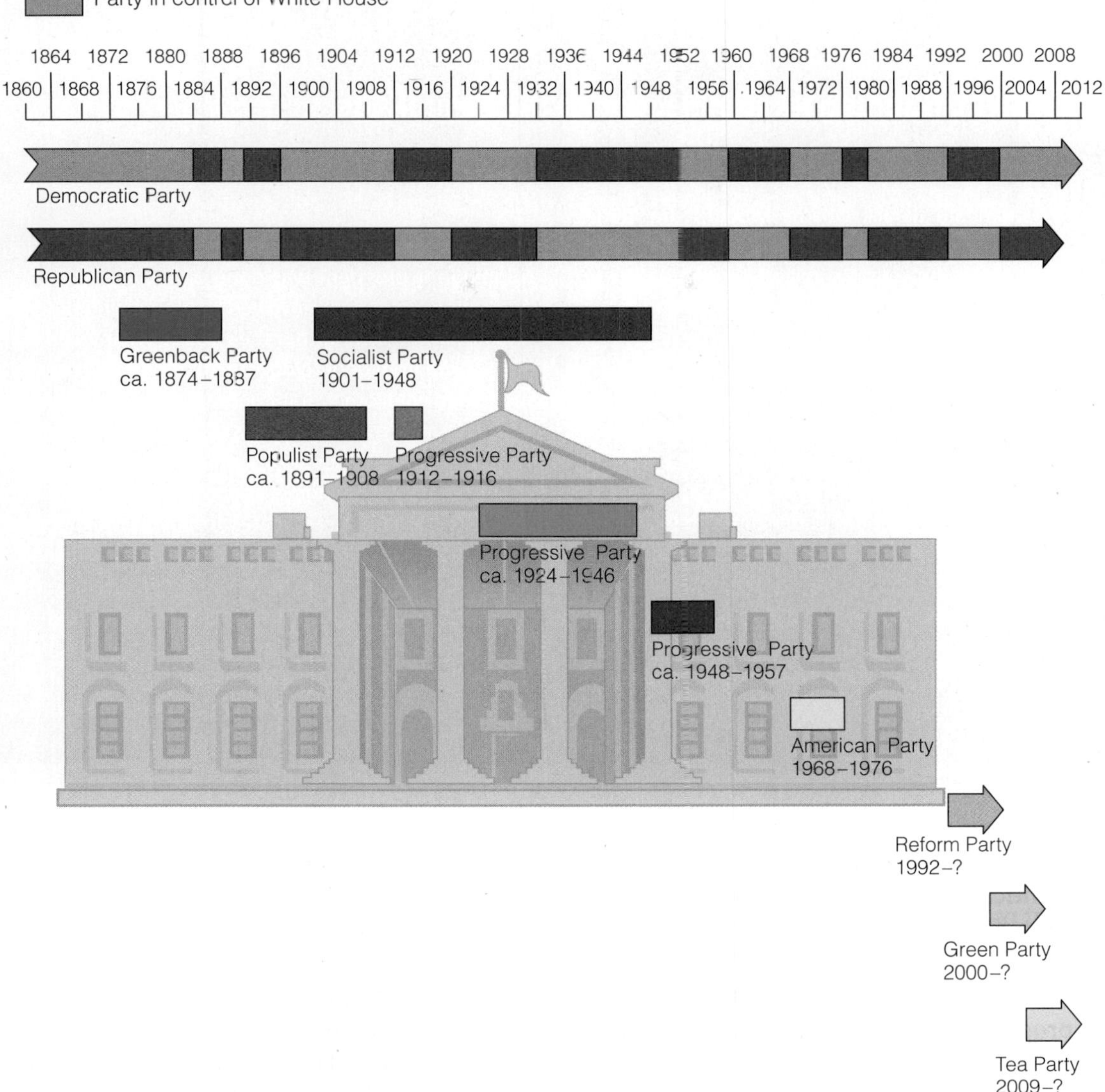

▲ **FIGURE 14.3 MAJOR U.S. POLITICAL PARTIES**

Despite recurring attempts by other groups to organize third parties, the Democratic and Republican parties have dominated national politics in the United States. Control of the presidency has alternated between these two parties since the Civil War.

Note: Three different "third parties" have used the name Progressive Party.

alternatives. Moreover, the two parties are oligarchies, dominated by active elites who hold views that are further from the center of the political spectrum than are those of a majority of members of their party. As a result, voters in primary elections (in which the nominees of political parties for most offices other than president and vice president are chosen) may select nominees whose views are closer to the center of the political spectrum and further away from the party's own platform. Likewise, party loyalties appear to be declining among voters, who may vote in one party's primary but then cast their ballot in general elections without total loyalty to that party, or cast a "split-ticket" ballot (voting for one party's candidate in one race and another party's candidate in the next one).

Finally, the media have replaced the party as a means of political communication. Often, the candidate who wins does so as a result of media presentation, not the political party's platform. Candidates no longer need political parties to carry their message to the people (see "Framing Politics in the Media").

Discontent with Current Political Parties

Although most individuals identify themselves as Republicans or Democrats, a growing number of

political party an organization whose purpose is to gain and hold legitimate control of government.

you can make a difference

Keeping an Eye on the Media

Do we get all of the news that we should about how our government operates and about the pressing social problems of our nation? Consider this list (shown by number) of the top five news stories that some media analysts believe were *not* adequately covered by the U.S. media in recent years:

1. Global Plans to Replace the Dollar: World leaders in China, Russia, and other nations in the Shanghai Cooperative Organization are taking steps to replace the dollar as the world's reserve currency. If successful, this would have a detrimental effect on the value of U.S. currency.
2. U.S. Department of Defense Is the Worst Polluter on the Planet: Uninhibited use of fossil fuels contributes many environmental problems, including the creation of greenhouse gases and extensive contamination of air, water, and soil.
3. Internet Privacy and Personal Access at Risk: Government surveillance of personal electronic communications such as the Internet and private computers puts many people at risk.
4. ICE Operates Secret Detention and Courts: U.S. Immigration and Customs Enforcement agents are holding and deporting tens of thousands of persons in secret court hearings.
5. Blackwater (Xe): The Secret US War in Pakistan: An elite branch of this private military firm is targeting assassinations of Taliban and al-Qaeda operatives in Pakistan.

According to Project Censored, an organization of students and professors who produce the annual "Top 25 Censored Stories" list at Sonoma State University's Sociology Department, many important stories are either missing from the news altogether or do not receive the attention they deserve. (To view the entire list, visit Project Censored's website at **www.projectcensored.org**.)

What should be the role of the media in keeping us informed? The media are referred to as the "Fourth Estate" or the "Fourth Branch of the Government" because they are supposed to provide people relevant information on important topics regarding how the government operates in a democratic society. This information can then be used by citizens to decide how they will vote on candidates and issues presented for their approval or disapproval on the election ballot.

The first step in keeping an eye on the news is to become more analytical about the "news" that we do receive. How can

of terrorism. War is an institution that involves *violence*—behavior intended to bring pain, physical injury, and/or psychological stress to people or to harm or destroy property. As such, war is a form of *collective violence* by people who are seeking to promote their cause or resist social policies or practices that they consider oppressive.

The direct effects of war are loss of human life and serious physical and psychological effects on some survivors. Although it is impossible to determine how many human lives have been lost in wars throughout human history, World War I took the lives of approximately 8 million combatants and 1 million civilians. In World War II, more than 50 million people (17 million combatants and 35 million civilians) lost their lives. During World War II, U.S. casualties alone totaled almost 300,000, and more than 600,000 were wounded. Later military actions in Korea, Vietnam, Iraq, Afghanistan, Libya, and other regions have brought about the deaths of millions of other military personnel and civilians, but estimates of those numbers range so widely that it is difficult, if not impossible, to come

▲ For many centuries, "warrior culture" has been the province of men. However, women now make up a growing percentage of soldiers in the U.S. armed services, and they have faced numerous obstacles in the process.

we evaluate the information we receive from the media? In *How to Watch TV News*, the media analysts Neil Postman and Steve Powers (1992: 160–168) suggest the following:

1. We should keep in mind that television news shows are called "shows" for a reason. They are not a public service or a public utility.
2. We should never underestimate the power of commercials, which tell us much about our society.
3. We should learn about the economic and political interests of those who run television stations or own a controlling interest in a media conglomerate.
4. We should pay attention to the *language* of newscasts, not just the visual imagery. For example, a *question* may reveal as much about the *questioner* as the person answering the question.

Becoming aware of the media's role in influencing people's opinions about how our government is run is the first step toward becoming an informed participant in the democratic political process. The second step in keeping an eye on the news is becoming aware of national and international events that should receive more coverage than they do or that might not be reported in a fair and unbiased manner. With traditional media, these steps are somewhat easier to follow; however, with newer social media, we must look even closer to distinguish what may be presented as nothing more than individual opinions from concrete information and facts. How do you think we might go about this endeavor?

Censored 2012: The Top Censored Stories and Media Analysis of 2010–2011, by Mickey Huff and Project Censored, published by Seven Stories Press

up with an accurate number. However, the number of casualties in conventional warfare would pale when compared to the potential loss of human life if all-out nuclear, biological, or chemical warfare were to occur. The devastation would be beyond description.

While conflicts continue in Afghanistan and Iraq (as of this writing, in 2011), U.S. armed forces are being pulled out in some regions while others are being sent for limited operations in other nations. No doubt, civilian and military casualties will continue to increase in various regions around the world as we have moved from a brief era of peacetime in the United States into what appears to be a protracted era of wartime in various regions of the world.

Politics and Government in the Future

Thinking about U.S. politics and government in the future is very much like the old story about optimists and pessimists. According to the story, an eight-ounce cup containing exactly four ounces of water is placed on a table. The optimist comes in, sees the cup, and says, "The cup is half full." The pessimist comes in, sees the cup, and says, "The cup is half empty." Clearly, both the pessimist and the optimist are looking at the same cup containing the same amount of water, but their perspective on what they see is quite different. For some analysts, looking at the future of the U.S. government and its political structure is very much like this.

Views of the future of politics and government relate to specific concerns about the United States:

- What will be the future of political parties? What have recent elections told us about the nature of these institutions?
- Are global corporate interests and the concerns of the wealthy in this nation and elsewhere overshadowing the needs and interests of everyday people?
- Is it possible to prevent future terrorist attacks in the United States (and other nations as

well) through tightening organizational intelligence and reorganizing some governmental bureaucracies?
- Will we be able to balance the need for national security with the individual's right to privacy and freedom of movement within this country?
- How will elected politicians and appointed government officials handle the challenges regarding ongoing budget crises and the changing demographics of the United States?
- Will immigration and employment policies be based on the best interests of the largest number of people, or will these policies be based on the best interests of elites and major transnational corporations that are major contributors to political campaigns?
- Do the media accurately report what is going on at all levels of government? To what extent can individuals and grassroots organizations influence the media and the political process? (See "You Can Make a Difference.")

In the second decade of the twenty-first century, these are a few of the many questions regarding politics and the government that face us and people in other nations. How we (and our elected officials) answer these (and related) questions will in large measure determine the future of politics and government in the United States. Our answers will also have a profound influence on people and governments in other countries, whether they are in high-, middle-, or low-income nations, around the world.

chapter review Q & A

Use these questions and answers to check how well you've achieved the learning objectives set out at the beginning of this chapter.

• What is power?

Power is the ability of persons or groups to carry out their will even when opposed by others.

• What is the relationship between power and politics?

Politics is the social institution through which power is acquired and exercised by some people or groups. A strong relationship between politics and power exists in all countries, and the military is closely tied to the political system.

• What are the three types of authority?

Max Weber identified these types of authority: traditional, charismatic, and rational–legal. Traditional authority is based on long-standing custom. Charismatic authority is power based on a leader's personal qualities. Rational–legal authority is based on law or written rules and regulations, as found in contemporary bureaucracies.

• What are the main types of political systems?

The main types of political systems are monarchies, authoritarian systems, totalitarian systems, and democratic systems. In a monarchy one person is the hereditary ruler of the nation. In authoritarian systems rulers tolerate little or no public opposition and generally cannot be removed from office by legal means. In totalitarian systems the state seeks to regulate all aspects of society and to monopolize all societal resources in order to exert complete control over both public and private life. In democratic systems the powers of government are derived from the consent of all the people.

• How do pluralist and power elite perspectives view power in the United States?

According to the pluralist (functionalist) model, power is widely dispersed throughout many competing interest groups. People influence policy by voting, joining special interest groups and political action campaigns, and forming new groups. According to the elite (conflict) model, power is concentrated in a small group of elites, whereas the masses are relatively powerless.

• Who makes up the power elite, and why are they important?

According to C. Wright Mills, the power elite is composed of influential business leaders, key government leaders, and the military. The elites possess greater resources than the masses, and public policy reflects their preferences.

• What is the military–industrial complex?

The military–industrial complex refers to a mutual interdependence of the government military establishment and private military contractors.

• What is terrorism?

Terrorism is the use of calculated, unlawful physical force or threats of violence against a government, organization, or individual to gain some political, religious, economic, or social objective.

key terms

authoritarianism 412
authority 409
charismatic authority 410
democracy 413
elite model 416
government 408
militarism 426
military–industrial complex 424
monarchy 412
pluralist model 414
political action committee 415
political party 418
political socialization 421
political sociology 408
politics 408
power 409
power elite 417
rational–legal authority 410
representative democracy 413
routinization of charisma 410
special interest group 414
state 408
terrorism 427
totalitarianism 412
traditional authority 409
war 427

questions for critical thinking

1. Who is ultimately responsible for decisions and policies that are made in a democracy such as the United States—the people or their elected representatives?
2. How would you design a research project to study the relationship between campaign contributions to elected representatives and their subsequent voting records? What would be your hypothesis? What kinds of data would you need to gather? How would you gather accurate data?
3. How does your school (or workplace) reflect a pluralist or elite model of power and decision making?
4. Can democracy survive in a context of rising concentrated power in the capitalist oligarchies? Why or why not?

turning to video

Watch the BBC video *Revolution in Egypt* (running time 1:54), available through **CengageBrain.com**. This video discusses the recent uprising in Egypt and features interviews with some of the people partaking in the rebellion. As you watch the video, think about your own reactions to the revolutions in the Middle East. After you watch the video, consider this question: Did the revolutions in Egypt and other Middle Eastern countries change your perception of the power you have as a citizen? Why or why not?

online study resources

Go to CENGAGE brain.com to access online study resources, including the Sociology CourseMate for this text as well as special features such as video, an interactive sociology time line and interactive maps, General Social Survey (GSS) data, and U.S. Census 2010 data.

CourseMate brings course concepts to life with interactive learning, study, and exam-preparation tools that support the printed textbook. A textbook-specific website, **Sociology CourseMate** includes an integrated interactive eBook and other interactive learning tools, including quizzes, flash cards, and videos.

Visit **www.cengagebrain.com** to access your account and purchase materials.

15 Families and Intimate Relationships

A lot of people wonder how Chinese parents raise such stereotypically successful kids. . . . Well, I can tell them, because I've done it. Here are some things my daughters were never allowed to do: attend a sleepover, have a play date, be in a school play, complain about not being in a school play, watch TV or play computer games, choose their own extracurricular activities, get any grade less than an A, not be the No. 1 student in every subject except gym and drama, play an instrument other than the piano or violin, and not play the piano or violin. . . . Even when Western parents think they're being strict, they usually don't come close to being Chinese mothers.

—Amy Chua (2011), a Chinese American law professor at Yale Law School who was born and raised in the United States, describing tough, intensive child-rearing practices in her best-selling book, *Battle Hymn of the Tiger Mother*

▲ **In 2011, when Amy Chua published *Battle Hymn of the Tiger Mother,* she set off a firestorm of controversy regarding the appropriate way to raise children.**

I didn't know you could [install a system to track driving]. I'm going to have a look into this. I wasn't familiar with that. . . . I trust [my daughter]. She's not a stupid girl. I guess she knows right from wrong. I trust her, but you know what kids are. Kids are kids. You can't just give the message that you're trusting them. You just have to watch them at all times. . . . She lied to me once, and she was punished for it, and I don't believe she'd do it again, but you never know.

—a white, working-class mother, interviewed by the sociologist Margaret K. Nelson (2010: 147)

Chapter Focus Question:

Why are many family-related issues—such as parenting, socialization, and supervision of children and young people—viewed as personal problems rather than as larger societal concerns?

The mother just quoted speaks out about distrusting her teenage daughter and feeling the need to keep her under control. Few family-related topics stir up as much controversy as how much control parents should exercise over their children. Media reports about "tiger moms" and "helicopter parents" call attention to the trend that many parents today are becoming so concerned about their children's safety and welfare that they are becoming control freaks who do not want to let their offspring out of their sight or beyond the range of a cell phone call. Parents are increasingly using the latest technology for "kiddy" surveillance, and the practice doesn't necessarily end when young people go to college: One survey found that parents interact with their college-age children with much greater frequently than parents did in the past. Cell phone conversations, text messages, and e-mail have all intensified interactions (both positively and negatively) between parents and children (National Survey of Student Engagement, 2007). However, the United States is a highly diverse nation in which most people socialize their children in a wide variety of ways, often on the basis of their socioeconomic status, race/ethnicity, religion, national origin, or other personal characteristics, attributes, or preferences, so our family interaction patterns still vary widely.

In this chapter we examine the increasing complexity of family life in the United States and other nations. Pressing family and societal issues such as communications, financial hardships, teenage pregnancy, divorce, and child-care issues will be used as examples of how families and intimate relationships continue to change over time and place. Before reading on, test your knowledge about trends in family life by taking the Sociology and Everyday Life quiz

Highlighted Learning Objectives

- Explain why it is difficult to define *family.*
- Compare marriage patterns around the world.
- Identify key assumptions of the functionalist, conflict/feminist, and symbolic interactionist perspectives on families.
- Describe significant trends that affect many U.S. families today.

sociology and everyday life

How Much Do You Know About Contemporary Trends in U.S. Family Life?

True	False	
T	F	1. Today, married-couple households account for nearly 75 percent of all U.S. households.
T	F	2. Men and women are waiting longer to get married.
T	F	3. Americans today are more likely than in the past to cohabit, divorce, marry later, or not marry at all.
T	F	4. Couples who cohabit before they get married are more likely to stay married.
T	F	5. The teenage birthrate has increased in recent years.
T	F	6. Stay-at-home moms are more likely to be younger, Latina (Hispanic), and foreign born than are mothers in the workforce.
T	F	7. The average U.S. household size is increasing each year in this country.
T	F	8. In the 2010s a larger percentage of U.S. men are married to women whose education and income exceed their own than in past decades.

Answers on page 436.

Families in Global Perspective

As the nature of family life has changed in high-, middle-, and low-income nations, the issue of what constitutes a "family" continues to be widely debated. In the "Universal Declaration of Human Rights," Article 16, adopted by the United Nations (1948), the family is defined as follows:

- Men and women of full age, without any limitation due to race, nationality, or religion, have the right to marry and to found a family. They are entitled to equal rights as to marriage, during marriage and at its dissolution.
- Marriage shall be entered into only with the free and full consent of the intending spouses.
- The family is the natural and fundamental group unit of society and is entitled to protection by society and the States.

According to this declaration, the social institution of family must be protected in all societies because family is the "natural" and "fundamental" group unit of society. Although families differ widely around the world, they also share certain common concerns in their everyday lives. Food, clothing, shelter, and child care are necessities important to all people.

In the United States the Census Bureau (2011a) defines a family as consisting of two or more people who are related by birth, marriage, or adoption, and residing in the same housing unit. (The Census Bureau specifies that one person in the household unit will be identified as the "householder.") For many years the standard sociological definition of *family* has been a group of people who are related to one another by bonds of blood, marriage, or adoption and who live together, form an economic unit, and bear and raise children. Many people believe that this definition should not be expanded—that social approval should not be extended to other relationships simply because the persons in those relationships wish to consider themselves to be a family. However, other people challenge this definition because it simply does not match the reality of family life in contemporary society (Lamanna and Riedmann, 2012).

Today's families include many types of living arrangements and relationships, including single-parent households, unmarried couples, lesbian and gay couples, and multiple generations (such as grandparent, parent, and child) living in the same household. To accurately reflect these changes in family life, some sociologists believe that we need a more encompassing definition of what constitutes a family. Accordingly, we will define ***families* as relationships in which people live together with commitment, form an economic unit and care for any young, and consider their identity to be significantly attached to the group.** Sexual expression and parent–child relationships are a part of most, but not all, family relationships.

How do sociologists approach the study of families? In our study of families we will use our sociological imagination to see how our personal experiences are related to the larger happenings in society. At the microlevel, each of us has a "biography," based on our experience within our family; at the macrolevel, our families are embedded in a specific social context that has a major effect on them. We will examine the institution of the family at both

© iofoto/iStockphoto.com

© Steven Rubin/The Image Works

▲ Whereas the relationship between a husband and wife is based on legal ties, relationships between parents and children may be established by either blood ties or legal ties.

of these levels, starting with family structure and characteristics.

Family Structure and Characteristics

In preindustrial societies the primary form of social organization is through kinship ties. ***Kinship* refers to a social network of people based on common ancestry, marriage, or adoption.** Through kinship networks, people cooperate so that they can acquire the basic necessities of life, including food and shelter. Kinship systems can also serve as a means by which property is transferred, goods are produced and distributed, and power is allocated.

In industrialized societies other social institutions fulfill some of the functions previously taken care of by the kinship network. For example, political systems provide structures of social control and authority, and economic systems are responsible for the production and distribution of goods and services. Consequently, families in industrialized societies serve fewer and more-specialized purposes than do families in preindustrial societies. Contemporary families are responsible primarily for regulating sexual activity, socializing children, and providing affection and companionship for family members.

Families of Orientation and Procreation

During our lifetime many of us will be members of two different types of families—a family of orientation and a family of procreation. **The *family of orientation* is the family into which a person is born and in which early socialization usually takes place.** Although most people are related to members of their family of orientation by blood ties, those who are adopted have a legal tie that is patterned after a blood relationship. **The *family of procreation* is the family that a person forms by having or adopting children.** Both legal and blood ties are found in most families of procreation. The relationship between a husband and wife is based on legal ties; however, the relationship between a parent and child may be based on either blood ties or legal ties, depending on whether the child has been adopted.

Some sociologists have emphasized that "family of orientation" and "family of procreation" do not encompass all types of contemporary families. Instead, many gay men and lesbians have *families we choose*—social arrangements that include intimate relationships between couples and close familial relationships among other couples and other adults and children. According to the sociologist Judy Root Aulette (1994), "families we choose" include blood ties and legal ties, but they also include *fictive kin*—persons who are not actually related by blood but who are accepted as family members.

Extended and Nuclear Families Sociologists distinguish between extended families and nuclear families based on the number of generations that live within a household. **An *extended family* is a family unit composed of relatives in addition to parents and children who live**

families relationships in which people live together with commitment, form an economic unit and care for any young, and consider their identity to be significantly attached to the group.

kinship a social network of people based on common ancestry, marriage, or adoption.

family of orientation the family into which a person is born and in which early socialization usually takes place.

family of procreation the family that a person forms by having or adopting children.

extended family a family unit composed of relatives in addition to parents and children who live in the same household.

sociology and everyday life

ANSWERS to the Sociology Quiz on Contemporary Trends in U.S. Family Life

1. False. U.S. Census Bureau data show that married households in the United States account for less than half of all households. Of 117,538 million households in 2010, only 58,410 million were made up of married couples.

2. True. Men and women are waiting longer to get married. The median age at first marriage has increased to 28.2 for men and 26.1 for women.

3. True. Americans are more likely than in the past to cohabit, divorce, marry later, or not marry at all. A variety of reasons have been identified for changes in these patterns.

4. False. Recent research indicates that couples who cohabit before they get married are less likely to stay married; however, their chances of remaining married improve if they were already engaged when they began living together.

5. False. The U.S. birthrate for teenagers in 2009 reached its lowest point in nearly 70 years. The rate was 39.1 births per 1,000 females age 15–19 years.

6. True. U.S. Census Bureau reports indicate that the 5 million stay-at-home mothers in 2010 were younger and more likely to be Latina (Hispanic) and foreign born than were mothers who were in the labor force.

7. False. The average U.S. household size declined to 2.59 persons in 2010, down from 2.62 people in 2000. The exception was households where the householder had less than a high school degree, and these increased in 2010 to an average of 2.87 people (up from 2.67 in 2001).

8. True. According to studies by the Pew Research Center and others, a larger share of men (22 percent) today are married to women whose income exceeds their own, compared with only 4 percent of married men in 1970.

Sources: Based on Fry and Cohn, 2010; Parker-Pope, 2010; Roberts, 2010b; U.S. Census Bureau 2011b; and Ventura and Hamilton, 2011.

in the same household. These families often include grandparents, uncles, aunts, or other relatives who live close to the parents and children, making it possible for family members to share resources. In horticultural and agricultural societies, extended families are extremely important; having a large number of family members participate in food production may be essential for survival. Today, extended-family patterns are found in Latin America, Africa, Asia, and some parts of Eastern and Southern Europe. With the advent of industrialization and urbanization, maintaining the extended-family pattern becomes more difficult. Increasingly, young people move from rural to urban areas in search of employment in the industrializing sector of the economy. At that time, some extended families remain, but the nuclear family typically becomes the predominant family type in the society.

A ***nuclear family* is a family composed of one or two parents and their dependent children, all of whom live apart from other relatives.** A traditional definition specifies that a nuclear family is made up of a "couple" and their dependent children; however, this definition became outdated when a significant shift occurred in the family structure. A comparison of Census Bureau data over the past four decades shows a significant decline in the percentage of U.S. households made up of a married couple who had children under age eighteen living with them (see "Census Profiles: Household Composition"). Conversely, there has been an increase in the percentage of households in which either a woman or a man lives alone.

Marriage Patterns

Across cultures, different forms of marriage characterize families. ***Marriage* is a legally recognized and/or socially approved arrangement between two or more individuals that carries certain rights and obligations and usually involves sexual activity.** In most societies, marriage involves a mutual commitment by each partner, and linkages between two individuals and families are publicly demonstrated.

In the United States the only legally sanctioned form of marriage is ***monogamy*—a marriage between two partners, usually a woman and a man.** For some people, marriage is a lifelong commitment that ends only with the death

Household Composition, 1970 and 2010

The Census Bureau asks a representative sample of the U.S. population about their marital status and also asks those individuals questions about other individuals residing in their household. Based on the most recent data, the Census Bureau reports that the percentage distribution of nonfamily and family households has changed substantially during the past four decades. The most noticeable trend is the decline in the number of married-couple households with their own children living with them, which decreased from about 40 percent of all households in 1970 to 21 percent in 2010. The distribution shown below reflects this significant change in family structure.

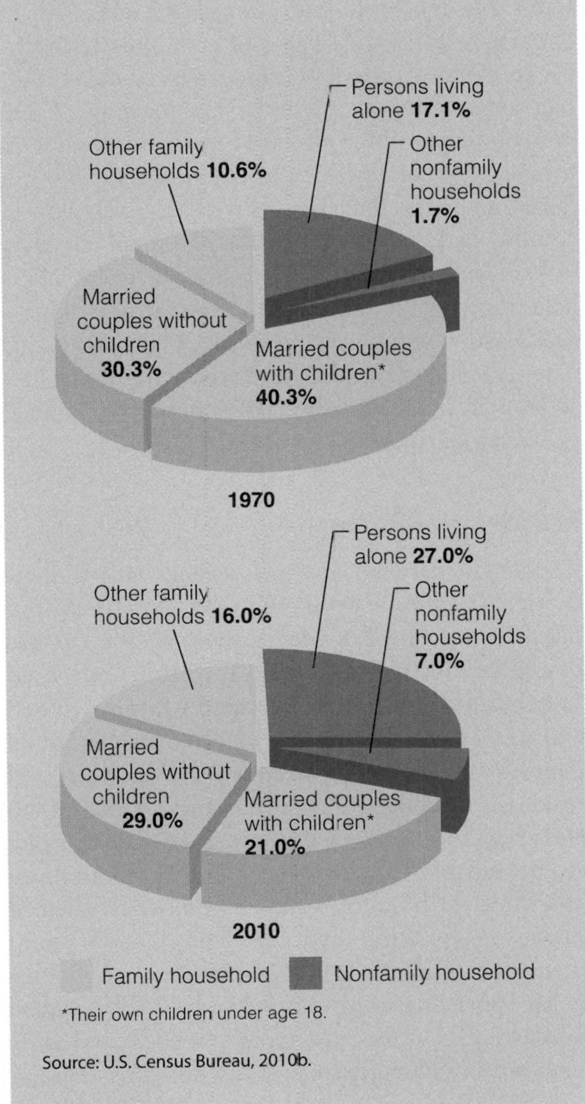

Source: U.S. Census Bureau, 2010b.

▲ Polygamy is the concurrent marriage of a person of one sex with two or more persons of the opposite sex. Although most people do not practice this pattern of marriage, some men are married to more than one wife.

of a partner. For others, marriage is a commitment of indefinite duration. Through a pattern of marriage, divorce, and remarriage, some people practice *serial monogamy*—a succession of marriages in which a person has several spouses over a lifetime but is legally married to only one person at a time.

***Polygamy* is the concurrent marriage of a person of one sex with two or more members of the opposite sex.** The most prevalent form of polygamy is ***polygyny*—the concurrent marriage of one man with two or more women.** Polygyny has been practiced in a number of societies, including parts of Europe until the Middle Ages. More recently, some marriages in Islamic societies in Africa and Asia have been polygynous; however, the cost of providing for multiple wives and numerous children makes the practice impossible for all but the wealthiest men. In addition, because roughly equal numbers of women and men live in these areas, this nearly balanced sex ratio tends to limit polygyny.

nuclear family a family composed of one or two parents and their dependent children, all of whom live apart from other relatives.

marriage a legally recognized and/or socially approved arrangement between two or more individuals that carries certain rights and obligations and usually involves sexual activity.

monogamy a marriage between two partners, usually a woman and a man.

polygamy the concurrent marriage of a person of one sex with two or more members of the opposite sex.

polygyny the concurrent marriage of one man with two or more women.

The second type of polygamy is ***polyandry*—the concurrent marriage of one woman with two or more men.** Polyandry is very rare; when it does occur, it is typically found in societies where men greatly outnumber women because of high rates of female infanticide.

Patterns of Descent and Inheritance

Even though a variety of marital patterns exist across cultures, virtually all forms of marriage establish a system of descent so that kinship can be determined and inheritance rights established. In preindustrial societies kinship is usually traced through one parent (unilineally). The most common pattern of unilineal descent is ***patrilineal descent*—a system of tracing descent through the father's side of the family.** Patrilineal systems are set up in such a manner that a legitimate son inherits his father's property and sometimes his position upon the father's death. In nations such as India, where boys are seen as permanent patrilineal family members but girls are seen as only temporary family members, girls tend to be considered more expendable than boys.

Even with the less common pattern of ***matrilineal descent*—a system of tracing descent through the mother's side of the family**—women may not control property. However, inheritance of property and position is usually traced from the maternal uncle (mother's brother) to his nephew (mother's son). In some cases mothers may pass on their property to daughters.

By contrast, kinship in industrial societies is usually traced through both parents (bilineally). The most common form is ***bilateral descent*—a system of tracing descent through both the mother's and father's sides of the family.** This pattern is used in the United States for the purpose of determining kinship and inheritance rights; however, children typically take the father's last name.

Power and Authority in Families

Descent and inheritance rights are intricately linked with patterns of power and authority in families. The most prevalent forms of familial power and authority are patriarchy, matriarchy, and egalitarianism. A ***patriarchal family* is a family structure in which authority is held by the eldest male (usually the father).** The male authority figure acts as the head of the household and holds power and authority over the women and children, as well as over the other males. A ***matriarchal family* is a family structure in which authority is held by the eldest female (usually the mother).** In this case the female authority figure acts as the head of the household. Although there has been a great deal of discussion about matriarchal families, scholars have found no historical evidence to indicate that true matriarchies ever existed.

The most prevalent pattern of power and authority in families is patriarchy. Across cultures, men are the primary (and often sole) decision makers regarding domestic, economic, and social concerns facing the family. The existence of patriarchy may give men a sense of power over their own lives, but it can also create an atmosphere in which some men feel greater freedom to abuse women and children.

An ***egalitarian family* is a family structure in which both partners share power and authority equally.** Recently, a trend toward more-egalitarian relationships has been evident in a number of countries as women have sought changes in their legal status and increased educational and employment opportunities. Some degree of economic independence makes it possible for women to delay marriage or to terminate a problematic marriage. However, one study of the effects of egalitarian values on the allocation and performance of domestic tasks in the family found that changes were relatively slow in coming: Fathers were more likely to share domestic tasks in nonconventional families where members held more-egalitarian values. Similarly, children's gender-role typing was more closely linked to their parents' egalitarian values and nonconventional lifestyles than to the domestic tasks they were assigned (Weisner, Garnier, and Loucky, 1994).

Residential Patterns

Residential patterns are interrelated with the authority structure and method of tracing descent in families. ***Patrilocal residence* refers to the custom of a married couple living in the same household (or community) as the husband's family.** Across cultures, patrilocal residency is the most common pattern. One example of contemporary patrilocal residency can be found in al-Barba, a lower-middle-class neighborhood in the Jordanian city of Irbid. According to researchers, the high cost of renting an apartment or building a new home has resulted in many sons building their own living quarters onto their parents' home, resulting in multifamily households consisting of an older married couple, their unmarried children, their married sons, and their sons' wives and children.

Few societies have residential patterns known as ***matrilocal residence*—the custom of a married couple living in the same household (or community) as the wife's parents.** In industrialized nations such as the United States, most couples hope to live in a ***neolocal residence*—the custom of a married couple living in their own residence apart from both the husband's and the wife's parents.**

To this point we have examined a variety of marriage and family patterns found around the world. Even with the diversity of these patterns, most people's behavior is shaped by cultural rules pertaining to endogamy and exogamy. ***Endogamy* is the practice of marrying within one's own social group or category.** In the United States, for example, most people practice endogamy: They marry people who come from the same social class, racial–ethnic group, religious affiliation, and other categories considered important within their own social group. ***Exogamy* is the practice of marrying outside one's own social group or category.** Depending on the circumstances, exogamy may not be noticed at all, or it may result in a person being ridiculed or ostracized by other members of the "in" group. The three most important sources of positive or negative sanctions for intermarriage are the family, the church, and the state. Participants in these social institutions may look unfavorably on the marriage of an in-group member to an "outsider" because of the belief that it diminishes social cohesion in the group. However, educational attainment is also a strong indicator of marital choice. Higher education emphasizes individual achievement, and college-educated people may be less likely than others to identify themselves with their social or cultural roots and thus more willing to marry outside their own social group or category if their potential partner shares a similar level of educational attainment.

Theoretical Perspectives on Families

The ***sociology of family* is the subdiscipline of sociology that attempts to describe and explain patterns of family life and variations in family structure.** Functionalist perspectives emphasize the functions that families perform at the macrolevel of society, whereas conflict and feminist perspectives focus on families as a primary source of social inequality. Symbolic interactionists examine microlevel interactions that are integral to the roles of different family members.

Functionalist Perspectives

Functionalists emphasize the importance of the family in maintaining the stability of society and the well-being of individuals. According to Emile Durkheim, marriage is a microcosmic replica of the larger society; both marriage and society involve a mental and moral fusion of physically distinct individuals. Durkheim also believed that a division of labor contributes to greater efficiency in all areas of life—including marriages and families—even though he acknowledged that this division imposes significant limitations on some people.

In the United States, Talcott Parsons was a key figure in developing a functionalist model of the family. According to Parsons (1955), the husband/father fulfills the *instrumental role* (meeting the family's economic needs, making important decisions, and providing leadership), whereas the wife/mother fulfills the *expressive role* (running the household,

polyandry the concurrent marriage of one woman with two or more men.

patrilineal descent a system of tracing descent through the father's side of the family.

matrilineal descent a system of tracing descent through the mother's side of the family.

bilateral descent a system of tracing descent through both the mother's and father's sides of the family.

patriarchal family a family structure in which authority is held by the eldest male (usually the father).

matriarchal family a family structure in which authority is held by the eldest female (usually the mother).

egalitarian family a family structure in which both partners share power and authority equally.

patrilocal residence the custom of a married couple living in the same household (or community) as the husband's family.

matrilocal residence the custom of a married couple living in the same household (or community) as the wife's parents.

neolocal residence the custom of a married couple living in their own residence apart from both the husband's and the wife's parents.

endogamy the practice of marrying within one's own social group or category.

exogamy the practice of marrying outside one's own social group or category.

sociology of family the subdiscipline of sociology that attempts to describe and explain patterns of family life and variations in family structure.

caring for children, and meeting the emotional needs of family members).

Contemporary functionalist perspectives on families derive their foundation from Durkheim. Division of labor makes it possible for families to fulfill a number of functions that no other institution can perform as effectively. In advanced industrial societies, families serve four key functions:

1. *Sexual regulation.* Families are expected to regulate the sexual activity of their members and thus control reproduction so that it occurs within specific boundaries. At the macrolevel, incest taboos prohibit sexual contact or marriage between certain relatives. For example, virtually all societies prohibit sexual relations between parents and their children and between brothers and sisters.
2. *Socialization.* Parents and other relatives are responsible for teaching children the necessary knowledge and skills to survive. The smallness and intimacy of families make them best suited for providing children with the initial learning experiences that they need.
3. *Economic and psychological support.* Families are responsible for providing economic and psychological support for members. In preindustrial societies families are economic production units; in industrial societies the economic security of families is tied to the workplace and to macrolevel economic systems. In recent years psychological support and emotional security have been increasingly important functions of the family.
4. *Provision of social status.* Families confer social status and reputation on their members. These statuses include the ascribed statuses with which individuals are born, such as race/ethnicity, nationality, social class, and sometimes religious affiliation. One of the most significant and compelling forms of social placement is the family's class position and the opportunities (or lack thereof) resulting from that position. Examples of class-related opportunities are access to quality health care, higher education, and a safe place to live.

© Getty Images/Jupiter Images

▲ Functionalist theorists believe that families serve a variety of functions that no other social institution can adequately fulfill. In contrast, conflict and feminist theorists believe that families may be a source of conflict over values, goals, and access to resources and power. Children in upper-class families have many advantages and opportunities that are not available to other children.

Conflict and Feminist Perspectives

Conflict and feminist analysts view functionalist perspectives on the role of the family in society as idealized and inadequate. Rather than operating harmoniously and for the benefit of all members, families are sources of social inequality and conflict over values, goals, and access to resources and power.

According to some conflict theorists, families in capitalist economies are similar to the work environment of a factory. Men in the home dominate women in the same manner that capitalists and managers in factories dominate workers (Engels, 1970/1884). Although childbearing and caring for family members in the home contribute to capitalism, these activities also reinforce the subordination of women through unpaid (and often devalued) labor. Other conflict analysts are concerned with the effect that class conflict has on the family. The exploitation of the lower classes by the upper classes contributes to family problems such as high rates of divorce and overall family instability.

Some feminist perspectives on inequality in families focus on patriarchy rather than class. From this viewpoint, men's domination over women existed long before capitalism and private ownership of property (Mann, 1994). Women's subordination is rooted in patriarchy and men's control over women's labor power (Hartmann, 1981). According to one scholar, "Male power in our society is expressed in economic terms even if it does not originate in property relations; women's activities in the home have been undervalued at the same time as their labor has been controlled by men" (Mann, 1994: 42). In addition, men have benefited from the privileges they derive from their status as family breadwinners.

Symbolic Interactionist Perspectives

Early symbolic interactionists such as Charles Horton Cooley and George Herbert Mead provided key insights into the roles we play as family members and how we modify or adapt our roles to

© Jon Feingersh/Blend Images/Jupiter Images

© Corbis Super RF/Alamy

© Pixland/Jupiter Images

▲ Marriage is a complicated process involving rituals and shared moments of happiness. When marriage is followed by divorce, couples must abandon a shared reality and then reestablish individual ones.

the expectations of others—especially significant others such as parents, grandparents, siblings, and other relatives. How does the family influence the individual's self-concept and identity? In order to answer questions such as this one, contemporary symbolic interactionists examine the roles of husbands, wives, and children as they act out their own parts and react to the actions of others. From such a perspective, what people think, as well as what they say and do, is very important in understanding family dynamics.

According to the sociologists Peter Berger and Hansfried Kellner (1964), interaction between marital partners contributes to a shared reality. Although newlyweds bring separate identities to a marriage, over time they construct a shared reality as a couple. In the process the partners redefine their past identities to be consistent with new realities. Development of a shared reality is a continuous process, taking place not only in the family but in any group in which the couple participates together. Divorce is the reverse of this process; couples may start with a shared reality and, in the process of uncoupling, gradually develop separate realities (Vaughan, 1985).

Symbolic interactionists explain family relationships in terms of the subjective meanings and everyday interpretations that people give to their lives. As the sociologist Jessie Bernard (1982/1973) pointed out, women and men experience marriage differently. Although the husband may see *his* marriage very positively, the wife may feel less positive about *her* marriage, and vice versa. Researchers have found that husbands and wives may give very different accounts of the same event and that their "two realities" frequently do not coincide (Safilios-Rothschild, 1969).

Postmodernist Perspectives

According to postmodern theories, we have experienced a significant decline in the influence of the family and other social institutions. As people have pursued individual freedom, they have been less inclined to accept the structural constraints imposed on them by institutions. Given this assumption, how might a postmodern perspective view contemporary family life? For example, how might this approach answer the question "How is family life different in the information age"? Social scientist David Elkind (1995) describes the postmodern family as *permeable*—capable of being diffused or invaded in such a manner that the family's original nature is modified or changed. According to Elkind (1995), if the nuclear family is a reflection of the age of modernity, the permeable family reflects the postmodern assumptions of difference and irregularity. This is evident in the fact that the nuclear family is now only one of many family forms. Similarly, under modernity the idea

[concept quick review]

Theoretical Perspectives on Families

	Focus	Key Points	Perspective on Family Problems
Functionalist	The role of families in maintaining stability of society and individuals' well-being.	In modern societies, families serve the functions of sexual regulation, socialization, economic and psychological support, and provision of social status.	Family problems are related to changes in social institutions such as the economy, religion, education, and law/government.
Conflict/Feminist	Families as sources of conflict and social inequality.	Families both mirror and help perpetuate social inequalities based on class and gender.	Family problems reflect social patterns of dominance and subordination.
Symbolic Interactionist	Family dynamics, including communication patterns and the subjective meanings that people assign to events.	Interactions within families create a shared reality.	How family problems are perceived and defined depends on patterns of communication, the meanings that people give to roles and events, and individuals' interpretations of family interactions.
Postmodernist	Permeability of families.	In postmodern societies, families are diverse and fragmented. Boundaries between workplace and home are blurred.	Family problems are related to cyberspace, consumerism, and the hyperreal in an age increasingly characterized by high-tech "haves" and "have-nots."

of romantic love has given way to the idea of consensual love: Individuals agree to have sexual relations with others whom they have no intention of marrying or, if they marry, do not necessarily see the marriage as having permanence. Maternal love has also been transformed into shared parenting, which includes not only mothers and fathers but also caregivers, who may either be relatives or nonrelatives (Elkind, 1995).

Urbanity is another characteristic of the postmodern family. The boundaries between the public sphere (the workplace) and the private sphere (the home) are becoming much more open and flexible. In fact, family life may be negatively affected by the decreasing distinction between what is work time and what is family time. As more people are becoming connected "24/7" (twenty-four hours a day, seven days a week), the boss who in the past would not call at 11:30 P.M. may send an e-mail asking for an immediate response to some question that has arisen while the person is away on vacation with family members (Leonard, 1999). According to some postmodern analysts, this is an example of the "power of the new communications technologies to integrate and control labour despite extensive dispersion and decentralization" (Haraway, 1994: 439).

The Concept Quick Review summarizes sociological perspectives on the family. Taken together, these perspectives on the social institution of families reflect various ways in which familial relationships may be viewed in contemporary societies. Now we shift our focus to love, marriage, intimate relationships, and family issues in the United States.

Developing Intimate Relationships and Establishing Families

The United States has been described as a "nation of lovers"; it has been said that we are "in love with love." Why is this so? Perhaps the answer lies in the fact that our ideal culture emphasizes *romantic love,* which refers to a deep emotion, the satisfaction of significant needs, a caring for and acceptance of the person we love, and involvement in an intimate relationship (Lamanna and Riedmann, 2012).

Love and Intimacy

In the late nineteenth century, during the Industrial Revolution, people came to view work and home as separate spheres in which different feelings and emotions were appropriate (Coontz, 1992). The public sphere of work—men's sphere—emphasized self-reliance and independence. By contrast, the private sphere of the home—women's sphere—emphasized the giving of services, the exchange of gifts, and love. Accordingly, love and emotions became the domain of women, and work and rationality became the domain of men (Lamanna and Riedmann, 2012). Although the roles of women and men changed dramatically in the twentieth century, women and men may still not share the same perceptions about romantic love today. According to the sociologist Francesca Cancian (1990), women tend to express their feelings verbally whereas men tend to express their love through nonverbal actions, such as running an errand for someone or repairing a child's broken toy.

Love and intimacy are closely intertwined. Intimacy may be psychic ("the sharing of minds"), sexual, or both. Although sexuality is an integral part of many intimate relationships, perceptions about sexual activities vary from one culture to the next and from one time period to another. For example, kissing is found primarily in Western cultures; many African and Asian cultures view kissing negatively (Reinisch, 1990).

© AP Images/FarmersOnly.com

▲ In the United States the notion of romantic love is deeply intertwined with our beliefs about how and why people develop intimate relationships and establish families. Not all societies share this concern with romantic love. However, in this country the number of opportunities for romance is sometimes increased by online matching services, such as this one, which caters to farmers and people who love nature.

For many years the work of the biologist Alfred C. Kinsey was considered to be the definitive research on human sexuality, even though some of his methodology had serious limitations. More recently, the work of Kinsey and his associates has been superseded by the National Health and Social Life Survey conducted by the National Opinion Research Center at the University of Chicago (see Laumann et al., 1994; Michael et al., 1994). Based on interviews with more than 3,400 men and women age 18 to 59, this random survey tended to reaffirm the significance of the dominant sexual ideologies. Most respondents reported that they engaged in heterosexual relationships, although 9 percent of the men said they had had at least one homosexual encounter resulting in orgasm. Although 6.2 percent of men and 4.4 percent of women said that they were at least somewhat attracted to others of the same gender, only 2.8 percent of men and 1.4 percent of women identified themselves as gay or lesbian. According to the study, persons who engaged in extramarital sex found their activities to be more thrilling than those with a marital partner, but they also felt more guilt. Persons in sustained relationships such as marriage or cohabitation found sexual activity to be the most satisfying emotionally and physically.

Cohabitation, Domestic Partnerships, and Same-Sex Marriage

Attitudes about cohabitation have changed dramatically over the past five decades. ***Cohabitation* refers to two people who live together, and think of themselves as a couple, without being legally married.** The number of unmarried couples sharing a household has grown rapidly, particularly among young people. It is estimated that more than 60 percent of first marriages are now preceded by living together. About 25 percent of unmarried women between the ages of 25 and 39 are currently living with a partner, and an additional 25 percent have lived with a partner at some time in the past (National Marriage Project, 2010). We do not have specific numbers of how many people actually cohabit because the Census Bureau does not ask about emotional or sexual involvement between unmarried individuals who share living quarters.

Recent studies have found that cohabitation is more common among people with lower educational and income levels. People who are most likely to cohabit are under age 45, have been married before, or are older individuals who do not want to lose financial benefits (such as retirement benefits) that are

cohabitation a situation in which two people live together, and think of themselves as a couple, without being legally married.

contingent upon not remarrying. Other factors associated with cohabitation include being less religious than one's peers, having been divorced previously, and having experienced parental divorce, lack of a father in the household, or high levels of marital discord while growing up (National Marriage Project, 2010).

Among heterosexual couples, many reasons exist for cohabitation; for gay and lesbian couples, however, no alternatives to cohabitation exist in most U.S. states. For that reason, many lesbians and gays seek recognition of their ***domestic partnerships*— household partnerships in which an unmarried couple lives together in a committed, sexually intimate relationship and is granted the same rights and benefits as those accorded to married heterosexual couples** (Aulette, 1994; Gerstel and Gross, 1995). However, this definition represents an ideal type; the reality of domestic partnerships varies from state to state. Although benefits such as health and life insurance coverage are extremely important to *all* couples, Gayle, a lesbian, points out that "It makes me angry that [heterosexuals] get insurance benefits and all the privileges, and Frances [her partner] and I take a beating financially. We both pay our insurance policies, but we don't get the discounts that other people get and that's not fair" (qtd. in Sherman, 1992: 197). Moreover, in some states with limited protection of domestic partnerships, some employers may choose to offer full benefits to domestic partners whereas other employers do not.

Over the past decade bitter controversy has emerged over the legal status of gay and lesbian couples, particularly those who seek to make their relationship a legally binding commitment through marriage. In 2004 Massachusetts became the first state to make same-sex marriage legal. For a five-month period in 2008, same-sex marriages were legally performed in California, but voters passed Proposition 8 to curtail this practice. As of 2011, same-sex marriage licenses are granted in Connecticut, Iowa, Massachusetts, New Hampshire, New York, Vermont, Washington, D.C., and the Coquille Tribe in Oregon, but a recent, major change at the federal level may bring about additional changes by the time you read this.

Here's why: For a number of years, same-sex marriage was recognized only at the state level because—for the purposes of federal law—the 1996 Defense of Marriage Act explicitly defined marriage as the union of one man and one woman. As a result, no act or agency of the federal government recognized same-sex marriage until President Obama's 2011 pronouncement that the Defense of Marriage Act is unconstitutional and his order that the U.S. Justice Department stop defending this law in court. This highly controversial decision was applauded by gay and lesbian rights advocates but sharply denounced by conservative political leaders. This decision raises many new questions about whether gay couples living in the states that already have legally recognized same-sex marriage may be discriminated against by the federal government, and it also brings to the forefront the more basic question about whether same-sex couples should have a legal right to marry. The status of same-sex marriages will no doubt remain in the public eye and before the courts for a number of years to come.

© David Butow/Redux

© ROBYN BECK/AFP/Getty Images

▲ Same-sex marriage was legal in California for a brief period of time, then was no longer an option after the passage of Proposition 8, a successful referendum that defined marriage on a heterosexual basis only. After challenges worked their way through the courts, a federal judge ruled in August 2010 that Proposition 8 violated the California Constitution. The state Supreme Court agreed to hear the case in September 2011, and you can learn the current status of this issue at **www.courts.ca.gov/supremecourt.htm**.

Marriage

Why do people get married? Couples get married for a variety of reasons. Some do so because they are "in love," desire companionship and sex, want to have children, feel social pressure, are attempting to escape from a bad situation in their parents' home, or believe that they will have more money or other resources if they get married. These factors notwithstanding, the selection of a marital partner is actually fairly predictable. As previously discussed, most people in the United States tend to choose marriage partners who are similar to themselves. ***Homogamy*** **refers to the pattern of individuals marrying those who have similar characteristics, such as race/ethnicity, religious background, age, education, or social class.** However, homogamy provides only the general framework within which people select their partners; people are also influenced by other factors. For example, some researchers claim that people want partners whose personalities match their own in significant ways. Thus, people who are outgoing and friendly may be attracted to other people with those same traits. However, other researchers claim that people look for partners whose personality traits differ from but complement their own.

© Ariel Skelley/Taxi/Getty Images

▲ Juggling housework, child care, and a job in the paid workforce is all part of the average day for many women. Why does sociologist Arlie Hochschild believe that many women work a "second shift"?

Housework and Child-Care Responsibilities

Today, women constitute more than half of all paid workers in the United States, and mothers are the primary breadwinner or co-breadwinner in nearly two-thirds of U.S. families. What this means is that more than 50 percent of all marriages in the United States are ***dual-earner marriages*****—marriages in which both spouses are in the labor force.** Over half of all employed women hold full-time, year-round jobs. Even when their children are very young, most working mothers work full time. For example, in 2010 more than 75 percent of employed mothers with children under age 6 worked full time (U.S. Census Bureau, 2010b). Also, many married women leave their paid employment at the end of the day and then go home to perform hours of housework and child care. Sociologist Arlie Hochschild (1989, 2003) refers to this as the ***second shift*****—the domestic work that employed women perform at home after they complete their workday on the job.** Thus, many married women contribute to the economic well-being of their families and also meet many, if not all, of the domestic needs of family members by cooking, cleaning, shopping, taking care of children, and managing household routines. According to Hochschild, the unpaid housework that women do on the second shift amounts to an extra month of work each year.

In recent years more husbands have shared some of the household and child-care responsibilities.

domestic partnerships household partnerships in which an unmarried couple lives together in a committed, sexually intimate relationship and is granted the same rights and benefits as those accorded to married heterosexual couples.

homogamy the pattern of individuals marrying those who have similar characteristics, such as race/ethnicity, religious background, age, education, or social class.

dual-earner marriages marriages in which both spouses are in the labor force.

second shift Arlie Hochschild's term for the domestic work that employed women perform at home after they complete their workday on the job.

However, studies continue to show that when husbands share some of the household responsibilities, they typically spend much less time in these activities than do their wives. Women and men perform different household tasks, and the deadlines for their work vary widely. Recurring tasks that have specific times for completion (such as bathing a child or cooking a meal) tend to be the women's responsibility; by contrast, men are more likely to do the periodic tasks that have no highly structured schedule (such as mowing the lawn or changing the oil in the car). Some gender reversals have occurred in many families, particularly in view of the national economic downturn, which has hurt the employment of men more than that of women: A larger share of men now perform some or all of the housework and child-care responsibilities at home while their wives are in the workplace.

In the United States millions of parents rely on help with child care from others so that they can work. Relatives take care of almost half of all children between the ages of birth and 4 who have full-time employed mothers. Among children between birth and 6 years of age (but not yet in kindergarten), 61 percent receive some form of nonparental child care on a regular basis (Federal Interagency Forum on Child and Family Statistics, 2010). The cost of such programs often makes it difficult for families to find high-quality child care.

Although organized after-school programs have become more numerous, the percentage of children staying home alone has remained steady in recent years. About 25 percent of school-age children stay alone after the school day ends until a parent returns home from work. Child-care specialists are concerned about this because children need productive and safe activities to engage in while their parents are working, but many home-alone children spend time eating junk food, watching television, talking on their cell phone, or playing computer and video games. Many of these children are under the supervision of an older brother or sister who may not be particularly interested in taking care of them (McClure, 2010).

Child-Related Family Issues and Parenting

Not all couples become parents, but everyone does not view this in the same way. Couples who decide not to have children typically consider themselves to be "child-free," whereas couples who do not have children through no choice of their own tend to consider themselves "childless" and sometimes feel stigmatized by others.

Deciding to Have Children

Cultural attitudes about having children and about the ideal family size began to change in the United States in the late 1950s. Women, on average, are now having 2.1 children each (see "Sociology Works!"). However, rates of fertility differ across racial and ethnic categories. In 2008, for example, the lowest fertility rate was for non-Hispanic white women (52 births per 1,000 women), whereas the rate for American Indian and Alaska Native women was among the highest (73 births per 1,000 women). African American women had 61 births per 1,000 women, and Asian American women had 63 births per 1,000. Latinas (Hispanic women) who were foreign born and not U.S. citizens had higher fertility rates than Latinas who were naturalized citizens or were born in the United States.

Advances in birth control techniques over the past four decades—including the birth control pill and contraceptive patches and shots—now make it possible for people to decide whether or not they want to have children, how many they wish to have, and to determine (at least somewhat) the spacing of their births. However, sociologists suggest that fertility is linked not only to reproductive technologies but also to women's beliefs that they do or do not have other opportunities in society that are viable alternatives to childbearing (Lamanna and Riedmann, 2012).

Today, the concept of reproductive freedom includes both the desire *to have* or *not to have* one or more children. According to the sociologists Leslie King and Madonna Harrington Meyer (1997), many U.S. women spend up to one-half of their life attempting to control their reproductivity. Other analysts have found that women, more often than men, are the first to choose a child-free lifestyle (Seccombe, 1991). However, the desire not to have children often comes in conflict with our society's *pronatalist bias,* which assumes that having children is the norm and can be taken for granted, whereas those who choose not to have children believe that they must justify their decision to others (Lamanna and Riedmann, 2012).

However, some couples experience the condition of *involuntary infertility,* whereby they want to have a child but find that they are physically unable to do so. *Infertility* is defined as an inability to conceive after a year of unprotected sexual relations. Today, infertility affects about 12 percent of couples in which the wife is between the ages of 15 and 44. Fertility problems originate in females in approximately 30–40 percent of the cases and with males in about 40 percent of the cases; in the other 20 percent

sociology works!

Social Factors Influencing Parenting: From the Housing Market to the Baby Nursery

One reason there are so few children in Italy is that housing is so hard to come by. Houses are bigger in the U.S. and generally more available. That may help explain why Americans have more babies.

—Robert Engelman, vice president for programs at the Worldwatch Institute, an environmental research organization, and author of *More: Population, Nature, and What Women Want* (qtd. in Leland, 2008: A12)

Social scientists have long traced a connection between housing and fertility. When homes are scarce or beyond the means of young couples, as in the 1930s, couples delay marriage or have fewer children. This tendency helps account for the relatively dismal birthrates of many developed nations. . . . (Leland, 2008: A12)

For many years, demographers and other sociologists who specialize in population trends have sought to identify how biological and social factors affect fertility rates in various nations. Although biological factors such as general health and levels of nutrition in a region clearly affect the number of children a couple may produce, social factors are also important in determining the estimated number of children a woman will have in her lifetime. A key social factor is the housing market where a couple lives: The ability to buy a house and having a relatively large home may influence a couple's decision about how many children to have. Early in the 2000s the housing market in some areas of the United States provided an opportune time for more young couples to purchase their own home. Mortgages were readily available, and it typically did not take much cash (in the way of a down payment) up front to sign the contract and move into one's dream home. The availability of such loans to people who really couldn't afford the home they were buying has since come back to haunt many underfunded homebuyers.

Sociological insights into the social aspects of fertility, which at first might appear to be primarily a biological phenomenon, have provided us with new information on why people decide to have children and how many children they might have. However, much remains unknown about the relationship between the housing market and the maternity ward, including how income and feelings of optimism or pessimism about the local and national economy might affect a couple's decisions regarding parenting.

Will the baby boomlet continue in the future? According to some social analysts, the boomlet may be short-lived because of a downturn in the economy and the housing market, which has culminated in fewer new homes being built and brought about more foreclosures—two key factors that may discourage couples from either having children or producing larger families.

reflect & analyze

How might sociological findings about factors that influence a couple's decision to have children be useful in your community? For example, why is information about the availability of housing and local fertility trends important to school board members and administrators when they are making enrollment projections or deciding where to build new schools in the future?

of the cases, the cause is impossible to determine. A leading cause of infertility is sexually transmitted diseases, especially those cases that develop into pelvic inflammatory disease. It is estimated that about half of infertile couples who seek treatments such as fertility drugs, artificial insemination, and surgery to unblock fallopian tubes can be helped; however, some are unable to conceive despite expensive treatments such as *in vitro fertilization,* which costs about $12,400 per attempt. Some health insurance policies cover this procedure; others do not. About 14 states require insurance companies to cover the diagnosis and treatment of infertility; however, this situation is subject to change because of the recent Affordable Health Care law.

How do people deal with being involuntarily childless? According to the sociologist Charlene Miall (1986), women who are involuntarily childless engage in "information management" to combat the social stigma associated with childlessness. Their tactics range from avoiding people who make them uncomfortable to revealing their infertility so that others will not think of them as "selfish" for being childless Some people who are involuntarily childless may choose surrogacy or adoption as an alternative way to become a parent (see "Sociology in Global Perspective").

Adoption

Adoption is a legal process through which the rights and duties of parenting are transferred from a child's biological and/or legal parents to a new legal parent or parents. This procedure gives the adopted child all the rights of a biological child. In most adoptions a new birth certificate is issued, and the child has no

framing teen pregnancy in the media

Sex Ed on TV? Mixed Messages on Teen Pregnancy

Chelsea, Jenelle, Kailyn, and Leah were like many other happy-go-lucky teenagers. Chelsea was a popular softball star until she became pregnant and dropped out of high school during her senior year to have her daughter. Jenelle was a carefree beach bunny until she gave birth to her son. Kailyn was a highly motivated high school senior until she became the mother of her son. Leah was a former cheerleader until she became the teen mother of twins.

Although this sounds like an episode of *Glee* or a soap opera, this is actually the cast of one season of MTV's *Teen Mom* franchise, where young girls are shown going through the process of finding out that they are pregnant, making adjustments to their daily lives, and entering the world of motherhood and complex relationships with their parents and boyfriends as their lives swiftly change around them. For some girls this means giving up their child for adoption and dealing with the consequences of that decision.

16 and Pregnant and the *Teen Mom* franchise have been very popular with many adolescent girls. Like *Teen Mom, 16 and Pregnant* also follows high school girls as they cope with being pregnant and becoming parents. These programs are so popular among some teens that rumors have spread that girls were becoming pregnant so that they could "star" on the shows. Although this rumor was never confirmed, it does brings up an interesting question: What kind of media message is being sent to young women about becoming pregnant while in high school?

A content analysis of these shows reveals several standard framing devices: (1) "What would it be like?" framing, (2) "Glad it's not me" framing, and (3) "cautionary tale" framing. In the series *16 and Pregnant,* viewers follow a different girl each week as she has her baby and moves into her new life. Viewers have an opportunity to see "what it would be like" because each girl describes her feelings and how she made the decisions she did along the way. Viewers see her

per 1,000 females in 2009 from 61.8 births per 1,000 females in 1991.

What are the primary reasons for the high rates of teenage pregnancy? At the microlevel, several issues are most important: (1) many sexually active teenagers do not use contraceptives; (2) teenagers—especially those from some low-income families and/or subordinate racial and ethnic groups—may receive little accurate information about the use of, and problems associated with, contraception; (3) some teenage males (because of a double standard based on the myth that sexual promiscuity is acceptable among males but not females) believe that females should be responsible for contraception; and (4) some teenagers view pregnancy as a sign of male prowess or as a way to gain adult status. At the macrolevel, structural factors also contribute to teenage pregnancy rates. Lack of education and employment opportunities in some central-city and rural areas may discourage young people's thoughts of upward mobility. Likewise, religious and political opposition has resulted in issues relating to reproductive responsibility not being dealt with as openly in the United States as in some other nations. Finally, advertising, films, television programming, magazines, music, and other forms of media often flaunt the idea of being sexually active without showing the possible consequences of such behavior. However, recent efforts have been made by television shows to fight teenage pregnancy, with varying degrees of success (see "Framing Teen Pregnancy in the Media").

Teen pregnancies are a concern because some teenage mothers may be less skilled at parenting, are less likely to complete high school than their counterparts without children, and possess few economic and social supports other than their relatives

© Fox Searchlight/Photofest

▲ In the 2007 film *Juno,* the title character, played by Ellen Page, becomes pregnant while still in high school. Although this film is a comedy, it takes a surprisingly straightforward look at the reality of teenage pregnancy.

friends, family, and sometimes the child's father as they respond to the situation. "Glad it's not me" framing comes into play in *16 and Pregnant* and *Teen Mom* when viewers see the many obstacles that a young mother faces as she tries to have a life of her own, parent an infant, and deal with interpersonal relationships. "Glad it's not me" framing provides a voyeuristic view of the often-troubled relationship between the new mother and the baby's father, and this is sometimes the focus of the show, more than other issues—such as the baby's future—that might be discussed.

Finally, "cautionary tale" framing of these shows is what many parents and educators like most about *16 and Pregnant* and *Teen Mom:* They believe that these programs open up the topic of teen pregnancy and parenting for discussion with adolescents who are rarely interested in talking about such matters with adults. It is the hope of some parents and educators that girls will be drawn into the lives of the young women featured in these reality shows and that they will come away with the message that they do not want to be in a similar situation. For example, some teachers ask their students to think about what it would be like to have schoolwork, babies who are ill, lack of sleep, and a boyfriend who does not help them out, all at the same time.

However, one of the major criticisms of this type of show as a cautionary tale is the fact that these shows do not provide useful sex education for young people. In some cases the shows may glamorize teen pregnancy and parenting for young women who have no other way of getting attention from their parents, peers, and boyfriends. Overall, though, shows such as these have been applauded for framing a pressing social concern in such a manner that it gains the attention of young women, the very audience that sex educators and parents have been trying to reach for many years (Hoffman, 2011).

reflect & analyze

Clearly, getting pregnant and having a child is a major milestone, but should teenage mothers share their thoughts and feeling about this event with millions of viewers? Do programs such as *Teen Mom* have the potential to generate constructive conversations about the choices teens make and how they deal with the consequences of their actions? What do you think?

(Maynard, 1996; Moore, Driscoll, and Lindberg, 1998). In addition, births among unmarried teenagers often have negative long-term consequences for both mothers and children, who have limited educational and employment opportunities and a high likelihood of living in poverty. Statistics also suggest that 43 percent of young women who first gave birth between the ages of 15 and 19 will have a second child within 3 years (Children's Defense Fund, 2008).

Teenage fathers have largely been left out of the picture. A number of myths exist regarding teenage fathers: (1) they are worldly wise "superstuds" who engage in sexual activity early and often, (2) they are "Don Juans" who sexually exploit unsuspecting females, (3) they have "macho" tendencies because they are psychologically inadequate and need to prove their masculinity, (4) they have few emotional feelings for the women they impregnate, and (5) they are "phantom fathers" who are rarely involved in caring for and rearing their children (Robinson, 1988). However, these myths overlook the fact that some teenage males work at being good fathers.

Single-Parent Households

In recent years there has been a significant increase in single-parent households because of divorce and births outside of marriage. About 25 percent of U.S. households are maintained by single parents who live with their own children under the age of 18. These figures exclude single parents who are living with an unmarried partner. ▶ Map 15.1 shows that most of the states with the largest percentages of single-parent households are in the South, with the exception of New Mexico, Michigan, Ohio, and New York. The area with the highest percentage of single-parent households is the District of Columbia (54 percent), whereas Utah (15 percent) is the state with the lowest percentage (Kreider and Elliott, 2009).

Even for a person with a stable income and a network of friends and family to help with child care, raising a child alone can be an emotional and financial burden. Single-parent households headed by women have been stereotyped by some journalists, politicians, and analysts as being problematic for children. One out of every two children in the United States will spend some time during childhood living in a household headed by a single mother who is divorced, separated, never married, or widowed. Statistics suggest that children from single-parent families are more likely than children in two-parent families to have poor academic achievement, higher school absentee and dropout rates, early marriage and parenthood, higher rates of divorce, and more drug and alcohol abuse. Does living in a one-parent family *cause* all of this? Certainly not! Many factors—including poverty, discrimination, unsafe neighborhoods, and high crime rates—contribute to these problems.

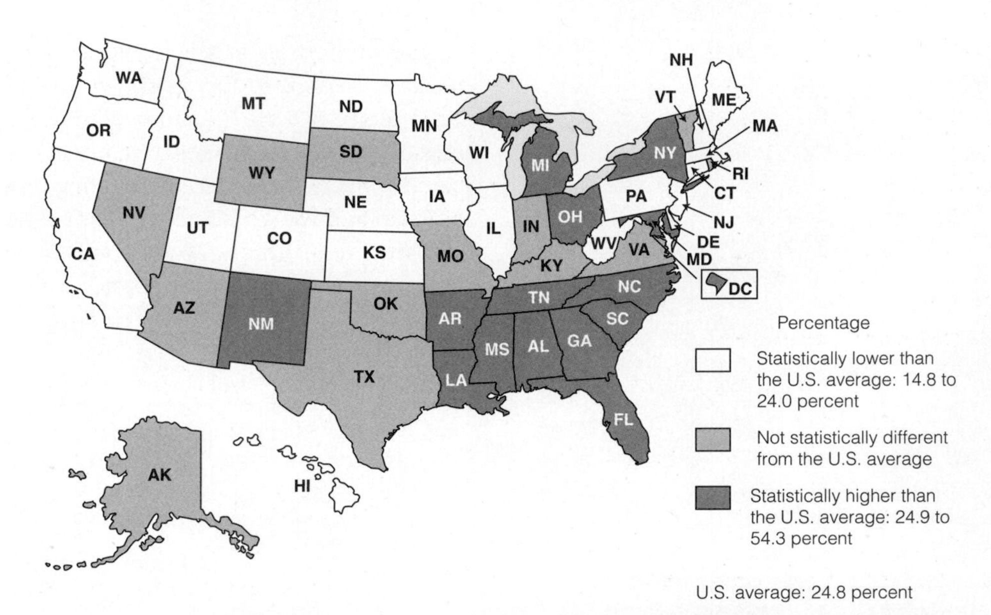

▲ **MAP 15.1 PERCENTAGE OF SINGLE-PARENT HOUSEHOLDS WITH CHILDREN UNDER 18 BY STATE: 2007**

Source: U.S. Census Bureau, 2009.

Single fathers who do not have custody of their children may play a relatively limited role in the lives of those children. Although many remain actively involved in their children's lives, others may become "Disneyland daddies" who take their children to recreational activities and buy them presents for special occasions but have a very small part in their children's day-to-day lives. Sometimes, this limited role is by choice, but more often it is caused by workplace demands on time and energy, the location of the ex-wife's residence, and limitations placed on visitation by custody arrangements.

Because of the nature of marriage laws in some states, lesbian mothers and gay fathers are counted in some studies as single parents even when they share parenting responsibilities with a same-sex partner. More research has been conducted in recent years on parenting by lesbian, gay, bisexual, and transgendered (LGBT) couples, and the results are favorable because living in these families appears in some respects to be better for children than living in heterosexual families. One of the areas that seem to be better is the division of parenting and household labor, which has a distinct pattern of equality and sharing among LGBT couples as compared with heterosexual parents. It also appears that lesbian and gay parents tend to be more responsive to their children and more child oriented in their outlook (Short et al., 2007).

Two-Parent Households

Since the 1970s the percentage of children living in two-parent households has dropped, while the percentage living with a single parent has increased (see ▶ Figure 15.1). In 2010 almost 70 percent of children lived with two parents, while nearly 23 percent lived with only their mother and 3.4 percent lived with only their father. When these statistics are computed, parents include not only biological parents but also stepparents who adopt their children. However, foster parents are considered nonrelatives.

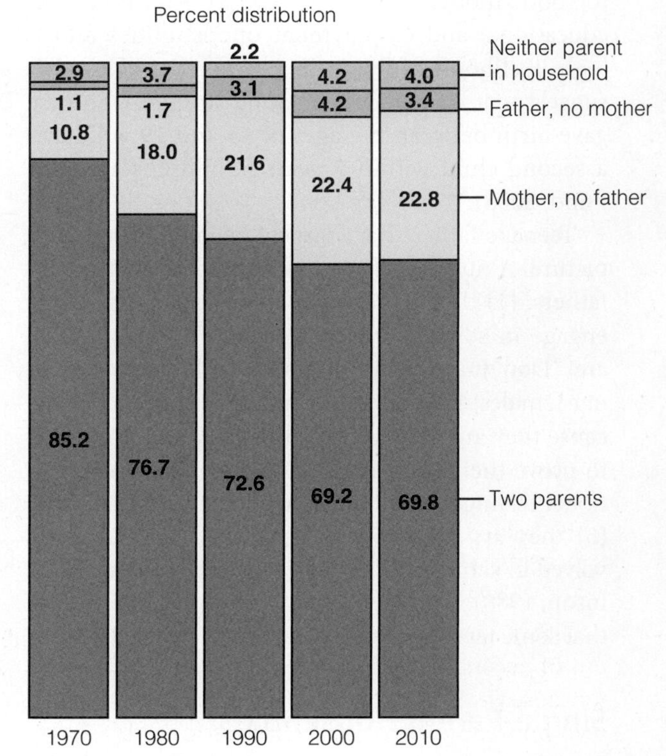

▲ **FIGURE 15.1 LIVING ARRANGEMENTS OF CHILDREN UNDER 18 YEARS OLD FOR SELECTED YEARS: 1970–2010**

Source: U.S. Census Bureau, 2010b.

© Yuri Arcurs/Shutterstock.com

▲ Mothers and fathers in single-parent households are confronted with the necessity of meeting most of their children's daily needs without help from others. However, even in two-parent households, children are not guaranteed a happy childhood simply because both parents live in the same household.

Children have two primary caregivers in families where a couple truly shares parenting. Some parents share parenting responsibilities by choice; others share out of necessity because both hold full-time jobs. Studies have found that men's taking an active part in raising the children is beneficial not only for mothers (who then have a little more time for other activities) but also for the men and the children. The men benefit through increased access to children and greater opportunity to be nurturing parents (Coltrane, 1989).

Remaining Single

Some never-married people remain single by choice. Reasons include opportunities for a career (especially for women), the availability of sexual partners without marriage, the belief that the single lifestyle is full of excitement, and the desire for self-sufficiency and freedom to change and experiment (Stein, 1976, 1981). Some scholars have concluded that individuals who prefer to remain single hold more-individualistic values and are less family oriented than those who choose to marry. Friends and personal growth tend to be valued more highly than marriage and children (Cargan and Melko, 1982; Alwin, Converse, and Martin, 1985).

Other never-married individuals remain single out of necessity. Being single is an economic reality for those who cannot afford to marry and set up their own household. Structural changes in the economy have limited the options of many working-class young people. Even some college graduates have found that they cannot earn enough money to set up a household separate from that of their parents.

The proportion of singles varies significantly by racial and ethnic group, as shown in ▶ Figure 15.2. Among persons age 15 and over, more than 40 percent of African Americans have never married, compared with about 32 percent of Latinos/as, 26 percent of Asian and Pacific Islander Americans, and 24 percent of whites. Among women age 20 and over, the difference is even more pronounced; almost twice as many African American women in this age category have never married, compared with U.S. women of the same age in general (U.S. Census Bureau, 2009).

Transitions and Problems in Families

Families go through many transitions and experience a wide variety of problems, ranging from high rates of divorce and teen pregnancy to domestic abuse and family violence. These all-too-common experiences highlight two important facts about families: (1) for good or ill, families are central to our existence, and (2) the reality of family life is far more complicated than the idealized image of families found in the media and in many political discussions. Family violence, children in foster care, divorce, and remarriage are four major issues in contemporary family life.

Family Violence

Family violence refers to various forms of abuse that take place among family members, including child abuse, spousal abuse, and elder abuse. We will primarily focus on domestic violence, also referred to as spousal abuse or intimate-partner violence. *Domestic violence* refers to any intentional act or series of acts—whether physical, emotional, or sexual—by one or both partners in an intimate relationship that causes injury to either person. An intimate relationship might include marriage or cohabitation, as well as people who are separated or living apart from a former partner or spouse.

There are numerous causes of domestic violence, and many factors are interrelated. Factors contributing to unequal power relations in families include social and economic inequality, legal and political sanctions that deny girls and women equal rights, and cultural sanctions that dictate appropriate sex roles and reinforce the belief that males are inherently superior to females. Cultural factors that perpetuate domestic violence include gender-specific socialization that establishes dominant–subordinate sex roles. Economic factors include poverty or limited financial resources within families that contribute to tension and sometimes to violence. Economic factors are intertwined with women's limited access to education, employment, and sufficient income

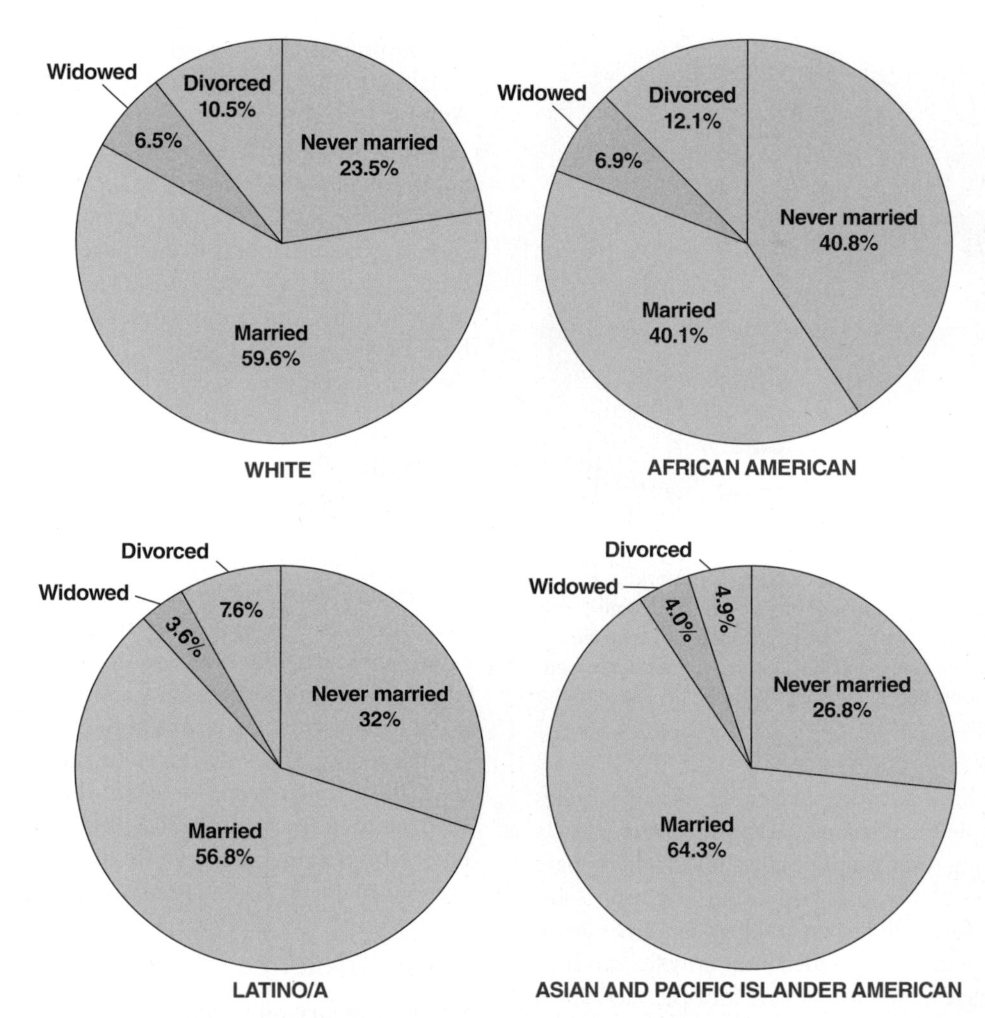

▲ FIGURE 15.2 MARITAL STATUS OF U.S. POPULATION AGE 15 AND OVER BY RACE/ETHNICITY

so that they could take care of themselves and their children. Regardless of the factors that contribute to domestic violence, control is central to all forms of abuse: *Gaining* and *maintaining control* over the victim is the key factor in abuse. As a result, family violence often involves a cycle of abuse that goes on for extended periods of time.

How much do we *really* know about family violence? Women, as compared with men, are more likely to be the victims of violence perpetrated by intimate partners. Recent statistics indicate that more than 25 million women in the United States experience domestic violence at some time during their lives. However, we cannot know the true extent of family violence because much of it is not reported to police. For example, it is estimated that only about one-half of the intimate-partner violence against women is reported to the police.

Although everyone in a household where family violence occurs is harmed psychologically, whether or not they are the victims of violence, children are especially affected by household violence. It is estimated that between three million and ten million children witness some form of domestic violence in their homes each year, and there is evidence to suggest that domestic violence and child maltreatment often take place in the same household (Children's Defense Fund, 2010).

In some situations, family violence can be reduced or eliminated through counseling, the removal of one parent from the household, or other steps that are taken either by the family or by social service or law enforcement officials. However, children who witness violence in the home may display certain emotional and behavioral problems that adversely affect their school life and communication with other people. In some families the problems of family violence are great enough that the children are removed from the household and placed in foster care.

Children in Foster Care

Not all of the children in foster care come from violent homes, but many foster children have been in dysfunctional homes where parents or other relatives lacked the ability to meet the children's daily needs. *Foster care* refers to institutional settings or residences where

you can make a difference

Providing Hope and Help for Children

I take it personally when I see kids mistreated. I just think they need an advocate to fight for them. . . . For me, it's very simple: The kids' needs come first. That's the bottom line at Hope Meadows. We make decisions as if these are our own children, and when you think that way, your decisions are different than if you are just trying to work within a bureaucratic system.

—sociologist Brenda Eheart describing why she founded Hope Meadows (qtd. in Smith, 2001: 22)

After five years of research into the adoptions of older children, the sociologist Brenda Eheart realized that foster families faced many problems when they tried to help children who had been moved from home to home. Thinking that she might be able to make a difference, Eheart developed the plan for Hope Meadows, a community established in 1994 on an abandoned air force base in Illinois. Hope Meadows is made up of a three-block-long series of ranch houses that provide multigenerational and multiracial housing for foster children, their temporary families, and older adults who live and work with the children. Older adults who interact with the children receive reduced rent in exchange for at least six hours per week of volunteer work. Foster families that reside at Hope Meadows gain a feeling of community as they work together to help children who have experienced severe abuse or neglect, have been exposed to drugs and numerous foster homes, and often have physical, emotional, and behavioral problems.

Since its inception, Hope Meadows has been largely successful in helping children get adopted. However, children are not the only beneficiaries of this community: Older residents gain the benefit of interacting with children and feeling that they can *make a positive contribution* to the lives of others (Barovick, 2001). Debbie Calhoun, a foster parent at Hope Meadows, has suggested things that children need the most when they come there, and we can make a difference by providing the children in our lives with these same things (based on Smith, 2001):

- *Understanding.* We need to gain an awareness of how children feel and why they say and do certain things.
- *Trust.* We need to help children to see us as people they can rely on and believe in as people they can trust.
- *Love.* We must show children that they are loved and that they will still be loved even when they make mistakes.
- *Compassion.* We must show children compassion because they must experience compassion in order to be able to show it to others.
- *Time.* We must give children time to be a part of our lives, and we must also give them time to adjust and to start over when they need to do so.
- *Security.* We must help children to feel secure in their surroundings and to believe that there is stability or permanency in their living arrangements.
- *Praise.* We must tell children when they are doing well and not always be critical of them.
- *Discipline.* We must let children know what behavior is acceptable and what behavior is not, all the while showing them that we love them, even when discipline is necessary.
- *Self-Esteem.* We must help children feel good about themselves.
- *Pride.* We should provide opportunities for children to learn to take pride in their accomplishments and in themselves.

If these suggestions are beneficial for children in foster care settings, then they are certainly useful ideas for each of us to implement in our own families and communities as well. What other ideas would you add to the list? Why?

adults other than a child's own parents or biological relatives serve as caregivers. States provide financial aid to foster parents, and the intent of such programs is that the children will either return to their own families or be adopted by other families. However, this is often not the case for "difficult to place" children, particularly those who are over 10 years of age, have illnesses or disabilities, or are perceived as suffering from "behavioral problems." Teenagers have a much more difficult time than younger children being placed for adoption and often spend their adolescent years moving from one foster home to another.

More than 500,000 children are in foster care at any given time, with about 75 percent of them being cared for by nonrelatives. About 60 percent of children in foster care are children of color (Children's Defense Fund, 2010). Many children in foster care have limited prospects for finding a permanent home; however, a few innovative programs have offered hope for children who previously had been moved from one foster care setting to another (see "You Can Make a Difference").

Sadly, recent studies have found that the problems of young people who have "aged out" of foster care continue into their midtwenties and beyond. Based on one study in Illinois, Iowa, and Wisconsin, about 60 percent of males previously in foster care had been convicted of at least one crime, and nearly 75 percent of females previously in foster care were now receiving public aid and struggling to raise their own children. And the picture was not better in regard to educational attainment: Only 6 percent (6 in 100) had finished two years of education beyond the high school diploma (Eckholm, 2010).

Divorce

Divorce is the legal process of dissolving a marriage that allows former spouses to remarry if they so choose. Although the U.S. divorce rate is nearly twice as high as it was in 1960, it has declined since hitting its highest point in the early 1980s. Most divorces today are granted on the grounds of *irreconcilable differences,* meaning that there has been a breakdown of the marital relationship for which neither partner is specifically blamed. Prior to the passage of more-lenient divorce laws, many states required that the partner seeking the divorce prove misconduct on the part of the other spouse. Under *no-fault divorce laws,* however, proof of "blameworthiness" is generally no longer necessary.

Over the past 100 years the U.S. divorce rate (number of divorces per 1,000 population) has varied from a low of 0.7 in 1900 to an all-time high of 5.3 in 1981; by 2009, it had decreased to 3.4 per 1,000 population (U.S. Census Bureau, 2010b). Divorce statistics may vary based on the source because some reporting organizations include annulments in their count, and a number of states no longer provide divorce statistics to national reporting agencies. ▶ Figure 15.3 shows the latest available U.S. divorce rates for each state.

Although many people believe that marriage should last for a lifetime, others believe that marriage is a commitment that may change over time. Financial stressors are a contributing factor to some divorces (National Marriage Project, 2010). During times of a national recession such as the one that United States experienced beginning in 2007–2008, some people decide to remain married until conditions change so that they can sell a house or have better financial stability when they part. However, others gain a deeper commitment to their marriage as they struggle through adversity (National Marriage Project, 2010).

We often hear that 50 percent of all marriages end in divorce. But this figure is misleading in some ways, even though it is true for the entire U.S. population as a whole. Although there is a 40 to 50 percent chance that a first marriage will end in either divorce or separation before one partner dies, divorce rates vary widely based on certain factors. There is a *decrease* in risk of divorce for people in the following categories:

- Making over $50,000 annually (as compared with under $25,000)
- Having graduated college (as opposed to not completing high school)
- Having a baby 7 months or more after marriage (as opposed to before marriage)
- Marrying over 25 years of age (as opposed to under 18)
- Coming from an intact family of origin (as opposed to having divorced parents)
- Having a religious affiliation (as opposed to none) (National Marriage Project, 2010)

Consequences of Divorce Divorce may have a dramatic economic and emotional impact on family members. An estimated 60 percent of divorcing couples have one or more children. By age sixteen, about one out of every three white children and two out of every three African American children will experience divorce within their families. As a result, most of them will remain with their mothers and live in a single-parent household for a period of time. Although divorce decrees provide for parental joint custody of hundreds of thousands of children annually, this arrangement may create unique problems for some children. Furthermore, some children experience more than one divorce during their childhood because one or both of their parents may remarry and subsequently divorce again.

But divorce does not have to be always negative. For some people, divorce may be an opportunity to terminate destructive relationships. For others, it may represent a means to achieve personal growth by managing their lives and social relationships and establishing their own social identity. Still others choose to remarry one or more times.

Remarriage

Most people who divorce get remarried. In recent years more than 40 percent of all marriages have been between previously married brides and/or grooms. Among individuals who divorce before age thirty-five, about half will remarry within three years of their first divorce (Bramlett and Mosher, 2001). Most divorced people remarry others who have been divorced. However, remarriage rates vary by gender and age. At all ages, a greater proportion of men than women

© Kayte Deioma/PhotoEdit

▲ Remarriage and blended families create new opportunities and challenges for parents and children alike.

	Rates per 1,000 population[a, b]			
	1990	2000	2005	2009
United States[c]	**4.7**	**4.2**	**3.6**	**3.4**
Alabama	6.1	5.4	4.9	4.4
Alaska	5.5	4.4	5.8	4.4
Arizona	6.9	4.4	4.1	3.5
Arkansas	6.9	6.9	6.0	5.7
California	4.3	n.a.	n.a.	n.a.
Colorado	5.5	n.a.	4.4	4.2
Connecticut	3.2	2.0	2.7	3.1
Delaware	4.4	4.2	3.9	3.6
District of Columbia	4.5	3.0	2.0	2.6
Florida	6.3	5.3	4.6	4.2
Georgia	5.5	3.9	n.a.	n.a.
Hawaii	4.6	3.9	n.a.	n.a.
Idaho	6.5	5.4	4.9	5.0
Illinois	3.8	3.2	2.5	2.5
Indiana	n.a.	n.a.	n.a.	n.a.
Iowa	3.9	3.3	2.7	2.4
Kansas	5.0	4.0	3.1	3.7
Kentucky	5.8	5.4	4.5	4.6
Louisiana	n.a.	n.a.	n.a.	n.a.
Maine	4.3	4.6	3.5	4.1
Maryland	3.4	3.3	3.1	2.8
Massachusetts	2.8	3.0	2.2	2.2
Michigan	4.3	4.0	3.4	3.3
Minnesota	3.5	3.1	n.a.	n.a.
Mississippi	5.5	5.2	4.5	4.1

	Rates per 1,000 population[a, b]			
	1990	2000	2005	2009
Missouri	5.1	4.8	3.6	3.7
Montana	5.1	2.4	3.8	4.1
Nebraska	4.0	3.8	3.4	3.4
Nevada	11.4	9.6	7.7	6.7
New Hampshire	4.7	5.8	3.3	3.7
New Jersey	3.0	3.1	2.9	2.8
New Mexico	4.9	5.3	4.6	4.0
New York	3.2	3.4	2.8	2.6
North Carolina	5.1	4.8	3.8	3.8
North Dakota	3.6	3.2	2.4	2.9
Ohio	4.7	4.4	3.6	3.3
Oklahoma	7.7	3.7	5.6	4.9
Oregon	5.5	5.0	4.3	3.9
Pennsylvania	3.3	3.2	2.3	2.7
Rhode Island	3.7	3.1	2.9	3.0
South Carolina	4.5	3.7	2.9	3.0
South Dakota	3.7	3.6	3.0	3.3
Tennessee	6.5	6.1	4.6	3.9
Texas	5.5	4.2	3.2	3.3
Utah	5.1	4.5	4.0	3.6
Vermont	4.5	8.6	3.3	3.5
Virginia	4.4	4.3	3.9	3.7
Washington	5.9	4.7	4.0	3.9
West Virginia	5.3	5.2	5.1	5.2
Wisconsin	3.6	3.3	3.0	3.0
Wyoming	6.6	5.9	5.3	5.2

▲ **FIGURE 15.3 U.S. DIVORCE RATES BY STATE, 1990–2009**

[a]Based on total population residing in area; population enumerated as of April 1 for 1990; estimated as of July 1 for all other years.

[b]Includes annulments.

[c]U.S. totals for the number of divorces are an estimate that includes states not reporting (California, Colorado, Indiana, and Louisiana).

Source: U.S. Census Bureau, 2010b.

remarry, often relatively soon after the divorce. Among women, the older a woman is at the time of divorce, the lower her likelihood of remarrying. Women who have not graduated from high school and who have young children tend to remarry relatively quickly; by contrast, women with a college degree and without children are less likely to remarry (Bramlett and Mosher, 2001).

As a result of divorce and remarriage, complex family relationships are often created. Some people become part of stepfamilies or ***blended families,* which consist of a husband and wife, children from previous marriages, and children (if any) from the new marriage.** At least initially, levels of family stress may be fairly high because of rivalry among the children and hostilities directed toward stepparents or babies born into the family. In spite of these problems, however, many blended families succeed. The family

blended family a family consisting of a husband and wife, children from previous marriages, and children (if any) from the new marriage.

that results from divorce and remarriage is typically a complex, binuclear family in which children may have a biological parent and a stepparent, biological siblings and stepsiblings, and an array of other relatives, including aunts, uncles, and cousins.

According to the sociologist Andrew Cherlin (1992), the norms governing divorce and remarriage are ambiguous. Because there are no clear-cut guidelines, people must make decisions about family life (such as whom to invite for a birthday celebration or wedding) based on their beliefs and feelings about the people involved.

Family Issues in the Future

As we have seen, families and intimate relationships changed dramatically during the twentieth century. Some people believe that the family as we know it is doomed. Others believe that a return to traditional family values will save this important social institution and create greater stability in society. However, the sociologist Lillian Rubin (1986: 89) suggests that clinging to a traditional image of families is hypocritical in light of our society's failure to support families: "We are after all, the only advanced industrial nation that has no public policy of support for the family whether with family allowances or decent publicly-sponsored childcare facilities." Some laws even have the effect of hurting children whose families do not fit the traditional model. For example, cutting back on government programs that provide food and medical care for pregnant women and infants will result in seriously ill children rather than model families (Aulette, 1994).

According to the psychologist Bernice Lott (1994: 155), people's perceptions about what constitutes a family will continue to change in the future:

> Persons on whom one can depend for emotional support, who are available in crises and emergencies, or who provide continuing affections, concern, and companionship can be said to make up a family. Members of such a group may live together in the same household or in separate households, alone or with others. They may be related by birth, marriage, or a chosen commitment to one another that has not been legally formalized.

Some of these changes are already becoming evident. For example, many men are attempting to take an active role in raising their children and helping with household chores. Many couples terminate abusive relationships and marriages.

Regardless of the problems facing families today, many people still demonstrate their faith in the future by getting married and having children. It will be interesting to see what people in the future decide to do about marriage and parenthood. How many children will people have? What will family life be like in 2020 or 2030? What will your own family be like?

chapter review Q & A

Use these questions and answers to check how well you've achieved the learning objectives set out at the beginning of this chapter.

- **What is the family?**

Today, families may be defined as relationships in which people live together with commitment, form an economic unit and care for any young, and consider their identity to be significantly attached to the group.

- **How does the family of orientation differ from the family of procreation?**

The family of orientation is the family into which a person is born; the family of procreation is the family a person forms by having or adopting children.

- **What pattern of marriage is legally sanctioned in the United States?**

In the United States, monogamy is the only form of marriage sanctioned by law. Monogamy is a marriage between two partners, usually a woman and a man.

- **What are the primary sociological perspectives on families?**

Functionalists emphasize the importance of the family in maintaining the stability of society and the well-being of individuals. Conflict and feminist perspectives view the family as a source of social inequality and an arena for conflict. Symbolic interactionists explain family relationships in terms of the subjective meanings and everyday interpretations that people give to their lives. Postmodern analysts view families as being permeable, capable of being diffused or invaded so that the original purpose is modified.

- **How are families in the United States changing?**

Families are changing dramatically in the United States. Cohabitation has increased significantly in the past three decades. With the increase in dual-earner marriages, women have become larger contributors to the financial well-being of their families, but some have become increasingly burdened by the second shift—the domestic work that employed women perform at home after they complete their workday on the job. Many single-parent families also exist today.

- **What is divorce, and what are some of its causes?**

Divorce is the legal process of dissolving a marriage. At the macrolevel, changes in social institutions may contribute to an increase in divorce rates; at the microlevel, factors contributing to divorce include age at marriage, length of acquaintanceship, economic resources, education level, and parental marital happiness.

key terms

bilateral descent 438
blended families 457
cohabitation 443
domestic partnerships 444
dual-earner marriages 445
egalitarian family 438
endogamy 439
exogamy 439
extended family 435
families 434
family of orientation 435
family of procreation 435
homogamy 445
kinship 435
marriage 436
matriarchal family 438
matrilineal descent 438
matrilocal residence 439
monogamy 436
neolocal residence 439
nuclear family 436
patriarchal family 438
patrilineal descent 438
patrilocal residence 438
polyandry 438
polygamy 437
polygyny 437
second shift 445
sociology of family 439

questions for critical thinking

1. In your opinion, what constitutes an ideal family? How might functionalist, conflict, feminist, and symbolic interactionist perspectives describe this family?
2. Suppose that you wanted to find out about women's and men's perceptions about love and marriage. What specific issues might you examine? What would be the best way to conduct your research?
3. You have been appointed to a presidential commission on child-care problems in the United States. How to provide high-quality child care at affordable prices is a key issue for the first meeting. What kinds of suggestions would you take to the meeting? How do you think your suggestions should be funded? How does the future look for children in high-, middle-, and low-income families in the United States?

turning to video

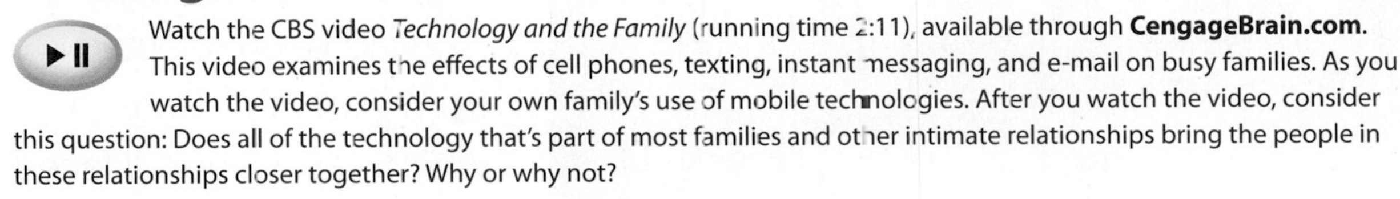

Watch the CBS video *Technology and the Family* (running time 2:11), available through **CengageBrain.com.** This video examines the effects of cell phones, texting, instant messaging, and e-mail on busy families. As you watch the video, consider your own family's use of mobile technologies. After you watch the video, consider this question: Does all of the technology that's part of most families and other intimate relationships bring the people in these relationships closer together? Why or why not?

online study resources

Go to CENGAGE brain.com to access online study resources, including the Sociology CourseMate for this text as well as special features such as video, an interactive sociology time line and interactive maps, General Social Survey (GSS) data, and U.S. Census 2010 data.

CourseMate brings course concepts to life with interactive learning, study, and exam-preparation tools that support the printed textbook. A textbook-specific website, **Sociology CourseMate** includes an integrated interactive eBook and other interactive learning tools, including quizzes, flash cards, and videos.

Visit **www.cengagebrain.com** to access your account and purchase materials.

16 Education

My parents came to the United States from Egypt 30 years ago. Today, they still can barely speak English and haven't been educated beyond high school. My parents, my three older brothers and I live in a two-bedroom apartment, which forces me to share a room with my parents. My household has always consisted of constant yelling, fighting, putting others down, and negative forces. . . . Slowly, I started to care less about my school work and failed my classes. . . . My only alternative was to spend time at my aunt's house. She and her husband are both college educated . . . [and] lived happily and comfortably in comparison to those who weren't educated (my parents), who lived a life of hard physical labor and financial troubles. It came down to one simple question: What kind of lifestyle do I want to live? . . .

As the end of my senior year [of high school] approached, I began to get response letters from the Cal State universities. . . . I had been rejected from every Cal State to which I applied. . . . I faced a crossroads. I could follow in my brothers' footsteps and begin working at a minimum-wage job instead of going to college, or I could prove my mother

Bob Daemmrich/PhotoEdit

▲ Achieving a college degree has long been a part of the American Dream. An increasingly popular way to start this process is to begin at the community college level.

wrong by starting at community college to begin the path of building the lifestyle I wanted in the future.

College it was! . . . [After completing two years of college,] I can't explain in words exactly how I feel inside because I don't think words can describe the level of happiness I now feel. . . . I have completely changed as a person, daughter, friend, leader, employee, and most importantly, as a student since I decided to attend Glendale Community College. If I had let those Cal State rejections stop me from this path, I wouldn't be as educated, well-rounded, happy and the Sally Morgan I am today. I graduate with my associate's degree in a few weeks, and I was accepted to several Cal States. . . . Remember that you have the power to live the lifestyle you want.

—Sally Morgan (2010) describing how completing a community college education changed her life

Chapter Focus Question:

How do race, class, gender, and country of origin affect people's access to and opportunities in education?

Sally Morgan is one of millions of people who have attended a community college to gain knowledge and skills that will benefit them and enhance their opportunities in the future. In fact, a substantial proportion of all college students start their higher education at a community college.

From prekindergarten through postgraduate studies, education is one of the most significant social institutions in the United States and other high-income nations. Although most social scientists agree that schools are supposed to be places where people acquire knowledge and skills, not all of them agree on how a large number of factors—including class, race, gender, national origin, age, religion, and family background—affect individuals' access to educational opportunities or to the differential rewards that accrue at various levels of academic achievement.

In this chapter we discuss education as a key social institution and analyze some of the problems that affect contemporary elementary, secondary, and higher education. Before reading on, test your knowledge about U.S. education by taking the Sociology and Everyday Life quiz.

Highlighted Learning Objectives

- Explain how educational goals differ in various nations.
- Identify the key assumptions of functionalist, conflict, and symbolic interactionist perspectives on education.
- Describe the major problems faced by U.S. schools today.
- Discuss ways in which problems in higher education are linked to problems in the nation and world.

sociology and everyday life

How Much Do You Know About U.S. Education?

True	False	
T	F	1. Public education in the United States dates back more than 160 years.
T	F	2. Equality of opportunity is a vital belief in the U.S. educational system.
T	F	3. Each year in the United States, men earn more doctoral degrees than women do.
T	F	4. Although there is much discussion about gangs in schools, very few students report that gangs are present at their school.
T	F	5. Young people who bully are more likely to smoke, drink alcohol, and get into fights.
T	F	6. Every school day more than 7,000 U.S. students drop out of high school or leave, never to return.
T	F	7. In public schools in the United States, core classes such as history and mathematics were never taught in any language other than English before the 1960s civil rights movement.
T	F	8. The federal government has only limited control over how funds are spent by individual school districts because most of the money comes from the state and local levels.

Answers on page 464.

An Overview of Education

***Education* is the social institution responsible for the systematic transmission of knowledge, skills, and cultural values within a formally organized structure.** As a social institution, education imparts values, beliefs, and knowledge considered essential to the social reproduction of individual personalities and entire cultures (Bourdieu and Passeron, 1990). Education grapples with issues of societal stability and social change, reflecting society even as it attempts to shape it. Education serves an important purpose in all societies. At the microlevel, people must acquire the basic knowledge and skills they need to survive in society. At the macrolevel, the social institution of education is an essential component in maintaining and perpetuating the culture of a society across generations. ***Cultural transmission*—the process by which children and recent immigrants become acquainted with the dominant cultural beliefs, values, norms, and accumulated knowledge of a society**—occurs through informal and formal education. However, the process of cultural transmission differs in preliterate, preindustrial, and industrial nations.

The earliest education in *preliterate societies,* which existed before the invention of reading and writing, was informal in nature. People acquired knowledge and skills through ***informal education*—learning that occurs in a spontaneous, unplanned way**—from parents and other group members who provided information on survival skills such as how to gather food, find shelter, make weapons and tools, and get along with others. Formal education for elites first came into being in *preindustrial societies,* where few people knew how to read and write. ***Formal education* is learning that takes place within an academic setting such as a school, which has a planned instructional process and teachers who convey specific knowledge, skills, and thinking processes to students.** Perhaps the earliest formal education occurred in ancient Greece and Rome, where philosophers such as Socrates, Plato, and Aristotle taught elite males the skills required to become thinkers and orators who could engage in the art of persuasion (Ballantine and Hammack, 2012). During the Middle Ages the first colleges and universities were developed under the auspices of the Catholic church. In the Renaissance era the focus of education shifted to the importance of developing well-rounded and liberally educated people. With the rapid growth of industrial capitalism and factories during the Industrial Revolution, it became necessary for workers to have basic skills in reading, writing, and arithmetic, and pressure to provide formal education for the masses increased significantly. In the United States, Horace Mann started the free public school movement in 1848 when he declared that education should be the "great equalizer." By the mid-1850s, the process of mass education had begun in the United States as all states established free, tax-supported elementary schools that were readily available to children throughout the country. ***Mass education* refers to providing free, public schooling for wide segments of a nation's population.** As industrialization and bureaucratization intensified, managers and business owners demanded that schools educate students beyond the third or fourth grade so that well-qualified workers would be available for rapidly emerging "white-collar" jobs in management and clerical work.

Today, schools attempt to meet the needs of industrial and postindustrial society by teaching a wide diversity of students a myriad of topics, including history and science, computer skills, how to balance a checkbook, and how to avoid contracting AIDS. According to sociologists, many functions performed by other social institutions in the past are now under the auspices of the public schools.

Sociological Perspectives on Education

Sociologists have divergent perspectives on education in contemporary society. Here, we examine functionalist, conflict, symbolic interactionist, and postmodernist approaches to analyzing schooling.

Functionalist Perspectives

Functionalists view education as one of the most important components of society. According to Emile Durkheim, education is crucial for promoting social solidarity and stability in society: Education is the "influence exercised by adult generations on those that are not yet ready for social life" (Durkheim, 1956: 28). Durkheim asserted that moral values are the foundation of a cohesive social order and that schools are responsible for teaching a commitment to the common morality. In analyzing the values and functions of education, sociologists using a functionalist framework distinguish between manifest functions and latent functions, which are compared in ▶ Figure 16.1.

Manifest Functions of Education Some functions of education are *manifest functions*—previously defined as open, stated, and intended goals or consequences of activities within an organization or institution. Education serves six major manifest functions in society:

1. *Socialization.* From kindergarten through college, schools teach students the student role, specific academic subjects, and political socialization.
2. *Transmission of culture.* Schools transmit cultural norms and values to each new generation and play an active part in the process of assimilation of recent immigrants.
3. *Multicultural education.* Schools promote awareness of and appreciation for cultural differences so that students can work and compete successfully in a diverse society and a global economy.
4. *Social control.* Schools teach values such as discipline, respect, obedience, punctuality, and perseverance. Schools teach conformity by encouraging young people to be good students, conscientious future workers, and law-abiding citizens (see "Sociology Works!").
5. *Social placement.* Schools identify the most-qualified people to fill the positions available in society. As a result, students are channeled into programs based on individual ability and academic achievement. Graduates receive the appropriate credentials to enter the paid labor force.

© Kayte M. Deioma/PhotoEdit

▲ Some early forms of mass education took place in one-room schoolhouses such as the one shown here, where children in various grades were all taught by the same teacher. How do changes in the larger society bring about changes in education?

education the social institution responsible for the systematic transmission of knowledge, skills, and cultural values within a formally organized structure.

cultural transmission the process by which children and recent immigrants become acquainted with the dominant cultural beliefs, values, norms, and accumulated knowledge of a society.

informal education learning that occurs in a spontaneous, unplanned way.

formal education learning that takes place within an academic setting such as a school, which has a planned instructional process and teachers who convey specific knowledge, skills, and thinking processes to students.

mass education the practice of providing free, public schooling for wide segments of a nation's population.

sociology and everyday life

ANSWERS to the Sociology Quiz on U.S. Education

1. True. As far back as 1848, free public education was believed to be important in the United States because of the high rates of immigration and the demand for literacy so that the country would have an informed citizenry that could function in a democracy.

2. True. Despite the fact that equality of educational opportunity has not been achieved, it remains a goal of many people in this country. A large number subscribe to the belief that this country's educational system provides equal educational opportunities and that it is up to each individual to make the most of them.

3. False. In recent years, women have earned slightly more of the doctoral degrees conferred. Of the approximately 69,600 doctoral degrees conferred in 2010, for example, women earned 36,300 as compared to 33,400 earned by men. However, men earn a higher percentage of the doctoral degrees conferred in computer and information sciences, engineering, and the physical sciences, while women earn a larger percentage of doctoral degrees in education, health professions, and psychology (*Chronicle of Higher Education*, 2010).

4. False. Approximately 23 percent of students report the presence of gangs at their school (National Center for Educational Statistics. 2010a).

5. True. Government data show that young people who bully *are* more likely to engage in other problematic behavior, such as smoking, drinking alcohol, and getting into fights.

6. True. On average, at least 7,000 students leave high school every school day and do not return (Whitaker, 2010).

7. False. Late in the nineteenth century and early in the twentieth century, some classes were conducted in Italian, Polish, German, and other languages of recent immigrants.

8. True. Most funding for public education comes from state and local property taxes, and similar sources of revenue.

Manifest functions—open, stated, and intended goals or consequences of activities within an organization or institution. In education, these are:

- socialization
- transmission of culture
- multicultural education
- social control
- social placement
- change and innovation

Latent functions—hidden, unstated, and sometimes unintended consequences of activities within an organization. In education, these include:

- matchmaking and production of social networks
- restricting some activities
- creating a generation gap

▲ **FIGURE 16.1 MANIFEST AND LATENT FUNCTIONS OF EDUCATION**

sociology works!

Bullying: Is Everyone Really Doing It?

Bullying, particularly in adolescence, is epidemic, not just in the USA but around the world.

—Marvin Berkowitz, professor of character education at the University of Missouri–St. Louis (qtd. in mindoh .com, 2007)

Bullying is aggressive behavior that is intentional and that involves an imbalance of power or strength. Bullying is often repeated, and the person who is routinely the object of the bullying often has a difficult time defending himself or herself (U.S. Department of Health and Human Services, 2008b).

How common is bullying? Some studies have found that as many as eight out of ten U.S. students in upper elementary, middle, and high schools (who have access to an anonymous online messaging service that provides them with the ability to communicate the problem to school counselors and administrators) report that they have been the victims of bullying (Gary, 2007). Other studies have shown that between 15 to 25 percent of U.S. students are bullied with some frequency and that 15 to 20 percent admit that they bully others with some frequency during the school year (U.S. Department of Health and Human Services, 2008b). Although the nature and extent of this problem are unknown, experts agree on one thing: Bullying can and must be addressed and reduced in schools.

What methods work best for eliminating negative behaviors and maintaining social control in schools? One method of reducing bullying is character education, which focuses on the reinforcement of acceptable character traits, such as respect and responsibility. From character-education courses and online studies, students learn problem-solving techniques, conflict-resolution approaches, and communication skills to help them interact with one another in a more positive manner.

However, an alternative approach has been suggested by the sociologist H. Wesley Perkins, who emphasizes that the best practice in bullying prevention and intervention is to focus on the *social environment* of the school. According to Perkins, to reduce bullying it is important to change the climate of the school and social norms regarding bullying so that it will become "uncool" for one individual to bully another person, but "cool" for other individuals to help out persons who are bullied. Based on previous research regarding college students' use of alcohol, Perkins developed a social norms approach to preventing bullying. From this approach, the first step to reducing the problem of bullying is to correct students' misperceptions about how prevalent this type of behavior really is: If students think that bullying is very common in their school and that "everybody is doing it," they may believe that the social norms support such behavior. If this misperception can be corrected, bullying will decrease as students come to see that the "nonbully status" is the social norm (Teicher, 2006).

Emphasizing the social environment of education and its norms is an important sociological contribution to our understanding of both positive and negative actions that take place in schools. It also makes us aware that social institutions fulfill a variety of important functions in society and that one of those functions for education is to maintain social control and to encourage civility among participants. Bullying isn't just a matter of "kids being kids": It is a serious sociological problem and must be dealt with as such.

▲ Sociologist H. Wesley Perkins sees the problem of bullying as an issue involving norms: If bullying becomes uncool, young people will be less likely to take part in it and more likely to help someone who is a victim of this very harmful activity. This scene from Fox television's hit show *Glee* illustrates Perkins's theory.

reflect & analyze

Have you encountered bullying at your school or elsewhere? How can sociology help us to reduce problems such as this?

framing education in the media

"A Bad Report Card"—How the Media Frame Stories About U.S. Schools

Sample Headlines:

- Test Scores Lag Behind Rising Grades
- Failing Science
- Most Students in Big Cities Lag Badly in Basic Science
- Students Ace State Tests, but Earn D's from U.S.

Headlines such as these are used by the media in reporting the results of national exams in the United States. Newspapers, Internet news sites, and television newscasts often use *report card framing* such as this in articles and news segments that describe students' academic achievement based on standardized tests in reading, science, and mathematics. Journalists use test scores to "grade" students and schools in much the same way that educators use report cards to inform students and their parents how the students are doing.

What is report card framing? As discussed in previous chapters, media framing refers to the manner in which reporters package information before it is presented to an audience. In regard to education reporting, journalists decide which key features or points to highlight in their coverage and which to minimize or leave out. For reporters on the "education beat," nationwide testing results are among the most important thing happening in schooling—and among the easiest subjects for an article—because journalists receive numerous press releases about the standardized tests and may be invited to a press conference by the Department of Education or other federal agencies to hear a discussion of the latest trends in test scores. For example, a downward trend in test scores is particularly bad news for schools, placing teachers and students under more pressure to bring up scores on future exams.

As compared to report card framing, which gives teachers, students, and schools "passing" or "failing" grades based on standardized test scores, the most frequent framing of media stories about education focuses on funding and wage-and-benefit disputes. If we borrow the title of a popular ABBA song to describe this type of approach, *"Money, Money, Money" framing* focuses on

6. *Change and innovation.* Schools are a source of change and innovation to meet societal needs. Faculty members are responsible for engaging in research and passing on their findings to students, colleagues, and the general public.

© Enigma/Alamy

▲ What values are these schoolchildren being taught? Is there a consensus about what today's schools should teach? Why or why not?

Latent Functions of Education Education serves at least three *latent functions,* which we have previously defined as hidden, unstated, and sometimes unintended consequences of activities within an organization or institution:

1. *Restricting some activities.* States have *mandatory education laws* that require children to attend school until they reach a specified age (usually age sixteen) or complete a minimum level of formal education (generally the eighth grade). Out of these laws grew one latent function of education: keeping students off the streets and out of the full-time job market until they are older.
2. *Matchmaking and production of social networks.* Because schools bring together people of similar ages, social class, and race/ethnicity, young people often meet future marriage partners and develop lasting social networks.
3. *Creation of a generation gap.* Students learn information and develop technological skills that may create a generation gap between them and their parents, particularly as the students come to embrace a newly acquired perspective.

Dysfunctions of Education Functionalists acknowledge that education has certain dysfunctions. Some analysts argue that U.S. education is not

how much more money is needed for education than is now available, where the money is going to come from, and who is most likely to benefit from revenues that are expended. Most reporters' sources for this type of framing are government officials or public-school-affiliated sources that have a vested interest in how the media report on these issues because such reports influence policy makers and public opinion regarding the appropriate allocation of money for public schools.

A final type of framing, *students-as-human-interest-story framing,* is one less frequently used by the media because it involves reporting on education from an inside-the-classroom view of what is actually going on in schools. About the closest that most reporters get to this type of coverage is highlighting a special student who has overcome great obstacles to accomplish a goal or gain a diploma. An example would be a student who overcame homelessness and did not drop out of school because parents, friends, and coaches inspired the student to finish high school.

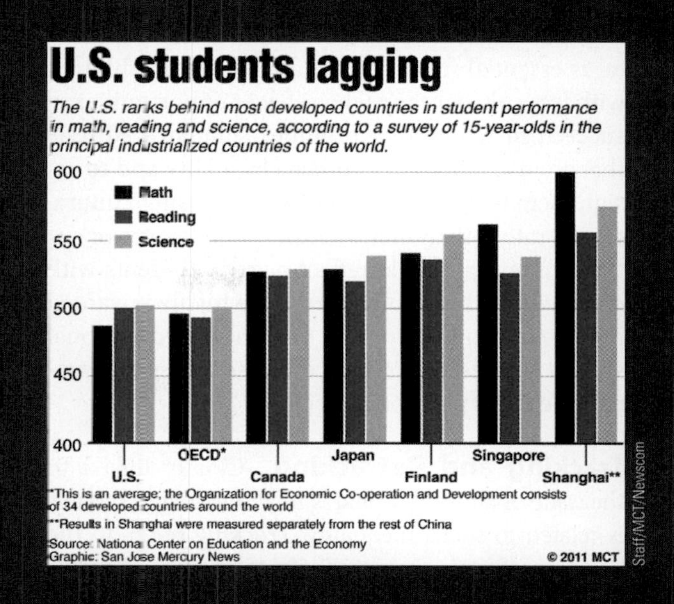

Staff/MCT/Newscom

reflect & analyze

Should the media expand their reporting on education in the United States? Do some reporters miss out on what's really happening in classrooms across the nation because they primarily rely on press releases from officials to set the agenda for reporting on education? What do you think?

promoting the high-level skills in reading, writing, science, and mathematics that are needed in the workplace and the global economy. For example, mathematics and science education in the United States does not compare favorably with that found in many other industrialized countries. Are U.S. schools dysfunctional as a result of lower test scores? Analysts do not agree on what the exam score differentials mean. For many functionalist thinkers, lagging test scores are a sign that dysfunctions exist in the nation's educational system. (To see how the media tend to treat the subject of standardized tests, see "Framing Education in the Media.") According to this approach, improvements will occur only when more-stringent academic requirements are implemented for students and when teachers receive sufficient training. Overall, functionalists typically advocate the importance of establishing a more rigorous academic environment in which students are required to learn the basics that will make them competitive in school and job markets.

Conflict Perspectives

Conflict theorists emphasize that schools solidify the privileged position of some groups at the expense of others by perpetuating class, racial–ethnic, and gender inequalities (Ballantine and Hammack, 2012). Contemporary conflict theorists also focus on how politics and corporate interests dominate schools, particularly higher education.

Cultural Capital and Class Reproduction

Although many factors—including intelligence, motivation, and previous accomplishments—are important in determining how much education a person will attain, conflict theorists argue that access to quality education is closely related to social class. From this approach, education is a vehicle for reproducing existing class relationships. According to the French sociologist Pierre Bourdieu, the school legitimates and reinforces the social elites by engaging in specific practices that uphold the patterns of behavior and the attitudes of the dominant class. Bourdieu asserts that students from diverse class backgrounds come to school with different amounts of ***cultural capital*—social assets that include values, beliefs, attitudes, and competencies in language and culture** (Bourdieu and Passeron, 1990). Cultural capital involves "proper" attitudes toward education, socially approved dress and manners, and knowledge about books, art, music, and

cultural capital Pierre Bourdieu's term for people's social assets, including values, beliefs, attitudes, and competencies in language and culture.

other forms of high and popular culture. Middle- and upper-income parents endow their children with more cultural capital than do working-class and poverty-level parents. Because cultural capital is essential for acquiring an education, children with less cultural capital have fewer opportunities to succeed in school. For example, standardized tests that are used to group students by ability and to assign them to classes often measure students' cultural capital rather than their "natural" intelligence or aptitude. Thus, a circular effect occurs: Students with dominant cultural values are more highly rewarded by the educational system. In turn, the educational system teaches and reinforces those values that sustain the elite's position in society.

Tracking and Detracking Closely linked to the issue of cultural capital is how tracking in schools is related to social inequality. ***Tracking* refers to the practice of assigning students to specific curriculum groups and courses on the basis of their test scores, previous grades, or other criteria.** Conflict theorists believe that tracking seriously affects many students' educational performance and their overall academic accomplishments. Tracking first came into practice in the early twentieth century, when a large influx of immigrant children entered U.S. schools for the first time and were sorted by ability and past performance. In elementary schools, tracking is often referred to *ability grouping* and is based on the assumption that it is easier to teach a group of students who have similar abilities. However, class-based factors also affect which children are most likely to be placed in "high," "middle," or "low" groups, often referred to by such innocuous terms as "Blue Birds," "Red Birds," and "Yellow Birds." This practice is described by the journalist Ruben Navarrette, Jr. (1997: 274–275), who tells us about his own experience with tracking:

> One fateful day, in the second grade, my teacher decided to teach her class more efficiently by dividing it into six groups of five students each. Each group was assigned a geometric symbol to differentiate it from the others. There were the Circles. There were the Squares. There were the Triangles and Rectangles.
>
> I remember being a Hexagon. . . . The Hexagons were the smartest kids in the class. These distinctions are not lost on a child of seven. . . . And on the day on which we were assigned our respective shapes, we knew that our teacher knew, too. As Hexagons, we would wait for her to call on us, then answer by hurrying to her with books and pencils in hand. We sat around a table in our "reading group," chattering excitedly to one another and basking in the intoxication of positive learning. We did not notice, did not care to notice, over our shoulders, the frustrated looks on the faces of Circles and Squares and Triangles who sat quietly at their desks, doodling on scratch paper or mumbling to one another. We knew also that, along with our geometric shapes, our books were different and that each group had different amounts of work to do. . . . The Circles had the easiest books and were assigned to read only a few pages at a time. . . . Not surprisingly, the Hexagons had the most difficult books of all, those with the biggest words and the fewest pictures, and we were expected to read the most pages.
>
> The result of all of this education by separation was exactly what the teacher had imagined that it would be: Students could, and did, learn at their own pace without being encumbered by one another. Some learned faster than others. Some, I realized only [later], did not learn at all.

As Navarrette suggests, tracking does make it possible for students to work together based on their perceived abilities and at their own pace; however, it also extracts a serious toll from students who are labeled as "underachievers" or "slow learners." Race, class,

▲ Children who are able to visit museums, libraries, and musical events may gain cultural capital that other children do not possess. What is cultural capital? Why is it important in the process of class reproduction?

© Gabe Palmer/Alamy

◀ As Ruben Navarrette, Jr., so powerfully describes, school is extremely tedious for underachieving students, who may find themselves "tracked" in such a way as to deny them upward mobility in the future.

language, gender, and many other social categories may determine the placement of children in elementary tracking systems as much as or more than their actual academic abilities and interests.

The practice of tracking continues in middle school/junior high and high school. Although schools in some communities bring together students from diverse economic and racial/ethnic backgrounds, the students do not necessarily take the same courses or move on the same academic career paths (Gilbert, 2010). Today, extreme forms of tracking are relatively rare. Most tracking involves grouping students by ability within subjects so that they are assigned to advanced, regular, or basic courses depending on their past performance.

The detracking movement has influenced some schools and teachers, stressing that students should be deliberately placed in classes of mixed ability. An important benefit of detracking is closing the achievement gap among students based on class or race; however, detracking is a major concern to parents of high-achieving students because they believe their children should have classes that maximize their potential, rather than hold them back with less-able or less-talented students. According to the sociologist Maureen Hallinan (2005), rather than tracking students, schools should provide more-engaging lessons for all students, alter teachers' assumptions about students, and raise students' performance requirements. Eventually, independent studies and technologies such as online learning may eliminate the "tracking" debate.

The Hidden Curriculum Another concern of conflict theorists is the ***hidden curriculum*—the transmission of cultural values and attitudes, such as conformity and obedience to authority, through implied demands found in the rules, routines, and regulations of schools** (Snyder, 1971). In other words, students pick up on subtle messages about attitudes, beliefs, values, and behavior that are either "appropriate" or "inappropriate" from teachers and other school personnel. These messages are not part of the official curriculum or the school's mission to educate students for the future.

© Spencer Grant/PhotoEdit

▲ Signs in this elementary classroom list the rules, rewards, and consequences of different types of student behavior. According to conflict theorists, schools impose rules on working-class and poverty-level students so that they will learn to follow orders and to be good employees in the workplace. How would functionalists and symbolic interactionists interpret these same signs?

tracking the assignment of students to specific curriculum groups and courses on the basis of their test scores, previous grades, or other criteria.

hidden curriculum the transmission of cultural values and attitudes, such as conformity and obedience to authority, through implied demands found in rules, routines, and regulations of schools.

Although all students are subjected to the hidden curriculum, students who are from low-income families and/or are African American or Hispanic (Latino/a) may be affected the most adversely by educational settings that have been established on the basis of upper- and middle-class white (non-Hispanic) values, attitudes, and behavior (AAUW, 2008). When teachers from middle- and upper-middle-class backgrounds instruct students from lower-income families, the teachers often establish a more structured classroom and a more controlling environment for students. These teachers may also have lower expectations for students' academic achievements. Schools with many students from low-income families often emphasize procedures and rote memorization without focusing on decision making and choice, or on providing explanations of why something is done a particular way. Schools for middle-class students stress the processes (such as figuring and decision making) involved in getting the right answer. Schools for affluent students focus on creative activities in which students express their own ideas and apply them to the subject under consideration, as well as building students' analytical and critical-thinking skills.

Over time, students become frustrated with the educational system and drop out or become very marginal students, making it even more difficult for them to attend college and gain the appropriate credentials for gaining better-paying jobs. Educational credentials are extremely important in a nation such as ours that emphasizes ***credentialism*—a process of social selection in which class advantage and social status are linked to the possession of academic qualifications** (Collins, 1979; Marshall, 1998). Credentialism is closely related to *meritocracy,* a social system in which status is assumed to be acquired through individual ability and effort (Young, 1994/1958). Persons who acquire the appropriate credentials for a job are assumed to have gained the position through what they know, not who they are or whom they know. According to conflict theorists, the hidden curriculum determines in advance that the most valued credentials will primarily stay in the hands of the middle and upper classes, so the United States is not actually as meritocratic as some might claim.

The hidden curriculum is also related to gender bias. For many years the focus in education was on how gender bias harmed girls and women: Reading materials, classroom activities, and treatment by teachers and peers contributed to a feeling among many girls and young women that they were less important than male students. The accepted wisdom was that, over time, differential treatment undermines females' self-esteem and discourages them from taking certain courses, such as math and science, that have been dominated by male teachers and students. In the early 1990s the American Association of University Women issued *The AAUW Report: How Schools Shortchange Girls,* which highlighted inequalities in women's education and started a national debate on gender equity (AAUW, 1995). Since this report was published, improvements have occurred in girls' educational achievement, as females have attended and graduated from high school and college at a higher rate than their male peers. More females have enrolled in advanced-placement or honors courses and in academic areas, such as math and science, where they had previously lagged (AAUW, 2008).

Ironically, after many years of discussion about how the hidden curriculum and other problems in schools served to disadvantage female students, the emphasis has now shifted to the question of whether girls' increasing accomplishments from elementary school to college and beyond have come at the expense of boys and young men. But this is not true, according to research by the AAUW (2008: 2): "Educational achievement is not a zero-sum game, in which a gain for one group results in a corresponding loss for the other. If girls' success comes at the expense of boys, one would expect to see boys' scores decline as girls' scores rise, but this has not been the case."

Regardless of gender, large differences remain in scores on academic tests among students by race/ethnicity. Studies have shown that white children are more likely to graduate from high school and college than are their African American and Hispanic peers. Likewise, children from higher-income families are more likely to graduate from high school than are children from lower-income families, who are also less likely to attend college (AAUW, 2008).

The conflict theorists' focus on the hidden curriculum calls our attention to the fact that students learn far more—both positively and negatively—than just the subject matter being taught in the classroom. Students are exposed to a wide range of beliefs, values, attitudes, and behavioral expectations that are not directly related to specific subject matter.

Symbolic Interactionist Perspectives

Unlike functionalist analysts, who focus on the functions and dysfunctions of education, and conflict theorists, who focus on the relationship between education and inequality, symbolic interactionists focus on classroom communication patterns and educational practices, such as labeling, which affect students' self-concept and aspirations.

Labeling and the Self-Fulfilling Prophecy

According to symbolic interactionists, the process of labeling is directly related to the power and status of those persons who do the labeling and those who are being labeled. Chapter 7 explains that *labeling* is the process whereby others identify a person as possessing a specific characteristic or exhibiting a certain pattern of behavior (such as being deviant). In schools, teachers and administrators are empowered to label children in various ways, including grades, written comments on classroom behavior, and placement in classes. For some students, labeling amounts to a *self-fulfilling prophecy*—an unsubstantiated belief or prediction resulting in behavior that makes the originally false belief come true (Merton, 1968).

A classic form of labeling and the self-fulfilling prophecy has occurred for many years through the use of various IQ (intelligence quotient) tests, which claim to measure a person's inherent intelligence, apart from any family or school influences on the individual. Schools have used IQ tests as one criterion in determining student placement in classes and ability groups (see ▶ Figure 16.2). The way in which IQ test scores may become a self-fulfilling prophecy was revealed in the 1960s when two social scientists conducted an experiment in an elementary school during which they intentionally misinformed teachers about the scores of students in their classes (Rosenthal and Jacobson, 1968). Although the students had no measurable differences in intelligence, the researchers informed the teachers that some of the students had extremely high IQ test scores whereas others had average to below-average scores. As the researchers observed, the teachers began to teach "exceptional" students in a different manner from other students. In turn, the "exceptional" students began to outperform their "average" peers and to excel in their classwork. This study called attention to the labeling effect of IQ scores.

Is IQ a good indicator of a person's potential? In their controversial book *The Bell Curve: Intelligence and Class Structure in American Life*, Richard J. Herrnstein and Charles Murray (1994) argue that intelligence is genetically inherited and that people cannot be "smarter" than they are born to be, regardless of their environment or education. According to Herrnstein and Murray, certain racial–ethnic groups differ in average IQ and are likely to differ in "intelligence genes" as well. Herrnstein and Murray claimed the people with lower intelligence are more likely to commit crimes, drop out of school, and live in poverty. In contrast, people with high intelligence are more likely to be successful. Many scholars disagree with Herrnstein and Murray's conclusions, pointing out that what these authors claim to be immutable intelligence is actually acquired skills (Weinstein, 1997).

In 2008 the British psychologist Richard Lynn's *The Global Bell Curve: Race, IQ, and Inequality Worldwide* expanded the ideas of Herrnstein and Murray to include the nations of the world. According to Lynn, in multiracial nations, people of Jewish and East Asian ancestry have the highest average IQ scores and socioeconomic positions, followed by whites, South Asians, Hispanics, and people of African descent. Lynn attributes people's positions in the socioeconomic hierarchy to differences in intelligence on the basis of race/ethnicity. Today, so-called IQ fundamentalists continue to label students and others on the basis of IQ tests, claiming that these tests measure some identifiable trait that predicts the quality of people's thinking and their ability to perform. Critics of IQ tests continue to argue that these exams measure a number of factors—including motivation, home environment, type of socialization at home, and quality of schooling—not intelligence alone (Yong, 2011).

Question 2: Consider the following two statements: all farmers who are also ranchers cannot come near town; and most of the ranchers who are also farmers cannot surf. Which of the following statements MUST be true?

- ❍ Most of the farmers who cannot come near town can surf
- ❍ Only some farmers who ranch can surf near town
- ❍ A surfer who ranches and farms cannot surf near town
- ❍ Some ranchers who farm can come to town to learn to surf
- ❍ Any farmer who cannot surf also ranches

▲ **FIGURE 16.2 IQ TEST SAMPLE QUESTION**
IQ tests containing items such as this are often used to place students in ability groups. Such placement can set the course of a person's entire education.

Postmodernist Perspectives

As discussed in Chapter 15, postmodern theories highlight *difference* and *irregularity* in society. How might a postmodern approach describe higher education? One of the major postmodern theorists is Jean-Francois Lyotard (1984), who described how knowledge has become a commodity that is exchanged between producers and consumers. "Knowledge" is now an automated database, and teaching and learning are primarily about data presentation, stripped of their former humanistic and spiritual associations.

In the postmodern era an emphasis in higher education is on how to make colleges and universities more efficient and how to bring these institutions into the service of business and industry. A major

credentialism a process of social selection in which class advantage and social status are linked to the possession of academic qualifications.

[concept quick review]

Sociological Perspectives on Education

	Key Points
Functionalist Perspectives	Education is one of the most important components of society: Schools teach students not only content but also to put group needs ahead of the individual's.
Conflict Perspectives	Schools perpetuate class, racial–ethnic, and gender inequalities through what they teach to whom.
Symbolic Interactionist Perspectives	Labeling and the self-fulfilling prophecy are an example of how students and teachers affect each other as they interpret their interactions.
Postmodernist Perspectives	Knowledge has become a commodity, and students and their parents are consumers of education in the twenty-first century.

objective is looking for the best way to transform these schools into corporate entities such as the "McUniversity." According to the sociologist George Ritzer (1998), "McUniversity" refers to a means of educational consumption that allows students to consume educational services and eventually obtain "goods" such as degrees and credentials:

> Students (and often, more importantly, their parents) are increasingly approaching the university as consumers; the university is fast becoming little more than another component of the consumer society. . . . Parents are, if anything, likely to be even more adept as consumers than their children and because of the burgeoning cost of higher education more apt to bring a consumerist mentality to it. (Ritzer, 1998: 151–152)

© Mike Booth/Alamy

▲ In the past, college cafeterias were simply places for students to eat. Today, however, at many schools the quality of the food has greatly improved, and the number of selections has increased. Campus restaurants and food courts have come to be seen as "amenities," something that can be used to help colleges recruit—and retain—new students.

Savvy college and university administrators are aware of the permeability of higher education and the "students-as-consumers" model. To attract new students and enhance current students' opportunities for consumption, most campuses have amenities such as food courts, ATMs, video games, Olympic-sized swimming pools, and massive rock-climbing walls. "High-tech" or "wired" campuses are also a major attraction for student consumers, and virtual classrooms make it possible for some students to earn college credit without having to look for a parking place at the traditional brick-and-mortar campus. Based on a postmodern approach, what do you believe will be the dominant means by which future students will consume educational services and goods at your college or university?

The Concept Quick Review summarizes the major theoretical perspectives on education.

Issues in Elementary and Secondary Schools

Education in kindergarten through high school is a microcosm of many of the concerns facing the United States. The issues we examine in this section include unequal funding of public schools, dropout rates, racial segregation and resegregation, equalizing opportunities for students with disabilities, schools vouchers, charter schools, home schooling, and school safety.

Unequal Funding of Public Schools

Why does unequal funding exist in public schools? Most educational funds come from state legislative appropriations and local property taxes (see ▶ Figure 16.3). As shown in Figure 16.3, state sources contribute less than half of public elementary–secondary school system revenue, and the rest comes from local sources and the federal

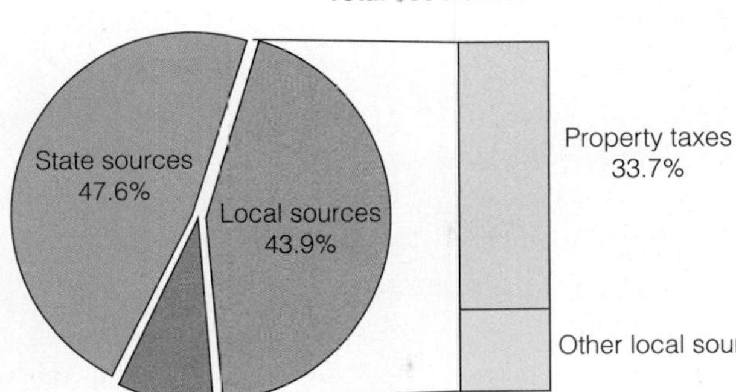

▶ **FIGURE 16.3 PERCENTAGE DISTRIBUTION OF TOTAL PUBLIC ELEMENTARY-SECONDARY SCHOOL SYSTEM REVENUE: 2006–2007**

Source: nces.ed.gov, 2010.

Dan Thornberg/iStockphoto.com

Steve Hebert/The New York Times/Redux Pictures

▲ "Rich" schools and "poor" schools are readily identifiable by their buildings and equipment. What are the long-term social consequences of unequal funding for schools?

government. In 2006–2007, for example, states contributed slightly less than 48 percent of total public school revenue, as compared to nearly 44 percent from local sources and 8.5 percent from the federal government. Much of the money from federal sources is earmarked for special programs for students who are disadvantaged (e.g., the Head Start program) or who have a disability. As shown in ▶ Map 16.1, expenditures per pupil for public and secondary public schools vary widely from state to state (Ballantine and Hammack, 2012).

School Dropouts

High dropout rates are a major problem facing contemporary schools. Dropout rates are computed in various ways, but one of the most telling is the *status dropout rate*—the percentage of people in a specific age range who are not currently enrolled in high school and who do not have a high school degree or its equivalent. In recent years slightly more than 3 million 16- to 24-year-olds were not enrolled in high school and had not earned a high school diploma or its equivalent.

Status dropout rates vary by gender, race/ethnicity, and region of the country (see ▶ Figure 16.4). Males (8.5 percent) have a higher status dropout rate than females (7.5 percent). Status dropout rates also vary by race/ethnicity: Hispanics/Latinos/as (18.3 percent), American Indian/Alaska native (14.6 percent), and blacks/African Americans (9.9 percent) have higher status dropout rates than whites (4.8 percent), Asian/Pacific Islanders (4.4 percent), and persons reporting two or more races (4.2 percent) (National Center for Educational Statistics, 2010b). Finally, region is also an issue in status dropout rates: The Northeastern United States has the lowest status dropout rates, while the South and West have the highest.

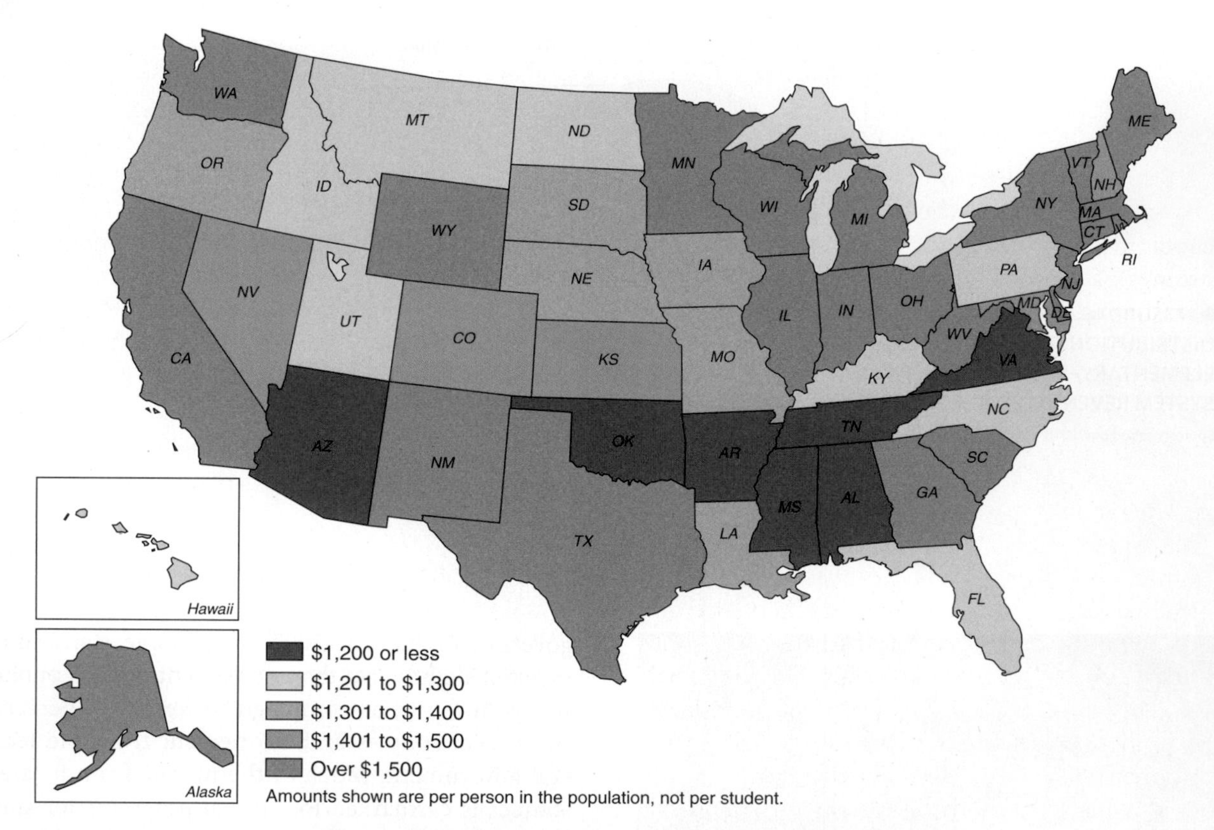

▲ **MAP 16.1 PER CAPITA PUBLIC ELEMENTARY AND SECONDARY SPENDING BY STATE**

Source: U.S. Census Bureau, 2008.

Using a second approach to determine dropout rates, the *event dropout rate*—which estimates the percentage of both public and private high school students who left high school between the beginning of one school year and the beginning of the next without earning a high school diploma or an alternative credential such as a GED—we find that every school day, at least 7,000 U.S. students (on average) leave high school and never return. What this means is that, on average, 3.5 percent of students who were enrolled in public or private high schools in October 2007 left school before October 2008 without completing a high school program. Perhaps the most telling statistic when using the event dropout rate is that students living in low-income families are about four-and-one-half times more likely to drop out in any given year than students living in high-income families (National Center for Educational Statistics, 2010b).

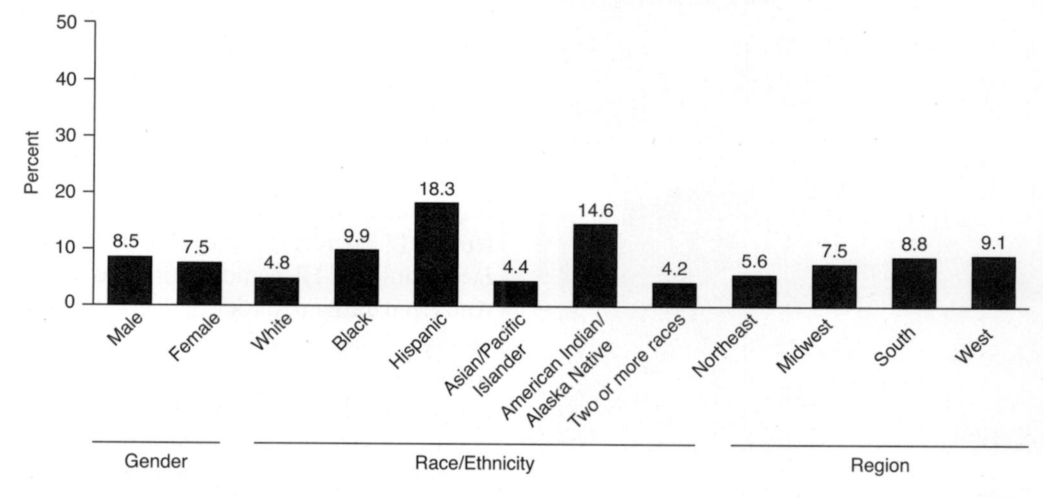

▲ **FIGURE 16.4 STATUS DROPOUT RATES FOR 16- to 24-YEAR-OLDS, BY RACE/ETHNICITY, GENDER, AND RELIGION**

Source: National Center for Educational Statistics, 2010b.

What is most important about dropping out is the economic and social consequences for individuals and nations. Young people without a high school education have a difficult time finding work that will pay a living wage: The median income of persons age 18 through 67 who had not completed high school (in 2008) was about $23,000, as compared to earnings of approximately $42,000 for those with at least a high school credential (including a GED certificate). This income gap adds up to a loss of approximately $630,000 over a lifetime. A higher percentage of dropouts (age 25 and over) are also unemployed or holding temporary or part-time jobs while seeking full-time work. Cities and states suffer because tax revenues are lower when many people are unemployed, and the societal costs for public assistance, crime control, and health care are higher (National Center for Educational Statistics, 2010b).

Why do students drop out of school? Some students believe that their classes are boring; others are skeptical about the value of schooling and think that completing high school will not increase their job opportunities. Upon leaving school, many dropouts have high hopes of making money and enjoying newfound freedom; however, many find that few jobs are available and that they do not have the minimum education required for any "good" jobs that exist.

Racial Segregation and Resegregation

Although some people believe that the issue of racial segregation has long been solved in America's schools, in many areas of the United States schools remain racially segregated or have become resegregated. In 1954 the U.S. Supreme Court ruled (in *Brown v. Board of Education of Topeka, Kansas*) that "separate but equal" segregated schools are unconstitutional because they are inherently unequal. Today, racial segregation remains a fact of life in education: Efforts to bring about *desegregation*—the abolition of legally sanctioned racial–ethnic segregation—or *integration*—the implementation of specific action to change the racial–ethnic and/or class composition of the student body—have failed in many districts. Some school systems have bused students across town to achieve racial integration. Others have changed school attendance boundaries or introduced magnet schools with specialized programs such as science or the fine arts to change the racial/ethnic composition of schools. But school segregation does not exist in isolation: Racially segregated housing patterns contribute to high rates of school segregation.

David Grossman/Alamy

▲ Although many people believe that the United States is a racially integrated nation, a look at schools throughout the country reveals that many of them remain segregated or have become largely resegregated in recent decades.

Resegregation is also an issue because some school districts have abandoned programs that had produced greater racial integration in local schools. Raleigh, North Carolina, is a case in point: A local school board decided to end consideration of race and socioeconomic status in determining school assignments and stopped the district's busing-for-diversity program. Those who opposed this return to the "neighborhood school" concept argued that resegregation would quickly occur throughout the district (Mooney, 2011).

How segregated are U.S. schools? Here are a few facts: More than half of all African American public school students in Illinois, Michigan, and New York State attend predominantly black schools. In Maryland, Alabama, Mississippi, Tennessee, Georgia, and Texas, approximately 30 percent of African American public school students attend schools that have at least a 95 percent black/African American population. Children of color now constitute more than half of public school students in a number of states because white (non-Hispanic) children are more often enrolled in charter schools, suburban school districts, or private schools with a high percentage of white students (Mack, 2010).

Equalizing Opportunities for Students with Disabilities

Another concern in education has been how to provide better educational opportunities for students with a disability—any physical and/or mental condition that limits students' access to, or full involvement in, school life. Along with other provisions, the Americans with Disabilities Act of 1990 requires schools to make their facilities, services, activities, and programs accessible to people with disabilities. Many schools have attempted to *mainstream* children with disabilities by providing *inclusion programs,* under which the special education curriculum is integrated with the regular education program and each child receives an *individualized education plan* that provides annual educational goals. *Inclusion* means that children with disabilities work with a wide variety of people; over the course of a day, children may interact with their regular education teacher, the special education teacher, a speech therapist, an occupational therapist, a physical therapist, and a resource teacher, depending on the child's individual needs. Although much remains to be done, recent measures to enhance education for children with disabilities have increased the inclusion of many young people who were formerly excluded or marginalized in the educational system.

School Choice and School Vouchers

School choice is a persistent issue in education. Much of the discussion about school choice focuses on school voucher programs in which public funds (tax dollars) are provided to parents so that they can pay their child's tuition at a private school of their choice. Many parents praise the voucher system because it provides them with options for schooling their children. Some political leaders applaud vouchers and other school choice policies for improving public school performance. However, voucher programs are controversial: Some critics believe that giving taxpayer money to parents so that they can spend it at private (often religious) schools violates constitutional requirements for the separation of church and state. However, the U.S. Supreme Court ruled in *Zelman v. Simmons-Harris,* a case involving a Cleveland, Ohio, school district, that voucher policies are constitutional because parents have a choice and are not required to send their children to church-affiliated schools. Other critics claim that voucher programs are less effective in educating children than public schools. According to studies in the District of Columbia, Milwaukee, and Cleveland, public school students outperformed voucher students in both reading and math on state proficiency tests; however, neither group reached state proficiency requirements (Ott, 2011). In sum, advocates like the choice factor in voucher programs while critics believe that vouchers undermine public education, lack accountability, and may contribute to the collapse of the public school system. The debate over school vouchers is further discussed in "Sociology and Social Policy."

Nivek Neslo/Taxi/Getty Images

▲ Computers make it possible for students with a disability to gain educational opportunities previously unavailable to them. Specialized software allows this hearing-impaired student to complete her assignment.

Charter Schools and "For Profit" Schools

Charter schools (or "schools of choice") are primary or secondary schools that receive public money but are free from some of the day-to-day bureaucracy of a larger school district that may limit classroom performance. These schools operate under a charter contract negotiated by the school's organizers (often parents or teachers) and a sponsor (usually a local school board, a state board of education, or a university) that oversees the provisions of the contract. Some school districts "contract out" by hiring for-profit companies on a contract basis to manage charter schools, but the schools themselves are nonprofit. Among the largest educational management organizations are Imagine Schools, National Heritage Academies, the Leona Group, EdisonLearning, White Hat Management, and Mosaica Education (Molnar, Miron, and Urschel, 2010).

Charter schools provide more autonomy for individual students and teachers, and can serve an important function in education: A large number of minority students receive a higher-quality education than they would in the public schools (see Hardy, 2010; Winters, 2010). Charter schools attempt to

sociology and social policy

Are Private School Vouchers a Good Idea?

One of the core American values is equality—at least equality of *opportunity*, an assumed equal opportunity to achieve success. If it is possible that a child could get a better education at a private school than at a public school, should the government provide that child's family with the resources to send the child to a private school in order to have the same opportunity as a child from a wealthier family? Should the government provide that funding even if the private school teaches specific religious beliefs as part of the academic curriculum?

The original idea for school choice was to provide a voucher (equal in value to the average amount spent per student by the local school district) to the family of any student who left the public schools to attend a private school. The private school could exchange the voucher for that amount of money and apply it to the student's tuition at the private school; the school district would save an equivalent amount of money as a result of the student not enrolling in the public schools. Between the origins of vouchers in 1955 and today, variations have been proposed, such as allowing vouchers to be used for transfers between public schools in the same school district, transfers from one public school district to another, transfers only to schools with no religious connections, and vouchers for use only by students from low-income families.

The Supreme Court's decision in the Cleveland, Ohio, case shifted the battle regarding voucher programs from a legal argument based on a constitutional question to a political debate involving social policy and politics. Advocates of the Cleveland program argued that the Court's decision was good because children from lower-income families could have a wider range of educational choices, a range similar to that available to children from middle-income families. They asserted that competition for students' vouchers would improve public education, force school administrators and teachers to perform at a higher level, and produce greater competencies in students. By contrast, opponents argued that public education would not be able to withstand the loss of funds and the "cream skimming" of the brightest students from public to private schools.

What major social policy issues surround voucher systems? Many people strongly believe that using public tax dollars for funding private, religiously based schools violates the separation of church and state. Despite the Court's decision, they will continue to make that argument, whether on constitutional grounds or on the basis of public policy. Ultimately, from a social policy standpoint, the future of school vouchers may depend on the willingness (or unwillingness) of political leaders to authorize vouchers and the willingness of voters to pay for them. At the bottom line, continuing debates about school choice and voucher programs distract us from real issues of education reform and questions of how to improve schools for students. How social issues are framed is a significant factor in shaping not only the future of vouchers but also of U.S. education.

reflect & analyze

Should public funds be used to send children to private schools in order for them to have the same educational opportunities as children from more-affluent families? Should public funds be used for vouchers even if the private school teaches specific religious beliefs as part of the academic curriculum?

maintain an organizational culture that motivates students and encourages achievement rather than having a negative school environment where minority students are ridiculed for "acting white" or making good grades. Some charter schools offer college-preparatory curriculums and help students of color achieve their goal of enrolling in the college or university of their choice (Hardy, 2010).

Charter schools have numerous challenges. Some schools have high turnover rates, perhaps partly because of family instability, students' socioeconomic status, or other factors not under the direct control of the schools. A number of charter-school officials have been accused of misappropriating school funds or other financial irregularities, but many analysts believe that the positives seem to outweigh the negatives when it comes to charter schools addressing the academic gap among minority students (Hardy, 2010).

Home Schooling

A final alternative, home schooling, has been chosen by some parents who hope to avoid the problems of public schools while providing a quality education for their children. It is estimated that about 1.5 million children are home-schooled in grades K through 12 (Kerkman, 2011). This is a significant increase from the estimated 1.1 million students who were home-schooled in 2003. The primary reasons that parents indicated for preferring to home-school their children are (1) concern about the school environment, (2) the desire to provide religious or moral instruction, and (3) dissatisfaction with the academic instruction available at traditional schools. Typically, the parents of home schoolers are better educated, on average, than other parents; however, their income is about the same. Researchers have found that boys and girls are equally likely to be home-schooled.

Parents who educate their children at home believe that their children are receiving a better education at home because instruction can be individualized to the needs and interests of their children. Some parents also indicate religious reasons for their decision to home-school their children. An association of home schoolers now provides communication links for parents and children, and technological advances in computers and the Internet have made it possible for home schoolers to gain information and communicate with one another. In some states parents organize athletic leagues, proms, and other social events so that their children will have an active social life without being part of a highly structured school setting. According to advocates, home-schooled students typically have high academic achievements and high rates of employment.

Critics of home schooling question how much parents know about school curricula and how competent they are to educate their own children at home, particularly in rapidly changing subjects such as science and computer technology. Some states have passed accountability laws that must be met by parents who teach their children at home. For example, Florida requires that parents register their children with their school districts or with umbrella schools that maintain and submit records to the state.

© AP Photo/Butch Dill

▲ The aftereffects of mass killings such as the recent ones at Virginia Tech and the University of Alabama at Huntsville can be devastating for a campus community. Here, a UAH student indicates both her grief and her hope.

School Safety and Violence at All Levels

Today, officials in schools from the elementary years to two-year colleges and four-year universities are focusing on how to reduce or eliminate violence. In many schools, teachers and counselors are instructed in anger management and peer mediation, and they are encouraged to develop classroom instruction that teaches values such as respect and responsibility (National Center for Educational Statistics, 2010a). Some schools create partnerships with local law enforcement agencies and social service organizations to link issues of school safety to larger concerns about safety in the community and the nation.

Clearly, some efforts to make schools a safe haven for students and teachers are paying off. Statistics related to school safety continue to show that U.S. schools are among the safest places for young people. According to "Indicators of School Crime and Safety," jointly released by the National Center for Education Statistics and the U.S. Department of Justice's Bureau of Justice Statistics, young people are more likely to be victims of violent crime at or near their home, on the streets, at commercial establishments, or at parks than they are at school (National Center for Educational Statistics, 2010a). However, these statistics do not keep many people from believing that schools are becoming more dangerous with each passing year and that all schools should have high-tech surveillance equipment to help maintain a safe environment.

© AP Images/Don Ryan

▲ Home schooling has grown in popularity in recent decades as parents have sought to have more control over their children's education. Although some home-school settings may resemble a regular classroom, other children learn in more informal settings, such as the family kitchen.

Even with all of these safety measures in place, violence and fear of violence continue to be pressing problems in schools throughout the United States. This concern extends from kindergarten through grade 12 because violent acts have resulted in deaths in communities such as Jonesboro, Arkansas; Springfield, Oregon; Littleton, Colorado; Santee, California; Red Lake, Minnesota; and an Amish schoolhouse in rural Pennsylvania. Similarly, college and university campuses are not immune to violence, as deranged individuals have engaged in acts of personal terrorism at the expense of students, professors, and other victims. University of Texas student Charles Whitman shot and killed sixteen people, seriously wounding thirty-one others, from the top of the university's tower in 1966. Decades later, Virginia Tech student Cho Seung-Hui killed thirty-two victims before taking his own life at the university. Less than a year later, a twenty-seven-year-old graduate of Northern Illinois University killed five students at that school. In 2010 Amy Bishop, a University of Alabama at Huntsville professor who was allegedly angry over being denied tenure, killed three faculty members and injured several others. The aftermath of each of these tragedies saw a massive outpouring of public sympathy and a call for greater campus security. Gun-control advocates called for greater control over the licensing and ownership of firearms and for heightened police security on college campuses, whereas pro-gun advocates argued that people should be allowed to carry firearms on campus for their own protection. Lawmakers in a number of states introduced measures seeking to relax concealed-weapons restrictions on college and university campuses. As of this writing, legislatures in a number of states were debating this issue, but many college administrators, faculty, and students expressed serious concerns about allowing concealed weapons on campus because having weapons readily available might make a bad situation even worse.

Opportunities and Challenges in Higher Education

Who attends college? What sort of college or university do they attend?

Community Colleges

One of the fastest-growing areas of U.S. higher education today is the community college; however, the history of two-year colleges goes back more than a century, with the establishment of Joliet Junior College in Illinois. Following World War II, the GI Bill of Rights provided the opportunity for more people to attend college, and in 1948 a presidential commission report called for the establishment of a network of public community colleges that would charge little or no tuition, serve as cultural centers, be comprehensive in their program offerings, and serve the area in which they were located.

Hundreds of community colleges were opened across the nation during the 1960s, and the number of such institutions has steadily increased since that time as community colleges have responded to the needs of their students and local communities. Community colleges offer a variety of courses, some of which are referred to as "transfer courses" in which students earn credits that are fully transferable to a four-year college or university. Other courses are in technical/occupational programs, which provide formal instruction in fields such as nursing, emergency medical technology, plumbing, carpentry, and computer information technology. Many community colleges are now also offering four-year degrees, sometimes in conjunction with four-year colleges located in the same state.

Community colleges educate about half of the nation's undergraduates. According to the American Association of Community Colleges (2011), the 1,167 community colleges (including public, private, and tribal colleges) in the United States enroll about 12.4 million students in credit and noncredit courses. Community college enrollment accounts for 44 percent of all U.S. undergraduate students. Women make up more than half (58 percent) of community college students. Community colleges are also important for underrepresented minority student enrollment: Fifty-five percent of all Native American college students attend a community college, as do 52 percent of all Hispanic students, 44 percent of African American students, and 45 percent of Asian American/Pacific Islanders (American Association of Community Colleges, 2011).

One of the greatest challenges facing community colleges today is money. Across the nation, state and local governments struggling to balance their budgets have slashed funding for community colleges. In a number of regions these cuts have been so severe that schools have been seriously limited in their ability to meet the needs of their students. In some cases colleges have terminated programs, slashed course offerings, reduced the number of faculty, and eliminated essential student services. Many people are hopeful that President Obama's "American Graduation Initiative" will strengthen community colleges and offer greater financial support for students. But this does not look promising because Congress, in an effort to make cuts in the federal budget, approved only a small portion of the $12 billion that President Obama called for to launch the initiative.

Four-Year Colleges and Universities

More than twelve million students attend public or private four-year colleges or universities in the United States. Four-year schools typically offer a general education curriculum that gives students exposure to multiple disciplines and ways of knowing, along with more in-depth study (known as a "major") in at least one area of concentration. However, many challenges are faced by four-year institutions, ranging from the cost of higher education to racial and ethnic differences in enrollment and lack of faculty diversity.

The High Cost of a College Education

What does a college education cost? According to the *Chronicle of Higher Education* (2010), the average tuition and fees per year at a public four-year institution are $6,319, as compared to $22,449 at a private four-year institution. Average tuition and fees per year at a public two-year institution are $2,137. Although public institutions such as community colleges and state colleges and universities typically have lower tuition and overall costs—because they are funded primarily by tax dollars—than private colleges have, the cost of attending public institutions has increased over the past decade.

According to some social analysts, a college education is a bargain. However, other analysts believe that the high cost of a college education reproduces the existing class system: Students who lack money may be denied access to higher education, and those who are able to attend college receive different types of education based on their ability to pay. For example, a community college student who receives an associate's degree or completes a certificate program may be prepared for a position in the middle of the occupational status range, such as a dental assistant, computer programmer, or auto mechanic. In contrast, university graduates with four-year degrees are more likely to find initial employment with firms where they stand a chance of being promoted to high-level management and executive positions.

Although higher education may be a source of upward mobility for talented young people from poor families, the U.S. system of higher education is sufficiently stratified that it may also reproduce the existing class structure. However, with the twenty-first-century global economy and a rapidly changing U.S. job market, many analysts suggest that a college education is worth what it costs because educational attainment plays an important role in income: "The higher their education level, the more that adults' household incomes have risen over the past four decades; within each level, married adults have seen larger gains than unmarried adults" (Fry and Cohn, 2010: 3). In other words, higher educational attainment is associated with higher earnings on average. In 2009 high school graduates had average earnings of $31,283, while those with a bachelor's degree earned about $58,613. Median

© Joliet Junior College

© Phil Roche/Miami Dade College

▲ Joliet Junior College (Illinois) is the oldest two-year college in the United States, having opened its doors in 1901. The bottom photo shows a scene from graduation day at the nation's largest two-year school, Miami Dade College (Florida), which recently expanded into offering four-year degrees, part of a national trend among two-year colleges. Today, Joliet, Miami Dade, and other schools like them fulfill many needs in the competitive world of higher education.

earnings for a worker with a high school diploma alone were about 53 percent of the median earnings of a worker with a bachelor's degree. Average earnings for those with an advanced degree totaled $83,144 (U.S. Census Bureau, 2010b).

Given the necessity of getting a college education in the twenty-first century, is any financial assistance on the way? In 2010 the Obama administration passed a student-loan bill to aid colleges and students. The legislation is designed to "cut out the middle person" by ending the bank-based system of distributing federally subsidized student loans and instead have the Department of Education give loan money directly to colleges and their students. With the savings from this approach, more money is to be put into the Pell Grant program. Unlike a loan, a federal Pell Grant does not have to be repaid. The maximum award for 2011–2012 was $5,500, but not all students were eligible for this amount. Some legislation also provides for additional assistance to historically black colleges in an effort to help more low-income students enroll and succeed in college (Basken, 2010). Many questions remain about student loans and the possible long-term effects of high student debt on individuals after they complete their college education.

Slashed Budgets at State Universities

The problem of increasing costs of higher education for students is compounded by state budget shortfalls, which have prompted slashed funding for public higher education. In states such as California, funding for the public higher education system has been cut by as much as 20 percent per year. Students and faculty have protested these cuts, but declining state support has become a major concern for colleges and universities not only in California but also throughout the nation. According to the *New York Times,* "The percentage of total spending at state universities provided by state tax revenue has been sinking for more than 20 years" (Folbre, 2009). In some states, major public universities have moved from being state supported to state assisted and now to nothing more than state located (Folbre, 2009). As a result, students and their families now contribute more than one-third of all educational revenue at many public institutions. As even larger federal and state budget deficits loom in the second decade of the twenty-first century, the problem of funding higher education will inevitably grow more problematic.

Bill Clark/Roll Call/Getty Images

▲ State universities and community colleges are largely funded by individual states and their counties. However, the recent economic downturn has led to cutbacks in services and in the number of students admitted. Here, members of the U.S. Student Association rally in the nation's capital to protest these changes.

Racial and Ethnic Differences in Enrollment

How does college enrollment differ by race and ethnicity? People of color (who are more likely than the average white student to be from lower-income families) are underrepresented in higher education. White Americans make up slightly more than 63 percent of all college students, as compared to African American enrollment at 13.5 percent and Hispanic/Latina/o enrollment at almost 12 percent (*Chronicle of Higher Education,* 2010). Native American/Alaska Native enrollment rates have remained stagnant at about 1.0 percent; however, tribal colleges on reservations have experienced growth in student enrollment. Founded to overcome racism experienced by Native American students in traditional four-year colleges and to shrink the high dropout rate among Native American college students, 37 colleges are now chartered and run by the Native American nations (American Association of Community Colleges, 2011). Unlike other community colleges, the tribal colleges receive no funding from state and local governments and, as a result, are often short of funds to fulfill their academic mission.

The proportionately low number of people of color enrolled in colleges and universities is reflected in the educational achievement of people age 25 and over, as shown in the "Census Profiles" feature. If we focus on persons who receive doctorate degrees, the underrepresentation of persons of color is even more striking. According to the *Chronicle of Higher Education* (2010), of the 84,960 doctoral degrees conferred in the 2007–2008 academic year, African Americans earned slightly less than 6 percent (4,766 degrees), Hispanics earned slightly less than 4 percent (3,199),

Educational Achievement of Persons Age 25 and Over

The Census Bureau asks people to indicate the highest degree or level of schooling they have completed. Sixteen categories, ranging from "no schooling completed" to "doctorate degree," are possible responses on the form that is used; however, we are looking only at the categories of high school graduate and above. As shown below, census data reflect that Asian Americans and others, followed by non-Hispanic white respondents, hold the highest levels of educational attainment. For these statistics to change significantly, greater educational opportunities and more-affordable higher education would need to be readily available to African Americans and Latinos/as, who historically have experienced racial discrimination and inadequately funded public schools with high dropout rates and low high school graduation rates.

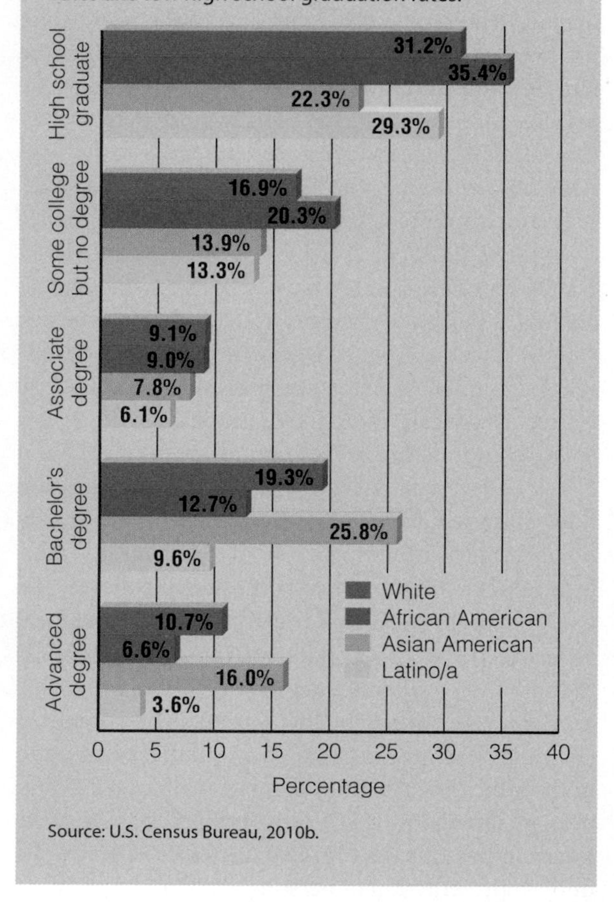

Source: U.S. Census Bureau, 2010b.

and Native Americans/American Indians earned .005 percent (432). By contrast, whites (non-Hispanic) earned 47,246 Ph.D. degrees, or 57 percent of the total number of degrees awarded.

Underrepresentation is not the only problem faced by students of color: Problems of prejudice and discrimination continue on some college campuses. Some problems are overt and highly visible; others are more covert and hidden from public view. Examples of overt racism include mocking Black History Month or a Latino celebration on campuses, referring to individuals by derogatory names, tying nooses on doorknobs of dorm rooms or faculty offices, and having "parties" where guests dress in outfits that ridicule people from different cultures or nations.

A study by the sociologists Leslie Houts Picca and Joe R. Feagin (2007) found that many blatant racist events, ranging from private jokes and conversations to violent incidents, occurred in the presence of 600 white students at 28 colleges and universities across the country who were asked to keep diaries and record any racist events that they observed. In addition to overt patterns of discrimination, other signs of racism included numerous conversations that took place "backstage" (in white-only spaces where no person of color was present) and involved derogatory comments, skits, or jokes about persons of color. According to Picca and Feagin, most of the racial events were directed at African Americans, but Latinos/as and Asian Americans were also objects of some negative comments.

E-mail and social networking sites offer additional avenues for college students and others to engage in backstage racism. However, racist statements made by white persons in "private" digital communications are sometimes made public and shared in the front stage by individuals who do not share their views. As a result, cyber racism brings embarrassment not only to the perpetrators but also to persons of color who experience emotional distress as a result of the behavior of others (Daniels, 2009, 2010).

The Lack of Faculty Diversity

Despite the widely held assumption that there has been a significant increase in the number of minority professors, the latest figures available (fall 2007) indicate that white Americans make up 78 percent of all full-time faculty members, as compared with 5.5 percent of African Americans, 3.5 percent of Hispanics, and 0.5 percent of Native Americans/American Indians. Overall, minority scholars are in the lowest tiers of the academic profession. For example, white Americans constitute 86 percent of full professors (with tenure), compared with only 3.4 percent of African Americans and 2.4 percent of Hispanics (*Chronicle of Higher Education*, 2010). By contrast, African Americans make up 8 percent of all part-time faculty and 5 percent of all lecturers.

Gender is also a factor in faculty diversity. In all ranks and racial and ethnic categories, men made up

59.2 percent of the full-time faculty in 2007, while women accounted for 41.8 percent. Across all racial and ethnic categories, women are underrepresented at the level of full professor and overrepresented at the lower, assistant professor and instructor levels. Although assistant professors may be on a "tenure track," neither of these last two ranks provides the security of tenured positions.

Faculty diversity along lines of race/ethnicity and gender is an important issue, and it is linked to another problem in higher education: Colleges and universities are experiencing a long-term trend toward more contingent faculty appointments. Data from 2009 indicate that part-time faculty and graduate student employees made up more than 75 percent of the total instructional staff in higher education. Between 2007 and 2009, the growth in full-time nontenure-track and part-time faculty positions outstripped the increase in tenure-line positions.

Affirmative Action

Affirmative action has been a controversial issue for many years. Why does affirmative action generate such a controversy among people? And what is affirmative action anyway? *Affirmative action* is a term that describes policies or procedures that are intended to promote equal opportunity for categories of people deemed to have been previously excluded from equality in education, employment, and other fields on the basis of characteristics such as race or ethnicity.

Education was one of the earliest targets of social policy pertaining to civil rights in the United States. Increased educational opportunity has been a goal of many subordinate-group members because of the widely held belief that education is the key to economic and social advancement. Beginning in the 1970s, most public and private colleges developed guidelines for admissions, financial aid, scholarships, and faculty hiring that took race, ethnicity, and gender into account. These affirmative action policies were challenged in a number of lawsuits, especially when the policies involved public colleges. Critics of affirmative action often assert that these policies amount to *reverse discrimination,* a term that describes a situation in which a person who is better qualified is denied enrollment in an educational program or employment in a specific position as a result of another person receiving preferential treatment as a result of affirmative action.

In 2003 the U.S. Supreme Court ruled in *Grutter v. Bollinger* (involving admissions policies of the University of Michigan's law school) and *Gratz v. Bollinger* (involving the undergraduate admissions policies of the same university) that race can be a factor for universities in shaping their admissions programs, but only within carefully defined limits. Colleges and universities continue to develop programs to produce more-diverse student bodies; however, most plans are displeasing to one group or another. One thing remains clear: Discussions regarding affirmative action and access to higher education—particularly regarding the way that access is influenced by income, race/ethnicity, gender, nationality, and other characteristics and attributes—are far from over as our country grows increasingly diverse in its population.

Future Trends in Education

What will the future of education be in the United States? The answer to this question depends on how successful elected officials are in getting their agendas through state legislatures or the U.S. Congress. In regard to public elementary and secondary schools in this country, the most important educational reform since compulsory attendance laws were implemented was the 2001 No Child Left Behind Act (NCLB) of the Bush administration. The Obama administration subsequently opposed this plan; however, the goals of this act remain in effect in the nation's schools. A primary purpose of NCLB was to close the achievement gap between rich and poor students by holding schools accountable for students' learning. The law required states to test every student's progress toward meeting specific standards that were established for each level of schooling. School districts

▲ The national trend of standardized testing has had some unforeseen consequences. Schools with low or failing outcomes may lose funding, so it is in the best interest of teachers to make sure that each child does as well as possible. In many instances, this situation has led to the phenomenon known as "teaching to the test."

you can make a difference

College Student Tutors Who Reach Out to Others

The students have grown attached to the fraternity members. They not only help them academically but also by mentoring them. . . . The main thing is that once you go back a couple of times, you form a connection with the kids, so we started making it more than just tutoring. We bought tickets to the Iowa State vs. Texas A&M basketball game and all went. We had a great time. I want to do more things like that 'cause we know the kids there.

—Miles Dunklin, then a senior business major at Texas A&M University, explaining why he and other members of his fraternity enjoyed tutoring elementary and middle school students in the Bryan–College Station community (qtd. in Getterman, 2011)

Tutoring is the highlight of my week. While going to a university, it is easy to lose a sense of curiosity and enthusiasm for learning because we get caught up in homework and grades. But, when I am at Martin Luther King Elementary, I feel refreshed and invigorated because of the excitement and curiosity with which young students approach their learning. . . . For this reason, I think that I get more out of this tutoring experience at times than the kids do.

—Zahra, then a University of Washington pre-med student, describing how it felt to be a tutor at a local public school (qtd. in Stickler, 2004)

Across the nation, college students like Miles and Zahra actively participate in tutoring and mentoring programs for kindergarten through grade twelve. College students tutoring younger students is a win–win situation for everyone. The students benefit from meeting and studying with college tutors because they have the opportunity to learn in a one-on-one situation and to ask questions about the subject matter that they otherwise might not ask their teacher. With the assistance of tutors, top students are able to achieve even more than they otherwise would, while middle- and lower-tier students are able to gain not only academic skills but also a greater feeling of confidence as they succeed in school.

For college students, tutoring provides an opportunity to make a difference in the lives of children and young people who are looking for role models closer to their own age with whom they can identify. Gaining knowledge of the real world through field experience and service activities is an important component of a college education. Tutoring also opens up new worlds of communication and social interaction with young people from diverse backgrounds. Even as college students provide a service for kids and local communities, the tutors also gain valuable experience and insights that may be useful throughout life.

Are you interested in making a difference in education? Would you like to learn more about children and young people and help them meet their educational needs? If so, tutoring might be a good avenue for you to begin to develop this interest, and you may wish to find out what tutoring programs are available at your college or university.

were required to report students' results to demonstrate that they were making progress toward meeting these standards. Schools that closed the education gap received additional federal dollars, but schools and districts that did not show adequate progress lost funding and pupils: In some school districts, parents were able to move their children from lower- to higher-performing schools. During the era of NCLB, some individuals, including many college students, began their own initiatives to improve education (see "You Can Make a Difference").

During the NCLB era, improvements occurred in fourth- and eighth-grade reading and math scores nationwide, particularly among African American and Hispanic students. However, critics believed that NCLB did not accurately define the main problem facing U.S. education: What the schools really needed was more money and more incentives for teaching and learning, not more testing of students or teachers.

In 2010 the Obama administration issued "A Blueprint for Reform" in an effort to reauthorize the Elementary and Secondary Education Act, an initiative that had been in place prior to the No Child Left Behind era. In his call for reform, President Obama stated that by 2020 the United States should once again lead all nations in college completion, rather than holding eleventh place. Four major priorities are included in this blueprint:

1. Improving teacher and principal effectiveness to ensure that every classroom has a great teacher and every school has a great leader.
2. Providing information to families to help them evaluate and improve their children's schools, and to educators to help them improve their students' learning.
3. Implementing college- and career-ready standards and developing improved assessments aligned with those standards.

4. Improving student learning and achievement in America's lowest-performing schools by providing intensive support and effective interventions. (U.S. Department of Education, 2010)

One goal of the blueprint is to increase the role of the federal government in raising standards for all students and providing them with a well-rounded education so that they can contribute as citizens to the process of democracy and learn to thrive in a global economy. Another goal is elevating the teaching profession and helping teachers and principals become more effective in their work. Overall, the plan's purpose is to produce greater equity and opportunity for all students. A "Race to the Top" in education, built on the American Recovery and Reinvestment Act of 2009, would provide additional competitive grants to expand innovations in education, support effective charter schools, promote public school choice, and provide assistance to magnet schools.

With many international problems and the national political climate heating up, it is difficult to determine if progress has been made in the implementation of Obama's plan. Critics claim that the plan cannot be adequately funded in difficult economic times and that it is far too ambitious, particularly the goal of leading the world in college completion rates by 2020. Will the Obama administration be able to rewrite the No Child Left Behind Act and reauthorize the Elementary and Secondary Education Act in such a manner that federal and state education funding is increased and genuine reform occurs in our schools? Or is this another set of unrealistic goals and expectations? We must wait to learn the answers to these questions.

Similarly, colleges and universities face many challenges today, particularly given the major financial constraints that many institutions must confront. President Obama's 2009 American Graduation Initiative was established to create a community-college challenge fund because these colleges are often underfunded and lack the basic resources they need to improve instruction, build ties with businesses, and adopt other reforms. Under the president's plan, community colleges will receive new competitive grants that will help them develop partnerships with businesses and create career pathways for workers. It will also help schools expand course offerings, provide dual enrollment at high schools and universities, and improve remedial and adult education programs (whitehouse.gov, 2009). If these measures are implemented, many more people will have an opportunity to acquire a two-year degree, complete a certificate program lasting six months to a year that provides specific credentials for middle-skill jobs that require more than a high school diploma but less than a four-year degree, or acquire sufficient college credits to transfer to a four-year college or university. Making college more affordable and significantly increasing the number of students enrolled in community colleges by 2020 are commendable goals for enhancing education in the United States. What is crucial is the approximately $12 billion price tag, which the president suggests it will cost, to bring about necessary changes in the current system.

At the university level, institutions continue to expand their focus while undergoing strenuous budget cuts coupled with increasing demands to meet the needs of widely diverse student populations. As previously discussed, budget cuts for higher education have been harsh because of the nation's economic downturn and the resulting drop in the state tax revenues that help fund public higher education. Tightening of the financial resources available to colleges and universities will lead to even more schools seeking alternative ways to fund their operations. Some will further increase tuition paid by students; others will seek different sources of funding. Some will move beyond the United States to find ways in which they can expand their base of operation. For example, some U.S. universities are expanding their educational operations to emerging nations where demand is high for certain kinds of curricula, such as advanced business and petroleum engineering courses in Qatar and other Middle Eastern countries. Experts suggest that "university globalization" is here to stay, with both the export of students from countries such as India and China to other countries to study and the development of top-tier research universities in countries, including China, Singapore, and Saudi Arabia, where students may study without living abroad. Increasing numbers of U.S. students may go to school in these countries if their institutions offer opportunities, possibly at a lower cost, than do colleges and universities in the United States (see Wildavsky, 2010). One of the major issues, present and future, in colleges and universities here and in other nations, is the part that higher education plays in reducing—or in maintaining and perpetuating—social inequality.

In conclusion, education is an important social institution that must be maintained and enhanced because we know that simply spending larger sums of money on education does not guarantee that the problems facing our schools will be resolved.

chapter review Q & A

Use these questions and answers to check how well you've achieved the learning objectives set out at the beginning of this chapter.

- **What is education?**

Education is the social institution responsible for the systematic transmission of knowledge, skills, and cultural values within a formally organized structure.

- **What is the functionalist perspective on education?**

According to functionalists, education has both manifest functions (socialization, transmission of culture, multicultural education, social control, social placement, and change and innovation) and latent functions (keeping young people off the streets and out of the job market, matchmaking and producing social networks, and creating a generation gap).

- **What is the conflict perspective on education?**

From a conflict perspective, education is used to perpetuate class, racial–ethnic, and gender inequalities through tracking, ability grouping, and a hidden curriculum that teaches subordinate groups conformity and obedience.

- **What is the symbolic interactionist perspective on education?**

Symbolic interactionists examine classroom dynamics and study ways in which practices such as labeling may become a self-fulfilling prophecy for some students, such that these students come to perform up—or down—to the expectations held for them by teachers.

- **How do postmodernist theorists view higher education?**

Some postmodernists suggest that in a consumer culture, education becomes a commodity that is bought by students and their parents. Colleges function as "McUniversities" that allow students to consume educational services and obtain goods such as degrees and credentials.

- **How are U.S. public schools funded?**

Most educational funds come from state legislative appropriations and local property taxes. In difficult economic times, this means that schools must do without the necessary funds to provide students with teachers, supplies, and the best educational environment for learning.

- **What issues are of concern in elementary and secondary education?**

Racial segregation and resegregation, and how to equalize educational opportunities for students are among the many pressing issues facing U.S. public education today.

- **What are the major problems in higher education?**

Among the most pressing problems are the high cost of a college education, the underrepresentation of minorities as students and faculty, and funding cuts that affect many schools and degree programs.

key terms

credentialism 470
cultural capital 467
cultural transmission 462
education 462
formal education 462
hidden curriculum 469
informal education 462
mass education 462
tracking 468

questions for critical thinking

1. What are the major functions of education for individuals and for societies?
2. Why do some theorists believe that education is a vehicle for decreasing social inequality whereas others believe that education reproduces existing class relationships?
3. Why does so much controversy exist over what should be taught in U.S. public schools?
4. How are the values and attitudes that you learned from your family reflected in your beliefs about education?

turning to video

Watch the CBS video *Teacher Shortage* (running time 1:46), available through **CengageBrain.com**. This video explores the effects of teacher layoffs and increases in class sizes that are happening across the country and the $100 billion in federal stimulus funds aimed at education. As you watch the video, think about the changes possibly taking place in the schools in your area. After you watch the video, consider this question: To what degree do you think the stimulus funds have helped to save teachers' jobs? Explain.

online study resources

Go to CENGAGE brain.com to access online study resources, including the Sociology CourseMate for this text as well as special features such as video, an interactive sociology time line and interactive maps, General Social Survey (GSS) data, and U.S. Census 2010 data.

CourseMate brings course concepts to life with interactive learning, study, and exam-preparation tools that support the printed textbook. A textbook-specific website, **Sociology CourseMate** includes an integrated interactive eBook and other interactive learning tools, including quizzes, flash cards, and videos.

Visit **www.cengagebrain.com** to access your account and purchase materials.

The Jewish people believe that they have a unique relationship with God, affirmed on the one hand by His covenant and on the other by His law. Judaism has three key components: God (the deity), Torah (God's teachings), and Israel (the community or holy nation). Although God guides human destiny, people are responsible for making their own ethical choices in keeping with His law; when they fail to act according to the law, they have committed a sin. Also fundamental to Judaism is the belief that one day the Messiah will come to Earth, ushering in an age of peace and justice for all.

Today, Jews worship in synagogues in congregations led by a *rabbi*—a teacher or ordained interpreter and leader of Judaism. The Sabbath is observed from sunset Friday to sunset Saturday, based on the story of Creation in Genesis, especially the belief that God rested on the seventh day after He had created the world. Worship services consist of readings from scripture, prayer, and singing. Jews celebrate a set of holidays distinct from U.S. dominant cultural religious celebrations. The most important holidays in the Jewish calendar are Rosh Hashana (New Year), Yom Kippur (Day of Atonement), Hanukkah (Festival of Lights), and Pesach (Passover).

Throughout their history, Jews have been the object of prejudice and discrimination. The Holocaust, which took place in Nazi Germany (and several other nations that the Germans occupied) between 1933 and 1945, remains one of the saddest eras in history. After the rise of Hitler in Germany in 1933 and the Nazi invasion of Poland, Jews were singled out with special registrations, passports, and clothing. Many of their families were separated by force, and some family members were sent to slave labor camps while others were sent to "resettlement." Eventually, many Jews were imprisoned in death camps, where six million lost their lives. Anti-Semitism has been a continuing problem in the United States. Like other forms of prejudice and discrimination, anti-Semitism has extracted a heavy toll on multiple generations of Jewish Americans.

Today, Judaism has three main branches—Orthodox, Reform, and Conservative. Orthodox Judaism follows the traditional practices and teachings, including eating only kosher foods prepared in a designated way, observing the traditional Sabbath, segregating women and men in religious services, and wearing traditional clothing. Reform Judaism, which began in Germany in the nineteenth century, is based on the belief that the Torah is binding only in its moral teachings and that adherents should no longer be required to follow all of the Talmud, the compilation of Jewish law setting forth the strict rabbinic teachings on practices such as food preparation, rituals, and dress. In some Reform congregations, gender-segregated seating is no longer required. In the United States, services are conducted almost entirely in English, a Sunday Sabbath is observed, and less emphasis is placed on traditional Jewish holidays (Albanese, 2007). Conservative Judaism emerged between 1880 and 1914 with the arrival of many Jewish immigrants in the United States from countries such as Russia, Poland, Rumania, and Austria. Seeking freedom and an escape from persecution, these new arrivals settled in major cities such as New York and Chicago, where they primarily became factory workers, artisans, and small shopkeepers. Conservative Judaism, which became a middle ground between Orthodox and Reform Judaism, teaches that the Torah and Talmud must be followed and that *Zionism*—the movement to establish and maintain a Jewish homeland in Israel—is crucial to the future of Judaism. In Conservative synagogues, worship services are typically performed in Hebrew. Men are expected to wear head coverings, and women have roles of leadership in the congregation; some may become ordained rabbis (Matthews, 2004). Despite centuries of religious hatred and discrimination, Judaism persists as one of the world's most influential religions.

Islam

Like Judaism, Islam is a religion in the Abrahamic tradition; both religions arise through sons of Abraham—Judaism through Isaac and Islam through Ishmael. Islam, whose followers are known as Muslims, is based on the teachings of its founder, Muhammad, who was born in Mecca (now in Saudi Arabia) in about 570 C.E. According to Muhammad, followers must adhere to the five Pillars of Islam: (1) believing that there is no god but Allah, (2) participating in five periods of prayer each day, (3) paying taxes to help support the needy, (4) fasting during the daylight hours in the month of Ramadan, and (5) making at least one pilgrimage to the Sacred House of Allah in Mecca (Matthews, 2004).

The Islamic faith is based on the Qur'an—the holy book of the Muslims—as revealed to the Prophet Muhammad through the Angel Gabriel at the command of God. According to the Qur'an, it is up to God, not humans, to determine which individuals are deserving of punishment and what kinds of violence are justified under various conditions.

The Islamic notion of *jihad*—meaning "struggle"—is a core belief. The Greater Jihad is believed to be the internal struggle against sin within a person's heart, whereas the Lesser Jihad is the external struggle that takes place in the world, including violence and war (Ferguson, 1977; Kurtz, 1995). The term *jihad* is typically associated with religious fundamentalism. Despite the fact that fundamentalism is found in most of the world's religions, some social analysts believe that Islamic fundamentalism is uniquely linked to the armed struggles of groups such as Hamas, an alleged terrorist organization, and the militant Islamic Jihad, which is believed to engage in continual conflict (see Barber, 1996).

Today, more than 22 percent of the world's population considers itself to be Muslim. Most of the more than 1.3 billion adherents of this religion reside in the Middle East; however, most people residing in Northern Africa and Western Asia also consider themselves to be Muslim. Other large populations of Muslims are located in Pakistan, India, Bangladesh, Indonesia, and the southern regions of the former Soviet Union.

Islam is one of the fastest-growing religions in the United States. Driven by recent waves of migration and a relatively high rate of conversion, there has been a significant increase in the number of Muslims in this country. Estimates place the adult Muslim population in the United States at about 1.4 million (Kosmin and Keysar, 2009). Recent Muslim arrivals in the United States typically have come from countries such as Pakistan, Iran, and Saudi Arabia. Most have settled in the Midwest or on the East Coast. The ten states with the largest Muslim populations are California, New York, Illinois, New Jersey, Indiana, Michigan, Virginia, Texas, Ohio, and Maryland. Muslims in the United States have experienced prejudice and discrimination because of fears that Muslims are terrorists. The 1993 bombing of New York's World Trade Center, the 2001 terrorist attacks on the United States, and other successful or attempted terrorist attacks around the world have intensified distrust regarding people appearing to be Muslim or from countries associated with Islam. Overall, however, many Muslims have fared well in the United States because of their level of education, professional status, and higher than average household income levels.

Christianity

Along with Judaism and Islam, Christianity follows the Abrahamic tradition, tracing its roots to Abraham and Sarah. Although Jews and Christians share common scriptures in the portions of the Bible known to Christians as the "Old Testament," they interpret them differently. The Christian teachings in the "New Testament" present a worldview in which the old covenant between God and humans, as found in the Old Testament, is obsolete in light of God's offer of a new covenant to the followers of Jesus, whom Christians believe to be God's only son (Matthews, 2004).

As described in the New Testament, Jesus was born to the virgin Mary and her husband, Joseph. After a period of youth in which He prepared himself for the ministry, Jesus appeared in public and went about teaching and preaching, including performing a series of miracles—events believed to be brought about by divine intervention—such as raising people from the dead.

The central themes in the teachings of Jesus are the kingdom of God and standards of personal conduct for adherents of Christianity. Jesus emphasized the importance of righteousness before God and of praying to the Supreme Being for guidance in the daily affairs of life (Matthews, 2004).

One of the central teachings of Christianity is linked to the unique circumstances surrounding the death of Jesus. Just prior to His death, Jesus and His disciples held a special supper, now referred to as "the last supper," which is commemorated in contemporary Christianity in the sacrament of Holy Communion. Afterward, Jesus was arrested by a group sent by the priests and scribes for claiming to be king of the Jews. After being condemned to death by political leaders, Jesus was executed by crucifixion, which made the cross a central symbol of the Christian religion. According to the New Testament, Jesus died, was placed in a tomb, and on the third day was resurrected—restored to life—establishing that He is the son of God. Jesus then remained on Earth for forty days, after which He ascended into heaven on a cloud. Two thousand years later, many Christian churches teach that one day Jesus will "come again in glory" and that His second coming will mark the end of the world as we know it.

▲ Throughout recorded history, churches and other religious bodies have provided people with a sense of belonging and of being part of something larger than themselves. Members of this congregation show their unity as they visit with one another.

[concept quick review]

Theoretical Perspectives on Religion

	Key Points
Functionalist Perspectives	Sacred beliefs and rituals bind people together and help maintain social control.
Conflict Perspectives	Religion may be used to justify the status quo (Marx) or to promote social change (Weber).
Symbolic Interactionist Perspectives	Religion may serve as a reference group for many people, but because of race, class, and gender, people may experience it differently.
Rational Choice Perspectives	Religious persons and organizations, interacting within a competitive market framework, offer a variety of religions and religious products to consumers, who shop around for religious theologies, practices, and communities that best suit them.

The major sociological perspectives (which are summarized in the Concept Quick Review) have different outlooks on the relationship between religion and society. Functionalists typically emphasize the ways in which religious beliefs and rituals can bind people together. Conflict explanations suggest that religion can be a source of false consciousness in society. Symbolic interactionists focus on the meanings that people give to religion in their everyday lives. Rational choice theorists view religion as a competitive marketplace in which religious organizations (suppliers) offer a variety of religions and religious products to potential followers (consumers), who shop around for the religious theologies, practices, and communities that best suit them.

Functionalist Perspectives on Religion

The functionalist perspective on religion finds its roots in the works of early sociologist Emile Durkheim, who emphasized that religion is essential to the maintenance of society. He suggested that religion is a cultural universal found in all societies because it meets basic human needs and serves important societal functions.

For Durkheim, the central feature of all religions is the presence of sacred beliefs and rituals that bind people together in a collectivity. In his studies of the religion of the Australian aborigines, Durkheim found that each clan had established its own sacred totem, which included kangaroos, trees, rivers, rock formations, and other animals or natural creations. To clan members, their totem was sacred; it symbolized some unique quality of their clan. People developed a feeling of unity by performing ritual dances around their totem, causing them to abandon individual self-interest. Durkheim suggested that the correct performance of the ritual gives rise to religious conviction. Religious beliefs and rituals are *collective representations*—group-held meanings that express something important about the group itself. Because of the intertwining of group consciousness and society, functionalists suggest that religion has three important functions in any society:

1. *Meaning and purpose.* Religion offers meaning for the human experience. Some events create a profound sense of loss on both an individual basis (such as injustice, suffering, or the death of a loved one) and a group basis (such as famine, earthquake, economic depression, or subjugation by an enemy). Inequality may cause people to wonder why their own situation is no better than it is. Most religions offer explanations for these concerns. Explanations may differ from one religion to another, yet each tells the individual or group that life is part of a larger system of order in the universe. Some (but not all) religions even offer hope of an afterlife for persons who follow the religion's tenets of morality in this life. Such beliefs help make injustices easier to endure.
2. *Social cohesion and a sense of belonging.* By emphasizing shared symbolism, religious teachings and practices help promote social cohesion. An example is the Christian ritual of communion, which not only commemorates a historical event but also allows followers to participate in the unity ("communion") of themselves with other believers. All religions have some form of shared experience that rekindles the group's consciousness of its own unity.
3. *Social control and support for the government.* All societies attempt to maintain social control through systems of rewards and punishments. Sacred symbols and beliefs establish powerful, pervasive, long-lasting motivations based on the concept of a general order of existence. In other words, if individuals consider

themselves to be part of a larger order that holds the ultimate meaning in life, they will feel bound to one another (and to past and future generations) in a way that might not be possible otherwise. Religion also helps maintain social control in society by conferring supernatural legitimacy on the norms and laws of a society. In some societies, social control occurs as a result of direct collusion between the dominant classes and the dominant religious organizations.

In the United States the separation of church and state reduces religious legitimation of political power. Nevertheless, political leaders often use religion to justify their decisions, stating that they have prayed for guidance in deciding what to do. This informal relationship between religion and the state has been referred to as ***civil religion*—the set of beliefs, rituals, and symbols that makes sacred the values of the society and places the nation in the context of the ultimate system of meaning.** Civil religion is not tied to any one denomination or religious group; it has an identity all its own. For example, many civil ceremonies in the United States have a marked religious quality. National values are celebrated on "high holy days" such as Memorial Day and the Fourth of July. Political inaugurations and courtroom trials often require people to place their hand on a Bible while swearing to do their duty or tell the truth, as the case may be. The U.S. flag is the primary sacred object of our civil religion, and the Pledge of Allegiance includes the phrase "one nation under God." U.S. currency bears the inscription "In God We Trust."

© AP Images

▲ Separation of church and state is a constitutional requirement that is often contested by people who believe that religion should be a part of public life. These workers are complying with a federal court order to remove a monument bearing the Ten Commandments from the rotunda of the Alabama State Judicial Building.

Some critics have attempted to eliminate all vestiges of civil religion from public life. However, sociologist Robert Bellah (1967), who has studied civil religion extensively, argues that civil religion is not the same thing as Christianity; rather, it is limited to affirmations of loyalty and patriotism that adherents of any religion can accept. However, Bellah's assertion does not resolve the problem for those who do not believe in the existence of God or for those who believe that *true* religion is trivialized by civil religion.

Conflict Perspectives on Religion

Many functionalists view religion, including civil religion, as serving positive functions in society, but some conflict theorists view religion negatively.

Karl Marx on Religion For Marx, *ideologies*—systematic views of the way the world ought to be—are embodied in religious doctrines and political values (Turner, Beeghley, and Powers, 2002). These ideologies serve to justify the status quo and restrict social change. The capitalist class uses religious ideology as a tool of domination to mislead the workers about their true interests. For this reason Marx wrote his now-famous statement that religion is the "opiate of the masses." People become complacent because they have been taught to believe in an afterlife in which they will be rewarded for their suffering and misery in this life. Although these religious teachings soothe the masses' distress, any relief is illusory. Religion unites people in a "false consciousness" that they share common interests with members of the dominant class.

Max Weber's Response to Marx Whereas Marx believed that religion restricts social change, Weber argued just the opposite. For Weber, religion could be a catalyst to produce social change. In *The Protestant Ethic and the Spirit of Capitalism* (1976/1904–1905), Weber asserted that the religious teachings of John Calvin are directly related to the rise of capitalism. Calvin emphasized the doctrine of *predestination*—the belief that even before they are born, all people are divided into two groups, the saved and the damned, and only God knows who will go to heaven (the elect) and

civil religion the set of beliefs, rituals, and symbols that makes sacred the values of the society and places the nation in the context of the ultimate system of meaning.

who will go to hell. Because people cannot know whether they will be saved, they tend to look for earthly signs that they are among the elect. According to the Protestant ethic, those who have faith, perform good works, and achieve economic success are more likely to be among the chosen of God. As a result, people work hard, save their money, and do not spend it on worldly frivolity; instead, they reinvest it in their land, equipment, and labor.

The spirit of capitalism grew in the fertile soil of the Protestant ethic. Even as people worked ever harder to prove their religious piety, structural conditions in Europe led to the Industrial Revolution, free markets, and the commercialization of the economy—developments that worked hand in hand with Calvinist religious teachings. From this viewpoint, wealth is an unintended consequence of religious piety and hard work. With the contemporary secularizing influence of wealth, people often think of wealth and material possessions as the major (or only) reason to work. Although the "Protestant ethic" is rarely invoked today, many people still refer to the "work ethic" in somewhat the same manner that Weber did.

Like Marx, Weber was acutely aware that religion could reinforce existing social arrangements, especially the stratification system. The wealthy can use religion to justify their power and privilege: It is a sign of God's approval of their hard work and morality. As for the poor, if they work hard and live a moral life, they will be richly rewarded in another life.

From a conflict perspective, religion tends to promote conflict between groups and societies. According to conflict theorists, conflict may be *between* religious groups (for example, anti-Semitism), *within* a religious group (for example, when a splinter group leaves an existing denomination), or between a religious group and the *larger society* (for example, the conflict over religion in the classroom). Conflict theorists assert that in attempting to provide meaning and purpose in life while at the same time promoting the status quo, religion is used by the dominant classes to impose their own control over society and its resources.

Feminist Perspectives on Religion Like other approaches in the conflict tradition, feminist perspectives focus on the relationship between religion and women's inequality. Some feminist perspectives highlight the patriarchal nature of religion and seek to reform religious language, symbols, and rituals to eliminate elements of patriarchy. As you will recall, *patriarchy* refers to a hierarchical system of social organization that is controlled by men. In virtually all religions, male members predominate in positions of power in the religious hierarchy, and women play subordinate roles in the hierarchy and in everyday life. For example, an Orthodox Jewish man may focus on *public* ritual roles and discussions of sacred texts, while an Orthodox Jewish woman may have few, if any, ritual duties and focus on her *private* responsibilities in the home. Orthodox Judaism does not permit women to become rabbis; however, more-liberal Jewish movements have granted more women this opportunity in recent years.

According to feminist theorists, religious symbolism and language consistently privilege men over women. Religious symbolism depicts the higher deities as male and the lower deities as female. Women are also more likely than men to be depicted as negative or evil spiritual forces in religion. In the Jewish and Christian traditions, for example, Eve in the Book of Genesis is a temptress who contributes to the Fall of Man. In the Hindu tradition, the goddess Kali represents men's eternal battle against the evils of materialism (Daly, 1973). Until recently, the language used in various religious texts made women virtually nonexistent. Phrases such as *for all men* in Catholic and Episcopal services have gradually been changed to *for all people;* however, some churches have retained the traditional liturgy, where all of the language is male-centered. As women have become increasingly aware of their subordination, more of them have fought to change existing rules and have a voice in their religious community. The fact that women make up between 56 percent and 58 percent of all adherents among Pentecostals, Baptists, and mainline Christians (such as Methodists, Lutherans, Presbyterians, and Episcopalians/Anglicans) may contribute to additional changes in the roles and statuses of women in organized religion (Kosmin and Keysar, 2009).

Symbolic Interactionist Perspectives on Religion

Thus far, we have been looking at religion primarily from a macrolevel perspective. Symbolic interactionists focus their attention on a microlevel analysis that examines the meanings people give to religion in their everyday lives.

Religion as a Reference Group For many people, religion serves as a reference group to help them define themselves. For example, religious symbols have meaning for large bodies of people. The

Star of David holds special significance for Jews, just as the crescent moon and star do for Muslims and the cross does for Christians. For individuals as well, a symbol may have a certain meaning beyond that shared by the group. For instance, a symbolic gift given to a child may have special meaning when he or she grows up and faces war or other crises. It may not only remind the adult of a religious belief but also create a feeling of closeness with a relative who is now deceased. It has been said that the symbolism of religion is so very powerful because it "expresses the essential facts of our human existence" (Collins, 1982: 37).

Religion and Social Meaning Religion provides social meaning for individuals as they learn about beliefs, rituals, and religious ideas from others. This learning process contributes to personal identity, which, in turn, helps people adjust to their surroundings. For example, children may learn appropriate conduct during the Christian sacrament of Communion or the Lord's Supper by attending instruction classes for prospective church members or by observing how their parents and other adults quietly and reverently participate in this ceremony.

Social meaning in religion emerges through the process of socialization and from interaction with others in a religious setting. This social meaning then develops into personal meaning that provides a religious identity for the individual. According to a symbolic interactionist approach, members of a religious group expect each member to perform certain normative religious behaviors and to adhere to certain normative religious beliefs. In other words, a specific kind of *role performance* is expected of members of a religious group, and this role performance contributes to religion becoming a *master status* for some people, despite the fact that each person performs numerous roles and has many identities in everyday life.

Rational Choice Perspectives on Religion

In terms of religion, *rational choice theory* is based on the assumption that religion is essentially a rational response to human needs; however, the theory does not claim that any particular religious belief is necessarily true or more rational than another. The rational choice perspective views religion as a competitive marketplace in which religious organizations (suppliers) offer a variety of religions and religious products to potential followers (consumers), who shop around for the religious theologies, practices, and communities that best suit them.

According to this approach, people need to know that life has a beneficial supernatural element, such as that there is meaning in life or that there is life after death, and they seek to find these *rewards* in various religious organizations. The rewards include explanations of the meaning of life and reassurances about overcoming death. However, because religious organizations cannot offer religious certainties, they instead offer *compensators*—a body of language and practices that compensate for some physical lack or frustrated goal. According to the sociologists Rodney Stark and William Bainbridge (1985), all religions offer compensators, such as a belief in heaven, personal fulfillment, and control over evil influences in the world, to offset the fact that they cannot offer certainty of an afterlife or other valued resources that potential followers and adherents might desire. Rational choice theory focuses on the process by which actors—individuals, groups, and communities—settle on one optimal outcome out of a range of possible choices (a cost–benefit analysis). These compensators provide a range of possible choices for people in the face of a limited (or nonexistent) supply of the choice (certainty, for example) that they truly desire.

Recently, sociologists of religion have applied rational choice theory to an examination of the very competitive U.S. religious marketplace and found that people are actively shopping around for beliefs, practices, and religious communities that best suit them. For example, in recent years some religious followers have been attracted to churches that preach the so-called *prosperity gospel,* which is based on the assumption that if you give your money to God, He will bless you with more money

Photo by Frank E. Lockwood/Lexington Herald-Leader/MCT via Getty Images

▲ One of the largest places of worship in the United States is Joel Osteen's Lakewood Church, in Houston, Texas. This church broadcasts its message worldwide.

and other material possessions (such as a larger house and a luxury vehicle) that you desire. Several megachurches, including Joel Osteen's Lakewood in Houston, T. D. Jakes's Potter's House in south Dallas, and Creflo Dollar's World Changers near Atlanta, are partly based on teaching that suggests God wants people to be prosperous if they are "right" with Him.

Based on the diverse teaching and practices of various religious bodies, adherents and prospective followers move among various religious organizations, with every major religious group simultaneously gaining and losing adherents. Although some find the religious home they seek, others decide to consider themselves unaffiliated with any specific faith (Pew Forum on Religion and Public Life, 2008). Based on the movement of possible adherents, religious groups challenge one another for followers, emphasize specific moral values, and create a civil society that offers followers a religious faith that does not unduly burden them (Stark and Tobin, 2008).

Types of Religious Organization

Religious groups vary widely in their organizational structure. Although some groups are large and somewhat bureaucratically organized, others are small and have a relatively informal authority structure. Some require total commitment of their members; others expect members to have only a partial commitment. (See ■ Table 17.2 for a summarized distinction of churches and sects.) Sociologists have developed typologies or ideal types of religious organization to enable them to study a wide variety of religious groups. The most common categorization sets forth four types: ecclesia, church, sect, and cult (also referred to as a "new religious movement").

Ecclesia

An ***ecclesia*** **is a religious organization that is so integrated into the dominant culture that it claims as its membership all members of a society.** Membership in the ecclesia occurs as a result of being born into the society, rather than by any conscious decision on the part of individual members. The linkages between the social institutions of religion and government are often very strong in such societies. Although no true ecclesia exists in the contemporary world, the Anglican church (the official church of England), the Lutheran church in Sweden and Denmark, the Roman Catholic church in Italy and Spain, and the Islamic mosques in Iran and Pakistan come fairly close.

Churches, Denominations, and Sects

Unlike an ecclesia, a church is not considered to be a state religion; however, it may still have a powerful influence on political and economic arrangements in society. A ***church*** **is a large, bureaucratically organized religious organization that tends to seek accommodation with the larger society in order to maintain some degree of control over it.** Church membership is largely based on birth; typically, children of church members are baptized as infants and become lifelong members of the church. Older children and adults may choose to join the church, but they are required to go through an extensive training program that culminates in a ceremony similar to the one that infants go through. Churches have a bureaucratic structure, and leadership is hierarchically arranged. Usually, the clergy have many years of formal education. Churches have very restrained services that appeal to the intellect rather than the emotions. Religious services are highly ritualized; they are led by clergy who wear robes, enter and exit in a formal processional, administer sacraments,

table 17.2
Characteristics of Churches and Sects

Characteristic	Church	Sect
Organization	Large, bureaucratic organization, led by a professional clergy	Small, faithful group, with high degree of lay participation
Membership	Open to all; members usually from upper and middle classes	Closely guarded membership, usually from lower classes
Type of Worship	Formal, orderly	Informal, spontaneous
Salvation	Granted by God, as administered by the church	Achieved by moral purity
Attitude Toward Other Institutions and Religions	Tolerant	Intolerant

table 17.3

Major U.S. Denominations That Self-Identify as Christian

Religious Body	Members	Churches
Roman Catholic Church	69,135,000	18,992
Southern Baptist Convention	16,270,000	43,669
United Methodist Church	8,075,000	34,660
Church of Jesus Christ of Latter-Day Saints	5,691,000	12,753
Church of God in Christ[a]	5,500,000	15,300
National Baptist Convention, U.S.A.[a]	5,000,000	9,000
Evangelical Lutheran Church in America	4,851,000	10,519
National Baptist Convention of America[a]	3,500,000	(N/A)
Presbyterian Church (U.S.A.)	3,099,000	10,960
Assemblies of God	2,831,000	12,298
African Methodist Episcopal Church[a]	2,500,000	4,174
National Missionary Baptist Convention of America[a]	2,500,000	(N/A)
Progressive National Baptist Convention[a]	2,500,000	2,000
Lutheran Church–Missouri Synod	2,441,000	6,144
Episcopal Church	2,248,000	7,200
Churches of Christ[a]	1,639,000	15,000
Greek Orthodox Church	1,500,000	566
Pentecostal Assemblies of the World[a]	1,500,000	1,750
African Methodist Episcopal Zion Church	1,440,000	3,260
American Baptist Churches in U.S.A.	1,397,000	5,740
United Church of Christ	1,244,000	5,567
Baptist Bible Fellowship, International[a]	1,200,000	4,500
Christian Churches and Churches of Christ[a]	1,072,000	5,579
Jehovah's Witnesses	1,046,000	12,384

[a]Current data not available; prior data used may no longer be comparable.

Source: U.S. Census Bureau, 2008.

and read services from a prayer book or other standardized liturgical format. The Lutheran church and the Episcopal church are two examples.

Midway between the church and the sect is the ***denomination*—a large organized religion characterized by accommodation to society but frequently lacking in ability or intention to dominate society** (Niebuhr, 1929). Denominations have a trained ministry, and although involvement by lay members is encouraged more than in the church, their participation is usually limited to particular activities, such as readings or prayers. Denominations tend to be more tolerant and are less likely than churches to expel or excommunicate members. This form of organization is most likely to thrive in societies characterized by *religious pluralism*—a situation in which many religious groups exist because they have a special appeal to specific segments of the population. Perhaps because of its diversity, the United States has more denominations than any other nation. ■ Table 17.3 shows this diversity in

ecclesia a religious organization that is so integrated into the dominant culture that it claims as its membership all members of a society.

church a large, bureaucratically organized religious organization that tends to seek accommodation with the larger society in order to maintain some degree of control over it.

denomination a large organized religion characterized by accommodation to society but frequently lacking in ability or intention to dominate society.

© Cindy Charles/PhotoEdit

▲ Christians around the world have been drawn to cathedrals such as the Basilica of Sacré Coeur in Paris (built between 1875 and 1914) to worship God and celebrate their religious beliefs.

Christian denominations. Today, denominations range from Baptists and members of the Church of Christ to Unitarians and Congregationalists.

To help compare religious organizations, Ernst Troeltsch (1960/1931) and his teacher, Max Weber (1963/1922), developed a typology that distinguishes between the characteristics of churches and sects (see Table 17.2). A ***sect* is a relatively small religious group that has broken away from another religious organization to renew what it views as the original version of the faith.** Unlike churches, sects offer members a more personal religion and an intimate relationship with a supreme being, depicted as taking an active interest in the individual's everyday life. Sects have informal prayers composed at the time they are given, whereas churches use formalized prayers, often from a prayer book. Typically, religious sects appeal to those who might be characterized as lower class, whereas denominations primarily appeal to the middle and upper-middle classes, and churches focus on the upper classes.

According to the church–sect typology, as members of a sect become more successful economically and socially, they tend to focus more on this world and less on the next. However, sect members who do not achieve financial success often believe that they are being left behind as the other members, and sometimes the minister, shift their priorities to things of this world. Eventually, this process weakens some religious groups, and the dissatisfied or downwardly mobile split off to create new, less worldly versions of the original group that will be more committed to "keeping the faith." Those who defect to form a new religious organization may start another sect or form a cult (Stark and Bainbridge, 1981).

Cults (New Religious Movements)

Previously, sociologists defined a ***cult* as a loosely organized religious group with practices and teachings outside the dominant cultural and religious traditions of a society.** Because the term *cult* has assumed a negative and sometimes offensive meaning because of the beliefs and actions of a few highly publicized cults, some researchers now use the term *new religious movement* (NRM) and point out that a number of major world religions (including Judaism, Islam, and Christianity) and some denominations (such as the Mormons) started as cults. Also, most cults or NRMs do not exhibit the bizarre behavior or have the unfortunate ending that a few notorious groups have had in the past. NRMs usually have a leader who exhibits *charismatic* characteristics (personal magnetism or mystical leadership) and possesses a unique ability to communicate and form attachments with people. An example is the Rev. Sun Myung Moon, a former Korean electrical engineer who founded the Unification church, or "Moonies," claiming God revealed to him that the Judgment Day was rapidly

AP Photo/Ken Cedeno

◀ This mass wedding ceremony of thousands of brides and grooms brought widespread media attention to the Rev. Sun Myung Moon and the Unification church, which many people view as a religious cult.

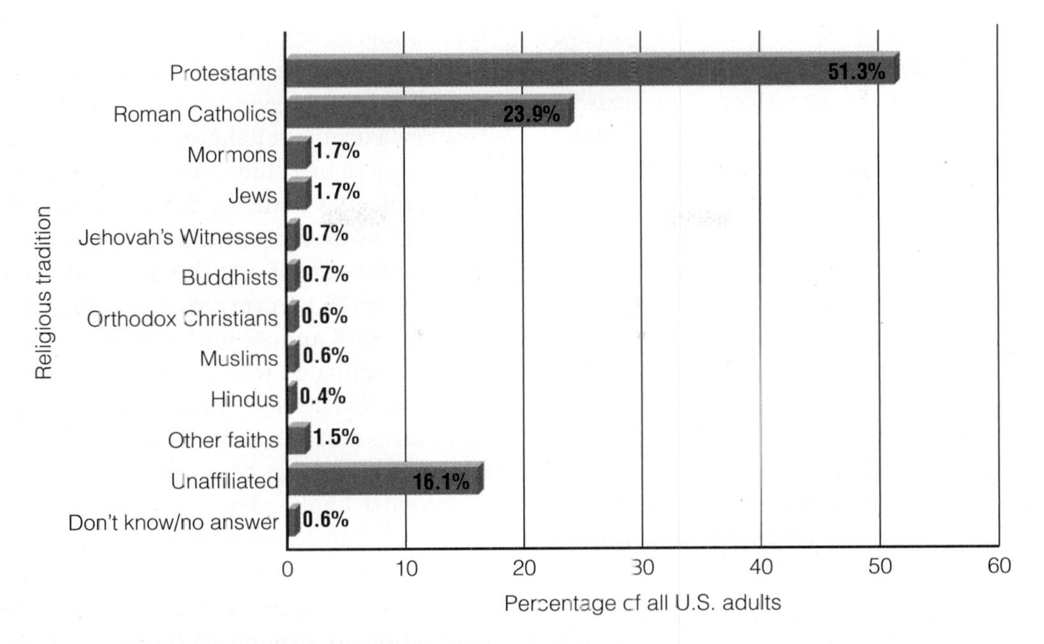

▲ **FIGURE 17.2 U.S. RELIGIOUS TRADITIONS' MEMBERSHIP**

Source: Pew Forum on Religion and Public Life, 2008.

approaching. Today, this former cult identifies itself as "comprised of families striving to embody the ideal of true love and to establish a world of peace and unity among all peoples, races, and religions as envisioned by Rev. Sun Myung Moon. Members of the Unification Church accept and follow Reverend Moon's particular religious teaching, the *Divine Principle*" (*Unification Church News*, 2011). Initially, this movement flourished because it recruited new members through their personal attachments to present members. In the twenty-first century, it has become more institutionalized. Other cult leaders have not fared so well, including Jim Jones, whose ill-fated cult ended up committing mass suicide in Guyana in 1978, and Marshall Herff Applewhite ("Do"), who led his 38 Heaven's Gate followers to commit mass suicide at their Rancho Santa Fe, California, mansion after convincing them that the comet Hale-Bopp, which swung by Earth in late March 1997, would be their celestial chariot taking them to a higher level (*Newsweek,* 1997). What eventually happens to cults or NRMs? Over time, some disappear; however, others gradually transform into other types of religious organizations, such as sects or denominations. An example is Mary Baker Eddy's Christian Science church, which started as a cult but became an established denomination with mainstream methods of outreach, such as Christian Science Reading Rooms placed in office buildings or shopping malls, where individuals can learn of the group's beliefs while going about their routine activities.

Trends in Religion in the United States

Religion in the United States is very diverse. Pluralism and religious freedom are among the cultural values most widely espoused, and no state church or single denomination predominates. As shown in ▶ Figure 17.2, Protestants constitute the largest religious body in the United States, followed by Roman Catholics, Mormons, Jews, Eastern churches, and others.

In this section we examine trends that have most influenced religion in the United States over the past two centuries. One of the most important has been the debate over secularization.

The Secularization Debate

During the Industrial Revolution and the ongoing process of modernization, scientific explanations began to compete with religious views of life. Rapid growth in scientific and technological knowledge gave rise to the idea that science would ultimately answer questions that had previously been in the realm of religion. Many scholars

sect a relatively small religious group that has broken away from another religious organization to renew what it views as the original version of the faith.

cult (also known as *new religious movement* or *NRM*) a religious group with practices and teachings outside the dominant cultural and religious traditions of a society.

believed that increases in scientific knowledge would result in ***secularization*—the process by which religious beliefs, practices, and institutions lose their significance in society and nonreligious values, principles, and institutions take their place.** Used in this context, secularization has two components: (1) a decline in religious values and institutions in everyday life and (2) a corresponding increase in nonreligious values or principles and greater significance given to secular institutions. The *secularization thesis* is a belief that as nations progress through various stages, such as modernization and rationalization, religion increasingly loses its authority in all aspects of social life and governance. However, secularization has been examined in a variety of ways in sociological research:

- *Decline of popular involvement in institutionalized religion.* What are the rates of church attendance and membership?
- *Level of prestige of religious institutions and symbols.* Is there a decline in the influence of religious organizations?
- *Extent of differentiation in social institutions such as the economy, politics, and religion.* Have these institutions become highly specialized and distinct from one another?
- *Focus on the things of this world rather than the spiritual world by religious organizations.* Are churches and other religious groups increasingly focused on the things of this world?

Although secularization has been widely studied and hotly debated, many scholars argue that levels of religiosity are not declining in the United States and other high-income nations. In fact, sociologist Peter Berger, who proposed the secularization theory, reconsidered his theory when he observed that modernity, rather than secularizing society, had contributed to a countersecularization movement. Berger also found that higher levels of religiosity accompanied modernity—the way people are influenced by religious beliefs and shape their social reality accordingly. Religiosity may also be referred to as religious commitment or "religiousness." Social scientists use various measures to determine religiosity, including the extent to which a person does one or more of the following: (1) believes in and "feels" or experiences certain aspects of religion, (2) becomes involved in religious activities such as attending church or reading sacred texts, (3) believes in the teachings of the church, and (4) lives in accordance with those teachings and beliefs.

Is secularization occurring in the twenty-first century? It is difficult for analysts to determine what patterns are occurring in the U.S. religious landscape because of a fluid and diverse pattern that is becoming evident in the ongoing research (see Pew Forum on Religion and Public Life, 2008). According to one study, about 28 percent of American adults have left the faith in which they were raised in favor of another religion or have no religious preference at all. This study also found that one in four (25 percent) of all respondents between the ages of 18 and 29 indicated that they were not currently affiliated with any religious organization. However, this does not necessarily mean that many Americans are losing their religion. Other researchers have found that how questions are worded on a survey strongly affects responses. For example, when individuals are asked to select their religious family or denomination from a list, they may not see one that matches up with how they view their connection to religion at the local level. A good example is Rick Warren's Saddleback Church in Southern California: Some who attend Saddleback may not associate it with the Southern Baptist denomination, with which it is affiliated, but think of their church instead as being "Saddleback" or "Rick Warren's church." As a result, some studies ask participants to provide the name and address of their place of worship to help researchers more closely identify the person's religious affiliation. Studies using this type of question typically have found that religion affiliation is not in a massive decline in the United States (Vara, 2006). However, attrition of followers from mainline churches and denominations has occurred, and more respondents in recent studies are rejecting all forms of organized religion (Kosmin and Keysar, 2009). Issues of secularization and the growing number of unaffiliated people in major national religion surveys are of concern to people who identify themselves as "fundamentalists."

The Rise of Religious Fundamentalism

***Fundamentalism* is a traditional religious doctrine that is conservative, is typically opposed to modernity, and rejects "worldly pleasures" in favor of otherworldly spirituality.** In the past, traditional fundamentalism primarily appealed to people from lower-income, rural, southern backgrounds; however, newer fundamentalist movements have had a much wider appeal to people from all socioeconomic levels, geographical areas, and occupations in the United States. One reason for the rise of fundamentalism has been a reaction against modernization and secularization. Around the world, those who adhere to fundamentalism—whether they are Muslims,

Christians, or followers of one of the other world religions—believe that sacred traditions must be revitalized. For example, public education in the United States has been the focus of some who follow the tenets of Christian fundamentalism. Various religious and political leaders vow to bring the Christian religion "back" into the public life of this country. They have been especially critical of educators who teach what they perceive to be *secular humanism*—the belief that human beings can become better through their own efforts rather than through belief in God and a religious conversion. According to some Christian fundamentalists, elementary schoolchildren do not receive a fair and balanced picture of the Christian religion, but instead are taught that their parents' religion is inferior and perhaps irrational (Carter, 1994). But how might students and teachers who come from the diverse religious heritages feel about Christian religious instruction or organized prayer in public schools? Many social analysts believe that such practices would cause conflict and perhaps discrimination on the basis of religion. The issue of whether religious or scientific explanations best explain various aspects of social life is not limited to the teachings of schools and religious organizations: Members of the contemporary media are also key players in the framing of religious and scientific debates, and journalists may influence how we view a number of these key issues (see "Framing Religion in the Media").

Trends in Race, Class, Gender, and Religion

One of the most comprehensive surveys on contemporary religion, the U.S. Religious Landscape Survey, conducted by the Pew Forum on Religion and Public Life (2008), identified a number of trends regarding religion and race, class, and gender. In regard to race, African Americans are the most likely of all the major racial and ethnic groups in the United States to report that they have a formal religious affiliation. Even among those who indicate that they are "unaffiliated," three in four state that religion is either somewhat or very important in their lives, as compared with slightly more than one-third of others who indicate that they are unaffiliated.

The linkage between race and religion has been very important throughout this nation's history. Most racial and cultural minorities who have felt overpowered by a lack of resources have been drawn to churches or other places of worship that help them establish a sense of dignity and self-worth that is otherwise missing in daily life.

In regard to religion and gender, there are more women than men in the U.S. population, partly because women live longer than men. Currently, the U.S. adult population is made up of about 49 males to every 52 females, and this pattern is reflected in religious adherents. Religious groups having more female adherents than men (between 56 and 58 percent) are Pentecostals, Baptists, and mainline Christians (Kosmin and Keysar, 2009). The most common trend in research on gender and religion is the finding that women consistently have higher rates of religiosity and are more involved in the life of the religious community than are men. Similarly, the Pew study found that men are significantly more likely than women to claim no religious affiliation: Nearly one in five men in the Pew study indicated that they had no formal religious affiliation, as compared to only 13 percent of women in the study. Family size is another gender-related issue: Mormons and Muslims have the largest families. More than one in five Mormon adults and 15 percent of Muslim adults in the United States have three or more children living at home (Pew Forum on Religion and Public Life, 2008). The presence of a larger number of children in the home may indicate a gendered division of labor in which the women have greater responsibility for care and socialization of children in the private sphere while men are more actively engaged in the public sphere of work and religious life.

Finally, social science research continues to identify significant patterns involving social class and religious affiliation. In the Pew study, Hindus and Jews in the United States reported higher incomes than people in other categories. Researchers attribute this higher income level to the fact that Hindus and Jews also have the highest levels of education. Historically, other religious groups with the most members making more than $100,000 per year have been the mainline liberal churches, such as the Episcopalians, the Presbyterians, and the Congregationalists. However, Mormons, Buddhists, and Orthodox Christians also tend to have higher income levels. But, looking at the other end

secularization the process by which religious beliefs, practices, and institutions lose their significance in society and nonreligious values, principles, and institutions take their place.

fundamentalism a traditional religious doctrine that is conservative, is typically opposed to modernity, and rejects "worldly pleasures" in favor of otherworldly spirituality.

you can make a difference

Understanding and Tolerating Religious and Cultural Differences

[The small strip of kente cloth that hangs on the marble altar and the flags of a dozen West Indies nations, including Trinidad, Jamaica, and Barbados,] represent all of the people in this parish. It's an opportunity for us to celebrate everyone's culture. You won't find your typical Episcopal church looking like that.

—Rev. Cannon Peter P. Q. Golden, rector of St. Paul's Episcopal Church in Brooklyn, New York (qtd. in Pierre-Pierre, 1997: A11)

One part of the history of St. Paul's Episcopal Church in Flatbush, Brooklyn, is etched into its huge stained-glass windows, which tell the story of the prominent descendants of English and Dutch settlers who founded the church. However, the church's more recent history is told in the faces of its parishioners—most of whom are Caribbean immigrants from the West Indies who labor as New York City's taxi drivers, factory workers, accountants, and medical professionals (Pierre-Pierre, 1997).

Traveling only a short distance, we find a growing enclave of Muslims in Brooklyn. It is their wish that people will recognize that Islam shares a great deal with Christianity and Judaism, including the fact that all three believe in one God, are rooted in the same part of the world, and share some holy sites (Sengupta, 1997). Some Muslim adherents also hope that more people in this country will learn greater tolerance toward those who have a different religion and celebrate different holidays. For example, in some years the Islamic holy season, Ramadan, falls at about the same time as Christmas and Hanukkah, but many Muslim children and adults are disparaged by their neighbors because they do not celebrate the same holidays

these factions primarily focus on issues that bring about discord, conflict, and sometimes violence to resolve perceived differences. It is important to consider ways in which the leaders and teachings of the world's religions can best help cope with the opportunities and challenges that we face in the twenty-first century. However, the challenge is greatest when we acknowledge that some peoples of the world may believe that their religion will be furthered only if they are able to remain aloof from, and perhaps even banish, Western civilization as we know it.

As to the future of religion in America, studies have shown that we have become more polarized religiously over the past fifty years, but this may not remain true in the future. According to the political scientists Robert Putnam and David Campbell (2010), the United States is not only religiously devout and diverse, but it is also more religiously tolerant than we initially might think. To support this point, they point to intermarriage across religious lines, as well as to the extent that people today are far more likely to change religions or list themselves as "unaffiliated" or "None" when asked about religious preference than they were fifty years ago. In the future, more people may turn away from organized religions, creating a drop in American religiosity, but that does not necessarily mean that these individuals have no religion at all: Many young people who indicate that they have no religious affiliation do state that they believe in God or provide other indicators of their religious beliefs. The extent to which religion is linked to conservative politics and/or discussions of sexual morality and rights regarding abortion, gay marriage, and other "hot topics" will play a central part in whether younger people become more (or less) involved in organized religion in the future.

In the twenty-first century, we are likely to see not only the creation of new religious forms but also a dramatic revitalization of traditional forms of religious life. Religious fundamentalism has found a new audience among young people between the ages of eighteen and twenty-four. Some are Christian fundamentalists who will continue to emphasize Biblical literalism, which means that every word of the Bible is literally true (inerrancy). Other Christians are creating their own modified version of what it means to be a Christian, or a fundamentalist, or an evangelical (a person actively involved in sharing the "Good News" of his or her religion with others). Ultimately, one thing we probably can expect is considerable religious tension between Americans with differing beliefs and people around the globe who do not share our worldview. All of these factors are going to make it imperative that we become more tolerant of others and hope that they will do likewise toward us.

As we look to the future, we can expect even more fluidity in religion and the expansion of religious pluralism. In the words of Putnam and Campbell (2010: 550),

> How has America solved the puzzle of religious pluralism—the coexistence of religious diversity

as others. As one person observed, "As Muslims, we have to respect all religions" (Sengupta, 1997: A12). Left unsaid was the belief that other people should do likewise. How can each of us—regardless of our race, color, creed, or national origin—help to bring about greater tolerance of religious diversity in this country? Here is one response to this question, from Huston Smith (1991: 389–390), a historian of religion:

> Whether religion is, for us, a good word or bad; whether (if on balance it is a good word) we side with a single religious tradition or to some degree open our arms to all: How do we comport ourselves in a pluralistic world that is riven by ideologies, some sacred, some profane? . . . We listen. . . . If one of the [world's religions] claims us, we begin by listening to it. Not uncritically, for new occasions teach new duties and everything finite is flawed in some respects. Still, we listen to it expectantly, knowing that it houses more truth than can be encompassed in a single lifetime.
>
> But we also listen to the faith of others, including the secularists. We listen first because . . . our times require it. The community today can be no single tradition; it is the planet. Daily, the world grows smaller, leaving understanding the only place where peace can find a home. . . . Those who listen work for peace, a peace built not on ecclesiastical or political hegemonies but on understanding and mutual concern.

Perhaps this is how each of us can make a difference—by *learning* more about our own beliefs and about the diverse denominations and world religions represented in the United States and around the globe, and *listening* to what other people have to say about their own beliefs and religious experiences.

Will religious tolerance increase in the United States? In the world? What steps can you take to help make a difference?

> and devotion? And how has it done so in the wake of growing religious polarization? By creating a web of interlocking personal relationships among people of many different faiths. This is America's grace.

This is an interesting and optimistic view of the future of religion. As we have seen in this chapter, the debate continues over what religion is, what it should do, and what its relationship to other social institutions such as education should be. It will be up to your generation to understand other religions and to work for greater understanding among the diverse people who make up our nation and the world (see "You Can Make a Difference").

chapter review Q & A

Use these questions and answers to check how well you've achieved the learning objectives set out at the beginning of this chapter.

- **What is religion, and what purpose does it serve in society?**

Religion is a system of beliefs, symbols, and rituals, based on some sacred or supernatural realm, that guides human behavior, gives meaning to life, and unites believers into a community.

- **What is the functionalist perspective on religion?**

According to functionalists, religion has three important functions in any society: (1) providing meaning and purpose to life, (2) promoting social cohesion and a sense of belonging, and (3) providing social control and support for the government.

- **What is the conflict perspective on religion?**

From a conflict perspective, religion can have negative consequences in that the capitalist class uses religion as a tool of domination to mislead workers about their true interests. However, Max Weber believed that religion could be a catalyst for social change.

- **What is the symbolic interactionist perspective on religion?**

Symbolic interactionists focus on a microlevel analysis of religion, examining the meanings that people give to religion and the meanings that they attach to religious symbols in their everyday life.

- **What is the rational choice perspective on religion?**

Religious persons and organizations, interacting within a competitive market framework, offer a variety of religions and religious products (compensators) to consumers who shop around for religious theologies, practices, and communities that best suit them. Religious organizations offer compensators (such as a belief in heaven) to help people deal with unanswerable questions such as "What happens when I die?"

- **What are the major types of religious organization?**

Religious organizations can be categorized as ecclesia, churches, denominations, sects, and cults (now frequently referred to as new religious movements or NRMs).

- **What are the major world religions?**

The major world religions are Buddhism, Christianity, Confucianism, Hinduism, Islam, and Judaism. More than 75 percent of the world's population is represented in one of these religions.

- **How has modernization contributed to the growth of fundamentalism?**

Fundamentalism has emerged in many religions because people do not like to see social changes taking place that affect their most treasured beliefs and values. In some situations, people have viewed modernity, especially science and new technologies, as a threat to their traditional beliefs and practices, which are important components of self-identity and group cohesion. As a result, fundamentalism opposes religious accommodation to the things of this world and demands higher standards of adherents, including a code of conduct and acceptance of specific beliefs.

- **Will religion continue as a major social institution?**

The short answer is "Yes." Religion meets many needs that no other institution can provide. Organized religion has grown and changed in many regions of the world because of modernization, democratization, and globalization. In the United States, religious organizations have become much more fluid and more diverse. Congregations in the future will hold an even wider variety of beliefs, and adherents will be drawn from many diverse cultures.

key terms

animism 493
church 508
civil religion 505
cult 510
denomination 509
ecclesia 508
faith 490
fundamentalism 512
monotheism 493
polytheism 493
profane 492
religion 490
rituals 492
sacred 491
sect 510
secularization 512
simple supernaturalism 493
spirituality 490
theism 493
transcendent idealism 493

questions for critical thinking

1. Why do some people who believe they have "no religion" subscribe to civil religion? How would you design a research project to study the effects of civil religion on everyday life?
2. How is religion a force for social stability? How is it a force for social change?
3. What is the relationship among race, class, gender, and religious beliefs?
4. If Durkheim, Marx, and Weber were engaged in a discussion about religion, on what topics might they agree? On what topics would they disagree?
5. How does the rational choice perspective differ from other theoretical explanations of religion in contemporary societies?

turning to video

Watch the BBC video *Christians Persecuted in Orissa, India* (running time 2:26), available through **CengageBrain.com**. This video explores the persecution of Christians by Hindu extremists in Orissa, India, where people have been killed and homes and churches have been burned down. As you watch the video, think about the area in which you live: How many different religions can be found in your area? Do conflicts arise among them? After you watch the video, try answering this question: Do you think people in the United States practice religious tolerance? Why or why not?

online study resources

Go to CENGAGE brain.com to access online study resources, including the Sociology CourseMate for this text as well as special features such as video, an interactive sociology time line and interactive maps, General Social Survey (GSS) data, and U.S. Census 2010 data.

CourseMate brings course concepts to life with interactive learning, study, and exam-preparation tools that support the printed textbook. A textbook-specific website, **Sociology CourseMate** includes an integrated interactive eBook and other interactive learning tools, including quizzes, flash cards, and videos.

Visit **www.cengagebrain.com** to access your account and purchase materials.

18 Health, Health Care, and Disability

Medicine is, I have found, a strange and in many ways disturbing business. The stakes are high, the liberties taken tremendous. We drug people, put needles and tubes into them, manipulate their chemistry, biology, and physics, lay them unconscious and open their bodies up to the world. We do so out of an abiding confidence in our know-how as a profession. What you find when you get in close, however—close enough to see the furrowed brows, the doubts and missteps, the failures as well as the successes—is how messy, uncertain, and also surprising medicine turns out to be.

The thing that still startles me is how fundamentally human an endeavor it is. Usually, when we think about medicine and its remarkable abilities, what comes to mind is the science and all it has given us to fight sickness and misery: the tests, the machines, the drugs, the procedures. And without question, these are at the center of virtually everything medicine achieves. But we rarely see how it all actually works. You have a cough that won't go away—and then? It's not science you call upon but a doctor. A doctor with good days and bad days. A doctor with a weird laugh and a bad haircut. A doctor with three other patients to see and, inevitably, gaps in what he knows and skills he's still trying to learn. . . .

ERIK JACOBS/The New York Times/Redux Pictures

▲ Dr. Atul Gawande (center) has written movingly about the difference between people's expectations of physicians and the medical establishment and the realities that they find in health care today. Sociologists study these contradictions to better understand a very complex and important part of U.S. social life.

We look for medicine to be an orderly field of knowledge and procedure. But it is not. It is an imperfect science, an enterprise of constantly changing knowledge, uncertain information, fallible individuals, and at the same time lives on the line. There is science in what we do, yes, but also habit, intuition, and sometimes plain old guessing. The gap between what we know and what we aim for persists. And this gap complicates everything we do.

—Atul Gawande, M.D. (2002: 4, 5, 7), a surgeon at Brigham and Women's Hospital in Boston, was a surgical resident when he wrote these words describing how he feels about the power and the limits of medicine.

Chapter Focus Question:

Why are health, health care, and disability significant concerns not only for individuals but also for entire societies?

The everyday life of a doctor like Atul Gawande is filled with its high points and low points: Some patients benefit from medical treatments they receive from physicians, whereas others have sustained injuries or have developed illnesses that are too severe or are beyond the scope of current knowledge and practice in the health care system to be successfully resolved. Physicians are human beings just like the patients they treat; however, much more is expected of them because of the availability of health care in the United States and other high-income nations and because the dominant role of doctors in modern high-tech medicine has led many individuals to believe that virtually anything should be possible when it comes to one's health and longevity. However, this assumption is often not an accurate reflection of how health, illness, and health care actually work.

In this chapter we will explore the dynamics of health, health care, and disability from a sociological perspective, as well as look at issues through the eyes of those who have experienced medical problems. Before reading on, test your knowledge about health, illness, and health care by taking the Sociology and Everyday Life quiz.

Highlighted Learning Objectives

- Discuss the relationship between the social environment and health and illness.
- Identify the major issues in U.S. health care.
- Explain how functionalist, conflict, and symbolic interactionist approaches differ in their analysis of health and health care.
- Discuss what is meant by the term *mental illness*, and explain why it is a difficult topic for sociological research.
- Describe some of the consequences of disability.

sociology and everyday life

How Much Do You Know About Health, Illness, and Health Care?

True	False	
T	F	1. The idea that everyone should have guaranteed health insurance coverage is accepted by nearly all Americans.
T	F	2. The field of epidemiology focuses primarily on how individuals acquire disease and bodily injury.
T	F	3. The primary reason that African Americans have shorter life expectancies than whites is the high rate of violence in central cities and the rural South.
T	F	4. Native Americans have shown dramatic improvement in their overall health level since the 1950s.
T	F	5. Health care in most high-income, developed nations is organized on a fee-for-service basis as it is in the United States.
T	F	6. The medical–industrial complex has operated in the United States with virtually no regulation, and allegations of health care fraud have largely been overlooked by federal and state governments.
T	F	7. Media coverage of chronic depression and other mental conditions focuses primarily on these problems as "women's illnesses."
T	F	8. It is extremely costly for employers to "mainstream" persons with disabilities in the workplace.

Answers on page 524.

What does the concept of health mean to you? At one time, health was considered to be simply the absence of disease. However, the World Health Organization (2011b) defines ***health*** **as a state of complete physical, mental, and social well-being.** According to this definition, health involves not only the absence of disease but also a positive sense of wellness. In other words, health is a multidimensional phenomenon: It includes physical, social, and psychological factors.

What is illness? Illness refers to an interference with health; like health, illness is socially defined and may change over time and between cultures. For example, in the United States and Canada, obesity is viewed as unhealthy, whereas in other times and places, obesity indicated that a person was prosperous and healthy.

What happens when a person is perceived to have an illness or disease? Healing involves both personal and institutional responses to perceived illness and disease. One aspect of institutional healing is health care and the health care delivery system in a society. ***Health care*** **is any activity intended to improve health.** When people experience illness, they often seek medical attention in hopes of having their health restored. A vital part of health care is ***medicine*****—an institutionalized system for the scientific diagnosis, treatment, and prevention of illness.**

Health in Global Perspective

Studying health and health care issues around the world offers insights into illness and how political and economic forces shape health care in nations. Disparities in health are glaringly apparent between high-income and low-income nations when we examine factors such as the prevalence of life-threatening diseases, rates of life expectancy and infant mortality, and access to health services. In regard to global health, for example, the number of people infected with HIV/AIDS more than doubled between 1990 and 2009 (from fewer than 15 million to more than 33 million) (UNAIDS, 2010). Estimates place the number of children living with HIV/AIDS at about 2.5 million, and the number of adults newly infected with HIV was approximately 2.2 million. ***Life expectancy*** **refers to an estimate of the average lifetime of people born in a specific year.** AIDS results in higher mortality rates in childhood and young adulthood, stages in the life course when mortality is otherwise low. However, AIDS is not the only disease reducing life expectancy in some nations. Most deaths in low- and middle-income nations are linked to infectious and parasitic diseases that are now rare in high-income, industrialized nations. Among these diseases are tuberculosis, polio, measles, diphtheria, meningitis, hepatitis, malaria, and leprosy. Although it is estimated that only 13 percent of U.S. citizens and 9 percent of Canadians will die prior to age 60, health experts estimate that more than 1.5 billion people around the world will die prior to age 60. This is particularly true in low-income nations such as Zambia, where 80 percent of the people are not expected to see their sixtieth birthday.

The ***infant mortality rate*** **is the number of deaths of infants under 1 year of age per 1,000**

live births in a given year. Among the highest infant mortality rates in low-income nations are these: 176 infants under 1 year of age die per 1,000 live births in Angola, 149 die in Afghanistan, 112 in Niger, and 111 in Mali (Central Intelligence Agency, 2011). Some analysts estimate that two-thirds of these infants die during the *first month* of life. There are several reasons for these differences in life expectancy and infant mortality. Many people in low-income countries have insufficient or contaminated food; lack access to pure, safe water; and do not have adequate sewage and trash disposal. Added to these hazards is a lack of information about how to maintain good health. Many of these nations also lack qualified physicians and health care facilities with up-to-date equipment and medical procedures.

Nevertheless, tremendous progress has been made in saving the lives of children and adults over the past 20 years. Life expectancy at birth has risen to more than 70 years in 146 countries, up from only 55 countries in 1990. In fact, 26 countries have a life expectancy of 80 or above. For comparison purposes, the United States has a life expectancy of 78.24. Although increases in life expectancy can be attributed to a number of factors, development of a safe water supply is an important advancement (United Nations Development Programme, 2009).

Despite the progress that has been made in life expectancy and medical care, many challenges remain. Consider these global health concerns:

- One billion people worldwide do not have access to health care.
- Cardiovascular diseases are the number-one group of conditions causing death, representing at least 30 percent of all global deaths. Over 80 percent of deaths from cardiovascular diseases occur in low- and middle-income countries.
- Each year, more than 12 million people die of infectious diseases, which is far more than the number of people killed in natural or technological disasters.
- Each year, 1.3 million people die of tuberculosis, and 9.4 million new cases are reported.
- Each year, pneumococcal diseases claim the lives of 1.6 million people, and more than half of these are children. (The pneumococcus is a bacterium that causes serious infections such as meningitis, pneumonia, and sepsis.) This is the number-one cause of death worldwide that is *vaccine-preventable.*
- Malaria causes more than 860,000 deaths and 243 million acute illnesses annually. (Shah, 2011)

With all of these global health care concerns, pressing issues remain about the costs and availability of advanced diagnostic and surgical technologies and lifesaving drugs around the world. Advanced technologies such the da Vinci surgery system (physician-directed robotic surgery) and MRI machines cost in the range of $1.5 milllion to $3 million, not including expenses involved in creating specially designed suites to house the equipment. Similarly, drugs to reduce pain and suffering or to save lives are very costly for individuals and medical institutions. The major pharmaceutical companies that hold patents on the most widely used drugs believe that their name-brand products need to be protected by law, whereas people in international human relief agencies believe that the most important concern is providing needed medication to the one-third of the world's population that does not have access to essential medicines (and this figure rises to one-half of the population in the poorest parts of Africa and Asia) (United Nations Development Programme, 2009). Transnational pharmaceutical companies fear that if they provide their name-brand drugs (if these are the only

© Lee Snider/The Image Works

▲ Singapore has the lowest infant mortality rate in the world. Why? Finding the answer to this question would benefit people from every nation.

health a state of complete physical, mental, and social well-being.

health care any activity intended to improve health.

medicine an institutionalized system for the scientific diagnosis, treatment, and prevention of illness.

life expectancy an estimate of the average lifetime of people born in a specific year.

infant mortality rate the number of deaths of infants under 1 year of age per 1,000 live births in a given year.

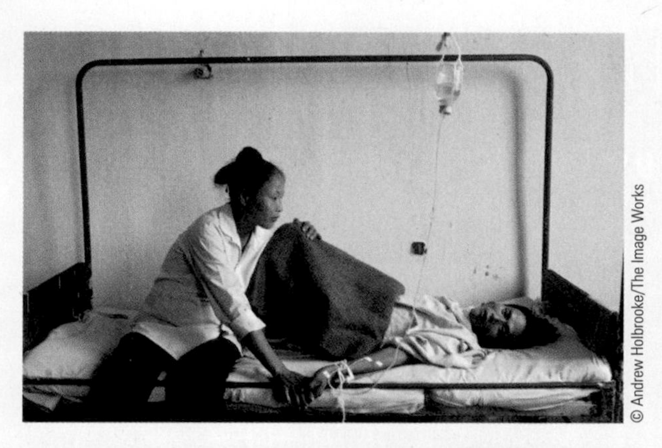

© Andrew Holbrooke/The Image Works

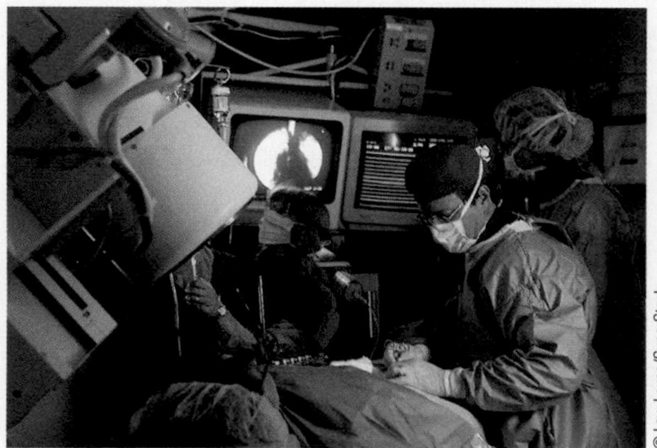

© Lew Lause/SuperStock

▲ Access to quality health care is much greater for some people than for others. The factors that are involved vary not only for people within one nation but also across the nations of the world.

Social Epidemiology

The field of social epidemiology attempts to answer questions such as these. ***Social epidemiology* is the study of the causes and distribution of health, disease, and impairment throughout a population** (Weiss and Lonnquist, 2009). Typically, the target of the investigation is disease agents, the environment, and the human host. *Disease agents* include biological agents such as insects, bacteria, and viruses that carry or cause disease; nutrient agents such as fats and carbohydrates; chemical agents such as gases and pollutants in the air; and physical agents such as temperature, humidity, and radiation. The *environment* includes the physical (geography and climate), biological (presence or absence of known disease agents), and social (socioeconomic status, occupation, and location of home) environments. The human *host* takes into account demographic factors (age, sex, and race/ethnicity), physical condition, habits and customs, and lifestyle (Weiss and Lonnquist, 2009). Let's look briefly at some of these factors.

Age Rates of illness and death are highest among the old and the young. Mortality rates drop shortly after birth and begin to rise significantly during middle age. After age 65, rates of chronic diseases and mortality increase rapidly. ***Chronic diseases* are illnesses that are long term or lifelong and that develop gradually or are present from birth;** in contrast, ***acute diseases* are illnesses that strike suddenly and cause dramatic incapacitation and sometimes death** (Weitz, 2010).

The fact that rates of chronic diseases increase rapidly after age 65 has obvious implications not only for people reaching that age (and their families) but also for society. The Census Bureau projects that about 20 percent of the U.S. population will be at least age 65 by the year 2050 and that the population of persons age 85 and over will have tripled from about 4 million (1.5 percent) in 2000 to about 12 million (5 percent). The cost of caring for many of these people—especially those who must be institutionalized—will increase in at least direct proportion to their numbers.

Sex Prior to the twentieth century, women had lower life expectancies than men because of high mortality rates during pregnancy and childbirth. Preventive measures have greatly reduced this cause of female mortality, and women now live longer than men. For babies born in the United States in 2010, for example, life expectancy at birth is estimated to be 78.3 years, with 75.7 years for males and 80.8 years for females. Females have a slight biological advantage over males in this regard from the beginning of life, as can be seen in the fact that they have lower mortality rates both in the prenatal stage and in the first month of life (Weiss and Lonnquist, 2009).

However, gender roles and gender socialization also contribute to the difference in life expectancy: Men are more likely to work in dangerous occupations such as commercial fishing, mining, construction, and public safety/firefighting. As a result of gender roles, males may be more likely than females to engage in risky behavior such as drinking alcohol, smoking cigarettes (there is more social pressure on women not to smoke), using drugs, driving dangerously, and engaging in fights. Two of the most common sources of chronic disease and premature death are tobacco use, which increases mortality among both smokers and people who breathe the tobacco smoke of others, and alcohol abuse, both of which are discussed later in this chapter. Finally, women are more likely to use the health care system, with the result that health problems are identified and treated earlier (while there is a better chance of a successful outcome), whereas many men are more reluctant to consult doctors.

© chuck kuhn photography/UpperCut Images/Getty Images

▲ Occupation and life expectancy may be related. Men are overrepresented in high-risk jobs, such as long-haul trucking, that may affect their life expectancy.

© David M. Grossman/The Image Works

▲ Can your neighborhood be bad for your health? According to recent research, it can indeed, especially if it predominantly contains fast-food restaurants, liquor stores, and similar lifestyle options.

Because women on average live longer than men, it is easy to jump to the conclusion that they are healthier than men. However, although men at all ages have higher rates of fatal diseases, women have higher rates of chronic illness.

Race/Ethnicity and Social Class Racial/ethnic differences are also visible in statistics pertaining to life expectancy. Projections for infants born in 2010 reveal this sobering fact: Life expectancy for African American males is estimated at 70.2 years as compared to 77.2 for African American females and 75.7 years for white males.

Although race/ethnicity and social class are related to issues of health and mortality, research continues to show that income and the neighborhood in which a person lives may be equally or more significant than race or ethnicity with respect to these issues. How is it possible that the neighborhood you live in may significantly affect your risk of dying during the next year? Numerous studies have found that people have a higher survival rate if they live in better-educated or wealthier neighborhoods than if the neighborhood is low income and has low levels of education. Among the reasons researchers believe that neighborhoods make a difference are the availability (or lack thereof) of safe areas to exercise, grocery stores with nutritious foods, and access to transportation, education, and good jobs. Many low-income neighborhoods are characterized by fast-food restaurants, liquor stores, and other facilities that do not afford residents healthy options.

As discussed in prior chapters, people of color are more likely to have incomes below the poverty line, and the poorest people typically receive less preventive care and less optimal management of chronic diseases than do other people. People living in central cities, where there are high levels of poverty and crime, or in remote rural areas generally have greater difficulty in getting health care because most doctors prefer to locate their practice in a "safe" area, particularly one with a patient base that will produce a high income. Although rural Americans make up 20 percent of the U.S. population, only about 10 percent of the nation's physicians practice in rural areas, and fewer specialists such as cardiologists are available in these areas.

Another factor is occupation. People with lower incomes are more likely to be employed in jobs that expose them to danger and illness—working in the construction industry or around heavy equipment in a factory, for example, or holding a job as a convenience store clerk or other position that exposes a person to the risk of armed robbery. Finally, people

social epidemiology the study of the causes and distribution of health, disease, and impairment throughout a population.

chronic diseases illnesses that are long term or lifelong and that develop gradually or are present from birth.

acute diseases illnesses that strike suddenly and cause dramatic incapacitation and sometimes death.

sociology in global perspective

Medical Crises in the Aftermath of Disasters: From Alabama to Japan

- Florence, Alabama (USA): Local Hospitals Remain on Alert (Singleton-Rickman, 2011)
- Kesennuma, Japan: Medical Crisis in Japan's Evacuation Shelters: Overworked Doctors Are Struggling to Provide Care to the Sick and Infirm (channelnewsasia.com, 2011)

Although Florence, Alabama, is thousands of miles away from Kesennuma, Japan, both of these cities had similar problems in 2011: how to take care of sick and injured patients quickly and effectively in the aftermath of deadly natural disasters. In Alabama and other U.S. states, devastating tornadoes ripped through cities and the countryside, leaving hundreds dead and many others injured. Hospitals were bombarded with injured patients whom rescue workers dug from the rubble or found in the streets. Although some of the injured had minor scrapes, bruises, or cuts, others had multiple fractures, contusions, and potentially life-threatening internal injuries that required surgery (Singleton-Rickman, 2011).

In Japan a powerful earthquake and tsunami devastated northern regions of the nation, leaving at least 15,000 people dead, 5,000 injured, 10,000 missing, and 300,000 homeless and in refugee status. Shortages of water, food, clothing, shelter, and medicine further contributed to the hardship that many people experienced. Fear of contamination from radiation leakage at earthquake-damaged nuclear power plants produced other health hazards, the short-term and long-term effects of which remain unknown at this time (Fackler, 2011). Adding to these problems was the fact that evacuation shelters were full of injured survivors, as well as individuals who were without their regular medications because they lost everything in the disaster, and patients who had been evacuated from hospitals to congested shelters that were ill-equipped to take care of their medical needs.

Given the horrific nature of major disasters such as these, what kind of national and global planning should be done to provide the best possible medical care for people? This question deserves careful consideration because the global

of color and poor people are more likely to live in areas that contain environmental hazards.

However, although Latinas/os are more likely than non-Latino/a whites to live below the poverty line, they have lower death rates from heart disease, cancer, accidents, and suicide, and an overall lower death rate. One explanation may be dietary factors and the strong family life and support networks found in many Latina/o families (Weiss and Lonnquist, 2009). Obviously, more research is needed on this point, for the answer might be beneficial to all people.

Health Effects of Disasters

When we hear about disease or impairment, most of us think about health problems associated with acute or chronic conditions, ranging from colds and flu to diabetes and coronary disease. However, disasters also have a detrimental effect on people's health and well-being, and they also contribute to higher rates of disability and mortality. The World Health Organization defines a *disaster* as a "sudden ecological phenomenon of sufficient magnitude to require external assistance" (Goolsby, 2011). This statement means that a disruption of such magnitude has occurred that the community, state, or nation is unable to return to a normal condition following the event without outside assistance. Disasters are commonly classified as natural or technological (human-made) disasters. Natural disasters include tornadoes, earthquakes, hurricanes, floods, volcano eruptions, tsunamis, and other potentially lethal conditions that originate in nature. By contrast, technological disasters include toxic spills, fires, and nuclear crises (such as Chernobyl and problems with nuclear power plants in Japan in the aftermath of a powerful 2011 earthquake and tsunami). Clearly, terrorist attacks and war also have a devastating effect on human life because many of the incidents have potential for mass casualties.

Both physical and mental health effects from disasters are important concerns. The risk of physical injury during and after natural and technological disasters is high. Individuals who sustain wound injuries are at a high risk for tetanus, a serious, often fatal toxic condition. However, this condition is virtually 100 percent preventable with the appropriate vaccination (Centers for Disease Control and Prevention, 2010). Similarly, infection is a potential problem with any wound or rash sustained in a disaster. For this reason, organizations such as the Centers for Disease Control and Prevention issue guidelines for health care professionals so that they will be aware of special precautions that

impact of natural disasters has taken a turn for the worse in recent years. In 2010 alone, 385 natural disasters worldwide killed more than 297.000 people and affected 217 million others (Canadian Broadcast Corporation, 2011).

The World Health Organization is one of the organizations leading the drive for more-effective disaster risk management for health-related concerns. According to this organization (2011a), disaster-related injuries, diseases, deaths, disabilities, and psychosocial problems can be avoided or reduced by effective disaster risk management. Unlike traditional approaches of the health sector that respond to emergencies only after they happen, disaster risk management offers a proactive approach that emphasizes prevention, or at least reduction, of the problem. Although many factors go into creating a health-related risk management plan, here are a few priorities:

1. Provide direction and support for disaster risk management at all levels, particularly local communities.
2. Assess potential risks to health and health systems, particularly from biological, natural, and technological sources, to enable early detection and warning to prompt action by the public and health workers.
3. Identify individuals, populations, infrastructure, and other community elements that are most vulnerable to harm in disasters and their aftermath (examples include young children, people over age 65, low-income persons, individuals with chronic illness or disability, and those who are socially isolated).
4. Evaluate the system's capacity to manage health risks when responding to, or recovering from, a major disaster.

Although these priorities are abstract, identifying them is the first step in disaster preparedness. It is important to manage health risks so that fewer lives will be lost and fewer injuries sustained in natural and technological disasters that are both hazardous to and frightening for all of us.

reflect & analyze

How do inequality and poverty contribute to some people's vulnerability in a major disaster? Can physicians and other health care providers become better prepared for disasters? How might they accomplish this goal?

are needed for emergency wound management. Often, the effects of disasters on mental health are not known for some period of time. The stress and trauma of survivors and those who are seriously injured may linger for extended periods. In fact, some medical specialists compare the psychological effects of being a survivor of a deadly disaster such as an earthquake, tornado, hurricane, or terrorist attack to the anxiety or post-traumatic stress disorders exhibited by some wartime combat survivors.

How does treatment differ in disasters as compared to medical emergencies? Emergency medical services typically provide *maximal resources* to a *small number* of people, while disaster medical services are designed to direct *limited resources* to the *greater number* of individuals. Although most of us do not like to think about these issues in regard to health and illness, in the United States and around the world, a disaster occurs somewhere almost daily, and some of these are of sufficient magnitude that individuals and nations are devastated for extended periods of time. For this reason, social epidemiologists have intensified research on this phenomenon, and the World Health Organization has created plans for disaster risk management for health (see "Sociology in Global Perspective").

Lifestyle Factors

In addition to looking at the effects of national and global disasters on heath, social epidemiologists also examine lifestyle choices as a factor in health, disease, and impairment. We will examine three lifestyle factors as they relate to health:

© Michael Newman/PhotoEdit

▲ Despite a variety of warnings from the U.S. surgeon general about the potentially harmful effects of smoking, many people continue to light up cigarettes. Even those who do not smoke may be affected by environmental tobacco smoke.

drugs, sexually transmitted diseases, and diet and exercise.

Drug Use and Abuse What is a drug? There are many different definitions, but for our purposes, a ***drug* is any substance—other than food and water—that, when taken into the body, alters its functioning in some way.** Drugs are used for either therapeutic or recreational purposes. *Therapeutic* use occurs when a person takes a drug for a specific purpose such as reducing a fever or controlling a cough. In contrast, *recreational* drug use occurs when a person takes a drug for no purpose other than achieving a pleasurable feeling or psychological state. Alcohol and tobacco are examples of drugs that are primarily used for recreational purposes; their use by people over a fixed age (which varies from time to time and place to place) is lawful. Other drugs—such as some antianxiety drugs or tranquilizers—may be used legally only if prescribed by a physician for therapeutic use but are frequently used illegally for recreational purposes.

Alcohol The use of alcohol is considered an accepted part of the dominant culture in the United States. Adults in this country consume an average of 2.5 gallons of wine, 21.8 gallons of beer, and 1.4 gallons of liquor per year (U.S. Census Bureau, 2009). In fact, adults consume more beer on average than milk or coffee. However, these statistics overlook the fact that among people who drink, 10 percent account for roughly half the total alcohol consumption in this country (Levinthal, 2010).

Although the negative short-term effects of alcohol are usually overcome, chronic heavy drinking or alcoholism can cause permanent damage to the brain or other parts of the body. For alcoholics, the long-term negative health effects include *nutritional deficiencies* resulting from poor eating habits (chronic heavy drinking contributes to high caloric consumption but low nutritional intake); *cardiovascular problems* such as inflammation and enlargement of the heart muscle, high blood pressure, and stroke; and eventually to *alcoholic cirrhosis*—a progressive development of scar tissue that chokes off blood vessels in the liver and destroys liver cells by interfering with their use of oxygen (Levinthal, 2010). Alcoholic cirrhosis is the ninth most frequent cause of death in the United States. The social consequences of heavy drinking are not always limited to the person doing the drinking. For example, abuse of alcohol and other drugs by a pregnant woman can damage the unborn fetus.

Can alcoholism be overcome? Many people get and stay sober. Some rely on organizations such as Alcoholics Anonymous or church support groups to help them overcome their drinking problem. Others undergo medical treatment and therapy sessions to learn more about the root causes of their problem. Caroline Knapp (1996: 242–243), who started drinking at age 14 and continued throughout college, describes how she felt after she regained sobriety and came to view herself as a recovering alcoholic:

> I still think about drinking, and not drinking, many many times each day, and sometimes I think I always will. We live in an alcohol-saturated world; it's simply impossible to avoid the stuff. When I read the papers now, I find myself scanning the pages for items about drink-related disasters, things that will reinforce my sense that I've made the right choice: what celebrity got pulled over for drunk driving; what college kid got drunk and plunged out of a five-story window; what alcohol-fueled argument between a couple fired up into a case of domestic violence. Evidence of the havoc liquor can wreak is there in black and white, almost every day, but it's not nearly as prevalent as the other messages: the liquor ads, the images of gaiety and romance, phrases like CHAMPAGNE BRUNCH: $19.95. At times, I've grumbled to friends about longing to return to Prohibition, a gripe that stems from the feeling, familiar among many alcoholics, that if *I* can't drink, no one should. But alcohol occupies a large role in the social world and it's important for me to remember that I have to come to terms with it, that I still have a relationship with liquor, even if the relationship now has the quality of a divorce rather than an active involvement.

Nicotine (Tobacco) The nicotine in tobacco is a toxic, dependency-producing psychoactive drug that is more addictive than heroin. It is classified as a stimulant because it stimulates central nervous system receptors and activates them to release adrenaline, which raises blood pressure, speeds up the heartbeat, and gives the user a temporary sense of alertness. Although the overall proportion of smokers in the general population has declined somewhat since the 1964 Surgeon General warning that smoking is linked to cancer and other serious diseases, tobacco is still responsible for about one in every five deaths in this country (National Institute on Drug Abuse, 2010). (▶ Figure 18.1 displays some of the nations and U.S. states that had instituted full or partial bans on public smoking as of

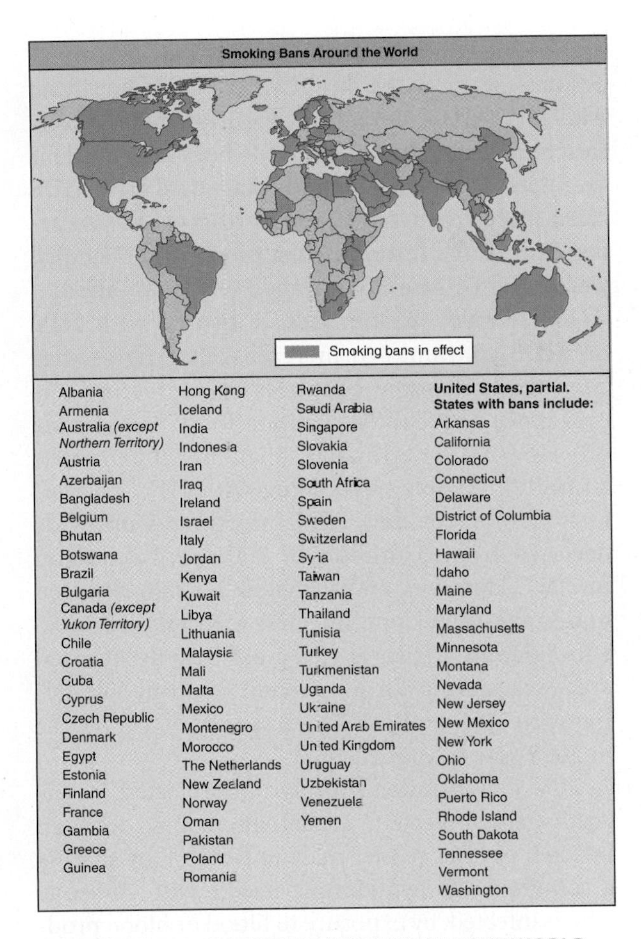

▲ **FIGURE 18.1 SMOKING BANS AROUND THE WORLD (2010)**

Sources: Based on Ray, 2007; and Wikipedia, 2011.

2010.) Even people who never light up a cigarette can be harmed by *environmental tobacco smoke*—the smoke in the air inhaled by nonsmokers as a result of other people's tobacco smoking and the residue of smoke on garments and furniture, for example (Levinthal, 2010). Researchers have found that environmental smoke is especially hazardous for nonsmokers who carpool or work with heavy smokers.

Illegal Drugs Marijuana is the most extensively used illegal drug in the United States. About one-third of all people over age 12 have tried marijuana at least once. Although most marijuana users are between the ages of 18 and 25, use by teenagers has more than doubled during the past decade. High doses of marijuana smoked during pregnancy can disrupt the development of a fetus and result in congenital abnormalities and neurological disturbances. Furthermore, some studies have found an increased risk of cancer and other lung problems associated with marijuana because its smokers are believed to inhale more deeply than tobacco users.

Another widely used illegal drug is cocaine: About 35.3 million people over the age of 12 in the United States report that they have used cocaine at least once, and about 2.4 million acknowledge having used it during the past month (National Survey on Drug Use and Health, 2006). Cocaine may be either inhaled, injected intravenously, or smoked ("crack cocaine"). People who use cocaine over extended periods of time have higher rates of infection, heart problems, internal bleeding, hypertension, stroke, and other neurological and cardiovascular disorders than do nonusers. Intravenous cocaine users who share contaminated needles are also at risk for contracting AIDS.

Each of the drugs discussed in these few paragraphs represents a lifestyle choice that affects health. Whereas age, race/ethnicity, sex, and—at least to some degree—social class are ascribed characteristics, taking drugs is a voluntary action on a person's part.

Sexually Transmitted Diseases The circumstances under which a person engages in sexual activity constitute another lifestyle choice with health implications. Although most people find sexual activity enjoyable, it can result in transmission of certain *sexually transmitted diseases* (STDs), including AIDS, gonorrhea, syphilis, and genital herpes. Prior to 1960, the incidence of STDs in this country had been reduced sharply by barrier-type contraceptives (e.g., condoms) and the use of penicillin as a cure. However, in the 1960s and 1970s the number of cases of STDs increased rapidly with the introduction of the birth control pill, which led to women having more sexual partners and couples being less likely to use barrier contraceptives.

Gonorrhea and Syphilis Until the 1960s, gonorrhea (today the second-most-common STD) and syphilis (which can be acquired not only by sexual intercourse but also by kissing or coming into intimate bodily contact with an infected person) were the principal STDs in this country. Today, however, they constitute less than 15 percent of all cases of STDs reported in U.S. clinics. Untreated gonorrhea may spread from the sexual organs to other parts of the body, among other things negatively affecting fertility; it can also spread to the brain or heart

drug any substance—other than food and water—that, when taken into the body, alters its functioning in some way.

and cause death. Untreated syphilis can, over time, cause cardiovascular problems, brain damage, or even death. Penicillin can cure most cases of either gonorrhea or syphilis as long as the disease has not spread.

Genital Herpes This sexually transmitted disease produces a painful rash on the genitals. Genital herpes cannot be cured: Once the virus enters the body, it stays there for the rest of a person's life, regardless of treatment. However, the earlier that treatment is received, the more likely it is that the severity of the symptoms will be reduced. About 40 percent of persons infected with genital herpes have only a first attack of symptoms of the disease; the remaining 60 percent may have attacks four or five times a year for several years.

AIDS Acquired immunodeficiency syndrome (AIDS), which is caused by HIV (human immunodeficiency virus), is among the most significant health problems that this nation—and the world—faces today. Although AIDS almost inevitably ends in death, no one actually dies *of* AIDS. Rather, AIDS reduces the body's ability to fight diseases, making a person vulnerable to many diseases—such as pneumonia—that result in death.

AIDS was first identified in 1981, and the total number of AIDS-related deaths in the United States through 1985 was only 12,493; however, the numbers rose rapidly and precipitously after that. In recent years the National Centers for Disease Control and Prevention has developed new estimates of HIV prevalence, or the total number of people living with HIV. Current estimates suggest that 1.1 million people are living with HIV/AIDS infection in the United States. Persons between the ages of 40 and 49 have accounted for the largest proportion of newly diagnosed HIV/AIDS cases; persons between 30 and 39 years of age have accounted for the second-largest proportion (National Centers for Disease Control and Prevention, 2009a).

Worldwide, the number of people with HIV or AIDS continues to increase, but progress has been made in some countries in addressing this epidemic. The United Nations Joint Programme on HIV/AIDS estimates that in 2007 about 33 million people were living with HIV. The annual number of new HIV infections worldwide declined from 3 million in 2001 to 2.7 million in 2007. However, an estimated 370,000 children under age 15 became infected with HIV in 2007. Sub-Saharan Africa is the most heavily affected area, accounting for 67 percent of all people living with HIV and for 72 percent of AIDS deaths in 2007 (see ▶ Map 18.1).

HIV is transmitted through unprotected (or inadequately protected) sexual intercourse with an infected partner (either male or female), by sharing a contaminated hypodermic needle with someone who is infected, by exposure to blood or blood products (usually from a transfusion), and by an infected woman who passes the virus on to her child during pregnancy, childbirth, or breast feeding. It is not transmitted by casual contact such as shaking hands.

Staying Healthy: Diet and Exercise Lifestyle choices also include positive actions such as a healthy

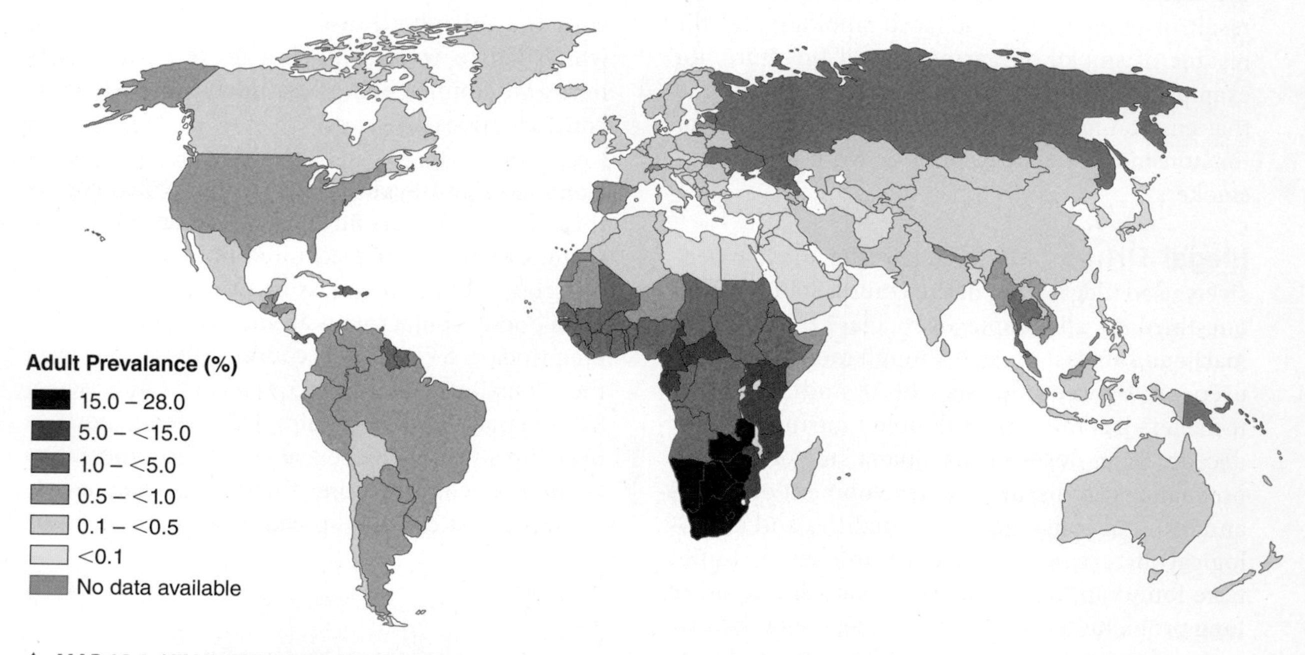

▲ MAP 18.1 HIV INFECTIONS WORLDWIDE, 2007

Source: UNAIDS, 2008.

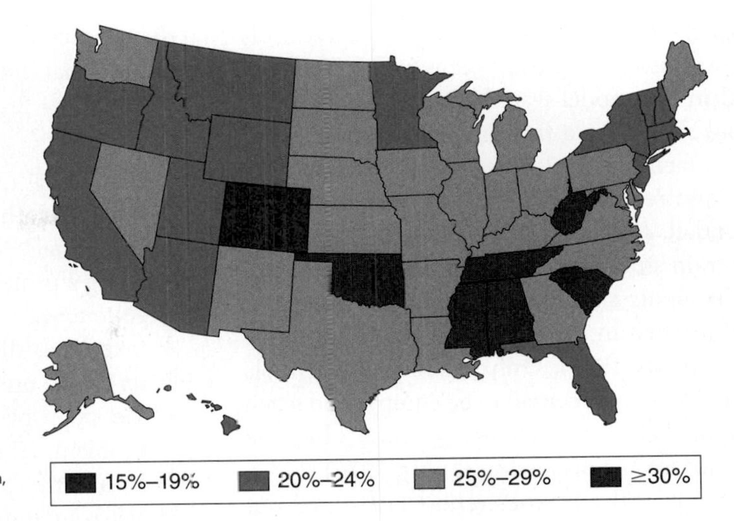

▶ **MAP 18.2 OBESITY IN THE UNITED STATES**

Source: National Centers for Disease Control and Prevention, 2008.

diet and good exercise. Over the past several decades a dramatic improvement in our understanding of food and diet has taken place, and many people in the United States have begun to improve their dietary habits. A significant portion of the population now eats larger amounts of vegetables, fruits, and cereals, substituting unsaturated fats and oils for saturated fats. These changes have contributed to a significant decrease in the incidence of heart disease and of some types of cancer. However, recent studies have raised concern that an increasing percentage of children and adults in this country are overweight or obese to an extent that may decrease their life expectancy. ▶ Map 18.2 shows obesity rates across the United States.

Exercise is another factor. Regular exercise (at least three times a week) keeps the heart, lungs, muscles, and bones in good health and slows the aging process. Studies have found that even moderate exercise can add years to people's lives, and it also contributes to how long people live a healthy life. Exercise helps reduce heart disease, obesity, and other health-related problems associated with living a sedentary lifestyle.

Health Care in the United States

Understanding health care as it exists in the United States today requires a brief examination of its history. During the nineteenth century, people became doctors in this country either through apprenticeships, purchasing a mail-order diploma, completing high school and attending a series of lectures, or obtaining bachelor's and M.D. degrees and studying abroad for a number of years. At that time, medical schools were largely proprietary institutions, and their officials were often more interested in acquiring students than in enforcing standards. The state licensing boards established to improve medical training and stop the proliferation of "irregular" practitioners failed to slow the growth of medical schools, and their number increased from 90 in 1880 to 160 in 1906. Medical school graduates were largely poor and frustrated because of the overabundance of doctors and quasi-medical practitioners, so doctors became highly competitive and anxious to limit the number of new practitioners. The obvious way to accomplish this was to reduce the number of medical schools and set up licensing laws to eliminate unqualified or irregular practitioners (Kendall, 1980).

The Rise of Scientific Medicine and Professionalism

Although medicine had been previously viewed more as an art than as a science, several significant discoveries during the nineteenth century in areas such as bacteriology and anesthesiology began to give medicine increasing credibility as a science (Nuland, 1997). At the same time that these discoveries were occurring, the ideology of science was being advocated in all areas of life, and people came to believe that almost any task could be done better if the appropriate scientific methods were used. To make medicine in the United States more scientific (and more profitable), the Carnegie Foundation (at the request of the American Medical Association and the forerunner of the Association of American Medical Colleges) commissioned an official study of medical education. The "Flexner report" that resulted from this study has been described as the catalyst of modern medical education but has also been criticized for its lack of objectivity.

The Flexner Report To conduct his study, Abraham Flexner met with the leading faculty at the Johns Hopkins University School of Medicine to develop a model of how medical education should take place; he next visited each of the 155 medical schools then in existence, comparing them with the model. Included in the model was the belief that a medical school should be a full-time, research-oriented laboratory facility that devoted all of its energies to teaching and research, not to the practice of medicine (Kendall, 1980). It should employ "laboratory men" to train students in the "science" of medicine, and the students should then apply the principles they had learned in the sciences to the illnesses of patients (Brown, 1979). Only a few of the schools Flexner visited were deemed to be equipped to teach scientific medicine; nonetheless, his model became the standard for the profession (Duffy, 1976).

As a result of the Flexner report (1910), all but two of the African American medical schools then in existence were closed, and only one of the medical schools for women survived. As a result, white women and people of color were largely excluded from medical education for the first half of the twentieth century. Until the civil rights movement and the women's movement of the 1960s and 1970s, virtually all physicians were white, male, and upper or upper-middle class.

The Professionalization of Medicine Despite its adverse effect on people of color and women who might desire a career in medicine, the Flexner report did help professionalize medicine. When we compare post-Flexner medicine with the characteristics of professions (see Chapter 13), we find that it meets those characteristics:

1. *Abstract, specialized knowledge.* Physicians undergo a rigorous education that results in a theoretical understanding of health, illness, and medicine. This education provides them with the credentials, skills, and training associated with being a professional.
2. *Autonomy.* Physicians are autonomous and (except as discussed subsequently in this chapter) rely on their own judgment in selecting the appropriate technique for dealing with a problem. They expect patients to respect that autonomy.
3. *Self-regulation.* Theoretically, physicians are self-regulating. They have licensing, accreditation, and regulatory boards and associations that set professional standards and require members to adhere to a code of ethics as a form of public accountability.
4. *Authority.* Because of their authority, physicians expect compliance with their directions and advice. They do not expect clients to argue about the advice rendered (or the price to be charged).
5. *Altruism.* Physicians perform a valuable service for society rather than acting solely in their own self-interest. Many physicians go beyond their self-interest or personal comfort so that they can help a patient.

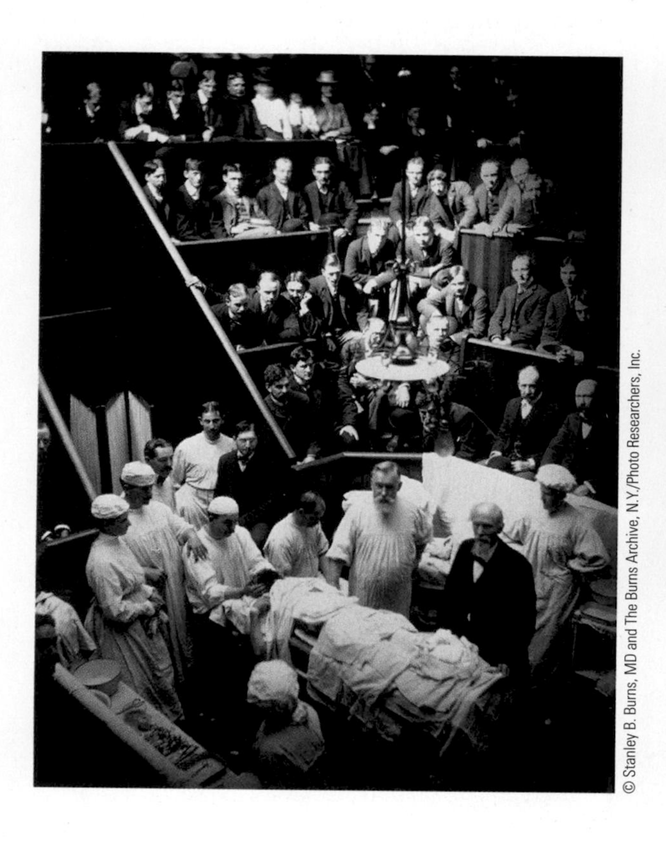

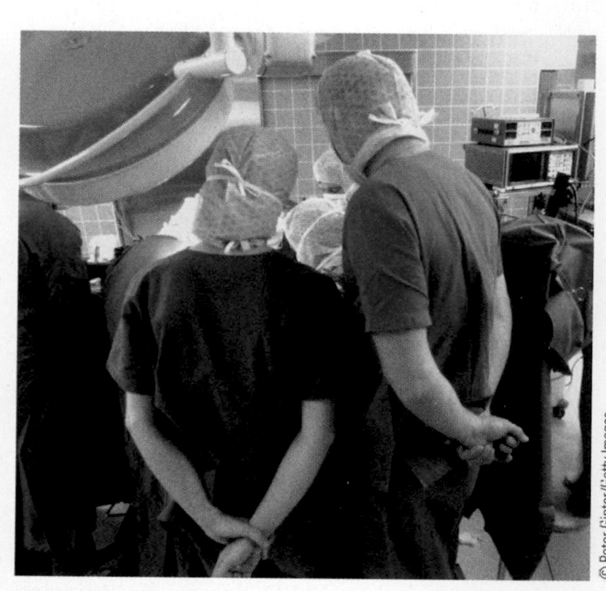

▲ A sight similar to the one on the left might have greeted Abraham Flexner as he prepared his report on medical education in the United States: students observing while their professor demonstrates surgical techniques. And although today's health care facilities look very different, many of the same teaching techniques are still employed.

However, with professionalization, licensed medical doctors gained control over the entire medical establishment, a situation that has continued until the present and—despite current efforts at cost control by insurance companies and others—may continue into the future.

Medicine Today

Throughout its history in the United States, medical care has been on a *fee-for-service* basis: Patients are billed individually for each service they receive, including treatment by doctors, laboratory work, hospital visits, prescriptions, and other health-related expenses. Fee for service is an expensive way to deliver health care because there are few restrictions on the fees charged by doctors, hospitals, and other medical providers.

There are both good and bad sides to the fee-for-service approach. The good side is that in the "true spirit" of capitalism, coupled with the hard work and scholarship of many people, this approach has resulted in remarkable advances in medicine. The bad side of fee-for-service medicine is its inequality of distribution. In effect, the United States has a two-tier system of medical care. Those who can afford it are able to get top-notch medical treatment. And where they receive it may not be much like the hospitals that most of us have visited:

> Every afternoon, between three and five, high above New York's Fifth Avenue, the usual quiet of Eleven West is broken by the soft rustle of white linen cloths and the clink of silver and china as high tea is served room by room. . . . Down the hall a concierge waits to take your dinner order, provide a video from a list of over 950 titles, arrange for a manicure or massage, or send up that magazine or best-seller that you wanted to read. No, this is not a hitherto unknown outpost of the Four Seasons or the Ritz, but a 19-room wing of the Mt. Sinai Medical Center, one of the nation's leading hospitals. . . .
>
> Here at Mt. Sinai and a few other top hospitals . . . sheets are 250-count cotton, the bathrooms are marble and stocked with toiletries, and there is ample room to accommodate a nice seating group of leather wing chairs and a brocade sofa. And here no call button is pressed in vain. Hospital personnel not only come when summoned but are eagerly waiting to cater to your every need. . . .
>
> On these select floors, multi-tiered food service carts and their clattering, plastic trays are gone. "Room service" is in full force. Meals are presented, often course by course, by bow-tied, black-jacketed waiters from rolling, linen-covered tables. (Winik, 1997)

The additional charges for rooms on floors such as those described above may run from $250 to $1,000 per night more than the cost of the standard private room (Winik, 1997). Obviously, this sort of medical care is not within the budget of most of us. However, the cost of health care per person in the United States rose from $141 in 1960 to $8,090 (more than 55 times as much) in 2009 (Plunkett Research, 2010) and is still increasing. ▶ Figure 18.2 reflects recent cost increases. In 2009 health care spending in the United States added up to

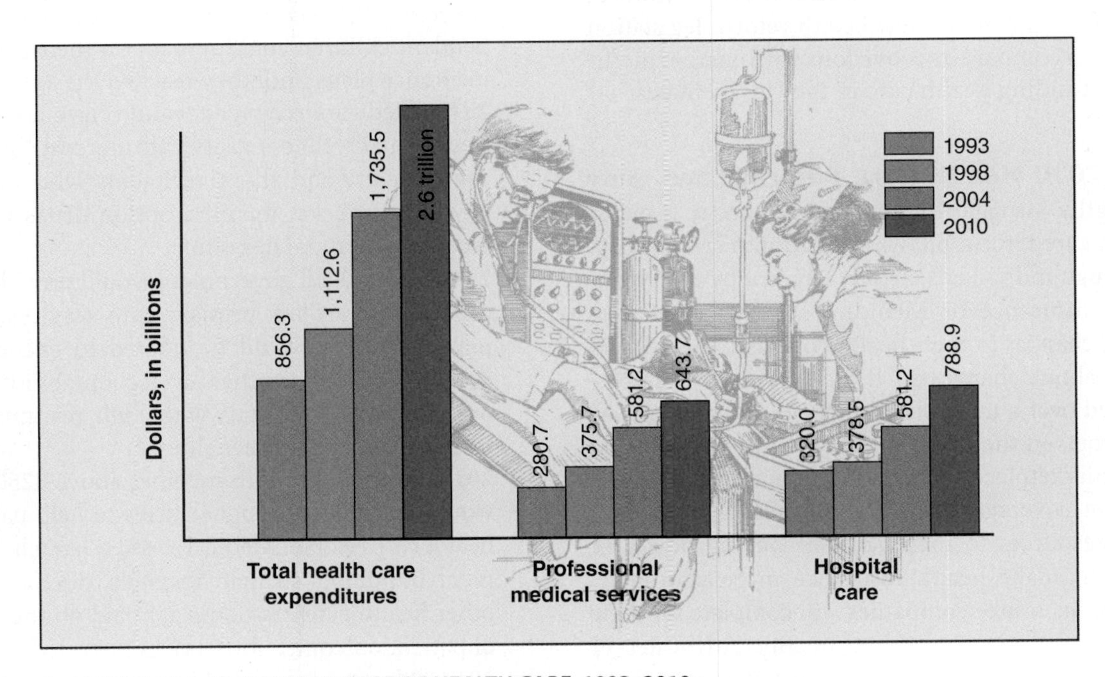

▲ FIGURE 18.2 INCREASE IN COST OF HEALTH CARE, 1993–2010

Source: Plunkett Research, 2010.

$2.5 trillion, which accounted for about 18 percent of the total economic output (GDP) of the nation (Pear, 2011). Health spending rose in 2009 over previous years but at the slowest rate of increase in 50 years. Spending went up even in a period of economic distress when more people lost their jobs, lost health insurance, and attempted to defer medical treatments as long as possible. Although there was a decline in visits to doctors' office and hospital admissions, this decrease was offset by a rapid increase in Medicaid spending, as an additional 3.5 million people were added to the program (Pear, 2011). Keeping in mind issues regarding income disparity in the United States, the questions to be considered at this point are "Who pays for medical care, and how?" and "What about the people who simply cannot afford medical care?"

© AP Photo/Gary Kazanjian

▲ The luxury services of the Mt. Sinai private wing are clearly intended for use by the wealthy, but for many Americans the crowded emergency room of the nearest hospital, such as the one shown here, is their only source of primary care.

Paying for Medical Care in the United States

Until recently the United States was the only high-income nation without some form of universal health coverage for all citizens. In 2010 the U.S. Congress passed a sweeping health care reform bill that was signed into law and will gradually bring about some changes in how health care is funded. Let's look first at the new health reform legislation and then compare its provisions with current methods of funding health care in the United States.

The 2010 Health Care Reform Law After a lengthy struggle in the U.S. Congress, a major health care reform bill was signed into law in 2010. Although individuals already covered by employer-based insurance or Medicare are unlikely to see major changes in their health coverage, the law will bring about changes in the future as it is implemented over a multiyear timetable. One of the central tenets in the law is the creation of a new insurance marketplace that lets individuals and families without coverage and small business owners pool their resources to increase their buying power in order to make health insurance more affordable. Private insurance companies will complete for their business based on cost and quality. Advocates of the new law believe that it is a first step in curbing abuses in the insurance industry.

Health care reform was scheduled to occur in the following stages:

- **2010:** Adults who had been unable to get coverage because of a preexisting condition could join a high-risk insurance pool (as a stopgap measure until the competitive health insurance marketplace begins in 2014). Insurance companies were required to cover children with preexisting conditions. Policies could not be revoked when people got sick. Preventive services were fully covered without co-payments or deductibles. Dependent children could remain on their parents' insurance plans until they reached the age of 26.
- **2011:** Medicare recipients would have access to free annual wellness visits with no cost for preventive care, and those recipients who had to pay out of pocket for prescription drugs would receive substantial discounts.
- **2012:** The federal government would provide additional money for primary-care services, and new incentives would be offered to encourage doctors to join together in accountability care organizations. Hospitals with high readmission rates would face stiff penalties.
- **2013:** Households with incomes above $250,000 would be subject to higher taxes to help pay for health care reform. Medicare would launch "payment bundling" so that hospitals, doctors, and other health care providers are paid on the basis of patient outcome, not services provided.
- **2014:** Most people would be required to buy health insurance or pay a penalty for not having

© KEVIN DIETSCH/UPI/Landov

▲ During the past century, many U.S. presidents, senators, and representatives have sought to reform health care. After a prolonged, intense debate in 2009 and 2010, the United States joined the other high-income nations in offering universal health care when President Barack Obama signed the historic legislation on March 23, 2010.

it. Insurance companies could not deny a policy to anyone based on health status, nor could they refuse to pay for treatment on the basis of preexisting health conditions. Annual limits on health care coverage would be abolished. Each state would be required to open a health insurance exchange, or marketplace, so that individuals and small businesses without coverage could comparatively shop for health packages. Tax credits would make insurance and health care more affordable for those who earned too much to qualify for Medicaid.

- **2018:** Insurance companies and plan administrators would pay a 40 percent excise tax on all family plans costing more than $27,500 a year.
- **2019:** The health reform law should have reduced the number of uninsured people by 32 million, leaving about 23 million uninsured. About one-third of the uninsured will be immigrants residing in the country without legal documentation.

These are a few of the highlights of the plan, which is a document of more than 2,000 pages in length. Examples of other provisions in the law include incentives to encourage doctors in training to pursue primary-care careers and to encourage more people to enter nursing. Some medical analysts believe that this reform measure will provide more cost control, such as competitive insurance exchanges that are supposed to lower premiums, and increase the quality of health care. Medicare pilot programs will be used to test innovative cost-reduction strategies, and greater emphasis is to be placed on the quality, not quantity, of services that health care providers deliver to their patients. At the time of this writing in 2011, several health care lawsuits are working their way through the courts because a number of states have challenged the constitutionality of the Affordable Care Act, particularly in regard to the "individual responsibility" provision, which requires everyone who can afford it to carry some form of health insurance. The Obama administration has staunchly defended the law, noting that it is constitutional and that further implementation will occur as these issues are resolved.

If this is what the future of health care funding looks like in the United States, how has it looked in the past? Let's look first at how private health insurance works.

Private Health Insurance Private health insurance is largely paid for by businesses and households. Beginning in the 1960s, medical insurance programs began to expand, and third-party providers (public and private insurers) began picking up large portions of doctor and hospital bills for insured patients. With third-party fee-for-service payment, patients pay premiums into a fund that in turn pays doctors and hospitals for each treatment the patient receives. Private health insurance premiums have continued to increase by about 6 percent per year, after a peak of 10.7 percent in 2002. Between 2002 and 2007, benefit payments slowed, from 9.4 percent to 6.6 percent, largely because of a decline in private health insurance spending growth on prescription drugs. During the same period, out-of-pocket spending (spending not reimbursed by a health insurance plan) increased by 5.3 percent because of increased out-of-pocket payments for prescription drugs, nursing home services, and nondurable medical supplies (Fritze, 2010).

Some believe that a third-party fee-for-service approach is the best and most cost-efficient method of delivering medical care. Others argue that fee for service is outrageously expensive and a very cost-ineffective way in which to provide for the medical needs of people in this country, particularly those who are without health insurance coverage. According to critics, third-party fee for service contributes greatly to medical inflation because it gives doctors and hospitals an incentive to increase medical services. In other words, the more services they provide, the more fees they charge, and the more money they make. Patients have no incentive to limit their visits to doctors or hospitals because they have already paid their premiums and feel entitled to medical care, regardless of the cost. This is one of the spiraling costs that advocates of the 2010 health care reform hope will be reduced when the new legislation is implemented.

Public Health Insurance Since the 1960s, the United States has had two nationwide public health insurance programs, Medicare and Medicaid. In 1965 Congress enacted Medicare, a federal program for people age 65 and over (who are covered by Social Security or railroad retirement insurance or who have been permanently and totally disabled for two years or more). This program was primarily funded through Social Security taxes paid by current workers. We refer to Medicare as an entitlement program because people who receive benefits under the plan must have paid something to be covered. Medicare Part A (hospital insurance) provides coverage for some inpatient hospital expenses, including critical-access hospitals and skilled-nursing facilities. It also helps cover hospice care and limited home health care. Part B (medical insurance) helps cover doctors' services and outpatient care. It covers some of the services of physical and occupational therapists, and some home health care. Most people pay a monthly premium for Part B. Beginning in 2006, Medicare prescription drug coverage became available to everyone covered under Medicare. Private companies provide the coverage, and beneficiaries choose the drug plan and pay a monthly premium.

Medicaid is the federal government's health care program for low-income and disabled persons and certain groups of seniors in nursing homes. Medicaid is jointly funded by federal/state/local monies, and various factors are taken into account when determining whether or not a person is eligible for Medicaid. Among these are age, disability, blindness, and pregnancy. Income and resources are also taken into consideration, as well as whether the person is a U.S. citizen or a lawfully admitted immigrant. Each state has its own rules regarding who may be covered under Medicaid, and some provide time-limited coverage for specific categories of individuals, such as uninsured women with breast or cervical cancer or those individuals diagnosed with TB (tuberculosis). As compared to Medicare, Medicaid has had a more tarnished image throughout its history. Unlike Medicare recipients, who are often seen as "worthy" of their health care benefits, Medicaid recipients have been stigmatized by politicians and media outlets for their participation in a "welfare program." Today, many physicians refuse to take Medicaid patients because the administrative paperwork is burdensome and reimbursements are low—typically less than one-half of what private insurance companies pay for the same services.

When the health care reform law passed in 2010, both the Medicaid program and the Medicare program were in financial difficulty. These two programs cost $760.6 billion annually and account for one-fourth (25 percent) of all federal spending. Medicare and Medicaid are growing more rapidly than the U.S. economy and the revenues that are used to finance them. For example, in 2007 Medicare spending grew 7.2 percent to $431.2 billion, which followed a growth of 18.5 percent in 2006, because of the one-time implementation of Medicare Part D (prescription coverage). Similarly, Medicaid spending grew 6.4 percent in 2007 to $329.4 billion (Fritze, 2010). Under the new health care reform law, eligibility for "free" Medicaid coverage will expand significantly, so it may be difficult to contain costs in this area. However, it is predicted that the new focus on preventive health care services will improve the health of recipients and reduce costs.

Health Maintenance Organizations (HMOs) Created in an effort to provide workers with health coverage by keeping costs down, ***health maintenance organizations*** (HMOs) **provide, for a set monthly fee, total care with an emphasis on**

AP Photo/Nick Ut

▲ Kaiser Permanente, a leading HMO in the United States, owns hospitals such as the one shown here.

prevention to avoid costly treatment later. The doctors do not work on a fee-for-service basis, and patients are encouraged to get regular checkups and to practice good health practices (e.g., exercise and eat right). However, research shows that preventive care is good for the individual's health but does not necessarily lower total costs. As long as patients use only the doctors and hospitals that are affiliated with their HMO, they pay no fees, or only small co-payments, beyond their insurance premiums.

Recent concerns about physicians being used as gatekeepers who might prevent some patients from obtaining referrals to specialists or from getting needed treatment have resulted in changes in the policies of some HMOs, which now allow patients to visit health care providers outside an HMO's network or to receive other previously unauthorized services by paying a higher co-payment. However, critics charge that those HMOs whose primary-care physicians are paid on a capitation basis—meaning that they receive only a fixed amount per patient whom they see, regardless of how long they spend with that patient—in effect encourage doctors to undertreat patients.

Managed Care Another approach to controlling health care costs in the United States is known as ***managed care*—any system of cost containment that closely monitors and controls health care providers' decisions about medical procedures, diagnostic tests, and other services that should be provided to patients.** One type of managed care in the United States is a *preferred provider organization* (PPO), which is an organization of medical doctors, hospitals, and other health care providers who enter into a contract with an insurer or a third-party administrator to provide health care at a reduced rate to patients who are covered under specific insurance plans. In most managed-care programs, patients choose a primary-care physician from a list of participating doctors. Unlike many of the HMOs, when a patient covered under a PPO plan needs medical services, he or she may contact any one of a number of primary-care physicians or specialists who are "in-network" providers. Like HMOs, most PPO plans do contain a precertification requirement in which scheduled (nonemergency) hospital admissions and certain kinds of procedures must be approved in advance. Through measures such as this, these insurance plans have sought unsuccessfully to curb the rapidly increasing costs of medical care and to reduce the extensive paperwork and bureaucracy involved in the typical medical visit. For the foreseeable future, HMOs and PPOs are supposed to remain somewhat the same. After the passage of the health care reform measure, the Obama administration widely publicized a statement that people in plans such as these would not be affected by the new law. Rather, they would have assurance that they could get health care coverage even if they lost their job, changed jobs, moved out of state, got divorced, or were diagnosed with a serious illness.

The Uninsured Despite existing public and private insurance programs, about one-third of all U.S. citizens were without health insurance or had difficulty getting or paying for medical care at some time in the last year. Under the new health care reform, this situation will not change significantly until about 2014, when each state will open a health insurance exchange, or marketplace, for individuals and small businesses without health insurance coverage. Prior to this, millions of people will remain uninsured. As shown on ▶ Map 18.3, the number of people not covered by health insurance varies from state to state. In 2009 the percentage of people without health insurance increased to 16.7 percent, as compared to 15.4 percent in 2008. The number of uninsured people increased to 50.7 million in 2009, up from 46.3 million in 2008 (DeNavas-Walt, Proctor, and Smith, 2010). Children made up 7.5 million of the uninsured in 2009, and this number represented about 10 percent of all children under age 18 in the United States (see ▶ Figure 18.3). Statistically

health maintenance organizations (HMOs) companies that provide, for a set monthly fee, total care with an emphasis on prevention to avoid costly treatment later.

managed care any system of cost containment that closely monitors and controls health care providers' decisions about medical procedures, diagnostic tests, and other services that should be provided to patients.

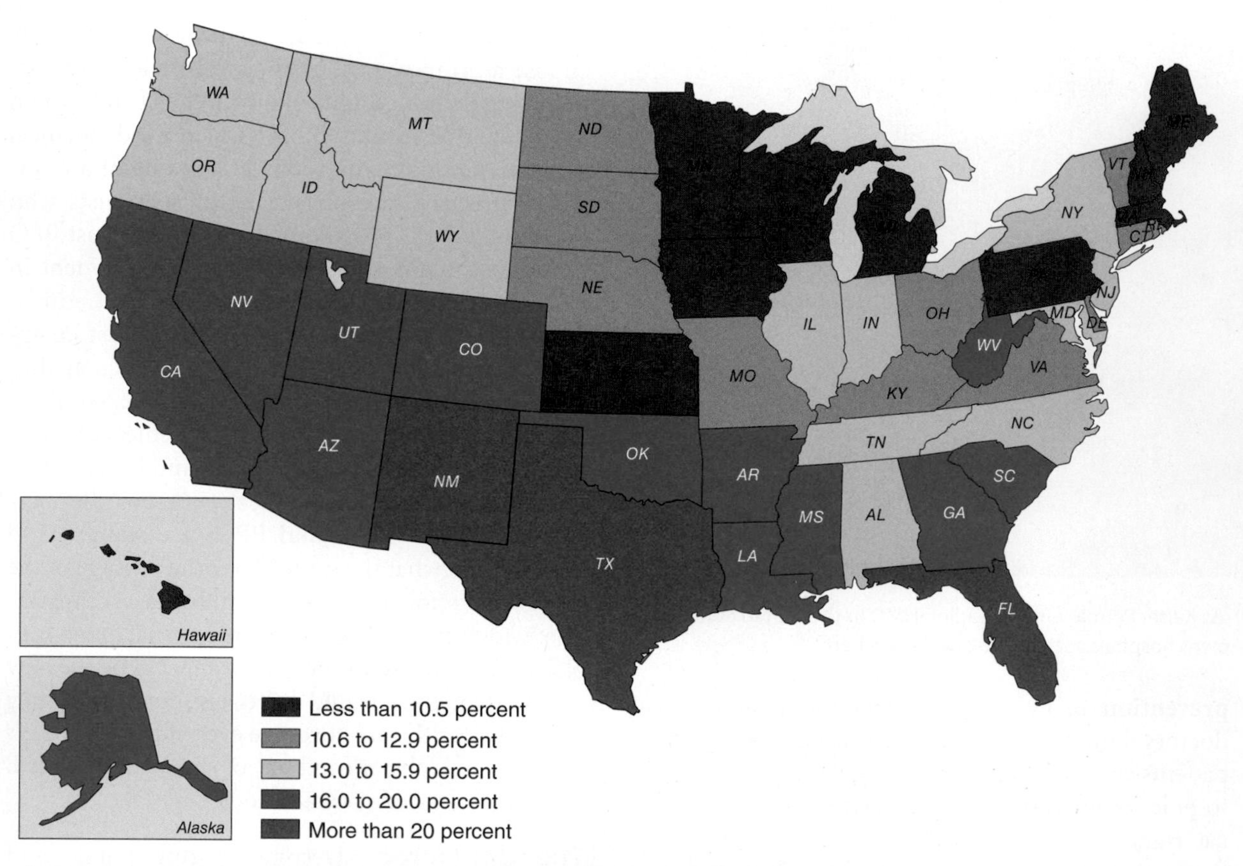

▲ **MAP 18.3 PERSONS NOT COVERED BY HEALTH INSURANCE, BY STATE**

Source: U.S. Census Bureau, 2008.

speaking, as this figure shows, children who are more likely to be uninsured are ones who live in poverty, are Hispanic (any race), and are not citizens of the United States. The rate of uninsured children would be even higher if there were not state-sponsored programs such as CHIP, a health insurance plan for children that is offered on a sliding scale with premiums (down to zero) based on family income.

Race and ethnicity are important factors in health insurance coverage. Although every racial and ethnic group is affected, Latinos/as and African Americans are more likely to be uninsured than are white (non-Hispanics) Americans. For Hispanic Americans the uninsured rate was 32.4 percent, as compared to African Americans at 21 percent, Asian Americans at 17.2 percent, and white (non-Hispanic) Americans at 12 percent.

Many people who are insured have coverage through their employer. An important negative trend in health insurance is a downward movement in the percentage of people covered by employment-based health insurance, from 58.5 percent in 2008 to 55.8 percent in 2009, as a result of job loss or private employers and large corporations cutting costs by reducing insurance plans and other employee benefits.

Paying for Medical Care in Other Nations

Other nations have various ways in which they provide health care for their citizens. Let's examine how other nations pay for health care.

Canada Health care in Canada is delivered through a publicly funded health care system. Services are provided by private entities and are mostly free to patients. Information remains private between physician and patient, and the government is not involved in patient care. Each citizen receives a health card, and all patients receive the same level of care. As long as a person's premiums are paid up, health coverage is not affected by losing or changing jobs.

How did the contemporary Canadian health care system get started? In 1962 the government of the province of Saskatchewan implemented a health insurance plan despite opposition from doctors, who went on strike to protest the program. The strike was not successful, as the vast majority of citizens supported the government, which maintained health services by importing doctors from Great Britain. The Saskatchewan program proved itself viable in the years following the strike, and by 1972 all Canadian provinces and territories had coverage for medical

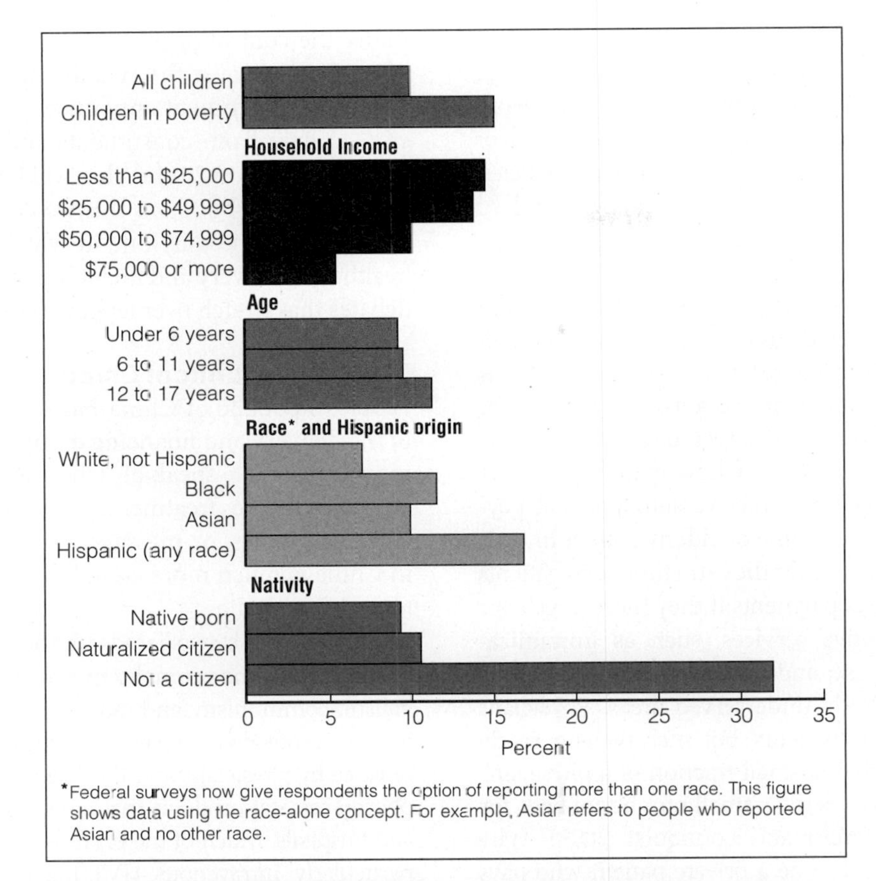

▲ **FIGURE 18.3 UNINSURED CHILDREN BY POVERTY STATUS, HOUSEHOLD INCOME, AGE, RACE, HISPANIC ORIGIN, AND NATIVITY: 2009**

Source: U.S. Census Bureau, Current Population Survey, 2010 Annual Social and Economic Supplement.

and hospital services (Kendall, Linden, and Murray, 2008). As a result, Canada has a ***universal health care* system—a health care system in which all citizens receive medical services paid for by tax revenues.** In Canada these revenues are supplemented by insurance premiums paid by all taxpaying citizens.

One major benefit of the Canadian system is a significant reduction in administrative costs. Whereas more than 20 percent of the U.S. health care dollar represents administrative costs, in Canada the corresponding figure is 10 percent (Weiss and Lonnquist, 2009). However, the system is not without its critics, who claim that it is costly and often wasteful. For example, Canadians are allowed unlimited trips to the doctor, and doctors can increase their income by ordering extensive tests and repeat visits, just as in the United States (Kendall, Linden, and Murray, 2008). Recent criticisms have emerged that many Canadians do not have a family physician and that a large number of acute-care beds in hospitals are filled with elderly patients who would be better served in long-term care facilities. This all adds up to massive problems in the nation's emergency rooms, where many patients seek medical attention that they otherwise cannot find, much like some of the complaints about the contemporary U.S. health care delivery system (MacQueen, 2011).

The Canadian health care system does not constitute what is referred to as ***socialized medicine*—a health care system in which the government owns the medical care facilities and employs the physicians.** Rather, Canada has maintained the private nature of the medical profession. Although the government pays most health care costs, the physicians are not government employees and have much greater autonomy than do physicians in the health care system in Great Britain.

Great Britain For many years, Britain has had a centralized, single-payer health care system that is funded by general revenues. Discussions are now under way to decentralize some aspects of health care in that nation, but let's look at the history of the socialized health care system there so that we can better understand what the issues are in the twenty-first century.

universal health care a health care system in which all citizens receive medical services paid for by tax revenues.

socialized medicine a health care system in which the government owns the medical care facilities and employs the physicians.

The National Health Service Act of 1946 provides for all health care services to be available at no charge to the entire population. Although physicians work out of offices or clinics—as in the United States or Canada—the government sets health care policies, raises funds and controls the medical care budget, owns health care facilities, and directly employs physicians and other health care personnel (Weiss and Lonnquist, 2009). Unlike the Canadian model, the health care system in Britain *does* constitute socialized medicine because physicians receive capitation payments from the government, which is a fixed annual fee for each patient in their practice regardless of how many times they see the patient or how many procedures they perform. They also receive supplemental payments for each low-income or elderly patient in their practice to compensate for the extra time such patients may require; bonus payments if they meet targets for providing preventive services, such as immunizations against disease; and financial incentives if they practice in medically underserved areas. Physicians may accept private patients, but such patients rarely constitute more than a small fraction of a physician's practice; hospitals reserve a small number of beds for private patients (Weiss and Lonnquist, 2009). Why would anyone want to be a private patient who pays for her or his own care or hospital bed? The answer is primarily found in the desire to avoid the long waits ("queues") that the general population encounters and the fact that private patients can enter the hospital for surgery at times convenient to the consumer rather than wait upon the convenience of the system.

When Britain's first coalition government in more than seventy years came into power in 2010, Prime Minister David Cameron, head of the Conservative Party, emphasized that deficit cutting was a top priority but that health care financing and delivery would be left alone. However, shortly thereafter a proposal was made to radically reorganize the National Health Service in England and to shift control of the $160 billion annual health budget from a centralized bureaucracy to physicians at the local level (Lyall, 2010). The goals of the new plan are to cut costs, avoid duplication of services, and make the system less cumbersome. Advocates of the changes in the Health and Social Care Bill 2011 believe that decision making should be shifted to physicians and patients, and away from 150 entities known as primary care trusts that previously made most of the decisions. Although this bill applies only to England because other parts of Britain have separate systems, both advocates and critics believe that it may have a long-term effect on all of Britain. Critics of the new plan are concerned that it may make matters worse, rather than better, to place the health care system and decisions about how it is financed and spends money under the control of physicians. Given the amount of controversy over the Health and Social Care Bill 2011, the government announced a "listening exercise" to hear from constituents and postponed the issue until later in 2011. What will happen in the future is unclear, but from this we can learn that our nation is not the only one in which issues such as health care delivery and financing create contentious debates that stretch over lengthy periods of time.

People's Republic of China In recent years the People's Republic of China has seen many changes in the delivery and financing of medical care. When most U.S. people speak of "Chinese medicine," they are referring to treatments such as acupuncture, herbal remedies, or massage. However, health care in China is much more complex than the use of alternative remedies.

During the past two decades the health care system in China has become a complex mix of market-driven capitalism, communism, and massive government spending. The profit-driven system is based on fee-for-service practice by physicians and the sale of highly expensive pharmaceutical products that are marketed by doctors and hospitals. Much of the profit is through drug sales, particularly intravenous (IV) treatments, which are used much more widely than in the United States. Until reforms occurred in about 2008–2009, pharmaceutical products were very costly; however, they have become somewhat more affordable since that time.

Ironically, as the Chinese government stepped up funding for health care, social disease epidemics, such as obesity and/or illnesses related to excessive use of tobacco, alcohol, and salt, took their toll on many in the population: Diabetes, hypertension, and environmentally induced heart and lung diseases have remained on the upswing for a number of years (Mills, 2010).

What is the history of health care delivery and funding in China? After a lengthy civil war, in 1949 the Communist Party won control of mainland China but found itself in charge of a vast nation with a population of one billion people, most of whom lived in poverty and misery. Malnutrition was prevalent, life expectancies were short, and infant and maternal mortality rates were high. In the cities only the elite could afford medical care; in the rural areas, where most of the population resided, Western-style health care barely existed (Weitz, 2010). With a lack of financial resources and not enough trained health care personnel, China adopted a policy to create a large number of *physician extenders* who could educate the public about health and the treatment of illness and disease. Referred to as *street doctors* in urban areas and *barefoot doctors* in the countryside, these individuals had little formal training and worked under the supervision of trained physicians (Weitz, 2010).

Over the past six decades, medical training has become more rigorous. All doctors receive training in both Western and traditional Chinese medicine. Most doctors who work in hospitals receive a salary; all other doctors now work on a fee-for-service basis. In urban areas about 94 percent of the working population has health insurance coverage paid for by employers, but individuals often pay large out-of-pocket fees because of gaps in health care coverage. Today, slightly more than half (approximately 53.4 percent) of the population of mainland China (excluding Hong Kong and Macau) live in rural areas, where large numbers of migrant farm workers must provide for their own health care. New government initiatives established in the 2010s call for the establishment of a clinic in each rural community; however, many more physicians will be needed to provide adequate care for the millions of people living in these areas.

In urban areas the latest reports show that it is difficult to get into a hospital even with insurance. Increasing demands for services have placed the already overburdened system under greater stress, and many people line up outside the more prestigious hospitals early in the morning on the day *before* they want treatment to get their name on lengthy waiting lists. Although local doctors might be able to take care of their needs, many Chinese patients want to be treated in the more prestigious hospitals, where they believe they will receive higher quality care. To help meet the growing demands on the health care delivery system, the Chinese government is attempting to improve thousands of medical centers and to focus on preventive care, especially for infants, children, pregnant women, and those in need of mental health care. Despite these efforts, both services and funding remain quite different when comparing urban and rural areas: The health care disparities remain great, and more money and medical personnel are needed throughout the system, but particularly in rural areas, to meet the needs of a rapidly growing population.

Regardless of which approach to health care delivery and financing that a nation uses, health care providers, hospitals, governmental agencies, political leaders, and the general public all face many difficult issues about the best way to meet patients' needs and not bankrupt the medical system and the nation. One area that may increase costs and generate complex concerns is advanced medical technology.

Social Implications of Advanced Medical Technology

Advances in medical technology are occurring at a speed that is almost unbelievable; however, sociologists and other social scientists have identified specific social implications of some of the new technologies (see Weiss and Lonnquist, 2009):

1. *Advanced technologies create options for people and for society, but options that alter human relationships.* An example is the ability of medical personnel to sustain a life that in earlier times would have ended as the result of disease or an accident. Although this can be beneficial, technologically advanced equipment (that can sustain life after consciousness is lost and there is no likelihood that the person will recover) can create

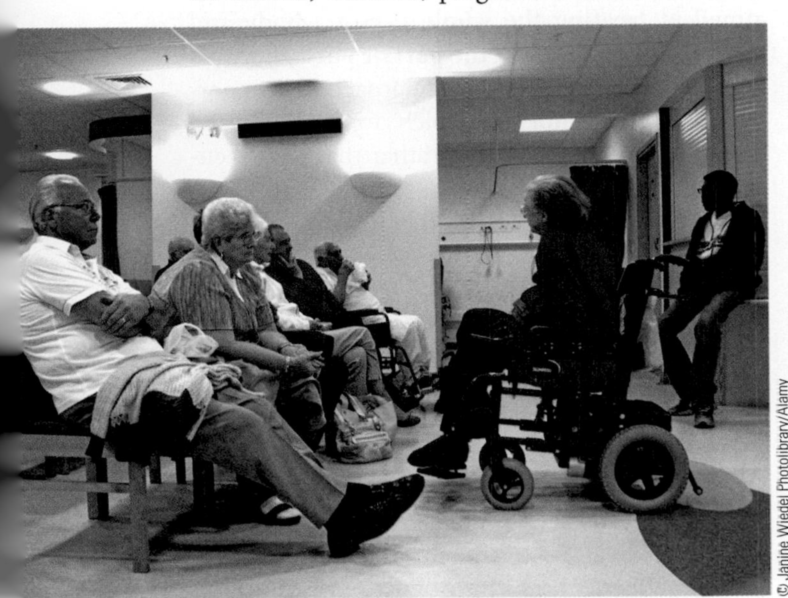

© Janine Wiedel Photolibrary/Alamy

▲ Socialized medicine is often criticized for the long waits that patients are forced to endure. Here, people at a British hospital patiently wait their turn. Will the new U.S. health care law bring about longer waits for medical care in this country?

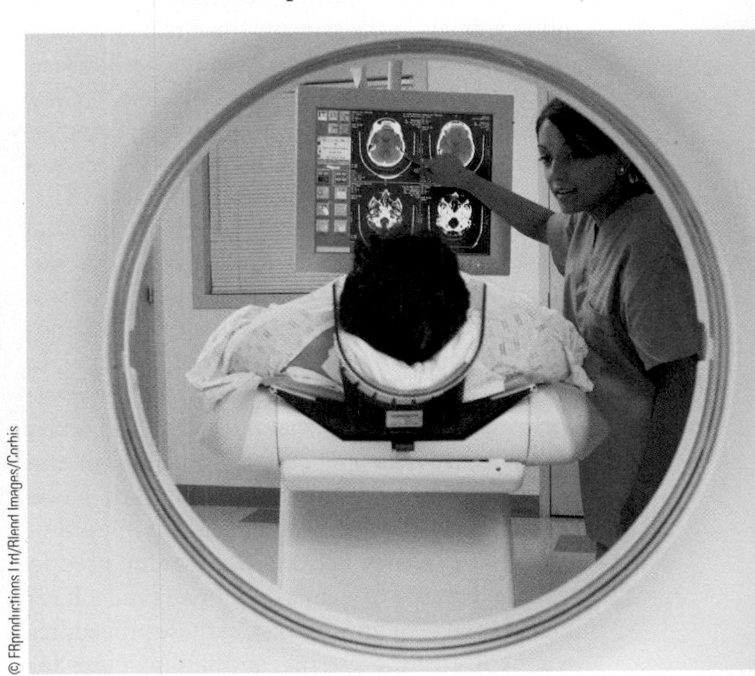

© FRproductions Ltd/Blend Images/Corbis

▲ High-technology medical imaging devices, such as the one shown here, are very expensive. Some analysts question whether expenditures of this size could be better used to provide other kinds of health care services for individuals.

a difficult decision for the family of that person if he or she has not left a *living will*—a document stating the person's wishes regarding the medical circumstances under which his or her life should be terminated. Federal law requires all hospitals and other medical facilities to honor the terms of a living will.

2. *Advanced technologies increase the cost of medical care.* For example, the computerized axial tomography (CT or CAT) scanner—which combines a computer with X-rays that are passed through the body at different angles—produces clear images of the interior of the body that are invaluable in investigating disease. The cost of such a scanner is around $1 million. Magnetic resonance imaging (MRI) equipment that allows pictures to be taken of internal organs ranges in cost from $1 million to $2.5 million. Can the United States afford such equipment in every hospital for every patient? The money available for health care is not unlimited, and when it is spent on high-tech equipment and treatment, it is being reallocated from other health care programs that might be of greater assistance to more people.
3. *Advanced technologies such as cloning and stem cell research raise provocative questions about the very nature of life.* In Chapter 15 we briefly discuss in vitro fertilization—a form of assisted reproductive technology. But during 1997, Dr. Ian Williams and his associates in Scotland took in vitro fertilization a step further: They cloned a lamb (that they named Dolly) from the DNA of an adult sheep. Subsequently, scientists have cloned other animals in the same manner, raising a number of profound questions: If scientists can duplicate mammals from adult DNA, is it possible to clone a perfect (whatever that may be) human being instead of taking a chance on a child that is born to a couple? If it is possible, would it be ethical?

   Like cloning, stem cell research has been an important and controversial issue in medicine. Stem cell research is important because these stem cells are building blocks that can be manipulated to perform the work of other cell types. In other words, stem cells can be used to generate virtually any type of specialized cell in the human body. Stems cells are accessible in the skin and through extraction from umbilical cord blood and human embryos. These cells are important to medical science because they can be used to replace diseased or damaged human tissue. It remains to be seen how successful these procedures will be in treating patients and finding cures for Alzheimer's, Parkinson's, diabetes, cancer, and numerous other diseases. These technological advances also produce ethical dilemmas and intense political and religious debates. Specifically, opponents of *embryonic* stem cell research believe that a human life is taken when a human embryo is destroyed in this research. Proponents of stem cell research respond that these studies do not always require the use of embryos. They also point out that this research may lead scientists to discoveries that they otherwise would never find. Research continues, even if on a small scale. In 2011, for example, Chicago medical researchers were testing the safety and usefulness of a treatment, derived from embryonic stem cells, on a patient paralyzed from the chest down as a result of a car crash. The purpose of this clinical study was to observe the effects of using these stem cells in humans and to determine if the stem cells might improve the patient's neuromuscular control or sensation (Japsen, 2011). As with most clinical trials, it will probably be decades before studies such as this make a contribution to our knowledge about how to best treat patients with certain medical conditions.

However, at the same time that advanced medical technologies are becoming an integral part of health care delivery, many people are turning to holistic medicine and alternative healing practices.

Holistic Medicine and Alternative Medicine

When examining the subject of medicine, it is easy to think only in terms of conventional (or mainstream) medical treatment. By contrast, ***holistic medicine* is an approach to health care that focuses on prevention of illness and disease and is aimed at treating the whole person—body and mind—rather than just the part or parts in which symptoms occur.** Under this approach, it is important that people not look solely to medicine and doctors for their health, but rather that people engage in health-promoting behavior. Likewise, medical professionals must not only treat illness and disease but also work with the patient to promote a healthy lifestyle and self-image.

Many practitioners of *alternative medicine*—healing practices inconsistent with dominant medical practice—take a holistic approach, and today many people are turning to alternative medicine either in addition to or in lieu of traditional medicine. However, some medical doctors are opposed to alternative medicine. In understanding the medical establishment's reaction to alternative medicine, it is important to keep in mind the philosophy of scientific medicine—that medicine is a science, not an art. Thus, to the extent to which alternative medicine is "nonscientific," it must be quackery and therefore something that is undoubtedly worthless and possibly harmful. Undoubtedly, self-interest is also involved in mainstream medicine's reaction to alternative medicine: If the public can be persuaded that scientific medicine is the only legitimate healing practice, fewer health care dollars will be spent on a form of medical treatment that is (at least to some extent) in competition with the medical establishment (Weiss and Lonnquist, 2009). But if all forms of alternative medicine (including chiropractic, massage, and spiritual) are taken into account, people spend more money on unconventional therapies than they do for all hospitalizations (Weiss and Lonnquist, 2009).

Just as the Internet and social networking sites have made people more aware of "mainstream" medical treatments, these global communications networks have informed millions of individuals about alternative medicines. Some sites provide valuable information to patients about how they can take better care of themselves or reduce the effects of a health care crisis. However, some medical professionals are concerned that the uninitiated (and perhaps gullible) may be misled by crafty individuals who want to sell bogus products and services that have no positive effect, but may have negative side effects, on a patient's condition. If people do spend more money on unconventional therapies than they do for all hospitalizations, many million and perhaps billions of dollars are spent on treatments that may not meet what is called the "gold standard" in medicine, which refers to the best-available test, treatment approach, or method that is accepted by the established medical community for dealing with a specific condition.

▲ The use of herbal therapies is a form of alternative medicine that is increasing in popularity in the United States. How does this approach to health care differ from a more traditional medical approach?

Sociological Perspectives on Health and Medicine

Functionalist, conflict, symbolic interactionist, and postmodernist perspectives focus on different aspects of health and medicine; each provides us with significant insights into the problems associated with these pressing social concerns.

A Functionalist Perspective: The Sick Role

According to the functionalist approach, if society is to function as a stable system, it is important for people to be healthy and to contribute to their society. Consequently, sickness is viewed as a form of deviant behavior that must be controlled by society. This view was initially set forth by the sociologist Talcott Parsons (1951) in his concept of the ***sick role*—the set of patterned expectations that defines the norms and values appropriate for individuals who are sick and for those who interact with them.** According to Parsons, the sick role has four primary characteristics:

1. People who are sick are not responsible for their condition. It is assumed that being sick is not a deliberate and knowing choice of the sick person.
2. People who assume the sick role are temporarily exempt from their normal roles and obligations. For example, people with illnesses are typically not expected to go to school or work.
3. People who are sick must want to get well. The sick role is considered to be a temporary one that people must relinquish as soon as their condition improves sufficiently. Those who do not return to their regular activities in a timely fashion may be labeled as hypochondriacs or malingerers.
4. People who are sick must seek competent help from a medical professional to hasten their recovery.

As these characteristics show, Parsons believed that illness is dysfunctional for both individuals and the larger society. Those who assume the sick role are unable to fulfill their necessary social roles, such as being parents or employees. Similarly, people who are ill lose days from their productive roles in

holistic medicine an approach to health care that focuses on prevention of illness and disease and is aimed at treating the whole person—body and mind—rather than just the part or parts in which symptoms occur.

sick role the set of patterned expectations that defines the norms and values appropriate for individuals who are sick and for those who interact with them.

society, thus weakening the ability of groups and organizations to fulfill their functions.

According to Parsons, it is important for society to maintain social control over people who enter the sick role. Physicians are empowered to determine who may enter this role and when patients are ready to exit it. Because physicians spend many years in training and have specialized knowledge about illness and its treatment, they are certified by the society to be "gatekeepers" of the sick role. When patients seek the advice of a physician, they enter into the patient–physician relationship, which does not contain equal power for both parties. The patient is expected to follow the "doctor's orders" by adhering to a treatment regime, recovering from the malady, and returning to a normal routine as soon as possible.

What are the major strengths and weaknesses of Parsons's model and, more generally, of the functionalist view of health and illness? Parsons's analysis of the sick role was pathbreaking when it was introduced. Some social analysts believe that Parsons made a major contribution to our knowledge of how society explains illness-related behavior and how physicians have attained their gatekeeper status. In contrast, other analysts believe that the sick-role model does not take into account racial–ethnic, class, and gender variations in the ways that people view illness and interpret this role. For example, this model does not take into account the fact that many individuals in the working class may choose not to accept the sick role unless they are seriously ill—because they cannot afford to miss time from work and lose a portion of their earnings. Moreover, people without health insurance may not have the option of assuming the sick role.

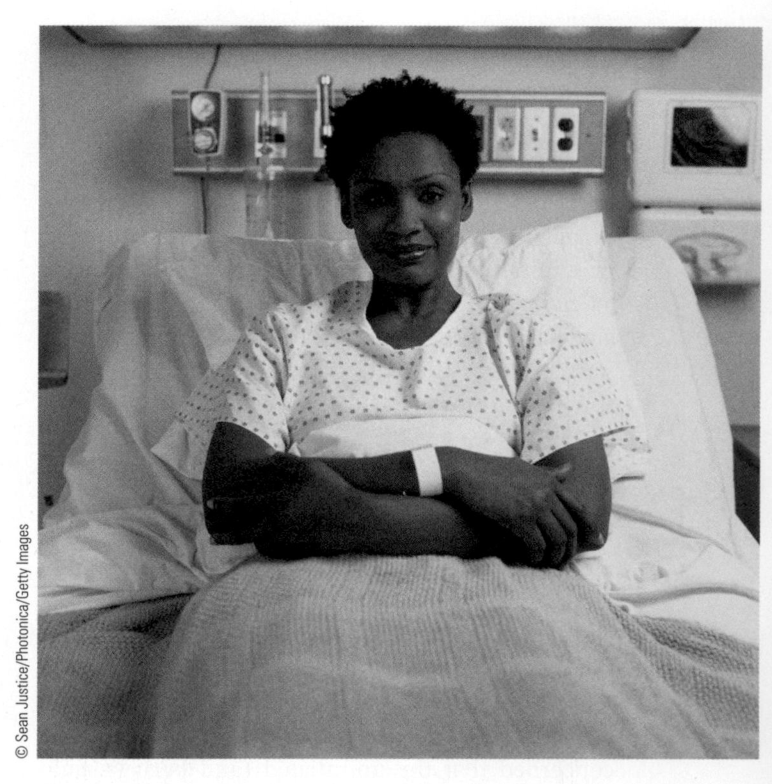

▲ According to the functionalist perspective, the sick role exempts the patient from routine activities for a period of time but assumes that the individual will seek appropriate medical attention and get well as soon as possible.

A Conflict Perspective: Inequalities in Health and Health Care

Unlike the functionalist approach, conflict theory emphasizes the political, economic, and social forces that affect health and the health care delivery system. Among the issues of concern to conflict theorists are the ability of all people to obtain health care; how race, class, and gender inequalities affect health and health care; power relationships between doctors and other health care workers; the dominance of the medical model of health care; and the role of profit in the health care system.

Who is responsible for problems in the U.S. health care system? According to many conflict theorists, problems in U.S. health care delivery are rooted in the capitalist economy, which views medicine as a commodity that is produced and sold by the medical–industrial complex. The ***medical–industrial complex*** **encompasses local physicians and hospitals as well as global health-related industries such as insurance companies and pharmaceutical and medical supply companies** (Relman, 1992).

The United States is one of the few industrialized nations that has relied almost exclusively on the medical–industrial complex for health care delivery and has not had any form of universal health coverage to provide some level of access to medical treatment for all people. Consequently, access to high-quality medical care has been linked to people's ability to pay and to their position within the class structure. Those who are affluent or have good medical insurance may receive high-quality, state-of-the-art care in the medical–industrial complex because of its elaborate technologies and treatments. However, people below the poverty level and those just above it have greater difficulty gaining access to medical care. Referred to as the *medically indigent,* these individuals do not earn enough to afford private medical care but earn just enough money to keep them from qualifying for Medicaid (Weiss and Lonnquist, 2009). In the profit-oriented capitalist economy, these individuals are said to "fall between the cracks" in the health care system.

Who benefits from the existing structure of medicine? According to conflict theorists, physicians—who hold a legal monopoly over medicine—benefit from the existing structure because they can charge inflated fees. Similarly, clinics, pharmacies, laboratories, hospitals, supply manufacturers, insurance companies, and many other corporations derive excessive profits from the existing system of payment in medicine. In recent years, large drug companies and profit-making hospital corporations have come to occupy a larger and larger part of health care delivery. As a result, medical costs have risen rapidly, and the federal government and many insurance companies have placed pressure for cost containment on other players in the medical–industrial complex.

Conflict theorists increase our awareness of inequalities of race, class, and gender as these statuses influence people's access to health care. They also inform us about the problems associated with health care becoming "big business." However, some analysts believe that the conflict approach is unduly pessimistic about the gains that have been made in health status and longevity—gains that are at least partially due to large investments in research and treatment by the medical–industrial complex.

A Symbolic Interactionist Perspective: The Social Construction of Illness

Symbolic interactionists attempt to understand the specific meanings and causes that we attribute to particular events. In studying health, symbolic interactionists focus on the meanings that social actors give their illness or disease and how these affect people's self-concept and relationships with others. According to symbolic interactionists, we socially construct "health" and "illness" and how both should be treated. For example, some people explain disease by blaming it on those who are ill. If we attribute cancer to the acts of a person, we can assume that we will be immune to that disease if we do not engage in the same behavior. Nonsmokers who learn that a lung cancer victim had a two-pack-a-day habit feel comforted that they are unlikely to suffer the same fate. Similarly, victims of AIDS are often blamed for promiscuous sexual conduct or intravenous drug use, regardless of how they contracted HIV. In this case the social definition of the illness leads to the stigmatization of individuals who suffer from the disease.

Although biological characteristics provide objective criteria for determining medical conditions such as heart disease, tuberculosis, or cancer, there is also a subjective component to how illness is defined. This subjective component is very important when we look at conditions such as childhood hyperactivity, mental illness, alcoholism, drug abuse, cigarette smoking, and overeating, all of which have been medicalized. The term ***medicalization*** **refers to the process whereby nonmedical problems become defined and treated as illnesses or disorders.** Medicalization may occur on three levels: (1) the conceptual level (e.g., the use of medical terminology to define the problem), (2) the institutional level (e.g., physicians are supervisors of treatment and gatekeepers to applying for benefits), and (3) the interactional level (e.g., when physicians treat patients' conditions as medical problems). For example, the sociologists Deborah Findlay and Leslie Miller (1994: 277) explain how gambling has been medicalized:

> Habitual gambling . . . has been regarded by a minority as a sin, and by most as a leisure pursuit—perhaps wasteful but a pastime nevertheless. Lately, however, we have seen gambling described as a psychological illness—"compulsive gambling." It is in the process of being medicalized. The consequences of this shift in discourse (that is, in the way of thinking and talking) about gambling are considerable for doctors, who now have in gamblers a new market for their services or "treatment"; perhaps for gambling halls, which may find themselves subject to new regulations, insofar as they are deemed to contribute to the "disease"; and not least, for gamblers themselves, who are no longer treated as sinners or wastrels, but as patients, with claims on our sympathy, and to our medical insurance plans as well.

Sociologists often refer to this form of medicalization as the *medicalization of deviance* because it gives physicians and other medical professionals greater authority to determine what should be considered "normal" and "acceptable" behavior and to establish the appropriate mechanisms for controlling "deviant behaviors."

According to symbolic interactionists, medicalization is a two-way process: Just as conditions can be medicalized, so can they be demedicalized. ***Demedicalization*** **refers to the process whereby a problem ceases to be defined as an illness or a disorder.** Examples include the removal of certain behaviors (such as homosexuality) from the list of mental disorders compiled by the American Psychiatric Association and the deinstitutionalization of mental health patients. The process of

medical–industrial complex local physicians, local hospitals, and global health-related industries such as insurance companies and pharmaceutical and medical supply companies that deliver health care today.

medicalization the process whereby nonmedical problems become defined and treated as illnesses or disorders.

demedicalization the process whereby a problem ceases to be defined as an illness or a disorder.

process exacerbated long-term problems associated with inadequate care for people with mental illness.

Admitting people to mental hospitals on an involuntary basis ("involuntary commitment") has always been controversial; however, it remains the primary method by which police officers, judges, social workers, and other officials deal with people—particularly the homeless—whom they have reason to believe are mentally ill and imminently dangerous to others or themselves if not detained (Monahan, 1992). State mental hospitals continue to provide most of the chronic inpatient care for poor people with mental illnesses; these institutions tend to serve as a revolving door to poverty-level board-and-care homes, nursing homes, or homelessness, as contrasted with the situation of patients who pay their bills at private psychiatric facilities through private insurance coverage or Medicare (Brown, 1985).

According to the sociologist Erving Goffman (1961a), mental hospitals are a classic example of a *total institution*, previously defined as a place where people are isolated from the rest of society for a period of time and come under the complete control of the officials who run the institution.

Disability

What is a disability? ***Disability* refers to a physical or mental impairment that substantially limits one or more major activities that a person would normally do at a given stage of life and that may result in stigmatization or discrimination against the person with a disability.** Some disabilities involve physical conditions, while others involve mental abilities. However, according to disability rights advocates, social attitudes and the social and physical environments in which people live are contributing factors to the extent to which a person may be considered "disabled." An example of a disabling environment is a school or office building in which elevator buttons and faucets on public restroom sinks are located beyond the reach of a person using a wheelchair. In such a setting, the person's disability derives from the fact that necessary objects in everyday life have been made inaccessible to some people. According to advocates, disability must be thought of in terms of how society causes or contributes to the problem—not in terms of what is "wrong" with the person with a disability (Albrecht, Seelman, and Bury, 2001).

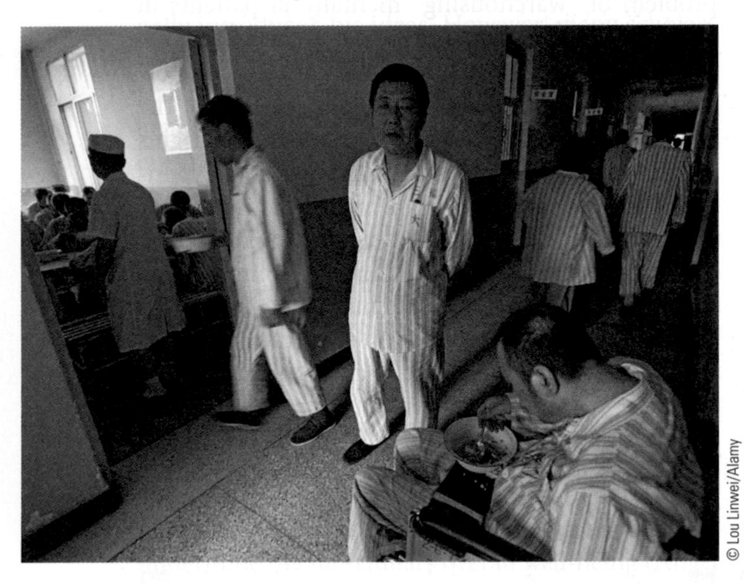

© Lou Linwei/Alamy

▲ According to some social analysts, psychiatric hospitals were created to treat people with mental illnesses; however, as total institutions these hospitals may also be used to silence political dissidents.

A second crucial factor in understanding disability is how a person with a disability is viewed and treated by other people. Many examples could be given of persons with a disability who have been made uncomfortable by "able-bodied" individuals who think they are "helping" the person with a disability when they make some comment about the disability or the perceived inadequacies it causes. However, Dr. Adrienne Asch, director of the Center for Ethics at Yeshiva University, speaks volumes about what her life has been like as an outstanding academic and as a person with a visual impairment—blindness:

> [A] growing number of professionals with disabilities, including myself, can point to professional recognition and the joys of doing work we love as well as its relative financial security and social status. Yet like . . . others, we have all-too-frequent reminders that we are unanticipated participants in workshops or conferences or unexpected guests at social gatherings. Sitting beside a stranger waiting for a lecture to begin at an academic conference, the stranger whispers loudly not "Hello, my name is Carol," but "Let me know how I can help you." What help do I need while waiting for the speaker to begin? Why not introduce herself, rather than assume that the only sociability I could possible want is her help? When I respond by saying that she can let me know if I can help her, she does not get the point and I am all too well aware that the point is subtle; instead she needs to be thanked for her offer and reassured that I will accept it—and then many pleasantries later perhaps we can discuss why we are at the lecture and whether we like it and what workshop we will attend that afternoon. (Asch, 2001)

Asch emphasizes that the incident she described is not an isolated one in her life: Many people who have known her for years do not feel comfortable treating her the same way they would a person without a disability. Even more important, countless other persons with a disability have shared similar experiences in which they been stigmatized or marginalized because of how another individual responded to them based on their disability.

Estimates suggest that about 54.4 million Americans (about 19 percent of the U.S. population) have some level of disability, and 35 million have a severe disability. Although anyone can become disabled, some people are more likely to be or to become disabled than others. People who work in hazardous settings have higher rates of disability than workers in seemingly safer occupations. African Americans have higher rates of disability than whites, especially more serious disabilities; persons with lower incomes also have higher rates of disability (Weitz, 2010). As age increases, the changes of disability also rise. But no one is immune to disability. Many people believe that most disabilities are caused by catastrophic events such as serious accidents; however, illness causes more than 90 percent of all disabilities.

Environment, lifestyle, and working conditions may all contribute to either temporary or chronic disability. For example, air pollution in automobile-clogged cities leads to a higher incidence of chronic respiratory disease and lung damage, which may result in severe disability for some people. Eating certain types of food and smoking cigarettes increase the risk for coronary and cardiovascular diseases. In contemporary industrial societies, workers in the second tier of the labor market (primarily recent immigrants, white women, and people of color) are at the greatest risk for certain health hazards and disabilities. Employees in data processing and service-oriented jobs may also be affected by work-related disabilities. The extensive use of computers has been shown to harm some workers' vision; to produce joint problems such as arthritis, low-back pain, and carpal tunnel syndrome; and to place employees under high levels of stress that may result in neuroses and other mental health problems. As shown in ■ Table 18.2, nearly one out of five people in the United States (21.3 percent) has a chronic health condition that, given the physical, attitudinal, and financial barriers built into the social system, makes it difficult to perform one or more activities generally considered appropriate for persons of their age.

For infants born with a disability, living with a disability is a long-term process for parents and child alike. Parents report that more than 200,000 children under age 3 have a disability, either a developmental delay or difficulty moving their arms or legs. In addition, about 475,000 children between the ages of 3 and 5 have a disability in which they experience a developmental delay or difficulty walking, running, or playing (Brault, 2008). According to a study by sociologist Meira Weiss (1994), an infant's appearance may determine how parents will view a child with a disability. Parents are more likely to be bothered by external, openly visible disabilities than by internal or disguised ones. According to Weiss, children born with internal (concealed) disabilities are at least initially more acceptable to parents because they do not violate the parents' perceived body images of their children. Although Weiss's study is about twenty years old, her findings remain important because they provide insight into the social significance that parents and others attach to congenital disabilities.

Among persons who acquire disabilities later in life, through disease or accidents, the social significance of their disability can be seen in how they initially respond to their symptoms and diagnosis, how they view the immediate situation and their future, and how the illness and disability affect their lives. When confronted with a disability, most people adopt one of two strategies—avoidance or

table 18.2

Percentage of U.S. Population with Disabilities

Characteristic	Percentage[a]
With a disability	21.3
Severe	14.2
Not severe	7.1
Has difficulty or is unable to:	
See	3.4
Hear	3.4
Have speech understood	1.1
Lift or carry ten pounds	6.9
Use stairs	9.4
Walk	9.8
Has difficulty or needs assistance with:	
Getting around inside the house	1.8
Getting into bed	2.3
Taking a bath or shower	2.2
Dressing	1.6
Eating	0.6
Getting to or using the toilet	1.0
Has difficulty or needs assistance with:	
Going outside the home alone	3.8
Managing money and bills	2.2
Preparing meals	2.2
Doing light housework	3.0
Using the phone	1.2

[a]Percentage of persons age 15 and older.

Source: Brault, 2008.

disability a physical or mental impairment that substantially limits one or more major activities that a person would normally do at a given stage of life and that may result in stigmatization or discrimination against the person with a disability.

likely to be educated and more likely to be malnourished and have inadequate access to health care—all of which contribute to risk of chronic illness, physical and mental disability, and the inability to participate in the labor force. As previously mentioned, the type of employment available to people with limited resources increases their chances of becoming disabled. They may work in hazardous places such as mines, factory assembly lines, and chemical plants, or in the construction industry, where the chance of becoming seriously disabled is much higher.

With ongoing efforts to enforce the Americans with Disabilities Act (ADA), as discussed in Chapter 13, and the ADA Amendments Act, which is designed to make it easier for an individual to establish in a lawsuit that he or she has a disability, more mentally disabled workers and individuals with impairments such as cancer, diabetes, and epilepsy may press for their rights in the future. Regardless of the outcome, disability concerns make us aware of how complicated equality and justice are in the distribution and financing of health care for all people.

Health Care in the Future

Central questions regarding the future of health care are how to provide coverage for the largest number of people and how to do this without bankrupting the entire nation. At the time of this writing, it is too early to evaluate how the Affordable Care Law will affect health care in this nation. This reform bill constitutes a form of universal health coverage and has widespread opposition from those who want to reduce taxes and keep the government out of their daily lives. Overall, our best hope for good medical care is that a payment system will be developed that pays for good patient results at a reasonable cost and thus brings about a transformation in the current U.S. health care system.

A key issue in the United States is how to prevent, reduce, and best treat epidemics that affect all Americans. If we do not have a specific illness or health condition, many of us do not see that disease as being "our problem." In the future, however, we must come to see problems such as HIV/AIDS as everyone's concern. Statistically, a person becomes infected with HIV every 9.5 minutes in the United States, and the epidemic so far has claimed more than 600,000 lives and left about 1.1 million more living with HIV. The Obama administration has developed a national HIV/AIDS strategy to help reduce the number of people who become infected with HIV, to increase access to care and improve health outcomes for people living with HIV, and to reduce HIV-related health disparities with more and better community-level approaches that integrate HIV prevention and care with other social service needs. Along with meeting these goals, it will be important to reduce stigma and the discrimination against people living with HIV. This HIV/AIDS national initiative is only one of many possible examples which show that in the future we must have a coordinated national response to preventive health care and to the equitable distribution of health care services throughout the country.

Another key issue in contemporary health care is the role that advanced technologies play in the rising costs of medical care and their usefulness as major tools for diagnosis and treatment. Technology is a major stimulus for social change, and the health care systems in high-income nations such as the United States reflect the rapid rate of technological innovation that has occurred in the last few decades. In the future, advanced health care technologies will no doubt provide even more accurate and quicker diagnosis, effective treatment techniques, and increased life expectancy.

However, technology alone cannot solve many of the problems confronting us in health and health care delivery. In fact, some aspects of technological innovation may be dysfunctional for individuals and society. As we have seen, some technological "advances" raise new ethical concerns, such as the moral and legal issues surrounding the cloning of human life. Some "advances" also may fail: A new prescription drug may be found to cause side effects that are more serious than the illness that it was supposed to remedy. Whether advanced technology succeeds or fails in some areas, it will probably continue to increase the cost of health care in the future. As a result, the gap between the rich and the poor in the United States will contribute to inequalities of access to vital medical services. On a global basis, new technologies may lower the death rate in some low-income countries, but it will primarily be the wealthy in those nations who will have access to the level of health care that many people in higher-income countries take for granted.

In the developing nations of the world, preventive health care and more effective and efficient health care delivery are crucial as the world's population continues to increase (see Chapter 19) and as some regions of the world are plagued by high rates of disease and poverty, a deadly combination in any setting. The concerns of the World Health Organization and other organizations must be heeded to prevent global *pandemics,* epidemics of infectious disease that spread through human populations across a wide region, country, continent, or the whole world. In the future it will be necessary to direct more money and attention toward preventing major health crises rather than trying to find some way to deal with them after they develop.

you can make a difference

Helping Others in the Fight Against Illness!

It's so nice to have someone come in and talk to you and do what they do. When you get kind of down, it definitely helps perk you up. I have visitors, but some of the people here don't. These ladies make us feel pampered.

—Lona, a patient recovering from colon cancer, describing how much she appreciated a group of volunteers who made her feel better in the hospital

She was depressed. Her doctor told her to put a little makeup on. She said, "Even if I had some, I wouldn't be motivated to do it." I saw a need for people to provide these things, especially with cancer, because their skin can get dry from treatments.

—Beverly Barnes explaining why she began Patient Pride to bring makeup and other cosmetic products to cancer patients (qtd. in Leptich, 2005)

Beverly Barnes did not intend to start the program now known as Patient Pride when she went to the hospital to visit a friend who was recuperating from surgery. However, by the time she left the hospital, having helped her friend put on makeup and freshen her appearance, Barnes realized how much seemingly small acts of kindness can mean to people who are ill or recuperating from surgery.

Patient Pride is only one of many examples of how everyday people can make a difference in the lives of people experiencing illness or disability. There are many Internet sources that you can check out to learn more about volunteering and participating in the fight against diseases such as cancer and cardiovascular disease:

- The American Cancer Society's home page (which has links to various volunteer activities): **www.cancer.org**
- The American Heart Association's home page: **http://www.heart.org/HEARTORG/**

© Jim West/Alamy

▲ One of the expected benefits of the 2010 health care reform legislation is more-efficient medical treatment and administration, ultimately lowering the overall costs of health care in the United States.

It goes without saying that health, illness, and health care will continue to change in the future. To a degree, health care in the future will be up to each of us. What measures will we take to safeguard ourselves against illness and disorders? How can we help others who are the victims of acute and chronic diseases or disabilities? Although we cannot change global or national health problems, there are some small (but not insignificant) things we can do to help others (see "You Can Make a Difference").

chapter review Q & A

Use these questions and answers to check how well you've achieved the learning objectives set out at the beginning of this chapter.

- **What is health, and why are sociologists interested in studying health and medicine?**

According to the World Health Organization, health is a state of complete physical, mental, and social well-being. In other words, health is not only a biological issue but also a social issue. For this reason, sociologists are interested in studying health and medicine. As a social institution, medicine is one of the most important components of quality of life.

- **How do acute and chronic diseases differ, and which is most closely linked to spiraling health care costs?**

Acute diseases are illnesses that strike suddenly and cause dramatic incapacitation and sometimes death. Chronic diseases, such as arthritis, diabetes, and heart disease, are long-term or lifelong illnesses that develop gradually or are present at birth. The treatment of chronic diseases is typically more costly because of the duration of these problems.

- **How is health care paid for in the United States?**

Throughout most of the past hundred years, medical care in the United States has been paid for on a fee-for-service basis. The term *fee for service* means that patients are billed individually for each service they receive. This approach to paying for medical services is expensive because few restrictions are placed on the fees that doctors, hospitals, and other medical providers can charge patients. Recently, there have been efforts at cost containment, and HMOs and managed care have produced both positive and negative results in the contemporary practice of medicine. Health maintenance organizations (HMOs) provide, for a set monthly fee, total care with an emphasis on prevention to avoid costly treatment later. Managed care refers to any system of cost containment that closely monitors and controls health care providers' decisions about medical procedures, diagnostic tests, and other services that should be provided to patients.

- **What is the functionalist perspective on health and illness?**

According to the functionalist approach, if society is to function as a stable system, it is important for people to be healthy and to contribute to their society. Consequently, sickness is viewed as a form of deviant behavior that must be controlled by society. Sociologist Talcott Parsons (1951) described the sick role—the set of patterned expectations that defines the norms and values appropriate for individuals who are sick and for those who interact with them. Although individuals are given permission to not perform their usual activities for a period of time, they are expected to seek medical attention and get well as soon as possible so that they can go about their normal routine.

- **What is the conflict perspective on health and illness?**

Conflict theory tends to emphasize the political, economic, and social forces that affect health and the health care delivery system. Among these issues are the ability of all people to obtain health care; how race, class, and gender inequalities affect health and health care; power relations between doctors and other health care workers; the dominance of the medical model of health care; and the role of profit in the health care system.

- **What is the symbolic interactionist perspective on health and illness?**

In studying health, symbolic interactionists focus on the fact that the meaning that social actors give their illness or disease will affect their self-concept and their relationships with others. According to symbolic interactionists, we socially construct "health" and "illness" and how both should be treated. Symbolic interactionists also examine medicalization—the process whereby nonmedical problems become defined and treated as illnesses or disorders.

- **What is the postmodernist perspective on health and illness?**

Postmodern theorists such as Michel Foucault argue that doctors and the medical establishment have gained control over illness and patients at least partly because of the physicians' clinical gaze, which replaces all other systems of knowledge. The myth of the wise doctor has also been supported by the development of disease classification systems and new tests.

- **How did deinstitutionalization change the way that mental illness is treated?**

Deinstitutionalization shifted many mental patients from hospital treatment to community- or family-based care. This was possible because of newer drugs and treatments; however, it also created new issues because of social stereotypes about the "mentally ill" and differences of opinion about various treatment options.

- **What is a disability, and how prevalent are disabilities in the United States?**

Disability is a physical or health condition that stigmatizes or causes discrimination. An estimated 54.4 million persons in the United States have one or more physical or mental disabilities. This number continues to increase for several reasons. First, with advances in medical technology, many people who once would have died from an accident or illness now survive, although with an impairment. Second, as more people live longer, they are more likely to experience diseases (such as arthritis) that may have disabling consequences. Third, persons born with serious disabilities are more likely to survive infancy because of medical technology. However, less than 15 percent of persons with a disability today were born with it; accidents, disease, and war account for most disabilities in this country.

key terms

acute diseases 526
chronic diseases 526
deinstitutionalization 551
demedicalization 547
disability 552
drug 530
health 522
health care 522
health maintenance organization (HMO) 538
holistic medicine 544
infant mortality rate 522
life expectancy 522
managed care 539
medical–industrial complex 546
medicalization 547
medicine 522
sick role 545
social epidemiology 526
socialized medicine 541
universal health care 541

questions for critical thinking

1. Why is it important to explain the social, as well as the biological, aspects of health and illness in societies?
2. In what ways are race, class, and gender intertwined with physical and mental disorders?
3. How would functionalists, conflict theorists, and symbolic interactionists suggest that health care delivery might be improved in the United States?
4. Based on this chapter, how do you think illness and disability will be handled in the United States in the near future? Are there things that we can learn from other nations regarding the delivery of health care? Why or why not?

turning to video

Watch the CBS video *Young and Uninsured* (running time 2:22), available through **CengageBrain.com.** This video examines the fact that young adults are the largest uninsured age group in the country, and it explores some of the ways in which the federal health care legislation passed into law in 2010 seeks to help. As you watch the video, think about you own health care situation: Did you have insurance before the new law started going in effect? Do you have it now? After you watch the video, try answering these questions: Should insuring young people, who generally have fewer health problems than older people, have been a priority of the new law? Why or why not?

online study resources

Go to CENGAGE brain.com to access online study resources, including the Sociology CourseMate for this text as well as special features such as video, an interactive sociology time line and interactive maps, General Social Survey (GSS) data, and U.S. Census 2010 data.

Course Mate brings course concepts to life with interactive learning, study, and exam-preparation tools that support the printed textbook. A textbook-specific website, **Sociology CourseMate** includes an integrated interactive eBook and other interactive learning tools, including quizzes, flash cards, and videos.

Visit **www.cengagebrain.com** to access your account and purchase materials.

have nothing to do with drug trafficking. They kill you just for having seen what they are doing.

—Vicente Burciaga, who escaped from El Porvenir, Mexico, to Fort Hancock, Texas, is seeking political asylum with his wife and infant son after gang members burned down five homes in their area and killed a neighbor in one of many episodes of drug violence in their community (qtd. in McKinley, 2010: A1).

Chapter Focus Question:

What effect does migration have on cities and on shifts in the global population?

Around the world, people move from one location to another for many reasons, and individuals and families that immigrate to the United States from other nations are no exception. As the interviews indicate, some persons move to the United States for economic opportunities while others move because they fear for their life. In the United States and many other high-income countries, there is a lack of consensus about the causes and consequences of immigration: Some are adamantly opposed to immigration; others believe that it is acceptable under certain circumstances.

Where do we stand in the United States in the second decade of the twenty-first century? As ...

the state of Arizona passed a stringent new law on illegal immigration in 2010, causing strife not only in that state but throughout the nation. The Arizona law required that immigrants meet federal requirements to carry necessary identity documents that legitimize their presence in the United States. Failure to carry immigration documents became an Arizona state crime, and police there were given broad power to detain any individual suspected of being in the country illegally (Archibold, 2010). In 2010 a district court judge is-

Highlighted Learning Objectives

- Explain how people are affected by population changes.
- Compare ecological/

19 Population and Urbanization

St. Louis, Missouri:

It was a good life, a really good ride. . . . I made a wonderful career out of understanding the cultures of Latin America and the culture of the United States and how to do business in both.

—Amparo Kollman-Moore, a naturalized U.S. citizen who migrated to this country from Columbia, describing how she felt about rising through the ranks of a medical supply company to become president of its Latin American division (qtd. in Preston, 2010: A1, A3)

The tomato-farming region of Immokalee, Florida:

We live here in fear. We fear Immigration will come, and many people just don't go out.

—Antonia Fuentes, a Mexican farmworker who has picked tomatoes for two years and would welcome even a guest worker program that would permit her to stay in the United States for a set period of time (qtd. in Goodnough and Steinhauer, 2006: A35)

© Ryan Rodrick Beiler/Shutterstock.com

▲ **The controversial 2010 Arizona legislation that cracked down on illegal immigration has had a perhaps unanticipated secondary effect: bringing the issue back into the national focus, where it had waned in the years after the 2008 real estate meltdown and subsequent economic recession.**

A migrant shelter in Nogales, Mexico:

It's hard to cross. But it's harder to see your children have little to eat.

—Raul Gonzalez, who turned himself in to U.S. authorities after he was robbed and his feet started bleeding from walking for five days to get across the border, states the common theme of most immigrants: We want a better life for ourselves and our children (qtd. in CNN.com, 2006).

Fort Hancock, Texas:

It's very hard over there [in El Porvenir, Mexico]. They are killing people over there who

© Chris Hondros/Getty Images

© David Young-Wolff/PhotoEdit

▲ Political unrest, violence, and war are "push" factors that encourage people to leave their country of origin. By contrast, job opportunities, such as construction work in the United States, are a major "pull" factor for people from low-income countries.

unauthorized immigration very difficult to maintain (Passel and Cohn, 2010).

country to escape oppression there in the early 2000s. Slavery is the most striking example of involuntary migration; for example, the 10 million to 20 million Africans forcibly transported to the Western Hemisphere prior to 1800 did not come by choice.

Population Composition

Changes in fertility, mortality, and migration affect the ***population composition*—the biological and social characteristics of a population, including age, sex, race, marital status, education, occupation, income, and size of household.**

One measure of population composition is the ***sex ratio*—the number of males for every hundred females in a given population.** A sex ratio of 100 indicates an equal number of males and females in the population. If the number is greater than 100, there are more males than females; if it is less than 100, there are more females than males. In the United States the estimated sex ratio for 2009 was 97, which means there were 97 males per 100 females. Although approximately 124 males are conceived for every 100 females, male fetuses miscarry at a higher rate. From birth to age 14, the sex ratio is 1.04; in the age 15–64 category, however, the ratio shifts to 1.0, and from 65 upward, women outnumber men. By age 65, the sex ratio is about 75—that is, there are 75 men for every 100 women.

For demographers, sex and age are significant population characteristics; they are key indicators of fertility and mortality rates. The age distribution of a population has a direct bearing on the demand for schooling, health, employment, housing, and pensions. The current distribution of a population can be depicted in a ***population pyramid*—a graphic rep-**

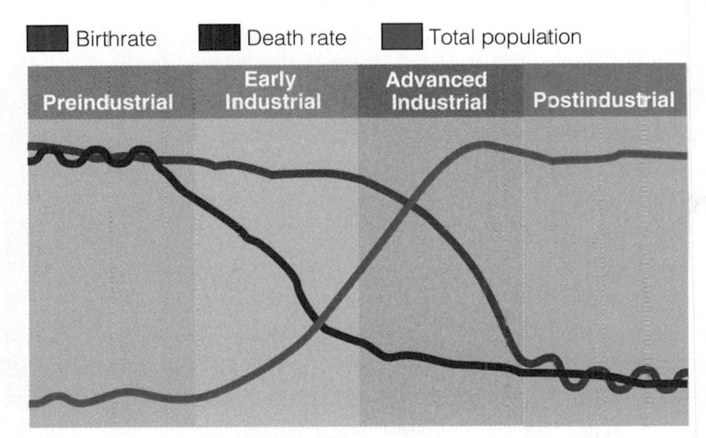

◀ **FIGURE 19.4 THE DEMOGRAPHIC TRANSITION**

by having only one or two children in order to bring about ***zero population growth*—the point at which no population increase occurs from year to year** because the number of births plus immigrants is equal to the number of deaths plus emigrants (Weeks, 2012).

Today, Ehrlich's proposal remains the same: "Adopt policies that gradually reduce birthrates and eventually start a global decline toward a human population size that is sustainable in the long run" (Ehrlich and Ehrlich, 2009).

Demographic Transition Theory

Some scholars who disagree with the neo-Malthusian viewpoint suggest that the theory of demographic transition offers a more accurate picture of future population growth. ***Demographic transition* is the process by which some societies have moved from high birth and death rates to relatively low birth and death rates as a result of technological development.** Demographic transition is linked to four stages of economic development (see ▶ Figure 19.4):

- *Stage 1: Preindustrial societies.* Little population growth occurs because high birthrates are offset by high death rates. Food shortages, poor sanitation, and lack of adequate medical care contribute to high rates of infant and child mortality.
- *Stage 2: Early industrialization.* Significant population growth occurs because birthrates are relatively high whereas death rates decline. Improvements in health, sanitation, and nutrition produce a substantial decline in infant mortality rates. Overpopulation is likely to occur because more people are alive than the society has the ability to support.
- *Stage 3: Advanced industrialization and urbanization.* Very little population growth occurs because both birthrates and death rates are low. The birthrate declines as couples control their fertility through contraceptives and become less likely to adhere to religious directives against their use. Children are not viewed as an economic asset; they consume income rather than produce it. Societies in this stage attain zero population growth, but the actual number of births per year may still rise because of an increased number of women of childbearing age.
- *Stage 4: Postindustrialization.* Birthrates continue to decline as more women gain full-time employment and the cost of raising children continues to increase. The population grows very slowly, if at all, because the decrease in birthrates is coupled with a stable death rate.

Debate continues as to whether this evolutionary model accurately explains the stages of population growth in all societies. Advocates note that demographic transition theory highlights the relationship between technological development and population growth, thus making Malthus's predictions obsolete. Scholars also point out that demographic transitions occur at a faster rate in now-low-income nations than they previously did in the nations that are already developed. For example, nations in the process of development have higher birthrates and death rates than the now-developed societies did when they were going through the transition. The death rates declined in the now-developed nations as a result of internal economic development—not, as is the case today, through improved methods of disease control (Weeks, 2012). Critics suggest that this theory best explains development in Western societies.

Other Perspectives on Population Change

In recent decades, other scholars have continued to develop theories about how and why changes in population growth patterns occur. Some have studied the relationship between economic development and a decline in fertility; others have focused on the process of *secularization*—the decline in the significance

zero population growth the point at which no population increase occurs from year to year.

demographic transition the process by which some societies have moved from high birthrates and death rates to relatively low birthrates and death rates as a result of technological development.

framing immigration in the media

Caught in the "Moat": Media Framing and Bipartisan Politics

In recent years, among the greatest impediments to [immigration] reform were questions about border security. . . . Well, over the past two years we have answered those concerns. . . . They wanted a fence. Well, that fence is now basically complete. And we've gone further. We tripled the number of intelligence analysts working the border [, etc.]. . . . But even though we've answered those concerns, I suspect there will be those who will try to move the goalposts one more time. They'll say we need to triple the border patrol. Or quadruple the border patrol. They'll say we need a higher fence to support reform.

Maybe they'll say we need a moat. Or alligators in the moat. They'll never be satisfied. And I understand that. That's politics. But the truth is, the measures we've put in place are getting results. . . .

—edited comments from President Barack Obama's speech in El Paso, Texas, where he advocated comprehensive immigration reform (qtd. in Foster, 2011)

This is a President who is more interested in trying out for *Saturday Night Live.* It seems like . . . the guy's a stand-up comic, and that was what he was playing to. The fact of the matter is, Americans don't want a stand-up comic for the President of the United States. . . . Anyone who knows what's happening on the border of Texas and Mexico, or for that matter the southern border of the United States with Mexico realizes this is not comedy, there are people's lives in jeopardy every day.

—Texas Governor Rick Perry chastising President Obama for his comment about Republicans wanting a moat filled with alligators to protect the border (qtd. in fireandreamitchell.com, 2011)

President Obama is riding my "moat-tails." This fits a pattern for Obama. He stole my idea for moat gators just like he stole President Bush's idea of killing bin Laden and George Washington's idea of being President.

—Stephen Colbert's statement on Comedy Central's *Colbert Report* in regard to his previous "jokes" about using a "gator-filled moat" along the U.S.–Mexico border to curtail illegal immigration (qtd. in TMPLiveWire.com, 2011)

Although intended as an ironic, or slightly humorous, comment in a longer speech on the need for immigration reform, President Obama's statement about protecting the border with alligator patrols received more comments from other politicians and media attention than did the rest of his speech. Obama was subjected to ribbing by the governor of Texas and by a late-night TV comedian, among many others.

Is it possible for the media to trivialize serious issues such as how we can best protect our national borders by what journalists and cable-channel satirists choose to emphasize and what they choose to ignore or downplay in coverage of an event such as the president's speech? When we discuss how the media "frame" immigration issues, it is important to recall that there is no completely neutral way to analyze and report on a political leader's ideologies and plans for action. In regard to President Obama's speech, for example, his comments may be represented in the media in several ways: (1) as a *bipartisan* policy statement highlighting key issues and goals, (2) as a *partisan* speech setting forth what one's own political party believes should be done, (3) as a *reelection bid* directed toward a specific constituency or groups of potential supporters, or (4) as a *neutral* commentary on a social problem.

If you were a member of the media, which, if any, of the above framing devices would you use to describe the president's speech? In regard to the El Paso speech, media sources took divergent stances. On *Fox & Friends,* for example, one host aired a clip of the president's remarks and stated, "I guess you're allowed to mock Republicans." The other host responded, "He's

of the sacred in daily life—and how a change from believing that otherworldly powers are responsible for one's life to a sense of responsibility for one's own well-being is linked to a decline in fertility. Based on this premise, some analysts argue that the processes of industrialization and economic development are typically accompanied by secularization but that the relationship between these factors is complex when it comes to changes in fertility.

Shifting from the macrolevel to the microlevel, education and social psychological factors also play into the decisions that individuals make about how many children to have. Family-planning information is more readily available to people with more years of formal education and may cause them to engage in decision making in accord with *rational choice theory,* which is based on the assumption that people make decisions based on a calculated cost–benefit analysis ("What do I gain and lose from a specific action?"). In low-income countries or other settings in which children are identified as an economic resource for their parents throughout life, fertility rates are higher than in higher-income countries. However, as modernization and urbanization occur in such societies, the positive economic effects of having more children may be offset by the cost of caring for those children and the lowered economic advantage gained from having children in an industrialized nation.

campaigning" and then staunchly criticized Obama's efforts at immigration reform. Throughout the *Fox & Friends* segment, the on-screen text below the hosts read, "Obama's Partisan Attack" (Johnson, 2011). This framing approach is a combination of numbers two and three above: The media commentators described the speech as partisan politics and also saw stated that it was the beginning of the president's reelection bid.

Another media report on this same incident highlighted something different. It seems that U.S. Representative Joe Walsh, a freshman Republican in Congress, responded to President Obama's speech with a "fake bipartisan" answer, the headlines for which stated: "GOP Rep. to Obama: I'm Game for Alligator-Filled Moat Along U.S.–Mexico Border" (dailycaller.com, 2011). According to Walsh, "I actually think a moat might be a very good idea and I'm wondering how many alligators it would take to secure the entire border" (qtd. in dailycaller.com, 2011). This is only one example of many in which media sources frame stories about an event with more emphasis on how an individual or group *responded* to a message or event than what the initial statement or event was about.

Finally, a more neutral framing of the El Paso immigration reform speech is found in this headline from the *New York Times:* "In Border City Talk, Obama Urges G.O.P. to Help Overhaul Immigration Law" (Calmes, 2011). The article first describes President Obama's efforts to deal with undocumented immigration and overhaul the nation's immigration laws for undocumented workers. Next, the article discusses how the Obama Administration tried to make the nation's borders safer and more secure. Finally, the article explains how Obama attempted to create a path so that some undocumented immigrants could become citizens, pay taxes, and learn English. Little attention was paid to the moat-and-alligator portion of the speech.

How should important concerns about immigration be represented in the media? Of course, no one answer can deal with all aspects of this question. In the past, undocumented workers and their families, particularly young children who worked in the fields or saw some of their family members deported, were shown in a sympathetic manner. Sympathetic framing refers to news writing that focuses on the human interest side of a story and shows that the individuals involved are caring people who are representative of a larger population. In this situation the focus would be on the issues and concerns of undocumented workers and why they feel a compelling need to come into the United States for work and what they perceive to be a "better quality of life."

In stark contrast to sympathetic framing are news reports that employ negative framing to describe recent immigrants from Central and South America. Negative framing sometimes describes "illegal" immigrants as a drain on education, health care, and the U.S. welfare system. Other negative framing in stories focuses on undocumented workers as a source of cheap labor that primarily benefits employers who do not have to pay a living wage. Negative framing of immigration in the media is not new in the United States. For many years, "immigrant, foreign labor" has been described as a threat to the livelihood of other workers and as a menace to public health and safety. As far back as 1904, the *San Francisco Chronicle* carried lengthy articles describing how Japanese laborers were taking jobs away from U.S. workers, reflecting a pattern of reporting that continues in the twenty-first century and screams the following: "Immigrants take jobs away from U.S. citizens!" (Puette, 1992).

How the media frame stories about immigration may influence social policy. Will we close our borders to immigration? Will we develop new programs to allow limited entry of immigrant workers? Not only are these important legal and social policy questions, but they also are issues that journalists and other news analysts must face when they frame stories on immigrants and the ways in which U.S. policy makers choose to either include or exclude them from this country.

reflect & analyze

What is your response to President Obama's comments about building a moat and stocking it with alligators? What point was he trying to make about partisan politics and how difficult it is to appease individuals on the other side of an argument? Do you believe that he successfully made his point with the moat-and-alligator comment? Why or why not?

As demographers have reformulated demographic transition theory, they have highlighted additional factors that are likely to be causes of fertility decline, and they have suggested that demographic transition is not just one process, but rather a set of intertwined transitions. One is the epidemiological transition—the shift from deaths at younger ages because of acute, communicable diseases. Another is the fertility transition—the shift from natural fertility to controlled fertility, resulting in a decrease in the fertility rate. Other transitions include the migration transition, the urban transition, the age transition, and the family and household transition, which occur as a result of lower fertility, longer life, an older age structure, and predominantly urban residence.

A Brief Glimpse at International Migration Theories

Why do people relocate from one nation to another? Several major theories have been developed in an attempt to explain international migration. The *neoclassical economic approach* assumes that migration patterns occur based on geographic differences in the supply of and demand for labor. The United States and other high-income countries that have had growing economies and a limited supply of workers for certain types of jobs have paid higher wages than are available in areas with a less-developed economy and a large labor force. As a result, people move to gain higher wages and sometimes better living conditions. They also may take jobs in other countries so that they can send money to their families in their country of origin. (See "Framing Immigration in the Media.")

Unlike the neoclassical explanation of migration, which focuses on individual decision making, the *new households economics of migration approach* emphasizes the part that entire families or households play in the migration process. From this approach, workers' temporary migration is examined not only from the perspective of the individual worker but also in terms of what the entire family gains from the process of having one or more migrant family members work in another country. By having a diversity of family income (originating from more than one source), the family is cushioned from the economic woes of the nation that most of the family members think of as "home."

Two conflict perspectives on migration add to our knowledge of why people migrate. Split-labor-market theory suggests that immigrants from low-income countries are often recruited for secondary labor market positions: dead-end jobs with low wages, unstable employment, and sometimes hazardous working conditions. By contrast, migrants from higher-income countries may migrate for primary-sector employment—jobs in which well-educated workers are paid high wages and receive benefits such as health insurance and a retirement plan. The global migration of some high-tech workers is an example of this process, whereas the migration of farmworkers and construction helpers is an example of secondary labor market migration.

Finally, world systems theory (discussed later in this chapter) views migration as linked to the problems caused by capitalist development around the world. As the natural resources, land, and workforce in low-income countries with little or no industrialization have come under the influence of international markets, there has been a corresponding flow of migrants from those nations to the highly industrialized, high-income countries, especially those with which the poorer nations have had the most economic, political, or military contact.

After flows of migration commence, the pattern may continue because potential migrants have personal ties with relatives and friends who now live in the country of destination and can serve as a source of stability when the potential migrants relocate to the new country. Known as *network theory,* this approach suggests that once migration has begun, it takes on a life of its own and that the migration pattern which ensues may be different from the original *push* or *pull* factors that produced the earlier migration. Another approach, *institutional theory,* suggests that migration may be fostered by groups—such as humanitarian aid organizations relocating refugees or smugglers bringing people into a country illegally—and that the actions of these groups may produce a larger stream of migrants than would otherwise be the case.

As you can see from these diverse approaches to explaining contemporary patterns of migration, the reasons that people migrate are numerous and complex, involving processes occurring at the individual, family, and societal levels. (See this chapter's Photo Essay.)

Urbanization in Global Perspective

Urban sociology is a subfield of sociology that examines social relationships and political and economic structures in the city. According to urban sociologists, a *city* is a relatively dense and permanent settlement of people who secure their livelihood primarily through nonagricultural activities. Although cities have existed for thousands of years, only about 3 percent of the world's population lived in cities 200 years ago, as compared with about 50 percent today. In 2011, more than 400 cities had a population of over 1 million, and 21 megacities had more than 10 million residents. A ***megacity* is a metropolitan area with a total population in excess of 10 million people.** Some megacities are a single metropolitan area, but others are two or more metropolitan areas that have converged. Examples include Tokyo, New York City, Mumbai, and Mexico City. ▶ Map 19.1 shows the percentage of the total world population living in urban areas as the process of urbanization continues on a global basis.

Emergence and Evolution of the City

Cities are a relatively recent innovation when compared with the length of human existence. The earliest humans are believed to have emerged anywhere from 40,000 to 1,000,000 years ago, and permanent human settlements are believed to have begun first about 8000 B.C.E. However, some scholars date the development of the first city between 3500 and 3100 B.C.E., depending largely on whether a formal writing system is considered as a requisite for city life (Sjoberg, 1965; Weeks, 2012).

According to the sociologist Gideon Sjoberg (1965), three preconditions must be present in order for a city to develop:

1. *A favorable physical environment,* including climate and soil favorable to the development of plant and animal life and an adequate water supply to sustain both.
2. *An advanced technology* (for that era) that could produce a social surplus in both agricultural and nonagricultural goods.
3. *A well-developed social organization,* including a power structure, in order to provide social stability to the economic system.

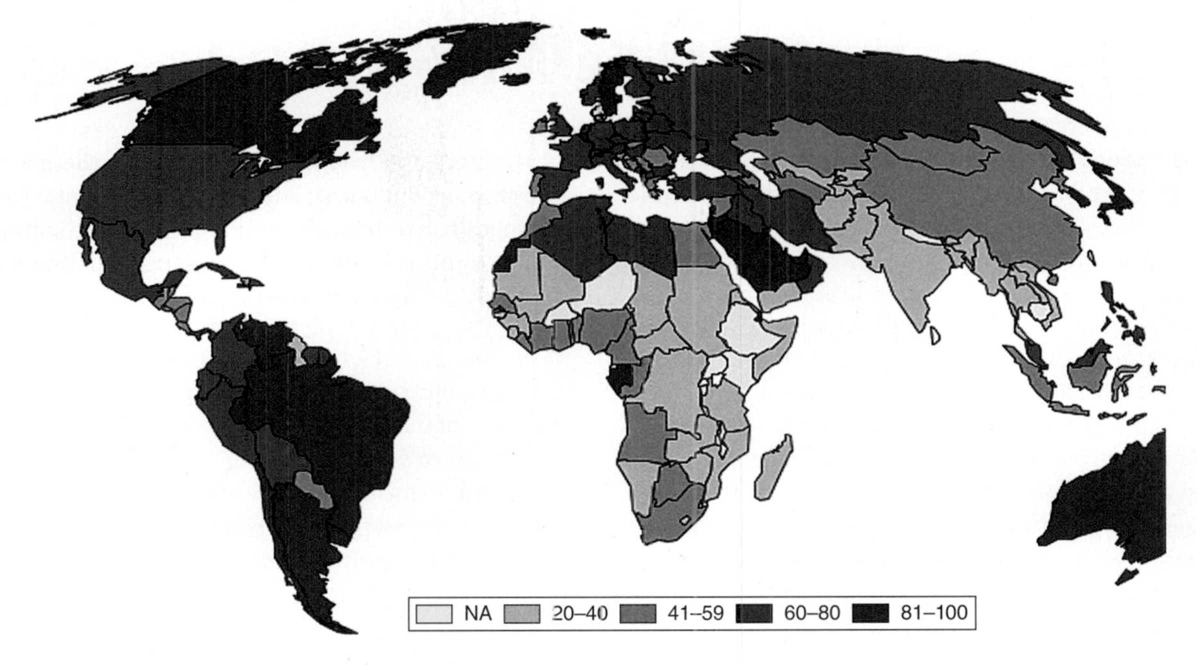

▲ **MAP 19.1 PERCENTAGE OF TOTAL POPULATION LIVING IN URBAN AREAS, 2009**

Source: Globalhealthfacts.org. Based on Population Reference Bureau, 2009. World Population Data Sheet.

Based on these prerequisites, Sjoberg places the first cities in the Middle Eastern region of Mesopotamia or in areas immediately adjacent to it at about 3500 B.C.E. However, not all scholars concur; some place the earliest city in Jericho (located in present-day Jordan) at about 8000 B.C.E., with a population of about 600 people (see Kenyon, 1957).

The earliest cities were not large by today's standards. The population of the larger Mesopotamian centers was between 5,000 and 10,000 (Sjoberg, 1965). The population of ancient Babylon (probably founded around 2200 B.C.E.) may have grown as large as 50,000 people; Athens may have held 80,000 people (Weeks, 2012). Four to five thousand years ago, cities with at least 50,000 people existed in the Middle East (in what today is Iraq and Egypt) and Asia (in what today is Pakistan and China), as well as in Europe. About 3,500 years ago, cities began to reach this size in Central and South America.

© AP Images/George Osodi

▲ An increasing proportion of the world's population lives in cities. How is this scene in Lagos, Nigeria, similar to and different from major U.S. cities?

Preindustrial Cities

The largest preindustrial city was Rome; by 100 C.E., it may have had a population of 650,000. With the fall of the Roman Empire in 476 C.E., the nature of European cities changed. Seeking protection and survival, those persons who lived in urban settings typically did so in walled cities containing no more than 25,000 people. For the next 600 years the urban population continued to live in walled enclaves as competing warlords battled for power and territory during the "dark ages." Slowly, as trade increased, cities began to tear down their walls.

Preindustrial cities were limited in size by a number of factors. For one thing, crowded housing

megacity a metropolitan area with a total population in excess of 10 million people.

photo essay Immigration and the Changing Face(s) of the United States

Throughout U.S. history, immigration has had a profound effect on our nation. Chances are very good that almost all of us can trace our heritage and our family roots to one or more other nations where our ancestors lived before coming to the United States. Immigration has also been a controversial topic at times throughout our nation's history, just as it is early in the twenty-first century.

When we look at the faces of the people around our country today, we see a wide diversity of human beings, most of whom are seeking to live their lives together positively and peacefully. Demographers and other social science researchers are interested in studying how people become part of the mainstream of a country to which they have immigrated while still maintaining their own unique cultural identity, and why some of them do not want to become part of that mainstream.

As you view the pictures on these three pages, think about how you and other members of your family came to view yourselves as Americans—residents of the United States of America—and what this means to you and to them in terms of what you think and do on a daily basis. Doing so helps us gain a better understanding of some of the sociological issues relating to immigration.

▲ At a march to show support for the Somali community in Lewiston, Maine, marchers urged people to "Love Thy Neighbor" rather than to attempt to keep new arrivals from other countries from moving into the city.

© AP Photo/Ann Heisenfelt

◀ **IMMIGRANTS AND EDUCATIONAL OPPORTUNITIES**

For recent immigrants, education is a key that may open the door to greater involvement in one's new country. Some educational experiences may be long term (such as completing a college degree), whereas others serve more immediate needs, such as the acculturation and parenting class that these recent Hmong immigrants have completed to help them adjust to their new life in the United States.

© AP Photo/Charles Krupa

▶ **POLITICS AND IMMIGRATION**

Voting and political participation help people have a voice in the U.S. democratic process. Here, four languages are used to encourage residents of South Boston to vote on "Super Tuesday" in 2008. How does politics influence our thinking about immigration? Does immigration influence our thinking about politics? Can our opinions change over time?

© Alexander Tamargo/Contributor/Getty Images

◀ **IMMIGRATION AND THE CHANGING FACE OF THE MEDIA**

As recent immigrants to the United States reach out to find media sources that reflect their culture and interests, executives at many media outlets strive to reach the large number of young people (typically between the ages of 18 and 24) who represent the future of the larger ethnic categories in this country. For example, Latinos and Latinas are a key target audience for both Hispanic and so-called mainstream media. On the Spanish-language cable network program *Acceso Maximo*, for instance, viewers vote for their favorite videos and artists by text messaging or voting online.

© Tony Freeman/PhotoEdit

◀ **IMMIGRANTS AND EMPLOYMENT**

Many people often view the world's immigrants and workers as being almost interchangeable because the primary purpose of much immigration is either to find work or for employers to have a larger pool of low-paid workers from which to hire new employees. Immigrant workers in the United States hold many jobs, ranging from agricultural and gardening positions to high-tech and health-care-related professions. As with many of our ancestors, these California landscape workers hope that their earnings will help their children have a more secure future in this country.

reflect & analyze

1. When did your ancestors immigrate to the United States? Or, if you are a Native American, what is the history of your group? Do some research to find out about your ancestors.
2. Do you believe that non-English-speaking immigrants to this country should be expected—or required—to read and speak English at school, work, and other public places? Why or why not?

▶ turning to video

Watch the ABC video *Illegal Immigrant Children* (running time 2:42), available through CengageBrain.com. According to the video, of the 5,000 immigrant and refugee children who are caught entering the United States illegally each year, about one-third of them end up in prison. As you watch the video, think about the photographs, commentary, and questions you encountered in this photo essay. After you've watched the video, consider another question: Do you think racism is a factor in not providing attorneys and guardians to undocumented children? Are these services that the government should provide for these children?

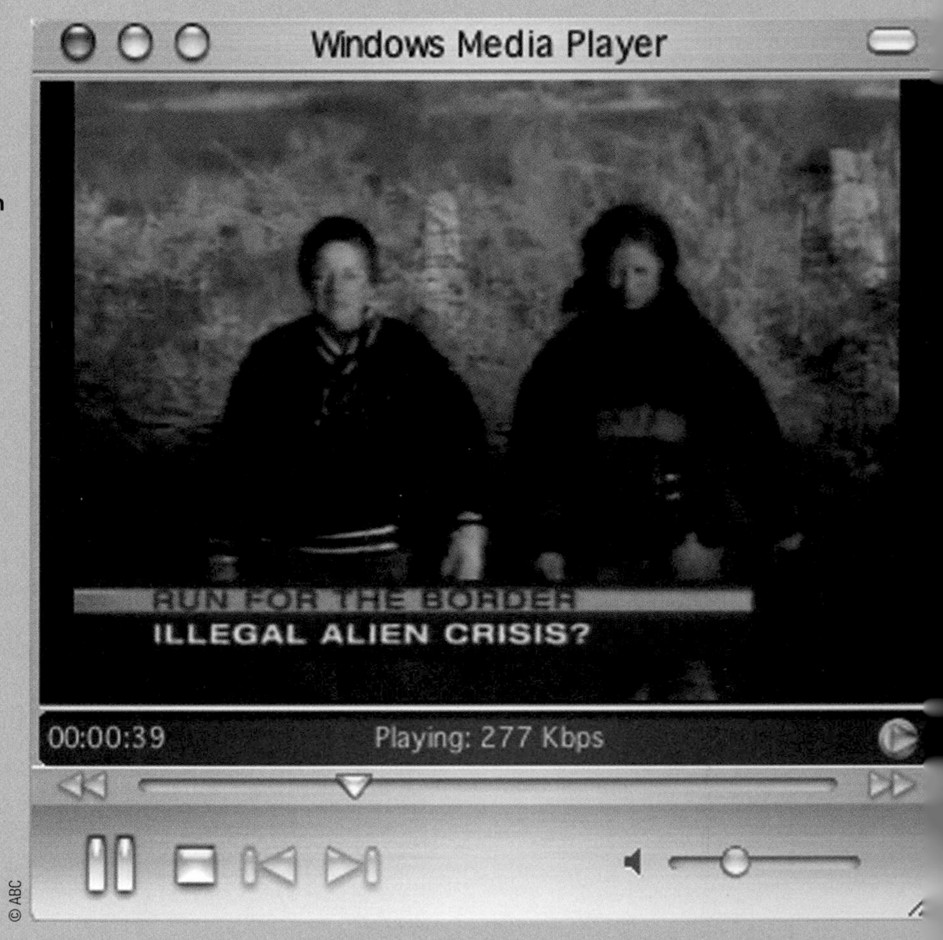

© ABC

conditions and a lack of adequate sewage facilities increased the hazards from plagues and fires, and death rates were high. For another, food supplies were limited. In order to generate food for each city resident, at least fifty farmers had to work in the fields, and animal power was the only means of bringing food to the city. Once foodstuffs arrived in the city, there was no effective way to preserve them. Finally, migration to the city was difficult. Many people were in serf, slave, and caste systems whereby they were bound to the land. Those able to escape such restrictions still faced several weeks of travel to reach the city, thus making it physically and financially impossible for many people to become city dwellers.

In spite of these problems, many preindustrial cities had a sense of *community*—a set of social relationships operating within given spatial boundaries or locations that provides people with a sense of identity and a feeling of belonging. The cities were full of people from all walks of life, both rich and poor, and they felt a high degree of social integration. You will recall that Ferdinand Tönnies (1940/1887) described such a community as *Gemeinschaft*—a society in which social relationships are based on personal bonds of friendship and kinship and on intergenerational stability, such that people have a commitment to the entire group and feel a sense of togetherness. By contrast, industrial cities were characterized by Tönnies as *Gesellschaft*—societies exhibiting impersonal and specialized relationships, with little long-term commitment to the group or consensus on values. In *Gesellschaft* societies, even neighbors are "strangers" who perceive that they have little in common with one another.

Industrial Cities

The Industrial Revolution changed the nature of the city. Factories sprang up rapidly as production shifted from the primary, agricultural sector to the secondary, manufacturing sector. With the advent of factories came many new employment opportunities not available to people in rural areas. Emergent technology, including new forms of transportation and agricultural production, made it easier for people to leave the countryside and move to the city. Between 1700 and 1900 the population of many European cities mushroomed. For example, the population of London increased from 550,000 to almost 6.5 million. Although the Industrial Revolution did not start in the United States until the mid-nineteenth century, the effect was similar. Between 1870 and 1910, for example, the population of New York City grew by 500 percent. In fact, New York City became the first U.S. *metropolis*—one or more central cities and their surrounding suburbs that dominate the economic and cultural life of a region. Nations such as Japan and Russia, which became industrialized after England and the United States, experienced a delayed pattern of urbanization, but this process moved quickly once it commenced in those countries.

© Rob Daemmrich/The Image Works

▲ Suburbia saw explosive growth after World War II, when more people possessed their own means of transportation, shrinking once-formidable work commutes to much more manageable trips.

Postindustrial Cities

Since the 1950s, postindustrial cities have emerged in nations such as the United States as their economies have gradually shifted from secondary (manufacturing) production to tertiary (service and information-processing) production. Postindustrial cities increasingly rely on an economic structure that is based on scientific knowledge rather than industrial production, and as a result, a class of professionals and technicians grows in size and influence. Postindustrial cities are dominated by "light" industry, such as software manufacturing; information-processing services, such as airline and hotel reservation services; educational complexes; medical centers; convention and entertainment centers; and retail trade centers and shopping malls. Most families do not live close to a central business district. Technological advances in communication and transportation make it possible for middle- and upper-income individuals and families to have more work options and to live greater distances from the workplace; however, these options are not often available to people of color and those at the lower end of the class structure.

On a global basis, cities such as New York, London, and Tokyo appear to fit the model of the postindustrial city. These cities have experienced a rapid growth in knowledge-based industries such as financial services. London, Tokyo, and New York have—at least until recently—experienced an increase in the number of highly paid professional jobs and more workers have been in high-income categories. Many people have benefited for a number of years from these high incomes and have created a lifestyle that is based on materialism and the gentrification of urban spaces. Meanwhile, those persons outside the growing professional categories have seen their own quality of life further deteriorate and their job opportunities become increasingly restricted to secondary labor markets in their respective "global" cities.

What is next for the postindustrial city? Various social analysts have proposed that cities are far from dead because of the many benefits they offer people. According to urban economist Edward Glaeser (2011), the future of nations and the world relies on cities that bring people together in a setting that is "healthier, greener, and richer" than urban myths would have us believe. From this perspective, incomes are higher in metropolitan areas; as well, cities are more energy efficient than suburban areas, where people commute great distances, often in heavy traffic congestion with stop-and-go traffic, between work and home. Cities are also centers of consumption; however, this benefit primarily accrues to the wealthy. However, the poor fare better than many think because they have inexpensive mass transit, the ability to "cram into small apartments in the outer boroughs," and "plenty of entry-level service-sector jobs with wages that beat those in Ghana or Guatemala" (Glaeser, 2011). People in the middle-income category often have a harder time because of the costs of housing in good neighborhoods and a quality education for their children (Glaeser, 2011).

A second perspective on the postmodern city has been offered by John D. Kasarda, who coined the term *aerotropolis* to describe a new urban pattern in which cities are built around airports rather than airports being built around cities. An aerotropolis is a combination of giant airport, planned city, shipping facility, and business hub. According to this approach, the pattern of the twentieth century was city in the center, airport on the periphery. However, this pattern has shifted, with the airport now in the center and the city on the periphery because of extensive growth in jet travel, 24/7 workdays, overnight shipping, and global business networks (Kasarda and Lindsay, 2011). Aerotropoli are now found in Seoul, Amsterdam, Dallas, Memphis, Washington, D.C., and other cities where globalization has forever changed the nature of urban life.

Perspectives on Urbanization and the Growth of Cities

Urban sociology follows in the tradition of early European sociological perspectives that compared social life with biological organisms or ecological processes. For example, Auguste Comte pointed out that cities are the "real organs" that make a society function. Emile Durkheim applied natural ecology to his analysis of *mechanical solidarity,* characterized by a simple division of labor and shared religious beliefs such as are found in small, agrarian societies, and *organic solidarity,* characterized by interdependence based on the elaborate division of labor found in large, urban societies. These early analyses became the foundation for ecological models/functionalist perspectives in urban sociology.

Functionalist Perspectives: Ecological Models

Functionalists examine the interrelations among the parts that make up the whole; therefore, in studying the growth of cities, they emphasize the life cycle of urban growth. Like the social philosophers and sociologists before him, the University of Chicago sociologist Robert Park (1915) based his analysis of the city on *human ecology*—the study of the relationship between people and their physical environment. According to Park (1936), economic competition produces certain regularities in land-use patterns and population distributions. Applying Park's idea to the study of urban land-use patterns, the sociologist Ernest W. Burgess (1925) developed the concentric zone model, an ideal construct that attempted to explain why some cities expand radially from a central business core.

© Comstock/Getty Images

▲ Despite an increase in telecommuting and more diverse employment opportunities in the high-tech economy, our highways have grown increasingly congested. Can we implement measures to reduce the problems of urban congestion and environmental pollution, or will these problems grow worse with each passing year?

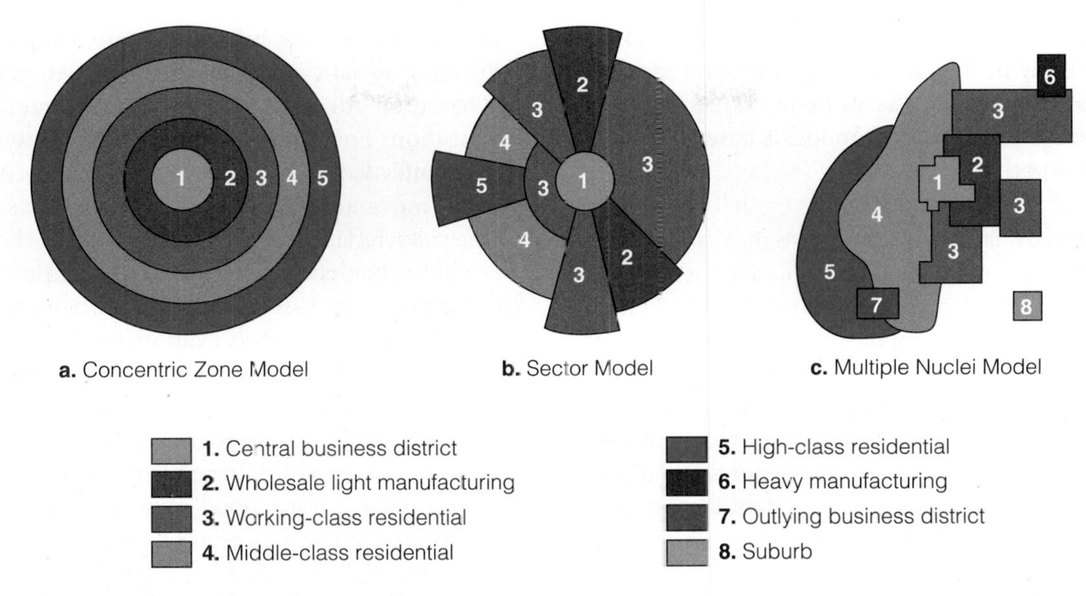

▲ **FIGURE 19.5 THREE MODELS OF THE CITY**

Source: Adapted from Harris and Ullman, 1945.

The Concentric Zone Model Burgess's *concentric zone model* is a description of the process of urban growth that views the city as a series of circular areas or zones, each characterized by a different type of land use, that developed from a central core (see ▶ Figure 19.5a). *Zone 1* is the central business district and cultural center. In *Zone 2* houses formerly occupied by wealthy families are divided into rooms and rented to recent immigrants and poor persons; this zone also contains light manufacturing and marginal businesses (such as secondhand stores, pawnshops, and taverns). *Zone 3* contains working-class residences and shops and ethnic enclaves. *Zone 4* comprises homes for affluent families, single-family residences of white-collar workers, and shopping centers. *Zone 5* is a ring of small cities and towns populated by persons who commute to the central city to work and by wealthy people living on estates.

Two important ecological processes are involved in the concentric zone theory: invasion and succession. ***Invasion* is the process by which a new category of people or type of land use arrives in an area previously occupied by another group or type of land use** (McKenzie, 1925). For example, Burgess noted that recent immigrants and low-income individuals "invaded" Zone 2, formerly occupied by wealthy families. ***Succession* is the process by which a new category of people or type of land use gradually predominates in an area formerly dominated by another group or activity** (McKenzie, 1925). In Zone 2, for example, when some of the single-family residences were sold and subsequently divided into multiple housing units, the remaining single-family owners moved out because the "old" neighborhood had changed. As a result of their move, the process of invasion was complete and succession had occurred.

Invasion and succession theoretically operate in an outward movement: Those who are unable to "move out" of the inner rings are those without upward social mobility, so the central zone ends up being primarily occupied by the poorest residents—except when gentrification occurs. ***Gentrification* is the process by which members of the middle and upper-middle classes, especially whites, move into the central-city area and renovate existing properties.** Centrally located, naturally attractive areas are the most likely candidates for gentrification. To urban ecologists, gentrification is the solution to revitalizing the central city. To conflict theorists, however, gentrification creates additional hardships for the poor by depleting the amount of affordable housing available and by "pushing" them out of the area.

The concentric zone model demonstrates how economic and political forces play an important part in the location of groups and activities, and it shows how a large urban area can have internal differentiation. However, the model is most applicable

invasion the process by which a new category of people or type of land use arrives in an area previously occupied by another group or type of land use.

succession the process by which a new category of people or type of land use gradually predominates in an area formerly dominated by another group or activity.

gentrification the process by which members of the middle and upper-middle classes, especially whites, move into a central-city area and renovate existing properties.

to older cities that experienced high levels of immigration early in the twentieth century and to a few midwestern cities such as St. Louis. No city, including Chicago (on which the model is based), entirely conforms to this model.

The Sector Model In an attempt to examine a wider range of settings, urban ecologist Homer Hoyt (1939) studied the configuration of 142 cities. Hoyt's *sector model* emphasizes the significance of terrain and the importance of transportation routes in the layout of cities. According to Hoyt, residences of a particular type and value tend to grow outward from the center of the city in wedge-shaped sectors, with the more-expensive residential neighborhoods located along the higher ground near lakes and rivers or along certain streets that stretch in one direction or another from the downtown area (see Figure 19.5b). By contrast, industrial areas tend to be located along river valleys and railroad lines. Middle-class residential zones exist on either side of the wealthier neighborhoods. Finally, lower-class residential areas occupy the remaining space, bordering the central business area and the industrial areas. Hoyt (1939) concluded that the sector model applied to cities such as Seattle, Minneapolis, San Francisco, Charleston (South Carolina), and Richmond (Virginia).

The Multiple Nuclei Model According to the *multiple nuclei model* developed by urban ecologists Chauncey Harris and Edward Ullman (1945), cities do not have one center from which all growth radiates, but rather have numerous centers of development based on specific urban needs or activities (see Figure 19.5c). As cities began to grow rapidly, they annexed formerly outlying and independent townships that had been communities in their own right. In addition to the central business district, other nuclei developed around entities such as an educational institution, a medical complex, or a government center. Residential neighborhoods may exist close to or far away from these nuclei. A wealthy residential enclave may be located near a high-priced shopping center, for instance, whereas less-expensive housing must locate closer to industrial and transitional areas of town. This model may be applicable to cities such as Boston. However, critics suggest that it does not provide insights into the uniformity of land-use patterns among cities and relies on an after-the-fact explanation of why certain entities are located where they are.

Contemporary Urban Ecology Urban ecologist Amos Hawley (1950) revitalized the ecological tradition by linking it more closely with functionalism. According to Hawley, urban areas are complex and expanding social systems in which growth patterns are based on advances in transportation and communication. For example, commuter railways and automobiles led to the decentralization of city life and the movement of industry from the central city to the suburbs (Hawley, 1981).

Other urban ecologists have continued to refine the methodology used to study the urban environment. *Social area analysis* examines urban populations in terms of economic status, family status, and ethnic classification (Shevky and Bell, 1966). For example, middle- and upper-middle-class parents with school-aged children tend to cluster together in "social areas" with a "good" school district; young single professionals may prefer to cluster in the central city for entertainment and nightlife.

The influence of human ecology on the field of urban sociology is still very strong today (see Frisbie and Kasarda, 1988). Contemporary research on European and North American urban patterns is often based on the assumption that spatial arrangements in cities conform to a common, most efficient design. However, some critics have noted that ecological models do not take into account the influence of powerful political and economic elites on the development process in urban areas (Feagin and Parker, 2002).

Conflict Perspectives: Political Economy Models

Conflict theorists argue that cities do not grow or decline by chance. Rather, they are the product of specific decisions made by members of the capitalist class and political elites. These far-reaching decisions regarding land use and urban development benefit the members of some groups at the expense of others (see Castells, 1977/1972). Karl Marx suggested that cities are the arenas in which the intertwined processes of class conflict and capital accumulation take place; class consciousness and worker revolt are more likely to develop when workers are concentrated in urban areas.

According to the sociologists Joe R. Feagin and Robert Parker (2002), three major themes prevail in political economy models of urban growth. First, both economic *and* political factors affect patterns of urban growth and decline. Economic factors include capitalistic investments in production, workers, workplaces, land, and buildings. Political factors include governmental protection of the right to own and dispose of privately held property as owners see fit and the role of government officials in promoting the interests of business elites and large corporations.

◀ **FIGURE 19.6 THE VALUE OF URBAN SPACE**

Second, urban space has both an exchange value and a use value. *Exchange value* refers to the profits that industrialists, developers, bankers, and others make from buying, selling, and developing land and buildings. By contrast, *use value* is the utility of space, land, and buildings for everyday life, family life, and neighborhood life. In other words, land has purposes other than simply for generating profits—for example, for homes, open spaces, and recreational areas (see ▶ Figure 19.6). Today, class conflict exists over the use of urban space, as is evident in battles over the rental costs, safety, and development of large-scale projects.

Third, both structure and agency are important in understanding how urban development takes place. *Structure* refers to institutions such as state bureaucracies and capital investment circuits that are involved in the urban development process. *Agency* refers to human actors, including developers, business elites, and activists protesting development, who are involved in decisions about land use.

Capitalism and Urban Growth in the United States According to political economy models, urban growth is influenced by capital investment decisions, power and resource inequality, class and class conflict, and government subsidy programs. Members of the capitalist class choose corporate locations, decide on sites for shopping centers and factories, and spread the population that can afford to purchase homes into sprawling suburbs located exactly where the capitalists think they should be located (Feagin and Parker, 2002).

Today, a few hundred financial institutions and developers finance and construct most major and many smaller urban development projects around the country, including skyscrapers, shopping malls, and suburban housing projects. These decision makers set limits on the individual choices of the ordinary citizen with regard to real estate, just as they do with regard to other choices (Feagin and Parker, 2002). They can make housing more affordable or totally unaffordable for many people. Ultimately, their motivation rests not in benefiting the community, but rather in making a profit; the cities that they produce reflect this mindset.

One of the major results of these urban development practices is *uneven development*—the tendency of some neighborhoods, cities, or regions to grow and prosper whereas others stagnate and decline. Conflict theorists argue that uneven development reflects inequalities of wealth and power in society. The problem not only affects areas in a state of decline but also produces external costs, even in "boom" areas, that are paid for by the entire community. Among these costs are increased pollution, increased traffic congestion, and rising rates of crime and violence. According to the sociologist Mark Gottdiener (1985: 214), these costs are "intrinsic to the very core of capitalism, and those who profit the most from development are not called upon to remedy its side effects."

The Gated Community in the Capitalist Economy The growth of *gated communities*—subdivisions or neighborhoods surrounded by barriers such as walls, fences, gates, or earth banks covered with bushes and shrubs, along with a secured entrance—is an example to many people of how developers, builders, and municipalities have encouraged an increasing division between public and private property in capitalist societies. Many gated communities are created by developers who hope to increase their profits by offering potential residents a semblance of safety, privacy, and luxury that they might not have in nongated residential areas. Other gated communities have been developed after the fact in established neighborhoods by adding walls, gates, and sometimes security-guard stations. For example, a recent study noted situations in which residents of elite residential enclaves, such as the River Oaks area of Houston or the "Old Enfield" area of Austin, Texas, were able to gain approval from the city to close certain streets and create cul-de-sacs, or to erect other barriers to discourage or prevent outsiders from driving through the neighborhood (Kendall, 2002). Gated communities for upper-middle-class and upper-class residents convey the idea of exclusivity and privilege, whereas similar communities for middle- and lower-income residents typically focus on such features as safety for children and the ability to share amenities such as a "community" swimming pool or recreational center with other residents.

© Andrew Holbrooke/The Image Works

▲ According to conflict theorists, exploitation by the capitalist class increasingly impoverishes poor whites and low-income subordinate-group members. Increasing rates of homelessness have made scenes such as this a recurring sight in many cities.

Regardless of the social and economic reasons given for the development of gated communities, many analysts agree that these communities reflect a growing divide between public and private space in urban areas. According to a study conducted by anthropologist Setha Low (2003), gated communities do more than simply restrict access to the residents' homes: They also limit the use of public spaces, making it impossible for others to use the roads, parks, and open space contained within the enclosed community. Low (2003) refers to this phenomenon as the "fortressing of America."

Gender Regimes in Cities Some feminist perspectives focus on urbanization as a reflection not only of the working of the political economy but also of patriarchy. According to the sociologist Lynn M. Appleton (1995), different kinds of cities have different *gender regimes*—prevailing ideologies of how women and men should think, feel, and act; how access to social positions and control of resources should be managed; and how relationships between men and women should be conducted. The higher density and greater diversity found in central cities such as New York City serve as a challenge to the private patriarchy found in the home and workplace in lower-density, homogeneous areas such as suburbs and rural areas. *Private patriarchy* is based on a strongly gendered division of labor in the home, gender-segregated paid employment, and women's dependence on men's income. At the same time, cities may foster *public patriarchy* in the form of women's increasing dependence on paid work and the state for income and their decreasing emotional interdependence with men. At this point, gender often intersects with class and race as a form of oppression because lower-income women of color often reside in central cities. Public patriarchy may be perpetuated by cities through policies that limit women's access to paid work and public transportation. However, such cities may also be a forum for challenging patriarchy; all residents who differ in marital status, paternity, sexual orientation, class, and/or race/ethnicity tend to live close to one another and may hold a common belief that both public and private patriarchy should be eliminated (Appleton, 1995).

Symbolic Interactionist Perspectives: The Experience of City Life

Symbolic interactionists examine the *experience* of urban life. How does city life affect the people who live in a city? Some analysts answer this question positively; others are cynical about the effects of urban living on the individual.

Simmel's View of City Life According to the German sociologist Georg Simmel (1950/1902–1917), urban life is highly stimulating, and it shapes people's thoughts and actions. Urban residents are influenced by the quick pace of the city and the pervasiveness of economic relations in everyday life. Because of the intensity of urban life, people have no choice but to become somewhat insensitive to events and individuals around them. Many urban residents avoid emotional involvement with one another and try to ignore events taking place around them. Urbanites feel wary toward other people because most interactions in the city are economic rather than social. Simmel suggests that attributes such as punctuality and exactness are rewarded but that friendliness and warmth in interpersonal relations are viewed as personal weaknesses. Some people act reserved to cloak their deeper feelings of distrust or dislike toward others. However, Simmel did not view city life as completely negative; he also pointed out that urban living could have a liberating effect on people because they had opportunities for individualism and autonomy.

Urbanism as a Way of Life Based on Simmel's observations on social relations in the city, the early Chicago School sociologist Louis Wirth (1938) suggested that urbanism is a "way of life." *Urbanism* refers to the distinctive social and psychological patterns of life typically found in the city. According to Wirth, the size, density, and heterogeneity of urban populations typically result in an elaborate division of labor and in spatial segregation of people by race/ethnicity, social class, religion, and/or lifestyle. In the city, primary-group ties are largely replaced by secondary relationships; social interaction is fragmented, impersonal, and often superficial. Even though people gain some degree of freedom and privacy by living in the city, they pay a price for their autonomy, losing the group support and reassurance that come from primary-group ties.

From Wirth's perspective, people who live in urban areas are alienated, powerless, and lonely. A sense of community is obliterated and replaced by the "mass society"—a large-scale, highly institutionalized society in which individuality is supplanted by mass messages, faceless bureaucrats, and corporate interests.

Gans's Urban Villagers Unlike Wirth's gloomy assessment of urban life, sociologist Herbert Gans (1982/1962) suggested that many people have positive experiences in cities. Based on research in the west end of Boston, Gans concluded that many residents develop strong loyalties and a sense of community in central-city areas that outsiders may view negatively. According to Gans, there are five major categories of adaptation among urban dwellers. *Cosmopolites* are students, artists, writers, musicians, entertainers, and professionals who choose to live in the city because they want to be close to its cultural facilities. *Unmarried people* and *childless couples* live in the city because they want to be close to work and entertainment. *Ethnic villagers* live in ethnically segregated

▲ Festive occasions such as this street fair provide opportunities for urban villagers to mingle with others, enjoying entertainment and social interaction.

sociology works!

Herbert Gans and Twenty-First-Century Urban Villagers

I moved to Austin because it's a high-tech city with a small-town feel. Kinda my own "urban village" where I can cycle around town when I want but still own a nice car to go out in. I chose my neighborhood because it's centrally located to downtown eating and live entertainment. Austin calls itself the "Live Music Capital of the World," and I have plenty of opportunities to hear the music I like. Overall, I'd say that I'm compatible with Austin, and Austin's compatible with me.

—"Brad," a twenty-four-year-old college graduate, explaining why he chose to become an "urban villager" in Austin, Texas (author's files, 2007)

Although urban sociologist Herbert J. Gans's research in Boston's west end took place a number of years ago, we find that many residents of cities still think of themselves as living in an "urban village." Today, many younger urban residents think of themselves as having strong loyalties to a specific segment of their community with which they share interests and common experiences. Although many contemporary studies of urban life have emphasized the problems of poverty, crime, racial and ethnic discrimination, inadequate health care, and poor schools in our nation's major cities, it is useful for us to examine positive aspects of urban life as well. We can also gain important insights from examining the experiences of middle- and upper-income residents of our cities.

Herbert Gans believed that the people he referred to as *cosmopolites* chose to live in the city so that they could be close to cultural facilities, while he thought that *unmarried people* and *childless couples* chose to live there because they wanted to be close to work and entertainment. Among some affluent residents in high-tech cities such as Austin, Texas, married people and families with children have increasingly joined the ranks of individuals who live in or near the city's downtown area. According to contemporary urban villagers such as "Brad," they can find other people who are like themselves, who participate in activities they enjoy, and who support one another much like the members of an extended family when they need friendship or assistance. From this perspective, Gans's ideas about the urban village work because they show us that an important way to understand city life is through the experiences of people who live there. In the final analysis, of course, all people—including lower-income individuals, who have been further disadvantaged or even displaced by gentrification, and the poor and homeless—must be included in any thorough sociological examination of city life in the twenty-first century. However, "urban villagers," as coined by Gans, has staying power as a concept because it encourages us to look at urban life as a kaleidoscope of diversity that includes the wealthy and the merely affluent, as well as those who are "just getting by" or who are poor, because they live close to one another as contemporary urban dwellers.

reflect & analyze

Can you identify categories of urban villagers in a city with which you are familiar? To what extent do people live in certain areas of the city based on personal choice? What factors appear to be beyond their control?

neighborhoods; some are recent immigrants who feel most comfortable within their own group. The *deprived* are poor individuals with dim future prospects; they have very limited education and few, if any, other resources. The *trapped* are urban dwellers who can find no escape from the city; this group includes persons left behind by the process of invasion and succession: downwardly mobile individuals who have lost their former position in society, older persons who have nowhere else to go, and individuals addicted to alcohol or other drugs. Gans concluded that the city is a pleasure and a challenge for some urban dwellers and an urban nightmare for others (see "Sociology Works!").

Gender and City Life In their everyday lives, do women and men experience city life differently? According to the scholar Elizabeth Wilson (1991), some men view the city as *sexual space* in which women, based on their sexual desirability and accessibility, are categorized as prostitutes, lesbians, temptresses, or virtuous women in need of protection. Wilson suggests that more-affluent, dominant-group women are more likely to be viewed as virtuous women in need of protection by their own men or police officers. Cities offer a paradox for women: On the one hand, cities offer more freedom than is found in comparatively isolated rural, suburban, and domestic settings; on the other, women may be in greater physical danger in the city. For Wilson, the answer to women's vulnerability in the city is not found in offering protection for them, but rather in changing people's perceptions that they can treat women as sexual objects because of the impersonality of city life (Wilson, 1991).

Cities and Persons with a Disability Chapter 18 describes how disability rights advocates believe that structural barriers create a "disabling" environment for many people, particularly in large urban settings. Many cities have made their streets and sidewalks more user-friendly for persons in wheelchairs and for individuals with visual disability by constructing concrete ramps with slide-proof surfaces at intersections or installing traffic lights with sounds designating when to "Walk." However, both urban and rural areas have a long way to go before many persons with disabilities will have the access to the things that they need to become productive members of the community: educational and employment opportunities. Some persons with disabilities cannot navigate the streets and sidewalks of their communities, and some face obstacles getting into buildings that marginally, at best, meet the accessibility standards of the Americans with Disabilities Act; thus, many persons with a disability are unemployed.

Political scientist Harlan Hahn (1997: 177) traces the problem of lack of access to the beginnings of industrialism:

> The rise of industrialism produced extensive changes in the lives of disabled as well as nondisabled people. As factories replaced private dwellings as the primary sites of production, routines and architectural configurations were standardized to suit nondisabled workers. Both the design of worksites and of the products that were manufactured gave virtually no attention to the needs of people with disabilities. As a result, patterns of aversion and avoidance toward disabled persons were embedded in the construction of commodities, landscapes, and buildings that would remain for centuries. . . .

© Andy Sacks/Stone/Getty Images

▲ Most U.S. cities are laid out for motorized vehicles, not people. Traveling urban areas via wheelchair can be a daunting proposition.

In addition to the effects of industrialism on public access, Hahn (1997: 178) also emphasizes the impact of urbanization on the plight of persons with a disability:

> The social and economic changes fostered by industrialization may have been exacerbated by the accompanying process of urbanization. As workers increasingly moved from farms and rural villages to live near the institutions of mass production, the character of community life appeared to shift perceptibly. Deviant or atypical personal characteristics that may have gradually become familiar in a small community seemed bizarre or disturbing in an urban milieu.

As Hahn's statement suggests, historical patterns in the dynamics of industrial capitalism contributed to discrimination against persons with disabilities, and this legacy remains evident in contemporary cities. Structural barriers are further intensified when other people do not respond favorably toward persons with disabilities. Scholar and disability rights advocate Sally French (1999: 25–26), who is visually disabled, describes her own experience:

> I have lived in the same house for 16 years and yet I cannot recognize my neighbors. I know nothing about them at all; which children belong to whom, who has come and gone, who is old or young, ill or well, black or white. . . . On moving to my present house I informed several neighbors that, because of my inability to recognize them, I would doubtless pass them by in the street without greeting them. One neighbor, who had previously seen me striding confidently down the road, refused to believe me, but the others said they understood and would talk to me if our paths crossed. For the first couple of weeks it worked and I was surprised how often we met, but after that their greetings rapidly decreased and then ceased altogether. Why this happened I am not sure, but I suspect that my lack of recognition strained the interaction and limited the social reward they received from the encounter.

The Concept Quick Review examines the multiple perspectives on urban growth and urban living.

Problems in Global Cities

Although some people have lived in cities for thousands of years, the time is rapidly approaching when more people worldwide will live in or near a city than live in a rural area. In the middle-income and

[concept quick review]

Perspectives on Urbanism and the Growth of Cities

Functionalist Perspectives: Ecological Models	Concentric zone model	Because of invasion, succession, and gentrification, cities are a series of circular zones, each characterized by a particular land use.
	Sector model	Cities consist of wedge-shaped sectors, based on terrain and transportation routes, with the most-expensive areas occupying the best terrain.
	Multiple nuclei model	Cities have more than one center of development, based on specific needs and activities.
Conflict Perspectives: Political Economy Models	Capitalism and urban growth	Members of the capitalist class choose locations for skyscrapers and housing projects, limiting individual choices by others.
	Gender regimes in cities	Different cities have different prevailing ideologies regarding access to social positions and resources for men and women.
	Global patterns of growth	Capital investment decisions by core nations result in uneven growth in peripheral and semiperipheral nations.
Symbolic Interactionist Perspectives: The Experience of City Life	Simmel's view of city life	Because of the intensity of city life, people become somewhat insensitive to individuals and events around them.
	Urbanism as a way of life	The size, density, and heterogeneity of urban population result in an elaborate division of labor and space.
	Gans's urban villagers	Five categories of adaptation occur among urban dwellers, ranging from cosmopolites to trapped city dwellers.
	Gender and city life	Cities offer women a paradox: more freedom than in more-isolated areas, yet greater potential danger.

low-income regions of the world, Latin America is becoming the most urbanized: Four megacities—Mexico City, Buenos Aires, Lima, and Santiago—already contain more than half of the region's population and continue to grow rapidly. Within the next ten years, Rio de Janeiro and São Paulo are expected to have a combined population of about 40 million people living in a 350-mile-long megalopolis.

Rapid population growth will have a major effect on cities throughout the world. Essential services such as health, education, transportation, and sanitation are already strained in many cities, and the problem will only grow worse as the world's population moves upward toward a projected 10 billion or more by the end of the twenty-first century. In China alone, it is estimated that 220 cities will have more than 1 million people by 2025, and more than 350 million rural residents will move to the cities from rural areas. Requirements for new and expanded infrastructure in these cities will be tremendous, particularly in the building of high-rise buildings, mass-transit systems, and other amenities that will be needed to support this colossal population shift (Roberts, 2008).

Today, some social analysts look beyond the city proper, which is defined as a locality with legally fixed boundaries and an administratively recognized urban status that is usually characterized by some form of local government, to see the larger picture of what takes place in urban agglomerations. An *urban agglomeration* is defined as comprising the city or town proper and also the suburban fringe or thickly settled territory lying outside of, but adjacent to, the city boundaries, as a more accurate reflection of population composition and density in a given region. ▶ Figure 19.7 shows the populations of the world's twelve largest urban agglomerations.

Natural increases in population (higher birthrates than death rates) account for two-thirds of new urban growth. In recent years, fewer deaths and more births have occurred than demographers had anticipated. High fertility brings about booming populations and a corresponding strain on food and other resources; however, low fertility contributes to an aging population and stress on social services.

The other component of new urban growth is rural-to-urban migration. Some people move from

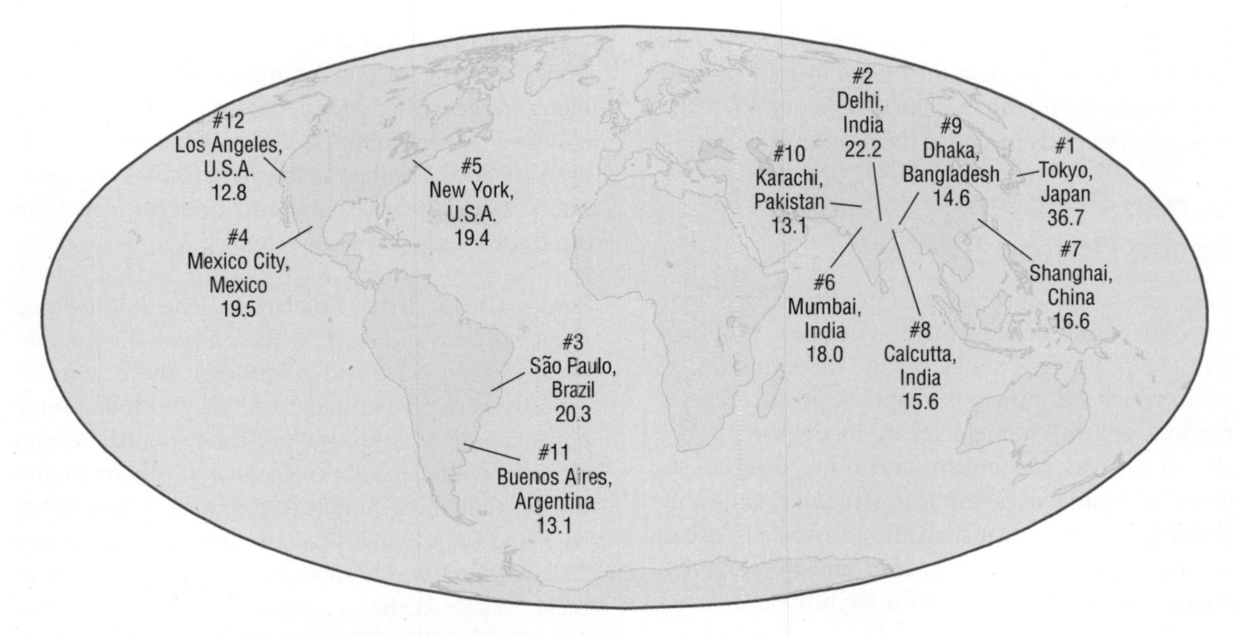

▲ **FIGURE 19.7 THE WORLD'S TWELVE LARGEST AGGLOMERATIONS**

Note: 2009 populations in millions.

Source: United Nations World Urbanization Prospects Report, 2009.

rural areas to urban areas because they have been displaced from their land. Others move because they are looking for a better life. No matter what the reason, migration has caused rapid growth in cities in China, sub-Saharan Africa, India, Algeria, and Egypt. At the same time that the population is growing rapidly, the amount of farmland available for growing crops to feed people is decreasing. In Egypt, for example, land that was previously used for growing crops is now used for petroleum refineries, food-processing plants, and other factories. Issues such as this contributed to the political uprising in 2011.

Rapid global population growth in Latin America and other regions is producing a wide variety of urban problems, including overcrowding, environmental pollution, and the disappearance of farmland. In fact, many cities in middle- and low-income nations are quickly reaching the point at which food, housing, and basic public services are available to only a limited segment of the population.

As global urbanization has continued to increase, differences in urban areas based on economic development at the national level have become apparent. Some cities in what Immanuel Wallerstein's (1984) world systems theory describes as core nations are referred to as *global cities*—interconnected urban areas that are centers of political, economic, and cultural activity. New York, Tokyo, and London are generally considered the largest global cities. These cities are the sites of new and innovative product development and marketing, and they are often the "command posts" for the world economy. But economic prosperity is not shared equally by all people in the core-nation global cities. Sometimes the living conditions of workers in low-wage service-sector jobs or in assembly production jobs more closely resemble the living conditions of workers in semiperipheral nations than they resemble the conditions of middle-class workers in their own country.

Many African countries and some countries in South America and the Caribbean are *peripheral* nations, previously defined as nations that depend on core nations for capital, have little or no industrialization (other than what may be brought in by core nations), and have uneven patterns of urbanization. According to Wallerstein (1984), the wealthy in peripheral nations support the exploitation of poor workers by core-nation capitalists in return for maintaining their own wealth and position. Poverty is thus perpetuated, and the problems worsen because of the unprecedented population growth in these countries. Like peripheral nations, semiperipheral nations—such as India, Iran, and Mexico—are confronted with unprecedented population growth. In addition, a steady flow of rural migrants to large cities is creating enormous urban problems.

More recently, Wallerstein suggested that a new world system is emerging that will retain some basic features of the existing one, but it will not be based on capitalism as we know it. According to Wallerstein, the new system will be either hierarchical and exploitative or relatively democratic and relatively egalitarian. In either scenario, national and local governments will be faced with difficult choices, including how to deal with major fiscal crises and immigration. In the United States and throughout Europe, for example, a growing cry demands that governments "do something" about evicting "foreigners"; however, the effects of such an action may create great turmoil and economic instability in cities. Wallerstein's original theory that work migrates from core nations to workers in semiperipheral and peripheral nations has shifted: Workers are now migrating to cities where they hope to find employment to support themselves and their families, and this process is not likely to stop in the near future.

Urban Problems in the United States

Even the most optimistic of observers tends to agree that cities in the United States have problems brought on by years of neglect and deterioration. As we have seen in previous chapters, poverty, crime, racism, sexism, homelessness, inadequate public school systems, alcoholism and other drug abuse, gangs and guns, decaying infrastructure, and other social problems are most visible and acute in urban settings. Issues of urban growth and development are intertwined with many of these problems.

Divided Interests: Cities, Suburbs, and Beyond

Since World War II, a dramatic population shift has occurred in this country as thousands of families have moved from cities to suburbs. Even though some people lived in suburban areas prior to the twentieth century, it took the involvement of the federal government and large-scale development to spur the dramatic shift that began in the 1950s (Palen, 2012). According to urban historian Kenneth T. Jackson (1985), postwar suburban growth was fueled by aggressive land developers, inexpensive real estate and construction methods, better transportation, abundant energy, government subsidies, and racial stress in the cities. However, the sociologist J. John Palen (2012) suggests that the Baby Boom following World War II and the liberalization of lending policies by federal agencies such as the Veterans Administration (VA) and the Federal Housing Authority (FHA) were significant factors in mass suburbanization.

Regardless of its causes, mass suburbanization has created a territorial division of interests between cities and suburban areas. Although many suburbanites rely on urban centers for their employment, entertainment, and other services, they pay their property taxes to suburban governments and school districts. Some affluent suburbs have state-of-the-art school districts, police and fire departments, libraries, and infrastructures (such as roads, sewers, and water-treatment plants). However, many suburban areas are not affluent and experience similar problems to those found in the urban core. Suburban inequality and poverty have continued to grow in the 2010s as more families have been hard hit by the economic recession, including higher rates of unemployment, steep rents, or lower property values on homes, combined with high interest on mortgages and high property taxes, as suburban communities attempt to meet budget shortfalls.

Similarly, central-city services and school districts languish for lack of funds. Affluent families living in "gentrified" properties typically send their children to elite private schools, whereas the children of poor families living in racially segregated public housing projects attend underfunded (and often substandard) public schools.

Race, Class, and Suburbs The intertwining impact of race and class is visible in the division between central cities and suburbs. In the past, most suburbs were predominantly white, and today some upper-middle-class and upper-class suburbs remain virtually white. Examples include northern Fulton County (adjoining Atlanta, Georgia) and the cities of Highland Park, University Park, and Plano (adjoining Dallas, Texas). In the suburbs, people of color (especially African Americans) often become resegregated. An example is the Detroit metropolitan area, where Census Bureau data show a 25 percent drop in the city's population over the last decade and a corresponding increase in minority population in the suburbs. However, as a study by historian and sociologist Thomas J. Sugrue (2011) shows, many African Americans are moving into so-called secondhand suburbs, defined as "established communities with deteriorating housing stock that are falling out of favor with younger white homebuyers." According to Sugrue (2011), if history holds true, these suburban areas will soon look like Detroit itself, with "resegregated schools, dwindling tax bases and decaying public services."

Some analysts argue that the location of one's residence is a matter of personal choice. African Americans and other persons of color have sought the same things in suburban properties that white Americans have, namely safe streets and low crime rates, the best housing they can afford, quality schools for their children, and the amenities of life outside the hubbub of the center city. However, what African Americans have often found are fewer job opportunities, poorer services, older houses, and run-down shopping districts (Sugrue, 2011).

Since the 1960s, other analysts have suggested that residential segregation reflects discriminatory practices by landlords, homeowners, and white realtors and their agents, who engage in *steering* people of color to different neighborhoods from those shown to their white counterparts. Lending practices of banks (including the *redlining* of certain properties so that acquiring a loan is virtually impossible) and the behavior of neighbors further intensify these problems. Some analysts suggest that African Americans are more likely to move to suburbs with declining tax bases because they have limited finances, because they are steered there by real estate agents, or because "white flight" occurs as African American homeowners move in, leaving a heavier tax burden and fewer of the benefits of "suburbia" for the newcomers and those who remain behind.

© megumi ito/Shutterstock.com

© AP Photo/Mark Duncan

▲ Affluent gated communities and enclaves of million-dollar homes stand in sharp contrast to low-income housing when we see them on the urban landscape. What sociological theories help us describe the disparity of lifestyles and life chances shown in these two settings?

Beyond the Suburbs: Edge Cities Urban fringes (referred to as *edge cities*) spring up beyond central cities and suburbs (Garreau, 1991). The Massachusetts Turnpike corridor west of Boston and the Perimeter area north of Atlanta are examples. Edge cities initially developed as residential areas; then retail establishments and office parks moved into the area. Commuters from the edge city are able to travel around (rather than in and out of) the metropolitan region's center and can avoid its rush-hour traffic quagmires. Many businesses and industries have moved physical plants and tax dollars to these areas: Land is cheaper, and utility rates and property taxes are lower.

The Continuing Fiscal Crises of the Cities

The largest cities in the United States have faced periodic fiscal crises for many years; however, in the 2010s, cities of all sizes are experiencing even greater financial problems partially linked to a major downturn in national and international economic trends. U.S. cities may experience a collective budget shortfall of more than $60 billion in the near future. Economic recoveries in cities take a number of years longer than a national recovery, which means that financial problems brought about by the Great Recession may last for years to come.

Why have national and international economic downturns hurt U.S. cities so drastically? What are cities doing about it? Cities have experienced extensive shortfalls in revenue because states have reduced the amount of money that they provide for cities, and the cities have had decreased revenue from sales taxes, corporate taxes, and personal income taxes. Funds from the federal government to states and cities have also been limited and are often earmarked for specific projects rather than for use in the general operating budget. These budget crises have forced states to cut funding to already cash-strapped cities. Vital services, including police, firefighting, and public works, have been cut drastically, and some analysts believe that the slashing of city budgets and programs will continue for some time in the future. As cities lose revenue, officials must decide to lay off or furlough employees, charge higher fees for services, and cancel major projects such as street repairs or infrastructure improvements (building a new water-treatment facility, for example).

City officials continue to urge leaders at the state and federal levels to create new programs that will help cities meet their residents' needs. Demands specifically are being made for more federal aid through job creation programs and other economic stimulus packages. Some analysts believe that inaction at the state and federal levels may create even greater financial chaos by forcing some cities into bankruptcy. Local officials emphasize that the state of America's cities continues to threaten the long-term national economic recovery (National League of Cities, 2010). It remains to be seen what the eventual effects of these continuing fiscal crises will be on various cities throughout the nation.

Rural Community Issues in the United States

Although most people think of the United States as highly urbanized, about 20 percent of the U.S. population resides in rural areas, identified as communities of 2,500 people or less by the U.S. Census Bureau. Sociologists typically identify *rural communities* as small, sparsely settled areas that have a relatively homogeneous population of people who primarily engage in agriculture. However, rural communities today are more diverse than this definition suggests.

Unlike the standard migration patterns from rural to urban places in the past, recently more people have moved from large urban areas and suburbs into rural areas. Many of those leaving urban areas today want to escape the high cost of living, crime, traffic congestion, and environmental pollution that make daily life difficult. Technological advances make it easier for people to move to outlying rural areas and still be connected to urban centers if they need to be. The proliferation of computers, cell phones, commuter airlines, and highway systems has made previously remote areas seem much more accessible to many people. However, many recent immigrants to rural areas do not face some traditional problems experienced by long-term rural residents, particularly farmers, small-business owners, teachers, doctors, and medical personnel in these rural communities.

For many people in rural areas who have made their livelihood through farming and other agricultural endeavors, recent decades have been very difficult, both financially and emotionally. Rural crises such as droughts, crop failures, and the loss of small businesses in the community have had a negative effect on many adults and their children. Like their urban counterparts, rural families have experienced problems of divorce, alcoholism, abuse, and other crises, but these issues have sometimes been exacerbated by such events as the loss of the family farm or business (Pitzer, 2003). Because home is also the center of work in farming families, the loss of the farm may also mean the loss of family and social life, and the loss of things dear to children such as their 4-H projects—often an animal that a child raises to show and sell (Pitzer, 2003). Some rural children and adolescents are also subject to injuries associated with farm work, such as livestock kicks or crushing, falling out of a tractor or pickup, and being injured while operating machinery designed for adults, that are not typically experienced by their urban counterparts (Schutske, 2002).

Economic opportunities are limited in many rural areas, and average salaries are typically lower than in urban areas, based on the assumption that a family can live on less money in rural communities than in cities. An example is rural teachers, who earn substantially less than their urban and

© D. Hurst/Alamy

▲ Areas with escalating unemployment frequently also have an increase in the use of illegal drugs. High unemployment in some rural areas of the United States has been accompanied by a huge increase in the production and use of methamphetamines (meth).

suburban counterparts. Some rural areas have lost many teachers and administrators to higher-paying districts in other cities.

Although many of the problems we have examined in this book are intensified in rural areas, one of the most pressing is the availability of health services and doctors. Recently, some medical schools have established clinics and practices in outlying rural regions of the states in which they are located in an effort to increase the number of physicians available to rural residents. Typically, physicians who have just started to practice medicine have chosen to work in large urban centers with accessible high-tech medical facilities. Because of the pressing time constraints of tending to patients with life-threatening problems, the availability of community clinics and hospitals in rural areas may be a life-or-death matter for some residents. Loss of these facilities can have a devastating effect on people's health and life chances.

In addition to the movement of some urban dwellers to rural areas, two other factors have changed the face of rural America in some regions. One is the proliferation of superstores, such as Wal-Mart, PetSmart, Lowe's, and Home Depot. In some cases these superstores have effectively put small businesses such as hardware stores and pet shops out of business because local merchants cannot meet the prices established by these large-volume discount chains. The development of superstores and outlet malls along the rural highways of this country has raised new concerns about environmental issues such as air and water pollution, and has brought about new questions regarding whether these stores benefit the rural communities where they are located.

A second factor that has changed the face of some rural areas (and is sometimes related to the growth of megastores and outlet malls) is an increase in tourism in rural America (Brown, 2003). According to one study, about 87 million people (nearly two-thirds of all U.S. adults) have taken a trip to a rural destination, usually for leisure purposes, over the past few years. Tourism produces jobs; however, many of the positions are for food servers, retail clerks, and hospitality workers, which are often low-paying, seasonal jobs that have few benefits. Tourism may improve a community's tax base, but this does not occur when the outlet malls, hotels, and fast-food restaurants are located outside of the rural community's taxing authority, as frequently occurs when developers decide where to locate malls and other tourist amenities.

Population and Urbanization in the Future

In the future, rapid global population growth is inevitable. Although death rates have declined in many low-income nations, there has not been a corresponding decrease in birthrates. Between 1985 and 2025, 93 percent of all global population growth will have occurred in Africa, Asia, and Latin America; 83 percent of the world's population will live in those regions by 2025. Perhaps even more amazing is the fact that in the five-year span between 1995 and 2000, 21 percent of the entire world's population increase occurred in two countries: China and India. ► Figure 19.8a shows the net annual additions to the populations of six countries during that five-year period. Figure 19.8b shows the growth of the world's population from 1927 to 1999 and the expected growth to eight billion by the year 2025.

In the future, low-income countries will have an increasing number of poor people. While the world's population will *double*, the urban population will *triple* as people migrate from rural to urban areas in search of food, water, and jobs.

By the 2020s, in a worst-case scenario, central cities and nearby suburbs in the United States will have

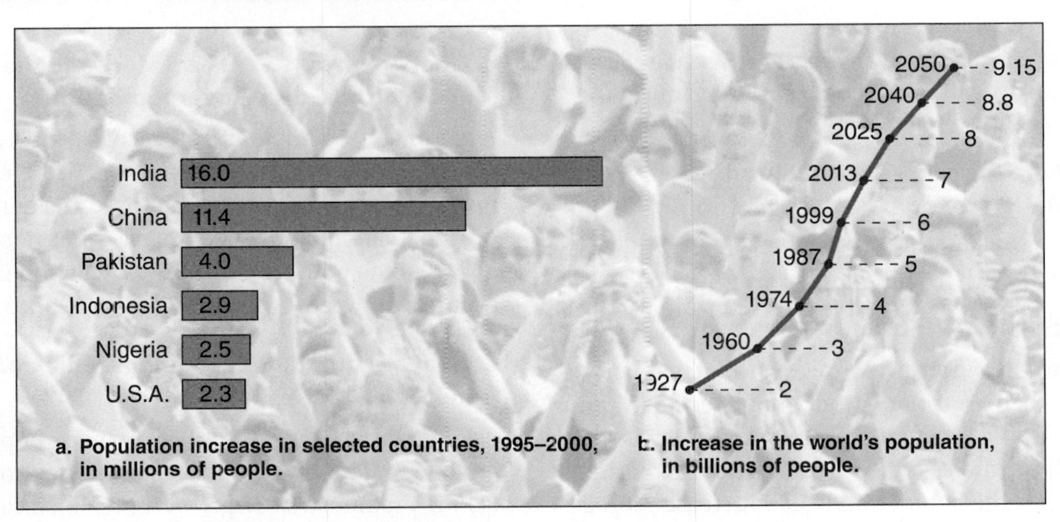

▲ FIGURE 19.8 GROWTH OF THE WORLD'S POPULATION

Sources: United Nations Population Division, 1999, 2008.

experienced bankruptcy exacerbated by sporadic race- and class-oriented violence. The infrastructure will be beyond the possibility of repair. Families and businesses with the ability to do so will have long since moved to "new cities," where they will inevitably diminish the quality of life that they originally sought there. Areas that we currently think of as being relatively free from such problems will be characterized by depletion of natural resources and by greater air and water pollution.

By contrast, in a best-case scenario, the problems brought about by rapid population growth in low-income nations will be remedied by new technologies that make goods readily available to people. International trade agreements are removing trade barriers and making it possible for all nations to fully engage in global trade. People in low-income nations will benefit by gaining jobs and opportunities to purchase goods at lower prices. Of course, the opposite may also occur: People may be exploited as inexpensive labor, and their country's natural resources may be depleted as transnational corporations buy up raw materials without contributing to the long-term economic stability of the nation.

In the United States a best-case scenario for the future might include improvements in how taxes are collected and spent. Some analysts suggest that regional governments should be developed that would be responsible for water, wastewater (sewage), transportation, schools, and other public services over a wider area. Others believe that we should put more effort into Smart Growth and/or green movements to improve our cities. Smart Growth is the name given to development decisions that cities are making across the United States. Examples of Smart Growth planning include working to conserve resources, preserving natural lands and critical environmental areas, protecting water and air quality, and reusing already developed land by reinvesting in existing infrastructure and reclaiming historic buildings. The following are Smart Growth principles (epa.gov, 2011):

- Employ mixed land use.
- Take advantage of compact building design.
- Create a range of housing opportunities and choices.
- Create walker-friendly neighborhoods.
- Foster distinctive, attractive communities with a strong sense of place.
- Preserve open spaces, farmland, natural beauty, and critical environmental areas.
- Strengthen and direct development toward existing communities.
- Provide a variety of transportation choices.
- Make development decisions predictable, fair, and cost effective.
- Encourage community and stakeholder collaboration in development decisions.

Although some critics argue that cities that have tried Smart Growth have not brought about profound changes through their efforts, goals such as these seem desirable for bringing about a better future for our communities. At the macrolevel, we may be able to do little about population and urbanization; however, at the microlevel, we may be able to exercise some degree of control over our communities and our lives. Reclaiming public space for daily life would be an important start, along with making neighborhoods more sustainable and helping inhabitants feel that they have a vested interest in their own community.

chapter review Q & A

Use these questions and answers to check how well you've achieved the learning objectives set out at the beginning of this chapter.

• What are the processes that produce population changes?

Populations change as the result of fertility (births), mortality (deaths), and migration.

• What is the Malthusian perspective?

Over two hundred years ago, Thomas Malthus warned that overpopulation would result in poverty, starvation, and other major problems that would limit the size of the population. According to Malthus, the population would increase geometrically while the food supply would increase only arithmetically, resulting in a critical food shortage and poverty.

• What are the views of Karl Marx and the neo-Malthusians on overpopulation?

According to Karl Marx, poverty is the result of capitalist greed, not overpopulation. More recently, neo-Malthusians have reemphasized the dangers of overpopulation and encouraged zero population growth—the point at which no population increase occurs from year to year.

• What are the stages in demographic transition theory?

Demographic transition theory links population growth to four stages of economic development: (1) the preindustrial stage, with high birthrates and death rates; (2) early industrialization, with relatively high birthrates and a decline in death rates; (3) advanced industrialization and urbanization, with low birthrates and death rates; and (4) postindustrialization, with additional decreases in the birthrate coupled with a stable death rate.

- **What are the three functionalist models of urban growth?**

Functionalists view urban growth in terms of ecological models. The concentric zone model sees the city as a series of circular areas, each characterized by a different type of land use. The sector model describes urban growth in terms of terrain and transportation routes. The multiple nuclei model views cities as having numerous centers of development from which growth radiates.

- **What is the political economy model/conflict perspective on urban growth?**

According to political economy models/conflict perspectives, urban growth is influenced by capital-investment decisions, class and class conflict, and government subsidy programs. At the global level, capitalism also influences the development of cities in core, peripheral, and semiperipheral nations.

- **How do symbolic interactionists view urban life?**

Symbolic interactionist perspectives focus on how people experience urban life. Some analysts view the urban experience positively; others believe that urban dwellers become insensitive to events and to people around them.

key terms

crude birthrate 563
crude death rate 564
demographic transition 571
demography 562
fertility 562
gentrification 581
invasion 581
megacity 574
migration 565
mortality 564
population composition 568
population pyramid 568
sex ratio 568
succession 581
zero population growth 571

questions for critical thinking

1. What impact might a high rate of immigration have on culture and personal identity in the United States?
2. If you were designing a study of growth patterns for the city where you live (or one you know well), which theoretical model(s) would provide the most useful framework for your analysis?
3. What do you think that everyday life in U.S. cities, suburbs, and rural areas will be like in 2020? Where would you prefer to live? What does your answer reflect about the future of U.S. cities?

turning to video

Watch the CBS video *Life Along the Mexican Border* (running time 7:35), available through **CengageBrain.com.** This video investigates the population growth along the border, reasons for the growth, and the effects on the environment. As you watch the video, don't think about carbon emissions from processed and burned fossil fuels; instead consider humans' efforts (such as offshore drilling) to maintain current supplies of these fuels. After you watch the video, answer these questions: What are the environmental impacts of having towns developed along the border? What are some ways the United States could alleviate these problems?

online study resources

Go to CENGAGE brain.com to access online study resources, including the Sociology CourseMate for this text as well as special features such as video, an interactive sociology time line and interactive maps, General Social Survey (GSS) data, and U.S. Census 2010 data.

CourseMate brings course concepts to life with interactive learning, study, and exam-preparation tools that support the printed textbook. A textbook-specific website, **Sociology CourseMate** includes an integrated interactive eBook and other interactive learning tools, including quizzes, flash cards, and videos.

Visit **www.cengagebrain.com** to access your account and purchase materials.

20 Collective Behavior, Social Movements, and Social Change

Along with my daughter Mariah and a team of human rights experts from the Robert F. Kennedy Center for Justice and Human Rights, I spent the last several days in Mississippi, Louisiana and Alabama speaking with commercial fishermen, deck hands, restaurateurs, ecologists, farmers, service providers, marina workers, hoteliers, kids and more whose lives are directly affected by BP's toxic tsunami swamping the Gulf Coast and wiping out the fishing and tourism industries which have been the mainstays of these communities for decades. "Oil will be all that's left," lamented one long-time resident. "And with the politicians in the pockets of the oil companies, there will be more pressure than ever to drill, baby, drill." . . . Eleven of us motored a small boat eight miles out from shore. Though far from shore, the water there appeared as though we had pulled up to a gargantuan gas dock, with a rainbow sheen covering the ocean, horizon to horizon. Our eyes stung, our throats closed and our heads ached despite the respirators we wore. . . . It may take decades for BP to make the Gulf "whole." In the aftermath of this oil tsunami, concrete actions that respect residents' right are the next step.

—Kerry Kennedy (2010), a prominent environmental activist, describing what she saw in the tragic aftermath of an explosion on the Deepwater Horizon rig that began spewing massive amounts of oil endlessly from a 5,000-foot-deep puncture in the Gulf of Mexico

▲ **Environmental activism is a powerful type of social movement that seeks to call public attention to pressing social concerns, such as the environmental impacts of harmful means of oil and gas exploration. Activists often stage public events, such as the one shown here, in an attempt to gain the attention of political leaders and everyday citizens.**

One of the by-products of having grown up alongside the Houston Ship Channel was very nearly becoming desensitized to the vast amounts of pollutants the oil and chemical industry poured into East Houston's air and waterways. I once fell in the Ship Channel while working on a crew that built launching pads for a new supertanker. The resulting kidney infection took nine months to heal. I urinated blood for three weeks. No one can tell me that the current state of global consumerism does not impact the world's climate adversely. To [people] who pooh pooh the notion of global warming, I say this: Go take a swim in the Houston Ship Channel.

—Grammy-winning songwriter and recording artist Rodney Crowel explaining why he joined the virtual march against global warming (qtd. in StopGlobalWarming.org, 2006)

When I circled the moon and looked back at Earth, my outlook on life and my viewpoint on Earth changed. You don't see Las Vegas, Boston or even New York. You don't see boundaries or people. No whites, blacks, French, Greeks, Christians or Jews. The Earth looks completely uninhabited, and you know that on Spaceship Earth, there live over six billion astronauts, all seeking the same things from life.

When viewed in total, Earth is a spaceship just like Apollo. We are all the crew of Spaceship Earth; and just like Apollo, the crew must learn to live and work together. We must learn to manage the resources of this world with new imagination. The future is up to you.

—Jim Lovell, a retired NASA astronaut, describing how his experience in space gave him a new perspective on environmental problems such as global warming—the process that occurs when carbon dioxide stays in the atmosphere and acts like a blanket that holds in the heat. Over time, global warming results in higher temperatures, rises in sea levels, and catastrophic weather such as powerful hurricanes (qtd. in StopGlobalWarming.org, 2006).

Highlighted Learning Objectives

- Define collective behavior and list factors that contribute to such behavior.
- Describe the most common types of crowd behavior.
- Identify and explain four explanations of crowd behavior.
- Explain how mass behavior differs from other forms of collective behavior.
- Identify and explain four types of social movements.
- Explain how people are drawn into social movements.
- Discuss factors that contribute to social change.

Chapter Focus Question:

How might collective behavior and social movements make people more aware of important social issues such as environmental destruction and global warming?

Sometimes it seems like environmental crises and people-made disasters occur routinely in the twenty-first century. Almost daily, the Internet, TV, and newspapers inform us of new or unresolved problems associated with environmental problems such as massive oil leaks, global warming, air pollution, or other disasters. As this chapter's opening narratives indicate, a number of well-known people such as Kerry Kennedy, Rodney Crowell, and Jim Lovell are deeply concerned about the crises that threaten our environment and want to encourage others to become part of the environmental movement. The message of the environmental movement is that we must act collectively and immediately to reduce environmental hazards before havoc comes to the Earth: Social change is essential. Sociologists define ***social change* as the alteration, modification, or transformation of public policy, culture, or social institutions over time.** Social change is usually brought about by collective behavior and social movements.

In this chapter we will examine collective behavior, social movements, and social change from a sociological perspective. We will use environmental activism as an example of how people may use social movements as a form of mass mobilization and social transformation. Before reading on, test your knowledge about collective behavior and environmental issues by taking the Sociology and Everyday Life quiz.

Collective Behavior

***Collective behavior* is voluntary, often spontaneous activity that is engaged in by a large number of people and typically violates dominant-group norms and values.** Unlike the *organizational behavior* found in corporations and voluntary associations (such as labor unions and environmental organizations), collective behavior lacks an official division of labor, hierarchy of authority, and established rules and procedures. Unlike *institutional behavior* (in education, religion, or politics, for example), it lacks institutionalized norms to govern behavior. Collective behavior can take various forms, including crowds, mobs, riots, panics, fads, fashions, and public opinion.

Early sociologists studied collective behavior because they lived in a world that was responding to the processes of modernization, including urbanization, industrialization, and the proletarianization of workers. Contemporary forms of collective behavior, particularly social protests, are variations on the themes that originated during the transition from feudalism to capitalism and the rise of modernity in Europe. Today, some forms of collective behavior and social movements are directed toward public issues such as air pollution, water pollution, and the exploitation of workers in global sweatshops by transnational corporations.

Tim Whitby/Stringer/Getty Images

▲ The flash mob is a recent type of collective behavior. As opposed to some other public displays of collective behavior, most flash mobs tend to form simply because people find this activity to be enjoyable. However, a few flash mobs have turned into unruly crowds with excessive physical contact and police intervention.

Conditions for Collective Behavior

Collective behavior occurs as a result of some common influence or stimulus that produces a response from a collectivity. A *collectivity* is a number of people who act together and may mutually transcend, bypass, or subvert established institutional patterns and structures. Three major factors contribute to the likelihood that collective behavior will occur: (1) structural factors that increase the chances of people responding in a particular way, (2) timing, and (3) a breakdown in social control mechanisms and a corresponding feeling of normlessness (McPhail, 1991; Turner and Killian, 1993).

A common stimulus is an important factor in collective behavior. For example, the publication of *Silent Spring* (1962) by former Fish and Wildlife Service biologist Rachel Carson is credited with triggering collective behavior directed at demanding a clean environment and questioning how much power large corporations should have. Carson described the dangers of pesticides such as DDT, which was then being promoted by the chemical

sociology and everyday life

How Much Do You Know About Collective Behavior and Environmental Issues?

True	False	
T	F	1. Scientists are forecasting a global warming of between 2 and 11 degrees Fahrenheit over the next century.
T	F	2. The environmental movement in the United States started in the 1960s.
T	F	3. People who hold strong attitudes regarding the environment are very likely to be involved in social movements to protect the environment.
T	F	4. Environmental groups may engage in civil disobedience or use symbolic gestures to call attention to their issue.
T	F	5. People are most likely to believe rumors when no other information is readily available on a topic.
T	F	6. Influencing public opinion is a very important activity for many social movements.
T	F	7. Most social movements in the United States seek to improve society by changing some specific aspect of the social structure.
T	F	8. Sociologists have found that people in a community respond very similarly to natural disasters and to disasters caused by technological failures.

Answers on page 600.

industry as the miracle that could give the United States the unchallenged position as food supplier to the world (Cronin and Kennedy, 1999). Carson's activism has been described in this way:

> Carson was not a wild-eyed reformer intent on bringing the industrial age to a grinding halt. She wasn't even opposed to pesticides per se. She was a careful scientist and brilliant writer whose painstaking research on pesticides proved that the "miraculous" bursts of agricultural productivity had long-term costs undisclosed in the chemical industry's exaggerated puffery. Americans were losing things—their health, many birds and fishes, and the purity of their waterways—that they should value more than modest savings at the grocery store. (Cronin and Kennedy, 1999: 151)

Timing is another significant factor in bringing about collective behavior. For example, in the 1960s smog had started staining the skies in this country; in Europe, birds and fish were dying from environmental pollution; and oil spills from tankers were provoking public outrage worldwide (Cronin and Kennedy, 1999). People in this country were ready to acknowledge that problems existed. By writing *Silent Spring,* Carson made people aware of the hazards of chemicals in their foods and the destruction of wildlife. However, that is not all she produced: As a consequence of her careful research and writing, she also produced anger in people at a time when they were beginning to wonder if they were being deceived by the very industries that they had entrusted with their lives and their resources. Once aroused to action, many people began demanding an honest, comprehensive accounting of where pollution was occurring and how it might be endangering public health and environmental resources. Public outcries also led to investigations in courts and legislatures throughout the United States as people began to demand legal recognition of the right to a clean environment (Cronin and Kennedy, 1999).

A breakdown in social control mechanisms has been a powerful force in triggering collective behavior regarding environmental protection and degradation. During the 1970s, people in the "Love Canal" area of Niagara Falls, New York, became aware that their neighborhood and their children's school had been built over a canal where tons of poisonous waste had been dumped by a chemical company between 1930 and 1950. After the company closed the site, covered it with soil, and sold it (for $1) to the city of Niagara Falls, homes and a school were built on the sixteen-acre site. Over the next two decades, an oily black substance began oozing into the homes in the area and killing the

social change the alteration, modification, or transformation of public policy, culture, or social institutions over time.

collective behavior voluntary, often spontaneous activity that is engaged in by a large number of people and typically violates dominant-group norms and values.

sociology and everyday life

ANSWERS to the Sociology Quiz on Collective Behavior and Environmental Issues

1. True. Global surface temperatures have increased about 0.4 degrees Fahrenheit during the past 25 years, and scientists believe that this trend will grow more pronounced during the next 100 years.

2. False. The environmental movement in the United States is the result of more than 100 years of collective action. The first environmental organization, the American Forestry Association (now American Forests), originated in 1875.

3. False. Since the 1980s, public opinion polls have shown that the majority of people in the United States have favorable attitudes regarding protection of the environment and banning nuclear weapons; however, far fewer individuals are actually involved in collective action to further these causes.

4. True. Environmental groups have held sit-ins, marches, boycotts, and strikes, which sometimes take the form of civil disobedience. Others have hanged political leaders in effigy or held officials hostage. Still others have dressed as grizzly bears to block traffic in Yellowstone National Park or created a symbolic "crack" (made of plastic) on the Glen Canyon Dam on the Colorado River to denounce development in the area.

5. True. Rumors are most likely to emerge and circulate when people have very little information on a topic that is important to them. For example, rumors abound in times of technological disasters, when people are fearful and often willing to believe a worst-case scenario.

6. True. Many social movements, including grassroots environmental activism, attempt to influence public opinion so that local decision makers will feel obliged to correct a specific problem through changes in public policy.

7. True. Most social movements are reform movements that focus on improving society by changing some specific aspect of the social structure. Examples include environmental movements and the disability rights movement.

8. False. Most sociological studies have found that people respond differently to natural disasters, which usually occur very suddenly, than to technological disasters, which may occur gradually. One of the major differences is the communal bonding that tends to occur following natural disasters, as compared to the extreme social conflict that may follow technological disasters.

trees and grass on the lots; schoolchildren reported mysterious illnesses and feelings of malaise. Tests indicated that the dump site contained more than two hundred different chemicals, many of which could cause cancer or other serious health problems. Upon learning this information, Lois Gibbs, a mother of one of the schoolchildren, began a grassroots campaign to force government officials to relocate community members injured by seepages from the chemical dump. The collective behavior of neighborhood volunteers was not only successful in eventually bringing about social change but also inspired others to engage in collective behavior regarding environmental problems in their own communities.

Dynamics of Collective Behavior

To better understand the dynamics of collective behavior, let's briefly examine several questions. First, how do people come to transcend, bypass, or subvert established institutional patterns and structures? Some environmental activists have found that they cannot get their point across unless they go outside established institutional patterns and organizations. For example, Lois Gibbs and other Love Canal residents initially tried to work within established means through the school administration and state health officials to clean up the problem. However, they quickly learned that their problems were not being solved through "official" channels. As the problem appeared to grow worse, organizational responses became more defensive and obscure. Accordingly, some residents began acting outside of established norms by holding protests and strikes (Gibbs, 1982). Some situations are more conducive to collective behavior than others. When people can communicate quickly and easily with one another, spontaneous behavior is more likely. The Internet and social networking sites have opened up a new world of instant communication that has made seemingly spontaneous behavior much easier than in the past. However, being in the same place at the same time is also important. For example, when people are gathered together in one general location (whether lining the streets or assembled in a massive stadium), they are more likely to respond to a common stimulus.

Second, how do people's actions compare with their attitudes? People's attitudes (as expressed in public opinion surveys, for instance) are not always reflected in their political and social behavior. Issues pertaining to the environment are no exception. For example, people may indicate in survey research that they believe the quality of the environment is very important, but the same people may not turn out on election day to support propositions that protect the environment or candidates who promise to focus on environmental issues. Likewise, individuals who indicate on a questionnaire that they are concerned about increases in ground-level ozone—the primary component of urban smog—often drive single-occupant, oversized vehicles which government studies have shown to be "gas guzzlers" that contribute to lowered air quality in urban areas. As a result, smog levels increase, contributing to human respiratory problems and dramatically reduced agricultural crop yields.

Third, why do people act collectively rather than singly? As sociologists Ralph H. Turner and Lewis M. Killian (1993: 12) note, people believe that there is strength in numbers: "[T]he rhythmic stamping of feet by hundreds of concert-goers in unison is different from isolated, individual cries of 'bravo.'" Likewise, people may act as a collectivity when they believe it is the only way to fight those with greater power and resources. Collective behavior is not just the sum total of a large number of individuals acting at the same time; rather, it reflects people's joint response to some common influence or stimulus.

Distinctions Regarding Collective Behavior

People engaging in collective behavior may be divided into crowds and masses. A ***crowd*** **is a relatively large number of people who are in one another's immediate vicinity.** Examples of crowds include the audience in a movie theater or people at a pep rally for a sporting event. By contrast, a ***mass*** **is a number of people who share an interest in a specific idea or issue but who are not in one another's immediate vicinity.** An example is the popularity of Facebook, Twitter, other social networking sites, and blogging on the Internet. Through these forms of instantaneous communication, people express their views on everyday life and on larger social issues such as the environment. Individuals who read what someone has posted and make comments in response usually share a common interest even if these individuals have not met in a face-to-face encounter.

Collective behavior may also be distinguished by the dominant emotion expressed. According to the sociologist John Lofland (1993: 72), the *dominant emotion* refers to the "publicly expressed feeling perceived by participants and observers as the most prominent in an episode of collective behavior." Lofland suggests that fear, hostility, and joy are three fundamental emotions found in collective behavior; however, grief, disgust, surprise, or shame may also predominate in some forms of collective behavior.

Types of Crowd Behavior

When we think of a crowd, many of us think of *aggregates*, previously defined as a collection of people who happen to be in the same place at the same time but who share little else in common. Think, for example, of people stranded in an airport when harsh weather or other conditions make it impossible for them to board flights and head for their destinations. Although stranded businesspeople and tourists are together in the airport, they do not necessarily share anything else in common with other weary air passengers. Moreover, the presence of a relatively

▲ The Love Canal area of Niagara Falls, New York, has been the site of protests and other forms of collective behavior because of hazardous environmental pollution. Original protests in the 1970s, demanding a cleanup of the site, were followed in the 1990s by new protests, this time over the proposed resettlement of the area.

crowd a relatively large number of people who are in one another's immediate vicinity.

mass a number of people who share an interest in a specific idea or issue but who are not in one another's immediate vicinity.

large number of people in the same location does not necessarily produce collective behavior. To help explain this phenomenon, sociologist Herbert Blumer (1946) developed a typology that divides crowds into four categories: casual, conventional, expressive, and acting. Other scholars have added a fifth category, protest crowds.

Casual and Conventional Crowds *Casual crowds* are relatively large gatherings of people who happen to be in the same place at the same time; if they interact at all, it is only briefly. People in a shopping mall or a subway car are examples of casual crowds. Other than sharing a momentary interest, such as a musician's performance or a small child's fall, a casual crowd has nothing in common. The casual crowd plays no active part in the event—such as the child's fall—which likely would have occurred whether or not the crowd was present; the crowd simply observes.

Conventional crowds are made up of people who come together for a scheduled event and thus share a common focus. Examples include religious services, graduation ceremonies, concerts, and college lectures. Each of these events has preestablished schedules and norms. Because these events occur regularly, interaction among participants is much more likely; in turn, the events would not occur without the crowd, which is essential to the event.

Expressive and Acting Crowds *Expressive crowds* provide opportunities for the expression of some strong emotion (such as joy, excitement, or grief). People release their pent-up emotions in conjunction with other persons experiencing similar emotions. Examples include worshippers at religious revival services; mourners lining the streets when a celebrity, public official, or religious leader has died; and revelers assembled at Mardi Gras or on New Year's Eve at Times Square in New York.

MCT via Getty Images

▲ Crowds of people come together for a variety of reasons. The people pictured here wanted to be near the front of the line to purchase an iPad on the first day the new device became available. How does a crowd such as this differ from other types of crowds?

Acting crowds are collectivities so intensely focused on a specific purpose or object that they may erupt into violent or destructive behavior. Mobs, riots, and panics are examples of acting crowds, but casual and conventional crowds may become acting crowds under some circumstances. A ***mob*** **is a highly emotional crowd whose members engage in, or are ready to engage in, violence against a specific target—a person, a category of people, or physical property.** Mob behavior in this country has included lynchings, fire bombings, effigy hangings, and hate crimes. Mob violence tends to dissipate relatively quickly once a target has been injured, killed, or destroyed. Sometimes, actions such as an effigy hanging are used symbolically by groups that are not otherwise violent. For example, Lois Gibbs and other Love Canal residents called attention to their problems with the chemical dump site by staging a protest in which they "burned in effigy" the governor and the health commissioner to emphasize their displeasure with the lack of response from these public officials. More recently, protesters seeking to call attention to social problems such as the environmental impact of offshore drilling, particularly in the aftermath of the explosion of the BP Gulf of Mexico oil rig, have held rallies in front of the White House and other public areas where they know that political leaders will observe them and take their demands for change into account.

Compared with mob actions, riots may be of somewhat longer duration. A ***riot*** **is violent crowd behavior that is fueled by deep-seated emotions but is not directed at one specific target.** Riots are often triggered by fear, anger, and hostility; however, not all riots are caused by deep-seated hostility and hatred—people may be expressing joy and exuberance when rioting occurs. Examples include celebrations after sports victories such as those that occurred in Montreal, Canada, following a Stanley Cup win and in Vancouver following a playoff victory (Kendall, Lothian Murray, and Linden, 2004).

A ***panic*** **is a form of crowd behavior that occurs when a large number of people react to a real or perceived threat with strong emotions and self-destructive behavior.** The most common type of panic occurs when people seek to escape from a perceived danger, fearing that few (if any) of them will be able to get away from that danger. Panics can also arise in response to events that people believe are beyond their control—such as a major disruption in the economy. Although panics are relatively rare, they receive extensive media coverage because they provoke strong feelings of fear in readers and viewers, and the number of casualties may be large. Examples of panics include soccer matches or other sporting events where large crowds gather at a stadium to cheer on their team, only to see that pandemonium has broken out at the end of the game as some fans brawl, throw drinks, and otherwise engage in physical violence that may quickly get out of control and cause others to panic as well. However, researchers have found that even in tragic situations such as the massive Rhode Island nightclub fire that killed 100 people and injured nearly 200 more, panic-like behaviors were not prevalent. Individuals tried to help each other evacuate the burning building until the very end, especially those who had strong social bonds with others in the crowd (Aguirre et al., 2011).

▲ In Thailand the Red Shirt movement was organized to protest what it saw as the antidemocratic policies of the current Thai government. As shown above, the group prefers to be seen as a nonviolent opposition, but many of its activities have escalated into bloody confrontations with government soldiers and police.

Protest Crowds *Protest crowds* engage in activities intended to achieve specific political goals. Examples include sit-ins, marches, boycotts, blockades, and strikes. Some protests take the form of ***civil disobedience*—nonviolent action that seeks to change a policy or law by refusing to comply with it.** Acts of civil disobedience may become violent, as in a confrontation between protesters and police officers; in this case, a protest crowd becomes an *acting crowd.* In the 1960s African American students and sympathetic whites used sit-ins to call attention to racial injustice and demand social change. Some of these protests can escalate into violent confrontations even when violence was not the intent of the organizers. Today, protest crowds show up in front of the White House or the U.S. Congress to express their approval or disapproval of certain actions of government officials, such as cutting funding for Social Security, Medicare, or Medicaid.

Explanations of Crowd Behavior

What causes people to act collectively? How do they determine what types of action to take? One of the earliest theorists to provide an answer to these questions was Gustave Le Bon, a French scholar who focused on crowd psychology in his contagion theory.

Contagion Theory *Contagion theory* focuses on the social–psychological aspects of collective behavior; it attempts to explain how moods, attitudes, and behavior are communicated rapidly and why they are accepted by others. Le Bon (1841–1931) argued that people are more likely to engage in antisocial behavior in a crowd because they are anonymous and feel invulnerable. Le Bon (1960/1895) suggested that a crowd takes on a life of its own that is larger than the beliefs or actions of any one person. Because of its anonymity, the crowd transforms individuals from rational beings into a single organism with a collective mind. In essence, Le Bon asserted that emotions such as fear and hate are contagious in crowds because people experience a decline in personal responsibility; they will do things as a collectivity that they would never do when acting alone.

Le Bon's theory is still used by many people to explain crowd behavior. However, critics argue that the "collective mind" has not been documented by systematic studies.

mob a highly emotional crowd whose members engage in, or are ready to engage in, violence against a specific target—a person, a category of people, or physical property.

riot violent crowd behavior that is fueled by deep-seated emotions but is not directed at one specific target.

panic a form of crowd behavior that occurs when a large number of people react to a real or perceived threat with strong emotions and self-destructive behavior.

civil disobedience nonviolent action that seeks to change a policy or law by refusing to comply with it.

Social Unrest and Circular Reaction Sociologist Robert E. Park was the first U.S. sociologist to investigate crowd behavior. Park believed that Le Bon's analysis of collective behavior lacked several important elements. Intrigued that people could break away from the powerful hold of culture and their established routines to develop a new social order, Park added the concepts of social unrest and circular reaction to contagion theory. According to Park, social unrest is transmitted by a process of *circular reaction*—the interactive communication between persons such that the discontent of one person is communicated to another, who, in turn, reflects the discontent back to the first person (Park and Burgess, 1921).

Convergence Theory *Convergence theory* focuses on the shared emotions, goals, and beliefs that many people may bring to crowd behavior. Because of their individual characteristics, many people have a predisposition to participate in certain types of activities (Turner and Killian, 1993). From this perspective, people with similar attributes find a collectivity of like-minded persons with whom they can express their underlying personal tendencies. Although people may reveal their "true selves" in crowds, their behavior is not irrational; it is highly predictable to those who share similar emotions or beliefs.

Convergence theory has been applied to a wide array of conduct, from lynch mobs to environmental movements. In social psychologist Hadley Cantril's (1941) study of one lynching, he found that the participants shared certain common attributes: They were poor and working-class whites who felt that their status was threatened by the presence of successful African Americans. Consequently, the characteristics of these individuals made them susceptible to joining a lynch mob even if they did not know the target of the lynching.

Convergence theory adds to our understanding of certain types of collective behavior by pointing out how individuals may have certain attributes—such as racial hatred or fear of environmental problems that directly threaten them—that initially bring them together. However, this theory does not explain how the attitudes and characteristics of individuals who take some collective action differ from those who do not.

Emergent Norm Theory Unlike contagion and convergence theories, *emergent norm theory* emphasizes the importance of social norms in shaping crowd behavior. Drawing on the symbolic interactionist perspective, the sociologists Ralph Turner and Lewis Killian (1993: 12) asserted that crowds develop their own definition of a situation and establish norms for behavior that fit the occasion:

> Some shared redefinition of right and wrong in a situation supplies the justification and coordinates the action in collective behavior. People do what they would not otherwise have done when they panic collectively, when they riot, when they engage in civil disobedience, or when they launch terrorist campaigns, because they find social support for the view that what they are doing is the right thing to do in the situation.

According to Turner and Killian (1993: 13), emergent norms occur when people define a new situation as highly unusual or see a long-standing situation in a new light.

Sociologists using the emergent norm approach seek to determine how individuals in a given collectivity develop an understanding of what is going on, how they construe these activities, and what type of norms are involved. For example, in a study of audience participation, the sociologist Steven E. Clayman (1993) found that members of an audience listening to a speech applaud promptly and independently but wait to coordinate their booing with other people; they do not wish to "boo" alone.

▲ Convergence theory is based on the assumption that crowd behavior involves shared emotions, goals, and beliefs, such as the importance of protecting the environment. An example is the Earth Day events that brought together these children carrying this banner to foster environmental causes.

Some emergent norms are permissive—that is, they give people a shared conviction that they may disregard ordinary rules, such as waiting in line, taking turns, or treating a speaker courteously. Collective activity such as mass looting may be defined (by participants) as taking what rightfully belongs to them. In the aftermath of the 2010 Haiti earthquake, when relief aid was slow in coming, looting was commonplace, but so too was "mob justice" for those who were caught stealing other people's possessions.

Emergent norm theory points out that crowds are not irrational. Rather, new norms are developed in a rational way to fit the immediate situation. However, critics note that proponents of this perspective fail to specify exactly what constitutes a norm, how new ones emerge, and how they are so quickly disseminated and accepted by a wide variety of participants. One variation of this theory suggests that no single dominant norm is accepted by everyone in a crowd; instead, norms are specific to the various categories of actors rather than to the collectivity as a whole (Snow, Zurcher, and Peters, 1981). For example, in a study of football victory celebrations, the sociologists David Snow, Louis Zurcher, and Robert Peters (1981) found that each week, behavioral patterns were changed in the postgame revelry, with some being modified, some added, and some deleted.

Mass Behavior

Not all collective behavior takes place in face-to-face collectivities. ***Mass behavior* is collective behavior that takes place when people (who often are geographically separated from one another) respond to the same event in much the same way.** For people to respond in the same way, they typically have common sources of information that provoke their collective behavior. The most frequent types of mass behavior are rumors, gossip, mass hysteria, public opinion, fashions, and fads. Under some circumstances, social movements constitute a form of mass behavior. However, we will examine social movements separately because they differ in some important ways from other types of dispersed collectivities.

Rumors and Gossip ***Rumors* are unsubstantiated reports on an issue or subject.** Whereas a rumor may spread through an assembled collectivity, rumors may also be transmitted among people who are dispersed geographically, including people spreading rumors on Twitter or posting messages on Facebook or talking by cell phone. Although rumors may initially contain a kernel of truth, they may be modified as they spread to serve the interests of those repeating them. Rumors thrive when tensions are high and when little authentic information is available on an issue of great concern. Once again, in the aftermath of the devastating Haiti earthquake, rumors quickly spread on Twitter that American Airlines would fly doctors and nurses to that country free of charge and that JetBlue was offering free flights and UPS was shipping packages for free (Griggs, 2010). These rumors were nothing more than a hoax, but they had many people inquiring because they wanted to help with disaster relief or perhaps because they saw the so-called offer as an opportunity for a free trip. Were people on Twitter trying to help the relief effort in Haiti but simply misinformed, or did they intentionally spread a rumor that was eventually proven to be untrue? Social networks such as Twitter provide people with an opportunity to spread "information," both substantiated and unsubstantiated, and the lack of validation can be particularly problematic during natural disasters and other times of crisis. In the case of a major disaster, it is important to know the reliability of information sources: The outcome might be much worse, or at least far different, than in the Haiti situation, where it created confusion and disappointment but was not necessarily life threatening.

As the example regarding free flights on airlines shows, people are willing to give rumors credence when no opposing information is available. Environmental issues are similar. For example, when residents of Love Canal waited for information from health department officials about their exposure to the toxic chemicals and from the government about possible relocation at state expense to another area, new waves of rumors spread through the community daily. By the time a meeting was called by health department officials to provide homeowners with the results of air-sample tests for hazardous chemicals (such as chloroform and benzene) performed on their homes, already fearful residents were ready to believe the worst, as Lois Gibbs (1982: 25) describes:

> Next to the names [of residents] were some numbers. But the numbers had no meaning. People stood there looking at the numbers, knowing nothing of what they meant but suspecting the worst.

mass behavior collective behavior that takes place when people (who often are geographically separated from one another) respond to the same event in much the same way.

rumor an unsubstantiated report on an issue or subject.

One woman, divorced and with three sick children, looked at the piece of paper with numbers and started crying hysterically: "No wonder my children are sick. Am I going to die? What's going to happen to my children?" No one could answer. . . .

The night was very warm and humid, and the air was stagnant. On a night like that, the smell of Love Canal is hard to describe. It's all around you. It's as though it were about to envelop you and smother you. By now, we were outside, standing in the parking lot. The woman's panic caught on, starting a chain reaction. Soon, many people there were hysterical.

Once a rumor begins to circulate, it seldom stops unless compelling information comes to the forefront that either proves the rumor false or makes it obsolete.

In industrialized societies with sophisticated technology, rumors come from a wide variety of sources and may be difficult to trace. Print media (newspapers and magazines) and electronic media (radio and television), fax machines, cellular networks, satellite systems, and the Internet aid the rapid movement of rumors around the globe. In addition, modern communications technology makes anonymity much easier. In a split second, messages (both factual and fictitious) can be disseminated to thousands of people through e-mail, computerized bulletin boards, and Internet newsgroups.

Whereas rumors deal with an issue or a subject, ***gossip* refers to rumors about the personal lives of individuals.** Charles Horton Cooley (1963/1909) viewed gossip as something that spread among a small group of individuals who personally knew the person who was the object of the rumor. Today, this is frequently not the case; many people enjoy gossiping about people whom they have never met. Tabloid newspapers and magazines such as the *National Enquirer* and *People,* along with television "news" programs that purport to provide "inside" information on the lives of celebrities, are sources of contemporary gossip, much of which has not been checked for authenticity.

Mass Hysteria and Panic *Mass hysteria* is a form of dispersed collective behavior that occurs when a large number of people react with strong emotions and self-destructive behavior to a real or perceived threat. Does mass hysteria actually occur? Although the term has been widely used, many sociologists believe that this behavior is best described as a panic with a dispersed audience.

An example of mass hysteria or a panic with a widely dispersed audience was actor Orson Welles's 1938 Halloween eve radio dramatization of H. G. Wells's science fiction classic *The War of the Worlds.* A CBS radio dance music program was interrupted suddenly by a news bulletin informing the audience that Martians had landed in New Jersey and were in the process of conquering Earth. Some listeners became extremely frightened even though an announcer had indicated before, during, and after the performance that the broadcast was a fictitious dramatization. According to some reports, as many as one million of the estimated ten million listeners believed that this astonishing event had occurred. Thousands were reported to have hidden in their storm cellars or to have gotten in their cars so that they could flee from the Martians (see Brown, 1954). In actuality, the program probably did not

NY Daily News via Getty Images

▲ When unexpected events such as the massive 2003 power outage in the United States and Canada occur, people frequently rely on rumors to help them know what is going on. Getting accurate information out quickly helped prevent people from panicking.

© AP Photo

▲ Although a spokesperson for CBS Radio stated to listeners that they were hearing a dramatization of a novel, the 1938 presentation of H. G. Wells's *The War of the Worlds,* as presented by Orson Welles and his Mercury Theatre, terrified untold numbers of people. Here Welles talks to interviewers the day after the event caused a nationwide panic.

generate mass hysteria, but rather a panic among gullible listeners. Others switched stations to determine if the same "news" was being broadcast elsewhere. When they discovered that it was not, they merely laughed at the joke being played on listeners by CBS. In 1988, on the fiftieth anniversary of the broadcast, a Portuguese radio station rebroadcast the program; once again, a panic ensued.

Fads and Fashions As you will recall from Chapter 3, a *fad* is a temporary but widely copied activity enthusiastically followed by large numbers of people. Fads can be embraced by widely dispersed collectivities; news networks such as CNN and Internet websites may bring the latest fad to the attention of audiences around the world.

Unlike fads, fashions tend to be longer lasting. In Chapter 3 *fashion* is defined as a currently valued style of behavior, thinking, or appearance. Fashion also applies to art, music, drama, literature, architecture, interior design, and automobiles, among other things. However, most sociological research on fashion has focused on clothing, especially women's apparel.

In preindustrial societies, clothing styles remained relatively unchanged. With the advent of industrialization, items of apparel became readily available at low prices because of mass production. Fashion became more important as people embraced the "modern" way of life and as advertising encouraged "conspicuous consumption."

Georg Simmel, Thorstein Veblen, and Pierre Bourdieu have all viewed fashion as a means of status differentiation among members of different social classes. Simmel (1957/1904) suggested a classic "trickle-down" theory (although he did not use those exact words) to describe the process by which members of the lower classes emulate the fashions of the upper class. As the fashions descend through the status hierarchy, they are watered down and "vulgarized" so that they are no longer recognizable to members of the upper class, who then regard them as unfashionable and in bad taste. Veblen (1967/1899) asserted that fashion serves mainly to institutionalize conspicuous consumption among the wealthy. Almost eighty years later, Bourdieu (1984) similarly (but more subtly) suggested that "matters of taste," including fashion sensibility, constitute a large share of the "cultural capital" possessed by members of the dominant class.

Herbert Blumer (1969) disagreed with the trickle-down approach, arguing that "collective selection" best explains fashion. Blumer suggested that people in the middle and lower classes follow fashion because it is *fashion*, not because they desire to emulate members of the elite class. Blumer thus shifted the focus on fashion to collective mood, tastes, and choices: "Tastes are themselves a product of experience. . . . They are formed in the context of social interaction, responding to the definitions and affirmation given by others. People thrown into areas of common interaction and having similar runs of experience develop common tastes" (qtd. in Davis, 1992: 116). Perhaps one of the best refutations of the trickle-down approach is the way in which fashion today often originates among people in the lower social classes and is mimicked by the elites. In the mid-1990s, the so-called grunge look was a prime example of this.

Jupiterimages/Getty Images

▲ In your opinion, are the fashions shown here ones that have "trickled down" from the elites to the masses or vice versa? What factors help shape your opinion?

Public Opinion ***Public opinion*** **consists of the attitudes and beliefs communicated by ordinary citizens to decision makers.** It is measured through polls and surveys, which use research methods such as interviews and questionnaires, as described in Chapter 2. Many people are not interested in all aspects of public policy but are concerned about issues that they believe are relevant to themselves. Even on a single topic, public opinion will vary widely based on race/ethnicity, religion, region, social class, education level, gender, age, and so on.

Scholars who examine public opinion are interested in the extent to which the public's attitudes are communicated to decision makers and the effect (if any) that public opinion has on policy making (Turner and Killian, 1993). Some political scientists argue that public opinion has a substantial effect on decisions at all levels of government; others strongly disagree.

Today, people attempt to influence elites, and vice versa. Consequently, a two-way process occurs with

gossip rumors about the personal lives of individuals.

public opinion the attitudes and beliefs communicated by ordinary citizens to decision makers.

the dissemination of ***propaganda*—information provided by individuals or groups that have a vested interest in furthering their own cause or damaging an opposing one.** Although many of us think of propaganda in negative terms, the information provided can be correct and can have a positive effect on decision making.

In recent decades, grassroots environmental activists (including the Love Canal residents) have attempted to influence public opinion. In a study of public opinion on environmental issues, the sociologist Riley E. Dunlap (1992) found that public awareness of the seriousness of environmental problems and public support for environmental protection increased dramatically between the late 1960s and the 1990s. However, it is less clear that public opinion translates into action by either decision makers in government and industry or by individuals (such as a willingness to adopt a more ecologically sound lifestyle).

Initially, most grassroots environmental activists attempt to influence public opinion so that local decision makers will feel the necessity of correcting a specific problem through changes in public policy. Although activists usually do not start out seeking broader social change, they often move in that direction when they become aware of how widespread the problem is in the larger society or on a global basis. One of two types of social movements often develops at this point—one focuses on NIMBY ("not in my backyard"), whereas the other focuses on NIABY ("not in anyone's backyard").

Social Movements

Although collective behavior is short-lived and relatively unorganized, social movements are longer lasting, are more organized, and have specific goals. A ***social movement* is an organized group that acts consciously to promote or resist change through collective action.** Because social movements have not become institutionalized and are outside the political mainstream, they offer "outsiders" an opportunity to have their voices heard.

Social movements are more likely to develop in industrialized societies than in preindustrial societies, where acceptance of traditional beliefs and practices makes such movements unlikely. Diversity and a lack of consensus (hallmarks of industrialized nations) contribute to demands for social change, and people who participate in social movements typically lack the power and other resources to bring about change without engaging in collective action. Social movements are most likely to spring up when people come to see their personal troubles as public issues that cannot be solved without a collective response.

Social movements make democracy more available to excluded groups. Historically, people in the United States have worked at the grassroots level to bring about changes even when elites sought to discourage activism. For example, the civil rights movement brought into its ranks African Americans in the South who had never before been allowed to participate in politics. The women's suffrage movement gave voice to women who had been denied the right to vote. Similarly, a grassroots environmental movement gave the working-class residents of Love Canal a way to "fight city hall" and Hooker Chemicals, as Lois Gibbs (1982: 38–40) explains:

> People were pretty upset. They were talking and stirring each other up. I was afraid there would be violence. We had a meeting at my house to try to put everything together [and] decided to form a homeowners' association. We got out the word as best we could and told everyone to come to the Frontier Fire Hall on 102d Street. . . . The firehouse was packed with people, and more were outside. . . .
>
> I was elected president. . . . I took over the meeting but I was scared to death. It was only the second time in my life I had been in front of a microphone or a crowd. . . . We set four goals right at the beginning—(1) get all the residents within the Love Canal area who wanted to be evacuated, evacuated and relocated, especially during the construction and repair of the canal; (2) do something about propping up property values; (3) get the canal fixed properly; and (4) have air sampling and soil and water testing done throughout the whole area, so we could tell how far the contamination had spread. . . .

Most social movements rely on volunteers like Lois Gibbs to carry out the work. Traditionally, women have been strongly represented in both the membership and the leadership of many grassroots movements.

The Love Canal activists set the stage for other movements that have grappled with the kind of issues that the sociologist Kai Erikson (1994) refers to as a "new species of trouble." Erikson describes the "new species" as environmental problems that "contaminate rather than merely damage . . . they pollute, befoul, taint, rather than just create wreckage . . . they penetrate human tissue indirectly rather than just wound the surfaces by assaults of a more straightforward kind. . . . And the evidence is growing that they scare human beings in new and special ways, that they elicit an uncanny fear in us" (Erikson, 1991: 15). The chaos that Erikson (1994: 141) describes is the result of technological disasters: "meaning everything that can go wrong when systems fail, humans err, designs prove faulty, engines misfire, and so on."

A recent example is the 2011 disaster in Japan where, in the aftermath of a major earthquake and massive flood, thousands of people living near Fukushima Daiichi and Daini nuclear power plants were ordered to evacuate because it was likely that small amounts of radioactive material would leak from the plants (Wald, 2011). The health effects of the radiation leak on residents and workers who sought to save the plants may not be known for years as problems such as this continue to be a "new species of trouble."

Social movements provide people who otherwise would not have the resources to enter the game of politics a chance to do so. We are most familiar with those movements that develop around public policy issues considered newsworthy by the media, ranging from abortion and women's rights to gun control and environmental justice. However, a number of other types of social movements exist as well.

Types of Social Movements

Social movements are difficult to classify; however, sociologists distinguish among movements on the basis of their *goals* and the *amount of change* they seek to produce (Blumer, 1974). Some movements seek to change people; others seek to change society.

Reform Movements Grassroots environmental movements are an example of *reform movements,* which seek to improve society by changing some specific aspect of the social structure. Members of reform movements usually work within the existing system to attempt to change existing public policy so that it more adequately reflects their own value system. Examples of reform movements (in addition to the environmental movement) include labor movements, animal rights movements, antinuclear movements, Mothers Against Drunk Driving, and the disability rights movement.

Some movements arise specifically to alter social responses to and definitions of stigmatized attributes. From this perspective, social movements may bring about changes in societal attitudes and practices while at the same time causing changes in participants' social emotions. For example, the civil rights and gay and lesbian rights movements helped replace shame with pride.

© AP Photo

▲ Martin Luther King, Jr., a leader of the civil rights movement in the 1950s and 1960s, advocated nonviolent protests that sometimes took the form of civil disobedience. Here he marches alongside his wife, Coretta Scott King, who for many years took over Dr. King's activities after he was assassinated.

Revolutionary Movements Movements seeking to bring about a total change in society are referred to as *revolutionary movements.* These movements usually do not attempt to work within the existing system; rather, they aim to remake the system by replacing existing institutions with new ones. Revolutionary movements range from utopian groups seeking to establish an ideal society to radical terrorists who use fear tactics to intimidate those with whom they disagree ideologically.

Movements based on terrorism often use tactics such as bombings, kidnappings, hostage taking, hijackings, and assassinations. A number of movements in the United States have engaged in terrorist activities or supported a policy of violence. However, the terrorist attacks in New York City and Washington, D.C., on September 11, 2001, and the events that followed those attacks proved to all of us that terrorism within this country can originate from the activities of revolutionary terrorists from outside the country as well.

Religious Movements Social movements that seek to produce radical change in individuals are typically based on spiritual or supernatural belief systems. Also referred to as *expressive movements, religious movements* are concerned with renovating or renewing people through "inner change." Fundamentalist religious groups seeking to convert nonbelievers to their belief system are an example of this type of movement. Some religious movements are *millenarian*—that is, they forecast that "the end is near" and assert that an immediate change in behavior is imperative. Relatively new

propaganda information provided by individuals or groups that have a vested interest in furthering their own cause or damaging an opposing one.

social movement an organized group that acts consciously to promote or resist change through collective action.

religious movements in industrialized Western societies have included Hare Krishnas, the Unification church, Scientology, and the Divine Light Mission, all of which tend to appeal to the psychological and social needs of young people seeking meaning in life that mainstream religions have not provided for them.

Alternative Movements Movements that seek limited change in some aspect of people's behavior are referred to as *alternative movements.* For example, early in the twentieth century the Women's Christian Temperance Union attempted to get people to abstain from drinking alcoholic beverages. Some analysts place "therapeutic social movements" such as Alcoholics Anonymous in this category; however, others do not, because of their belief that people must change their lives completely in order to overcome alcohol abuse. More recently, a variety of "New Age" movements have directed people's behavior by emphasizing spiritual consciousness combined with a belief in reincarnation and astrology. Such practices as vegetarianism, meditation, and holistic medicine are often included in the self-improvement category. Some alternative movements have included the practice of yoga (usually without its traditional background in the Hindu religion) as a means by which the self can be liberated and union can be achieved with the supreme spirit or universal soul.

Resistance Movements Also referred to as *regressive movements, resistance movements* seek to prevent change or to undo change that has already occurred. Virtually all of the social movements previously discussed face resistance from one or more reactive movements that hold opposing viewpoints and want to foster public policies that reflect their own beliefs. Examples of resistance movements are groups organized since the 1950s to oppose school integration, civil rights and affirmative action legislation, and domestic partnership initiatives. However, perhaps the most widely known resistance movement includes many who label themselves "pro-life" advocates—such as Prolife America and Operation Rescue, which seek to close abortion clinics and make abortion illegal under all circumstances. Protests by some radical antiabortion groups have grown violent, resulting in the deaths of several doctors and clinic workers, and creating fear among health professionals and patients seeking abortions.

Stages in Social Movements

Do all social movements go through similar stages? Not necessarily, but there appear to be identifiable stages in virtually all movements that succeed beyond their initial phase of development.

In the *preliminary* (or *incipiency*) *stage,* widespread unrest is present as people begin to become aware of a problem. At this stage, leaders emerge to agitate others into taking action. In the *coalescence stage,* people begin to organize and to publicize the problem. At this stage, some movements become formally organized at local and regional levels. In the *institutionalization* (or *bureaucratization*) *stage,* an organizational structure develops, and a paid staff (rather than volunteers) begins to lead the group. When the movement reaches this stage, the initial zeal and idealism of members may diminish as administrators take over management of the organization. Early grassroots supporters may become disillusioned and drop out; they may also start another movement to address some as-yet-unsolved aspect of the original problem. For example, some national environmental organizations—such as the Sierra Club, the

▲ The term *netroots,* an update of *grassroots,* used to describe locally grown movements, is now commonly used in reference to Internet-based activism. How do Web-based research and information organizations such as Media Matters for America support or undermine the efforts of activist organizations such as MoveOn.org and Americans for Prosperity, which rely heavily on the Internet for promoting their causes?

▲ Yoga has become an increasingly popular activity in recent years as many people have turned to alternative social movements derived from Asian traditions.

Dallas Events Inc./Shutterstock.com

National Audubon Society, and the National Parks and Conservation Association—that started as grassroots conservation movements are currently viewed by many people as being unresponsive to local environmental problems. As a result, new movements have arisen.

Social Movement Theories

What conditions are most likely to produce social movements? Why are people drawn to these movements? Sociologists have developed several theories to answer these questions.

Relative Deprivation Theory

According to relative deprivation theory, people who are satisfied with their present condition are less likely to seek social change. Social movements arise as a response to people's perception that they have been deprived of what they consider to be their fair share. Thus, people who suffer relative deprivation are more likely to feel that change is necessary and to join a social movement in order to bring about that change. *Relative deprivation* refers to the discontent that people may feel when they compare their achievements with those of similarly situated persons and find that they have less than they think they deserve. Karl Marx captured the idea of relative deprivation in this description: "A house may be large or small; as long as the surrounding houses are small it satisfies all social demands for a dwelling. But let a palace arise beside the little house, and it shrinks from a little house to a hut" (qtd. in Ladd, 1966: 24). Movements based on relative deprivation are most likely to occur when an upswing in the standard of living is followed by a period of decline, such that people have *unfulfilled rising expectations*—newly raised hopes of a better lifestyle that are not fulfilled as rapidly as the people expected or are not realized at all.

Although most of us can relate to relative deprivation theory, it does not fully account for why people experience social discontent but fail to join a social movement. Even though discontent and feelings of deprivation may be necessary to produce certain types of social movements, they are not sufficient to bring movements into existence.

Value-Added Theory

The value-added theory developed by sociologist Neil Smelser (1963) is based on the assumption that certain conditions are necessary for the development of a social movement. Smelser called his theory the "value-added" approach based on the concept (borrowed from the field of economics) that each step in the production process adds something to the finished product. For example, in the process of converting iron ore into automobiles, each stage "adds value" to the final product (Smelser, 1963). Similarly, Smelser asserted, six conditions are necessary and sufficient to produce social movements when they combine or interact in a particular situation:

1. *Structural conduciveness.* People must become aware of a significant problem and have the opportunity to engage in collective action. Movements are more likely to occur when a person, class, or agency can be singled out as the source of the problem; when channels for expressing grievances either are not available or fail; and when the aggrieved have a chance to communicate among themselves.
2. *Structural strain.* When a society or community is unable to meet people's expectations that something should be done about a problem, strain occurs in the system. The ensuing tension and conflict contribute to the development of a social movement based on people's belief that the problem would not exist if authorities had done what they were supposed to do.
3. *Spread of a generalized belief.* For a movement to develop, there must be a clear statement of the problem and a shared view of its cause, effects, and possible solution.
4. *Precipitating factors.* To reinforce the existing generalized belief, an inciting incident or

sociology in global perspective

China: A Nation of Environmental Woes and Emergent Social Activism

News Bulletin:

Estimated number of premature deaths in China each year that are caused by pollution:

- outdoor air pollution: 350,000 to 400,000 people
- indoor air pollution: 300,000 people
- water pollution: 60,000 people

—World Bank data (Kahn and Yardley, 2008: A1)

China is frequently in the international news these days because of its rapid industrial growth and swift rise as a global economic power. However, accompanying this nation's double-digit growth rate has been an unprecedented pollution problem. According to some social analysts, "China is choking on its own success" (Kahn and Yardley, 2008: A1). Although the economy has grown rapidly, much of this growth is related to a vast expansion of industry and rapid patterns of urbanization. For this kind of growth to be possible, staggering amounts of energy are needed, and China derives almost all of its energy from coal, one of the dirtiest sources of energy.

If this is China's problem, why should those of us who live in the United States be concerned? For humanitarian reasons, we must be concerned about the effects of deadly pollution on the residents of China. But we must also be concerned about the effects of such environmental degradation because "What happens in China *does not* stay in China." China's pollution problems are not just national problems; they are global problems. According to the *Journal of Geophysical Research*, "Sulfur dioxide and nitrogen oxides spewed by China's coal-fired power plants fall as acid rain on Seoul, South Korea, and Tokyo. Much of the particulate pollution over Los Angeles originates in China" (qtd. in Kahn and Yardley, 2008: A6). Yes, that is correct: Some of the pollution found in Los Angeles, California, may be attributed to what happens in China!

Can anything be done about the problem? Are activists and environmental movements trying to bring about environmental conservation in China? Some environmental activists are indeed attempting to highlight the causes and consequences of the various forms of pollution that are assaulting their nation. For example, environmental activist Wu Lihong repeatedly warned public officials in Wuxi, China, that pollution was strangling Lake Tai, but little attention was paid to his concerns until after the city was forced to shut off its drinking water because a deadly algae bloom was growing rapidly on the lake. Environmental researchers partly attributed this algae bloom to heavy pollution in the area. However, rather than praising Wu Lihong for his efforts to mobilize people and raise awareness of the problem, public officials had him arrested on blackmail and extortion charges, claiming that he demanded money from businesses

dramatic event must occur. With regard to technological disasters, some (including Love Canal) gradually emerge from a long-standing environmental threat, whereas others (including the Japanese nuclear power plants) involve a suddenly imposed problem.

5. *Mobilization for action.* At this stage, leaders emerge to organize others and give them a sense of direction.
6. *Social control factors.* If there is a high level of social control on the part of law enforcement officials, political leaders, and others, it becomes more difficult to develop a social movement or engage in certain types of collective action.

Value-added theory takes into account the complexity of social movements and makes it possible to test Smelser's assertions regarding the necessary and sufficient conditions that produce such movements. However, critics note that the approach is rooted in the functionalist tradition and views structural strains as disruptive to society.

Resource Mobilization Theory

Smelser's value-added theory tends to underemphasize the importance of resources in social movements. By contrast, *resource mobilization theory* focuses on the ability of members of a social movement to acquire resources and mobilize people in order to advance their cause (McCarthy and Zald, 1977). Resources include money, people's time and skills, access to the media, and material goods, such as property and equipment. Assistance from outsiders is essential for social movements. For example, reform movements are more likely to succeed when they gain the support of political and economic elites.

Resource mobilization theory is based on the assumption that participants in social movements are rational people. According to the sociologist Charles Tilly (1973, 1978), movements are formed and dissolved, mobilized and deactivated, based on rational decisions about the goals of the group, available resources, and the cost of mobilization and collective action. Resource mobilization theory also assumes that participants must have some degree

Although some people believe that the Chinese government sends mixed messages about care for the environment, China has joined other countries in an effort to limited the use of plastic shopping bags. Other alternatives are shown here.

by threatening that he would expose them for illegal pollution (Bodeen, 2007). Other social movement organizers in China have also found that they risk arrest and prosecution if they publicize their concerns and try to gather resources and mobilize others for environmental causes. As a result, some organizers are hesitant to take action because they fear the consequences.

Indeed, it appears that "China is choking on its own success" and that social movements so far have made few, if any, inroads on addressing the problem. According to the environmental researcher Wang Jinnan, "It is a very awkward situation for the country because our greatest achievement is also our biggest burden. There is pressure for change, but many people refuse to accept that we need a new approach so soon" (qtd. in Kahn and Yardley, 2008: A1, A6).

What will the future hold for environmental protection in China? According to resource mobilization theory, widespread discontent alone cannot produce a social movement: Adequate resources and motivated people are essential for any concerned social action. Some analysts believe that environmental leaders in China will eventually be able to mobilize people for change because more-affluent Chinese residents are becoming very concerned about quality-of-life issues and because the Internet, Facebook, and other social media are making it possible for people to organize quickly and demand governmental action on pressing problems such as this one.

reflect & analyze

Do you believe that environmental movements in China might be organized like the most successful ones in the United States? Why or why not?

of economic and political resources to make the movement a success. In other words, widespread discontent alone cannot produce a social movement: Adequate resources and motivated people are essential to any concerted social action. Based on an analysis of U.S. social protest groups (ranging from labor unions to peace movements) between 1800 and 1945, one study concluded that the organization and tactics of a movement strongly influence its chances of success. However, critics note that this theory fails to account for social changes brought about by groups with limited resources.

In the twenty-first century, scholars continue to modify resource mobilization theory and to develop new approaches for investigating the diversity of movements (see Buechler, 2000). For example, newer perspectives based on resource mobilization theory emphasize the ideology and legitimacy of movements as well as material resources (Zald and McCarthy, 1987). Additional perspectives are also needed on social movements in other nations to determine how activists in those countries acquire resources and mobilize people to advance causes such as environmental protection (see "Sociology in Global Perspective").

Social Constructionist Theory: Frame Analysis

Theories based on a symbolic interactionist perspective focus on the importance of the symbolic presentation of a problem to both participants and the general public (see Snow et al., 1986; Capek, 1993). Social constructionist theory is based on the assumption that a social movement is an interactive, symbolically defined, and negotiated process that involves participants, opponents, and bystanders (Buechler, 2000).

Research based on this perspective often investigates how problems are framed and what names they are given. This approach reflects the influence of the sociologist Erving Goffman's *Frame Analysis* (1974), in which he suggests that our interpretation of the particulars of events and activities is dependent on the framework from which we perceive them. According to Goffman (1974: 10), the purpose of frame analysis is "to try to isolate some

© AP Photo/Rich Pedroncelli

Chip Somodevilla/Getty Images

▲ How is the issue of immigration framed in these photos? Research based on frame analysis often investigates how social issues are framed and what names they are given.

of the basic frameworks of understanding available in our society for making sense out of events and to analyze the special vulnerabilities to which these frames of reference are subject." In other words, various "realities" may be simultaneously occurring among participants engaged in the same set of activities. Sociologist Steven M. Buechler (2000: 41) explains the relationship between frame analysis and social movement theory:

> Framing means focusing attention on some bounded phenomenon by imparting meaning and significance to elements within the frame and setting them apart from what is outside the frame. In the context of social movements, framing refers to the interactive, collective ways that movement actors assign meanings to their activities in the conduct of social movement activism. The concept of framing is designed for discussing the social construction of grievances as a fluid and variable process of social interaction—and hence a much more important explanatory tool than resource mobilization theory has maintained.

Sociologists have identified at least three ways in which grievances are framed. First, *diagnostic framing* identifies a problem and attributes blame or causality to some group or entity so that the social movement has a target for its actions. Second, *prognostic framing* pinpoints possible solutions or remedies, based on the target previously identified. Third, *motivational framing* provides a vocabulary of motives that compel people to take action (Benford, 1993; Snow and Benford, 1988). When successful framing occurs, the individual's vague dissatisfactions are turned into well-defined grievances, and people are compelled to join the movement in an effort to reduce or eliminate those grievances (Buechler, 2000).

Beyond motivational framing, additional frame-alignment processes are necessary in order to supply a continuing sense of urgency to the movement. *Frame alignment* is the linking together of interpretive orientations of individuals and social movement organizations so that there is congruence between individuals' interests, beliefs, and values and the movement's ideologies, goals, and activities (Snow et al., 1986). Four distinct frame-alignment processes occur in social movements: (1) *frame bridging* is the process by which movement organizations reach individuals who already share the same worldview as the organization, (2) *frame amplification* occurs when movements appeal to deeply held values and beliefs in the general population and link those to movement issues so that people's preexisting value commitments serve as a "hook" that can be used to recruit them, (3) *frame extension* occurs when movements enlarge the boundaries of an initial frame to incorporate other issues that appear to be of importance to potential participants, and (4) *frame transformation* refers to the process whereby the creation and maintenance of new values, beliefs, and meanings induce movement participation by redefining activities and events in such a manner that people believe they must become involved in collective action (Buechler, 2000). Some or all of these frame-alignment processes are used by social movements as they seek to define grievances and recruit participants.

Frame analysis provides new insights into how social movements emerge and grow when people are faced with problems such as technological disasters, about which greater ambiguity typically exists, and when people are attempting to "name" the problems associated with things such as nuclear or chemical contamination. However, frame analysis has been criticized for its "ideational biases" (McAdam, 1996). According to the sociologist Doug McAdam (1996), frame analyses of social movements have looked almost exclusively at ideas and their formal expression, whereas little attention has been paid to other significant factors, such as movement tactics, mobilizing structures, and changing political opportunities that influence the signifying work of movements. In this context, *political opportunities* means government structure, public policy, and political conditions that set the boundaries for change and political action. These boundaries are crucial variables in explaining why various social movements have different outcomes.

Political Opportunity Theory

Why do social protests occur? According to political opportunity theorists, the origins of social protests cannot be explained solely by the fact that people possess a variety of grievances or that they have resources available for mobilization. Instead, social protests are directly related to the political opportunities that potential protesters and movement organizers believe exist within the political system at any given point in time. Political opportunity theory is based on the assumption that social protests that take place *outside* of mainstream political institutions are deeply intertwined with more conventional political activities that take place *inside* these institutions. As used in this context, *opportunity* refers to "options for collective action, with chances and risks attached to them that depend on factors outside the mobilizing group" (Koopmans, 1999: 97). Political opportunity theory states that people will choose those options for collective action that are most readily available to them and those options that will produce the most favorable outcome for their cause.

What are some specific applications of political action theory? Urban sociologists and social movement analysts have found that those cities that provided opportunities for people's protests to be heard within urban governments were less likely to have extensive protests or riots in their communities because aggrieved people could use more conventional means to make their claims known. By contrast, urban riots were more likely to occur when activists believed that all conventional routes to protest were blocked. Changes in demography, migration, and the political economy in the United States (factors that were seemingly external to the civil rights movement) all contributed to a belief on the part of African Americans in the late 1960s and early 1970s that they could organize collective action and that their claims regarding the need for racial justice might be more readily heard by government officials.

Political opportunity theory has grown in popularity among sociologists who study social movements because this approach highlights the interplay of opportunity, mobilization, and political influence in determining when certain types of behavior may occur. However, like other perspectives, this theory has certain limitations, including the fact that social movement organizations may not always be not completely distinct from, or external to, the existing political system. For example, it is difficult to classify the Tea Party movement, which emerged in the aftermath of the election of President Barack Obama. Some supporters were outside the political mainstream and felt like they had no voice in what was happening in Washington. Keli Carender, who is credited with being one of the first Tea Party campaigners, complained that she tried to call her senators to urge them to vote against the $787 billion stimulus bill but constantly found that their mailboxes were full. As a result, she decided to protest against "porkulus"; in her words, "I basically thought to myself: 'I have two courses. I can give up, go home, crawl into bed and be really depressed and let it happen, or I can do something different, and I can find a new avenue to have my voice get out'" (qtd. in Zernike, 2010: A1). By contrast, other active supporters of the Tea Party movement are players in the mainstream political process. An example is Sarah Palin, the former Alaska governor and Republican vice presidential candidate, who is a frequent spokesperson at Tea Party rallies across the United States. In political movements, social activists typically *create* their own opportunities rather than wait for them to emerge, and activists often are political entrepreneurs in their own right, much like the state and federal legislators and other governmental officials whom they seek to influence on behalf of their social cause. Political opportunity theory calls our attention to how important the degree of openness of a political system is to the goals and tactics of social movements' organizers.

New Social Movement Theory

New social movement theory looks at a diverse array of collective actions and the manner in which those actions are based on politics, ideology, and culture. It also incorporates factors of identity, including race, class, gender, and sexuality, as sources of collective action and social movements. Examples of "new social movements" include ecofeminism and environmental justice movements.

Ecofeminism emerged in the late 1970s and early 1980s out of the feminist, peace, and ecology movements. Prompted by the near-meltdown at the Three Mile Island nuclear power plant, ecofeminists established World Women in Defense of the Environment. *Ecofeminism* is based on the belief that patriarchy is a root cause of environmental problems. According to ecofeminists, patriarchy not only results in the domination of women by men but also contributes to a belief that nature is to be possessed and dominated, rather than treated as a partner.

sociology works!

Fine-Tuning Theories and Data Gathering on Environmental Racism

Throughout *Sociology in Our Times,* we have examined social theories that help us understand the interplay of factors such as race, class, and gender in the everyday lives of millions of people. In the "Sociology Works!" feature, we have focused on specific theories and how applications of those theories can help us understand the world and sometimes make it a better place in which to live.

In this chapter we have looked at the work of new social movement theorists who have demonstrated the intersection of environmental justice with race and class: the belief that hazardous-waste treatment, storage, and disposal facilities are more likely to be located near low-income, nonwhite neighborhoods than to higher-income, predominantly white neighborhoods. This is an important issue because of the potential health risks that such sites may pose for people who live nearby. However, critics have scoffed at the suggestion that race- or class-based discrimination is involved in decisions about where hazardous-waste-materials facilities are located. Can more accurate data be gathered to help determine the nature and extent to which environmental racism exists?

During the 1980s and 1990s, the most frequently used method employed in national-level studies documenting the location of waste sites and other polluting industrial facilities was referred to as "unit-hazard coincidence" methodology. Based on this approach, researchers selected a predefined geographic unit (such as certain ZIP code areas or census tracts). Then they identified subsets of the units (areas located within a specific ZIP code or census tract) that had, or did not have, the hazard present. The researchers then compared the demographic characteristics of people living within each of the subsets to see if a larger minority population was present near the hazardous facility (see Mohai and Saha, 2007). Unit-hazard coincidence methodology assumes that the people who live within the predefined geographic units included in a study are located closer to the hazard than those individuals who do not live in those geographic units (Mohai and Saha, 2007). The problem with this approach is that the hazardous site is usually not located at the center of the ZIP code

Another "new social movement" focuses on environmental justice and the intersection of race and class in the environmental struggle (see "Sociology Works!"). Sociologist Stella M. Capek (1993) investigated a contaminated landfill in the Carver Terrace neighborhood of Texarkana, Texas, and found that residents were able to mobilize for change and win a federal buyout and relocation by symbolically linking their issue to a larger *environmental justice* framework. Since the 1980s, the emerging environmental

© Andrew Lichtenstein/The Image Works

▲ Referred to as "Cancer Alley," this area of Baton Rouge, Louisiana, is home to a predominantly African American population and also to many refineries that heavily pollute the region. Sociologists suggest that environmental racism is a significant problem in the United States and other nations. What do you think?

or census tract and that the geographic area being examined may be large or small, making it difficult to know for sure the racial and class characteristics of the people who live closest to the waste facility.

In recent years, sociologists and other social scientists have begun to use other methods such as GIS (a computer system for capturing, storing, checking, integrating, manipulating, analyzing, and displaying data related to positions on the Earth's surface) to more adequately determine the distance between environmentally hazardous sites and nearby populations. By using distance-based methods to control for proximity around environmentally hazardous sites, those researchers have demonstrated that nonwhites, who made up about 25 percent of the nation's population in 1990, constituted over 40 percent of the population living within one mile of hazardous-waste facilities, meaning that racial disparities in the distribution of hazardous sites are much greater than what had been previously reported. According to social scientists Paul Mohai and Robin Saha (2007: 343), "We [find that] these disparities persist even when controlling for economic and sociopolitical variables, suggesting that factors uniquely associated with race, such as racial targeting, housing discrimination, or other race-related factors, are associated with the location of the nation's hazardous waste facilities."

Sociological theories and research pertaining to environmental racism have raised public awareness that the location of hazardous facilities is not purely coincidental in communities throughout our nation. Clearly, proximity to hazardous sites is related to the cost of the land on which the facilities are located, but the issue of proximity based on the racial/ethnic composition of residents raises an even more challenging social and ethical dilemma. But it is also clear that the vast quantity of data available today—and the methods for obtaining those data—make it possible for us to fine-tune previous theories and obtain a better understanding of the social world in which we live.

reflect & analyze

New technology can cause problems for society—but it can also improve people's lives. Can you think of another way that current technology could be used to help correct a social problem in your community?

justice movement has focused on the issue of ***environmental racism*****—the belief that a disproportionate number of hazardous facilities (including industries such as waste disposal/treatment and chemical plants) are placed in low-income areas populated primarily by people of color** (Bullard and Wright, 1992). These areas have been left out of most of the environmental cleanup that has taken place in the last two decades. Capek concludes that linking Carver Terrace with environmental justice led to it being designated as a cleanup site. She also views this as an important turning point in new social movements: "Carver Terrace is significant not only as a federal buyout and relocation of a minority community, but also as a marker of the emergence of environmental racism as a major new component of environmental social movements in the United States" (Capek, 1993: 21).

Sociologist Steven M. Buechler (2000) has argued that theories pertaining to twenty-first-century social movements should be oriented toward the structural, macrolevel contexts in which movements arise. These theories should incorporate both political and cultural dimensions of social activism:

> Social movements are historical products of the age of modernity. They arose as part of a sweeping social, political, and intellectual change that led a significant number of people to view society as a social construction that was susceptible to social reconstruction through concerted collective effort. Thus, from their inception, social movements have had a dual focus. Reflecting the political, they have always involved some form of challenge to prevailing forms of authority. Reflecting the cultural, they have always operated as symbolic laboratories in which reflexive actors pose questions of meaning, purpose, identity, and change. (Buechler, 2000: 211)

This chapter's Concept Quick Review summarizes the main theories of social movements.

As we have seen, social movements may be an important source of social change. Throughout this text, we have examined a variety of social problems that have been the focus of one or more social movements. For this reason, many groups focus on preserving their gains while simultaneously fighting for changes that they believe are still necessary.

environmental racism the belief that a disproportionate number of hazardous facilities (including industries such as waste disposal/treatment and chemical plants) are placed in low-income areas populated primarily by people of color.

[concept quick review]

Social Movement Theories

	Key Components
Relative Deprivation	People who are discontent when they compare their achievements with those of others consider themselves relatively deprived and join social movements in order to get what they view as their "fair share," especially when there is an upswing in the economy followed by a decline.
Value-Added	Certain conditions are necessary for a social movement to develop: (1) structural conduciveness, such that people are aware of a problem and have the opportunity to engage in collective action; (2) structural strain, such that society or the community cannot meet people's expectations for taking care of the problem; (3) growth and spread of a generalized belief as to causes and effects of and possible solutions to the problem; (4) precipitating factors, or events that reinforce the beliefs; (5) mobilization of participants for action; and (6) social control factors, such that society decides to allow the movement to take action.
Resource Mobilization	A variety of resources (money, members, access to media, and material goods such as equipment) are necessary for a social movement; people participate only when they feel the movement has access to these resources.
Social Construction Theory: Frame Analysis	Based on the assumption that social movements are an interactive, symbolically defined, and negotiated process involving participants, opponents, and bystanders, frame analysis is used to determine how people assign meaning to activities and processes in social movements.
Political Opportunity	People will choose the options for collective action (i.e., "opportunities") that are most readily available to them and those options that will produce the most favorable outcome for their cause.
New Social Movement	The focus is on sources of social movements, including politics, ideology, and culture. Race, class, gender, sexuality, and other sources of identity are also factors in movements such as ecofeminism and environmental justice.

Social Change in the Future

In this chapter, we have focused on collective behavior and social movements as potential forces for social change in contemporary societies. A number of other factors also contribute to social change, including the physical environment, population trends, technological development, and social institutions.

The Physical Environment and Change

Changes in the physical environment often produce changes in the lives of people; in turn, people can make dramatic changes in the physical environment, over which we have only limited control. Throughout history, natural disasters have taken their toll on individuals and societies. Major natural disasters—including hurricanes, floods, and tornadoes—can devastate an entire population. In September 2005, the United States experienced the worst natural disaster in its history when Hurricane Katrina left a wide path of death and destruction through Louisiana, Mississippi, and Alabama. However, damage from the hurricane itself was just the beginning of how the physical environment abruptly changed, how this disaster altered the lives of millions of people, and how it raised serious questions about our national priorities and the future of the environment. More recent earthquakes, hurricanes, and tornadoes, though not as devastating as Katrina, have changed the lives of many people not only in terms of property loss but also in regard to the long-term trauma that people experience after a major disaster (see Erikson, 1976, 1994).

Some natural disasters are exacerbated by human decisions. For example, floods are viewed as natural disasters, but excessive development may contribute to a flood's severity. As office buildings, shopping malls, industrial plants, residential areas, and highways are developed, less land remains as groundcover to absorb rainfall. When heavier-than-usual rains occur, flooding becomes inevitable; some regions of the United States—such as in and around New Orleans—have remained under water for days or even weeks in recent years. Clearly, humans cannot control the rain, but human decisions can worsen the consequences. If Hurricane Katrina's first wave was the storm itself, the second wave was a *human-made disaster* resulting in part from decisions relating to planning and budgetary priorities, allocation of funds for maintaining infrastructure, and the importance of emergency preparedness. *Infrastructure* refers to a framework of systems, such

Jocelyn Augustino/FEMA

© AP Photo/Bill Haber

▲ Natural disasters such as Hurricane Katrina produce flooding and devastation in cities such as New Orleans and in less populated areas, but they also make us acutely aware of vast racial and economic inequalities that persist in the United States and other nations. Volunteers did a significant part of the recovery work in New Orleans.

as transportation and utilities, that makes it possible to have specific land uses (commercial, residential, and recreational, for example) and a built environment (buildings, houses, and highways) that support people's daily activities and the nation's economy. It takes money and commitment to make sure that the components of the infrastructure remain strong so that cities can withstand natural disasters and other concerns such as climate change. Consider, for example, that the city of Chicago is taking a proactive stance on dealing with future increases in temperature and climate conditions that will make Chicago's weather feel more like Baton Rouge, Louisiana, than a northern city. To cope with this change, Chicago city officials are already repaving alleyways with water-permeable materials and planting swamp oak and sweet gum trees from the South rather than the more indigenous white oak, which is the state tree of Illinois (Kaufman, 2011). Long-range planning such as this to cope with changes in the physical environment may seem far-fetched, but when the time comes, between fifty and a hundred years from now, such endeavors may seem farsighted instead. In the words of one Chicago city official, "Cities adapt or they go away" (qtd. in Kaufman, 2011).

The changing environment is one of many reasons why experts are also concerned about availability of water in the future. Water is a finite resource that is necessary for both human survival and the production of goods. However, water is being wasted and polluted, and the supply of *potable* (drinkable) water is limited. People are causing—or at least contributing to—that problem (see "Sociology and Social Policy").

People also contribute to changes in the Earth's physical condition. Through soil erosion and other degradation of grazing land, often at the hands of people, more than 25 billion tons of topsoil is lost annually. As people clear forests to create farmland and pastures and to acquire lumber and firewood, the Earth's tree cover continues to diminish. As millions of people drive motor vehicles, the amount of carbon dioxide in the environment continues to rise each year, contributing to global warming.

Just as people contribute to changes in the physical environment, human activities must also be adapted to changes in the environment. For example, we are being warned to stay out of the sunlight because of increases in ultraviolet rays, a cause of skin cancer, as a result of the accelerating depletion of the ozone layer. If the ozone warnings are accurate, the change in the physical environment will dramatically affect those who work or spend their leisure time outside.

Population and Change

Changes in population size, distribution, and composition affect the culture and social structure of a society and change the relationships among nations. As discussed in Chapter 19, the countries experiencing the most rapid increases in population have a less-developed infrastructure to deal with those

sociology and social policy

The Fight Over Water Rights

Who controls water rights—the rights to the world's finite supply of potable (drinkable) water? (Figure 20.1 shows how finite that supply is.) Is the water supply something that individual people or governments should be able to *own*? Answers to that question may vary depending on whether an individual is one of the "haves" or the "have-nots" with regard to water. Court decisions speak of water rights in terms of *sovereignty* with regard to governmental actions and *riparian rights* with regard to individual or group water rights. *Riparian* refers to the rights of a person (or group) to water by virtue of owning or occupying the bank of a river

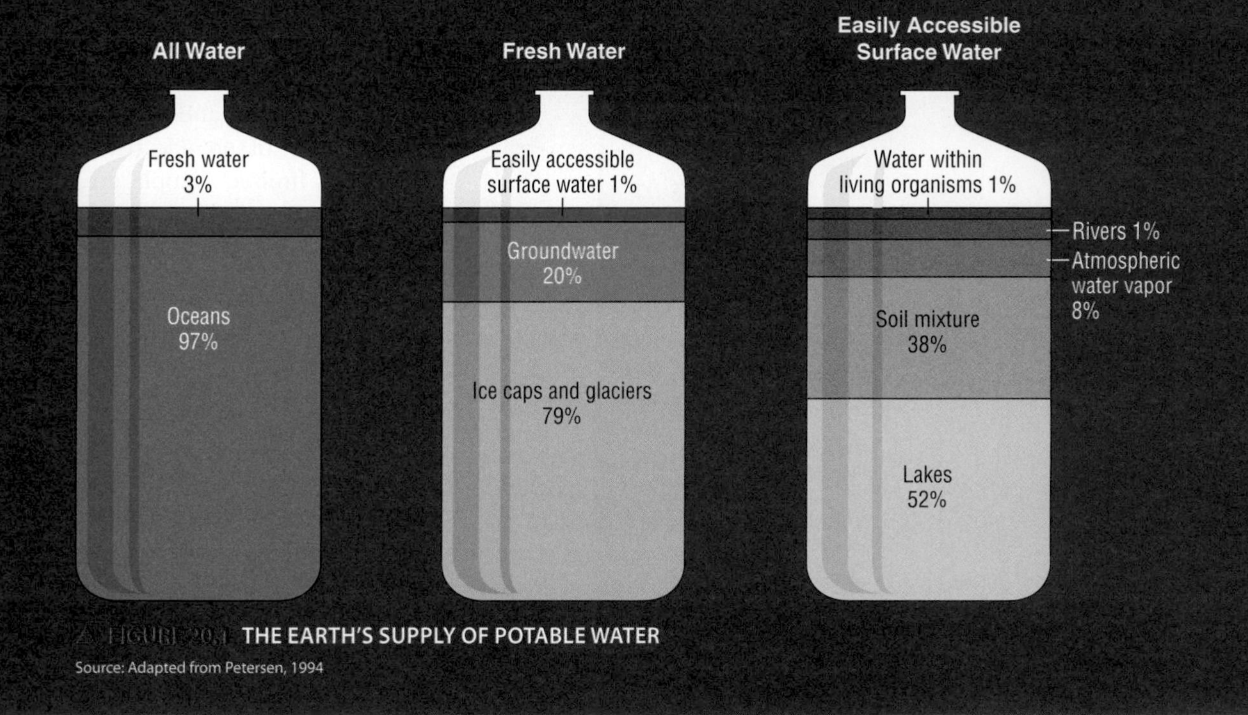

FIGURE 20.1 THE EARTH'S SUPPLY OF POTABLE WATER

Source: Adapted from Petersen, 1994

changes. How will nations of the world deal with population growth as the global population continues to move toward seven billion? Only time will provide a response to this question.

In the United States a shift in population distribution from central cities to suburban and exurban areas has produced other dramatic changes. Central cities have experienced a shrinking tax base as middle-income and upper-middle-income residents and businesses have moved to suburban and outlying areas. Some suburban areas have high rates of hidden poverty and are experiencing the same problems of central cities, including decaying infrastructure, low-performing schools, inadequate transportation, and rising rates of crime, to name only a few concerns. In the urban centers, schools and public services have declined in many areas, leaving those people with the greatest needs with the fewest public resources and essential services. The changing composition of the U.S. population has resulted in children from more-diverse cultural backgrounds entering school, producing a demand for new programs and changes in curricula. An increase in the birthrate has created a need for more child care; an increase in the older population has created a need for services such as medical care and placed greater demand on programs such as Social Security.

Population growth and the movement of people to urban areas have brought profound changes to many regions and intensified existing social problems. Among other factors, growth in the global population will be one of the most significant driving forces in the future.

Technology and Change

Technology is an important force for change; in some ways, technological development has made our lives much easier. Advances in communication and transportation have made instantaneous worldwide communication possible but have also brought old belief systems and the status quo into question as never before. Today, we are increasingly moving information

or lake. Historically, if a river passed through your property, you had the right to take and use as much of its water as you wanted or needed, without regard to the effect this had on people farther down the river. Accordingly, those who lived higher up a river could build a dam and divert water into lavish agricultural irrigation projects even when this resulted in water rationing for people farther down the river. Often, untreated wastewater was intentionally discharged back into the river, again without regard to the effect such action had on those farther down the river.

In the United States—as well as in the rest of the world—the assertion of riparian rights is being challenged by those whose water supply is threatened, whether by dwindling supplies or by pollution. A nation, a state, or a city may assert absolute sovereignty over the natural resources within its territory (such as its water supply), whereas nations, states, or cities farther downstream may assert another principle—governmental integrity, or the right to a supply that is adequate (in terms of both quantity and quality) to meet their own survival requirements.

One example of this conflict can be found in the Edwards Aquifer debate in Texas (an *aquifer* is an underground water supply). The Edwards Aquifer reaches from Austin to San Antonio and beyond. San Antonio relies on the aquifer for its potable water supply, but other—smaller—cities and many rural businesses also depend on the aquifer, including farmers who need the water to grow their crops. When the supply in the aquifer drops, the farmers must compete with the cities for water; attorneys representing both groups often go before the courts and regulatory agencies to argue over how much of the water each should be entitled to receive and use.

Referred to as the "western water wars," water policies have angered many people in Nevada and California. Over the past century in Nevada, billions of tax dollars have been used to finance irrigation projects that siphoned off water from Pyramid Lake and diverted it to alfalfa farms and cattle ranches in the middle of the high desert east of Reno. However, the U.S. government recently started buying back much of that water and giving it to the Pyramid Lake Paiute Indians to restore fish runs and wetlands. The owners of the irrigated farms in the desert are extremely frustrated at this change in policy. In California, large volumes of water have been transferred from the big farms in the Central Valley to help depleted fish runs in the San Joaquin and Sacramento rivers, and an area "sucked dry" of its water by Los Angeles demanded its water back to no avail.

What will the future hold in regard to water? Many people agree that water policies are necessary; they just do not agree on what those policies should be. However, most of us believe that taking care of the environment—for today and for the future—is a worthy and necessary goal. Maintaining an adequate and unpolluted supply of water is an integral part of that task; however, as you can see, social policy in this area often produces a great deal of conflict, and it is likely to produce even greater conflict in the future.

reflect & analyze

The average person in the United States uses eighty to one hundred gallons of water per day, and this figure does not include the amount consumed outside the home—for example, water used in restaurants for preparing food and cleaning up after meals. Can you think of some easy, basic ways that individuals could conserve water during their daily lives?

▲ As the very large Baby Boom generation enters retirement age, the health care needs of people in this age bracket will strain the capabilities of federal programs such as Medicare.

instead of people—and doing it almost instantly. Advances in science and medicine have made significant changes in people's lives in high-income countries.

Scientific advances will continue to affect our lives, from the foods we eat to our reproductive capabilities. Genetically engineered plants have been developed and marketed in recent years, and biochemists are creating potatoes, rice, and cassava with the same protein value as meat. Advances in medicine have made it possible for those formerly unable to have children to procreate; women well beyond menopause are now able to become pregnant with the

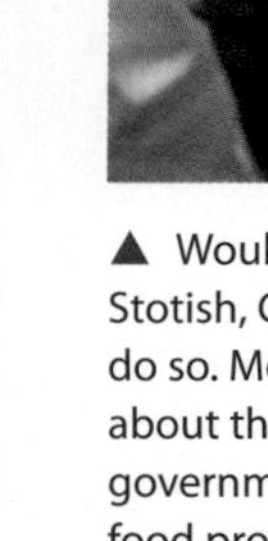

AP Photo/Charles Dharapak

▲ Would you like to eat a genetically modified salmon? Ron Stotish, CEO of AquaBounty, thinks that you might want to do so. More importantly, would you like to know beforehand about the genetic modification? The federal and state governments are considering how much information that food producers should have to disclose to consumers.

assistance of medical technology. Advances in medicine have also increased the human life span, especially for white and middle- or upper-class individuals in high-income nations; medical advances have also contributed to the declining death rate in low-income nations, where birthrates have not yet been curbed.

Just as technology has brought about improvements in the quality and length of life for many, it has also created the potential for new disasters, ranging from global warfare to localized technological disasters at toxic waste sites. As the sociologist William Ogburn (1966) suggested, when a change in the material culture occurs in society, a period of *cultural lag* follows in which the nonmaterial (ideological) culture has not yet caught up with material development. The rate of technological advance at the level of material culture today is mind-boggling. Many of us can never hope to understand technological advances in the areas of artificial intelligence, holography, virtual reality, biotechnology, cold fusion, and robotics.

One of the ironies of twenty-first-century high technology is the increased vulnerability that results from the increasing complexity of such systems. We have already seen this in situations ranging from jetliners that are used as terrorist weapons to identity theft and fraud on the Internet.

Social Institutions and Change

Many changes have occurred in the family, religion, education, the economy, and the political system over the last century. As discussed in Chapter 15, the size and composition of families in the United States changed with the dramatic increase in the number of single-person and single-parent households. Changes in families produced changes in the socialization of children, many of whom now spend much of their time playing video games, texting friends, posting their daily activities on Facebook or Twitter, or spending time in a child-care facility outside their own home.

Public education changed dramatically in the United States during the last century. This country was one of the first to provide "universal" education for students regardless of their ability to pay. As a result, at least until recently the United States has had one of the most highly educated populations in the world. Today, the United States still has one of the best public education systems in the world for the top 15 percent of the students, but it badly fails the bottom 25 percent. As the nature of the economy changes, schools almost inevitably will have to change, if for no other reason than demands from leaders in business and industry for an educated workforce that allows U.S. companies to compete in a global economic environment. Many business and political leaders believe that education is the single most important factor in the future of the United States; however, in difficult economic times when local, state, and federal budgets are strained, public education is one of the first institutions to undergo the axe as teachers are let go, school buildings are allowed to further decay, and students are not provided with the necessary physical setting and learning tools.

Although we have examined changes in the physical environment, population, technology, and social institutions separately, they all operate together in a complex relationship, sometimes producing large, unanticipated consequences. As we move further into the twenty-first century, we need new ways of conceptualizing social life at both the macrolevel and the microlevel. The sociological imagination helps us think about how personal troubles—regardless of our race, class, gender, age, sexual orientation, or physical abilities and disabilities—are intertwined with the public issues of our society and the global community of which we are a part. As one analyst noted regarding Lois Gibbs and Love Canal,

> If Love Canal has taught Lois Gibbs—and the rest of us—anything, it is that ordinary people become very smart very quickly when their lives are threatened. They become adept at detecting absurdity, even when it is concealed in bureaucratese and scientific jargon. Lois Gibbs learned that one cannot always rely on government to act in the best interests of ordinary citizens—at least, not without considerable prodding. She determined that she would prod them until her objectives were attained. She led one of the most successful, single-purpose grass roots efforts of our time. (Levine, 1982: xv)

Taking care of the environment is an example of something that government and each of us as individuals can do to help (see "You Can Make a Difference").

you can make a difference

College Students Are Taking the Lead in the "Go-Green" Movement

We feel like we have a social responsibility not to leave the next generation's environment in a worse way than it is.

—Adam Yarnell, a Brown University student, discussing why sustainability is an important issue to many college students (qtd. in Riley, 2009)

There is a huge youth movement, at least in my school, Colorado College. And I have witnessed it across a lot of schools, [the] back to earth movement of organic farming.

—Sophia Maravell, a student whose father owns an organic farm in Maryland, describing how students are becoming involved in green movements, ranging from organic farming to the development of clean energy (qtd. in Palacio, 2009)

These comments are supported by recent studies that show college students are leading the way in the green movement (Palacio, 2009). A National Wildlife Federation study, "Generation E: Students Leading for a Sustainable, Clean Energy Future," looked at 160 college campuses where students are active in the greening of campus operations and concluded that students frequently are at the forefront, encouraging faculty and administrators to organize green gatherings and help their school become a green campus.

Would you like to become involved in making your school a green campus? Here are suggestions from students working toward eco-friendly campuses:

- Recycle everything, especially paper! Look for recycling bins; if they're not available, encourage officials to have them placed near dorms, classrooms, and cafeterias and food courts.
- Use your printer wisely. Ask professors if printing on both sides of the paper is acceptable for class assignments. Use low-quality print settings for rough drafts, and limit the number of items printed from the Internet if you are unlikely to reuse them.
- Limit the use of disposable cups, plates, and napkins.
- Recycle cans, bottles, plastic bags, newspapers, and other items (recycling helps diminish waste-disposal problems by reducing the amount of waste hauled off to landfills).
- Use compact fluorescent light bulbs. Turn off lights, televisions, and computers when you are not using them.
- Walk, bike, and limit the use of your car.
- Buy green products that have been recycled, such as paper goods and cleaning products.
- Use refillable binders instead of notebooks, or use your laptop to take notes.
- Carry a water bottle and refill it at water fountains or drink dispensers.
- Buy used clothing. (based on CollegeUniversity.suite101.com, 2010)

If you are already involved in the green movement and are helping to reduce global warming, keep up the good work! If you would like additional information on the green movement or global warming, many websites are available to help you take the first steps toward making a difference on your campus and in the world where you live. A famous statement by the author Mark Twain is often quoted: "Everybody talks about the weather, but nobody ever does anything about it." In the case of environmental issues, if nobody does anything, we may face dire consequences now, and future generations may truly be imperiled.

© Jeff Greenberg/PhotoEdit

▲ Pollution of lakes, rivers, and other bodies of water has an adverse effect on food supplies, air quality, and the entire environment. What influence does a "business as usual" approach have on environmental quality in your area?

And it is vitally important that we all do everything that we can in order to protect the environment.

A Few Final Thoughts

In this text we have covered a substantial amount of material, examined different perspectives on a wide variety of social issues, and suggested different methods by which to deal with them. The purpose of this text is not to encourage you to take any particular point of view; rather, it is to allow you to understand different viewpoints and ways in which they may be helpful to you and to society in dealing with the issues of the twenty-first century. Possessing that understanding, we can hope that the future will be something we can all look forward to—producing a better way of life, not only in this country but worldwide as well.

chapter review Q & A

Use these questions and answers to check how well you've achieved the learning objectives set out at the beginning of this chapter.

- **What is the relationship between social change and collective behavior?**

Social change—the alteration, modification, or transformation of public policy, culture, or social institutions over time—is usually brought about by collective behavior, which is defined as a relatively spontaneous, unstructured activity that typically violates established social norms.

- **When is collective behavior likely to occur?**

Collective behavior occurs when some common influence or stimulus produces a response from a relatively large number of people.

- **What is a crowd?**

A crowd is a relatively large number of people in one another's immediate presence. Sociologist Herbert Blumer divided crowds into four categories: (1) casual crowds, (2) conventional crowds, (3) expressive crowds, and (4) acting crowds (including mobs, riots, and panics). A fifth type of crowd is a protest crowd.

- **What causes crowd behavior?**

Social scientists have developed several theories to explain crowd behavior. Contagion theory asserts that a crowd takes on a life of its own as people are transformed from rational beings into part of an organism that acts on its own. A variation on this is social unrest and circular reaction—people express their discontent to others, who communicate back similar feelings, resulting in a conscious effort to engage in the crowd's behavior. Convergence theory asserts that people with similar attributes find other like-minded persons with whom they can release underlying personal tendencies. Emergent norm theory asserts that as a crowd develops, it comes up with its own norms that replace more-conventional norms of behavior.

- **What are the primary forms of mass behavior?**

Mass behavior is collective behavior that occurs when people respond to the same event in the same way even if they are not in geographic proximity to one another. Rumors, gossip, mass hysteria, fads and fashions, and public opinion are forms of mass behavior.

- **What are the major types of social movements, and what are their goals?**

A social movement is an organized group that acts consciously to promote or resist change through collective action. Reform, revolutionary, religious, and alternative movements are the major types identified by sociologists. Reform movements seek to improve society by changing some specific aspect of the social structure. Revolutionary movements seek to bring about a total change in society—sometimes by the use of terrorism. Religious movements seek to produce radical change in individuals based on spiritual or supernatural belief systems. Alternative movements seek limited change of some aspect of people's behavior. Resistance movements seek to prevent change or to undo change that has already occurred.

- **How do social movements develop?**

Social movements typically go through three stages: (1) a preliminary stage (unrest results from a perceived problem), (2) coalescence (people begin to organize), and (3) institutionalization (an organization is developed, and paid staff replaces volunteers in leadership positions).

- **How do relative deprivation theory, value-added theory, and resource mobilization theory explain social movements?**

Relative deprivation theory asserts that if people are discontented when they compare their accomplishments with those of others similarly situated, they are more likely to join a social movement than are people who are relatively content with their status. According to value-added theory, six conditions are required for a social movement: (1) a perceived problem, (2) a perception that the authorities are not resolving the problem, (3) a spread of the belief to an adequate number of people, (4) a precipitating incident, (5) mobilization of other people by leaders, and (6) a lack of social control. By contrast, resource mobilization theory asserts that successful social movements can occur only when they gain the support of political and economic elites, who provide access to the resources necessary to maintain the movement.

- **What is the primary focus on research based on frame analysis, political opportunity theory, and new social movement theory?**

Research based on frame analysis often highlights the social construction of grievances through the process of social interaction. Various types of framing occur as problems are identified, remedies are sought, and people feel compelled to take action. Political opportunity theory focuses on how social protests are directly related to the political opportunities that potential protesters and movement organizers believe exist within the political system at any given point in time. Research based on new social movement theory has examined factors of identity, such as race, class, gender and sexuality, as sources of collective action and social movements (for example, environmental racism).

key terms

civil disobedience 603
collective behavior 598
crowd 601
environmental racism 617
gossip 606
mass 601
mass behavior 605
mob 602
panic 602
propaganda 608
public opinion 607
riot 602
rumor 605
social change 598
social movement 608

questions for critical thinking

1. What types of collective behavior in the United States do you believe are influenced by inequalities based on race/ethnicity, class, gender, age, or disabilities? Why?
2. Which of the four explanations of crowd behavior (contagion theory, social unrest and circular reaction, convergence theory, and emergent norm theory) do you believe best explains crowd behavior? Why?
3. In the text the Love Canal environmental movement is analyzed in terms of the value-added theory. How would you analyze that movement under (a) the relative deprivation theory and (b) the resource mobilization theory?
4. Using the sociological imagination that you have gained in this course, what are some positive steps that you believe might be taken in the United States to make our society a better place for everyone? What types of collective behavior and/or social movements might be required in order to take those steps?

turning to video

Watch the CBS video *Powered by Coal* (running time 11:12), available through **CengageBrain.com**. This video investigates the United States' significant dependency on coal and other fossil fuels and the enormous task of regulating and cleaning up carbon emissions. As you watch the video, think about what coal mining does to the landscape and to the lives of people who live near coal mines. After you watch the video, answer these questions: Should the development and use of alternative forms of energy such as solar and wind be a backup plan for the United States—or the main plan? Which, and why?

online study resources

Go to CENGAGE brain.com to access online study resources, including the Sociology CourseMate for this text as well as special features such as video, an interactive sociology time line and interactive maps, General Social Survey (GSS) data, and U.S. Census 2010 data.

CourseMate brings course concepts to life with interactive learning, study, and exam-preparation tools that support the printed textbook. A textbook-specific website, **Sociology CourseMate** includes an integrated interactive eBook and other interactive learning tools, including quizzes, flash cards, and videos.

Visit **www.cengagebrain.com** to access your account and purchase materials.

glossary

absolute poverty a level of economic deprivation that exists when people do not have the means to secure the most basic necessities of life.

achieved status a social position that a person assumes voluntarily as a result of personal choice, merit, or direct effort.

activity theory the proposition that people tend to shift gears in late middle age and find substitutes for previous statuses, roles, and activities.

acute diseases illnesses that strike suddenly and cause dramatic incapacitation and sometimes death.

age stratification inequalities, differences, segregation, or conflict between age groups.

ageism prejudice and discrimination against people on the basis of age, particularly against older persons.

agents of socialization the persons, groups, or institutions that teach us what we need to know in order to participate in society.

aggregate a collection of people who happen to be in the same place at the same time but share little else in common.

aging the physical, psychological, and social processes associated with growing older.

agrarian societies societies that use the technology of large-scale farming, including animal-drawn or energy-powered plows and equipment, to produce their food supply.

alienation a feeling of powerlessness and estrangement from other people and from oneself.

animism the belief that plants, animals, or other elements of the natural world are endowed with spirits or life forces that have an effect on events in society.

anomie Emile Durkheim's designation for a condition in which social control becomes ineffective as a result of the loss of shared values and of a sense of purpose in society.

anticipatory socialization the process by which knowledge and skills are learned for future roles.

ascribed status a social position conferred at birth or received involuntarily later in life, based on attributes over which the individual has little or no control, such as race/ethnicity, age, and gender.

assimilation a process by which members of subordinate racial and ethnic groups become absorbed into the dominant culture.

authoritarian leaders people who make all major group decisions and assign tasks to members.

authoritarian personality a personality type characterized by excessive conformity, submissiveness to authority, intolerance, insecurity, a high level of superstition, and rigid, stereotypic thinking.

authoritarianism a political system controlled by rulers who deny popular participation in government.

authority power that people accept as legitimate rather than coercive.

beliefs the mental acceptance or conviction that certain things are true or real.

bilateral descent a system of tracing descent through both the mother's and father's sides of the family.

blended family a family consisting of a husband and wife, children from previous marriages, and children (if any) from the new marriage.

body consciousness a term that describes how a person perceives and feels about his or her body.

bureaucracy an organizational model characterized by a hierarchy of authority, a clear division of labor, explicit rules and procedures, and impersonality in personnel matters.

bureaucratic personality a psychological construct that describes those workers who are more concerned with following correct procedures than they are with getting the job done correctly.

capitalism an economic system characterized by private ownership of the means of production, from which personal profits can be derived through market competition and without government intervention.

capitalist class (or **bourgeoisie**) Karl Marx's term for the class that consists of those who own and control the means of production.

caste system a system of social inequality in which people's status is permanently determined at birth based on their parents' ascribed characteristics.

category a number of people who may never have met one another but share a similar characteristic (such as education level, age, race, or gender).

charismatic authority power legitimized on the basis of a leader's exceptional personal qualities or the demonstration of extraordinary insight and accomplishment that inspire loyalty and obedience from followers.

chronic diseases illnesses that are long term or lifelong and that develop gradually or are present from birth.

chronological age a person's age based on date of birth.

church a large, bureaucratically organized religious organization that tends to seek accommodation with the larger society in order to maintain some degree of control over it.

civil disobedience nonviolent action that seeks to change a policy or law by refusing to comply with it.

civil religion the set of beliefs, rituals, and symbols that makes sacred the values of the society and places the nation in the context of the ultimate system of meaning.

class conflict Karl Marx's term for the struggle between the capitalist class and the working class.

class system a type of stratification based on the ownership and control of resources and on the type of work that people do.

cohabitation a situation in which two people live together, and think of themselves as a couple, without being legally married.

cohort a group of people born within a specified period of time.

collective behavior voluntary, often spontaneous activity that is engaged in by a large number of people and typically violates dominant-group norms and values.

comparable worth (or **pay equity**) the belief that wages ought to reflect the worth of a job, not the gender or race of the worker.

conflict perspectives the sociological approach that views groups in society as engaged in a continuous power struggle for control of scarce resources.

conformity the process of maintaining or changing behavior to comply with the norms established by a society, subculture, or other group.

conglomerate a combination of businesses in different commercial areas, all of which are owned by one holding company.

content analysis the systematic examination of cultural artifacts or various forms of communication to extract thematic data and draw conclusions about social life.

contingent work part-time work, temporary work, or subcontracted work that offers advantages to employers but that can be detrimental to the welfare of workers.

control group in an experiment, the group containing the subjects who are not exposed to the independent variable.

core nations according to world systems theory, dominant capitalist centers characterized by high levels of industrialization and urbanization.

corporate crime illegal acts committed by corporate employees on behalf of the corporation and with its support.

corporations large-scale organizations that have legal powers, such as the ability to enter into contracts and buy and sell property, separate from their individual owners.

correlation a relationship that exists when two variables are associated more frequently than could be expected by chance.

counterculture a group that strongly rejects dominant societal values and norms and seeks alternative lifestyles.

credentialism a process of social selection in which class advantage and social status are linked to the possession of academic qualifications.

crime a behavior that violates criminal law and is punishable with fines, jail terms, and/or other negative sanctions.

criminal justice system the more than 55,000 local, state, and federal agencies that enforce laws, adjudicate crimes, and treat and rehabilitate criminals.

criminology the systematic study of crime and the criminal justice system, including the police, courts, and prisons.

cross-dresser a male who dresses like a woman or a women who dresses like a man but does not alter the genitalia.

crowd a relatively large number of people who are in one another's immediate vicinity.

crude birthrate the number of live births per 1,000 people in a population in a given year.

crude death rate the number of deaths per 1,000 people in a population in a given year.

cult (also known as *new religious movement* or *NRM*) a religious group with practices and teachings outside the dominant cultural and religious traditions of a society.

cultural capital Pierre Bourdieu's term for people's social assets, including values, beliefs, attitudes, and competencies in language and culture.

cultural imperialism the extensive infusion of one nation's culture into other nations.

cultural lag William Ogburn's term for a gap between the technical development of a society and its moral and legal institutions.

cultural relativism the belief that the behaviors and customs of any culture must be viewed and analyzed by the culture's own standards.

cultural transmission the process by which children and recent immigrants become acquainted with the dominant cultural beliefs, values, norms, and accumulated knowledge of a society.

cultural universals customs and practices that occur across all societies.

culture the knowledge, language, values, customs, and material objects that are passed from person to person and from one generation to the next in a human group or society.

culture shock the disorientation that people feel when they encounter cultures radically different from their own and believe that they cannot depend on their own taken-for-granted assumptions about life.

deinstitutionalization the practice of rapidly discharging patients from mental hospitals into the community.

demedicalization the process whereby a problem ceases to be defined as an illness or a disorder.

democracy a political system in which the people hold the ruling power either directly or through elected representatives.

democratic leaders leaders who encourage group discussion and decision making through consensus building.

democratic socialism an economic and political system that combines private ownership of some of the means of production, governmental distribution of some essential goods and services, and free elections.

demographic transition the process by which some societies have moved from high birthrates and death rates to relatively low birthrates and death rates as a result of technological development.

demography a subfield of sociology that examines population size, composition, and distribution.

denomination a large organized religion characterized by accommodation to society but frequently lacking in ability or intention to dominate society.

dependency theory the belief that global poverty can at least partially be attributed to the fact that the high-income countries have exploited the low-income countries.

dependent variable a variable that is assumed to depend on or be caused by one or more other (independent) variables.

deviance any behavior, belief, or condition that violates significant social norms in the society or group in which it occurs.

differential association theory the proposition that people have a greater tendency to deviate from societal norms when they frequently associate with individuals who are more favorable toward deviance than conformity.

disability a physical or mental impairment that substantially limits one or more major activities that a person would normally do at a given stage of life and that may result in stigmatization or discrimination against the person with a disability.

discrimination actions or practices of dominant-group members (or their representatives) that have a harmful effect on members of a subordinate group.

disengagement theory the proposition that older persons make a normal and healthy adjustment to aging when they detach themselves from their social roles and prepare for their eventual death.

domestic partnerships household partnerships in which an unmarried couple lives together in a committed, sexually intimate relationship and is granted the same rights and benefits as those accorded to married heterosexual couples.

dominant group a group that is advantaged and has superior resources and rights in a society.

dramaturgical analysis the study of social interaction that compares everyday life to a theatrical presentation.

drug any substance—other than food and water—that, when taken into the body, alters its functioning in some way.

dual-earner marriages marriages in which both spouses are in the labor force.

dyad a group composed of two members.

ecclesia a religious organization that is so integrated into the dominant culture that it claims as its membership all members of a society.

economy the social institution that ensures the maintenance of society through the production, distribution, and consumption of goods and services.

education the social institution responsible for the systematic transmission of knowledge, skills, and cultural values within a formally organized structure.

egalitarian family a family structure in which both partners share power and authority equally.

ego Sigmund Freud's term for the rational, reality-oriented component of personality that imposes restrictions on the innate pleasure-seeking drives of the id.

elder abuse a term used to describe physical abuse, psychological abuse, financial exploitation, and medical abuse or neglect of people age 65 or older.

elite model a view of society that sees power in political systems as being concentrated in the hands of a small group of elites whereas the masses are relatively powerless.

endogamy the practice of marrying within one's own social group or category.

entitlements certain benefit payments made by the government.

environmental racism the belief that a disproportionate number of hazardous facilities (including industries such as waste disposal/treatment and chemical plants) are placed in low-income areas populated primarily by people of color.

ethnic group a collection of people distinguished, by others or by themselves, primarily on the basis of cultural or nationality characteristics.

ethnic pluralism the coexistence of a variety of distinct racial and ethnic groups within one society.

ethnocentrism the practice of judging all other cultures by one's own culture.

ethnography a detailed study of the life and activities of a group of people by researchers who may live with that group over a period of years.

ethnomethodology the study of the commonsense knowledge that people use to understand the situations in which they find themselves.

exogamy the practice of marrying outside one's own social group or category.

experiment a research method involving a carefully designed situation in which the researcher studies the impact of certain variables on subjects' attitudes or behavior.

experimental group in an experiment, the group that contains the subjects who are exposed to an independent variable (the experimental condition) to study its effect on them.

expressive leadership an approach to leadership that provides emotional support for members.

extended family a family unit composed of relatives in addition to parents and children who live in the same household.

faith a confident belief that cannot be proven or disproven but is accepted as true.

families relationships in which people live together with commitment, form an economic unit and care for any young, and consider their identity to be significantly attached to the group.

family of orientation the family into which a person is born and in which early socialization usually takes place.

family of procreation the family that a person forms by having or adopting children.

feminism the belief that women and men are equal and should be valued equally and have equal rights.

feminization of poverty the trend in which women are disproportionately represented among individuals living in poverty.

fertility the actual level of childbearing for an individual or a population.

field research the study of social life in its natural setting: observing and interviewing people where they live, work, and play.

folkways informal norms or everyday customs that may be violated without serious consequences within a particular culture.

formal education learning that takes place within an academic setting such as a school, which has a planned instructional process and teachers who convey specific knowledge, skills, and thinking processes to students.

formal organization a highly structured group formed for the purpose of completing certain tasks or achieving specific goals.

functional age a term used to describe observable individual attributes such as physical appearance, mobility, strength, coordination, and mental capacity that are used to assign people to age categories.

functionalist perspectives the sociological approach that views society as a stable, orderly system.

fundamentalism a traditional religious doctrine that is conservative, is typically opposed to modernity, and rejects "worldly pleasures" in favor of otherworldly spirituality.

Gemeinschaft (guh-MINE-shoft) a traditional society in which social relationships are based on personal bonds of friendship and kinship and on intergenerational stability.

gender the culturally and socially constructed differences between females and males found in the meanings, beliefs, and practices associated with "femininity" and "masculinity."

gender bias behavior that shows favoritism toward one gender over the other.

gender identity a person's perception of the self as female or male.

gender role the attitudes, behavior, and activities that are socially defined as appropriate for each sex and are learned through the socialization process.

gender socialization the aspect of socialization that contains specific messages and practices concerning the nature of being female or male in a specific group or society.

gendered racism the interactive effect of racism and sexism on the exploitation of women of color.

generalized other George Herbert Mead's term for the child's awareness of the demands and expectations of the society as a whole or of the child's subculture.

genocide the deliberate, systematic killing of an entire people or nation.

gentrification the process by which members of the middle and upper-middle classes, especially whites, move into a central-city area and renovate existing properties.

gerontology the study of aging and older people.

Gesellschaft (guh-ZELL-shoft) a large, urban society in which social bonds are based on impersonal and specialized relationships, with little long-term commitment to the group or consensus on values.

global stratification the unequal distribution of wealth, power, and prestige on a global basis, resulting in people having vastly different lifestyles and life chances both within and among the nations of the world.

goal displacement a process that occurs in organizations when the rules become an end in themselves rather than a means to an end, and organizational survival becomes more important than achievement of goals.

gossip rumors about the personal lives of individuals.

government the formal organization that has the legal and political authority to regulate the relationships among members of a society and between the society and those outside its borders.

groupthink the process by which members of a cohesive group arrive at a decision that many individual members privately believe is unwise.

Hawthorne effect a phenomenon in which changes in a subject's behavior are caused by the researcher's presence or by the subject's awareness of being studied.

health a state of complete physical, mental, and social well-being.

health care any activity intended to improve health.

health maintenance organizations (HMOs) companies that provide, for a set monthly fee, total care with an emphasis on prevention to avoid costly treatment later.

hidden curriculum the transmission of cultural values and attitudes, such as conformity and obedience to authority, through implied demands found in rules, routines, and regulations of schools.

high culture classical music, opera, ballet, live theater, and other activities usually patronized by elite audiences.

high-income countries (sometimes referred to as **industrial countries**) nations with highly industrialized economies; technologically advanced industrial, administrative, and service occupations; and relatively high levels of national and personal income.

holistic medicine an approach to health care that focuses on prevention of illness and disease and is aimed at treating the whole person—body and mind—rather than just the part or parts in which symptoms occur.

homogamy the pattern of individuals marrying those who have similar characteristics, such as race/ethnicity, religious background, age, education, or social class.

homophobia extreme prejudice and sometimes discriminatory actions directed at gays, lesbians, bisexuals, and others who are perceived as not being heterosexual.

horticultural societies societies based on technology that supports the cultivation of plants to provide food.

hospice an organization that provides a homelike facility or home-based care (or both) for people who are terminally ill.

hunting and gathering societies societies that use simple technology for hunting animals and gathering vegetation.

hypothesis in research studies, a tentative statement of the relationship between two or more concepts.

id Sigmund Freud's term for the component of personality that includes all of the individual's basic biological drives and needs that demand immediate gratification.

ideal type an abstract model that describes the recurring characteristics of some phenomenon (such as bureaucracy).

illegitimate opportunity structures circumstances that provide an opportunity for people to acquire through illegitimate activities what they cannot achieve through legitimate channels.

impression management (presentation of self) Erving Goffman's term for people's efforts to present themselves to others in ways that are most favorable to their own interests or image.

income the economic gain derived from wages, salaries, income transfers (governmental aid), and ownership of property.

independent variable a variable that is presumed to cause or determine a dependent variable.

individual discrimination behavior consisting of one-on-one acts by members of the dominant group that harm members of the subordinate group or their property.

industrial societies societies based on technology that mechanizes production.

industrialization the process by which societies are transformed from dependence on agriculture and handmade products to an emphasis on manufacturing and related industries.

infant mortality rate the number of deaths of infants under 1 year of age per 1,000 live births in a given year.

informal education learning that occurs in a spontaneous, unplanned way.

informal side of a bureaucracy those aspects of participants' day-to-day activities and interactions that ignore, bypass, or do not correspond with the official rules and procedures of the bureaucracy.

ingroup a group to which a person belongs and with which the person feels a sense of identity.

institutional discrimination the day-to-day practices of organizations and institutions that have a harmful impact on members of subordinate groups.

instrumental leadership goal- or task-oriented leadership.

intergenerational mobility the social movement experienced by family members from one generation to the next.

interlocking corporate directorates members of the board of directors of one corporation who also sit on the board(s) of other corporations.

internal colonialism according to conflict theorists, a practice that occurs when members of a racial or ethnic group are conquered or colonized and forcibly placed under the economic and political control of the dominant group.

intersexed person an individual who is born with a reproductive or sexual anatomy that does not correspond to typical definitions of male or female; in other words, the person's sexual differentiation is ambiguous.

interview a research method using a data-collection encounter in which an interviewer asks the respondent questions and records the answers.

intragenerational mobility the social movement of individuals within their own lifetime.

invasion the process by which a new category of people or type of land use arrives in an area previously occupied by another group or type of land use.

iron law of oligarchy according to Robert Michels, the tendency of bureaucracies to be ruled by a few people.

job deskilling a reduction in the proficiency needed to perform a specific job that may lead to a corresponding reduction in the wages for that job.

juvenile delinquency a violation of law or the commission of a status offense by young people.

kinship a social network of people based on common ancestry, marriage, or adoption.

labeling theory the proposition that deviance is a socially constructed process in which social control agencies designate certain people as deviants, and they, in turn, come to accept the label placed upon them and begin to act accordingly.

labor union a group of employees who join together to bargain with an employer or a group of employers over wages, benefits, and working conditions.

laissez-faire leaders leaders who are only minimally involved in decision making and who encourage group members to make their own decisions.

language a set of symbols that expresses ideas and enables people to think and communicate with one another.

latent functions unintended functions that are hidden and remain unacknowledged by participants.

laws formal, standardized norms that have been enacted by legislatures and are enforced by formal sanctions.

life chances Max Weber's term for the extent to which individuals have access to important societal resources such as food, clothing, shelter, education, and health care.

life expectancy an estimate of the average lifetime of people born in a specific year.

looking-glass self Charles Horton Cooley's term for the way in which a person's sense of self is derived from the perceptions of others.

low-income countries (sometimes referred to as **underdeveloped countries**) primarily agrarian nations with little industrialization and low levels of national and personal income.

macrolevel analysis an approach that examines whole societies, large-scale social structures, and social systems.

managed care any system of cost containment that closely monitors and controls health care providers' decisions about medical procedures, diagnostic tests, and other services that should be provided to patients.

manifest functions functions that are intended and/or overtly recognized by the participants in a social unit

marginal jobs jobs that differ from the employment norms of the society in which they are located.

marriage a legally recognized and/or socially approved arrangement between two or more individuals that carries certain rights and obligations and usually involves sexual activity.

mass a number of people who share an interest in a specific idea or issue but who are not in one another's immediate vicinity.

mass behavior collective behavior that takes place when people (who often are geographically separated from one another) respond to the same event in much the same way.

mass education the practice of providing free, public schooling for wide segments of a nation's population.

mass media organizations that use print, analog electronic, and digital electronic means to communicate with large numbers of people.

master status the most important status that a person occupies.

material culture a component of culture that consists of the physical or tangible creations that members of a society make, use, and share.

matriarchal family a family structure in which authority is held by the eldest female (usually the mother).

matriarchy a hierarchical system of social organization in which women control cultural, political, and economic structures.

matrilineal descent a system of tracing descent through the mother's side of the family.

matrilocal residence the custom of a married couple living in the same household (or community) as the wife's parents.

mechanical solidarity Emile Durkheim's term for the social cohesion of preindustrial societies, in which there is minimal division of labor and people feel united by shared values and common social bonds.

medical–industrial complex local physicians, local hospitals, and global health-related industries such as insurance companies and pharmaceutical and medical supply companies that deliver health care today.

medicalization the process whereby nonmedical problems become defined and treated as illnesses or disorders.

medicine an institutionalized system for the scientific diagnosis, treatment, and prevention of illness.

megacity a metropolitan area with a total population in excess of 10 million people.

meritocracy a hierarchy in which all positions are rewarded based on people's ability and credentials.

microlevel analysis sociological theory and research that focus on small groups rather than on large-scale social structures.

middle-income countries (sometimes referred to as **developing countries**) nations with industrializing economies and moderate levels of national and personal income.

migration the movement of people from one geographic area to another for the purpose of changing residency.

militarism a societal focus on military ideals and an aggressive preparedness for war.

military–industrial complex the mutual interdependence of the military establishment and private military contractors.

mixed economy an economic system that combines elements of a market economy (capitalism) with elements of a command economy (socialism).

mob a highly emotional crowd whose members engage in, or are ready to engage in, violence against a specific target—a person, a category of people, or physical property.

modernization theory a perspective that links global inequality to different levels of economic development and suggests that low-income economies can move to middle- and high-income economies by achieving self-sustained economic growth.

monarchy a political system in which power resides in one person or family and is passed from generation to generation through lines of inheritance.

monogamy a marriage between two partners, usually a woman and a man.

monotheism a belief in a single, supreme being or god who is responsible for significant events such as the creation of the world.

mores strongly held norms with moral and ethical connotations that may not be violated without serious consequences in a particular culture.

mortality the incidence of death in a population.

neolocal residence the custom of a married couple living in their own residence apart from both the husband's and the wife's parents.

network a web of social relationships that links one person with other people and, through them, with other people they know.

nonmaterial culture a component of culture that consists of the abstract or intangible human creations of society that influence people's behavior.

nonverbal communication the transfer of information between persons without the use of words.

norms established rules of behavior or standards of conduct.

nuclear family a family composed of one or two parents and their dependent children, all of whom live apart from other relatives.

occupational (white-collar) crime illegal activities committed by people in the course of their employment or financial affairs.

occupations categories of jobs that involve similar activities at different work sites.

official poverty line the federal income standard that is based on what is considered to be the minimum amount of money required for living at a subsistence level.

oligopoly a condition that exists when several companies overwhelmingly control an entire industry.

organic solidarity Emile Durkheim's term for the social cohesion found in industrial (and perhaps postindustrial) societies, in which people perform very specialized tasks and feel united by their mutual dependence.

organized crime a business operation that supplies illegal goods and services for profit.

outgroup a group to which a person does not belong and toward which the person may feel a sense of competitiveness or hostility.

panic a form of crowd behavior that occurs when a large number of people react to a real or perceived threat with strong emotions and self-destructive behavior.

participant observation a research method in which researchers collect data while being part of the activities of the group being studied.

pastoral societies societies based on technology that supports the domestication of large animals to provide food.

patriarchal family a family structure in which authority is held by the eldest male (usually the father).

patriarchy a hierarchical system of social organization in which men control cultural, political, and economic structures.

patrilineal descent a system of tracing descent through the father's side of the family.

patrilocal residence the custom of a married couple living in the same household (or community) as the husband's family.

pay gap a term used to describe the disparity between women's and men's earnings.

peer group a group of people who are linked by common interests, equal social position, and (usually) similar age.

peripheral nations according to world systems theory, nations that are dependent on core nations for capital, have little or no industrialization (other than what may be brought in by core nations), and have uneven patterns of urbanization.

personal space the immediate area surrounding a person that the person claims as private.

pink-collar occupations relatively low-paying, nonmanual, semiskilled positions primarily held by women.

pluralist model an analysis of political systems that views power as widely dispersed throughout many competing interest groups.

political action committees (PACs) organizations of special interest groups that solicit contributions from donors and fund campaigns to help elect (or defeat) candidates based on their stances on specific issues.

political crime illegal or unethical acts involving the usurpation of power by government officials, or illegal/unethical acts perpetrated against the government by outsiders seeking to make a political statement, undermine the government, or overthrow it.

political party an organization whose purpose is to gain and hold legitimate control of government.

political socialization the process by which people learn political attitudes, values, and behavior.

political sociology the area of sociology that examines the nature and consequences of power within or between societies, as well as the social and political conflicts that lead to changes in the allocation of power.

politics the social institution through which power is acquired and exercised by some people and groups.

polyandry the concurrent marriage of one woman with two or more men.

polygamy the concurrent marriage of a person of one sex with two or more members of the opposite sex.

polygyny the concurrent marriage of one man with two or more women.

polytheism a belief in more than one god.

popular culture activities, products, and services that are assumed to appeal primarily to members of the middle and working classes.

population composition the biological and social characteristics of a population, including age, sex, race, marital status, education, occupation, income, and size of household.

population pyramid a graphic representation of the distribution of a population by sex and age.

positivism a term describing Auguste Comte's belief that the world can best be understood through scientific inquiry.

postindustrial societies societies in which technology supports a service- and information-based economy.

postmodern perspectives the sociological approach that attempts to explain social life in modern societies that are characterized by postindustrialization, consumerism, and global communications.

power according to Max Weber, the ability of people or groups to achieve their goals despite opposition from others.

power elite C. Wright Mills's term for the group made up of leaders at the top of business, the executive branch of the federal government, and the military.

prejudice a negative attitude based on faulty generalizations about members of specific racial, ethnic, or other groups.

prestige the respect or regard with which a person or status position is regarded by others.

primary deviance the initial act of rule-breaking.

primary group Charles Horton Cooley's term for a small, less specialized group in which members engage in face-to-face, emotion-based interactions over an extended period of time.

primary labor market the sector of the labor market that consists of high-paying jobs with good benefits that have some degree of security and the possibility of future advancement.

primary sector production the sector of the economy that extracts raw materials and natural resources from the environment.

primary sex characteristics the genitalia used in the reproductive process.

probability sampling choosing participants for a study on the basis of specific characteristics, possibly including such factors as age, sex, race/ethnicity, and educational attainment.

profane the everyday, secular, or "worldly" aspects of life that we know through our senses.

professions high-status, knowledge-based occupations.

propaganda information provided by individuals or groups that have a vested interest in furthering their own cause or damaging an opposing one.

property crimes burglary (breaking into private property to commit a serious crime), motor vehicle theft, larceny-theft (theft of property worth $50 or more), and arson.

public opinion the attitudes and beliefs communicated by ordinary citizens to decision makers.

punishment any action designed to deprive a person of things of value (including liberty) because of some offense the person is thought to have committed.

questionnaire a printed research instrument containing a series of items to which subjects respond.

race a category of people who have been singled out as inferior or superior, often on the basis of real or alleged physical characteristics such as skin color, hair texture, eye shape, or other subjectively selected attributes.

racial socialization the aspect of socialization that contains specific messages and practices concerning the nature of one's racial or ethnic status.

racism a set of attitudes, beliefs, and practices that is used to justify the superior treatment of one racial or ethnic group and the inferior treatment of another racial or ethnic group.

random sampling a study approach in which every member of an entire population being studied has the same chance of being selected.

rational choice theory of deviance the proposition that deviant behavior occurs when a person weighs the costs and benefits of nonconventional or criminal behavior and determines that the benefits will outweigh the risks involved in such actions.

rational–legal authority power legitimized by law or written rules and procedures. Also referred to as *bureaucratic authority.*

rationality the process by which traditional methods of social organization, characterized by informality and spontaneity, are gradually replaced by efficiently administered formal rules and procedures.

reference group a group that strongly influences a person's behavior and social attitudes, regardless of whether that individual is an actual member.

relative poverty a condition that exists when people may be able to afford basic necessities but are still unable to maintain an average standard of living.

religion a social institution composed of a unified system of beliefs, symbols, and rituals—based on some sacred or supernatural realm—that guides human behavior, gives meaning to life, and unites believers into a community.

representative democracy a form of democracy whereby citizens elect representatives to serve as bridges between themselves and the government.

research methods specific strategies or techniques for systematically conducting research.

resocialization the process of learning a new and different set of attitudes, values, and behaviors from those in one's background and previous experience.

respondents persons who provide data for analysis through interviews or questionnaires.

riot violent crowd behavior that is fueled by deep-seated emotions but is not directed at one specific target.

rituals regularly repeated and carefully prescribed forms of behaviors that symbolize a cherished value or belief.

role a set of behavioral expectations associated with a given status.

role conflict a situation in which incompatible role demands are placed on a person by two or more statuses held at the same time.

role exit a situation in which people disengage from social roles that have been central to their self-identity.

role expectation a group's or society's definition of the way that a specific role *ought* to be played.

role performance how a person *actually* plays a role.

role strain a condition that occurs when incompatible demands are built into a single status that a person occupies.

role-taking the process by which a person mentally assumes the role of another person or group in order to understand the world from that person's or group's point of view.

routinization of charisma the process by which charismatic authority is succeeded by a bureaucracy controlled by a rationally established authority or by a combination of traditional and bureaucratic authority.

rumor an unsubstantiated report on an issue or subject.

sacred those aspects of life that exist beyond the everyday, natural world that we cannot experience with our senses.

sanctions rewards for appropriate behavior or penalties for inappropriate behavior.

Sapir–Whorf hypothesis the proposition that language shapes the view of reality of its speakers.

scapegoat a person or group that is incapable of offering resistance to the hostility or aggression of others.

second shift Arlie Hochschild's term for the domestic work that employed women perform at home after they complete their workday on the job.

secondary analysis a research method in which researchers use existing material and analyze data that were originally collected by others.

secondary deviance the process that occurs when a person who has been labeled a deviant accepts that new identity and continues the deviant behavior.

secondary group a larger, more specialized group in which members engage in more-impersonal, goal-oriented relationships for a limited period of time.

secondary labor market the sector of the labor market that consists of low-paying jobs with few benefits and very little job security or possibility for future advancement.

secondary sector production the sector of the economy that processes raw materials (from the primary sector) into finished goods.

secondary sex characteristics the physical traits (other than reproductive organs) that identify an individual's sex.

sect a relatively small religious group that has broken away from another religious organization to renew what it views as the original version of the faith.

secularization the process by which religious beliefs, practices, and institutions lose their significance in society and nonreligious values, principles, and institutions take their place.

segregation the spatial and social separation of categories of people by race, ethnicity, class, gender, and/or religion.

self-concept the totality of our beliefs and feelings about ourselves.

self-fulfilling prophecy the situation in which a false belief or prediction produces behavior that makes the originally false belief come true.

semiperipheral nations according to world systems theory, nations that are more developed than peripheral nations but less developed than core nations.

sex the biological and anatomical differences between females and males.

sex ratio the number of males for every hundred females in a given population.

sexism the subordination of one sex, usually female, based on the assumed superiority of the other sex.

sexual orientation a person's preference for emotional–sexual relationships with members of the opposite sex (heterosexuality), the same sex (homosexuality), or both (bisexuality).

shared monopoly a condition that exists when four or fewer companies supply 50 percent or more of a particular market.

sick role the set of patterned expectations that defines the norms and values appropriate for individuals who are sick and for those who interact with them.

significant others those persons whose care, affection, and approval are especially desired and who are most important in the development of the self.

simple supernaturalism the belief that supernatural forces affect people's lives either positively or negatively.

slavery an extreme form of stratification in which some people are owned by others.

small group a collectivity small enough for all members to be acquainted with one another and to interact simultaneously.

social bond theory the proposition that the probability of deviant behavior increases when a person's ties to society are weakened or broken.

social change the alteration, modification, or transformation of public policy, culture, or social institutions over time.

social construction of reality the process by which our perception of reality is largely shaped by the subjective meaning that we give to an experience.

social control systematic practices developed by social groups to encourage conformity to norms, rules, and laws and to discourage deviance.

social Darwinism Herbert Spencer's belief that those species of animals, including human beings, best adapted to their environment survive and prosper, whereas those poorly adapted die out.

social devaluation a situation in which a person or group is considered to have less social value than other individuals or groups.

social epidemiology the study of the causes and distribution of health, disease, and impairment throughout a population.

social facts Emile Durkheim's term for patterned ways of acting, thinking, and feeling that exist *outside* any one individual but that exert social control over each person.

social group a group that consists of two or more people who interact frequently and share a common identity and a feeling of interdependence.

social institution a set of organized beliefs and rules that establishes how a society will attempt to meet its basic social needs.

social interaction the process by which people act toward or respond to other people.

social mobility the movement of individuals or groups from one level in a stratification system to another.

social movement an organized group that acts consciously to promote or resist change through collective action.

social stratification the hierarchical arrangement of large social groups based on their control over basic resources.

social structure the complex framework of societal institutions (such as the economy, politics, and religion) and the social practices (such as rules and social roles) that make up a society and that organize and establish limits on people's behavior.

socialism an economic system characterized by public ownership of the means of production, the pursuit of collective goals, and centralized decision making.

socialization the lifelong process of social interaction through which individuals acquire a self-identity and the physical, mental, and social skills needed for survival in society.

socialized medicine a health care system in which the government owns the medical care facilities and employs the physicians.

society a large social grouping that shares the same geographical territory and is subject to the same political authority and dominant cultural expectations.

sociobiology the systematic study of how biology affects social behavior.

socioeconomic status (SES) a combined measure that, in order to determine class location, attempts to classify individuals, families, or households in terms of factors such as income, occupation, and education.

sociological imagination C. Wright Mills's term for the ability to see the relationship between individual experiences and the larger society.

sociology the systematic study of human society and social interaction.

sociology of family the subdiscipline of sociology that attempts to describe and explain patterns of family life and variations in family structure.

special interest groups political coalitions composed of individuals or groups that share a specific interest that they wish to protect or advance with the help of the political system.

spirituality the relationship between the individual and something larger than oneself, such as a broader sense of connection with the surrounding world.

split labor market a term used to describe the division of the economy into two areas of employment: a primary sector or upper tier, composed of higher-paid (usually dominant-group) workers in more-secure jobs, and a secondary sector or lower tier, composed of lower-paid (often subordinate-group) workers in jobs with little security and hazardous working conditions.

state the political entity that possesses a legitimate monopoly over the use of force within its territory to achieve its goals.

status a socially defined position in a group or society characterized by certain expectations, rights, and duties.

status symbol a material sign that informs others of a person's specific status.

stereotypes overgeneralizations about the appearance, behavior, or other characteristics of members of particular categories.

strain theory the proposition that people feel strain when they are exposed to cultural goals that they are unable to obtain because they do not have access to culturally approved means of achieving those goals.

subcontracting an agreement in which a corporation contracts with other (usually smaller) firms to provide specialized components, products, or services to the larger corporation.

subculture a category of people who share distinguishing attributes, beliefs, values, and/or norms that set them apart in some significant manner from the dominant culture.

subordinate group a group whose members, because of physical or cultural characteristics, are disadvantaged and subjected to unequal treatment by the dominant group and who regard themselves as objects of collective discrimination.

succession the process by which a new category of people or type of land use gradually predominates in an area formerly dominated by another group or activity.

superego Sigmund Freud's term for the conscience, consisting of the moral and ethical aspects of personality.

survey a poll in which the researcher gathers facts or attempts to determine the relationships among facts.

symbol anything that meaningfully represents something else.

symbolic interactionist perspectives the sociological approach that views society as the sum of the interactions of individuals and groups.

taboos mores so strong that their violation is considered to be extremely offensive and even unmentionable.

technology the knowledge, techniques, and tools that allow people to transform resources into a usable form and the knowledge and skills required to use what is developed.

terrorism the calculated, unlawful use of physical force or threats of violence against persons or property in order to intimidate or coerce a government, organization, or individual for the purpose of gaining some political, religious, economic, or social objective.

tertiary deviance deviance that occurs when a person who has been labeled a deviant seeks to normalize the behavior by relabeling it as nondeviant.

tertiary sector production the sector of the economy that is involved in the provision of services rather than goods.

theism a belief in a god or gods who shape human affairs.

theory a set of logically interrelated statements that attempts to describe, explain, and (occasionally) predict social events.

theory of racial formation the idea that actions of the government substantially define racial and ethnic relations in the United States.

total institution Erving Goffman's term for a place where people are isolated from the rest of society for a set period of time and come under the control of the officials who run the institution.

totalitarianism a political system in which the state seeks to regulate all aspects of people's public and private lives.

tracking the assignment of students to specific curriculum groups and courses on the basis of their test scores, previous grades, or other criteria.

traditional authority power that is legitimized on the basis of long-standing custom.

transcendent idealism a belief that each individual has the freedom to function as an autonomous source in regard to principles of thought and conduct.

transgendered person an individual whose gender identity (self-identification as woman, man, neither, or both) does not match the person's assigned sex (identification by others as male, female, or intersex based on physical/genetic sex).

transnational corporations large corporations that are headquartered in one or a few countries but sell and produce goods and services in many countries.

triad a group composed of three members.

underclass those who are poor, seldom employed, and caught in long-term deprivation that results from low levels of education and income and high rates of unemployment.

unemployment rate the percentage of unemployed persons in the labor force actively seeking jobs.

universal health care a health care system in which all citizens receive medical services paid for by tax revenues.

unstructured interview an extended, open-ended interaction between an interviewer and an interviewee.

urbanization the process by which an increasing proportion of a population lives in cities rather than in rural areas.

value contradictions values that conflict with one another or are mutually exclusive.

values collective ideas about what is right or wrong, good or bad, and desirable or undesirable in a particular culture.

victimless crimes crimes involving a willing exchange of illegal goods or services among adults.

violent crime actions—murder, forcible rape, robbery, and aggravated assault—involving force or the threat of force against others.

war organized, armed conflict between nations or distinct political factions.

wealth the value of all of a person's or family's economic assets, including income, personal property, and income-producing property.

welfare state a state in which there is extensive government action to provide support and services to the citizens.

working class (or **proletariat**) those who must sell their labor to the owners in order to earn enough money to survive.

zero population growth the point at which no population increase occurs from year to year.

references

AAUW (American Association of University Women). 1995. *How Schools Shortchange Girls/ The AAUW Report: A Study of Major Findings on Girls and Education.* New York: Marlowe.

—. 2008. "Where the Girls Are: The Facts About Gender Equity in Education." Retrieved Apr. 2, 2010. Online: www.aauw.org/research/upload/whereGirlsARe.pdf

Abaya, Carol. 2011. "Welcome to the Sandwich Generation." Retrieved Mar. 20 2011. Online: www.sandwichgeneration.com

ABC Family. 2010. "Greek: Recaps." Retrieved Dec. 22, 2010. Online: www.abcfamily.go.com/shows/greek/recaps

ABC.net.au. 2008. "Community Urged to End Homelessness Stigma." Retrieved Feb. 23, 2010. Online: www.abc.net.au/news/stories/2008/05/22/2252087.htm

ABCNews.com. 2006. "Kids Getting 'Thinspiration' from Dangerously Skinny Stars." *Good Morning America* (July 12). Retrieved June 15, 2011. Online: http://abcnews.go.com/GMA/Diet/story?id=2182068&page=1

Aberle, D. F., A. K. Cohen, A. K. Davis, M. J. Leng, Jr., and F. N. Sutton. 1950. "The Functional Prerequisites of Society." *Ethics,* 60 (January): 100–111.

Adler, Jerry. 1999. "The Truth About High School." *Newsweek* (May 10): 56–58.

Adler, Patricia A., and Peter Adler. 1998. *Peer Power: Preadolescent Culture and Identity.* New Brunswick, NJ: Rutgers University Press.

—. 2003. *Constructions of Deviance: Social Power, Context, and Interaction* (4th ed.). Belmont, CA: Wadsworth.

Adorno, Theodor W., Else Frenkel-Brunswick, Daniel J. Levinson, and R. Nevitt Sanford. 1950. *The Authoritarian Personality.* New York: Harper & Row.

Agency for Healthcare Research and Quality. 2000. "Hospitalization in the United States, 1997." Retrieved Aug. 7, 2003. Online: www.ahcpr.gov/data/hcup/factbk1/hcupfbk1.pdf

Agger, Ben. 1993. *Gender, Culture, and Power: Toward a Feminist Postmodern Critical Theory.* Westport, CT: Praeger.

Agnew, Robert. 1985. "Social Control Theory and Delinquency: A Longitudinal Test." *Criminology,* 23: 47–61.

Agonafir, Rebecca. 2002. "Workplace Surveillance of Internet and E-mail Usage." Retrieved July 6, 2002. Online: http://firstclass.wellesley.edu/~ragonafi/cs100.rp1.html

AGS Foundation for Health in Aging. 2007. "Aging in the Know: Nursing Home Care." Retrieved Mar. 29, 2007. Online: www.healthinaging.org/agingintheknow/chapters_ch_trial.asp?ch=15

Aguirre, B. E., Manuel R. Torres, Kimberly B. Gill, and H. Lawrence Hotchkiss. 2011. "Normative Collective Behavior in the Station Building Fire." *Social Science Quarterly* (Mar.): 100–118.

Aiello, John R., and S. E. Jones. 1971. "Field Study of Proxemic Behavior of Young School Children in Three Subcultural Groups." *Journal of Personality and Social Psychology,* 19: 351–356.

Akers, Ronald L. 1998. *Social Learning and Social Structure: A General Theory of Crime and Deviance.* Boston: Northeastern University Press.

Albanese, Catherine L. 2007. *America: Religions and Religion* (4th ed.). Belmont, CA: Wadsworth.

Albas, Daniel, and Cheryl Albas. 1988. "Aces and Bombers: The Post-Exam Impression Management Strategies of Students." *Symbolic Interaction,* 11 (Fall): 289–302.

Albrecht, Gary L., Katherine D. Seelman, and Michael Bury (Eds.). 2001. *The Handbook of Disability Studies.* Thousand Oaks, CA: Sage.

Alden, William. 2010. "Goldman Sachs Sued for Sex Discrimination." Huffingtonpost.com (Sept. 15). Retrieved Mar. 11, 2011. Online: www.huffingtonpost.com/2010/09/15/goldman-sachs-sex-discrimination_n_717938.html

Allport, Gordon. 1958. *The Nature of Prejudice* (abridged ed.). New York: Doubleday/Anchor.

Altemeyer, Bob. 1981. *Right-Wing Authoritarianism.* Winnipeg: University of Manitoba Press.

—. 1988. *Enemies of Freedom: Understanding Right-Wing Authoritarianism.* San Francisco: Jossey-Bass.

Alwin, Duane, Philip Converse, and Steven Martin. 1985. "Living Arrangements and Social Integration." *Journal of Marriage and the Family,* 47: 319–334.

Alzheimer's Association. 2011. "2011 Alzheimer's Disease Facts and Figures." Retrieved Mar. 19, 2010. Online: www.alz.org/downloads/Facts_Figures_2011.pdf

Amelie, Mdmse. 2010. "Tips for Surviving Your First Day of Higher Education." Retrieved Jan. 27, 2010. Online: www.helium.com/items/1094810-first-day-college-first-day-university-higher-education-success-good-impression

American Academy of Child and Adolescent Psychiatry. 1997. "Children and Watching TV." Retrieved June 22, 1999. Online: www.aacap.org/factsfam/tv.htm

American Anthropological Association. 2007. "Race: Sports Quiz—'White Men Can't Jump' and Other Assumptions About Sports and Race." Retrieved Mar. 7, 2011. Online: www.understandingrace.org/lived/sports/index.html

American Association of Community Colleges. 2011. "2011 Fact Sheet." Retrieved Apr. 18, 2011. Online: www.aacc.nche.edu/AboutCC/Documents/FactSheet2011.pdf

American Association of Suicidology. 2009. "Suicide in the U.S.A. Based on Current (2006) Statistics." Retrieved Jan. 24, 2010. Online: www.suicidology.org/web/guest/stats-and-tools/fact-sheets

American Bar Association, 2010. "Privacy: How Will You Maintain Your Privacy?" Retrieved Nov. 23, 2010. Online: www.safeshopping.org/privacy.shtml.

American Management Association. 2008. "2007 Electronic Monitoring and Surveillance Survey." Retrieved Feb. 7, 2011. Online: http://press.amanet.org/press-releases/177/2007-electronic-monitoring-surveillance-survey

American Psychiatric Association. 1994. *Diagnostic and Statistical Manual of Mental Disorders IV.* Washington, DC: American Psychiatric Association.

American Sociological Association. 1997. *Code of Ethics.* Washington, DC: American Sociological Association (orig. pub. 1971).

Amira, Dan. 2011. "This Week in Fake Presidential Candidates." *New York* (Apr. 8). Retrieved Mar. 11, 2011. Online: http://nymag.com/daily/intel/2011/04/fake_candidates.html

Amott, Teresa, and Julie Matthaei. 1996. *Race, Gender, and Work: A Multicultural Economic History of Women in the United States* (rev. ed.). Boston: South End.

Andersen, Margaret L., and Patricia Hill Collins (Eds.). 1998. *Race, Class, and Gender: An Anthology* (3rd ed.). Belmont, CA: Wadsworth.

Anderson, Elijah. 1990. *Streetwise: Race, Class, and Change in an Urban Community.* Chicago: University of Chicago Press.

Angel, Ronald J., and Jacqueline L. Angel. 1993. *Painful Inheritance: Health and the New Generation of Fatherless Families.* Madison: University of Wisconsin Press.

Angelotti, Amanda. 2006. "Confessions of a Beauty Pageant Drop-Out." *Campus Progress: A Project of Center for American Progress* (Jan. 25). Retrieved Jan. 27, 2008. Online: www.campusprogress.org/features/727/confessions-of-a-beauty-pageant-drop-out

Angier, Natalie. 1993. "'Stopit!' She Said. 'Nomore!'" *New York Times Book Review* (Apr. 25): 12.

APA Online. 2000. "Psychiatric Effects of Violence." Public Information: APA Fact Sheet Series. Washington, DC: American Psychological Association. Retrieved Apr. 5, 2000. Online: www.psych.org/psych/htdocs/public_info/media_violence.html

Appleton, Lynn M. 1995. "The Gender Regimes in American Cities." In Judith A. Garber and Robyne S. Turner (Eds.), *Gender in Urban Research.* Thousand Oaks, CA: Sage, pp. 44–59.

Archibold, Randal C. 2010. "Arizona Enacts Stringent Law on Immigration." *New York Times* (Apr. 23): A1.

Arendt, Hannah. 1973. *On Revolution.* London: Penguin.

Arenson, Karen W. 2004. "Suicide of N.Y.U. Student, 19, Brings Sadness and Questions." *New York Times* (Mar. 10). Retrieved Jan. 8, 2006. Online: http://query.nytimes.com/gst/fullpage.html?sec=health&res=9904EFDF153EF933A25750C0A9629C8B63

Arnold, Regina A. 1990. "Processes of Victimization and Criminalization of Black Women." *Social Justice,* 17 (3): 153–166.

Asch, Adrienne. 2001. "Critical Race Theory, Feminism, and Disability: Reflections on Social Justice and Personal Identity." *Ohio State Law Journal* (62). Retrieved May 14, 2011. Online: http://moritzlaw.osu.edu/lawjournal/issues/volume62/number1/asch.pdf

Asch, Solomon E. 1955. "Opinions and Social Pressure." *Scientific American,* 193 (5): 31–35.

—. 1956. "Studies of Independence and Conformity: A Minority of One Against a Unanimous Majority." *Psychological Monographs,* 70(9) (Whole No. 416).

Ashe, Arthur R., Jr. 1988. *A Hard Road to Glory: A History of the African-American Athlete.* New York: Warner.

Aulette, Judy Root. 1994. *Changing Families.* Belmont, CA: Wadsworth.

Axtell, Roger E. 1991. *Gestures: The Do's and Taboos of Body Language Around the World.* New York: Wiley.
Aydelotte, Becky. 2010. "The Normal Paranormal." *Baylor Arts & Sciences* (Fall): 6–7.

Babbie, Earl. 2010. *The Practice of Social Research* (12th ed.). Belmont, CA: Cengage/Wadsworth.
Bader, Christopher, F. Carson Mencken, and Joseph Baker. 2010. *Paranormal America: Ghost Encounters, UFO Sightings, Bigfoot Hunts, and Other Curiosities in Religion and Culture.* New York: NYU Press.
Baker, Robert. 1993. "'Pricks' and 'Chicks': A Plea for 'Persons.'" In Anne Minas (Ed.), *Gender Basics: Feminist Perspectives on Women and Men.* Belmont, CA: Wadsworth, pp. 66–68.
Ballantine, Jeanne H., and Floyd M. Hammack. 2009. *The Sociology of Education: A Systematic Analysis* (6th ed.). Englewood Cliffs, NJ: Prentice Hall.
—. 2012. *The Sociology of Education: A Systematic Analysis* (7th ed.). Upper Saddle River, NJ: Prentice-Hall/Pearson.
Banet-Weiser, Sarah. 1999. *The Most Beautiful Girl in the World: Beauty Pageants and National Identity.* Berkeley: University of California Press.
Banner, Lois W. 1993. *In Full Flower: Aging Women, Power, and Sexuality.* New York: Vintage.
Barber, Benjamin R. 1996. *Jihad vs. McWorld: How Globalism and Tribalism Are Reshaping the World.* New York: Ballantine.
Barber, Brian. 2010. "Spartan College Students to Help Erase Graffiti." *Tulsa World* (July 1). Retrieved Feb. 9, 2011. Online: www.tulsaworld.com/news/article.aspx?subjectid=334&articleid=20100701_11_A12_Thomas624356&rss_lnk=11
Bardwell, Jill R., Samuel W. Cochran, and Sharon Walker. 1986. "Relationship of Parental Education, Race, and Gender to Sex Role Stereotyping in Five-Year-Old Kindergarteners." *Sex Roles,* 15: 275–281.
Barlow, Hugh D., and David Kauzlarich. 2002. *Introduction to Criminology* (8th ed.). Upper Saddle River, NJ: Prentice Hall.
Barna, George. 1996. *Index of Leading Spiritual Indicators.* Dallas: Word.
Barnard, Chester. 1938. *The Functions of the Executive.* Cambridge, MA: Harvard University Press.
Barnett, Harold. 1979. "Wealth, Crime and Capital Accumulation." *Contemporary Crises,* 3: 171–186.
Baron, Dennis. 1986. *Grammar and Gender.* New Haven, CT: Yale University Press.
Barovick, Harriet. 2001. "Hope in the Heartland." *Time* (special edition, July): G1–G3.
Basken, Paul. 2010. "Final Student-Loan Bill Offers Aid to Colleges and Students." *Chronicle of Higher Education* (Mar. 19). Retrieved Apr. 5, 2010. Online: http://chronicle.com/article/Final-Student-Loan-Bill-Offers/64769
Basow, Susan A. 1992. *Gender Stereotypes and Roles* (3rd ed.). Pacific Grove, CA: Brooks/Cole.
Baudrillard, Jean. 1983. *Simulations.* New York: Semiotext.
—. 1998. *The Consumer Society: Myths and Structures.* London: Sage (orig. pub. 1970).
Baxter, J. 1970. "Interpersonal Spacing in Natural Settings." *Sociology,* 36 (3): 444–456.
BBC News. 2005. "Forced Labour: Global Problem." *Slavery in the 21st Century.* Retrieved Mar. 14, 2010. Online: http://news.bbc.co.uk/2/shared/spl/hi/world/05/slavery/html/1.stm
Beard, Stephen. 2011. "Jordan's Elite Stifles Economic Opportunity." National Public Radio: Marketplace (Feb. 9). Retrieved May 19, 2011. Online: http://marketplace.publicradio.org/display/web/2011/02/10/pm-poverty-fuels-protests-in-the-middle-east
Becker, Bernie. 2010. "A Revised Contract for America, Minus 'with' and Newt." *New York Times* (Apr. 15): A17.
Becker, Howard S. 1963. *Outsiders: Studies in the Sociology of Deviance.* New York: Free Press.
Belkin, Lisa. 2006. "Life's Work: The Best Part Comes in the Third Act." *New York Times* (July 2). Retrieved Mar. 25, 2007. Online: http://select.nytimes.com/search/restricted/article?res=FB0E10F83E540C718CDDAE0894DE404482
Bell, Inge Powell. 1989. "The Double Standard: Age." In Jo Freeman (Ed.), *Women: A Feminist Perspective* (4th ed.). Mountain View, CA: Mayfield, pp. 236–244.
Bellah, Robert N. 1967. "Civil Religion." *Daedalus,* 96: 1–21.
Benford, Robert D. 1993. "'You Could Be the Hundredth Monkey': Collective Action Frames and Vocabularies of Motive Within the Nuclear Disarmament Movement." *Sociological Quarterly,* 34: 195–216.
Bengston, Vern L., Elisabeth O. Burgess, and Tonya M. Parrott. 1997. "Theory, Explanation, and a Third Generation of Theoretical Development in Social Gerontology." *Journal of Gerontology* (52B): 572–588.
Benjamin, Lois. 1991. *The Black Elite: Facing the Color Line in the Twilight of the Twentieth Century.* Chicago: Nelson-Hall.
Bennahum, David S. 1999. "For Kosovars, an On-Line Phone Directory of a People in Exile." *New York Times* (July 15): D7.
Berg, Bruce L. 1998. *Qualitative Research Methods for the Social Sciences.* Boston: Allyn & Bacon.
Berger, Peter. 1963. *Invitation to Sociology: A Humanistic Perspective.* New York: Anchor.
—. 1967. *The Sacred Canopy: Elements of a Sociological Theory of Religion.* New York: Doubleday.
Berger, Peter, and Hansfried Kellner. 1964. "Marriage and the Construction of Reality." *Diogenes,* 46: 1–32.
Berger, Peter, and Thomas Luckmann. 1967. *The Social Construction of Reality: A Treatise in the Sociology of Knowledge.* Garden City, NY: Anchor.
Bernard, Jessie. 1982. *The Future of Marriage.* New Haven, CT: Yale University Press (orig. pub. 1973).
Better Health Channel. 2007. "Food and Celebrations." Retrieved Feb. 11, 2007. Online: www.betterhealth.vic.gov.au
Biblarz, Arturo, R. Michael Brown, Dolores Noonan Biblarz, Mary Pilgram, and Brent F. Baldree. 1991. "Media Influence on Attitudes Toward Suicide." *Suicide and Life-Threatening Behavior,* 21 (4): 374–385.
Bishaw, Alemajehu, and Trudi J. Renwick. 2009. *Poverty: 2007 and 2008, American Community Survey.* Retrieved May 18, 2011. Online: www.census.gov/prod/2009pubs/acsbr08-1.pdf
Blau, Peter M., and Otis Dudley Duncan. 1967. *The American Occupational Structure.* New York: Wiley.
Blau, Peter M., and Marshall W. Meyer. 1987. *Bureaucracy in Modern Society* (3rd ed.). New York: Random House.
Blauner, Robert. 1972. *Racial Oppression in America.* New York: Harper & Row.
Block, Fred, Anna C. Korteweg, and Kerry Woodward, with Zach Schiller and Imrul Mazid. 2008. "The Compassion Gap in American Poverty Policy." In Jeff Goodwin and James M. Jasper (Eds.), *The Contexts Reader.* New York: Norton, pp. 166–175. Originally appeared in *Contexts,* a journal published by the American Sociological Association (Spring 2006).
Blumer, Herbert G. 1946. "Collective Behavior." In Alfred McClung Lee (Ed.), *A New Outline of the Principles of Sociology.* New York: Barnes & Noble, pp. 167–219.
—. 1969. *Symbolic Interactionism: Perspective and Method.* Englewood Cliffs, NJ: Prentice Hall.
—. 1974. "Social Movements." In R. Serge Denisoff (Ed.), *The Sociology of Dissent.* New York: Harcourt, pp. 74–90.
—. 1986. *Symbolic Interactionism: Perspective and Method.* Berkeley: University of California Press (orig. pub. 1969).
Bodeen, Christopher. 2007. "China's Woes Vindicate Whistle-Blower, But Court Case Against Him Continues." *International Business Times* (June 9). Retrieved Mar. 8, 2008. Online: www.ibtimes.com/articles/20070609/china-whistleblower-wu-lihong_all.htm
Bonacich, Edna. 1972. "A Theory of Ethnic Antagonism: The Split Labor Market." *American Sociological Review,* 37: 547–549.
—. 1976. "Advanced Capitalism and Black–White Relations in the United States: A Split Labor Market Interpretation." *American Sociological Review,* 41: 34–51.
Bonanno, George A. 2009. "Grief Does Not Come in Stages and It's Not the Same for Everyone." *Psychology Today* (Oct. 26). Retrieved Mar. 25, 2011. Online: www.psychologytoday.com/blog/thriving-in-the-face-trauma/200910/grief-does-not-come-in-stages-and-its-not-the-same-everyone
—. 2010. *The Other Side of Sadness: What the New Science of Bereavement Tells Us About Life After Loss.* New York: Basic.
Bonvillain, Nancy. 2001. *Women & Men: Cultural Constructs of Gender* (3rd ed.). Upper Saddle River, NJ: Prentice Hall.
Bordo, Susan. 2004. *Unbearable Weight: Feminism, Western Culture, and the Body* (10th anniversary edition). Berkeley: University of California Press.
Bourdieu, Pierre. 1984. *Distinction: A Social Critique of the Judgement of Taste.* Trans. Richard Nice. Cambridge, MA: Harvard University Press.
Bourdieu, Pierre, and Jean-Claude Passeron. 1990. *Reproduction in Education, Society and Culture.* Newbury Park, CA: Sage.
Bourdon, Karen H., Donald S. Rae, Ben Z. Locke, William E. Narrow, and Darrel A. Regier. 1992. "Estimating the Prevalence of Mental Disorders in U.S. Adults from the Epidemiological Catchment Area Survey." *Public Health Reports,* 107: 663–668.
Bowles, Samuel, and Herbert Gintis. 1976. *Schooling in Capitalist America: Education and the Contradictions of Economic Life.* New York: Basic.
Boyes, William, and Michael Melvin. 2011. *Economics* (8th ed.). Stamford, CT: Cengage.
Bramlett, Matthew D., and William D. Mosher. 2001. "First Marriage Dissolution, Divorce, and Remarriage: United States." DHHS publication no. 2001–1250 01–0384 (5/01). Hyattsville, MD: Department of Health and Human Services.
Brandl, Bonnie, and Loree Cook-Daniels. 2002. "Domestic Abuse in Later Life." Retrieved July 26, 2003. Online: www.elderabusecenter.org/pdf/research/abusers.pdf
Brault, Matthew W. 2008. "Americans with Disabilities: 2005." U.S. Census Bureau. Retrieved May 15, 2011. Online: www.census.gov/prod/2008pubs/p70-117.pdf
Braun, Ralph. 2009. "Disabled Workers: Employer Fears Are Groundless." *Bloomberg Businessweek* (Oct. 2). Retrieved Apr. 3, 2011. Online: www.businessweek.com/managing/content/oct2009/ca2009102_029034.htm
Breault, K. D. 1986. "Suicide in America: A Test of Durkheim's Theory of Religious and Family

Integration, 1933–1980." *American Journal of Sociology,* 92 (3): 628–656.

Broder, John M. 2003. "Debris Is Now Leading Suspect in Shuttle Catastrophe." *New York Times* (Feb. 4): A1–A25.

Brody, Alan. 2002. "Students Say Hollywood Fails to Portray College Realistically." Clarksonintegrator.com (Mar. 4). Retrieved Dec. 21, 2010. Online: http://media.www.clarksonintegrator.com/media/storage/paper280/news/2002/03/04/Opinion/Students.Say.Hollywood.Fails.To.Portray.College.Realistically-204729.shtml

Brooks, Robert A. 2011. *Cheaper by the Hour: Temporary Lawyers and the Deprofessionalization of the Law.* Philadelphia: Temple University Press.

Brooks-Gunn, Jeanne. 1986. "The Relationship of Maternal Beliefs About Sex Typing to Maternal and Young Children's Behavior." *Sex Roles,* 14: 21–35.

Brown, Dennis M. 2003. "Rural Tourism: An Annotated Bibliography." United States Department of Agriculture. Retrieved Aug. 9, 2003. Online: www.nal.usda.gov/ric/ricpubs/rural_tourism.html#summary

Brown, E. Richard. 1979. *Rockefeller Medicine Men.* Berkeley: University of California Press.

Brown, Phil. 1985. *The Transfer of Care: Psychiatric Deinstitutionalization and Its Aftermath.* Boston: Routledge & Kegan Paul.

Brown, Robbie. 2011. "Georgia Gives Police Added Power to Seek Out Illegal Immigrants." *New York Times* (May 13). Retrieved May 15, 2011. Online: www.nytimes.com/2011/05/14/us/14georgia.html

Brown, Robert W. 1954. "Mass Phenomena." In Gardner Lindzey (Ed.), *Handbook of Social Psychology* (vol. 2). Reading, MA: Addison-Wesley, pp. 833–873.

Brustad, Robert J. 1996. "Attraction to Physical Activity in Urban Schoolchildren: Parental Socialization and Gender Influence." *Research Quarterly for Exercise and Sport,* 67: 316–324.

Buechler, Steven M. 2000. *Social Movements in Advanced Capitalism: The Political Economy and Cultural Construction of Social Activism.* New York: Oxford University Press.

Bullard, Robert B., and Beverly H. Wright. 1992. "The Quest for Environmental Equity: Mobilizing the African-American Community for Social Change." In Riley E. Dunlap and Angela G. Mertig (Eds.), *American Environmentalism: The U.S. Environmental Movement, 1970–1990.* New York: Taylor & Francis, pp. 39–49.

Burgess, Ernest W. 1925. "The Growth of the City." In Robert E. Park and Ernest W. Burgess (Eds.), *The City.* Chicago: University of Chicago Press, pp. 47–62.

Burnham, Walter Dean. 1983. *Democracy in the Making: American Government and Politics.* Englewood Cliffs, NJ: Prentice Hall.

Burros, Marian. 1994. "Despite Awareness of Risks, More in U.S. Are Getting Fat." *New York Times* (July 17): 1, 8.

—. 1996. "Eating Well: A Law to Encourage Sharing in a Land of Plenty." *New York Times* (Dec. 11): B6.

Buvinić, Mayra. 1997. "Women in Poverty: A New Global Underclass." *Foreign Policy* (Fall): 38–53.

Calmes, Jackie. 2011. "In Border City Talk, Obama Urges G.O.P. to Help Overhaul Immigration Law." *New York Times* (May 10): A1.

Campaign for Disability Employment. 2011. "What Can Employers Do?" Retrieved July 8, 2011. Online: www.whatcanyoudocampaign.org/blog/index.php/what-can-employers-do

Campbell, Anne. 1984. *The Girls in the Gang* (2nd ed.). Cambridge, MA: Basil Blackwell.

Campus Times. 2008. "Survey Analyzes Students' Stress Levels" (Aug. 10). Retrieved Feb. 8, 2010. Online: www.campustimes.org/2.4981/survey-analyzes-students-stress-levels-1.492369

Canadian Broadcast Corporation. 2011. "Global Disaster Impact Rises in 2010" (May 10). *CBC News.* Retrieved May 11, 2011. Online: www.cbc.ca/news/world/story/2011/05/10/disaster-report.html?sms_ss=facebook&at_xt=4dcac2ba4422a2a7,0

Cancian, Francesca M. 1990. "The Feminization of Love." In C. Carlson (Ed.), *Perspectives on the Family: History, Class, and Feminism.* Belmont, CA: Wadsworth, pp. 171–185.

—. 1992. "Feminist Science: Methodologies That Challenge Inequality." *Gender & Society,* 6 (4): 623–642.

Canetto, Silvia Sara. 1992. "She Died for Love and He for Glory: Gender Myths of Suicidal Behavior." *OMEGA,* 26 (1): 1–17.

Canter, R. J., and S. S. Ageton. 1984. "The Epidemiology of Adolescent Sex-Role Attitudes." *Sex Roles,* 11: 657–676.

Cantor, Muriel G. 1980. *Prime-Time Television: Content and Control.* Newbury Park, CA: Sage.

—. 1987. "Popular Culture and the Portrayal of Women: Content and Control." In Beth B. Hess and Myra Marx Ferree (Eds.), *Analyzing Gender: A Handbook of Social Science Research.* Newbury Park, CA: Sage, pp. 190–214.

Cantril, Hadley. 1941. *The Psychology of Social Movements.* New York: Wiley.

Capek, Stella M. 1993. "The 'Environmental Justice' Frame: A Conceptual Discussion and Application." *Social Problems,* 40 (1): 5–23.

Cargan, Leonard, and Matthew Melko. 1982. *Singles: Myths and Realities.* Newbury Park, CA: Sage.

Carson, Rachel. 1962. *Silent Spring.* Boston: Houghton Mifflin.

Carter, Stephen L. 1994. *The Culture of Disbelief: How American Law and Politics Trivializes Religious Devotion.* New York: Anchor/Doubleday.

Cashmore, E. Ellis. 1996. *Dictionary of Race and Ethnic Relations* (4th ed.). London: Routledge.

Castells, Manuel. 1977. *The Urban Question.* London: Edward Arnold (orig. pub. 1972 as *La Question Urbaine,* Paris).

—. 1998. *End of Millennium.* Malden, MA: Blackwell.

Cattaneo, Olivier, Gary Gereffi, and Cornelia Staritz. 2010. "Global Value Chains in a Postcrisis World: Reliance, Consolidation, and Shifting End Markets." Washington, DC: World Bank. Retrieved Mar. 6, 2011. Online: www.cggc.duke.edu/pdfs/Gereffi_GVCs_in_the_Postcrisis_World_Book.pdf

Cavender, Gray. 1995. "Alternative Theory: Labeling and Critical Perspectives." In Joseph F. Sheley (Ed.), *Criminology: A Contemporary Handbook* (2nd ed.). Belmont, CA: Wadsworth, pp. 349–371.

Cawthorne, Aleandra. 2008. "The Straight Facts on Women in Poverty." Center for American Progress (Oct. 8). Retrieved Feb. 21, 2011. Online: www.americanprogress.org/issues/2008/10/women_poverty.html

CBS News. 2007. "Outsourced 'Wombs-for-Rent' in India." Retrieved Feb. 9, 2008. Online: www.cbsnews.com/stories/2007/12/31/health/main3658750.shtml

CDC Morbidity and Mortality Weekly Report. 2011. "CDC Health Disparities and Inequalities Report—United States, 2011." Retrieved May 23, 2011. Online: www.cdc.gov/mmwr/pdf/other/su6001.pdf

census.gov. 2008. "Public Education Finances: 2006." Retrieved Aug. 26, 2009. Online: www2.census.gov/govs/school/06f33pub.pdf

—. 2010. "Median Household Income by State, 2009." Retrieved May 18, 2011. Online: www.census.gov/hhes.www/income/data/statemedian/index.html

—. 2010a. "Census Bureau Reports Hispanic-Owned Businesses Increase at More Than Double the National Rate" (Sept. 21). Retrieved Feb. 20, 2011. Online: www.census.gov/newsroom/releases/archives/business_ownership/cb10-145.html

—. 2010b. "Census Bureau Reports Women-Owned Firms Numbered 7.8 Million in 2007, Generated Receipts of $1.2 Trillion" (Dec. 7). Retrieved Feb. 20, 2011. Online: www.census.gov/newsroom/releases/archives/business_ownership/cb10-184.html

—. 2011. "Census Bureau Reports the Number of Black-Owned Business Increased at Triple the National Rate" (Feb. 8). Retrieved Feb. 20, 2011. Online: www.census.gov/newsroom/releases/archives/business_ownership/cb11-24.html

Center on Budget and Policy Priorities. 2010. "Policy Basics: Top Ten Facts About Social Security on the Program's 75th Anniversary" (Aug.). Retrieved Mar. 21, 2011. Online: www.cbpp.org/cms/index.cfm?fa=view&id=3261

Central Intelligence Agency. 2011. *World Factbook.* Retrieved May 12, 2011. Online: www.cia.gov/library/publications/the-world-factbook/rankorder/2091rank.html

Chafetz, Janet Saltzman. 1984. *Sex and Advantage: A Comparative, Macro-Structural Theory of Sex Stratification.* Totowa, NJ: Rowman & Allanheld.

Chagnon, Napoleon A. 1992. *Yanomamo: The Last Days of Eden.* New York: Harcourt (rev. from 4th ed., *Yanomamo: The Fierce People,* published by Holt, Rinehart & Winston).

Chalfant, H. Paul, Robert E. Beckley, and C. Eddie Palmer. 1994. *Religion in Contemporary Society* (3rd ed.). Itasca, IL: Peacock.

Chambliss, William J. 1973. "The Saints and the Roughnecks." *Society,* 11: 24–31.

channelnewsasia.com. 2011. "Medical Crisis in Japan's Evacuation Shelters." Retrieved May 11, 2011. Online: www.channelnewsasia.com/stories/afp_asiapacific/view/1117487/1/.html

Cherlin, Andrew J. 1992. *Marriage, Divorce, Remarriage.* Cambridge, MA: Harvard University Press.

Chesney-Lind, Meda. 1989. "Girls' Crime and Woman's Place: Toward a Feminist Model of Female Delinquency." *Crime and Delinquency,* 35 (1): 5–29.

—. 1997. *The Female Offender.* Thousand Oaks, CA: Sage.

chiefexecutive.com. 2010. "Changing Management Practice for the 21st Century." Retrieved Mar. 8, 2010. Online: www.the-chiefexecutive.com/features/feature54442

Children's Defense Fund, 2001. *The State of America's Children: Yearbook 2001.* Boston: Beacon.

—. 2008. *The State of America's Children: Yearbook 2008.* Retrieved July 28, 2009. Online: www.childrensdefense.org/child-research-data-publications/data/state-of-americas-children-highlights.html

—. 2009. *Kinship Care Resource Kit.* Retrieved Feb. 15, 2010. Online: www.childrensdefense.org/child-research-data-publications/data/kinship-care-resource-kit.pdf

—. 2010. *The State of America's Children 2010.* Retrieved May 18, 2011. Online: www.childrensdefense.org/child-research-data-publications/data/state-of-americas-children-2010-report-child-poverty.pdf

—. 2011. "Children in the United States: Fact Sheet." Retrieved Feb. 21, 2011. Online: www.childrensdefense.org/child-research-data-publications/data/state-data-repository/

cits/2011/children-in-the-states-2011-united-states.pdf
Christoffersen, John. 2010. "Conn. Man's Trial to Open in Fatal Home Invasion." msnbc.com (Sept. 12). Retrieved Jan. 15, 2011. Online: www.msnbc.msn.com/id/3913475?ocid=twitter
Chronicle of Higher Education. 2009. "Almanac Issue: 2009–2010" (Aug. 28).
—. 2010. "Data Points: From Community College to Ph.D." (Jan. 29): A4.
Chua, Amy. 2011. *Battle Hymn of the Tiger Mother.* New York: Penguin.
CIA. 2010. *World Fact Book: 2010.* Washington, D.C.: Central Intelligence Agency. Retrieved May 19, 2011. Online: www.cia.gov/library/publications/the-world-factbook/fields/2172.html
cia.gov. 2011. "The World Factbook." Retrieved May 15, 2011. Online: www.cia.gov/library/publications/the-world-factbook/fields/print_2054.html
Cinquemani, Anthony M. 1997. "C. Wright Mills, Media, and Mass: Only More So." *Educational Change* (Spring): 88–90.
Clayman, Steven E. 1993. "Booing: The Anatomy of a Disaffiliative Response." *American Sociological Review,* 58 (1): 110–131.
Clinard, Marshall B., and Peter C. Yeager. 1980. *Corporate Crime.* New York: Free Press.
Clines, Francis X. 1996. "A Chef's Training Program That Feeds Hope as Well as Hunger." *New York Times* (Dec. 11): B1, B6.
Clinton, Hillary Rodham. 2011. "Empowering Women Helps Fuel Global Economic Growth." Bloomberg.com (Mar. 8). Retrieved Mar. 14, 2011. Online: www.bloomberg.com/news/print/2011-03-08/empowering-women-helps-global-growth-commentary-by-hillary-rodham-clinton.html
Cloward, Richard A., and Lloyd E. Ohlin. 1960. *Delinquency and Opportunity: A Theory of Delinquent Gangs.* New York: Free Press.
CNN.com. 2006. "Immigrant Hopefuls Gather Near U.S. Border." (Apr. 12). Retrieved Apr. 29, 2006. Online: www.cnn.com/2006/WORLD/americas/04/12/mexico.border.ap/index.html
—. 2011. "Teen Pregnancy Rate Lowest in Two Decades." CNN Health (Apr. 5). Retrieved Apr. 16, 2011. Online: http://thechart.blogs.cnn.com/2011/04/05/teen-pregnancy-rate-lowest-in-two-decades
CNNmoney.com. 2009. "Life on Unemployment." Retrieved July 8, 2011. Online: http://money.cnn.com/galleries/2009/news/0903/gallery.living_on_unemployment/3.html
—. 2010. "Global 500: Our Annual Rankings of the World's Largest Corporations." Retrieved Mar. 31, 2011. Online: http://money.cnn.com/magazines/fortune/global500/2010/full_list
Coakley, Jay J. 2004. *Sport in Society: Issues and Controversies* (8th ed.). New York: McGraw-Hill.
Coasports.org. 2009. "Estimated Probability of Competing in Athletics Beyond the High-School Interscholastic Level." Online: www.coasports.org/pdf/odds%20of%20becoming%20a%20pro.pdf
Coburn, Andrew F., and Elise J. Bolda. 1999. "The Rural Elderly and Long-Term Care." In Thomas C. Ricketts, III (Ed.), *Rural Health in the United States.* New York: Oxford University Press, pp. 179–189.
Cock, Jacklyn. 1994. "Women and the Military: Implications for Demilitarization in the 1990s in South Africa." *Gender & Society,* 8 (2): 152–169.
Cockerham, William C. 2009. *Medical Sociology* (11th ed.). Upper Saddle River, NJ: Prentice Hall.
Cohen, Adam. 1999. "A Curse of Cliques." *Time* (May 3): 44–45.
Cole, David. 2000. *No Equal Justice: Race and Class in the American Criminal Justice System.* New York: New Press.
Cole, George F., and Christopher E. Smith. 2004. *The American System of Criminal Justice* (10th ed.). Belmont, CA: Wadsworth.
Coleman, Richard P., and Lee Rainwater. 1978. *Social Standing in America: New Dimensions of Class.* New York: Basic.
collegefund.org. 2011. "Tribal Colleges and Universities Map." Retrieved May 23, 2011. Online: www.collegefund.org/userfiles/TribalColleges-map.pdf
CollegeUniversity.suite11.com. 2010. "10 Ways to Make Your School a Green Campus" (by mrubin) (Mar. 15). Retrieved Apr. 20, 2010. Online: www.greenstudentu.com/green_campus/10_ways_to_make_your_school_a_green_campus.aspx
Collins, Patricia Hill. 1990. *Black Feminist Thought: Knowledge, Consciousness, and the Politics of Empowerment.* London: HarperCollins Academic.
—. 1991. "The Meaning of Motherhood in Black Culture." In Robert Staples (Ed.), *The Black Family: Essays and Studies.* Belmont, CA: Wadsworth, pp. 169–178. Orig. pub. in *SAGE: A Scholarly Journal on Black Women,* 4 (Fall 1987): 3–10.
Collins, Randall. 1971. "A Conflict Theory of Sexual Stratification." *Social Problems,* 19 (1): 3–21.
—. 1979. *The Credential Society: An Historical Sociology of Education.* New York: Academic Press.
—. 1982. *Sociological Insight: An Introduction to Non-Obvious Sociology.* New York: Oxford University Press.
—. 1987. "Interaction Ritual Chains, Power, and Property: The Micro–Macro Connection as an Empirically Based Theoretical Problem." In Jeffrey C. Alexander et al. (Eds.), *The Micro–Macro Link.* Berkeley: University of California Press, pp. 193–206.
—. 1994. *Four Sociological Traditions.* New York: Oxford University Press.
Coltrane, Scott. 1989. "Household Labor and the Routine Production of Gender." *Social Problems,* 36: 473–490.
Commonwealth of Massachusetts. 2011. "Labor and Workforce Development: Job Search Tips." Retrieved Apr. 3, 2011. Online: www.mass.gov/?pageID=elwdterminal&L=5&L0=Home&L1=Workers+and+Unions&L2=Job+Seekers&L3=Special+Programs&L4=Connecting+Disabled+Workers+and+Employers&sid=Elwd&b=terminalcontent&f=dcs_cc_services_job_search_tips_disabled&csid=Elwd
Condry, Sandra McConnell, John C. Condry, Jr., and Lee Wolfram Pogatshnik. 1983. "Sex Differences: A Study of the Ear of the Beholder." *Sex Roles,* 9: 697–704.
Conrad, Peter. 1996. "Medicalization and Social Control." In Phil Brown (Ed.), *Perspectives in Medical Sociology* (2nd ed.). Prospect Heights, IL: Waveland, pp. 137–162.
Cook, Sherburn F. 1973. "The Significance of Disease in the Extinction of the New England Indians." *Human Biology,* 45: 485–508.
Cookson, Peter W., Jr., and Caroline Hodges Persell. 1985. *Preparing for Power: America's Elite Boarding Schools.* New York: Basic.
Cooley, Charles Horton. 1963. *Social Organization: A Study of the Larger Mind.* New York: Schocken (orig. pub. 1909).
—. 1998. "The Social Self—the Meaning of 'I.'" In Hans-Joachim Schubert (Ed.), *On Self and Social Organization—Charles Horton Cooley.* Chicago: University of Chicago Press, pp. 155–175. Reprinted from Charles Horton Cooley, *Human Nature and the Social Order.* New York: Schocken, 1902.
Coontz, Stephanie. 1992. *The Way We Never Were: American Families and the Nostalgia Trap.* New York: Basic.
Cooper, Elizabeth. 2011. "Citizen's Voice: Why Intelligent Design Doesn't Qualify as Science." Knoxnews.com (Apr. 9). Retrieved May 2, 2011. Online: www.knoxnews.com/news/2011/apr/09/why-intelligent-design-doesnt-qualify-as-science/?print=1
Corr, Charles A., Clyde M. Nabe, and Donald M. Corr. 2003. *Death and Dying, Life and Living* (4th ed.). Pacific Grove, CA: Brooks/Cole.
Corsaro, William A. 1985. *Friendship and Peer Culture in the Early Years.* Norwood, NJ: Ablex.
—. 1992. "Interpretive Reproduction in Children's Peer Cultures." *Social Psychology Quarterly,* 55 (2): 160–177.
—. 1997. *Sociology of Childhood.* Thousand Oaks, CA: Pine Forge.
Corsbie-Massay, Charisse L. 2005. "Beauty Pageants and Television: A Perfect Marriage." Retrieved Jan. 27, 2008. Online: http://alum.mit.edu/www/charisse
Cortese, Anthony J. 2004. *Provocateur: Images of Women and Minorities in Advertising* (2nd ed.). Latham, MD: Rowman and Littlefield.
Coser, Lewis A. 1956. *The Functions of Social Conflict.* Glencoe, IL: Free Press.
Coughlin, Ellen K. 1993. "Author of a Noted Study on Black Ghetto Life Returns with a Portrait of Homeless Women." *Chronicle of Higher Education* (Mar. 31): A7–A8.
Cox, Harvey. 1995. "Christianity." In Arvind Sharma (Ed.), *Our Religions.* San Francisco: HarperCollins, pp. 359–423.
Cox, Oliver C. 1948. *Caste, Class, and Race.* Garden City, NY: Doubleday.
Craig, Steve. 1992. "Considering Men and the Media." In Steve Craig (Ed.), *Men, Masculinity, and the Media.* Newbury Park, CA: Sage, pp. 1–7.
Cressey, Donald. 1969. *Theft of the Nation.* New York: Harper & Row.
Creswell, John W. 1998. *Qualitative Inquiry and Research Design: Choosing Among Five Traditions.* Thousand Oaks, CA: Sage.
Creswell, Julie. 2011. "Even Funds That Lagged Paid Richly." *New York Times* (Apr. 1): B1, B6.
Crinnion, Walter J. 1995. "Are Organic Foods Really Healthier for You?" Retrieved Feb. 17, 2007. Online: http://lookwayup.com/free/organic.htm
Cronin, John, and Robert F. Kennedy, Jr. 1999. *The Riverkeepers: Two Activists Fight to Reclaim Our Environment as a Basic Human Right.* New York: Touchstone.
Crosnoe, Robert. 2006. *Mexican Roots, American Schools: Helping Mexican Immigrant Children Succeed.* Stanford, CA: Stanford University Press.
Cumming, Elaine C., and William E. Henry. 1961. *Growing Old: The Process of Disengagement.* New York: Basic.
Currie, Elliott. 1998. *Crime and Punishment in America.* New York: Metropolitan.
Curry, Pat. 2009. "How Credit Card Scores Work, How a Score Is Calculated." *Bankrate.com.* Retrieved Apr. 25, 2011. Online: www.bankrate.com/brm/news/credit-scoring/20031104a1.asp
Curtiss, Susan. 1977. *Genie: A Psycholinguistic Study of a Modern Day "Wild Child."* New York: Academic Press.
Cyrus, Virginia. 1993. *Experiencing Race, Class, and Gender in the United States.* Mountain View, CA: Mayfield.

Dahl, Robert A. 1961. *Who Governs?* New Haven, CT: Yale University Press.

Dahrendorf, Ralf. 1959. *Class and Class Conflict in an Industrial Society.* Stanford, CA: Stanford University Press.

Dailycaller.com. 2011. "GOP Rep. to Obama: 'I'm Game' for Alligator-Filled Moat Along U.S.–Mexico Border" (May 12). Retrieved May 16, 2011. Online: http://dailycaller.com/2011/05/12/gop-rep-to-obama-im-game-for-alligator-filled-moat-along-u-s-mexico-border

Dailytroll.com. 2006. "Food as Bonding." Retrieved Dec. 29, 2007. Online: http://dailytroll.com/?p=890

Daly, Kathleen, and Meda Chesney-Lind. 1988. "Feminism and Criminology." *Justice Quarterly,* 5: 497–533.

Daly, Mary. 1973. *Beyond God the Father.* Boston: Beacon.

Daniels, Jessie. 2009. *Cyber Racism: White Supremacy Online and the New Attack on Civil Rights.* Lanham, MD: Rowman & Littlefield.

—. 2010. "Cyber Racism on College Campuses." *Race-Talk* (May 4). Retrieved Apr. 29, 2011. Online: http://blogs.alternet.org/speakeasy/2010/05/04/cyber-racism-on-college-campuses

Daniels, Roger. 1993. *Prisoners Without Trial: Japanese-Americans in World War II.* New York: Hill & Wang.

Darley, John M., and Thomas R. Shultz. 1990. "Moral Rules: Their Content and Acquisition." *Annual Review of Psychology,* 41: 525–556.

Dart, Bob. 1999. "Kids Get More Screen Time Than School Time." *Austin American-Statesman* (June 28): A1, A5.

Davis, F. James. 1991. *Who Is Black?* University Park: Pennsylvania State University Press.

Davis, Fred. 1992. *Fashion, Culture, and Identity.* Chicago: University of Chicago Press.

Davis, James A., Tom W. Smith, and Peter V. Marsden. 2007. *General Social Survey, 1972–2006: Cumulative Codebook.* Chicago: National Opinion Research Center.

Davis, Kingsley. 1940. "Extreme Social Isolation of a Child." *American Journal of Sociology,* 45 (4): 554–565.

Davis, Kingsley, and Wilbert Moore. 1945. "Some Principles of Stratification." *American Sociological Review,* 7 (April): 242–249.

Dean, L. M., F. N. Willis, and J. N. la Rocco. 1976. "Invasion of Personal Space as a Function of Age, Sex and Race." *Psychological Reports,* 38 (3) (pt. 1): 959–965.

Death Penalty Information Center. 2010. "Facts About the Death Penalty" (Dec. 17). Retrieved Jan. 14, 2011. Online: www.deathpenaltyinfo.org/documents/FactSheet.pdf

Deegan, Mary Jo. 1988. *Jane Addams and the Men of the Chicago School, 1892–1918.* New Brunswick, NJ: Transaction.

Degher, Douglas, and Gerald Hughes. 1991. "The Identity Change Process: A Field Study of Obesity." *Deviant Behavior,* 12: 385–402.

Delgado, Richard. 1995. "Introduction." In Richard Delgado (Ed.), *Critical Race Theory: The Cutting Edge.* Philadelphia: Temple University Press, pp. xiii–xvi.

DeNavas-Walt, Carmen, Robert W. Cleveland, and Bruce H. Webster, Jr. 2003. "Income in the United States: 2002." U.S. Census Bureau, Current Population Reports, P60–221. Washington, DC: U.S. Government Printing Office.

DeNavas-Walt, Carmen, Bernadette D. Proctor, and Jessica C. Smith. 2010. *Income, Poverty, and Health Insurance Coverage in the United States: 2009.* Current Population Report P60-238. Washington, DC: U.S. Census Bureau.

DeParle, Jason. 2007. "A Global Trek to Poor Nations, from Poorer Ones." *New York Times* (Dec. 27): A1, A16.

DeParle, Jason, and Robert Gebeloff. 2009. "Food Stamp Use Soars Across U.S., and Stigma Fades." *New York Times* (Nov. 29): A1, A25.

—. 2010. "Once Stigmatized, Food Stamps Find New Users and Acceptance." *New York Times* (Feb. 11): A1, A20.

Derber, Charles. 1983. *The Pursuit of Attention: Power and Individualism in Everyday Life.* New York: Oxford University Press.

Dohan, Daniel, and Martin Sanchez-Jankowski. 1998. "Using Computers to Analyze Ethnographic Field Data: Theoretical and Practical Considerations." *Annual Review of Sociology,* 24: 477–499.

Dollard, John, Neal E. Miller, Leonard W. Doob, O. H. Mowrer, and Robert R. Sears. 1939. *Frustration and Aggression.* New Haven, CT: Yale University Press

Dolnick, Sam. 2008. "Surrogate Business Makes Birth the Latest Job Outsourced to India." *Austin American-Statesman* (Jan. 1): A10.

—. 2010. "The Obesity–Hunger Paradox." *New York Times* (Mar. 14): A27.

Domhoff, G. William. 1978. *The Powers That Be: Processes of Ruling Class Domination in America.* New York: Random House.

—. 1983. *Who Rules America Now? A View for the '80s.* Englewood Cliffs, NJ: Prentice Hall.

—. 2002. *Who Rules America? Power and Politics* (4th ed.). New York: McGraw-Hill.

—. 2011. "Wealth, Income, and Power. Who Rules America?" (Sept. 2005, updated Jan. 2011). Retrieved Feb. 20, 2011. Online: http://sociology.ucsc.edu/whorulesamerica/power/wealth.html

Driskell, Robyn Bateman, and Larry Lyon. 2002. "Are Virtual Communities True Communities? Examining the Environments and Elements of Community." *City & Community,* 1 (4): 1–18.

Du Bois, W. E. B. 1967. *The Philadelphia Negro: A Social Study.* New York: Schocken (orig. pub. 1899).

Dubowitz, Howard, Maureen Black, Raymond H. Starr, Jr., and Susan Zuravin. 1993. "A Conceptual Definition of Child Neglect." *Criminal Justice and Behavior,* 20 (1): 8–26.

Duffy, John. 1976. *The Healers.* New York: McGraw-Hill.

Dunbar, Polly. 2007. "Wombs to Rent: Childless British Couples Pay Indian Women to Carry Their Babies." *The Daily Mail* (Dec. 8). Retrieved Feb. 9, 2008. Online: www.dailymail.co.uk/pages/live/articles/news/worldnews.html?in_article_id=500601

Dunlap, Riley E. 1992. "Trends in Public Opinion Toward Environmental Issues: 1965–1990." In Riley E. Dunlap and Angela G. Mertig (Eds.), *American Environmentalism: The U.S. Environmental Movement, 1970–1990.* New York: Taylor & Francis, pp. 89–113.

Dupre, Roslyn, and Paul Gains. 1997. "Fundamental Differences." *Women's Sports & Fitness* (October): 63–68.

Durkheim, Emile. 1933. *The Division of Labor in Society.* Trans. George Simpson. New York: Free Press (orig. pub. 1893).

—. 1956. *Education and Sociology.* Trans. Sherwood D. Fox. Glencoe, IL: Free Press.

—. 1964a. *The Rules of Sociological Method.* Trans. Sarah A. Solovay and John H. Mueller. New York: Free Press (orig. pub. 1895).

—. 1964b. *Suicide.* Trans. John A. Spaulding and George Simpson. New York: Free Press (orig. pub. 1897).

—. 1995. *The Elementary Forms of Religious Life.* Trans. Karen E. Fields. New York: Free Press (orig. pub. 1912).

Dye, Thomas R., Harmon Zeigler, and Louis Schubert. 2012. *The Irony of Democracy: An Uncommon Introduction to American Politics* (15th ed.). Boston: Wadsworth.

Early, Kevin E. 1992. *Religion and Suicide in the African-American Community.* Westport, CT: Greenwood.

Ebaugh, Helen Rose Fuchs. 1988. *Becoming an EX: The Process of Role Exit.* Chicago: University of Chicago Press.

Eckholm, Erik. 2009. "Gang Violence Grows on an Indian Reservation." *New York Times* (Dec. 14): A13.

—. 2010. "Study Finds More Woes Following Foster Care." *New York Times* (Apr. 7). Retrieved Apr. 18, 2011. Online: www.nytimes.com/2010/04/07/us/07foster.html

Egan, Mary Ellen. 2010. "Bank of America and Merrill Lynch Sex Discrimination Lawsuit." Forbes.com (Mar. 31). Retrieved Mar. 11, 2011. Online: http://blogs.forbes.com/work-in-progress/2010/03/31/bank-of-america-and-merrill-lynch-sex-discrimination-lawsuit

Egley, Arlen, Jr., James C. Howell, and John P. Moore. 2010. "Highlights of the 2008 National Youth Gang Survey." *OJJDP Fact Sheet* (Mar.). Retrieved Dec. 31, 2010. Online: www.ncjrs.gov/pdffiles1/ojjdp/229249.pdf

Ehrenreich, Barbara. 2001. *Nickel and Dimed: On (Not) Getting by in America.* New York: Metropolitan.

—. 2008. "The Communist Manifesto Hits 160." *Current Affairs* (Oct. 10). Retrieved July 4, 2009. Online: http://ehrenreich.blogs.com/barbaras_blog/2008/10/the-communist-manifesto-hits-160.html

Ehrlich, Paul R., and Anne H. Ehrlich. 2009. "The Population Bomb Revisited." *Electronic Journal of Sustainable Development* 1 (3). Retrieved May 22, 2011. Online: http://fragette.free.fr/demography/The_Population_Bomb_Revisited.pdf

Eighner, Lars. 1993. *Travels with Lizbeth.* New York: St. Martin's.

Eisenstein, Zillah R. 1994. *The Color of Gender: Reimaging Democracy.* Berkeley: University of California Press.

Elkin, Frederick, and Gerald Handel. 1989. *The Child and Society: The Process of Socialization* (5th ed.). New York: Random House.

Elkind, David. 1995. "School and Family in the Postmodern World." *Phi Delta Kappan* (September): 8–21.

Elster, Jon. 1989. *Nuts and Bolts for the Social Sciences.* Cambridge, England: Cambridge University Press.

Emling, Shelley. 1997. "Haiti Held in Grip of Another Drought." *Austin American-Statesman* (Sept. 19): A17, A18.

Engels, Friedrich. 1970. *The Origins of the Family, Private Property, and the State.* New York: International (orig. pub. 1884).

Engerman, Stanley L. 1995. "The Extent of Slavery and Freedom Throughout the World as a Whole and in Major Subareas." In Julian L. Simon (Ed.), *The State of Humanity.* Cambridge, MA: Blackwell, pp. 171–177.

Enloe, Cynthia H. 1987. "Feminists Thinking About War, Militarism, and Peace." In Beth H. Hess and Myra Marx Ferree (Eds.), *Analyzing Gender: A Handbook of Social Science Research.* Newbury Park, CA: Sage, pp. 526–547.

epa.gov. 2011. "Smart Growth." Retrieved May 23, 2011. Online: www.epa.gov/smartgrowth/about_sg.htm

Epstein, Cynthia Fuchs. 1988. *Deceptive Distinctions: Sex, Gender, and the Social Order.* New Haven, CT: Yale University Press.

Epstein, Jennifer, and Glenn Thrush. 2011. "Obama Launches Reelection Campaign." Politico.com (Apr. 4). Retrieved Apr. 4, 2011. Online: www.politico.com/news/stories/0411/52457.html

Erikson, Erik H. 1963. *Childhood and Society.* New York: Norton.

Erikson, Kai T. 1962. "Notes on the Sociology of Deviance." *Social Problems,* 9: 307–314.

—. 1964. "Notes on the Sociology of Deviance." In Howard S. Becker (Ed.), *The Other Side: Perspectives on Deviance.* New York: Free Press, pp. 9–21.

—. 1976. *Everything in Its Path: Destruction of Community in the Buffalo Creek Flood.* New York: Simon & Schuster.

—. 1994. *A New Species of Trouble: Explorations in Disaster, Trauma, and Community.* New York: Norton.

Espiritu, Yen Le. 1995. *Filipino American Lives.* Philadelphia: Temple University Press.

Essed, Philomena. 1991. *Understanding Everyday Racism.* Newbury Park, CA: Sage.

Esterberg, Kristin G. 1997. *Lesbian and Bisexual Identities: Constructing Communities, Constructing Self.* Philadelphia: Temple University Press.

Etzioni, Amitai. 1975. *A Comparative Analysis of Complex Organizations: On Power, Involvement, and Their Correlates* (rev. ed.). New York: Free Press.

Evans, Glen, and Norman L. Farberow. 1988. *The Encyclopedia of Suicide.* New York: Facts on File.

Evans, Peter B., and John D. Stephens. 1988. "Development and the World Economy." In Neil J. Smelser (Ed.), *Handbook of Sociology.* Newbury Park, CA: Sage, pp. 739–773.

Fackler, Martin. 2011. "Powerful Quake and Tsunami Devastate Northern Japan." *New York Times* (Mar. 11). Retrieved May 11, 2011. Online: www.nytimes.com/2011/03/12/world/asia/12japan.html?_r=1&pagewanted=print

Fagan, Kevin. 2003. "Shame of the City: Homeless Island." *San Francisco Chronicle* (Nov. 30). Retrieved Apr. 11, 2004. Online: www.sfgate.com

Fallon, Patricia, Melanie A. Katzman, and Susan C. Wooley. 1994. *Feminist Perspectives on Eating Disorders.* New York: Guilford.

Farb, Peter. 1973. *Word Play: What Happens When People Talk.* New York: Knopf.

farenet.org. 2011. "Star Tells of Sadness at Racist Taunts." FARE. Retrieved May 23, 2011. Online: www.farenet.org/default.asp?intPageID=7&intArticleID=714

Farley, John E. 2000. *Majority–Minority Relations* (4th ed.). Englewood Cliffs, NJ: Prentice Hall.

Fausto-Sterling, Anne. 1985. *Myths of Gender: Biological Theories About Women and Men.* New York: Basic.

Feagin, Joe R. 1991. "The Continuing Significance of Race: Antiblack Discrimination in Public Places." *American Sociological Review,* 56 (February): 101–116.

Feagin, Joe R., David B. Baker, and Clairece B. Feagin. 2006. *Social Problems: A Critical Power-Conflict Perspective* (6th ed.). Englewood Cliffs. NJ: Prentice Hall.

Feagin, Joe R., and Clairece Booher Feagin. 1994. *Social Problems: A Critical Power–Conflict Perspective* (4th ed.). Englewood Cliffs, NJ: Prentice Hall.

—. 2003. *Racial and Ethnic Relations* (7th ed.). Upper Saddle River, NJ: Prentice Hall.

—. 2008. *Racial and Ethnic Relations* (8th ed.). Upper Saddle River, NJ: Prentice Hall.

—. 2011. *Racial and Ethnic Relations* (9th ed.). Upper Saddle River, NJ: Prentice Hall.

Feagin, Joe R., Anthony M. Orum, and Gideon Sjoberg (Eds.). 1991. *A Case for the Case Study.* Chapel Hill: University of North Carolina Press.

Feagin, Joe R., and Robert Parker. 2002. *Building American Cities: The Urban Real Estate Game* (2nd ed.). Hopkins, MN: Beard.

Feagin, Joe R., and Hernán Vera. 1995. *White Racism: The Basics.* New York: Routledge.

Federal Bureau of Investigation (FBI). 2009. *Crime in the United States: 2008.* Retrieved Mar. 12, 2010. Online: www.fbi.gov/ucr/cius2008/index.html

—. 2010. *Crime in the United States, 2009.* Retrieved Dec. 31, 2010. Online: www2.fbi.gov/ucr/cius2009/data/table_29.html

Federal Interagency Forum on Aging Related Statistics. 2010. "Older Americans 2010: Key Indicators of Well-Being." Retrieved June 14, 2011. Online: www.agingstats.gov/agingstatsdotnet/Main_Site/Data/2010_Documents/Docs/OA_2010.pdf

Federal Interagency Forum on Child and Family Statistics. 2010. "America's Children in Brief: Key National Indicators of Well-Being, 2010." Retrieved Apr. 16, 2012. Online: http://childstats.gov/pdf/ac2010/ac_10.pdf

Feifel, H., and W. T. Nagy. 1981. "Another Look at Fear of Death." *Journal of Consulting and Clinical Psychology,* 49: 278–286. Cited in N. Hooyman and H. A. Kiyak, 2002.

Fenstermaker, Sarah, and Candace West (Eds.). 2002. *Doing Gender, Doing Difference: Inequality, Power, and Institutional Change.* New York: Routledge.

Ferguson, John. 1977. *War and Peace in the World's Religions.* New York: Oxford University Press.

Ferraro, Gary. 1992. *Cultural Anthropology: An Applied Perspective.* St. Paul, MN: West.

Ferrell, Bruce. 2008. "Gangs and the Internet." *United States Attorneys' Bulletin* (July). Retrieved Mar. 10, 2010. Online: www.nationalgangcenter.gov/Content/Documents/Gang-Issues-July-2008.pdf

Findlay, Deborah A., and Leslie J. Miller. 1994. "Through Medical Eyes: The Medicalization of Women's Bodies and Women's Lives." In B. Singh Bolaria and Harley D. Dickinson (Eds.), *Health, Illness, and Health Care in Canada* (2nd ed.). Toronto: Harcourt, pp. 276–306.

Fine, Michelle, and Lois Weis. 1998. *The Unknown City: The Lives of Poor and Working-Class Young People.* Boston: Beacon.

Finklea, Kristin M. 2009. "Organized Crime in the United States: Trends and Issues for Congress." *CRS Report for Congress.* Washington, DC: Congressional Research Service. Retrieved May 25, 2009. Online: www.fas.org/sgp/crs/misc/R40525.pdf

Finley, M. I. 1980. *Ancient Slavery and Modern Ideology.* New York: Viking.

Finn Paradis, Leonora, and Scott B. Cummings. 1986. "The Evolution of Hospice in America Toward Organizational Homogeneity." *Journal of Health and Social Behavior,* 27: 370–386.

fireandreamitchell.com. 2011. "Texas Governor Rick Perry Destroys Obama Over Moat Comment—Obama Is Just a Standup Comic Trying Out for Saturday Night Live." Retrieved May 16, 2011. Online: www.fireandreamitchell.com/2011/05/11/texas-governor-rick-perry-destroys-obama-over-moat-comment-obama-is-just-a-stand-up-comic-trying-out-for-saturday-night-live

Firestone, Shulamith. 1970. *The Dialectic of Sex.* New York: Morrow.

Fisher-Thompson, Donna. 1990. "Adult Sex-Typing of Children's Toys." *Sex Roles,* 23: 291–303.

Fjellman, Stephen M. 1992. *Vinyl Leaves: Walt Disney World & America.* Boulder, CO: Westview.

Flexner, Abraham. 1910. *Medical Education in the United States and Canada.* New York: Carnegie Foundation.

Foderaro, Lisa W. 2010. "Private Moment Made Public, Then a Fatal Jump." *New York Times* (Sept. 30): A1, A4.

Folbre, Nancy. 2009. "When State Universities Lose State Support." *New York Times* (Oct. 5). Retrieved Apr. 29. Online: http://economix.blogs.nytimes.com/2009/10/05/when-state-universities-lose-state-support

Fong-Torres, Ben. 2007. "Hungry Heart." *New York Times Book Review* (Feb. 4): 11.

Forbes. 2010. "The World's Billionaires." Retrieved Mar. 15, 2010. Online: www.forbes.com/2010/03/10/worlds-richest-people-slim-gates-buffett-billionaires-2010_land.html

Foster, Daniel. 2011. "Obama's Immigration Speech: Republicans Want a 'Moat with Alligators.'" *National Review Online* (May 10). Retrieved May 16, 2011. Online: www.nationalreview.com/blogs/print/266896

Foucault, Michel. 1979. *Discipline and Punish: The Birth of the Prison.* New York: Vintage.

—. 1988. *Madness and Civilization: A History of Insanity in the Age of Reason.* New York: Vintage (orig. pub. 1961).

—. 1994. *The Birth of the Clinic: An Archeology of Medical Perception.* New York: Vintage (orig. pub. 1963).

Foxnews.com. 2011. "Thousands March in Los Angeles for Organized Labor." Foxnews.com (Mar. 27). Retrieved Mar. 29, 2011. Online: www.foxnews.com/us/2011/03/27/thousands-march-los-angeles-organized-labor

Frank, Robert H. 1999. *Luxury Fever: Why Money Fails to Satisfy in an Era of Excess.* New York: Free Press.

Frankenberg, Ruth. 1993. *White Women, Race Matters: The Social Construction of Whiteness.* Minneapolis: University of Minnesota Press.

Franklin, John Hope. 1980. *From Slavery to Freedom: A History of Negro Americans.* New York: Vintage.

Freidson, Eliot. 1965. "Disability as Social Deviance." In Marvin B. Sussman (Ed.), *Sociology and Rehabilitation.* Washington, DC: American Sociology Association, pp. 71–99.

—. 1970. *Profession of Medicine.* New York: Dodd, Mead.

—. 1986. *Professional Powers.* Chicago: University of Chicago Press.

French, Howard W. 2008. "Lives of Grinding Poverty, Untouched by China's Boom." *New York Times* (Jan. 13): YT4.

French, Sally. 1999. "The Wind Gets in My Way." In Mairian Corker and Sally French (Eds.), *Disability Discourse.* Buckingham, England: Open University Press, pp. 21–27.

Freud, Sigmund. 1924. *A General Introduction to Psychoanalysis* (2nd ed.). New York: Boni & Liveright.

Freudenheim, Milt. 2007. "Showdown Looms in Congress Over Drug Advertising on TV." *New York Times* (Jan. 22): A1.

Friedman, Emily. 2010. "Victim of Secret Dorm Sex Tape Posts Facebook Goodbye, Jumps to Death." abcnews.com (Sept. 29). Retrieved Nov. 26, 2010. Online: http://abcnews.go.com/US/victim-secret-dorm-sex-tape-commits-suicide/story?id=11758716

Friedman, Thomas L. 2005. "It's a Flat World, After All." *New York Times Magazine* (Apr. 3): 33ff.

—. 2007. *The World Is Flat 3.0: A Brief History of the Twenty-First Century.* New York: Picador.

Frisbie, W. Parker, and John D. Kasarda. 1988. "Spatial Processes." In Neil Smelser (Ed.), *The Handbook of Sociology*. Newbury Park, CA: Sage, pp. 629–666.

Fritze, John. 2010. "Medical Expenses Have 'Very Steep' Rate of Growth." *USA Today* (Feb. 4). Retrieved June 7, 2010. Online: www.usatoday.com/cleanprint/?1275957329928

Fry, Richard, and D'Vera Cohn. 2010. *Women, Men and the New Economics of Marriage*. Pew Research Center. Retrieved Apr. 2, 2010. Online: http://pewsocialtrends.org/assets/pdf/new-economics-of-marriage.pdf

Galbraith, John Kenneth. 1985. *The New Industrial State* (4th ed.). Boston: Houghton Mifflin.

Gallardo, Roberto. 2010. "Rural America in the 2000s: Age." Retrieved Mar. 21, 2011. Online: www.dailyyonder.com/print/2849

Gambino, Richard. 1975. *Blood of My Blood*. New York: Doubleday/Anchor.

Gandara, Ricardo. 1995. "*Dichos de la Vida*: Homespun Proverbs Link Hispanic Culture's Past with the Present." *Austin American-Statesman* (Jan. 21): E1, E10.

Gans, Herbert. 1982. *The Urban Villagers: Group and Class in the Life of Italian Americans* (updated and expanded ed.; orig. pub. 1962). New York: Free Press.

—. 1999. *Popular Culture and High Culture: An Analysis and Evaluation of Taste* (revised and updated ed.). New York: Basic.

Gardner, Carol Brooks. 1989. "Analyzing Gender in Public Places: Rethinking Goffman's Vision of Everyday Life." *American Sociologist*, 20 (Spring): 42–56.

Garfinkel, Harold. 1967. *Studies in Ethnomethodology*. Englewood Cliffs, NJ: Prentice Hall.

Garreau, Joel. 1991. *Edge City: Life on the New Frontier*. New York: Doubleday.

Gary, Keahn. 2007. "New Report Reveals Top Ten Problems Facing U.S. Students." Retrieved Feb. 11, 2008. Online: www.nbc26.com/news/trends/8000952.html

Gawande, Atul. 2002. *Complications: A Surgeon's Notes on an Imperfect Science*. New York: Picador.

Gaylin, Willard. 1992. *The Male Ego*. New York: Viking/Penguin.

Geertz, Clifford. 1966. "Religion as a Cultural System." In Michael Banton (Ed.), *Anthropological Approaches to the Study of Religion*. London: Tavistock, pp. 1–46.

General Motors. 2010. "2010 Annual Report." Retrieved July 8, 2011. Online: https://materials.proxyvote.com/Approved/37045V/20110408/AR_87685/HTML1/general_motors-ar2010_0019.htm

Gerbner, George, Larry Gross, Michael Morton, and Nancy Signorielli. 1987. "Charting the Mainstream: Television's Contributions to Political Orientations." In Donald Lazere (Ed.), *American Media and Mass Culture: Left Perspectives*. Berkeley: University of California Press, pp. 441–464.

Gereffi, Gary. 1994. "The International Economy and Economic Development." In Neil J. Smelser and Richard Swedberg (Eds.), *The Handbook of Economic Sociology*. Princeton, NJ: Princeton University Press, pp. 206–233.

Gerstel, Naomi, and Harriet Engel Gross. 1995. "Gender and Families in the United States: The Reality of Economic Dependence." In Jo Freeman (Ed.), *Women: A Feminist Perspective* (5th ed.). Mountain View, CA: Mayfield, pp. 92–127.

Gerth, Hans H., and C. Wright Mills. 1946. *From Max Weber: Essays in Sociology*. New York: Oxford University Press.

Getterman, Rosalee. 2011. "Beyond Tutoring." *Battalion* (Apr. 25). Retrieved Apr. 30, 2011. Online: www.thebatt.com/news/beyond-tutoring-1.2198477

Gibbs, Lois Marie, as told to Murray Levine. 1982. *Love Canal: My Story*. Albany: SUNY Press.

Gilbert, Dennis. 2010. *The American Class Structure in an Age of Growing Inequality* (8th ed.). Newbury Park, CA: Pine Forge/Sage.

Gilkey, Langdon. 1993. "Theories in Science and Religion." In James Huchingson (Ed.) *Religion and the Natural Sciences: The Range of Engagement*. Fort Worth, TX: Harcourt, pp. 61–65.

Gilligan, Carol. 1982. *In a Different Voice: Psychological Theory and Women's Development*. Cambridge, MA: Harvard University Press.

Gilmore, David D. 1990. *Manhood in the Making: Cultural Concepts of Masculinity*. New Haven, CT: Yale University Press.

Gilson, Dave. 2009. "Black and White and Dead All Over." *Mother Jones* (July/Aug.). Retrieved Apr. 12, 2010. Online: http://motherjones.com/print/24631

Giuliano, Traci A., Kathryn E. Popp, and Jennifer L. Knight. 2000. "Footballs Versus Barbies: Childhood Play Activities as Predictors of Sports Participation by Women." *Sex Roles*, 42: 159–181.

Glaeser, Edward. 2011. *Triumph of the City*. New York: Penguin.

Glanz, James, and Edward Wong. 2003. "'97 Report Warned of Foam Damaging Tiles." *New York Times* (Feb. 4): A1–A26.

Glaser, Barney, and Anselm Strauss. 1967. *Discovery of Grounded Theory: Strategies for Qualitative Research*. Chicago: Aldine.

—. 1968. *Time for Dying*. Chicago: Aldine.

Glastris, Paul. 1990. "The New Way to Get Rich." *U.S. News & World Report* (May 7): 25–36.

Goffman, Erving. 1956. "The Nature of Deference and Demeanor." *American Anthropologist*, 58: 473–502.

—. 1959. *The Presentation of Self in Everyday Life*. Garden City, NY: Doubleday.

—. 1961a. *Asylums: Essays on the Social Situation of Mental Patients and Other Inmates*. Chicago: Aldine.

—. 1961b. *Encounters: Two Studies in the Sociology of Interaction*. London: Routledge and Kegan Paul.

—. 1963a. *Behavior in Public Places: Notes on the Social Structure of Gatherings*. New York: Free Press.

—. 1963b. *Stigma: Notes on the Management of Spoiled Identity*. Englewood Cliffs, NJ: Prentice Hall.

—. 1967. *Interaction Ritual: Essays on Face to Face Behavior*. Garden City, NY: Anchor.

—. 1974. *Frame Analysis: An Essay on the Organization of Experience*. Boston: Northeastern University Press.

Goode, Erich. 1996. "The Stigma of Obesity." In Erich Goode (Ed.), *Social Deviance*. Boston: Allyn & Bacon, pp. 332–340.

Goode, William J. 1960. "A Theory of Role Strain." *American Sociological Review*, 25: 483–496.

Goodnough, Abby, and Jennifer Steinhauer. 2006. "Senate's Failure to Agree on Immigration Plan Angers Workers and Employers Alike." *New York Times* (Apr. 9): A35.

Goolsby, Craig A. 2011. "Disaster Planning: The Scope and Nature of the Problem" (Mar. 18). Retrieved May 11, 2011. Online: http://emedicine.medscape.com/article/765495-overview.pdf

Gordon, Milton. 1964. *Assimilation in American Life: The Role of Race, Religion, and National Origins*. New York: Oxford University Press.

Gordon, Philip L. 2001. "Federal Judge's Victory Just the First Shot in the Battle Over Workplace Monitoring." Retrieved July 8, 2002. Online: www.privacyfoundation.org/workplace/law/law_show.asp?id=75&action=0

Gottdiener, Mark. 1985. *The Social Production of Urban Space*. Austin: University of Texas Press.

Gouldner, Alvin W. 1970. *The Coming Crisis of Western Sociology*. New York: Basic.

Governors Highway Safety Association. 2011. "Mature Drivers." Retrieved Mar. 20, 2011. Online: www.ghsa.org/html/issues/olderdriver.html

Green, Donald E. 1977. *The Politics of Indian Removal: Creek Government and Society in Crisis*. Lincoln: University of Nebraska Press.

Green, Erica L. 2011. "Prayer Service at City School Called Improper." *Baltimore Sun* (Mar. 13). Retrieved May 8, 2011. Online: http://articles.baltimoresun.com/2011-03-13/news/bs-md-ci-msa-prayer-service-20110313_1_prayer-service-prayer-in-public-schools-voluntary-student-prayer

Greenberg, Edward S., and Benjamin I. Page. 1993. *The Struggle for Democracy*. New York: HarperCollins.

Greendorfer, Susan L. 1993. "Gender Role Stereotypes and Early Childhood Socialization." *Psychology of Women Quarterly*, 18: 85–104.

Greenstone, Michael, and Adam Looney. 2010. "The Long-Term Effects of the Great Recession for America's Youth." Brookings (Sept. 3). Retrieved Mar. 26, 2011. Online: www.brookings.edu/opinions/2010/0903_jobs_greenstone_looney.aspx?p=1

Griggs, Brandon. 2010. "Twitter Hoax Spreads Rumors of Airlines' Free Flights to Haiti." CNN.com (Jan. 14). Retrieved Apr. 18, 2010. Online: http://edition.cnn.com/2010/TECH/01/14/twitter.hoax.haiti

Guha, Ramachandra. 2004. "The Sociology of Suicide." Retrieved Dec. 20, 2007. Online: www.indiatogether.org/2004/aug/rgh-suicide.htm

Hadden, Richard W. 1997. *Sociological Theory: An Introduction to the Classical Tradition*. Peterborough, Ontario: Broadview.

Hahn, Harlan. 1997. "Advertising the Acceptably Employable Image." In Lennard J. Davis (Ed.), *The Disability Studies Reader*. New York: Routledge, pp. 172–186.

Haines, Valerie A. 1997. "Spencer and His Critics." In Charles Camic (Ed.), *Reclaiming the Sociological Classics: The State of the Scholarship*. Malden, MA: Blackwell, pp. 81–111.

Halberstadt, Amy G., and Martha B. Saitta. 1987. "Gender, Nonverbal Behavior, and Perceived Dominance: A Test of the Theory." *Journal of Personality and Social Psychology*, 53: 257–272.

Hale-Benson, Janice E. 1986. *Black Children: Their Roots, Culture and Learning Styles* (rev. ed.). Provo, UT: Brigham Young University Press.

Hall, Edward. 1966. *The Hidden Dimension*. New York: Anchor/Doubleday.

Hallinan, Maureen. 2005. "Should Your School Eliminate Tracking? The History of Tracking and Detracking in America's Schools." *Education Matters*, 1–2.

Haraway, Donna. 1994. "A Cyborgo Manifesto: Science, Technology, and Socialist-Feminism in the Late Twentieth Century." In Anne C. Herrmann and Abigail J. Stewart (Eds.), *Theorizing Feminism: Parallel Trends in the Humanities and Social Sciences*. Boulder, CO: Westview, pp. 424–457.

Hardy, Jackie. 2010. "Charter Schools Closing the Achievement Gap Among Minority Students." *North Dallas Gazette* (Mar. 3, 2010). Retrieved Apr. 29, 2011. Online: http://northdallasgazette.com/2010/03/charter-schools-closing-the-achievement-gap-among-minority-students

Harlow, Harry F., and Margaret Kuenne Harlow. 1962. "Social Deprivation in Monkeys." *Scientific American,* 207 (5): 137–146.

—. 1977. "Effects of Various Mother–Infant Relationships on Rhesus Monkey Behaviors." In Brian M. Foss (Ed.), *Determinants of Infant Behavior* (vol. 4). London: Methuen, pp. 15–36.

Harrell, Erika. 2007. "Black Victims of Violent Crime." Bureau of Justice Statistics Special Report (Aug.). Retrieved Mar. 10, 2010. Online: http://bjs.ojp.usdoj.gov/content/pub/pdf/bvvc.pdf

—. 2009. "Asian, Native Hawaiian, and Pacific Islander Victims of Crime." Bureau of Justice Statistics Special Report (Mar.). Retrieved Mar. 14, 2010. Online: http://bjs.ojp.usdoj.gov/content/pub/pdf/anhpivc.pdf

Harrington, Michael. 1985. *The New American Poverty.* New York: Viking/Penguin.

Harrington Meyer, Madonna. 1990. "Family Status and Poverty Among Older Women: The Gendered Distribution of Retirement Income in the United States." *Social Problems,* 37: 551–563.

Harris, Chauncey D., and Edward L. Ullman. 1945. "The Nature of Cities." *Annals of the Academy of Political and Social Sciences* (November): 7–17.

Harris, Marvin. 1974. *Cows, Pigs, Wars, and Witches.* New York: Random House.

—. 1985. *Good to Eat: Riddles of Food and Culture.* New York: Simon & Schuster.

Harrison, Algea O., Melvin N. Wilson, Charles J. Pine, Samuel Q. Chan, and Raymond Buriel. 1990. "Family Ecologies of Ethnic Minority Children." *Child Development,* 61 (2): 347–362.

Hartmann, Heidi. 1976. "Capitalism, Patriarchy, and Job Segregation by Sex." *Signs: Journal of Women in Culture and Society,* 1 (Spring): 137–169.

—. 1981. "The Unhappy Marriage of Marxism and Feminism." In Lydia Sargent (Ed.), *Women and Revolution.* Boston: South End.

Havighurst, Robert J., Bernice L. Neugarten, and Sheldon S. Tobin. 1968. "Patterns of Aging." In Bernice L. Neugarten (Ed.), *Middle Age and Aging.* Chicago: University of Chicago Press, pp. 161–172.

Haviland, William A. 1993. *Cultural Anthropology* (7th ed.). Orlando, FL: Harcourt.

—. 1999. *Cultural Anthropology* (9th ed.). Orlando, FL: Harcourt.

Hawley, Amos. 1950. *Human Ecology.* New York: Ronald.

—. 1981. *Urban Society* (2nd ed.). New York: Wiley.

Hays, Tom. 2011. "More Than 120 Busted in Northeast Mafia Crackdown." *Yahoo! News* (Jan. 20). Retrieved Jan. 20, 2011. Online: http://news.yahoo.com/s/ap/us_mob_arrests/print

Headley, Bernard. 1985. "The Atlanta Establishment and the Atlanta Tragedy." *Phylon,* 46(4): 333–340.

Heard, Amy. 2011. "Point of View: Campus Kitchen Is Worth Being Excited About." Baylor Lariat.com (Feb. 8). Retrieved Feb. 11, 2011. Online: http://baylorlariat.com/2011/02/08/2310

Helpguide.org. 2011. "Adult Day Care Centers." Retrieved Mar. 24, 2011. Online: www.helpguide.org/elder/adult_day_care_centers.htm

Henley, Nancy. 1977. *Body Politics: Power, Sex, and Nonverbal Communication.* Englewood Cliffs, NJ: Prentice Hall.

Heritage, John. 1984. *Garfinkel and Ethnomethodology.* Cambridge, MA: Polity.

Herrnstein, Richard J., and Charles Murray. 1994. *The Bell Curve: Intelligence and Class Structure in American Life.* New York: Free Press.

Heshka, Stanley, and Yona Nelson. 1972. "Interpersonal Speaking Distances as a Function of Age, Sex, and Relationship." *Sociometry,* 35 (4): 491–498.

Hesse-Biber, Sharlene, and Gregg Lee Carter. 2000. *Working Women in America: Split Dreams.* New York: Oxford University Press.

Hibbard, David R., and Duane Buhrmester. 1998. "The Role of Peers in the Socialization of Gender-Related Social Interaction Styles." *Sex Roles,* 39: 185–203.

Hiebert, Paul G. 1983. *Cultural Anthropology* (2nd ed.). Grand Rapids, MI: Baker.

Higginbotham, Elizabeth. 1994. "Black Professional Women: Job Ceilings and Employment Sectors." In Maxine Baca Zinn and Bonnie Thornton Dill (Eds.), *Women of Color in U.S. Society.* Philadelphia: Temple University Press, pp. 113–131.

Hirschi, Travis. 1969. *Causes of Delinquency.* Berkeley: University of California Press.

Hochschild, Arlie Russell. 1983. *The Managed Heart: Commercialization of Human Feeling.* Berkeley: University of California Press.

—. 1997. *The Time Bind: When Work Becomes Home and Home Becomes Work.* New York: Metropolitan.

—. 2003. *The Commercialization of Intimate Life: Notes from Home and Work.* Berkeley: University of California Press.

Hochschild, Arlie Russell, with Ann Machung. 1989. *The Second Shift: Working Parents and the Revolution at Home.* New York: Viking/Penguin.

Hoecker-Drysdale, Susan. 1992. *Harriet Martineau: First Woman Sociologist.* Oxford, England: Berg.

Hoffman, Jan. 2009. "Binging on Celebrity Weight Battles." *New York Times* (May 31). Retrieved Mar. 11, 2011. Online: www.nytimes.com/2009/05/31/fashion/31fat.html?_r=2&pagewanted=1

—. 2011. "Fighting Teenage Pregnancy with MTV Stars as Exhibit A." *New York Times* (Apr. 10): ST1, ST11.

Holland, Dorothy C., and Margaret A. Eisenhart. 1990. *Educated in Romance: Women, Achievement, and College Culture.* Chicago: University of Chicago Press.

Hoose, Phillip M. 1989. *Necessities: Racial Barriers in American Sports.* New York: Random House.

Hoover, Kenneth R. 1992. *The Elements of Social Scientific Thinking.* New York: St. Martin's.

Hooyman, Nancy, and H. Asuman Kiyak. 2011. *Social Gerontology: A Multidisciplinary Perspective* (9th ed.). Upper Saddle River, NJ: Pearson.

Horan, Patrick M. 1978. "Is Status Attainment Research Atheoretical?" *American Sociological Review,* 43: 534–541.

Horovitz, Brude. 2006. "More University Students Call for Organic, 'Sustainable' Food." *USA Today* (Sept. 26). Retrieved Feb. 17, 2007. Online: www.usatoday.com/money/industries/food/2006-09-26-college-food-usat_x.htm

Housing Assistance Council. 2003. "Rural Seniors and Their Homes." Retrieved Mar. 21, 2011. Online: www.ruralhome.org/storage/documents/ruralseniors.pdf

Howie, John R. R. 2010. "A Final Word." *Blog-Abroad.* StudyAbroad.com. Retrieved Feb. 12, 2010. Online: www.studyabroad.com/blog-abroad/howie

Hoyt, Homer. 1939. *The Structure and Growth of Residential Neighborhoods in American Cities.* Washington, DC: Federal Housing Administration.

Hughes, Everett C. 1945. "Dilemmas and Contradictions of Status." *American Journal of Sociology,* 50: 353–359.

Hull, Gloria T., Patricia Bell-Scott, and Barbara Smith. 1982. *All the Women Are White, All the Blacks Are Men, But Some of Us Are Brave.* Old Westbury, NY: Feminist.

Humes, Karen R., Nicholas A. Jones, and Roberto R. Ramirez. 2011. "Overview of Race and Hispanic Origin: 2010." U.S. Census Bureau. Retrieved June 2, 2011. Online: www.census.gov/prod/cen2010/briefs/c2010br-02.pdf

Humphreys, Laud. 1970. *Tearoom Trade: Impersonal Sex in Public Places.* Chicago: Aldine.

Hurtado, Aida. 1996. *The Color of Privilege: Three Blasphemies on Race and Feminism.* Ann Arbor: University of Michigan Press.

Huston, Aletha C. 1985. "The Development of Sex Typing: Themes from Recent Research." *Developmental Review,* 5: 2–17.

Huyssen, Andreas. 1984. *After the Great Divide.* Bloomington: Indiana University Press.

Idaho Association of Soil Conservation Districts. 2004. "Organic Pest Control." Retrieved Feb. 17, 2007. Online: www.oneplan.org/Crop/OrganicPestCtrl.shtml

IFAD. 2002. "IFAD in China: The Rural Poor Speak." Rome, Italy: International Fund for Agricultural Development. Retrieved Jan. 22, 2008. Online: www.ifad.org/media/success/China.pdf

Inciardi, James A., Ruth Horowitz, and Anne E. Pottieger. 1993. *Street Kids, Street Drugs, Street Crime: An Examination of Drug Use and Serious Delinquency in Miami.* Belmont, CA: Wadsworth.

Institute of International Education. 2010. "Open Doors 2009 Report on International Educational Exchange." Retrieved Feb. 12, 2010. Online: http://opendoors.iienetwork.org/?p=150651

International Foundation for Electoral Systems. 2011. "Gender Issues." Retrieved Mar. 12, 2011. Online: www.ifes.org/Content/Topics/Gender-Issues.aspx

International Longevity Center—Japan. 2010. "A Profile of Older Japanese" (Mar. 31). Retrieved Mar. 15, 2011. Online: www.ilcjapan.org/agingE/POJ10.html

International Monetary Fund. 2010. "World Economic Outlook" (Oct.). Retrieved Mar. 31, 2011. Online: www.imf.org/external/pubs/ft/weo/2010/02/weodata/weorept.aspx?

Intersex Society of North America. 2011. "What Is Intersex?" Retrieved Mar. 11, 2011. Online: www.isna.org/faq/what_is_intersex

Ishikawa, Kaoru. 1984. "Quality Control in Japan." In Naoto Sasaki and David Hutchins (Eds.), *The Japanese Approach to Product Quality: Its Applicability to the West.* Oxford: Permagon, pp. 1–5.

Isidore, Chris. 2010. "7.9 Million Jobs Lost—Many Forever." CNNMoney.com (July 2). Retrieved Mar. 25, 2011. Online: http://money.cnn.com/2010/07/02/news/economy/jobs_gone_forever/index.htm

Jackson, Kenneth T. 1985. *Crabgrass Frontier: The Suburbanization of the United States.* New York: Oxford University Press.

Jacobs, Andrew. 2008. "One-Child Policy Lifted for Quake Victims' Parents." *New York Times* (May 27): A1.

—. 2011. "Where 'Jasmine' Means Tea, Not a Revolt." *New York Times* (Apr. 3): WK4.

Jameson, Fredric. 1984. "Postmodernism, or, The Cultural Logic of Late Capitalism." *New Left Review,* 146: 59–92.

Janis, Irving. 1972. *Victims of Groupthink.* Boston: Houghton Mifflin.

—. 1989. *Crucial Decisions: Leadership in Policymaking and Crisis Management.* New York: Free Press.

Jankowski, Martin Sanchez. 1991. *Islands in the Street: Gangs and American Urban Society.* Berkeley: University of California Press.

Japsen, Bruce. 2011. "Chicago Has 2nd Patient in Landmark Embryonic Stem Cell Trial." Chicago Tribune.com (May 10). Retrieved May 12, 2011. Online: http://articles.chicagotribune.com/2011-05-10/business/ct-biz-0511-stem-cells-geron-20110510_1_cell-trial-cell-types-oligodendrocyte

Jaramillo, P. T., and Jesse T. Zapata. 1987. "Roles and Alliances Within Mexican-American and Anglo Families." *Journal of Marriage and the Family,* 49 (November): 727–735.

Jary, David, and Julia Jary. 1991. *The Harper Collins Dictionary of Sociology.* New York: HarperPerennial.

Jewell, K. Sue. 1993. *From Mammy to Miss America and Beyond: Cultural Images and the Shaping of US Social Policy.* New York: Routledge.

John Curley Center for Sports Journalism at Penn State. 2007. "Jason Whitlock and the New Racism." *Sports, Media and Society* (Oct. 24). Retrieved Mar. 8, 2010. Online: http://sportsmediasociety.blogspot.com/2007/10/jason-whitlock-and-new-racism.html

Johnson, Earvin "Magic," with William Novak. 1992. *My Life.* New York: Fawcett Crest.

Johnson, Guy. 2006. On the Move: A Longitudinal Study of Pathways in and out of Homelessness. Ph.D. Dissertation, submitted to the School of Global Studies, Social Science and Planning, RMIT University, Australia (May). Retrieved Feb. 26, 2010. Online: www.ahuri.edu/au/downloads/Research_Training/PhDs/Johnson_Onthemove.pdf

Johnson, Jenna. 2010. "First Lady Pays Off on Challenge to Serve with GWU Commencement Speech." *Washington Post* (May 17). Retrieved Dec. 31, 2010. Online: www.washingtonpost.com/wp-dyn/content/article/2010/05/16/AR2010051601114.html

Johnson, Melody. 2011. "They Don't Get It: Unable to Take a Joke, Right-Wing Media Attack Obama Over 'Moat' Comment." *Media Matters for America* (May 11). Retrieved May 16, 2011. Online: http://mediamatters.org/print/research/201010010027

Joint Center for Political and Economic Studies. 2003. "Black Elected Officials: A Statistical Summary 2001." Retrieved Mar. 17, 2007. Online: www.jointcenter.org/publications1/publication-PDFs/BEO-pdfs/2001-BEO.pdf

Jolly, David, and Matthew Saltmarsh. 2009. "After Suicides: France Wrestles with Worker Stress." *New York Times* (Oct. 1): B3.

Kabagarama, Daisy. 1993. *Breaking the Ice: A Guide to Understanding People from Other Cultures.* Boston: Allyn & Bacon.

Kahn, Joseph, and Jim Yardley. 2008. "As China Roars, Pollution Reaches Deadly Extremes." *New York Times* (Aug. 26): A1, A6–7.

Kansas State University. 2010. "Timeline for Transition." Retrieved Dec. 22, 2010. Online: www.k-state.edu/parentsandfamily/resources/timeline.htm

Kanter, Rosabeth Moss. 1983. *The Change Masters: Innovation and Entrepreneurship in the American Corporation.* New York: Simon & Schuster.

—. 1985. "All That Is Entrepreneural Is Not Gold." *Wall Street Journal* (July 22): 18.

—. 1993. *Men and Women of the Corporation.* New York: Basic (orig. pub. 1977).

Karnowski, Steve. 2011. "Dalai Lama Says bin Laden's Killing Understandable." Associated Press (May 8). Retrieved May 8, 2011. Online: http://minnesota.publicradio.org/display/web/2011/05/08/dalai-lama-minnesota/?refid=0

Kasarda, John D., and Greg Lindsay. 2011. *Aerotropolis: The Way We'll Live Next.* New York: Farrar, Straus and Giroux.

Kaspar, Anne S. 1986. "Consciousness Re-evaluated: Interpretive Theory and Feminist Scholarship." *Sociological Inquiry,* 56 (1): 30–49.

Katzer, Jeffrey, Kenneth H. Cook, and Wayne W. Crouch. 1991. *Evaluating Information: A Guide for Users of Social Science Research.* New York: McGraw-Hill.

Kaufman, Gayle. 1999. "The Portrayal of Men's Family Roles in Television Commercials." *Sex Roles,* 313: 439–451.

Kaufman, Leslie. 2011. "City Prepares for a Warm Long-Term Forecast." *New York Times* (May 23): A1, A14.

Kaufman, Sharon R. 1994. "The Social Construction of Frailty: An Anthropological Perspective." *Journal of Aging Studies* (Spring): 45–58.

Kaufman, Tracy L. 1996. *Out of Reach: Can America Pay the Rent?* Washington, DC: National Low Income Housing Coalition.

Keep America Beautiful. 2009. "Graffiti Hurts® e-News" (Aug.). Retrieved Feb. 9, 2011. Online: www.kab.org/site/MessageViewer?dlv_id=0&em_id=2181.0

Kemp, Alice Abel. 1994. *Women's Work: Degraded and Devalued.* Englewood Cliffs, NJ: Prentice Hall.

Kendall, Diana. 1980. Square Pegs in Round Holes: Non-Traditional Students in Medical Schools. Unpublished doctoral dissertation, Department of Sociology, the University of Texas at Austin.

—. 2002. *The Power of Good Deeds: Privileged Women and the Social Reproduction of the Upper Class.* Lanham, MD: Rowman & Littlefield.

—. 2004. *Social Problems in a Diverse Society* (3rd ed.). Boston: Allyn & Bacon.

—. 2008. *Members Only: Elite Clubs and the Process of Exclusion.* Lanham, MD: Rowman & Littlefield.

—. 2011. *Framing Class: Media Representations of Wealth and Poverty in the United States* (2nd ed.). Lanham, MD: Rowman & Littlefield.

Kendall, Diana, Rick Linden, and Jane Murray. 2008. *Sociology in Our Times: The Essentials* (4th Canadian edition). Scarborough, Ontario: Nelson Thomson Learning.

Kendall, Diana, Jane Lothian Murray, and Rick Linden. 2004. *Sociology in Our Times* (3rd Canadian edition). Scarborough, Ontario: Nelson Thomson Learning.

Kennedy, Kerry. 2010. "Gulf Needs Concrete Actions That Respect Residents' Rights." *Huffington Post* (June 11). Retrieved June 12, 2010. Online: www.huffingtonpost.com/kerry-kennedy/gulf-needs-concrete-actions_b_609827.html

Kenyon, Kathleen. 1957. *Digging Up Jericho.* London: Benn.

Kerbo, Harold R. 2000. *Social Stratification and Inequality: Class Conflict in Historical, Comparative, and Global Perspective* (4th ed.). New York:McGraw-Hill.

Kerkman, Maggie. 2011. "Educating Our Children: The Evolution of Home Schooling." Foxnews.com (Feb. 9). Retrieved Apr. 29, 2011. Online: www.foxnews.com/us/2011/02/09/educating-children-evolution-home-schooling

Kessler, Ronald C. 1994. "Lifetime and 12-Month Prevalence of DSM-III-R Psychiatric Disorders in the United States: Results of the National Comorbidity Survey." *JAMA, the Journal of the American Medical Association,* 271 (Mar. 2): 654E.

Khan, Mahmood Hasan. 2001. "Rural Poverty in Developing Countries: Implications for Public Policy: Economic Issues No. 26." Washington, DC: International Monetary Fund. Retrieved Jan. 21, 2008. Online: www.imf.org/external/pubs/ft/issues/issues26/index.htm

Kilbourne, Jean. 1999. *Deadly Persuasion: The Addictive Power of Advertising.* New York: Simon & Schuster.

Kimmell, Michael S., and Michael A. Messner. 2004. *Men's Lives* (6th ed.). Boston: Allyn & Bacon.

King, Gary, Robert O. Keohane, and Sidney Verba. 1994. *Designing Social Inquiry: Scientific Inference in Qualitative Research.* Princeton, NJ: Princeton University Press.

King, Leslie, and Madonna Harrington Meyer. 1997. "The Politics of Reproductive Benefits: U.S. Insurance Coverage of Contraceptive and Infertility Treatments." *Gender and Society,* 11 (1): 8–30.

Kitsuse, John I. 1980. "Coming Out All Over: Deviance and the Politics of Social Problems." *Social Problems,* 28: 1–13.

Klein, Alan M. 1993. *Little Big Men: Bodybuilding Subculture and Gender Construction.* Albany: SUNY Press.

Knapp, Caroline. 1996. *Drinking: A Love Story.* New York: Dial.

Knox, Paul L., and Peter J. Taylor (Eds.). 1995. *World Cities in a World-System.* Cambridge, England: Cambridge University Press.

Knudsen, Dean D. 1992. *Child Maltreatment: Emerging Perspectives.* Dix Hills, NY: General Hall.

Kocieniewski, David. 2011. "G.E.'s Strategies Let It Avoid Taxes Altogether." *New York Times* (Mar. 24). Retrieved July 8, 2011. Online: www.nytimes.com/2011/03/25/business/economy/25tax.html?_r=1&ref=davidkocieniewski

Kohl, Beth. 2007. "On Indian Surrogates." *The Huffington Post* (Oct. 30). Retrieved Feb. 9, 2008. Online: www.huffingtonpost.com/beth-kohl/on-indian-surrogates_b_70425.html

Kohlberg, Lawrence. 1969. "Stage and Sequence: The Cognitive–Developmental Approach to Socialization." In David A. Goslin (Ed.), *Handbook of Socialization Theory and Research.* Chicago: Rand McNally, pp. 347–480.

—. 1981. *The Philosophy of Moral Development: Moral Stages and the Idea of Justice,* vol. 1: *Essays on Moral Development.* San Francisco: Harper & Row.

Kohn, Alfie. 2001. "Five Reasons to Stop Saying 'Good Job!'" Retrieved Jan. 3, 2008. Online: www.alfiekohn.org/parenting/gj.htm

Kohn, Melvin L. 1977. *Class and Conformity: A Study in Values* (2nd ed.). Homewood, IL: Dorsey.

Kohn, Melvin L., Atsushi Naoi, Carrie Schoenbach, Carmi Schooler, and Kazimierz M. Slomczynski. 1990. "Position in the Class Structure and Psychological Functioning in the United States, Japan, and Poland." *American Journal of Sociology,* 95: 964–1008.

Kolata, Gina. 1993. "Fear of Fatness: Living Large in a Slimfast World." *Austin American-Statesman* (Jan. 3): C1, C6.

Koopmans, Ruud. 1999. "Political. Opportunity. Structure. Some Splitting to Balance the Lumping." *Sociological Forum* (Mar.): 93–105.

Korsmeyer, Carolyn. 1981. "The Hidden Joke: Generic Uses of Masculine Terminology." In Mary Vetterling-Braggin (Ed.), *Sexist Language: A Modern Philosophical Analysis.* Totowa, NJ: Littlefield, Adams, pp. 116–131.

Korten, David C. 1996. *When Corporations Rule the World.* West Hartford, CT: Kumarian.
Kosmin, Barry A., and Ariela Keysar. 2009. "American Religious Identification Survey 2008: Summary Report" (Mar.). Retrieved May 8, 2011. Online: http://b27.cc.trincoll.edu/weblogs/AmericanReligionSurvey-ARIS/reports/ARIS_Report_2008.pdf
Kovacs, M., and A. T. Beck. 1977. "The Wish to Live and the Wish to Die in Attempted Suicides." *Journal of Clinical Psychology,* 33: 361–365.
Kozol, Jonathan. 1991. *Savage Inequalities: Children in America's Schools.* New York: Crown.
Kramer, Andrew E. 2010. "Philip Morris Is Said to Benefit from Child Labor." *New York Times* (July 14). Retrieved Mar. 6, 2011. Online: www.nytimes.com/2010/07/14/business/global/14smoke.html?_r=2&ref=childlabor
—. 2011. "Russia Cashes in on Anxiety Over Supply of Middle East Oil." *New York Times* (Mar. 7). Retrieved Mar. 28, 2011. Online: www.nytimes.com/2011/03/08/business/global/08oil.html
Kreider, Rose M., and Diana B. Elliott. 2009. *America's Families and Living Arrangements: 2007.* U.S. Census Bureau. Retrieved Apr. 2, 2010. Online: www.census.gov/population/www/socdemo/hh-fam/p20-561.pdf
Kristof, Nicholas D. 2006. "Looking for Islam's Luthers." *New York Times* (Oct. 15): A22.
Kübler-Ross, Elisabeth. 1969. *On Death and Dying.* New York: Macmillan.
Kumar, V. Phani. 2011. "China Reportedly Considering 'Two-Child' Policy." Marketwatch.com (Mar. 8). Retrieved May 15, 2011. Online: www.marketwatch.com/story/china-reportedly-considering-two-child-policy-2011-03-07
Kurtz, Lester. 1995. *Gods in the Global Village: The World's Religions in Sociological Perspective.* Thousand Oaks, CA: Sage.
Kvale, Steinar. 1996. *Interviews: An Introduction to Qualitative Research Interviewing.* Thousand Oaks, CA: Sage.

Ladd, E. C., Jr. 1966. *Negro Political Leadership in the South.* Ithaca, NY: Cornell University Press.
Lamanna, Mary Ann, and Agnes Riedmann. 2012. *Marriages, Families, and Relationships: Making Choices in a Diverse Society* (11th ed.). Belmont, CA: Wadsworth/Cengage.
Lapchick, Richard. 2010. "The 2010 Racial and Gender Report Card: College Sport." University of Central Florida: The Institute for Diversity and Ethics in Sport. Retrieved Mar. 7, 2011. Online: www.tidesport.org/RGRC/2010/2010_College_RGRC_FINAL.pdf
Lapham, Lewis H. 1988. *Money and Class in America: Notes and Observations on Our Civil Religion.* New York: Weidenfeld & Nicolson.
Lapsley, Daniel K., 1990. "Continuity and Discontinuity in Adolescent Social Cognitive Development." In Raymond Montemayor, Gerald R. Adams, and Thomas P. Gullota (Eds.), *From Childhood to Adolescence: A Transitional Period?* (*Advances in Adolescent Development,* vol. 2). Newbury Park, CA: Sage.
Larson, Magali Sarfatti. 1977. *The Rise of Professionalism: A Sociological Analysis.* Berkeley: University of California Press.
Lash, Scott, and John Urry. 1994. *Economies of Signs and Space.* London: Sage.
Latino Legends in Sports. 2007. "Sports News." Retrieved Mar. 17, 2007. Online: www.latinosportslegends.com
Laumann, Edward O., John H. Gagnon, Robert T. Michael, and Stuart Michaels. 1994. *The Social Organization of Sexuality.* Chicago: University of Chicago Press.
Le Bon, Gustave. 1960. *The Crowd: A Study of the Popular Mind.* New York: Viking (orig. pub. 1895).
Leenaars, Antoon A. 1988. *Suicide Notes: Predictive Clues and Patterns.* New York: Human Sciences Press.
Lefrançois, Guy R. 1996. *The Lifespan* (5th ed.). Belmont, CA: Wadsworth.
Leland, John. 2008. "From the Housing Market to the Maternity Ward." *New York Times* (Feb. 1): A12.
Lemert, Charles. 1997. *Postmodernism Is Not What You Think.* Malden, MA: Blackwell.
Lemert, Edwin M. 1951. *Social Pathology.* New York: McGraw-Hill.
Lengermann, Patricia Madoo, and Jill Niebrugge-Brantley. 1998. *The Women Founders: Sociology and Social Theory, 1830–1930.* New York: McGraw-Hill.
Leonard, Andrew. 1999. "We've Got Mail—Always." *Newsweek* (Sept. 20): 58–61.
Leonhardt, David. 2011. "In Wreckage of Lost Jobs, Lost Power." *New York Times* (Jan. 19). Retrieved Mar. 29, 2011. Online: www.nytimes.com/2011/01/19/business/economy/19leonhardt.html?_r=2&partner=rssnyt&emc=rss
Leptich, John. 2005. "Volunteers Pamper, Perk Up Cancer Patients." East Valley Tribune.com (Dec. 26). Retrieved Apr. 16, 2010. Online: www.eastvalleytribune.com/story/55921
Lerner, Gerda. 1986. *The Creation of Patriarchy.* New York: Oxford University Press.
Lester, David. 1992. *Why People Kill Themselves: A 1990s Summary of Research Findings of Suicidal Behavior* (3rd ed.). Springfield, IL: Thomas.
Levey, Hilary. 2007. "Here She Is . . . and There She Goes?" *Contexts* (Summer): 70–72.
Levin, William C. 1988. "Age Stereotyping: College Student Evaluations." *Research on Aging,* 10 (1): 134–148.
Levine, Peter. 1992. *Ellis Island to Ebbets Field: Sport and the American Jewish Experience.* New York: Oxford University Press.
Levinthal, Charles F. 2010. *Drugs, Behavior, and Modern Society* (6th ed.). Boston: Allyn & Bacon.
Levy, Becca R. 2009. "Stereotype Embodiment: A Psychosocial Approach to Aging." *Current Directions in Psychological Science,* 18: 332–336.
Lewis, Thabiti. 2010. *Ballers of the New School: Race and Sports in America.* Chicago: Third World.
Liebow, Elliot. 1993. *Tell Them Who I Am: The Lives of Homeless Women.* New York: Free Press.
Lindblom, Charles. 1977. *Politics and Markets.* New York: Basic.
Linton, Ralph. 1936. *The Study of Man.* New York: Appleton-Century-Crofts.
Lips, Hilary M. 2001. *Sex and Gender: An Introduction* (4th ed.). New York: McGraw-Hill.
Liptak, Adam. 2008. "Inmate Count in U.S. Dwarfs Other Nations." *New York Times* (Apr. 23). Retrieved Mar. 13, 2010. Online: www.nytimes.com/2008/04/23/us/23prison.html
Littal, Robert. 2011. "Why Is This Site Called BlackSportsOnline?" (Jan. 3). Retrieved Mar. 8, 2011. Online: http://blacksportsonline.com/home/why-is-the-site-called-blacksportsonline
Loeb, Paul Rogat. 1994. *Generation at the Crossroads: Apathy and Action on the American Campus.* New Brunswick, NJ: Rutgers University Press.
Lofland, John. 1993. "Collective Behavior: The Elementary Forms." In Russell L. Curtis, Jr., and Benigno E. Aguirre (Eds.), *Collective Behavior and Social Movements.* Boston: Allyn & Bacon, pp. 70–75.
Longman, Jere. 2009. "Alleging Racism, Soccer Star Seeks 'Moral Compensation.'" *New York Times* (June 14): Y1, Y8.
Lorber, Judith. 1994. *Paradoxes of Gender.* New Haven, CT: Yale University Press.
Lorber, Judith (Ed.). 2005. *Gender Inequality: Feminist Theories and Politics* (3rd ed.). Los Angeles: Roxbury.
Lott, Bernice. 1994. *Women's Lives: Themes and Variations in Gender Learning* (2nd ed.). Pacific Grove, CA: Brooks/Cole.
Low, Setha. 2003. *Behind the Gates: Life, Security, and the Pursuit of Happiness in Fortress America.* New York: Routledge.
Lummis, C. Douglas. 1992. "Equality." In Wolfgang Sachs (Ed.), *The Development Dictionary.* Atlantic Highlands, NJ: Zed, pp. 38–52.
Lundberg, Ferdinand. 1988. *The Rich and the Super-Rich: A Study in the Power of Money Today.* Secaucus, NJ: Lyle Stuart.
Lupton, Deborah. 1997. "Foucault and the Medicalisation Critique." In Alan Petersen and Robin Bunton (Eds.), *Foucault: Health and Medicine.* London: Routledge, pp. 94–110.
Lyall, Sarah. 2010. "Britain Plans to Decentralize Health Care." *New York Times* (July 24). Retrieved May 13, 2011. Online: www.nytimes.com/2010/07/25/world/europe/25britain.html?pagewanted=print
Lynd, Robert S., and Helen M. Lynd. 1929. *Middletown.* New York: Harcourt.
—. 1937. *Middletown in Transition.* New York: Harcourt.
Lynn, Kelci. 2010. "College Life: How to Reduce Stress While in College." Retrieved Feb. 3, 2010. Online: http://collegelife.about.com/od/healthwellness/ht/Stress.htm
Lynn, Richard. 2008. *The Global Bell Curve: Race, IQ, and Inequality Worldwide.* Augusta, GA: Washington Summit.
Lyotard, Jean-Francois. 1984. *The Postmodern Condition.* Minneapolis: University of Minnesota Press.

Maccoby, Eleanor E., and Carol Nagy Jacklin. 1987. "Gender Segregation in Childhood." *Advances in Child Development and Behavior,* 20: 239–287.
Mack, Julie. 2010. "Charter Schools Split Along Racial Lines: New Study Finds Parents' Choices Accelerate Resegregation." Mlive.com (Feb. 15). Retrieved Apr. 25, 2011. Online: http://blog.mlive.com/kzgazette_impact/print.html?entry=/2010/02/charter_schools_split_along_ra.html
Mack, Raymond W., and Calvin P. Bradford. 1979. *Transforming America: Patterns of Social Change* (2nd ed.). New York: Random House.
MacQueen, Ken. 2011. "Our Health Care Delusion." *Maclean's* (Jan. 25). Retrieved May 11, 2011. Online: www2.macleans.ca/2011/01/25/our-health-care-delusion
Madden, Mary. 2010. "Older Adults and Social Media." Pew Internet and American Life Project (Apr. 27). Retrieved Mar. 19, 2011. Online: http://pewinternet.org/Reports/2010/Older-Adults-and-Social-Media.aspx
Mahapatra, Rajesh. 2007. "Outsourced Jobs Take Toll on Indians' Health." *Austin American-Statesman* (Dec. 30): H1, H6.
Mahler, Sarah J. 1995. *American Dreaming: Immigrant Life on the Margins.* Princeton, NJ: Princeton University Press.
Malinowski, Bronislaw. 1922. *Argonauts of the Western Pacific.* New York: Dutton.
Malthus, Thomas R. 1965. *An Essay on Population.* New York: Augustus Kelley (orig. pub. 1798).

Mangione, Jerre, and Ben Morreale. 1992. *La Storia: Five Centuries of the Italian American Experience.* New York: HarperPerennial.

Mann, Coramae Richey. 1993. *Unequal Justice: A Question of Color.* Bloomington: Indiana University Press.

Mann, Patricia S. 1994. *Micro-Politics: Agency in a Postfeminist Era.* Minneapolis: University of Minnesota Press.

Manning, P. K., and B. Cullum-Swan. 1994. "Narrative, Content, and Semiotic Analysis." In Norman K. Denzin and Y. S. Lincoln (Eds.), *Handbook of Qualitative Research.* Thousand Oaks, CA: Sage.

Manning, Robert D. 2000. *Credit Card Nation: The Consequences of America's Addiction to Credit.* New York: Basic.

Mantsios, Gregory. 2003. "Media Magic: Making Class Invisible." In Michael S. Kimmel and Abby L. Ferber (Eds.), *Privilege: A Reader.* Boulder, CO: Westview, pp. 99–109.

Marger, Martin N. 1994. *Race and Ethnic Relations: American and Global Perspectives.* Belmont, CA: Wadsworth.

—. 2009. *Race and Ethnic Relations: American and Global Perspectives* (8th ed.). Belmont, CA: Wadsworth, Cengage Learning.

Marshall, Gordon. 1998. *A Dictionary of Sociology* (2nd ed.). New York: Oxford University Press.

Martin, Carol L. 1989. "Children's Use of Gender-Related Information in Making Social Judgments." *Developmental Psychology,* 25: 80–88.

Martin, Linda, and Kevin Kinsella. 1994. "Research in the Demography of Aging in Developing Countries." In Linda Martin and Samuel Preston (Eds.), *Demography of Aging.* Washington, DC: National Academic Press.

Martin, Philip, and Elizabeth Midgley. 2010. "Immigration in America 2010" (June). Population Reference Bureau. Retrieved May 15, 2011. Online: www.prb.org/Publications/PopulationBulletins/2010/immigrationupdate1.aspx

Martin, Susan Ehrlich, and Nancy C. Jurik. 1996. *Doing Justice, Doing Gender.* Thousand Oaks, CA: Sage.

Martineau, Harriet. 1962. *Society in America* (edited, abridged). Garden City, NY: Doubleday (orig. pub. 1837).

Marx, Karl. 1967. *Capital: A Critique of Political Economy.* Ed. Friedrich Engels. New York: International (orig. pub. 1867).

Marx, Karl, and Friedrich Engels. 1967. *The Communist Manifesto.* New York: Pantheon (orig. pub. 1848).

—. 1970. *The German Ideology,* Part 1. Ed. C. J. Arthur. New York: International (orig. pub. 1845–1846).

Matthews, Warren. 2004. *World Religions* (4th ed.). Belmont, CA: Wadsworth.

Maynard, R. A. 1996. *Kids Having Kids: A Robin Hood Foundation Special Report on the Costs of Adolescent Childbearing.* New York: Robin Hood Foundation.

McAdam, Doug. 1996. "Conceptual Origins, Current Problems, Future Directions." In Doug McAdam, John D. McCarthy, and Meyer N. Zald (Eds.), *Comparative Perspectives on Social Movements.* New York: Cambridge University Press, pp. 23–40.

McCarthy, John D., and Mayer N. Zald. 1977. "Resource Mobilization and Social Movements: A Partial Theory." *American Journal of Sociology,* 82: 1212–1241.

McClure, Robin. 2010. "Afterschool Child Care—Number of Kids Home Alone After School Has Risen." Retrieved Apr. 2, 2010. Online: http://childcare.about.com/od/schoolagetopics/a/homealone.htm

McDonnell, Janet A. 1991. *The Dispossession of the American Indian, 1887–1934.* Bloomington: Indiana University Press.

McHale, Susan M., Ann C. Crouter, and C. Jack Tucker. 1999. "Family Context and Gender Role Socialization in Middle Childhood: Comparing Girls to Boys and Sisters to Brothers." *Child Development,* 70: 990–1004.

McKenzie, Roderick D. 1925. "The Ecological Approach to the Study of the Human Community." In Robert Park, Ernest Burgess, and Roderick D. McKenzie, *The City.* Chicago: University of Chicago Press.

McKinley, James C., Jr. 2010. "Fleeing Extreme Drug Violence, Mexican Families Pour into U.S." *New York Times* (Apr. 18): A1, A14.

McPhail, Clark. 1991. *The Myth of the Maddening Crowd.* New York: Aldine de Gruyter.

McPherson, Barry D., James E. Curtis, and John W. Loy. 1989. *The Social Significance of Sport: An Introduction to the Sociology of Sport.* Champaign, IL: Human Kinetics.

Mead, George Herbert. 1934. *Mind, Self, and Society.* Chicago: University of Chicago Press.

Meals on Wheels Association of America. 2011. "Take Action: Volunteer." Retrieved June 15, 2011. Online: www.mowaa.org/page.aspx?pid=396

Medicare.gov. 2007. "Nursing Homes." Retrieved Mar. 29, 2007. Online: www.medicare.gov/nursing/overview.asp

Mennell, Stephen. 1996. *All Manners of Food: Eating and Taste in England and France from the Middle Ages to the Present.* Urbana: University of Illinois Press.

Mennell, Stephen, Anne Murcott, and Anneke H. van Otterloo. 1993. *The Sociology of Food: Eating, Diet and Culture.* Thousand Oaks, CA: Sage.

Merton, Robert King. 1938. "Social Structure and Anomie." *American Sociological Review,* 3 (6): 672–682.

—. 1949. "Discrimination and the American Creed." In Robert M. MacIver (Ed.), *Discrimination and National Welfare.* New York: Harper & Row, pp. 99–126.

—. 1968. *Social Theory and Social Structure* (enlarged ed.). New York: Free Press.

Messenger, David. 2009. "Studies Show People More Stressed as Students Than at Other Stages of Life Due to Work, Relationships." Retrieved Feb. 8, 2010. Online: www.studlife.com/news/2009/10/30/studies-shows-people-more-stressed-as-students-than-at-other-stages-of-life-due-to-work-relationships

Miall, Charlene. 1986. "The Stigma of Involuntary Childlessness." *Social Problems,* 33 (4): 268–282.

Michael, Robert T., John H. Gagnon, Edward O. Laumann, and Gina Kolata. 1994. *Sex in America. Boston: Little, Brown.*

Michels, Robert. 1949. Political Parties. Glencoe, IL: Free Press (orig. pub. 1911).

Miethe, Terance, and Charles Moore. 1987. "Racial Differences in Criminal Processing: The Consequences of Model Selection on Conclusions About Differential Treatment." *Sociological Quarterly,* 27: 217–237.

Milgram, Stanley. 1963. "Behavioral Study of Obedience." *Journal of Abnormal and Social Psychology,* 67: 371–378.

—. 1967. "The Small World Problem." *Psychology Today,* 1: 61–67.

—. 1974. *Obedience to Authority.* New York: Harper & Row.

Miller, Casey, and Kate Swift. 1991. *Words and Women: New Language in New Times* (updated ed.). New York: HarperCollins.

Miller, Dan E. 1986. "Milgram Redux: Obedience and Disobedience in Authority Relations." In Norman K. Denzin (Ed.), *Studies in Symbolic Interaction.* Greenwich, CT: JAI, pp. 77–106.

Mills, C. Wright. 1956. *White Collar.* New York: Oxford University Press.

—. 1959a. *The Power Elite.* Fair Lawn, NJ: Oxford University Press.

—. 1959b. *The Sociological Imagination.* London: Oxford University Press.

—. 1976. *The Causes of World War Three. Westport, CT: Greenwood.*

Mills, Iain. 2010. "China Puts Healthcare Cart Before the Horse." *Asia Times Online* (Apr. 21). Retrieved May 13, 2011. Online: www.atimes.com/atimes/China/LD21Ad01.html

Min, Eungjun (Ed.). 1999. *Reading the Homeless: The Media's Image of Homeless Culture.* Westport, CT: Praeger.

Min, Pyong Gap. 1988. "The Korean American Family." In Charles H. Mindel, Robert W. Habenstein, and Roosevelt Wright, Jr. (Eds.), *Ethnic Families in America: Patterns and Variations* (3rd ed.). New York: Elsevier, pp. 199–229.

Mindlin, Alex. 2009. "Children Watch More TV Than Ever." *New York Times* (Nov. 2): B3.

mindoh.com. 2007. "I Wish I Knew What to Do?!" Retrieved Feb. 15, 2008. Online: www.mindoh.docs/Bullyingbook_excerpt_noCCC.pdf

Mishler, Elliot G. 1984. *The Discourse of Medicine: Dialectics of Medical Interviews.* Norwood, NJ: Ablex.

—. 2005. "The Struggle Between the Voice of Medicine and the Voice of the Lifeworld." In Peter Conrad (Ed.), *The Sociology of Health and Illness: Critical Perspectives* (7th ed.). New York: Worth, pp. 319–330.

MIT 21st Century Manifesto Working Group. 1999. "What Do We Really Want? A Manifesto for the Organizations of the 21st Century." Sloan School of Management, Massachusetts Institute of Technology (Nov.). Retrieved Mar. 8, 2010. Online: http://ccs.mit.edu/papers/pdf/wp032manifesto21C.pdf

Mixedfolks.com. 2011. "Well Known Mixed Athletes." Retrieved Mar. 23, 2011. Online: www.mixedfolks.com/mfc/Athletes.html

Mohai, Paul, and Robin Saha. 2007. "Racial Inequality in the Distribution of Hazardous Waste: A National-Level Reassessment." *Social Problems* (August): 343–370.

Molnar, Alex, Gary Miron, and Jessica L. Urschel. 2010. "Profiles of For-Profit Education Management Organizations, 2009–2010." National Education Policy Center, University of Colorado at Boulder. Retrieved Apr. 29, 2011. Online: http://nepc.colorado.edu/files/EMO-FP-09-10.pdf

Monahan, John. 1992. "Mental Disorder and Violent Behavior: Perceptions and Evidence." *American Psychologist,* 47: 511–521.

Mooney, Karen. 2011. "Raleigh Marchers Fight 'Re-segregation' Plan." abc.news.com (Feb. 12). Retrieved Apr. 25, 2011. Online: http://abcnews.go.com/Politics/raleigh-nc-marchers-fight-segregation-plan/story?id=12904201

Moore, K. A., A. K. Driscoll, and L. D. Lindberg. 1998. *A Statistical Portrait of Adolescent Sex, Contraception, and Childbearing.* Washington, DC: National Campaign to Prevent Teen Pregnancy.

Moore, Kristin Anderson, Zakia Redd, Mary Burkhauser, Kassim Mbwana, and Ashleigh Collins. 2009. "Children in Poverty: Trends, Consequences, and Policy Options." *Trends: Child Research Brief* (Publication #2009-11). Retrieved Feb. 21, 2011. Online: www.childtrends.org/files/child_trends-2009_04_07_rb_childreninpoverty.pdf

Moore, Patricia, with C. P. Conn. 1985. *Disguised.* Waco, TX: Word.

Morello, Carol. 2010. "Before Downturn, Opportunities Boomed; Census: Firms Owned by Women, Minorities Were on the Rise." *Washington Post* (July 14). Retrieved Feb. 20, 2011. Online: www.washingtonpost.com/wp-dyn/content/article/2010/07/13/AR2010071302389.html

Morgan, Leslie, and Suzanne Kunkel. 1998. *Aging: The Social Context.* Thousand Oaks, CA: Pine Forge.

Morgan, Sally. 2010. "Community College Changed My Life." BetterGrads.org (May 25). Retrieved Apr. 23, 2011. Online: http://bettergrads.org/blog/2010/05/25/why-college-part-7-%E2%80%93-community-college-changed-my-life

Morselli, Henry. 1975. *Suicide: An Essay on Comparative Moral Statistics.* New York: Arno (orig. pub. 1881).

msnbc.com. 2011. "Japanese Earthquake: Gripping Survivor Stories" (Mar. 13). Retrieved Mar. 15, 2011. Online: http://bltwy.msnbc.msn.com/politics/japan-earthquake-gripping-survivor-stories-1683413.story

Murdock, George P. 1945. "The Common Denominator of Cultures." In Ralph Linton (Ed.), *The Science of Man in the World Crisis.* New York: Columbia University Press, pp. 123–142.

Mydans, Seth. 1993. "A New Tide of Immigration Brings Hostility to the Surface, Poll Finds." *New York Times* (June 27): A1.

Naffine, Ngaire. 1987. *Female Crime: The Construction of Women in Criminology.* Boston: Allen & Unwin.

National Center for Educational Statistics. 2010a. "Indicators of School Crime and Safety: 2010." Retrieved Apr. 18, 2011. Online: http://nces.ed.gov/programs/crimeindicators/crimeindicators2010/ind_11.asp

—. 2010b. "Trends in High School Dropout and Completion Rates in the United States: 1972–2008." Retrieved Apr. 24, 2011. Online: http://nces.ed.gov/pubs2011/dropout08/findings1.asp

National Center for Health Statistics. 2008. "National Suicide Statistics at a Glance." Retrieved Apr. 20, 2011. Online: www.cdc.gov/violenceprevention/suicide/statistics/rates02.html

—. 2010. *Health, United States, 2010.* Hyattsville, MD. Retrieved May 18, 2011. Online: www.cdc.gov/nchs .data/hus/hus10.pdf

National Center for Injury Prevention and Control. 2006. "2004 United States Suicide Injury Deaths and Rates per 100,000." Retrieved Jan. 28, 2007. Online: http://webapp.cdc.gov/sasweb/ncipc/mortrate10_sy.html

National Center for Victims of Crime. 2011. "Elder Abuse." Retrieved Mar. 21, 2010. Online: www.ncvc.org/ncvc/main.aspx?dbName=DocumentViewer&DocumentID=32350#2

National Center on Elder Abuse. 2010. "What Is Abuse? Why Should I Care About Elder Abuse?" Retrieved Mar. 24, 2011. Online: www.ncea.aoa.gov/Ncearoot/Main_Site/pdf/publication/NCEA_WhatIsAbuse-2010.pdf

National Centers for Disease Control and Prevention. 2008. "Understanding School Violence: Fact Sheet." Retrieved Aug. 26, 2009. Online: www.cdc.gov/Violence Prevention/pdf/SchoolViolence_FactSheet-a.pdf

—. 2009a. "FastStats." Retrieved July 21, 2009. Online: www.cdc.gov/nchs/fastats.htm

—. 2009b. "National Suicide Statistics at a Glance." Retrieved Jan. 24, 2010. Online: www.cdc.gov/violenceprevention/suicide/statistics/rates02.html.

—. 2010. "Emergency Wound Management for Healthcare Professionals" (Jan. 15). Retrieved May 11, 2011. Online: http://emergency.cdc.gov/disasters/emergwoundhcp.asp

National Coalition on Health Care. 2009. "Facts on Health Care Costs." Retrieved July 21, 2009. Online: www.nchc.org/documents/Cost%20Fact%20Sheet-2009.pdf

National Council on Crime and Delinquency. 1969. *The Infiltration into Legitimate Business by Organized Crime.* Washington, DC: National Council on Crime and Delinquency.

National Crime Prevention Council. 2011. "Graffiti Is Everybody's Problem." Retrieved Feb. 9, 2011. Online: www.ncpc.org/cms/cms-upload/ncpc/files/graffiti_r.pdf

National Institute of Justice. 2011. "Transnational Organized Crime." Retrieved Feb. 9, 2011. Online: www.ojp.usdoj.gov/nij/topics/crime/transnational-organized-crime/welcome.htm

National Institute on Drug Abuse. 2010. "NIDA InfoFacts: Cigarettes and Other Tobacco Products." Retrieved Apr. 16, 2010. Online: www.drugabuse.gov/Infofacts/Tobacco.html

National Law Center on Homelessness and Poverty. 2004. "Homelessness and Poverty in America." Retrieved Jan. 21, 2006. Online: www.nlchp.org/FA%5FHAPIA

National League of Cities. 2010. "Significant Budget Shortfalls Could Mean More Job Losses" (May 24). National League of Cities. Retrieved June 10, 2010. Online: www.ncl.org.PRESSROOM/PRESSRELEASEITEMS/SoACJobsEcon5.10.aspx

National Marriage Project. 2010. "When Marriage Disappears: The New Middle America." Retrieved Apr. 18, 2011. Online: www.virginia.edu/marriageproject/pdfs/Union_11_12_10.pdf

National Science Foundation. 2010. "Science and Engineering Indicators, 2010: Current Expenditures per Pupil for Elementary and Secondary Public Schools." Retrieved July 11, 2011. Online: www.nsf.gov/statistics/seind10/c8/c8s1o11.htm

National Survey of Student Engagement. 2007. "Annual Report 2007: Experiences That Matter: Enhancing Student Learning and Success." Retrieved Apr. 15, 2011. Online: http://nsse.iub.edu/NSSE_2007_Annual_Report/docs/withhold/NSSE_2007_Annual_Report.pdf

National Survey on Drug Use and Health. 2006. "Trends in Alcohol, Tobacco and other Drug Use." Retrieved Apr. 13, 2010. Online: www.oas.samhsa.gov/trends.htm#Cocaine

National Vital Statistics System. 2010. "Deaths: Final Data for 2007" (May 20). Retrieved May 15, 2011. Online: www.cdc.gov/NCHS/data/nvsr/nvsr58/nvsr58_19.pdf

Navarrette, Ruben, Jr. 1997. "A Darker Shade of Crimson." In Diana Kendall (Ed.), *Race, Class, and Gender in a Diverse Society.* Boston: Allyn & Bacon, 1997: 274–279. Reprinted from Ruben Navarrette, Jr., *A Darker Shade of Crimson.* New York: Bantam, 1993.

NCDC. 2007. "About NCDC: Northwestern Community Development Corps." Retrieved Mar. 31, 2007. Online: http://groups.northwestern.edu/ncdc/about.html

nces.ed.gov. 2008. "Revenues and Expenditures for Public Elementary and Secondary Education." Retrieved July 11, 2011. Online: http://nces.ed.gov/pubs2008/expenditures/figures/figure_01.asp

—. 2010. "The Condition of Education: 2010: Indicator 33, Public School Revenue Sources." Retrieved Apr. 30, 2011. Online: http://nces.ed.gov/programs/coe/2010/pdf/33_2010.pdf

NCSL. 2010. "Homeless and Runaway Youth" (Apr.). Retrieved June 12, 2010. Online: www.ncsl.org/?tabid=18275

Nelson, Margaret K. 2010. *Parenting Out of Control: Anxious Parents in Uncertain Times.* New York: New York University Press.

Nelson, Margaret K., and Joan Smith. 1999. *Working Hard and Making Do: Surviving in Small Town America.* Berkeley: University of California Press.

New York Post. 2004. "Death Plunge No. 4: NYU's Grief." (Mar. 10): 1.

New York State Troopers. 2011. "Crime Prevention: Graffiti Is Everybody's Problem." Retrieved Feb. 9, 2011. Online: www.troopers.state.ny.us/Crime_Prevention/Juvenile_Crime/Graffiti

New York Times. 2008. "President Map—Election Results." Online: http://elections.nytimes.com/2008/results/president/map.html

Newfoundation.org. 2011. "Love Your Body: On Campus, in Community." Washington, D.C.: National Organization of Women. Retrieved June 12, 2011. Online: http://loveyourbody.nowfoundation.org/oncampus.html

Newman, Katherine S. 1988. *Falling from Grace: The Experience of Downward Mobility in the American Middle Class.* New York: Free Press.

—. 1993. *Declining Fortunes: The Withering of the American Dream.* New York: Basic.

—. 1999. *No Shame in My Game: The Working Poor in the Inner City.* New York: Knopf and the Russell Sage Foundation.

Newsweek. 1997. "Cult: Now on the Next Level." (Dec. 29/Jan. 5): 17.

Nguyen, Bich Minh. 2007. *Stealing Buddha's Dinner: A Memoir.* New York: Viking.

Niebuhr, H. Richard. 1929. *The Social Sources of Denominationalism.* New York: Meridian.

Nielsen, Joyce McCarl. 1990. *Sex and Gender in Society: Perspectives on Stratification* (2nd ed.). Prospects Heights, IL: Waveland.

Noel, Donald L. 1972. *The Origins of American Slavery and Racism.* Columbus, OH: Merrill.

Nolan, Patrick, and Gerhard E. Lenski. 1999. *Human Societies: An Introduction to Macrosociology* (8th ed.). New York: McGraw-Hill.

Norris, Floyd. 2011. "Japan's Meltdown and the Global Economy's." *New York Times* (Mar. 17). Retrieved Apr. 2, 2011. Online: www.nytimes.com/2011/03/18/business/18norris.html

Nowrasteh, Alex. 2011. "The Unhappy Anniversary of Arizona's Anti-Immigration Law." Forbes.com (Apr. 27). Retrieved May 15, 2011. Online: www.forbes.com/2011/04/26/immigration-law-anniversary.html

Nuland, Sherwin B. 1997. "Heroes of Medicine." *Time* (Fall special edition): 6–10.

nytimes.com. 2011. Multimedia: "A Horrible Feeling" (audio clip of interview with Maricela Aguilar, Feb. 9). Retrieved Feb. 10, 2010. Online: www.nytimes.com/2011/02/09/us/09immigration.html

Odendahl, Teresa. 1990. *Charity Begins at Home: Generosity and Self-Interest Among the Philanthropic Elite.* New York: Basic.

Ogburn, William F. 1966. *Social Change with Respect to Culture and Original Nature.* New York: Dell (orig. pub. 1922).

Ohio State University. 2007. "Cultural Diversity: Eating in America." Ohio State University Extension Fact Sheet, Family and Consumer Sciences Series. Retrieved Feb. 11, 2007. Online: www.ohioline.ag.ohio-state.edu

Omaar, Rageh. 2007. "The World of Modern Child Slavery." *BBC News* (Mar. 27). Retrieved Mar. 14, 2010. Online: http://news.bbc.co.uk/go/pr/fr/-/2/hi/programmes/this_world/6458377.stm

Omi, Michael, and Howard Winant. 1994. *Racial Formation in the United States: From the 1960s to the 1990s.* New York: Routledge.

Orr, Matthew, and Susan Saulny. 2011. nytimes.com (Jan. 29): Video: "Young and Mixed in America." Retrieved Mar. 7, 2011. Online: www.nytimes.com

Ott, Thomas. 2011. "Cleveland Students Hold Their Own with Voucher Students on State Tests." Cleveland.com (Feb. 22). Retrieved Apr. 27, 2011. Online: http://blog.cleveland.com/metro/2011/02/cleveland_students_hold_own_wi.html

Oxendine, Joseph B. 2003. *American Indian Sports Heritage* (rev. ed.). Lincoln: University of Nebraska Press.

Padilla, Felix M. 1993. *The Gang as an American Enterprise.* New Brunswick, NJ: Rutgers University Press.

Page, Charles H. 1946. "Bureaucracy's Other Face." *Social Forces,* 25 (October): 89–94.

Palacio, Zulima. 2009. "Study Shows U.S. College Students Active in Promoting Sustainability." Voanews.com (December). Retrieved Apr. 20, 2010. Online: www1.voanews.com/English/news/environment/Study-Shows-US-College-Students-Active-in-Promoting-Sustainability-78675327.html

Palen, J. John. 2012. *The Urban World* (9th ed.). Boulder, CO: Paradigm.

Parenti, Michael. 1996. *Democracy for the Few* (5th ed.). New York: St. Martin's.

—. 1998. *America Besieged.* San Francisco: City Lights.

Park, Robert E. 1915. "The City: Suggestions for the Investigation of Human Behavior in the City." *American Journal of Sociology,* 20: 577–612.

—. 1928. "Human Migration and the Marginal Man." *American Journal of Sociology,* 33.

—. 1936. "Human Ecology." *American Journal of Sociology,* 42: 1–15.

Park, Robert E., and Ernest W. Burgess. 1921. *Human Ecology.* Chicago: University of Chicago Press.

Parker-Pope, Tara. 2010. "She Works. They're Happy." *New York Times* (Jan. 24): ST1.

Parrish, Dee Anna. 1990. *Abused: A Guide to Recovery for Adult Survivors of Emotional/Physical Child Abuse.* Barrytown, NY: Station Hill.

Parsons, Talcott. 1951. *The Social System.* Glencoe, IL: Free Press.

—. 1955. "The American Family: Its Relations to Personality and to the Social Structure." In Talcott Parsons and Robert F. Bales (Eds.), *Family, Socialization and Interaction Process.* Glencoe, IL: Free Press, pp. 3–33.

—. 1960. "Toward a Healthy Maturity." *Journal of Health and Social Behavior,* 1: 163–173.

Passel, Jeffrey S., and D'Vera Cohn. 2010. "U.S. Unauthorized Immigration Flows Are Down Sharply Since Mid-Decade" (Sept. 1). Retrieved May 15, 2011. Online: http://pewhispanic.org/files/reports/126.pdf

PBS. 2005a. "The Meaning of Food: Food & Culture." Retrieved Feb. 11, 2007. Online: www.pbs.org/opb/meaningoffood

—. 2005b. "Online NewsHour: The Schiavo Case Receives Strong Media Coverage" (Mar. 24). Retrieved Apr. 10, 2005. Online: www.pbs.org/newshour/bb/media/jan-june05/schiavo_3-24.html

—. 2008. "Facts About Global Poverty and Microcredit." Retrieved Jan. 19, 2008. Online: www.pbs.org/toourcredit/facts_one.htm

Pear, Robert. 2011. "Health Spending Rose in '09, but at Low Rate." *New York Times* (Jan. 5). Retrieved May 12, 2011. Online: www.nytimes.com/2011/01/06/health/06health.html?_r=2&sq=health law&st=cse&scp=3&pagewanted=print

Pearce, Diana. 1978. "The Feminization of Poverty: Women, Work, and Welfare." *Urban and Social Change Review,* 11 (1/2): 28–36.

Pearson, Judy C. 1985. *Gender and Communication.* Dubuque, IA: Brown.

People. 2010. "Tormented to Death?" (Oct. 6): 56–58.

PerfSpot.com. 2011. "The Enjoyment of Chat Groups." Retrieved Feb. 7, 2011. Online: www.perfspot.com/articles/groups_for_chat.asp

Perry, Steven W. 2004. "American Indians and Crime." U.S. Bureau of Justice Statistics. Retrieved Mar. 19, 2005. Online: www.ojp.usdoj.gov/bjs/pub/pdf/aic02.pdf

Petersen, John L. 1994. *The Road to 2015: Profiles of the Future.* Corte Madera, CA: Waite Group.

Peterson, Robert. 1992. *Only the Ball Was White: A History of Legendary Black Players and All-Black Professional Teams.* New York: Oxford University Press (orig. pub. 1970).

Pew Charitable Trusts. 2007. "Economic Mobility: Is the American Dream Alive and Well?" Retrieved Dec. 20, 2007. Online: www.economicmobility.org

Pew Forum on Religion and Public Life. 2008. "U.S. Religious Landscape Survey: Religious Affiliation—Diverse and Dynamic." Retrieved May 8, 2011. Online: http://religions.pewforum.org/pdf/report-religious-landscape-study-full.pdf

Pew Internet & American Life Project. 2010. "Teens and Mobile Phones" (Apr. 20). Retrieved Mar. 19, 2011. Online: www.pewinternet.org/~/media//Files/Reports/2010/PIP-Teens-and-Mobile-2010-with-topline.pdf

Pew Research Center. 2008. "Inside the Middle Class: Bad Times Hit the Good Life" (Apr. 9). Retrieved May 18, 2011. Online: http://pewsocialtrends.org/files/2010/10/MC-Middle-class-report1.pdf

—. 2011. "State of the News Media: 2011." Retrieved Apr. 8, 2011. Online: http://pewresearch.org/pubs/1924/state-of-the-news-media-2011

Phillips, John C. 1993. *Sociology of Sport.* Boston: Allyn & Bacon.

Phillips, Sarah. 2009. "Ivy Bean: The Oldest Tweeter in Town." *Guardian* (Oct. 23). Retrieved Mar. 21, 2011. Online: www.guardian.co.uk/technology/2009/oct/23/ivy-bean-oldest-tweeter/print

—. 2010. "Ivy Bean Obituary." *Guardian* (July 30). Retrieved Mar. 21, 2011. Online: www.guardian.co.uk/technology/2010/jul/30/ivy-bean-obituary/print

Piaget, Jean. 1954. *The Construction of Reality in the Child.* Trans. Margaret Cook. New York: Basic.

Picca, Leslie Houts, and Joe R. Feagin. 2007. *Two-Faced Racism: Whites in the Backstage and Frontstage.* New York: Routledge.

Pierre-Pierre, Garry. 1997. "Traditional Church's New Life." *New York Times* (Nov. 15): A11.

Pillemer, Karl A. 1985. "The Dangers of Dependency: New Findings on Domestic Violence Against the Elderly." *Social Problems,* 33 (December): 146–158.

Pinderhughes, Dianne M. 1986. "Political Choices: A Realignment in Partisanship Among Black Voters?" In James D. Williams (Ed.), *The State of Black America 1986.* New York: National Urban League, pp. 85–113.

Pines, Maya. 1981. "The Civilizing of Genie." *Psychology Today,* 15 (September): 28–29, 31–32, 34.

Pitzer, Ronald. 2003. "Rural Children Under Stress." University of Minnesota Extension Service. Retrieved Aug. 9, 2003. Online: www.extension.umn.edu/distribution/familydevelopment/components/7269cm.html

Plunkett Research. 2010. "Health Care Trends." Retrieved Apr. 15, 2010. Online: www.plunkettresearch.com/Industries/HealthCare/HealthCareTrends/tabid/294/Default.aspx

Polanyi, Karl. 1944. *The Great Transformation: The Political and Economic Origins of Our Time.* New York: Beacon.

Postman, Joel. 2011. "MLK: From a Thing-Oriented Society to a Person-Oriented One." *Social Kapital.* Retrieved Apr. 20, 2011. Online: http://socialized.tumblr.com/post/71619428/mlk-from-a-thing-oriented-society-to-a-person-oriented

Postman, Neil, and Steve Powers. 1992. *How to Watch TV News.* New York: Penguin.

Powell, Michael. 2004. "Evolution Shares a Desk with 'Intelligent Design.'" *Washington Post* (Dec. 26): A1.

Powers, Ron. 2010. "Obama Shocked, Saddened by Tyler Clementi, Other Youth Suicides." DailyRecord.com (Oct. 22). Retrieved Nov. 26, 2010. Online: www.dailyrecord.com/article/20101022/UPDATES01/101022002/1005/RSS

Preston, Julia. 2010. "Immigrants in Work Force: Study Belies Image." *New York Times* (Apr. 16): A1, A3.

Project for Excellence in Journalism. 2010. "The State of the News Media." Pew Internet and American Life Project. Retrieved Apr. 12, 2010. Online: www.stateofthemedia.org/2010/printable_online_survey_chapter.htm

Prus, Robert. 1996. *Symbolic Interaction and Ethnographic Research: Intersubjectivity and the Study of Human Lived Experience.* Albany: State University of New York Press.

Psychology Today. 2007. "Teen Spirit: Give and Let Live" (Apr. 29). Retrieved Mar. 20, 2011. Online: www.psychologytoday.com/print/23810

Puette, William J. 1992. *Through Jaundiced Eyes: How the Media View Organized Labor.* Ithaca, NY: ILR.

Puffer, J. Adams. 1912. *The Boy and His Gang.* Boston: Houghton Mifflin.

Putnam, Robert D., and David E. Campbell. 2010. *American Grace: How Religion Divides and Unites Us.* New York: Simon & Schuster.

Quinney, Richard. 2001. *Critique of the Legal Order.* Piscataway, NJ: Transaction (orig. pub. 1974).

Qvortrup, Jens. 1990. *Childhood as a Social Phenomenon.* Vienna: European Centre for Social Welfare Policy and Research.

Raby, Rosalind. 2010. "Community College Study Abroad: Making Study Abroad Accessible to All Students." Institute of International Education (IIE Network). Retrieved Feb. 12, 2010. Online: www.iienetwork.org/page/91081

Radcliffe-Brown, A. R. 1952. *Structure and Function in Primitive Society.* New York: Free Press.

Raffaelli, Marcela, and Lenna L. Ontai. 2004. "Gender Socialization in Latino/a Families: Results from Two Retrospective Studies." *Sex Roles,* 50: 287–299.

Ramirez, Marc. 1999. "A Portrait of a Local Muslim Family." *Seattle Times* (Jan. 24). Retrieved Aug. 16, 1999. Online: http://archives.seattletimes.com/cgi-bin/texis.mummy/web/vortex/display?storyID=36d4d218

Rand, Michael R. 2009. "Criminal Victimization, 2008." Bureau of Justice Statistics, U.S. Department of Justice (Sept.). Retrieved Mar. 14, 2010. Online: http://bjs.ojp.usdoj.gov/content/pub/pdf/cv08.pdf

Ratha, Dilip, and William Shaw. 2007. "South–South Migration and Remittances." New York:

World Bank Development Prospects Group. Retrieved Jan. 21, 2008. Online: http://siteresources.worldbank.org/INTPROSPECTS/Resources/SouthSouthMigrationandRemittances.pdf

Ray, Matt. 2007. *Environmental Health Perspectives,* 115 (8) (Oct.).

Reckless, Walter C. 1967. *The Crime Problem.* New York: Meredith.

Reich, Robert. 1993. "Why the Rich Are Getting Richer and the Poor Poorer." In Paul J. Baker, Louis E. Anderson, and Dean S. Dorn (Eds.), *Social Problems: A Critical Thinking Approach* (2nd ed.). Belmont, CA: Wadsworth, pp. 145–149. Adapted from *The New Republic,* May 1, 1989.

Reiman, Jeffrey. 1998. *The Rich Get Richer and the Poor Get Prison: Ideology, Class, and Criminal Justice* (5th ed.). Boston: Allyn & Bacon.

Reiner, Robert. 2007. "Media Made Criminality: The Representation of Crime in the Mass Media." In Mike Maguire, Rodney Morgan, and Robert Reiner (Eds.), *The Oxford Handbook of Criminology.* New York: Oxford University Press, pp. 376–416.

Reinharz, Shulamit. 1992. *Feminist Methods in Social Research.* New York: Oxford University Press.

Reinisch, June. 1990. *The Kinsey Institute New Report on Sex: What You Must Know to Be Sexually Literate.* New York: St. Martin's.

religioustolerance.org. 2011. "Science and/versus Religion." Retrieved May 10, 2011. Online: www.religioustolerance.org/scirelintro.htm

Relman, Arnold S. 1992. "Self-Referral—What's at Stake?" *New England Journal of Medicine,* 327 (Nov. 19): 1522–1524.

Reuters News Service. 2008. "New mtvU and Associated Press Poll Shows How Stress, War, the Economy, and Other Factors Are Affecting College Students' Mental Health." Retrieved Feb. 8, 2010. Online: www.reuters.com/article/idUS173716+19-Mar-2008+PRN20080319

Rich, Motoko. 2011. "Many Jobs Seen as Failing to Meet the Basics." *New York Times* (Apr. 1): B1, B6.

Richardson, Laurel. 1993. "Inequalities of Power, Property, and Prestige." In Virginia Cyrus (Ed.), *Experiencing Race, Class, and Gender in the United States.* Mountain View, CA: Mayfield, pp. 229–236.

Rigler, David. 1993. "Letters: A Psychologist Portrayed in a Book About an Abused Child Speaks Out for the First Time in 22 Years." *New York Times Book Review* (June 13): 35.

Riley, Dylan. 2009. "New Report Shows How Students Can Go Green." mainecampus.com (Nov. 19). Retrieved Apr. 20, 2010. Online: http://mainecampus.com/2009/11/19/new-report-shows-how-students-can-go-green

Ritzer, George. 1995. *Expressing America: A Critique of the Global Credit Card Society.* Thousand Oaks, CA: Pine Forge.

—. 1997. *Postmodern Society Theory.* New York: McGraw-Hill.

—. 1998. *The McDonaldization Thesis.* London: Sage.

—. 1999. *Enchanting a Disenchanted World: Revolutionizing the Means of Consumption.* Thousand Oaks, CA: Pine Forge.

—. 2000a. *The McDonaldization of Society.* Thousand Oaks, CA: Pine Forge.

—. 2000b. *Modern Sociological Theory* (5th ed.). New York: McGraw-Hill.

—. 2011. *Sociological Theory* (8th ed.). New York: McGraw-Hill.

Rizzo, Thomas A., and William A. Corsaro. 1995. "Social Support Processes in Early Childhood Friendships: A Comparative Study of Ecological Congruences in Enacted Support." *American Journal of Community Psychology,* 23: 389–418.

Robbins, Alexandra. 2004. *Pledged: The Secret Life of Sororities.* New York: Hyperion.

Roberts, Dexter. 2008. "China Prepares for Urban Revolution." *Bloomberg Businessweek* (Nov. 13). Retrieved May 22, 2011. Online: www.businessweek.com/print/globalbiz/content/nov2008/gb20081113_305364.htm

Roberts, Sam. 2010a. "Census Figures Challenge View of Race and Ethnicity." *New York Times* (Jan. 22): A13.

—. 2010b. "Study Finds Cohabiting Doesn't Make a Union Last." *New York Times* (Mar. 3): A14.

Robertson, Ann. 1990. "The Politics of Alzheimer's Disease: A Case Study in Apocalyptic Demography." *International Journal of Health Services,* 20 (3): 429–442.

Robinson, Brian E. 1988. *Teenage Fathers.* Lexington, MA: Lexington.

Rocca, Mo. 2007. "TV Drug Ads' Side Effects." CBS News *Sunday Morning* (Oct. 14). Retrieved Feb. 24, 2008. Online: www.cbsnews.com/stories/2007/10/14/sunday/main3365346.shtml

Rodriguez, Clara E. 1989. *Puerto Ricans: Born in the U.S.A.* New York: Unwin Hyman.

Roethlisberger, Fritz J., and William J. Dickson. 1939. *Management and the Worker.* Cambridge, MA: Harvard University Press.

Rogers, Dexter. 2011. "White Heterosexual Men Dominate the Sports Journalism Arena but Are They Objective About Race Issues?" *Colorlines Magazine* (Jan. 3). Retrieved Mar. 8, 2010. Online: http://gayblackcanadianman.com/2011/01/03/colorlines-magazine-article-white-heterosexual-men-dominate-the-sports-journalism-arena-but-are-they-objective-about-race-issues

Rollins, Judith. 1985. *Between Women: Domestics and Their Employers.* Philadelphia: Temple University Press.

Ropers, Richard H. 1991. *Persistent Poverty: The American Dream Turned Nightmare.* New York: Plenum.

Rosenthal, Robert, and Lenore Jacobson. 1968. *Pygmalion in the Classroom: Teacher Expectation and Student's Intellectual Development.* New York: Holt, Rinehart, and Winston.

Ross, Dorothy. 1991. *The Origins of American Social Science.* Cambridge, England: Cambridge University Press.

Ross, Shannon E., Bradley C. Niebling, and Teresa M. Heckert. 1999. "Sources of Stress Among College Students." *College Student Journal* (June): 312.

Rossides, Daniel W. 1986. *The American Class System: An Introduction to Social Stratification.* Boston: Houghton Mifflin.

Rostow, Walt W. 1971. *The Stages of Economic Growth: A Non-Communist Manifesto* (2nd ed.). Cambridge: Cambridge University Press (orig. pub. 1960).

—. 1978. *The World Economy: History and Prospect.* Austin: University of Texas Press.

Roth, Guenther. 1988. "Marianne Weber and Her Circle." In Marianne Weber, *Max Weber.* New Brunswick, NJ: Transaction, p. xv.

Rothchild, John. 1995. "Wealth: Static Wages, Except for the Rich." *Time* (Jan. 30): 60–61.

Rothman, Ellen Lerner. 1999. *White Coat: Becoming a Doctor at Harvard Medical School.* New York: Morrow.

Rousseau, Ann Marie. 1981. *Shopping Bag Ladies: Homeless Women Speak About Their Lives.* New York: Pilgrim.

Rubin, Lillian B. 1986. "A Feminist Response to Lasch." *Tikkun,* 1 (2): 89–91.

Rural Policy Research Institute. 2004. "Place Matters: Addressing Rural Poverty." Rural Poverty Research Institute. Retrieved Jan. 21, 2008. Online: www.rprconline.org/synthesis.pdf

Rymer, Russ. 1993. *Genie: An Abused Child's Flight from Silence.* New York: HarperCollins.

Sadker, David, and Myra Sadker. 1985. "Is the OK Classroom OK?" *Phi Delta Kappan,* 55: 358–367.

—. 1986. "Sexism in the Classroom: From Grade School to Graduate School." *Phi Delta Kappan,* 68: 512–515.

Sadker, Myra, and David Sadker. 1994. *Failing at Fairness: How America's Schools Cheat Girls.* New York: Scribner.

Safilios-Rothschild, Constantina. 1969. "Family Sociology or Wives' Family Sociology? A Cross-Cultural Examination of Decision-Making." *Journal of Marriage and the Family,* 31 (2): 290–301.

SallieMae.com. 2009. "Study Finds Rising Number of College Students Using Credit Cards for Tuition." SallieMae.com. Retrieved Nov. 13, 2010. Online: www.salliemae.com/about/news_info/newsreleases/041309.html

Salt Lake City Sheriff's Department. 2007. "Graffiti: That Writing on the Wall." Retrieved Mar. 10, 2007. Online: www.slsheriff.org/html/org/metrogang/graffiti.html

Samovar, Larry A., and Richard E. Porter. 1991. *Communication Between Cultures.* Belmont, CA: Wadsworth.

Sandals, Leah. 2007. "'Public Space Protection'—But for Which 'Public'?" Retrieved Feb. 24, 2007. Online: http://spacing.ca/wire/?p=1466

Sapir, Edward. 1961. *Culture, Language and Personality.* Berkeley: University of California Press.

Sargent, Margaret. 1987. *Sociology for Australians* (2nd ed.). Melbourne, Australia: Longman Cheshire.

SAT. 2010. "2010 College-Bound Seniors: Total Group Profile Report." College Board. Retrieved July 8, 2011. Online: http://professionals.collegeboard.com/profdownload/2010-total-group-profile-report-cbs.pdf

Saulny, Susan. 2011a. "Black? White? Asian? More Young Americans Choose All of the Above." *New York Times* (Jan. 29). Retrieved Mar. 7, 2011. Online: www.nytimes.com/2011/01/30/us/30mixed.html

—. 2011b. "In a Multiracial Nation, Many Ways to Tally." *New York Times* (Feb. 10): A1, A17.

Savin-Williams, Ritch C. 2004. "Memories of Same-Sex Attractions." In Michael S. Kimmel and Michael A. Messner (Eds.), *Men's Lives* (6th ed.). Boston: Allyn & Bacon, pp. 116–132.

Schaefer, James. 2005. "Reporting Complexity: Science and Religion." In Claire Hoertz Badaracco (Ed.), *Quoting God: How Media Shape Ideas About Religion and Culture.* Waco, TX: Baylor University Press, pp. 211–224.

Schaefer, Richard T., and William W. Zellner. 2007. *Extraordinary Groups: An Examination of Unconventional Lifestyles* (8th ed.). New York: Worth.

Schama, Simon. 1989. *Citizens: A Chronicle of the French Revolution.* New York: Knopf.

Schneider, Donna. 1995. *American Childhood: Risks and Realities.* New Brunswick, NJ: Rutgers University Press.

Scholastic Parent & Child. 2007. "How and When to Praise." Retrieved Jan. 3, 2008. Online: http://content.scholastic.com/browse/article.jsp?id=2064

Schor, Juliet B. 1999. *The Overspent American: Upscaling, Downshifting, and the New Consumer.* New York: HarperPerennial.

Schubert, Hans-Joachim (Ed.). 1998. "Introduction." In *On Self and Social Organization—Charles Horton Cooley.* Chicago: University of Chicago Press, pp. 1–31.

Schur, Edwin M. 1983. *Labeling Women Deviant: Gender, Stigma, and Social Control.* Philadelphia: Temple University Press.

Schutske, John. 2002. "Keeping Farm Children Safe." University of Minnesota Extension Service. Retrieved Aug. 9, 2003. Online: www.extension.umn.edu/distribution/youth development/DA6188.html

Schwartz, John. 2003. "Too Much Information, Not Enough Knowledge." *New York Times* (June 9): WK5.

—. 2010. "Bullying, Suicide, Punishment." *New York Times* (Oct. 3): WK 1, 3.

Schwartz, John, and John M. Broder. 2003. "Engineer Warned of Consequences of Liftoff Damage." *New York Times* (Feb. 13): A1–A29.

Schwartz, John (with Matthew L. Wald). 2003. "Costs and Risk Clouding Plans to Fix Shuttles." *New York Times* (June 8): A1–A20.

Schwartz, John, and Matthew L. Wald. 2003. "'Groupthink' Is 30 Years Old, and Still Going Strong." *New York Times* (Mar. 9): WK3.

Schwarz, John E., and Thomas J. Volgy. 1992. *The Forgotten Americans.* New York: Norton.

ScienceDaily. 2007. "Study Examines Video Game Play Among Adolescents." *Science Daily* (July 4). Retrieved Dec. 21, 2010. Online: www.sciencedaily.com/releases/2007/07/070702161141.htm

Scott, Alan. 1990. *Ideology and the New Social Movements.* Boston: Unwin & Hyman.

Seccombe, Karen. 1991. "Assessing the Costs and Benefits of Children: Gender Comparisons Among Childfree Husbands and Wives." *Journal of Marriage and the Family,* 53 (1): 191–202.

Seegmiller, B. R., B. Suter, and N. Duviant. 1980. *Personal, Socioeconomic, and Sibling Influences on Sex-Role Differentiation.* Urbana: ERIC Clearinghouse of Elementary and Early Childhood Education, ED 176 895, College of Education, University of Illinois.

Seid, Roberta P. 1994. "Too 'Close to the Bone': The Historical Context for Women's Obsession with Slenderness." In Patricia Fallon, Melanie A. Katzman, and Susan C. Wooley (Eds.), *Feminist Perspectives on Eating Disorders.* New York: Guilford, pp. 3–16.

Semple, Kirk. 2011. "Illegal Immigrants' Children Face Hardships, Study Says." *New York Times* (May 21): A15.

Sengoku, Tamotsu. 1985. *Willing Workers: The Work Ethic in Japan, England, and the United States.* Westport, CT: Quorum.

Sengupta, Somini. 1997. "At Holidays, Test of Patience of Muslims." *New York Times* (Dec. 25): A12.

Senna, Joseph J., and Larry J. Siegel. 2002. *Introduction to Criminal Justice* (9th ed.). Belmont, CA: Wadsworth.

Serbin, Lisa A., Phyllis Zelkowitz, Anna-Beth Doyle, Dolores Gold, and Bill Wheaton. 1990. "The Socialization of Sex-Differentiated Skills and Academic Performance: A Mediational Model." *Sex Roles,* 23: 613–628.

Shah, Anup. 2011. "Health Issues" (Apr. 9). Retrieved May 10, 2011. Online: www.globalissues.org/print/issue/587

Shapin, Steven. 2006. "Paradise Sold: What Are You Buying When You Buy Organic?" *New Yorker* (May 15). Retrieved Feb. 17, 2007. Online: www.newyorker.com/critics/atlarge/articles/060515crat_atlarge

Shawver, Lois. 1998. "Notes on Reading Foucault's *The Birth of the Clinic.*" Retrieved Oct. 2, 1999. Online: www.california.com/~rathbone/foucbc.htm

Sheen, Fulton J. 1995. *From the Angel's Blackboard: The Best of Fulton J. Sheen.* Liguori, MO: Triumph.

Sherman, Suzanne (Ed.). 1992. "Frances Fuchs and Gayle Remick." In *Lesbian and Gay Marriage: Private Commitments, Public Ceremonies.* Philadelphia: Temple University Press, pp. 189–201.

Shevky, Eshref, and Wendell Bell. 1966. *Social Area Analysis: Theory, Illustrative Application and Computational Procedures.* Westport, CT: Greenwood.

Short, Elizabeth, Damien W. Riggs, Amaryll Perlesz, Rhonda Brown, and Graeme Kane. 2007. "Lesbian, Gay, Bisexual and Transgender (LGBT) Parented Families" (Aug.). Australian Psychological Society. Retrieved Apr. 17, 2011. Online: www.psychology.org.au/Assets/Files/LGBT-Families-Lit-Review.pdf

Shum, Tedd. 1997. "Olympic Gymnast Chow Makes Impact on All Americans." Retrieved Aug. 15, 1999. Online: www.dailybruin.ucla.edu/DB/issues/97/0530/view.shum.html

Siegel, Larry J. 2006. *Criminology* (9th ed.). Belmont, CA: Wadsworth.

—. 2007. *Criminology: Theories, Patterns, and Typologies* (9th ed.). Belmont, CA: Wadsworth.

Simmel, Georg. 1950. *The Sociology of Georg Simmel.* Trans. Kurt Wolff. Glencoe, IL: Free Press (orig. written in 1902–1917).

—. 1957. "Fashion." *American Journal of Sociology,* 62 (May 1957): 541–558 (orig. pub. 1904).

—. 1990. *The Philosophy of Money.* Ed. David Frisby. New York: Routledge (orig. pub. 1907).

Simon, David R. 2008. *Elite Deviance* (9th ed.). Boston: Allyn & Bacon.

Simpson, Sally S. 1989. "Feminist Theory, Crime, and Justice." *Criminology,* 27: 605–632.

Singleton-Rickman, Lisa. 2011. "Local Hospitals Remain on Alert" *Times Daily* (Florence, Ala., Apr. 29). Retrieved May 11, 2011. Online: www.timesdaily.com/article/20110429/news/110429780&tc=yahoo

Sjoberg, Gideon. 1965. *The Preindustrial City: Past and Present.* New York: Free Press.

Skinner, E. Benjamin. 2008. *A Crime So Monstrous: Face-to-Face with Modern-Day Slavery.* New York: Free Press.

Smelser, Neil J. 1963. *Theory of Collective Behavior.* New York: Free Press.

—. 1988. "Social Structure." In Neil J. Smelser (Ed.), *Handbook of Sociology.* Newbury Park, CA: Sage, pp. 103–129.

Smith, Adam. 1976. *An Inquiry into the Nature and Causes of the Wealth of Nations.* Ed. Roy H. Campbell and Andrew S. Skinner. Oxford, England: Clarendon (orig. pub. 1776).

Smith, Allen C., III, and Sheryl Kleinman. 1989. "Managing Emotions in Medical School: Students' Contacts with the Living and the Dead." *Social Science Quarterly,* 52 (1): 56–69.

Smith, Douglas, Christy Visher, and Laura Davidson. 1984. "Equity and Discretionary Justice: The Influence of Race on Police Arrest Decisions." *Journal of Criminal Law and Criminology,* 75: 234–249.

Smith, Huston. 1991. *The World's Religions.* San Francisco: HarperSanFrancisco.

Smith, Wes. 2001. *Hope Meadows: Real-Life Stories of Healing and Caring from an Inspiring Community.* New York: Berkley.

Smolkin, Rachel. 2007. "What the Mainstream Media Can Learn from Jon Stewart." *American Journalism Review* (June/July 2007). Retrieved Feb. 16, 2008. Online: www.ajr.org/Article.asp?id=4329

Snow, David A., and Leon Anderson. 1991. "Researching the Homeless: The Characteristic Features and Virtues of the Case Study." In Joe R. Feagin, Anthony M. Orum, and Gideon Sjoberg (Eds.), *A Case for the Case Study.* Chapel Hill: University of North Carolina Press, pp. 148–173.

—. 1993. *Down on Their Luck: A Case Study of Homeless Street People.* Berkeley: University of California Press.

Snow, David A., and Robert Benford. 1988. "Ideology, Frame Resonance, and Participant Mobilization." In Bert Klandermans, Hanspeter Kriesi, and Sidney Tarrow (Eds.), *International Social Movement Research,* Vol. 1, *From Structure to Action.* Greenwich, CT: JAI, pp. 133–155.

Snow, David A., E. Burke Rochford, Jr., Steven K. Worden, and Robert D. Benford. 1986. "Frame Alignment Processes, Micromobilization, and Movement Participation." *American Sociological Review,* 51: 464–481.

Snow, David A., Louis A. Zurcher, and Robert Peters. 1981. "Victory Celebrations as Theater: A Dramaturgical Approach to Crowd Behavior." *Symbolic Interaction,* 4 (1): 21–41.

Snyder, Benson R. 1971. *The Hidden Curriculum.* New York: Knopf.

Snyder, Michael. 2010. "The Middle Class in America Is Radically Shrinking. Here Are the Stats to Prove It." *Business Insider* (July 15). Retrieved July 24, 2010. Online: http://finance.yahoo.com/ttech-ticker/the-u.s.-middle-class-is-being-wiped-out-here%27s-the-stats-to-prove-it-520657.html

Society for Human Resource Management. 2010. "Workers with Disabilities Face Steep Occupational Obstacles" (Dec. 8). Retrieved Apr. 2, 2011. Online: www.shrm.org/hrdisciplines/Diversity/Articles/Pages/FaceSteepOccupationalObstacles.aspx

Solem, Per Erik. 2008. "Age Changes in Subjective Work Ability." *International Journal of Aging and Later Life* 3 (2): 43–70.

Stack, Steven. 1998. "Gender, Marriage, and Suicide Acceptability: A Comparative Analysis." *Sex Roles,* 38: 501–521.

Stack, Steven, and I. Wasserman. 1995. "The Effect of Marriage, Family, and Religious Ties on African American Suicide Ideology." *Journal of Marriage and the Family,* 57: 215–222.

Stake, Robert E. 1995. *The Art of Case Study Research.* Thousand Oaks, CA: Sage.

Stanley, Alessandra. 2004. "Old-Time Sexism Suffuses New Season." *New York Times* (Oct. 1): B1–B22.

—. 2009. "A Wink at Colleges and a Nod to Cliches." *New York Times* (Sept. 17). Retrieved Dec. 21, 2010. Online: http://tv.nytimes.com/2009/09/17/arts/television/17community.html

Stannard, David E. 1992. *American Holocaust: Columbus and the Conquest of the New World.* New York: Oxford University Press.

Stapleton-Paff, Katie. 2007. "College Students Prefer 'The Daily Show' to Real News." *The Daily of the University of Washington* (May 21, 2007): 1.

Stark, Rodney, and William Sims Bainbridge. 1981. "American-Born Sects: Initial Findings." *Journal for the Scientific Study of Religion,* 20: 130–149.

—. 1985. *The Future of Religion: Secularization, Revival and Cult Formation.* Berkeley, CA: University of California Press.

Stark, Rodney, and Gary Tobin. 2008. "Guest Commentary: Competition and the American Religious Marketplace." JCR: Institute for Jewish and Community Research (Mar.). Retrieved

May 2, 2011. Online: www.jewishresearch.org/v2/2008/articles/demography/03_08.htm
Steffensmeier, Darrell, and Emilie Allan. 2000. "Looking for Patterns: Gender, Age, and Crime." In Joseph F. Sheley (Ed.), *Criminology: A Contemporary Handbook* (3rd ed.). Belmont, CA: Wadsworth, pp. 85–128.
Stein, Peter J. 1976. *Single*. Englewood Cliffs, NJ: Prentice Hall.
Stein, Peter J. (Ed.). 1981. *Single Life: Unmarried Adults in Social Context*. New York: St. Martin's.
Stevenson, Mary Huff. 1988. "Some Economic Approaches to the Persistence of Wage Differences Between Men and Women." In Ann H. Stromberg and Shirley Harkess (Eds.), *Women Working: Theories and Facts in Perspective* (2nd ed.). Mountain View, CA: Mayfield, pp. 87–100.
Stewart, Abigail J. 1994. "Toward a Feminist Strategy for Studying Women's Lives." In Carol E. Franz and Abigail J. Stewart (Eds.), *Women Creating Lives: Identities, Resilience, and Resistance*. Boulder, CO: Westview, pp. 11–35.
Stickler, Christine. 2004. "One Response to Special Needs in the Classroom: Utilizing College Students as an Untapped Resource." *New Horizons for Learning* (September). Retrieved Apr. 17, 2005. Online: www.newhorizons.org/lifelong/higher_ed/stickler.htm
St. John, Warren. 2007. "A Laboratory for Getting Along." *New York Times* (Dec. 25): A1, A14.
StopGlobalWarming.org. 2006. "Marchers." Retrieved May 6, 2006. Online: www.stopglobalwarming.org
Sugrue, Thomas J. 2011. "A Dream Still Deferred." *New York Times* (Mar. 26). Retrieved May 23, 2011. Online: www.nytimes.com/2011/03/27/opinion/27Sugrue.html?ref=michigan&pagewanted=print
Suhr, Jim. 2007. "Police: Elderly Driver in School Crash Was Bound for Driving School." *Chicago Tribune* (Jan. 30). Retrieved Mar. 25, 2007. Online: www.chicagotribune.com/news/local/illinois/chi-ap-il-carhitsschool,1,3832602.story?coll=chi-newsap_il-hed&ctrack=1&cset=true
Sumner, William G. 1959. *Folkways*. New York: Dover (orig. pub. 1906).
Surowiecki, James. 2011. "State of the Unions." *New Yorker* (Jan. 17). Retrieved Mar. 26, 2011. Online: www.newyorker.com/talk/financial/2011/01/17/110117ta_talk_surowiecki?printable=true&mbid=social_release
Sutherland, Edwin H. 1939. *Principles of Criminology*. Philadelphia: Lippincott.
—. 1949. *White Collar Crime*. New York: Dryden.
Swartz, Mimi. 2010. "The Lost Girls." *Texas Monthly* (Apr.): 104–109, 194–199.
Swidler, Ann. 1986. "Culture in Action: Symbols and Strategies." *American Sociological Review*, 51 (April): 273–286.
Szasz, Thomas S. 1984. *The Myth of Mental Illness: Foundations of a Theory of Personal Conduct*. New York: HarperCollins.

Takaki, Ronald. 1993. *A Different Mirror: A History of Multicultural America*. Boston: Little, Brown.
Tannen, Deborah. 1993. "Commencement Address, State University of New York at Binghamton." Reprinted in *Chronicle of Higher Education* (June 9): B5.
Tarbell, Ida M. 1925. *The History of Standard Oil Company*. New York: Macmillan (orig. pub. 1904).
Tau, Byron. 2011. "Obama Faces Brave New Web World in 2012." Politico.com (Apr. 4). Retrieved Apr. 4, 2011. Online: www.politico.com/news/stories/0311/51594.html
Tavris, Carol. 1993. *The Mismeasure of Woman*. New York: Touchstone.
Tax Policy Center. 2009. "Historical Number of Households, Average Pretax and After-Tax Income and Shares, and Minimum Income." Retrieved May 28, 2009. Online: www.taxpolicy-center.org/UploadedPDF/901006_taxpolicy.pdf
Taylor, Steve. 1982. *Durkheim and the Study of Suicide*. New York: St. Martin's.
Teicher, Stacy A. 2006. "Researchers Say That Middle-School Bullying Could Be Curbed by Showing That It's Not Normal." *Christian Science Monitor* (Aug. 17). Retrieved Feb. 12, 2008. Online: www.csmonitor.com/2006/0817/p15s02-legn.html
Terkel, Studs. 1990. *Working: People Talk About What They Do All Day and How They Feel About What They Do*. New York: Ballantine (orig. pub. 1972).
That, Sovanny. 2007. "Refugee Women's Alliance." Retrieved Mar. 31, 2007. Online: http://students.washington.edu/sovannyt/communityservice.htm
theadventuresofiman.com. 2007. "The Adventures of Iman." Retrieved Mar. 18, 2007. Online: www.theadventuresofiman.com/AboutIman.asp
Thompson, Becky W. 1994. *A Hunger So Wide and So Deep: American Women Speak Out on Eating Problems*. Minneapolis: University of Minnesota Press.
Thorne, Barrie. 1993. *Gender Play: Girls and Boys in School*. New Brunswick, NJ: Rutgers University Press.
Thorne, Barrie, Cheris Kramarae, and Nancy Henley. 1983. *Language, Gender, and Society*. Rowley, MA: Newbury.
Thornton, Russell. 1984. "Cherokee Population Losses During the Trail of Tears: A New Perspective and a New Estimate." *Ethnohistory*, 31: 289–300.
Tilly, Charles. 1973. "Collective Action and Conflict in Large-Scale Social Change: Research Plans, 1974–78." Center for Research on Social Organization. Ann Arbor: University of Michigan, October.
—. 1975. *The Formation of National States in Western Europe*. Princeton, NJ: Princeton University Press.
—. 1978. *From Mobilization to Revolution*. Reading, MA: Addison-Wesley.
Tiryakian, Edward A. 1978. "Emile Durkheim." In Tom Bottomore and Robert Nisbet (Eds.), *A History of Sociological Analysis*. New York: Basic, pp. 187–236.
TMPLiveWire. 2011. "Colbert: Obama Stole My Awesome Idea for a Gator-Filled Border Moat." Retrieved May 16, 2011. Online: http://tpmlivewire.talkingpointsmemo.com/2011/05/colbert-obama-stole-my-awesome-idea-for-a-gator-filled-border-moat-video.php
Toft, Monica Duffy, Daniel Philpott, and Timothy Samuel Shah. 2011. "God's Partisans Are Back." *Chronicle Review* (Apr. 22): B4, B5.
Tong, Rosemarie. 1989. *Feminist Thought: A Comprehensive Introduction*. Boulder, CO: Westview.
Tönnies, Ferdinand. 1940. *Fundamental Concepts of Sociology* (Gemeinschaft *und* Gesellschaft). Trans. Charles P. Loomis. New York: American Book Company (orig. pub. 1887).
—. 1963. *Community and Society* (Gemeinschaft *and* Gesellschaft). New York: Harper & Row (orig. pub. 1887).
Tower, Cynthia Crosson. 1996. *Child Abuse and Neglect* (3rd ed.). Boston: Allyn & Bacon.
Toobin, Jeffrey. 2011. "Money Talks." *New Yorker* (Apr. 11). Retrieved Mar. 9, 2011. Online: www.newyorker.com/talk/comment/2011/04/11/110411taco_talk_toobin
Tracy, C. 1980. "Race, Crime and Social Policy: The Chinese in Oregon, 1871–1885." *Crime and Social Justice*, 14: 11–25.
Troeltsch, Ernst. 1960. *The Social Teachings of the Christian Churches*, vols. 1 and 2. Trans. O. Wyon. New York: Harper & Row (orig. pub. 1931).
Tumin, Melvin. 1953. "Some Principles of Stratification: A Critical Analysis." *American Sociological Review*, 18 (August): 387–393.
Turk, Austin. 1969. *Criminality and Legal Order*. Chicago: Rand McNally.
—. 1977. "Class, Conflict and Criminology." *Sociological Focus*, 10: 209–220.
Turkle, Sherry. 2011. *Alone Together: Why We Expect More from Technology and Less from Each Other*. New York: Basic.
Turner, Jonathan, Leonard Beeghley, and Charles H. Powers. 2002. *The Emergence of Sociological Theory* (5th ed.). Belmont, CA: Wadsworth.
—. 2007. *The Emergence of Sociological Theory* (6th ed.). Belmont, CA: Wadsworth.
Turner, Jonathan H., Royce Singleton, Jr., and David Musick. 1984. *Oppression: A Socio-History of Black–White Relations in America*. Chicago: Nelson-Hall (reprinted 1987).
Turner, Ralph H., and Lewis M. Killian. 1993. "The Field of Collective Behavior." In Russell L. Curtis, Jr., and Benigno E. Aguirre (Eds.), *Collective Behavior and Social Movements*. Boston: Allyn & Bacon, pp. 5–20.
Turner, Sherry L. 1997. "The Influence of Fashion Magazines on the Body Image Satisfaction of College Women: An Exploratory Analysis." *Adolescence* (Fall). Retrieved Feb. 2, 2008. Online: http://findarticles.com/p/articles/mim2248/isn127v32ai20413253
Twitchell, James B. 1996. *ADCULTusa: The Triumph of Advertising in American Culture*. New York: Columbia University Press.
—. 1999. *Lead Us into Temptation: The Triumph of American Materialism*. New York: Columbia University Press.

uefa.com. 2011. "Tackling Racism in Club Football." Retrieved May 23, 2011. Online: www.uefa.com/newsfiles/459063.pdf
UNAIDS. 2008. "2008 Report on the Global AIDS Epidemic." Retrieved Apr. 12, 2010. Online: http://www.unaids.org/en/KnowledgeCentre/HIVData/GlobalReport/2008/2008_Global_report.asp
—. 2010. "UNAIDS Report on the Global AIDS Epidemic." Retrieved May 15, 2011. Online: www.unaids.org/globalreport
UNAIDS/WHO. 2000. "Report on the Global HIV/AIDS Epidemic—June 2000." Retrieved Aug. 5, 2000. Online: www.unaids.org/epidemic_update/report/glo_estim.pdf
unicef.org. 2011. "Child Protection from Violence, Exploitation and Abuse: Child Labour." Retrieved Mar. 6, 2011. Online: www.unicef.org/protection/index_childlabour.html
Unification Church News. 2011. "Who We Are." Retrieved May 10, 2011. Online: www.familyfed.org/about
United Nations. 1948. "The Universal Declaration of Human Rights." Retrieved Apr. 14, 2011. Online: www.un.org/en/documents/udhr/index.shtml
—. 2011. "World Population to Reach 10 Billion by 2100 If Fertility in All Countries Converges to Replacement Level." *World Population Prospects: The 2010 Revision* (May 3). Retrieved May 22, 2011. Online: http://esa.un.org/unpd/wpp/Other-Information/Press_Release_WPP2010.pdf
United Nations Development Programme. 2008. *Human Development Report 2007/2008*. Retrieved

June 6, 2009. Online: http://hdrstats.undp.org/en/indicators/2.html
—. 2009. *Human Development Report: 2009.* New York: Oxford University Press.
—. 2010. *Human Development Report: 2010.* Retrieved Feb. 25, 2010. Online: http://hdr.undp.org/en/media/HDR_2010_EN_Complete_reprint.pdf
United Nations Population Division. 1999. Retrieved Oct. 17, 1999. Online: gopher:/gopher.undp.org/00/ungophers/popin/wdtrends
—. 2008. "World Population Prospects." Retrieved June 7, 2010. Online: www.un.org/esa/population/publications/wpp2008/wpp2008_highlights.pdf
United Nations World Urbanization Prospects Report. 2009. "World's Largest Urban Agglomerations." Retrieved Apr. 19, 2010. Online: http://esa.un.org/wup2009/unup/index.asp?panel=2
United Press International. 2007. "Indian Women Carry Children for Foreigners." *United Press International.com* (Nov. 11). Retrieved Feb. 9, 2008. Online: www.upi.com/NewsTrack/Science/2007/11/11/indian_women_carry_children_for_foreigners/2909
UPI.com. 2010. "Philip Morris Vows to End Child Labor." UPI.com (July 14). Retrieved Mar. 6, 2011. Online: www.upi.com/Top_News/US/2010/07/14/Philip-Morris-vows-to-end-child-labor/UPI-61821279121448
U.S. Bureau of Justice Statistics. 2008. "Criminal Victimization, 2007." Retrieved July 9, 2009. Online: www.ojp.usdoj.gov/bjs/abstract/cv07.htm
U.S. Bureau of Labor Statistics. 2007. "Labor Force Statistics from the Current Population Survey." Retrieved Jan. 12, 2008. Online: http://data.bls.gov/cgi-bin/surveymost?ln
—. 2010a. "Consumer Expenditure Survey, 2006–2007." Retrieved Nov. 24, 2010. Online: www.bls.gov/cex/twoyear/200607/csxtwoyr.pdf
—. 2010b. "Highlights of Women's Earnings in 2009" (June). Retrieved Mar. 12, 2010. Online: www.bls.gov/cps/cpswom2009.pdf
—. 2011a. "The Employment Situation—February 2011." Retrieved Mar. 25, 2011. Online: www.bls.gov/news.release/pdf/empsit.pdf
—. 2011b. "Employment Status of the Civilian Population by Sex, Age, and Disability Status" (Apr. 1). Retrieved Apr. 2, 2011. Online: www.bls.gov/news.release/empsit.t06.htm
—. 2011c. "Major Work Stoppages in 2010" (Feb. 8). Retrieved Mar. 25, 2011. Online: www.bls.gov/news.release/pdf/wkstp.pdf
—. 2011d. "Union Members—2010" (Jan. 21). Retrieved Mar. 25, 2011. Online: www.bls.gov/news.release/pdf/union2.pdf
U.S. Census Bureau. 2007. *Statistical Abstract of the United States: 2007* (126th ed.). Washington, DC: U.S. Government Printing Office.
—. 2008. *Statistical Abstract of the United States* (127th ed.). Washington, DC: U.S. Government Printing Office.
—. 2009. *Statistical Abstract of the United States: 2009* (128th ed.). Washington, DC: U.S. Government Printing Office.
—. 2010a. "Current Population Survey, October, 2009" (released Feb. 2010). Retrieved Dec. 27, 2010. Online: www.census.gov/population/www/socdemo/computer.html
—. 2010b. *Statistical Abstract of the United States* (129th ed.) Washington, DC: U.S. Government Printing Office.
—. 2010c. "USA QuickFacts." Retrieved Apr. 20, 2011. Online: http://quickfacts.census.gov/qfd/states/00000.html
—. 2011a. "Current Population Survey: Definitions and Explanations." Retrieved Apr. 14, 2011. Online: www.census.gov/population/www/cps/cpsdef.html
—. 2011b. "Mother's Day: May 8, 2011." Retrieved Apr. 14, 2011. Online: www.census.gov/newsroom/releases/archives/facts_for_features_special_editions/cb11-ff07.html
U.S. Census Bureau, American Community Survey. 2009. "Percentage of People in Poverty." Retrieved Feb. 20, 2010. Online: www.census.gov/prod/2010pubs/acsbr09-1.pdf
U.S. Census Bureau International Data Base. 2011. "World Vital Events Per Time Unit: 2011." Retrieved May 15, 2011. Online: www.census.gov/cgi-bin/ipc/pcwe
U.S. Census Bureau Population Estimates. 2010. "Annual Estimates of the Resident Population by Age and Sex for States and for Puerto Rico: April 1, 2000 to July 1, 2009." Retrieved May 14, 2011. Online: www.census.gov/popest/states/asrh/SC-EST2009-02.html
U.S. Conference of Mayors, 2010. "Hunger and Homelessness Survey" (Dec.). Retrieved Jan. 25, 2011. Online: www.usmayors.org/pressreleases/uploads/2010HungerHomelessnessReportfinalDec212010.pdf
U.S. Congress, Joint Economic Committee. 1986. *The Concentration of Wealth in the United States: Trends in the Distribution of Wealth Among American Families.* Washington, DC: U.S. Government Printing Office.
U.S. Department of Education. 2009. "Indicators of School Crime and Safety: 2009." Retrieved Mar. 12, 2010. Online: http://nces.ed.gov/programs/crimeindicators/crimeindicators2009/figures/figure_01_1.asp?referrer=report
—. 2010. "A Blueprint for Reform: The Reauthorization of the Elementary and Secondary Education Act." Office of Planning, Evaluation and Policy Development. Retrieved Apr. 29, 2011. Online: www2.ed.gov/policy/elsec/leg/blueprint/blueprint.pdf
U.S. Department of Health and Human Services. 2008a. "Adoption Experiences of Women and Men and Demand for Children to Adopt by Women 18–44 Years of Age in the United States, 2002." National Centers for Disease Control and Prevention. Retrieved Apr. 2, 2010. Online www.cdc.gov/nchs/data/series/sr_23/sr23_027.pdf
U.S. Department of Health and Human Services. 2008b. "Stop Bullying Now." Retrieved Feb. 12, 2008. Online: www.stopbullyingnow.hrsa.gov
U.S. Department of Labor. 2011. "Myths and Facts About Workers with Disabilities." Retrieved Apr. 3, 2011. Online: www.doleta.gov/disability/htmldocs/myths.cfm
U.S. English. 2011. "Official English: States with Official English Laws." Retrieved Mar. 21, 2011. Online: http://us-english.org/view/13
U.S. Office of Personnel Management. 2008. "Profile of Federal Civilian Non-Postal Employees." Retrieved Aug. 21, 2009. Online: www.opm.gov/feddata/html/prof0908.asp

Van Ausdale, Debra, and Joe R. Feagin. 2001. *The First R: How Children Learn Race and Racism.* Lanham, MD: Rowman & Littlefield
Van Horn, Carl, and Cliff Zukin. 2009. "'What a Difference a Decade Makes:' The Declining Job Satisfaction of the American Worker" (Dec. 15). John H. Heldrich Center for Workplace Development, Rutgers, The State University of New Jersey. Retrieved Apr. 3, 2011. Online: www.heldrich.rutgers.edu/sites/default/files/content/Work_Trends_Dec._2009.pdf
Vara, Richard. 2006. "Baylor Study Finds Religion Thriving in the U.S." *Houston Chronicle* (Sept. 11). Retrieved May 9, 2011. Online: www.chron.com/disp/story.mpl/front/4177637.html
Vaughan, Diane. 1985. "Uncoupling: The Social Construction of Divorce." In James M. Henslin (Ed.), *Marriage and Family in a Changing Society* (2nd ed.). New York: Free Press, pp. 429–439.
Veblen, Thorstein. 1967. *The Theory of the Leisure Class.* New York: Viking (orig. pub. 1899).
Venkatesh, Sudhir Alladi. 2006. *Off the Books: The Underground Economy of the Urban Poor.* Cambridge: Harvard University Press.
—. 2009. "Feeling Too Down to Rise Up." *New York Times* (Mar. 29): WK 10.
Ventura, Stephanie J., and Brady E. Hamilton. 2011. "U.S. Teenage Birth Rate Resumes Decline." National Centers for Disease Control and Prevention. NCHS Data Brief, #58 (Feb.). Retrieved Apr. 16, 2011. Online: www.cdc.gov/nchs/data/databriefs/db58.pdf
Vincent, Isabel. 2004. "Women Bear Brunt of 'Sandwich' Caregiving." *National Post* (Sept. 29). Retrieved Apr. 16, 2005. Online: www.albertacaregiversassociation.org/natpost-sandwichcaregiving.htm
Vito, Gennaro F., and Ronald M. Holmes. 1994. *Criminology: Theory, Research and Policy.* Belmont, CA: Wadsworth.
Volti, Rudi. 1995. *Society and Technological Change* (3rd ed.). New York: St. Martin's.
VolunteerMatch.org. 2010. "Volunteer Spotlight: Christine French, the George Washington University" (June 16). Retrieved Dec. 29, 2010. Online: www.volunteermatch.org/volunteers/stories/spotlight.jsp?id=48

Wadud, Amina. 2002. "A'ishah's Legacy: Amina Wadud Looks at the Struggle for Women's Rights Within Islam." *New Internationalist* (May). Retrieved Mar. 18, 2007. Online: http://newint.org/features/2002/05/01/aishahs-legacy
Wagner, Elvin, and Allen E. Stearn. 1945. *The Effects of Smallpox on the Destiny of the American Indian.* Boston: Bruce Humphries.
Wald, Matthew L. 2011. "Japan Orders Evacuation Near 2nd Nuclear Plant." *New York Times* (Mar. 11). Retrieved May 23, 2011. Online: www.nytimes.com/2011/03/12/world/asia/12nuclear.html?_r=1&scp=3&sq=evaculation%20near%20nuclear%20reaction%20in%20Japan?&st=cse
Wallace, Walter L. 1971. *The Logic of Science in Sociology.* New York: Aldine de Gruyter.
Wallerstein, Immanuel. 1979. *The Capitalist World-Economy.* Cambridge, England: Cambridge University Press.
—. 1984. *The Politics of the World Economy.* Cambridge, England: Cambridge University Press.
—. 1991. *Unthinking Social Science: The Limits of Nineteenth-Century Paradigms.* Cambridge, England: Polity.
Walmart Corporate. 2010. "Walmart Fact Sheet" (Mar. 2010). Retrieved Nov. 14, 2010. Online: http://walmartstores.com/pressroom/FactSheets
Warner, W. Lloyd, and Paul S. Lunt. 1941. *The Social Life of a Modern Community.* New Haven, CT: Yale University Press.
Warr, Mark. 1993. "Age, Peers, and Delinquency." *Criminology,* 31 (1): 17–40.
Waters, Malcolm. 1995. *Globalization.* London and New York: Routledge.
Watson, Elwood, and Darcy Martin (Eds). 2004. *"There She Is, Miss America": The Politics of Sex, Beauty, and Race in America's Most Famous Pageant.* New York: Palgrave Macmillan.
Watson, Tracey. 1987. "Women Athletes and Athletic Women: The Dilemmas and Contradictions

of Managing Incongruent Identities." *Sociological Inquiry,* 57 (Fall): 431–446.

Weber, Max. 1963. *The Sociology of Religion.* Trans. E. Fischoff. Boston: Beacon (orig. pub. 1922).

—. 1968. *Economy and Society: An Outline of Interpretive Sociology.* Trans. G. Roth and G. Wittich. New York: Bedminster (orig. pub. 1922).

—. 1976. *The Protestant Ethic and the Spirit of Capitalism.* Trans. Talcott Parsons. Introduction by Anthony Giddens. New York: Scribner (orig. pub. 1904–1905).

Weeks, John R. 2005. *Population: An Introduction to Concepts and Issues* (9th ed.). Belmont, CA: Wadsworth.

—. 2012. *Population: An Introduction to Concepts and Issues* (11th ed.). Belmont, CA: Wadsworth.

Weigel, Russell H., and P. W. Howes. 1985. "Conceptions of Racial Prejudice: Symbolic Racism Revisited." *Journal of Social Issues,* 41: 124–132.

Wei-ming, Tu. 1995. "Confucianism." In Arvind Sharma (Ed.), *Our Religions.* San Francisco: HarperCollins, pp. 141–227.

Weinstein, Michael M. 1997. "'The Bell Curve,' Revisited by Scholars." *New York Times* (Oct. 11): A20.

Weisner, Thomas S., Helen Garnier, and James Loucky. 1994. "Domestic Tasks, Gender Egalitarian Values and Children's Gender Typing in Conventional and Nonconventional Families." *Sex Roles* (January): 23–55.

Weiss, Gregory L., and Lynne E. Lonnquist. 2009. *The Sociology of Health, Healing, and Illness* (6th ed.). Upper Saddle River, NJ: Prentice Hall.

Weiss, Jeffrey. 2004. "Beliefs in UFOs and Ritual Abuse Are Studied as New Religious Movements." *Dallas Morning News* (Aug. 28). Retrieved Jan. 11, 2008. Online: www.dallasnews.com

Weiss, Meira. 1994. *Conditional Love: Attitudes Toward Handicapped Children.* Westport, CT: Bergin & Garvey.

Weitz, Rose. 2004. *The Sociology of Health, Illness, and Health Care* (3rd ed.). Belmont, CA: Wadsworth.

—. *The Sociology of Health, Illness, and Health Care* (5th ed.). Belmont, CA: Wadsworth/Cengage.

Wellhousen, Karyn, and Zenong Yin. 1997. "Peter Pan Isn't a Girls' Part: An Investigation of Gender Bias in a Kindergarten Classroom." *Women and Language,* 20: 35–40.

Wellman, Barry. 2001. "Physical Place and Cyberplace: The Rise of Personalized Networking." *International Journal of Urban and Regional Research,* 22 (2): 227–252.

West, Candace, and Don H. Zimmerman. 1987. "Doing Gender." *Gender & Society* (June): 125–151.

Westrum, Ron. 1991. *Technologies and Society: The Shaping of People and Things.* Belmont, CA: Wadsworth.

Wharton, Amy S. 2004. *The Sociology of Gender: An Introduction to Theory and Research.* London: Blackwell.

whitehouse.gov. 2009. "Excerpts of the President's Remarks in Warren, Michigan" (July 14). Retrieved July 20, 2009. Online: www.whitehouse.gov/the_press_office/Excerpts-of-the-Presidents-remarks-in-Warren-Michigan-and-fact-sheet-on-the-American-Graduation-Initiative

Whitaker, Bill. 2010. "High School Dropouts Costly for American Economy." cbsnews.com (May 28). Retrieved Apr. 23, 2011. Online: www.cbsnews.com/stories/2010/05/28/eveningnews/main6528227.shtml

Whitman, Neal A. 1985. "Student Stress: Effects and Solutions." Retrieved Feb. 8, 2010. Online: www.ericdigests.org/pre-926/stress.htm

Whitney, Craig R. 1997. "Jeanne Calment, World's Elder, Dies at 122." *New York Times* (Aug. 5): B8.

Whorf, Benjamin Lee. 1956. *Language, Thought and Reality.* Ed. John B. Carroll. Cambridge, MA: MIT Press.

Whyte, William Foote. 1989. "Advancing Scientific Knowledge Through Participatory Action Research." *Sociological Forum,* 4: 367–386.

Wikipedia. 2011. "List of Smoking Bans Around the World, 2010." Retrieved July 12, 2011. Online: http://en.wikipedia.org/wiki/List_of_smoking_bans

Wildavsky, Ben. 2010. "University Globalization Is Here to Stay." *Chronicle of Higher Education* (Aug. 27): 44.

Wilkerson, Jamie. 1996. "Thoughts of Internet Friendships." Retrieved July 5, 2003. Online: http://web.cetlink.net/~parrothd/Jamie/Poetry1.htm

Williams, Christine. 2004. "The Glass Escalator: Hidden Advantages for Men in the 'Female' Professions." In Michael S. Kimmel and Michael S. Messner (Eds.), *Men's Lives* (6th ed.). Boston: Allyn & Bacon.

Williams, Geoff. 2011. "Made in the USA: The Weight-Loss Industry Makes Huge Gains." Smallbusiness.aol.com (Feb. 16). Retrieved Feb. 27, 2011. Online: http://smallbusiness.aol.com/2011/02/16/made-in-the-usa-the-weight-loss-industry-makes-huge-gains

Williams, Robin M., Jr. 1970. *American Society: A Sociological Interpretation* (3rd ed.). New York: Knopf.

Williamson, Robert C., Alice Duffy Rinehart, and Thomas O. Blank. 1992. *Early Retirement: Promises and Pitfalls.* New York: Plenum.

Wilson, David (Ed.). 1997. "Globalization and the Changing U.S. City." *Annals of the American Academy of Political and Social Sciences,* 551 (May special issue).

Wilson, Edward O. 1975. *Sociobiology: A New Synthesis.* Cambridge, MA: Harvard University Press.

Wilson, Elizabeth. 1991. *The Sphinx in the City: Urban Life, the Control of Disorder, and Women.* Berkeley: University of California Press.

Wilson, James Q. 1996. "Foreword." In George L. Kelling and Catherine M. Coles (Eds.), *Fixing Broken Windows: Restoring Order and Reducing Crime in Our Communities.* New York: Touchstone, pp. xiii–xvi.

Wilson, William Julius. 1978. *The Declining Significance of Race: Blacks and Changing American Institutions.* Chicago: University of Chicago Press.

—. 1996. *When Work Disappears: The World of the New Urban Poor.* New York: Knopf.

Winik, Lyric Wallwork. 1997. "Oh Nurse, More Beluga Please." *Forbes FYI: The Good Life* (Winter): 157–166.

Winn, Maria. 1985. *The Plug-in Drug: Television, Children, and the Family.* New York: Viking.

Winters, Marcus A. 2010. "For Minorities, a Charter-School Boost." *New York Post* (Apr. 27). Online: www.manhattan-institute.org/cgi-bin/apMI/print.cgi

Wirth, Louis. 1938. "Urbanism as a Way of Life." *American Journal of Sociology,* 40: 1–24.

Wischnowsky, Dave. 2005. "Small Town a Bastion of Bigfoot Belief." *Chicago Tribune* (Oct. 10): 2.

Wiseman, Jacqueline. 1970. *Stations of the Lost: The Treatment of Skid Row Alcoholics.* Chicago: University of Chicago Press.

Wollstonecraft, Mary. 1974. *A Vindication of the Rights of Woman.* New York: Garland (orig. pub. 1797).

Women's Research & Education Institute. 2009. "Breaking Through the Brass Ceiling." Retrieved July 9, 2009. Online: www.wrei.org/News.html

Wonders, Nancy. 1996. "Determinate Sentencing: A Feminist and Postmodern Story." *Justice Quarterly,* 13: 610–648.

Wood, Daniel B. 2002. "As Homelessness Grows, Even Havens Toughen Up." *Christian Science Monitor* (Nov. 21). Retrieved July 1, 2003. Online: www.csmonitor.com/2002/1121/p01s04-ussc.htm

Wood, Julia T. 1994. *Gendered Lives: Communication, Gender, and Culture.* Belmont, CA: Wadsworth.

—. 1999. *Gendered Lives: Communication, Gender, and Culture* (3rd ed.). Belmont, CA: Wadsworth.

Woods, Rebecca Bowman. 2010. "Slavery in the Twenty-First Century." DisciplesWorld.com. Retrieved Mar. 14, 2010. Online: www.disciples-world.com/article_print.html?id=1523

Wooley, Susan C. 1994. "Sexual Abuse and Eating Disorders: The Concealed Debate." In Patricia Fallon, Melanie A. Katzman, and Susan C. Wooley (Eds.), *Feminist Perspectives on Eating Disorders.* New York: Guilford, pp. 171–211.

Woolsey, Ben, and Matt Schulz. 2010. "Credit Card Statistics, Industry Facts, Debt Statistics." CreditCards.com. Retrieved Nov. 13, 2010. Online: www.creditcards.com/credit-card-news/credit-card-industry-facts-personal-debt-statistics-1276.php

World Bank. 2007. "Poverty in China: What Do the Numbers Say?" Washington, DC: World Bank. Retrieved Jan. 20, 2008. Online: http://web.worldbank.org

—. 2009. *World Development Indicators 2009.* Retrieved June 6, 2009. Online: http://siteresources.worldbank.org/DATASTATISTICS/Resources/wdi09introch1.pdf

—. 2010. *World Development Indicators: 2010.* Retrieved Feb. 25, 2010. Online: http://data.worldbank.org/sites/default/files/wdi-final.pdf

World Health Organization, 2009. "Global Summary of the AIDS Epidemic." Retrieved May 19, 2011. Online: www.who.int/hiv/data/2009_global_summary.png

—. 2011a. "Disaster Risk Management for Health" (May). Retrieved May 11, 2011. Online: www.who.int/hac/events/drm_fact_sheet_

—. 2011b. "Frequently Asked Questions: What Is the WHO Definition of Health?" Retrieved May 12, 2011. Online: www.who.int/suggestions/faq/en

Wouters, Cas. 1989. "The Sociology of Emotions and Flight Attendants: Hochschild's Managed Heart." *Theory, Culture & Society,* 6: 95–123.

Wright, Erik Olin. 1978. "Race, Class, and Income Inequality." *American Journal of Sociology,* 83 (6): 1397.

—. 1979. *Class Structure and Income Determination.* New York: Academic Press.

—. 1985. *Class.* London: Verso.

—. 1997. *Class Counts: Comparative Studies in Class Analysis.* Cambridge, England: Cambridge University Press.

Wuthnow, Robert. 1992. *Rediscovering the Sacred: Perspectives on Religion in Contemporary Society.* Grand Rapids, MI: Eerdmans. Prepared for Religion on Line by William E. Chapman. Retrieved July 12, 2009. Online: www.religion-online.org/showchapter.asp?title=1509&C=1350

Yablonsky, Lewis. 1997. *Gangsters: Fifty Years of Madness, Drugs, and Death on the Streets of America.* New York: New York University Press.

Yinger, J. Milton. 1960. "Contraculture and Subculture." *American Sociological Review,* 25 (October): 625–635.

—. 1982. *Countercultures: The Promise and Peril of a World Turned Upside Down.* New York: Free Press.

Yong, Ed. 2011. "New Evidence That IQ Is Not Set in Stone." cbsnews.com (Apr. 26). Retrieved Apr. 28, 2011. Online: www.cbsnews.com/stories/2011/04/26/scitech/main20057536.shtml

Yoshikawa, Hirokazu. 2011. *Immigrants Raising Citizens: Undocumented Parents and Their Young Children.* New York: Russell Sage Foundation.

Young, Michael Dunlap. 1994. *The Rise of the Meritocracy.* New Brunswick, NJ: Transaction (orig. pub. 1958).

YouTube. 2011. "Credit Card Debt: A Student's Story." TG (Texas Guaranteed Student Loan Corporation). Retrieved Apr. 21, 2011. Online: www.youtube.com/watch?v=7U6pmkTC8i0

Ypulse.com. 2009. "MTV Surveys College Students on What Stresses Them Out." Retrieved Feb. 8, 2010. Online: www.ypulse.com/wordpress/wordpress/mtv-surveys-college-students-on-what-stresses-them-out

Zald, Mayer N., and John D. McCarthy (Eds.). 1987. *Social Movements in an Organizational Society.* New Brunswick, NJ: Transaction.

Zavella, Patricia. 1987. *Women's Work and Chicano Families: Cannery Workers of the Santa Clara Valley.* Ithaca, NY: Cornell University Press.

Zellner, William M. 1978. Vehicular Suicide: In Search of Incidence. Unpublished M.A. thesis, Western Illinois University, Macomb. Quoted in Richard T. Schaefer and Robert P. Lamm. 1992. *Sociology* (4th ed.). New York: McGraw-Hill, pp. 54–55.

Zephyr. 2009. "Would You Wear . . . Designer Knockoffs?" *College Fashion.* Retrieved July 8, 2011. Online: www.collegefashion.net/would-you-wear/would-you-wear-designer-2knockoffs

Zernike, Kate. 2010. "A Young and Unlikely Activist Who Got to the Tea Party Early." *New York Times* (Feb. 28): A1, A19.

Zernike, Kate, and Megan Thee-Brenan. 2010. "Discontent's Demography: Who Backs the Tea Party?" *New York Times* (Apr. 15): A1, A17.

Zipp, John F. 1985. "Perceived Representativeness and Voting: An Assessment of the Impact of 'Choices' vs. 'Echoes.'" *American Political Science Review,* 60 (3): 738–759.

Zuboff, Shoshana. 1988. *In the Age of the Smart Machine: The Future of Work and Power.* New York: Basic.

name index

subject index